TIC DISORDERS AND TOURETTE SYNDROME

HANDBOOK OF CLINICAL NEUROLOGY

Series Editors

MICHAEL J. AMINOFF, FRANÇOIS BOLLER, AND DICK F. SWAAB

VOLUME 215

ELSEVIER

TIC DISORDERS AND TOURETTE SYNDROME

Series Editors

MICHAEL J. AMINOFF, FRANÇOIS BOLLER, AND DICK F. SWAAB

Volume Editors

CHRISTOS GANOS AND YULIA WORBE

VOLUME 215

3rd Series

ELSEVIER

Elsevier
Radarweg 29, PO Box 211, 1000 AE Amsterdam, Netherlands
125 London Wall, London EC2Y 5AS, United Kingdom
50 Hampshire Street, 5th Floor, Cambridge, MA 02139, United States

ISBN: 978-0-443-13554-5

For Information on all Elsevier publications
visit our website at https://www.elsevier.com/books-and-journals

Publisher: Megan Ball
Acquisitions Editor: Megan McManus
Editorial Project Manager: Himanshi Yadav
Production Project Manager: Punithavathy Govindaradjane
Cover Designer: Gopalakrishnan Venkatraman

Typeset by MPS Limited, Chennai, India

Printed in Great Britain by CPI Group (UK) Ltd, Croydon CR0 4YY
Last digit is the print number: 10 9 8 7 6 5 4 3

Handbook of Clinical Neurology, 3rd Series

Available titles

Vol. 79, The human hypothalamus: Basic and clinical aspects, Part I, D.F. Swaab, ed. ISBN 9780444513571
Vol. 80, The human hypothalamus: Basic and clinical aspects, Part II, D.F. Swaab, ed. ISBN 9780444514905
Vol. 81, Pain, F. Cervero and T.S. Jensen, eds. ISBN 9780444519016
Vol. 82, Motor neurone disorders and related diseases, A.A. Eisen and P.J. Shaw, eds. ISBN 9780444518941
Vol. 83, Parkinson's disease and related disorders, Part I, W.C. Koller and E. Melamed, eds. ISBN 9780444519009
Vol. 84, Parkinson's disease and related disorders, Part II, W.C. Koller and E. Melamed, eds. ISBN 9780444528933
Vol. 85, HIV/AIDS and the nervous system, P. Portegies and J. Berger, eds. ISBN 9780444520104
Vol. 86, Myopathies, F.L. Mastaglia and D. Hilton Jones, eds. ISBN 9780444518996
Vol. 87, Malformations of the nervous system, H.B. Sarnat and P. Curatolo, eds. ISBN 9780444518965
Vol. 88, Neuropsychology and behavioural neurology, G. Goldenberg and B.C. Miller, eds. ISBN 9780444518972
Vol. 89, Dementias, C. Duyckaerts and I. Litvan, eds. ISBN 9780444518989
Vol. 90, Disorders of consciousness, G.B. Young and E.F.M. Wijdicks, eds. ISBN 9780444518958
Vol. 91, Neuromuscular junction disorders, A.G. Engel, ed. ISBN 9780444520081
Vol. 92, Stroke – Part I: Basic and epidemiological aspects, M. Fisher, ed. ISBN 9780444520036
Vol. 93, Stroke – Part II: Clinical manifestations and pathogenesis, M. Fisher, ed. ISBN 9780444520043
Vol. 94, Stroke – Part III: Investigations and management, M. Fisher, ed. ISBN 9780444520050
Vol. 95, History of neurology, S. Finger, F. Boller and K.L. Tyler, eds. ISBN 9780444520081
Vol. 96, Bacterial infections of the central nervous system, K.L. Roos and A.R. Tunkel, eds. ISBN 9780444520159
Vol. 97, Headache, G. Nappi and M.A. Moskowitz, eds. ISBN 9780444521392
Vol. 98, Sleep disorders Part I, P. Montagna and S. Chokroverty, eds. ISBN 9780444520067
Vol. 99, Sleep disorders Part II, P. Montagna and S. Chokroverty, eds. ISBN 9780444520074
Vol. 100, Hyperkinetic movement disorders, W.J. Weiner and E. Tolosa, eds. ISBN 9780444520142
Vol. 101, Muscular dystrophies, A. Amato and R.C. Griggs, eds. ISBN 9780080450315
Vol. 102, Neuro-ophthalmology, C. Kennard and R.J. Leigh, eds. ISBN 9780444529039
Vol. 103, Ataxic disorders, S.H. Subramony and A. Durr, eds. ISBN 9780444518927
Vol. 104, Neuro-oncology Part I, W. Grisold and R. Sofietti, eds. ISBN 9780444521385
Vol. 105, Neuro-oncology Part II, W. Grisold and R. Sofietti, eds. ISBN 9780444535023
Vol. 106, Neurobiology of psychiatric disorders, T. Schlaepfer and C.B. Nemeroff, eds. ISBN 9780444520029
Vol. 107, Epilepsy Part I, H. Stefan and W.H. Theodore, eds. ISBN 9780444528988
Vol. 108, Epilepsy Part II, H. Stefan and W.H. Theodore, eds. ISBN 9780444528995
Vol. 109, Spinal cord injury, J. Verhaagen and J.W. McDonald III, eds. ISBN 9780444521378
Vol. 110, Neurological rehabilitation, M. Barnes and D.C. Good, eds. ISBN 9780444529015
Vol. 111, Pediatric neurology Part I, O. Dulac, M. Lassonde and H.B. Sarnat, eds. ISBN 9780444528919
Vol. 112, Pediatric neurology Part II, O. Dulac, M. Lassonde and H.B. Sarnat, eds. ISBN 9780444529107
Vol. 113, Pediatric neurology Part III, O. Dulac, M. Lassonde and H.B. Sarnat, eds. ISBN 9780444595652
Vol. 114, Neuroparasitology and tropical neurology, H.H. Garcia, H.B. Tanowitz and O.H. Del Brutto, eds. ISBN 9780444534903
Vol. 115, Peripheral nerve disorders, G. Said and C. Krarup, eds. ISBN 9780444529022
Vol. 116, Brain stimulation, A.M. Lozano and M. Hallett, eds. ISBN 9780444534972
Vol. 117, Autonomic nervous system, R.M. Buijs and D.F. Swaab, eds. ISBN 9780444534910
Vol. 118, Ethical and legal issues in neurology, J.L. Bernat and H.R. Beresford, eds. ISBN 9780444535016
Vol. 119, Neurologic aspects of systemic disease Part I, J. Biller and J.M. Ferro, eds. ISBN 9780702040863
Vol. 120, Neurologic aspects of systemic disease Part II, J. Biller and J.M. Ferro, eds. ISBN 9780702040870
Vol. 121, Neurologic aspects of systemic disease Part III, J. Biller and J.M. Ferro, eds. ISBN 9780702040887
Vol. 122, Multiple sclerosis and related disorders, D.S. Goodin, ed. ISBN 9780444520012
Vol. 123, Neurovirology, A.C. Tselis and J. Booss, eds. ISBN 9780444534880
Vol. 124, Clinical neuroendocrinology, E. Fliers, M. Korbonits and J.A. Romijn, eds. ISBN 9780444596024
Vol. 125, Alcohol and the nervous system, E.V. Sullivan and A. Pfefferbaum, eds. ISBN 9780444626196
Vol. 126, Diabetes and the nervous system, D.W. Zochodne and R.A. Malik, eds. ISBN 9780444534804
Vol. 127, Traumatic brain injury Part I, J.H. Grafman and A.M. Salazar, eds. ISBN 9780444528926
Vol. 128, Traumatic brain injury Part II, J.H. Grafman and A.M. Salazar, eds. ISBN 9780444635211
Vol. 129, The human auditory system: Fundamental organization and clinical disorders, G.G. Celesia and G. Hickok, eds. ISBN 9780444626301
Vol. 130, Neurology of sexual and bladder disorders, D.B. Vodušek and F. Boller, eds. ISBN 9780444632470
Vol. 131, Occupational neurology, M. Lotti and M.L. Bleecker, eds. ISBN 9780444626271

All volumes in the Third Series of the **Handbook of Clinical Neurology** are published electronically,
on ScienceDirect: http://www.sciencedirect.com/science/handbooks/00729752.

Foreword

The boundary between normality and pathology is always tenuous. A prominent child neurologist, Dr Paul Rosman, used to recall that he had been asked to see a child with "pathologically abnormal tics." It turned out that the child was perfectly normal: he just amused himself and those around him by "making faces." Actually, the diagnosis of the syndrome named after Gilles de la Tourette (Tourette syndrome [TS]) is based on the presence of multiple motor tics plus at least one vocal tic, with onset before the age of 18 years and a chronic course. The syndrome is now well recognized and is even part of general culture, including some recent cinematographic productions. In recent years, there have been numbers of publications dedicated to its pathogenesis and therapy. Several points mentioned by Olivier Walusinski in the introductory chapter of this volume are worth reiterating here. The syndrome had of course occurred long before the original description by Itard and the Salpêtrière group, but it was confused with chorea. As is the case with several other syndromes, Charcot himself deserves much of the credit for having isolated and described the syndrome. In attributing the disease to one of his favorite pupils, he reportedly remarked that it was "such a beautiful name for such a loathsome condition." Finally, it needs to be emphasized that Gilles de la Tourette is a last name and not a first name followed by a last name (the man's first name is Georges). However, as per many recent publications and the International Movement Disorders Society, this volume is titled "Tourette syndrome."

We felt it important to include a volume on TS in the current series of the *Handbook of Clinical Neurology*, and we were delighted that two important specialists in the field agreed to serve as volume editors: Christos Ganos, Associate Professor of Neurology at the University of Toronto, as well as Yulia Worbe, Professor of Clinical Neurophysiology at Sorbonne University in Paris. Together they have planned an exciting and comprehensive volume, with an outstanding group of contributors. As series editors, we reviewed all the chapters and made suggestions for improvement, but we were delighted that the volume editors and chapter authors had produced such scholarly and comprehensive accounts of different aspects of the topic.

The volume is divided into five sections following an introductory chapter on the history of the syndrome, with Section 1 dedicated to the clinical diagnosis of tics and Section 2 to the comorbidities in primary tic disorders. Section 3 deals with the etiology and pathophysiology of Tourette syndrome while Section 4 is dedicated to treatment. Section 5 covers the lived experience of the disorder and also discusses patient organizations and other useful resources.

In addition to the printed version, the volume will be available electronically on Elsevier's ScienceDirect website, as are all of the volumes in the present series of the *Handbook*. This website is popular with readers as it increases the book's accessibility and facilitates searches for specific information.

As always, it is a pleasure to thank Elsevier, our publisher, and in particular Kristi Anderson in San Diego, Megan McManus in Salt Lake City, a well as Abdus Salam Mazumder and Punithavathy Govindaradjane at Elsevier Global Book Production in Chennai, for their assistance in the development and production of the *Handbook of Clinical Neurology*.

Michael J. Aminoff
François Boller
Dick F. Swaab

Preface

Tic disorders, including their most recognized form – Gilles de la Tourette syndrome – occupy a unique position in medical science. On the one hand, diagnostic manuals state that the mere presence of tics is sufficient for the designation of a disorder. On the other hand, many individuals with tics are unaware of them, do not consider them pathologic, and may never seek medical attention. Thus, the point at which a tic, present as part of everyday behavior without strong awareness, becomes a disorder remains blurred.

Moreover, the classification of tics also remains ambiguous: tics do not conform easily to standard categories of movement disorders or psychiatric syndromes. They are described as either involuntary, unwanted, or unavoidable responses to excessive bodily signals, yet their expression is modulated through attention and can be volitionally suppressed. Moreover, tics are amenable to behavioral training. These ambiguities have spurred ongoing debates about the exact pathophysiologic basis of tics, their clinical boundaries, and therapeutic approaches. For example, neurologists may view tics primarily as movement disorders, while psychiatrists may emphasize their association with comorbid conditions such as obsessive-compulsive disorder (OCD) or attention-deficit/hyperactivity disorder (ADHD). Behavioral psychologists may examine tics as expressions of aberrantly reinforced behavioral patterns, while neurosurgeons may become involved in select cases to provide interventions such as deep brain stimulation based on growing knowledge of brain network dysfunctions associated with tics. The rise in functional tic-like behaviors, particularly among adolescents during the COVID-19 pandemic, has only sharpened these questions and drawn renewed attention to how we define and recognize tics.

We are honored to present the first volume on tic disorders for the *Handbook of Clinical Neurology* and present a collection of articles that we believe reflects the breadth and complexity of tic disorders as they are currently understood. In addition to providing an up-to-date overview of clinical, pathophysiologic, and therapeutic advances, we are also pleased to expand the scope of this volume by including contributions that broaden its clinical and scientific relevance. These include first-person accounts, parent perspectives, and a practical overview of patient and caregiver support associations. Together, these perspectives underscore a central aim of this work: that scientific progress ultimately serves to improve care and support for those living with tics.

We recognize that several excellent books on tic disorders have already contributed foundational knowledge to the field. This volume does not attempt to replace them but consolidate the expansion of knowledge in this exciting field.

Tics are still rarely the focus of sustained scientific inquiry, despite their relevance to some of the most fundamental questions in neuroscience—how voluntary and involuntary actions are generated, how motor control is organized, and how intentions to act are formed and expressed. Their study provides unique insights into the development of the motor system, as well as broader processes of brain plasticity and compensation. It also offers a window into the nature of human volition and forces us to reconsider where we draw the line between normal variation and disorder.

We are indebted to the many contributors to this volume whose expertise, critical thinking, and commitment have shaped its content and made this effort possible.

Christos Ganos
Yulia Worbe

List of contributors

E. Abi-Jaoude
Department of Psychiatry, Hospital for Sick Children; Department of Psychiatry, Toronto Western Hospital – University Health Network, University of Toronto, Toronto, ON, Canada

B. Al-Fatly
Movement Disorders and Neuromodulation Unit, Department of Neurology with Experimental Neurology, Charité–Universitätsmedizin Berlin, Corporate Member of Freie Universität Berlin and Humboldt-Universität zu Berlin, Berlin, Germany

A. Alruwaita
Edmond J. Safra Program in Parkinson's Disease and the Morton and Gloria Shulman Movement Disorders Clinic, Toronto Western Hospital, Toronto, ON, Canada; Neurology Department, Prince Sultan Military Medical City, Riyadh, Saudi Arabia

T. Amelung
Institute of Sexology and Sexual Medicine, Charité - University Medicine Berlin, Corporate Member of Free University of Berlin and Humboldt University of Berlin; Department of Psychology, Medical School Berlin, Berlin, Germany

G. Amorelli
Division of Neurology, Department of Medicine, University of Ottawa; Ottawa Hospital Research Institute, Ottawa Hospital, Ottawa, ON, Canada

S. Anderson
Villers Sainte Gertrude, Luxembourg, Belgium

B. Angeloni
Edmond J. Safra Program in Parkinson's Disease and the Morton and Gloria Shulman Movement Disorders Clinic, Toronto Western Hospital; Division of Neurology, University of Toronto, Toronto, ON, Canada; Department of Neurology, Catholic University of Sacred Heart, Rome, Italy

E. Antelmi
Neurology Unit, Parkinson Disease and Movement Disorders Division, AOUI; DIMI, Department of Engineering and Medicine of Innovation, University of Verona, Verona, Italy

J.F. Baizabal-Carvallo
Parkinson Disease Center and Movement Disorders Clinic, Department of Neurology, Baylor College of Medicine, Houston, TX, United States; Department of Sciences and Engineering, University of Guanajuato, León, Mexico

I. Bar-Gad
Leslie & Susan Goldschmied Gonda Multidisciplinary Brain Research Center, Bar-Ilan University, Ramat-Gan, Israel

K. Belelovsky
Leslie & Susan Goldschmied Gonda Multidisciplinary Brain Research Center, Bar-Ilan University, Ramat-Gan, Israel

V. Brandt
School of Psychology, Centre for Innovation in Mental Health, University of Southampton, Southampton, United Kingdom

B. Brown
Sanford, NC, United States

C. Budman
Department of Psychiatry, Zucker School of Medicine at Hofstra/Northwell Health, Hempstead, NY, United States

P. Burbaud
Department of Clinical Neurophysiology, Pellegin Hospital; Institute for Neurodegenerative Disorders, CNRS – UMR 5293, Bordeaux, France

A. Cavanna
Department of Neuropsychiatry, BSMHFT and University of Birmingham, Birmingham, United Kingdom; Department of Child Neuropsychiatry, School of Medicine and Surgery, University of Milano-Bicocca, Milan, Italy

E. Courtin
Department of Clinical Neurophysiology, Pellegin Hospital; Institute for Neurodegenerative Disorders, CNRS – UMR 5293, Bordeaux, France

I.A. Di Vico
Neurology Unit, Parkinson Disease and Movement Disorders Division, AOUI, Verona, Italy

K. Dyke
School of Psychology, University of Nottingham, Nottingham, United Kingdom

C. Ganos
Edmond J. Safra Program in Parkinson's Disease and the Morton and Gloria Shulman Movement Disorders Clinic, Toronto Western Hospital; Division of Neurology, University of Toronto; Krembil Brain Institute, Toronto, ON, Canada

D.L. Gilbert
Division of Neurology, Cincinnati Children's Hospital Medical Center; Department of Pediatrics, University of Cincinnati Collecge of Medicine, Cincinnati, OH, United States

D.A. Gorman
Department of Psychiatry, Hospital for Sick Children, University of Toronto, Toronto, ON, Canada

E. Greenberg
Department of Psychiatry, Massachusetts General Hospital, Harvard Medical School, Boston, MA, United States

A. Gunduz
Norman Fixel Institute for Neurological Diseases; Department of Neurology; J. Crayton Pruitt Family Department of Biomedical Engineering, University of Florida, Gainesville, FL, United States

T. Hedderly
Tic and Neurodevelopmental Movements Service (TANDeM), Evelina London Children's Hospital, Guy's and St Thomas' NHS Foundation Trust; School of Life Sciences and Medicine, King's College London, London, United Kingdom

M.B. Himle
Department of Psychology, University of Utah, Salt Lake City, UT, United States

M.S. Houlgreave
School of Psychology, University of Nottingham, Nottingham, United Kingdom

S.R. Jackson
School of Psychology, University of Nottingham, Nottingham, United Kingdom

P. Jain
Department of Biological Sciences, Purdue University, West Lafayette, IN, United States

J. Jankovic
Parkinson Disease Center and Movement Disorders Clinic, Department of Neurology, Baylor College of Medicine, Houston, TX, United States

K.A. Johnson
Norman Fixel Institute for Neurological Diseases; Department of Neurology, University of Florida, Gainesville, FL, United States

A.E. Lang
Edmond J. Safra Program in Parkinson's Disease and the Morton and Gloria Shulman Movement Disorders Clinic, Toronto Western Hospital; Division of Neurology, University of Toronto; Krembil Brain Institute, Toronto, ON, Canada

A. Latorre
Department of Clinical and Movement Neurosciences, UCL Queen Square Institute of Neurology, University College London, London, United Kingdom

F. Lazaroaie
Department of Psychiatry, Zucker School of Medicine at Hofstra/Northwell Health, Hempstead, NY, United States

Y. Loewenstern
Leslie & Susan Goldschmied Gonda Multidisciplinary Brain Research Center, Bar-Ilan University, Ramat-Gan, Israel

D. Macias-Garcia
Servicio de Neurología y Neurofisiología Clínica, Unidad de Trastornos del Movimiento, Instituto de Biomedicina de Sevilla, (IBiS), Hospital Universitario Virgen del Rocío/CSIC/Universidad de Sevilla, Seville; Centro de Investigación Biomédica en Red sobre Enfermedades Neurodegenerativas (CIBERNED), Madrid, Spain

S. Mahate
Department of Neurology, Johns Hopkins University School of Medicine, Baltimore, MD, United States

T. Mainka
Department of Neurology with Experimental Neurology, Charité - University Medicine Berlin, Berlin, Germany

O. Malik
Tic and Neurodevelopmental Movements Service (TANDeM), Evelina London Children's Hospital, Guy's and St Thomas' NHS Foundation Trust; South London and Maudsley NHS Foundation Mental Health Trust, London, United Kingdom

M. Marklin
Department of Psychology, University of Utah, Salt Lake City, UT, United States

D. Martino
Department of Clinical Neurosciences; Hotchkiss Brain Institute, University of Calgary, Calgary, AB, Canada

J.F. Martin-Rodriguez
Servicio de Neurología y Neurofisiología Clínica, Unidad de Trastornos del Movimiento, Instituto de Biomedicina de Sevilla, (IBiS), Hospital Universitario Virgen del Rocío/CSIC/Universidad de Sevilla, Seville; Centro de Investigación Biomédica en Red sobre Enfermedades Neurodegenerativas (CIBERNED), Madrid; Departamento de Psicología Experimental, Facultad de Psicología, Universidad de Sevilla, Sevilla, Spain

M. Meloni
Department of Medical Sciences and Public Health, Institute of Neurology, University Hospital of Cagliari, Cagliari, Italy

G.P. Mingolla
Department of Neurosciences, Biomedicine and Movement Sciences, University of Verona, Verona, Italy

P. Mir
Servicio de Neurología y Neurofisiología Clínica, Unidad de Trastornos del Movimiento, Instituto de Biomedicina de Sevilla, (IBiS), Hospital Universitario Virgen del Rocío/CSIC/ Universidad de Sevilla, Seville; Centro de Investigación Biomédica en Red sobre Enfermedades Neurodegenerativas (CIBERNED), Madrid; Departamento de Medicina, Facultad de Medicina, Universidad de Sevilla, Seville, Spain

S. Morand-Beaulieu
Department of Psychology, McGill University; Department of Psychology, Université du Québec à Montréal, Montréal, QC, Canada

C. Morris-Berry
Department of Neurology, Johns Hopkins University School of Medicine; Department of Neurology, Kennedy Krieger Institute, Baltimore, MD, United States

M.S. Okun
Norman Fixel Institute for Neurological Diseases; Department of Neurology, University of Florida, Gainesville, FL, United States

P. Paschou
Department of Biological Sciences, Purdue University, West Lafayette, IN, United States

B. Pasquereau
Institute of Cognitive Sciences–Marc Jeannerod, UMR-5229 CNRS, Bron, France

N.A. Perroud
Division of Psychiatric Specialties, Department of Psychiatry, University Hospitals of Geneva, Geneva, Switzerland

K.J. Plessen
Division of Child and Adolescent Psychiatry, Department of Psychiatry, Lausanne University Hospital (CHUV) and University of Lausanne, Lausanne, Switzerland

R. Rizzo
Department of Clinical and Experimental Medicine, Child and Adolescent Neurology and Psychiatry, Catania University, Catania, Italy

L. Rocchi
Department of Medical Sciences and Public Health, University of Cagliari, Cagliari, Italy

P. Sandor
Department of Psychiatry, Toronto Western Hospital–University Health Network, University of Toronto, Toronto, ON, Canada

A. Santer
Department of Neurology, Medical University of Vienna; Department of Neurology, Klinik Ottakring, Vienna, Austria

H.S. Singer
Department of Neurology, Johns Hopkins University School of Medicine; Department of Neurology, Kennedy Krieger Institute, Baltimore, MD, United States

J. Stewart
Toronto, ON, Canada

O. Tahary
Leslie & Susan Goldschmied Gonda Multidisciplinary Brain Research Center, Bar-Ilan University, Ramat-Gan, Israel

A. Topaloudi
Department of Biological Sciences, Purdue University, West Lafayette, IN, United States

O. Walusinski
Lauréat de l'Académie de Médecine (Paris), Brou, France

B. Wellen
Department of Psychiatry and Behavioral Sciences, University of Minnesota, Minneapolis, MN, United States

Y. Worbe
Department of Clinical Neurophysiology, Sorbonne University, APHP-Sorbonne University – Hospital Saint-Antoine; Paris Brain Institute (ICM), Sorbonne University, Inserm U1127, CNRS UMR7225, UM75, Movement Investigation and Therapeutics Team, Paris, France

Contents

Handbook of Clinical Neurology, Vol. 215 (3rd series)
Tic Disorders and Tourette Syndrome
C. Ganos and Y. Worbe, Editors
https://doi.org/10.1016/B978-0-443-13554-5.00009-2

Chapter 1

Georges Gilles de la Tourette in the history of tics

OLIVIER WALUSINSKI*

Lauréat de l'Académie de Médecine (Paris), Brou, France

Abstract

After a year spent working for Jean-Martin Charcot, Georges Gilles de la Tourette published an article in 1885 that would become a seminal work, reporting on nine cases of tic disease. However, only four observations meet the current criteria of the eponym, including echolalia and coprolalia, the latter term coined by Gilles de la Tourette. In 1886, Charcot asked Georges Guinon to complete the clinical picture by adding obsessional disorders temporary control of tics, and the premonitory sensation leading to their occurrence. In 1888, Charcot presided over the jury for Grégoire Breitman's thesis, then in 1890 for Jacques Catrou's thesis; both theses covered Gilles de la Tourette disease and were prepared with the namesake's help. Gilles de la Tourette put the finishing touch on the disease's description in 1900, which implicitly credited Guinon with completing the clinical picture and naming the disease at the time: "la maladie des tics convulsifs" (convulsive tic disease). That said, Gilles de la Tourette did not give much importance to his own nosographic contribution to neurology. After having been nearly forgotten, it was not until the 1960s that this pathology was definitively resurrected and validated by Arthur and Elaine Shapiro under the name of Gilles de la Tourette syndrome. This was a testament to its significant prevalence and its treatment possibilities.

FROM ÉTIENNE BOUTEILLE TO JEAN-GASPARD ITARD AND THE MARQUISE DE DAMPIERRE

In 1884 Jean-Martin Charcot (1825–93) asked his resident Georges Gilles de la Tourette (1857–1904) to study tics. At that time, the medical nosography was inaccurate and tics were not distinguished from chorea (Walusinski, 2022). In his treatise on chorea in 1810, Étienne Michel Bouteille (1732–1916) was the first to distinguish certain forms he called "pseudo-chorea" or "false chorea." Although he never used the word "tic," he described the case of a 10-year-old child: "The facial features are not simply altered; they express through their contortion—such as the grimacing of the lips and the impetuous rolling of the eyes—the most ridiculous and bizarre figures" (Bouteille, 1810). This description of abnormal movements already existed in 1790, authored by Joseph Sumeire (1735–1806), a physician in Marignane, Provence: "She continually agitated primarily her head; her eyes, lips, and facial muscles were in uninterrupted, convulsive motion" (Sumeire, 1790).

After citing Bouteille in his article, Gilles de la Tourette proposed as his first demonstrative clinical case, the observation made by Jean-Gaspard Itard (1775–1838) in 1825 (Itard, 1825), involving the famous case of the Marquise de Dampierre (1800–84), whose origins were recently discovered (Walusinski and Féray, 2020). She was in fact "Madame La Comtesse Picot de Dampierre," that is, Ernestine Émilie Prondre de Guermantes, born on August 22, 1800 ("4 Fructidor an VIII" according to the revolutionary calendar) at the Guermantes family castle in Seine et Marne, east of Paris. Ernestine was a source of worry for her mother, Eulalie de Brisay-Tholozan (1779–1866). Itard described her behavioral problems as follows: "At age 7, she was overcome with convulsive contractions in the muscles of her hands and arms, especially when she was practicing her writing; she would abruptly move her hand away from the letters she was forming. After shifting her hand, its movements became regular and could be controlled voluntarily, until another jolt once again interrupted her work. At first this was considered a sign of vivacity and harmless mischief, but since the behavior was repeated over and over, adults around the girl began to respond with reprimands and punishment. Soon, however, they became convinced that these movements were involuntary and convulsive, involving the shoulder, neck, and facial muscles. The result was extraordinary contortions and grimaces. The disease continued to progress, and when the spasms spread to the organs of the voice and speech, this youngster would utter

*Correspondence to: Olivier Walusinski, Private Practice, Lauréat de l'Académie de Médecine (Paris), 20 rue de Chartres, Brou, 28160, France. Tel: 33952470349, E-mail: walusinski@baillement.com.

bizarre cries and words devoid of meaning, but never in a state of delirium, without any disturbance of the mental faculties." Her physicians held out hope that the symptoms would improve with puberty, but in vain. They advised Ernestine's mother to find a husband for her, since marriage was an ancestral treatment for hysteria. Eulalie selected a first cousin, Louis Augustin Picot de Dampierre (1780–1841), as a potential son-in-law capable of accepting this young woman as his wife, despite her natural defects. Since he was Ernestine's third-degree relative, he had to apply for a dispensation for the marriage from the Bishop of Meaux, requested in these terms: "The most pressing reason is this young woman's health: for several years, she has suffered from an acute nervous and convulsive disease. All the medical remedies used by physicians in Paris, Brittany, and Switzerland have not improved her constitution. Her doctors agree that only a change in her civil status can bring an end to the fits she experiences. In this situation, it is improbable that she could rapidly find a husband appropriate to her rank and birth, and the idea of a delay in the consummation of the union proposed to her and approved by her, or the idea of some other obstacle, may worsen her condition" (Archives de Seine-et-Marne, 1817). Authorization was granted and the marriage contract approved by King Louis XVIII. The marriage was celebrated at the Guermantes castle on December 7, 1817 when she had just turned 17.

Itard was especially struck by the verbal tics: "Among the continual and chaotic movements resulting from these morbid contractions, those affecting the organs of the voice and speech are the only ones worthy of our attention, as they represent an exceptional phenomenon, and constitute a most disagreeable inconvenience that deprives the patient of all of the joys of society; the disturbance they create is due to the pleasure the subject takes in social interaction. For example, in the middle of a conversation in which she is most interested, she suddenly, without being able to restrain herself, interrupts what she is saying or hearing with bizarre cries and words that are even more extraordinary and contrast deplorably with her intellect and her distinguished manners. These words are mostly vulgar curses, obscene epithets, and, what is no less embarrassing for her and others listening, blunt expressions of judgment or unfavorable opinions of those present in society." During his Tuesday lesson on December 13, 1887, Charcot explained: "In Parisian high society, there was a person of the most aristocratic circles who was known for uttering vulgar words. I did not have the honor of knowing her; I only crossed paths with her one day on my way up the stairs from the Salon. I was surprised to hear her suddenly say 'shit' and 'fucking pig'" (Charcot, 1887). Gilles de la Tourette himself wrote: "On several occasions, Professor Charcot saw this patient who, even when she was quite elderly, continued to suffer from muscular incoordination and to utter obscene words in spite of herself, even in public places", as Mr. Charcot was able to witness. The political journals announced her death, which occurred in July or August 1884, and some of them treated their readers to a list of the obscene words.

Although virtually nothing is known about the course of Ernestine's disorder from 1825 until her death in 1884, we can appreciate that some of Ernestine's reported behaviors would currently be regarded as not only coprolalic, but also corresponding to what has been described as nonobscene, socially inappropriate symptoms (NOSISs) (Kurlan et al., 1996). There are significant social and emotional sequelae to living with tics and NOSISs, which can adversely affect quality of life (Eddy and Cavanna, 2013). However, these inconveniences do not seem to have hindered Ernestine's social life; for example, she was at an art exhibition when Charcot heard her swear (Fig. 1.1).

Fig. 1.1. Georges Gilles de la Tourette around 1900 (Collection OW).

CLINICAL CASES COMPILED BY GILLES DE LA TOURETTE

The second case discussed by Gilles de la Tourette involved a 17-year-old boy, hospitalized at La Salpêtrière for "sudden, involuntary movements" in the right leg and arm, accompanied by "a soft, inarticulate cry, a sort of hmm! or ahh! uttered loud enough to be perfectly audible for people around the boy." This condition had developed four years prior (Gilles de la Tourette, 1885a). These movements, which Gilles de la Tourette only qualified as tics after discussing them at length, secondarily spread to the face, then to the left limbs. They occurred in irregular fits and ceased during sleep. In addition, "the patient echoed words and even brief phrases he had heard uttered," without being able to restrain himself. Moreover, "the vulgar nature of the word or phrase accompanying the movement is constant … He often couples his contortions with the word 'shit'." The treatments tested were not effective, but his condition improved spontaneously after he returned home. The patient "seems to have imperceptibly lost the habit of uttering vulgar words, but he remains echolalic."

The third case was that of an 11-year-old boy who, since the age of 8, had exhibited "rapid movements of flection and extension of the head and neck. Soon thereafter, these jolts spread throughout his body, his facial muscles took on various grimaces, and his arms moved, with alternating lifting of the shoulders." Then he began to utter "shit" and "damn" and became "echolalic." Follow-up until the end of 1884 showed the persistence of "frequent tics in the face with an echolalic tendency."

The fourth patient was 28 years old and complained of verbal phobic ruminations. Gilles de la Tourette did not use the word "tic" to describe his jumping movements. The absence of true echolalia and coprolalia makes it doubtful that the clinical picture he described was indeed the disease of interest, suggesting instead a case of hysteria. The fifth patient, aged 14, had facial and shoulder tics that were isolated and few in number, calling to mind transient benign tics of childhood. The sixth patient was 11 years old and had "small convulsive tics," including ocular and shoulder tics, sometimes accompanied by a few verbal tics that were only transient. The seventh patient was 21 years old. Since having been traumatized by the bombings of Paris in 1870, he had a few isolated ocular and shoulder tics that were spontaneously in the process of disappearing. Once again, this man, an anxious adult, did not really match the clinical picture for the disease. The eighth case was that of a 15-year-old girl with tics in her face and limbs. She involuntarily uttered "get away from me, get away from me you imbecile" soon replaced by "in the name of God, fuck, shit," typical of coprolalia, which she could briefly contain by inarticulate grunting. Albert Pitres (1848–1928), a former resident of Charcot who sent this patient to La Salpêtrière Hospital, reported echolalia involving dog barks and imitation of the movements and postures the patient observed around her. The ninth and last observation compiled by Gilles de la Tourette was that of a 23-year-old man with infantilism. At age 20, he developed "convulsive phenomena" consisting of "rhythmic movements" of the limbs, without echolalia or coprolalia. Once again, these symptoms had nothing in common with the disease Gilles de la Tourette was attempting to introduce into the neurologic nosography.

In total, only four observations out of the nine presented by Gilles de la Tourette correspond to the symptom complex of the disease, including the age of onset, motor and verbal tics, echolalia, and coprolalia. The others are either transient benign tics of childhood or hysterical-phobic disorders.

GILLES DE LA TOURETTE'S INITIAL INTERPRETATION

After presenting his clinical cases, Gilles de la Tourette wrote a long discussion in which he argued that the symptoms described in 1878 by the American alienist George M. Beard (1839–83), referring to the "Jumpers of Maine" (Beard, 1878), were similar. The translation of Beard's publications by Gilles de la Tourette, published in the *Archives of Neurologie* when he was a first-year resident (Walusinski, 2019), undoubtedly captured Charcot's attention and led the Maître to choose him to study these symptoms (Beard and Gilles de la Tourette, 1881). However, the Jumpers of Maine did not resemble the patients observed at La Salpêtrière (Gilles de la Tourette, 1884). Gilles de la Tourette's desire to underscore the similarity of these two disorders led him to avoid using the word "tic" in several contexts and in the title of his article. He instead employed the term "motor incoordination," unsuitable because the muscular actions involved in tics are perfectly coordinated. He saw a predominance of these tics in boys and was right to suggest that they were hereditary. He likened the suddenness and rapidity of the jumping to limb tics. Gilles de la Tourette pointed out the variability of tics over time, and how at times they could disappear totally for a while, in what can be called inexplicable "periods of remission." He also pointed out that a voluntary effort could block the appearance of tics, followed by an intense rebound. He underscored what he considered to be pathognomonic elements: "The patient cries out inarticulately," then "the cry becomes articulate and the word the patient pronounces, a variable word, is at times an echo, the subject becoming echolalic." This involuntary echolalia is never controlled by the patient. And beyond that, Gilles de la Tourette spoke of "echolalia of gestures or actions," what Charcot would name "echokinesia." In order to specifically identify "the obscene expression that their upbringing and mental state should have protected them against," he coined the word "coprolalia." "This coprolalia is not found in all subjects, in the same way that echolalia is not always present." Lacking sufficient perspective, Gilles de la Tourette wondered whether the disease could gradually disappear. He had never seen it disappear completely, but considered the prognosis to be good, with no impact on life expectancy. That said, he noted, "There is no denying that [this condition] is a deplorable companion." For him, the chronic progression "results in lazy habits and prevents intellectual development by eliminating work or resulting in serious obstacles to working." For Gilles de la Tourette, the differential diagnoses were easy to make out, especially true chorea. Facial motor tics without pain may be a way into the disease but may also spontaneously disappear. The complete clinical picture must combine "motor incoordination with articulate cries, … the utterance of articulate words with echolalia and coprolalia." Treatment was yet to be found: "How many curative procedures have been unsuccessfully tried; all sedatives of the nervous systems have failed." He recommended, not very strongly, isolation, iron, hydrotherapy, static electricity – the therapies used at La Salpêtrière to treat hysteria! As he lacked anatomic-pathologic data, he hoped that psychology would know how to interpret these disorders, and he recommended the book of the director of the *La Revue philosophique*, Théodule Ribot (1839–1916): *Les maladies de la volonté* (illnesses of the will) (Ribot, 1883) (Fig. 1.2).

Fig. 1.2. Georges Guinon in 1890 (Collection OW).

Fig. 1.3. Jean-Martin Charcot around 1881 (Collection OW).

DECISIVE ROLE OF GEORGES GUINON

Charcot quickly realized the shortcomings in the initial article by Gilles de la Tourette. Georges Guinon (1859–1932) succeeded him in 1885 as one of Charcot's residents (Walusinski, 2021). Since Charcot saw two new cases in 1885, brought back by his Italian student Giulio Melotti (1857–?) from Italy the following year (Charcot and Melotti, 1887; Kushner et al., 1999), Guinon published them in 1886 (Guinon, 1886): "Mr. Charcot has called our attention to the existence in our patients of a series of psychic phenomena not noted in other, similar observations and which, if we refer to the few cases we have been able to study, must be quite frequent in the serious forms of the tic disease. These phenomena are *idées fixes* [fixed ideas]." Also cited were the *folie du pourquoi* (compulsion to ask the reasons underlying the most insignificant things), *folie du doute avec délire du toucher* (compulsion to doubt leading to delirium), arithmomania (compulsion to calculate), and onomatomania (anxiety-producing need to find exactly the right word), recently described by Charcot and Valentin Magnan (1835–1916) (Charcot and Magnan, 1885). It was thus Guinon who added the "idea tic" WHAT IS THAT? to the clinical picture of motor tics, specifically the obsessional and compulsive disturbances that Gilles de la Tourette did not initially mention. Moreover, Guinon strengthened the overlooked significance of his contribution, highlighting two other specific criteria: the possibility for the patient to temporarily inhibit the exteriorization of motor tics and the premonitory sensation of their occurrence. Finally, he contested, and rightly so, the title Gilles de la Tourette gave his article, notably the term "motor incoordination," underscoring that tics were involuntary movements but perfectly coordinated. In addition, in 1887, Guinon authored the "Tics" chapter in the *Dictionnaire Encyclopédique des Sciences Médicales* directed by Amédée Dechambre (1812–86). In this chapter, he did a better job than Gilles de la Tourette at explaining the differential diagnoses, notably for dyskinesia, dystonia, and chorea. He noted the disease's course and its fluctuating nature with the possibility of periods of remission with few symptoms. However, and in contrast to Charcot, Guinon thought "convulsive tic disease" to be a form of hysteria. It is noteworthy that he suggested keeping the name "convulsive tic disease" whereas Charcot has coined the eponym "Gilles de la Tourette disease" to mask the approximation of the initial title, but also perhaps for the harmonious sound of the name and his affection for his unique student. Was Guinon concealing acrimony? Perhaps, but it should be noted that when Rose Kamper, a deranged patient, shot Gilles de la Tourette in 1893, Guinon was the first to help "his friend," as he warmly recounted in the article in *Progrès médical* covering this episode a few days later (Guinon, 1893) (Fig. 1.3).

In his Tuesday lesson on October 23, 1888, Charcot clearly contradicted Gilles de la Tourette: "The movements of ticcers, however complex and bizarre they may be, are not always and essentially disorganized, uncoordinated, or contradictory, as is too often believed. On the contrary, they are generally systematic, to the extent that they always recur in the same way in a given subject. Moreover, very often they reproduce, albeit exaggeratedly, certain automatic physiologic movements that have a purpose." It is during this lesson that Charcot credited Gilles de la Tourette with the coinage of the word "coprolalia": "This is the remarkable phenomenon, among all the others, that Gilles de la Tourette, in his interesting work on the tic disease, ingeniously designated using the term 'coprolalia'" (Charcot, 1889).

The last Tuesday lesson during which Charcot mentioned "convulsive tic disease" took place on June 4, 1889. At the time, he refined the semiologic description: "The complex tic movement is not absurd in itself. It is absurd and illogical because it is executed outside of a context, without an apparent reason. The act of scratching oneself occurs without an itch, the eyes blink in the absence of any foreign object, and so forth. In addition, tic movements are abrupt, rapid, momentary, and do not have, for example, the slowness of choreic gesticulations. They are not continuous but supervene by fits repeated at greater or lesser frequencies and lasting for shorter or longer periods, fits that patients can often, for a time, stop by an effort of will. It is also often the case that while jerking and grimacing, patients utter exclamations and entire words, quite often vulgar words. You will accumulate such specific clinical characteristics that will enable you to distinguish with certainty between the tic disease and all other types of convulsive conditions" (Charcot, 1889). Charcot was once again teaching the difference between tics and chorea by presenting the students with a ticcer alongside a chorea patient (Huntington chorea), both female. By way of introduction, he declared: "You can see to what extent the electric jerks of the ticcer differ profoundly from the slow and permanent gesticulations of subjects with Sydenham's chorea. Between tics and chorea, the distance is vast. Do not forget this; these conditions are sometimes wrongly given the same name and the prognosis is quite different." Charcot showed the opposition between the speed of the movements and highlighted the constancy of choreic writhing, whereas tics stopped intermittently. The chorea patient made no sounds, whereas the ticcer "makes, from time to time, expressive laryngeal noises." The movements of the latter were chaotic and unpredictable, whereas in the former, everything was rhythmic, regular, and systematic. The ticcer started having tics at the age of 12, whereas the other patient was "over 51" at the time, with an age of onset at 33.

BREITMAN INTRODUCES DEGENERATION THEORY

Grégoire (Gregor) Breitman (1859–1914) was a physician from the city of Ananyiv, a part of Moldavia at the time, currently in the Ukraine (160 km north of Odessa). On November 28, 1888, he defended the first thesis on Gilles de la Tourette syndrome since the seminal article's publication, with Charcot presiding over his jury (Breitman, 1888). Breitman, a student alienist at the Vaucluse asylum near Paris (currently the Hôpital de santé mentale de Perray-Vaucluse, in Epinay sur Orge), wanted to study the heredity of involuntary imitation symptoms to clarify the role of "degeneration" in their genesis. This had been the accepted etiology of neuropsychiatric diseases since the work of Bénédicte-Augustin Morel (1809–73): "As the semiology of hereditary degenerates grows more detailed, exact, and manifest, we are relegating the search for the original flaw in the various forms of alienation to a secondary position"; given that "the aim is to find out whether, by studying the semiology of each psychic state and by grouping the signs into syndromes, we can, by deduction, demonstrate the manifestly degenerative hereditary origin" (Morel, 1857).

On March 6, 1890, with Charcot presiding over the jury, Jacques Catrou (1865–?), a former nonresident medical student, defended his thesis on convulsive tic disorder (Catrou, 1890), in which he included Jumping, Latah, and Myriachit, previously described by Gilles de la Tourette (Gilles de la Tourette, 1884). He started by summarizing (nearly copying) the entire beginning of the seminal article, then cited Charcot's lesson recorded by Melotti. Catrou added 28 unpublished observations that Charcot and Gilles de la Tourette had sent him. He then proceeded to analyze all of the 45 observations published to date. His thesis took a highly original turn when he argued against the theories attributing the etiology to degeneration: "We do not subscribe in any way to the current opinion whereby all nervous pathology falls under mental degeneration, aside from conditions where a physical lesion has been demonstrated."

GILLES DE LA TOURETTE'S FINAL CONTRIBUTION

Aside from the observations he gave Catrou, Gilles de la Tourette never published any other data on the disease named for him during Charcot's lifetime. It was not until 1899 that the transcript of one of his lessons at Saint-Antoine Hospital was published (Gilles de la Tourette, 1899). A 22-year-old woman came "to ask for treatment for chorea, which seemed to explain, at least in appearance, the agitation of the muscles of her face and the left side of her body." The abnormal movements began when she was 8 years old. Periods lasting a few weeks to a few months alternated with calm periods, without "muscular incoordination," but with a few residual tics in her face. Gilles de la Tourette once again employed this awkward expression, as if not to admit defeat, all while rejecting it in the lines that followed! All the physicians consulted had diagnosed chorea and prescribed arsenic and antipyrine. In addition to her motor tics, the patient had phonic tics (hum! hum! or oh ah!). "When she grows irritated with her little boy, who is very agitated, her jerks worsen and she calls him 'pig, ass', often uttering the word of Cambronne [shit] in her interjections. I was now sure; the unanimous diagnosis of chorea should be replaced by a diagnosis of convulsive tic disease with coprolalia." This young woman also exhibited phobias. In his commentary, Gilles de la Tourette recognized the contributions of Guinon and Charcot to his initial description and gave this recommendation: "If you wish to read a good survey of the question, I urge you to consult the thesis of Mr. Catrou, my student, prepared in 1890 on the convulsive tic disease, based on twenty-six unpublished observations I sent to him." Gilles de la Tourette admitted at this point that the expression "motor incoordination" was unsuitable and acknowledged the validity of Guinon's "convulsive tics": "Jerks, or tics, to use what has become the established expression." There followed a long section on coprolalia and echolalia, "which can truly be deemed psychic stigmata." Fourteen years after the initial article, this clarification in 1899 presented the disease's clinical picture and its course in a comprehensive fashion that remains relevant today. It was an occasion for Gilles de la Tourette to repeat the aphorism: "Once a ticcer, always a ticcer."

CONCLUSION

The introduction of tics into the neurologic nosography owes in large part to the La Salpêtrière School. Once he had obtained the Chair of Nervous System Diseases, created especially for him in 1881, Charcot no longer published new descriptions of neurologic diseases in his own name, instead promoting the publications of his students. The convulsive tic disease is a demonstrative example of this strategy, which played an important role in the renown of the La Salpêtrière Master. Throughout the conceptual and observational development of Gilles de la Tourette syndrome, Charcot was the real mastermind, always in the background, but constantly vigilant that the work published by his students was relevant and thorough, as in the case of Gilles de la Tourette then Guinon, the latter being the one that the history of this creation forgot.

Gilles de la Tourette was one of Charcot's favorite disciples, and the publications that established his fame while he was alive are all linked to hysteria (Gilles de la Tourette, 1891–1895), or with the diagnosis of gait disturbances in neurology (Gilles de la Tourette, 1885b), the subject of his thesis. By contrast, those that have ensured his current fame were only considered as modestly original and innovative in his day. He himself seems to have attached little importance to them. Gilles de la Tourette disease disappeared almost totally from the medical literature from the early 20th century until the late 1950s. In the 1960s, the disease became a syndrome, firmly reassociated with Gilles de la Tourette's name after Arthur K. Shapiro (1923–95) and his wife Elaine Shapiro-Schlaffer (1925–2014), both American psychiatrists, successfully replicated Jean-Noël Seignot (1925-2022)'s observation that the disease could be treated by haloperidol (Seignot, 1961). In their article, which American medical journals refused to publish, they convincingly demonstrated the efficacy of haloperidol in calming the symptomatology of Gilles de la Tourette syndrome (Shapiro and Shapiro, 1968). Their work had a large impact, first in the community of American psychiatrists, and later among American neurologists. Shapiro thus became the instigator of a radical paradigm shift. This disease was henceforth seen as organic; psychotherapy, and to an even greater degree psychoanalysis, had no role in its treatment, aside from helping patients and their families tolerate its psycho-affective repercussions (Kushner, 1999).

Both Itard and Gilles de la Tourette "expressed their observations in beautiful, clear language, above all, in narratives, because one requires a narrative, as opposed to an itemization or checklist, to convey the full complexity and vicissitudes of a life" (Sacks, 1993). We believe that by publishing their work, Gilles de la Tourette and Charcot have pioneered a specialty still in the making, the neuroanthropology, studying not just an individual, but a society, the attitudes and reactions of which get internalized in each patient.

We conclude by noting that Gilles de la Tourette is a last name, and not a first name followed by a last name. As a result, the expression "Tourette syndrome" used in English-language articles, should be replaced with "Gilles de la Tourette syndrome."

References

Archives de Seine-et-Marne (1817). Chartrier du château de Guermantes, Dossier 159 J 28, pièce «À Monseigneur l'évêque de Meaux».

Beard GM (1878). Remarks upon "Jumpers or Jumping Frenchmen". J Nerva Ment Dis 5: 526.

Beard GM, Gilles de la Tourette G (1881). Les auteur's du Maine (États-Unis). Arch Neurol 2: 146–150.

Bouteille EM (1810). Traité de la chorée ou danse de St Guy. Paris: Chez Vinçard.

Breitman G (1888). Contribution à l'étude de l'écholalie, de la coprolalie et de l'imitation des gestes chez les dégénérés et les aliénés. Paris: Ollier-Henrythèse n°439.

Catrou J (1890). Étude sur la maladie des tics convulsifs (Jumping, Latah, Myriachit). Paris: H. Jouvethèse n°129.

Charcot JM (1887). Leçons du Mardi à La Salpêtrière. Policlinique 1887-1888. Paris: Bureaux du Progrès Médical et A. Delahaye & E. Lecrosnier.

Charcot JM (1889). Leçons du Mardi à La Salpêtrière. Policliniques 1888-1889. Paris: Bureaux du Progrès Médical et E Lecrosnier & Babé.

Charcot JM, Magnan V (1885). De l'onomatomanie. Arch Neurol 10: 157–168.

Charcot JM, Melotti G (1887). Lezione quattordicesima: intorno ad alcuni casi di tic convulsivo con coprolalia ed ecolalia. Analogia col Jumping di Beard, il Latah della Malesia ed il Miriachit di Hammond. Milano: Dottor Francesco Vallardi.

Eddy CM, Cavanna AE (2013). On being your own worst enemy: an investigation of socially inappropriate symptoms in Tourette syndrome. J Psychiatr Res 47: 1259–1263.

Gilles de la Tourette G (1884). Jumping, Latah, Myriachit. Arch Neurol 8: 68–74.

Gilles de la Tourette G (1885). Étude sur une affection, nerveuse caractérisée par l'incoordination motrice et accompagnée d'écholalie et de coprolalie. Arch Neurol 919-42/158-200.

Gilles de la Tourette G (1885). Études cliniques et physiologiques sur la marche, la marche dans les maladies du système nerveux étudiée par la méthode des empreintes (avec 31 figures). Paris: Société Typographiethèse n°84.

Gilles de la Tourette G (1891–1895). Traité clinique et thérapeutique de l'hystérie d'après l'enseignement de la Salpêtrière. Paris: E. Plon, Nourrit et Cie.

Gilles de la Tourette G (1899). La maladie des tics convulsifs. Sem Méd 19: 153–156.

Guinon G (1886). Sur la maladie des tics convulsifs. Rev Méd 6: 50–80.

Guinon G (1893). Attentat contre le Dr Gilles de la Tourette. Prog Méd 18: 446.

Itard JG (1825). Mémoire sur quelques fonctions involontaires des appareils de la locomotion, de la préhension et de la voix. Arch Gén Méd 3: 385–407.

Kurlan R, Daragjati C, Como PG et al. (1996). Non obscene complex socially inappropriate behavior in Tourette's syndrome. J Neuropsychiatry Clin Neurosci 8: 311–317.

Kushner H (1999). A cursing brain? The histories of Tourette syndrome. Cambridge: Harvard University Press.

Kushner HI, Luzzatti C, Finger S (1999). A perplexing document in the early history of Gilles de la Tourette Syndrome: Melotti's rendition of a "Lecture of Charcot" (including a complete translation from the Italian with commentary). J Hist Neurosci 8: 5–20.

Morel BA (1857). Traité des dégénérescence physiques, intellectuelles et morales de l'espèce humaine et des causes qui produisent ces variétés maladives. Paris: JB Baillière.

Ribot TH (1883). Les maladies de la volonté. Paris: Germer Baillière.

Sacks O (1993). Tourette's syndrome: a human condition. In: Kurlan R. Handbook of Tourette's syndrome and related tic and behavioral disorders. New York: Marcel Dekker.

Seignot JN (1961). Un cas de maladie de Gilles de la Tourette guéri par le R 1625. Ann Med Psychol (Paris) 119: 578–579.

Shapiro AK, Shapiro E (1968). Treatment of Gilles de la Tourette's Syndrome with haloperidol. Br J Psychiatry 114: 345–350.

Sumeire J (1790). Sur la danse de Saint-With. J Méd Chir Pharm 85: 23–25.

Walusinski O (2019). Georges Gilles de la Tourette, beyond the eponym. New York: Oxford University Press.

Walusinski O Georges Guinon (1859–1832). In: Les 33 internes de Jean-Martin Charcot à La Salpêtrière de 1862 à 1893. Brou (France): Oscitatio.

Walusinski O (2022). Historical background of the Gilles de la Tourette syndrome. Int Rev Mov Disord 3: 3–67.

Walusinski O, Féray JC (2020). The Marquise de Dampierre identified at last, the first described clinical case of Gilles de la Tourette syndrome. Rev Neurol (Paris) 176: 754–762.

Section I

Clinical diagnosis of tics

Handbook of Clinical Neurology, Vol. 215 (3rd series)
Tic Disorders and Tourette Syndrome
C. Ganos and Y. Worbe, Editors
https://doi.org/10.1016/B978-0-443-13554-5.00013-4

Chapter 2

Clinical presentation of tics and Gilles de la Tourette syndrome

ANDREA E. CAVANNA[1,2,*]

[1]*Department of Neuropsychiatry, BSMHFT and University of Birmingham, Birmingham, United Kingdom*

[2]*Department of Child Neuropsychiatry, School of Medicine and Surgery, University of Milano-Bicocca, Milan, Italy*

Abstract

Tics are the most common hyperkinetic manifestations during development. The clinical phenomenology of motor tics ranges from mild twitches affecting a single facial muscle to orchestrated contractions of different muscular districts resembling purposeful behaviors. Likewise, the repertoire of vocal tics (also called phonic tics) covers the whole spectrum between isolated grunting noises and meaningful strings of words. Simple and complex tics arguably sit on a continuum of symptom severity and respond to the same treatment interventions. The diagnosis of Gilles de la Tourette syndrome (GTS) is based on the presence of multiple motor tics plus at least one vocal tic, with onset before the age of 18 years and chronic course. It has been argued that the different tic disorders belong to a spectrum of increasing complexity, from the transient form (provisional tic disorder), through persistent motor or vocal tic disorder, to GTS. However, the clinical phenotype of GTS stands out because of the frequent association with specific behavioral problems, ranging from tic-related obsessive-compulsive disorder to other neurodevelopmental conditions. The diagnosis of tic disorders is based on clinical observation and requires expertise. The recent outbreak of functional tics, documented across several countries during the COVID-19 pandemic, introduced unprecedented challenges to the differential diagnosis of neurodevelopmental tics.

INTRODUCTION

It could be argued that tics are literally "here, there, and everywhere," as they are currently recognized as a relatively common variant of motor development. According to the Diagnostic and Statistical Manual of Mental Disorders, Fifth Edition, Text Revision (DSM-5-TR), tics are defined as sudden, rapid, recurrent, nonrhythmic movements or vocalizations (American Psychiatric Association, 2022). Tics have been documented in virtually all cultures, irrespective of ethnicity (Robertson et al., 2009), and affect males more frequently than females, with a ratio of 3:1 to 4:1 (Cavanna and Seri, 2013; Robertson and Eapen, 2014; Garcia-Delgar et al., 2022). Neurodevelopmental tics are the hallmark of primary tic disorders: Gilles de la Tourette syndrome (GTS), persistent (chronic) motor or vocal tic disorder, and provisional tic disorder. Taken together, tic disorders are the most common hyperkinetic movement disorders during the developmental age. According to a recent systematic review and meta-analysis covering the period between 1986 and 2022, the estimated global prevalence of GTS is 0.5% (0.7% in the population of children and adolescents) (Jafari et al., 2022). These findings are in line with the results of two similar meta-analyses conducted in the previous decade (Knight et al., 2012; Scharf et al., 2015). Epidemiologic studies that included transient tics have shown that up to 1 in 5 children (6–17 years old) exhibit tics at some time during their childhood (Kurlan et al., 2001). Finally, the prevalence of tics is higher (22%–27%) in children with special education needs related to autistic spectrum disorder or other neurodevelopmental conditions (Canitano and Vivanti, 2007).

HISTORICAL NOTE

Retrospective diagnoses of tic disorders have resulted in a number of reports, based on historical records (Pearce, 1994; Monaco et al., 2009; Murad, 2010; Cavanna et al., 2014) and fictional works (Cavanna et al., 2010b; Voss, 2012; Calder-Sprackman et al., 2014; Cavanna and Cavanna, 2014). Early descriptions of patients with tics in the medical literature date back to the 19th century. In particular, three contributions

*Correspondence to: Prof. Andrea E. Cavanna, MD, PhD, FRCP, FANPA, SFHEA, Department of Neuropsychiatry, National Centre for Mental Health, 25 Vincent Drive, Birmingham B15 2FG, United Kingdom. Tel: +44 121 3012317, E-mail: a.e.cavanna@bham.ac.uk.

marked the beginning of the modern scientific discourse on tic disorders. In 1825, Jean-Marc Gaspard Itard published the case report of the Marquise de Dampierre, a French noblewoman whose tic repertoire included obscene utterances (coprolalia) (Itard, 1825; Walusinski and Féray, 2020). In 1873, Armand Trousseau completed his description of the "tic non douloureux," which included the hereditary pattern – among other features that later became hallmarks of tics (Trousseau, 1873; Rickards et al., 2010). Finally, in 1885 Georges Gilles de la Tourette accomplished his mentor Jean-Martin Charcot's request by publishing a case series of nine patients affected by the "maladie des tics" (Gilles de la Tourette, 1885; Yorston and Hindley, 1998). The first case description in Gilles de la Tourette's series was actually a summary of Itard's previous report of the "cursing marquise" (Cavanna and Seri, 2019; Walusinski and Féray, 2020) (see Chapter 1).

In 1902, in Meige and Feindel's "Les tics et leur traitement," it was still possible to read that the "response to the question whether tics are hysterical in origin is a direct negative": tics were attributed to "hereditary weakness of the will" (Meige and Feindel, 1902). This view was soon replaced by psychoanalytic theories that stemmed from Freud's growing influence (Kushner, 1998). In 1921, Sándor Ferenczi proposed that tics were "a stereotyped expression of an onanistic equivalent" (Ferenczi, 1921). It was another Hungarian-born clinician, the pediatrician Margaret Mahler, who refined the psychoanalytic view of tics by updating Ferenczi's explanation for a new generation of American psychiatrists. In 1943, Mahler was already working in New York when she described the case of an 11-year-old boy with GTS: according to her, this patient's tics were the "expression of an emotional conflict, representing, on the one hand, partial gratification of repressed instinctual impulses, and on the other, defenses against these impulses" (Mahler and Rangell, 1943). Between the lines, it is possible to read a subtext of frustration about the widely popular psychoanalytic account of tic disorders, as 30 months of intensive psychotherapy ameliorated the patient's condition without fundamentally changing it. Interestingly, Mahler presented evidence "in support of the idea that there is an underlying substratum of organic disease of the central nervous system in the etiology of the disease in this case," raising the possibility that the organic basis of the patient's "incontinence of emotions" made the condition only partly accessible to psychotherapy (Mahler and Rangell, 1943).

The disappointing outcome of psychotherapy was only one of the many contributors to a "paradigm shift" in the medical perception and understanding of tic disorders (Kushner, 1999). With the therapeutic breakthrough marked by the use of Haloperidol in the treatment of tics, which was popularized by Arthur and Elaine Shapiro in 1968 (Shapiro and Shapiro, 1968), the time was ripe for psychoanalytic approaches to be replaced by neurobiologic models. Throughout the second half of the 20th century, tics were increasingly less likely to be seen as the conversion of suppressed instinctual impulses in an overprotective familial environment. Instead, progress in neuropharmacology and clinical genetics prompted a reappraisal of the concept of tic disorder as a relatively common hyperkinetic movement disorder with a complex heritability pattern (genetic heterogeneity). Since the turn of the millennium, important contributions have shaped the characterization of the neural correlates of both tic generation (insula, anterior cingulate cortex, supplementary motor area) and tic expression (cortico-striato-thalamo-cortical networks) (Cavanna et al., 2017; Ganos et al., 2022). An outdated psychoanalytic view of tic disorders has been replaced by a more promising, yet incomplete neurobiologic picture, meaning that clinical acumen and expertise remain the cornerstones of effective care pathways.

PHENOMENOLOGY OF MOTOR AND VOCAL TICS

Tics are listed in the group of hyperkinetic movement disorders, alongside tremor, myoclonus, chorea, and dystonia. Any single movement or sound that the human body is capable of making has the potential to become a tic. Intriguingly, it has been documented that people with tic disorders are sometimes able to control specific muscle contractions as part of their tic behaviors to a degree that most people without tics typically cannot (Ganos et al., 2021). Consequently, tics are characterized by a wide phenomenological variability, reflecting the range of normal voluntary motor behavior. Moreover, the lifetime tic repertoire is typically variable (the same individual will have different tics over time). From the neurophysiologic perspective, tics can be indistinguishable from voluntary actions (Hallett, 2001; Paszek et al., 2010). Indeed, most tics can be easily mimicked, and external observers often develop the misconception that they are produced "on purpose." However, a few subtle features can separate tics from goal-directed actions: these include their sudden and unexpected nature, that can excite surprise, their lack of flexibility ("stereotyped quality"), and their exaggerated character in both intensity and frequency. Moreover, voluntary actions are context-bound, whereas tics can be described as fragments of normal motor behavior that appear without any logical reason (misplaced in context and time) (Jankovic and Fahn, 1986; Ganos et al., 2014b).

Tics are best characterized based on their anatomic location, frequency, intensity or "forcefulness," complexity, and impact on the individual's day-to-day life. The Yale Global Tic Severity Scale, the most widely used instrument for the rating of tic severity (Leckman et al., 1989; Martino et al., 2017b; McGuire et al., 2018), includes separate ratings of the number, frequency, intensity, complexity, and interference associated with motor and vocal tics, as well as an overall impairment rating. The Tourette Syndrome Diagnostic Confidence Index (DCI) is another clinician-rated instrument that documents a number of characteristic features of tics. Specifically, the DCI focuses on those characteristics of tics, such as their complexity, their changeable nature, their subjective correlates, and the temporal features of their expression, that increase the likelihood of correctly diagnosing a primary tic disorder (Robertson et al., 1999).

Motor tics are often classed as either simple or complex. Simple motor tics are sudden, meaningless movements that involve single muscles or isolated muscle groups. Common examples include eye blinking, nose twitching, facial grimacing, shoulder shrugging, mouth movements, and head jerking. Fig. 2.1 presents a schematic representation of the most frequently reported motor tics, according to body location.

Complex motor tics are sudden, coordinated movements resembling goal-directed actions or gestures that appear inappropriately timed and intense (Jankovic, 1997). Examples are countless and include repetitive touching, jumping, twisting, and whole-body movements. Complex motor tics are rarely seen in the absence of simple motor tics. Overall, motor tics can also be described in terms of their duration. Although the majority of motor tics are brief ("clonic tics"), a few patients report isometric contractions, such as tensing of the abdominal muscles ("tonic tics") or slower, twisting movements that result in abnormal postures ("dystonic tics") (Jankovic and Stone, 1991; Jankovic, 1997). A minority of patients exhibit specific complex motor tics that have also been referred to as tic-related symptoms. Palipraxia denotes the repetition of certain movements or actions, usually for a given number of times or until they have been performed "just right" (Cath et al., 2001a). Echopraxia, the imitation of another person's actions, encompasses elements of early normal motor and social development that can pathologically persist in neurodevelopmental conditions (Ganos et al., 2012b). Copropraxia is the inappropriate and out-of-context making of obscene and often insulting gestures, depleted of intent (Freeman et al., 2009).

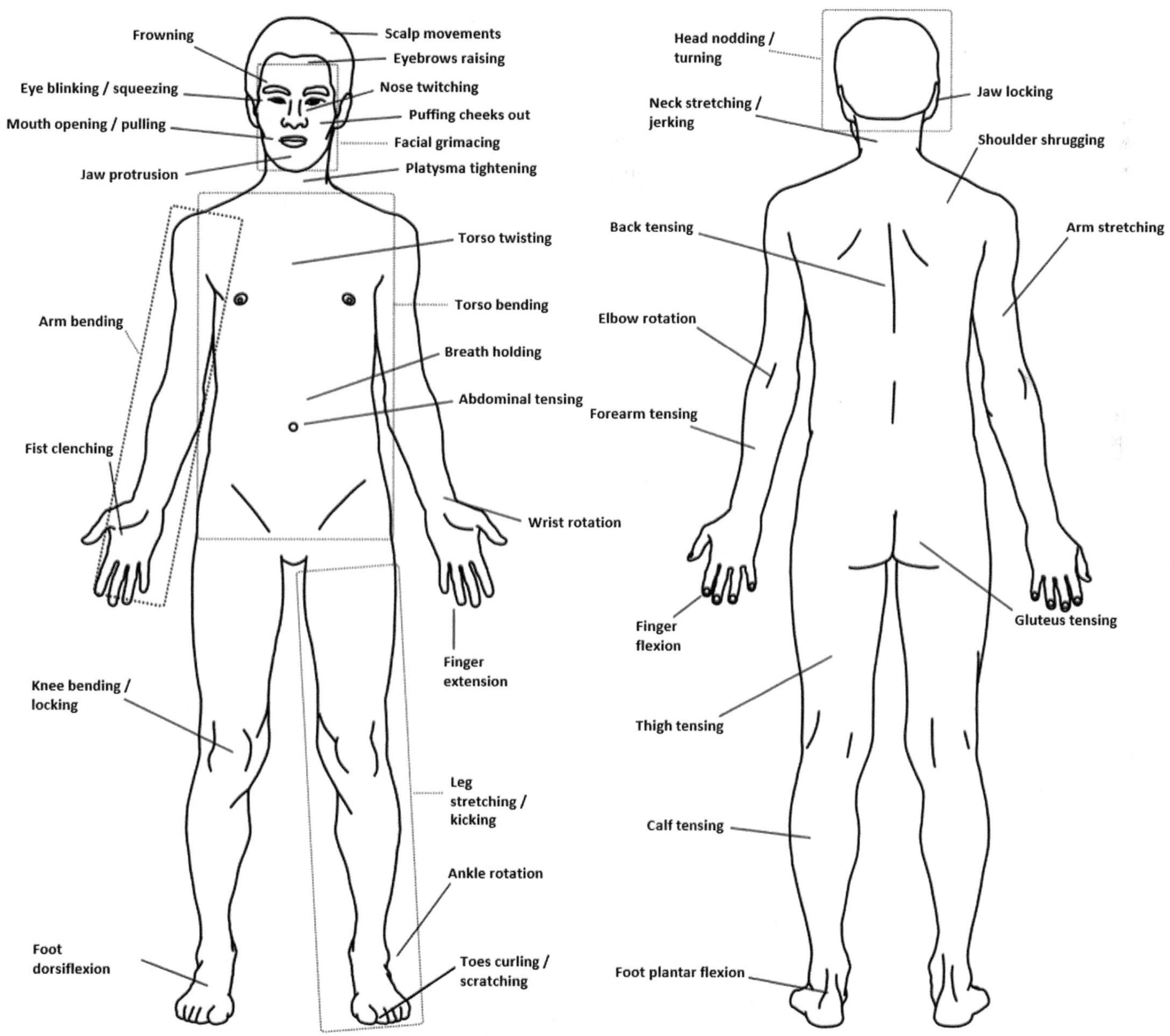

Fig. 2.1. Schematic representation of some of the most frequently reported motor tics, according to body location.

Pain and injuries as a result of severe motor tics are not rare, albeit often underreported (Stafford and Cavanna, 2020; Baizabal-Carvallo et al., 2022). The repetitiveness of tics is an important contributor to their potential for wear and tear at the level of the musculoskeletal system. Overall, self-injurious behaviors of different degrees of severity, sometimes overlapping with body-focused repetitive behaviors, can be present in up to 60% of patients with tic disorders (Mathews et al., 2004). Self-hitting, poking, biting, and other forms of tic-related self-injurious behaviors can also result in permanent physical injuries, including spinal cord damage and blindness from retinal detachment (Isung et al., 2021; Mito et al., 2023). More severe forms of tic-related self-injurious behavior have been found to correlate with measures of impulsivity (Mathews et al., 2004). Extreme cases of self-injurious tics leading to hospitalization, loss of bodily function, or permanent disfigurement have been documented (Stafford and Cavanna, 2020).

The classical distinction between motor and vocal tics has been questioned as somewhat arbitrary, because vocal tics are sounds caused by diaphragmatic and oropharyngeal muscle contractions (Jankovic, 1997). Simple vocal tics are abrupt, meaningless sounds or noises that are more appropriately referred to as phonic tics, as they are often produced without the involvement of the vocal chords. Throat clearing, grunting, coughing, sniffing, snorting, and squeaking are common simple phonic symptoms. Table 2.1 presents a brief compendium of some of the most common vocal tics.

Complex vocal tics are meaningful vocalizations ranging from syllables, through words, to entire phrases. Sudden changes in the rate, volume, intonation, and/or rhythm, among other odd patterns of speech, have also been classed as complex vocal tics. Pathologic laughter, the occurrence of episodic and contextually inappropriate outbursts of laughter, is a rare complex vocal tic that is understandably described as distressing and socially disabling (Cavanna et al., 2010a). It has to be noted that complex vocal tics are rarely, if ever, present in the absence of simple vocal tics and motor tics of any kind.

Table 2.1

Selection of some of the most commonly reported vocal tics.

Simple vocal tics	Complex vocal tics
Animal sounds (e.g., dog, pig)	Coprolalia
Car sounds (e.g., hooting, beeping)	Echolalia
Clicking	Isolated random words
Consonant sounds (e.g., "sh sh," "t t")	Muttering
Coughing	Nonobscene socially inappropriate words
Grunting	Palilalia
Gulping	Talking to oneself
Humming	
Moaning	
Shouting	
Sniffing	
Snorting	
Squeaking	
Throat clearing	
Vowel sounds (e.g., "ah," "ee")	

More complex vocal tics – sometimes referred to as tic-related symptoms – are palilalia (the repetition of words or phrases said by one's self) and echolalia (the repetition of words or phrases said by another person) (Ganos et al., 2012b). Finally, coprolalia defines inappropriate and out-of-context swearing. Coprolalia has traditionally received much attention in popular culture; however, it is relatively uncommon, affecting up to one-third of patients with tic disorders (Freeman et al., 2009; Eddy and Cavanna, 2013a). The presence of coprolalia correlates with tic severity and number of psychiatric comorbidities, suggesting that it is a feature of more severe forms of tic disorders (Eapen and Robertson, 2015).

Coprophenomena result in the gestural (copropraxia) or verbal (coprolalia) expression of obscenities, whereas nonobscene socially inappropriate behaviors (NOSIBs) involve the expression of other kinds of socially inappropriate gestures, words, or phrases (Kurlan et al., 1996; Eddy and Cavanna, 2013c). Engaging in socially inappropriate comments toward family members, familiar individuals, or strangers, is a common example of NOSIB. A particular type of tic-related behaviors that share the inability to refrain from acting out undesired behaviors, despite awareness of negative implications, has been recently described and named intrusive-destructive behaviors (IDBs) (Greenberg et al., 2021). IDBs are diagnosed when one feels simultaneously compelled to and unable to stop oneself from performing the "worst" possible action in a particular situation. IDBs are conceptualized as a hybrid of complex tics and obsessive-compulsive symptoms atop an overlay of disinhibition and impulsivity (reward-seeking). For example, shouting "bomb" in a loud, explosive manner while in sensitive environments such as airports would be a NOSIB that fulfils the more stringent criteria for IDB. Of note, IDBs expand on the concept of NOSIBs by also including acts directed solely toward the self ("self-sabotage behaviors"), such as turning earphones to maximum volume. Over the last few years it has become apparent that such phenomena might also be seen in the context of functional disorders and it is difficult to establish phenomenologic boundaries between the different etiologies.

Tics that lead to disruption of ongoing action (motor tics) or interfere with communication (vocal tics) are labeled as "blocking" (Ganos et al., 2014b). Blocking tics are a particular type of complex tics reported by a minority of patients and tend to be associated with the presence of dystonic tics (Baizabal-Carvallo and Jankovic, 2023) and with more severe forms of tic disorders (Kaczyńska and Janik, 2021). Finally, it is a common clinical finding that multiple types of tics frequently coexist and can occur in paroxysmal orchestrated sequences, with displays of motor and vocal tics (e.g., facial grimacing followed by blowing, hopping accompanied by squeaking sounds).

PREMONITORY URGES

Individuals are not always consciously aware of all their tics (Schubert et al., 2021). Conscious awareness is facilitated by a distressing somatic sensation that precedes tic expression, known as "premonitory urge" (Kwak et al., 2003; Cavanna and Nani, 2013; Rajagopal et al., 2013; Eddy and Cavanna, 2014a; Brandt et al., 2023). The vast majority of patients with tic disorders report specific premonitory urges in association with their tics; however, tics can occasionally be perceived as coming "out of the blue," especially in younger patients, as awareness of the presence of premonitory urges increases with age (Patel et al., 2014; Cox et al., 2018). It is not uncommon for patients with tic disorders to attribute voluntariness to tic generation in responses to urge (Lang, 1991). In fact, tics have been described as resulting from a willful capitulation to an irresistible urge (Bliss, 1980). Tic expression usually relieves the urge for a limited amount of time, after which the urge starts building up again ("tic cycle"). A few patients report generalized urges that are best captured by a sense of mounting inner tension. However, it has been found that premonitory urges are often focal in distribution and localized to specific anatomic regions (Leckman et al., 1993). Fig. 2.2 presents a body map depicting the anatomic location and density of the premonitory urges according to the two largest studies conducted to date (Leckman et al., 1993; Essing et al., 2022). The localization of urges does not appear to be random: they are most frequent in the head and shoulders and tend to be located in the same body areas as, or in direct proximity to, the corresponding tics (Leckman et al., 1993; Essing et al., 2022).

In addition to physical tension, pressure, and energy that needs to be released, premonitory urges are sometimes described in terms of feelings of incompleteness or not-just-right experiences (Prado et al., 2008). These latter are driven by a need to have movements, actions, objects, and sounds that feel a certain "just right" way. Of note, not-just-right experiences overlap with compulsions, although they seem to be more closely related to tic disorders than to obsessive-compulsive and related disorders (Worbe et al., 2010; Neal and Cavanna, 2013a). Based on the presence of the premonitory urge, it has been suggested that tics are semi-voluntary or "unvoluntary" movements, as opposed to either voluntary or involuntary movements (Jankovic, 1997). Unvoluntary movements have been defined as "actions that are perceived as voluntary responses to uncontrolled and involuntary urges to move," thus bearing more resemblance to a cough or a sneeze than to voluntary actions (Cavanna and Nani, 2013).

Of note, premonitory urges are not the only bothersome sensory phenomena experienced by patients with tic disorders, as over 80% of patients with GTS report subjectively enhanced perception of various sensory stimuli (sensory hypersensitivity) (Isaacs and Riordan, 2020). Sensory hypersensitivity is defined as a constant, heightened awareness of external and/or internal stimuli. Among the different sensory modalities, altered perception of specific auditory stimuli has been investigated in the context of tic disorders, with reports of patients with GTS affected by distressing symptoms of misophonia (selective sound sensitivity syndrome) (Neal and Cavanna, 2013b; Cavanna, 2014b; Cavanna and Seri, 2015; Robinson et al., 2018).

TIC-MODULATING FACTORS AND NATURAL COURSE

Tics naturally tend to occur in bouts and are characterized by spontaneous fluctuations over time ("waxing and waning course") (Peterson and Leckman, 1998). Moreover, tic severity can be modulated by a number of internal and environmental factors (Barnea et al., 2016). Relaxed concentration – focused

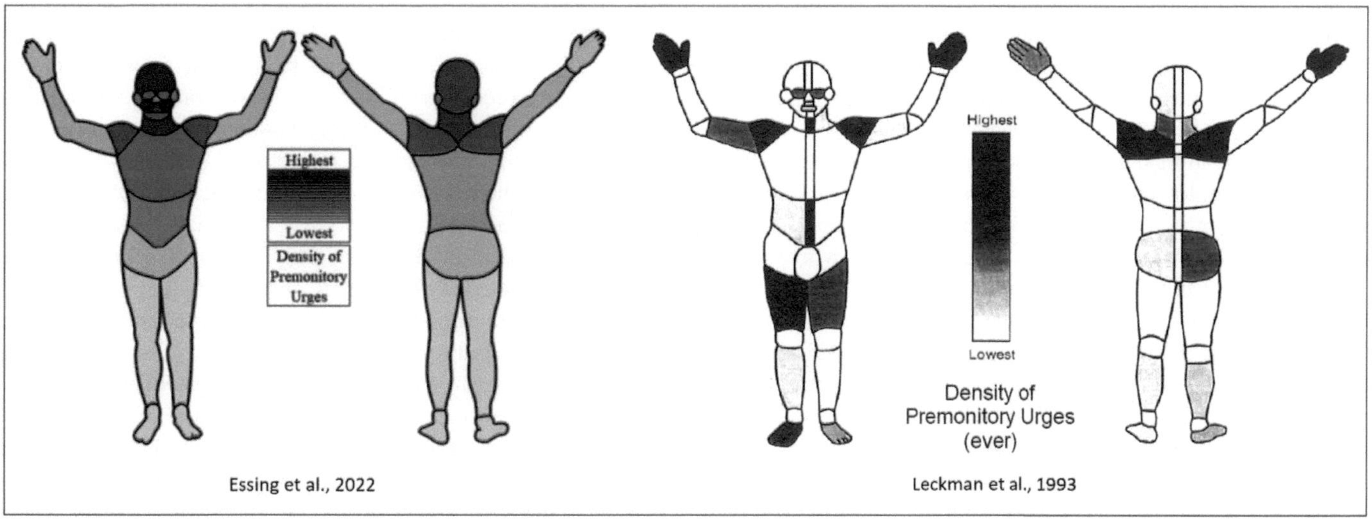

Fig. 2.2. Density of premonitory urges according to body location in patients with tic disorders: findings of the studies by Essing et al. (2022) (291 adults) and Leckman et al. (1993) (135 children and adults).

attention when the individual is engaged in nonstressful mental or physical activity – seems to be an alleviating factor, whereas stress, self-consciousness, tiredness, and emotional excitement have all been reported as common exacerbating factors (Conelea and Woods, 2008; Iverson and Black, 2022) Overall, tics are typically distractible and suggestible: discussing tics or seeing someone else's tics can induce the same tics (Leckman, 2003). Finally, tics can often be suppressed – at least partially – for variable lengths of time. The ability to suppress tics varies between individuals, and between different situations within the same individual (Ganos et al., 2012a; Conelea et al., 2018). The same patients who struggle to suppress their tics in social situations might easily succeed while engrossed in demanding tasks within their comfort zone. Moreover, the ability to suppress tics is not homogeneous, but varies between different body parts: tics in more commonly represented anatomic locations, such as the eyes, are apparently less likely to be amenable to effective suppression (Ganos et al., 2015). Individuals often report that voluntary tic suppression results in a "buildup" of inner tension, followed by a sense of relief when the tics are finally performed with increased frequency and/or severity ("tic rebound"); however, these observations lack experimental confirmation (Specht et al., 2013).

In most cases, the onset of neurodevelopmental tics is between 5 and 7 years of age, with motor tics generally appearing earlier than vocal tics (Freeman et al., 2000; Khalifa and von Knorring 2003; Cavanna and Seri, 2013; Ganos et al., 2014b). The most common tics at onset are simple motor tics affecting the eyes and face. Specifically, eye blinking is often the first tic to emerge (Martino et al., 2012). In general, eye tics are extremely common, as ocular involvement has been estimated to occur in over 95% of individuals with tic disorders (Martino et al., 2012). Other common motor tics at onset affect the cranial (e.g., facial grimacing, nose twitching) and cervical (e.g., neck jerks) regions (Leckman et al., 1998). The appearance of additional tics typically progresses according to a rostro-caudal pattern, with facial and head tics appearing first, followed by trunk and limb tics, and subsequently by simple vocal tics such as grunting, throat clearing, coughing, sniffing, and squeaking (Leckman et al., 1998). Vocal tics typically occur a few years after the onset of motor tics (Leckman, 2003). With age, both motor and vocal tics tend to be more complex, although not all individuals with tic disorders develop complex tics (Bloch and Leckman, 2009). Tics usually worsen for several years until they reach their maximum severity by early adolescence (Bloch and Leckman, 2009). Most patients report a gradual improvement starting from late adolescence or early adulthood. About one-third of patients experience complete remission (Kompoliti, 2016). The remaining two-thirds of individuals continue to have tics in adulthood, often with mild severity. However, in a subset of patients (up to 20%), tic severity may not improve. These patients continue to have significant symptoms throughout their adult life (Pappert et al., 2003; Groth et al., 2017).

Relatively little is known about the predictors of tic persistence and tic severity in adulthood (Hassan and Cavanna, 2012) (see Chapter 13). Higher tic severity, poorer fine motor skills, lower caudate volumes, and higher levels of psychosocial stress, as well as the presence of psychiatric comorbidities during childhood, have all been found to be predictive of poorer prognosis (Bloch et al., 2005, 2006; Lin et al., 2007; Bloch et al., 2011; Rizzo et al., 2012). Preliminary data suggest that youngsters with severe tics associated with characteristic premonitory urges and a family history of tic disorders are at higher risk for poorer health-related quality of life as adults (Cavanna et al., 2012). However, the complexities of the clinical phenotypes and their variable presentation throughout the lifespan, particularly during the transition from childhood to early adulthood, do not allow reliable predictions of future tic severity (Silvestri et al., 2016).

DIFFERENTIAL DIAGNOSIS

In most cases, the diagnosis of a neurodevelopmental tic disorder can be confirmed by skilled clinicians after observing the phenomenology and obtaining a history focused on the characteristic features of tics (including possible family history) and comorbid conditions. In order to establish the diagnosis of a tic disorder, it is important to avoid potential diagnostic confounders, including normal behaviors (e.g., habits), pathologic behaviors (e.g., compulsions), and other hyperkinetic movement disorders (e.g., stereotypies) that can be mistaken for tics, especially during childhood (Sanger et al., 2010). Diagnostic uncertainty is more likely to arise with a few particular tics. For instance, among motor tics, eye-rolling and blinking tics can appear similar to certain seizures. Among vocal tics, sniffing tics may be misinterpreted as reactions to allergies, whereas coughing tics may be misattributed to gastric reflux, as well as allergies. Tics that affect the limbs and occur preferentially at bedtime before falling asleep can be mistaken for restless leg syndrome, periodic limb movements, frontal lobe epilepsy, or sleep myoclonus (Voderholzer et al., 1997; Kompoliti and Goetz, 1998).

Specific clinical features can be helpful in the differential diagnosis between tics and other types of abnormal movements. Clonic tics often include brief, sudden, jerky, or shock-like movements that resemble myoclonus ("myoclonic tics"). Despite the phenomenologic similarities, myoclonic tics are unlikely to share pathophysiology with other forms of myoclonus. Crucially, myoclonus is typically not suppressible and may be associated with epilepsy, especially in young patients. Likewise, although "dystonic tics" involve briefly sustained postures, it is unlikely that these rarer types of tics share pathophysiologic aspects with other forms of dystonia. The differential diagnosis between a dystonic tic and dystonia can be challenging in case of facial involvement and in the context of specific disorders such as neuroacanthocytosis, where tics and dystonia can co-occur. Focal types of dystonia can be differentiated by the absence of premonitory urges and suppressibility (Kompoliti and Goetz, 1998; Ganos et al., 2014b). Dystonia is usually triggered by and interferes with voluntary movement, whereas tics can be suppressed by and do not necessarily interfere with voluntary movement. Motor tics differ from chorea and athetosis because of the lack of continuity of the movement and the presence of intervening periods of normal movement. Moreover, most tics cause less interference with ongoing tasks. Tics can also be distinguished from chorea based on their predictability and repeatability (stereotyped nature), in addition to their association with premonitory urges and suppressibility. Motor tics can be distinguished from tremor and stereotypies by the clear initiation and termination of each individual tic, and by the lack of rhythmicity (although certain tics may include rhythmic shaking). Moreover, some of the most common stereotypies such as arm flapping or truncal rocking are situation-specific, self-gratifying repetitive movements that are not driven by premonitory urges. The differential diagnosis between tics and stereotypies can be particularly challenging in individuals with multiple neurodevelopmental conditions, including both autism spectrum disorders and tic disorders (Ubhi et al., 2020; Termine et al., 2021). Finally, secondary movement disorders can be mistaken for tics when they affect patients with primary tic disorders (Ganos et al., 2014a). In a large videotape-based study of patients referred to a tertiary movement disorders clinic for suspected exacerbation of their neurodevelopmental tics, about one patient in 20 was eventually diagnosed with a nontic movement disorder; the majority were secondary disorders associated with use of first-generation antidopaminergic agents, mainly akathisia, tardive dystonia, and acute dystonic reaction (Kompoliti and Goetz, 1998).

FUNCTIONAL TICS

During the COVID-19 pandemic, there has been a worldwide increase in adolescents and young adults presenting acutely or subacutely with functional neurologic symptoms resembling tics. Before the pandemic, functional tics were considered to be relatively rare, compared to other clinical manifestations within the broader spectrum of functional movement disorders, such as functional dystonia and functional tremor (Baizabal-Carvallo and Jankovic, 2014; Baizabal-Carvallo and Fekete, 2015; Demartini et al., 2015; Ganos et al., 2016, 2019; Cavanna et al., 2023a) (see Chapter 5). Interestingly, a fictional description of functional tics can be traced back to 1884, the year before Gilles de la Tourette's original report of the eponymous condition characterized by multiple neurodevelopmental tics (Cavanna, 2022). However, since 2020 an unprecedented number of referrals of patients with functional tics have been received by health services, with a strong preponderance of young females. Specifically, a large population-based study conducted in England documented a more than fourfold increase in teenage girls who developed tic-like behaviors, often in association with anxiety and/or other psychiatric comorbidities (Jack et al., 2023). A significant female preponderance was consistently documented across the majority of case series (Hull et al., 2021; Buts et al., 2022; Cavanna et al., 2023c) and controlled studies (Paulus et al., 2021; Pringsheim and Martino, 2021; Pringsheim et al., 2021; Han et al., 2022; Trau et al., 2022; Andersen et al., 2023; Baizabal-Carvallo et al., 2023; Cavanna et al., 2023b) conducted in specialist clinics for tic disorders. The existence of specific demographic and phenomenologic differences between functional tics and neurodevelopmental tics was confirmed by the clinical data from ten specialist centers across North America, Australia, and Europe, collated in an international registry established during the COVID-19 pandemic (Martino et al., 2023). In addition to the female gender predominance, functional tics are more likely to have an abrupt and explosive presentation of severe symptoms, later age of onset, and a higher rate of obscene and socially inappropriate symptoms (Amorelli et al., 2022; Fremer et al., 2023). Compared to patients with neurodevelopmental tics, patients with functional tics are more likely to have a lower rate of comorbid neurodevelopmental disorders, such as attention-deficit and hyperactivity disorder, as well as lower rates of comorbid obsessive-compulsive disorder (Amorelli et al., 2022; Fremer et al., 2023). High rates of comorbid anxiety and depression can be found in both patient populations, although patients with functional tics are more likely to report previous traumatic experiences (Cavanna et al., 2022b). Based on these findings, a group of experts in tic disorders from the European Society for the Study of Tourette Syndrome published a set of criteria for the clinical diagnosis of functional tics (Pringsheim et al., 2023) (Table 2.3).

In clinical practice, the differential diagnosis between neurodevelopmental tics and functional tics can pose considerable challenges, even to experienced clinicians (Rigas et al., 2023). The diagnostic process can be complicated by similarities in the clinical features and associated behavioral symptoms (Kurvits et al., 2021), as well as by the possibility that both types of tics coexist in the same individual, either simultaneously or at different times (Cavanna et al., 2022a). Little is known about the cause(s) of the recently observed global increase in functional tics. The direct and indirect effects of the COVID-19 pandemic, including increased exposure to social media content by influencers portraying tic-like behaviors, have been listed as possible precipitating factors for what has been referred to as a "pandemic within a pandemic" (Hull and Parnes, 2021; Olvera et al., 2021; Frey et al., 2022; Martindale and Mink, 2022; Nilles et al., 2022; Janiri et al., 2023). The term "mass social media-induced illness" has been proposed for the cases of young individuals who present with a repertoire of functional tics that are similar or identical to the tic-like behaviors portrayed by popular social media influencers (Müller-Vahl et al., 2022; Fremer et al., 2023). According to this model, functional tics can be regarded as a modern form of the well-known motor variant of mass sociogenic illness (Müller-Vahl et al., 2022).

DIAGNOSTIC CRITERIA FOR TIC DISORDERS

There is no laboratory or imaging test available to diagnose tic disorders: the diagnosis is clinical and rests on the implementation of validated criteria. Tic disorders are listed in the current edition of the American Psychiatric Association's Diagnostic and Statistical Manual for Mental Disorders (DSM-5-TR) among the motor disorders of developmental origin, alongside stereotypic movement disorder and developmental coordination disorder (American Psychiatric Association, 2022). In addition to motor disorders, the DSM-5-TR classification of neurodevelopmental disorders includes autism spectrum disorder, attention-deficit and hyperactivity disorder, intellectual disability, and communication disorder.

In continuity with its previous version, the DSM-5-TR lists three main types of tic disorders: GTS, persistent motor or vocal tic disorder, and provisional tic disorder (Cavanna, 2014a; American Psychiatric Association, 2022). The diagnostic criteria for GTS require multiple motor tics plus at least one vocal tic to be present at some time during the illness (but not necessarily concurrently). Moreover, the criteria stipulate that tic onset occurs before the age of 18 years and tics persist for at least one year despite fluctuations. Finally, tics are not explained by the effects of a substance or another medical condition. The diagnosis of persistent motor or vocal tic disorder requires single or multiple vocal or motor tics (but not both). As in GTS, the criteria stipulate that tic onset occurs before the age of 18 years, tics persist for at least one year despite fluctuations and are not explained by the effects of a substance or another medical condition. Provisional tic disorder, previously known as transient tic disorder, is defined by childhood onset of one or more motor or vocal tics that are present for less than 12 months in a row and are not attributed to another medical condition. The use of the term provisional reflects the working nature of the diagnosis, whereas a tic disorder can be classed as transient only in retrospect, when the tics cease. For both persistent and provisional tic disorders, there is the requirement that the patient does not meet diagnostic criteria for GTS at any time. The current diagnostic criteria for neurodevelopmental tic disorders are summarized in Tables 2.2 and 2.3.

Recent findings support the hypothesis that persistent motor or vocal tic disorder and GTS occur along a clinical continuum, in which GTS is a more severe condition and persistent motor or vocal tic disorder a less severe manifestation of a continuous neurodevelopmental tic spectrum disorder. A retrospective chart review of a large sample of patients with chronic tic disorders of all ages showed that the demographic and clinical characteristics of patients with a diagnosis of persistent motor tic disorder were similar to those of patients with GTS, with few significant differences (Müller-Vahl et al., 2019). Patients with persistent motor tic disorder had lower tic severity, lower prevalence of complex motor tics (including copropraxia and echopraxia), and fewer comorbid conditions, compared to patients with GTS. Based on these findings, the term "tic spectrum disorders" was introduced (Müller-Vahl et al., 2019). The hypothesis that persistent motor or vocal tic disorder is a *forme fruste* of GTS was further supported by the findings of a comprehensive study conducted by the Tourette Association of America International Consortium for Genetics and the OCD Collaborative Genetics Association (Claudio-Campos et al., 2021). Specifically, there is evidence that age of tic onset is a few years later, and tic severity is lower in persistent motor or vocal tic disorder than in GTS. Moreover, consistent findings indicate that rates of comorbid obsessive-compulsive disorder and attention-deficit and hyperactivity disorder are lower in persistent motor or vocal tic disorder than in GTS. In turn, tic-related obsessive-compulsive disorder and attention-deficit and hyperactivity disorder have been found to increase the odds of developing comorbid anxiety and affective symptoms, as well as substance misuse and disruptive behaviors (Robertson et al., 2015). These clinical observations account for the observed differences between persistent motor or vocal tic disorder and GTS.

BEHAVIORAL COMORBIDITIES: THE "GILLES DE LA TOURETTE SYNDROME SPECTRUM"

GTS is often referred to as a quintessentially neuropsychiatric disorder because, in addition to tics, comorbid behavioral symptoms are present in up to 90% of patients (Hirschtritt et al., 2015; Martino et al., 2017a; Cavanna, 2018a, 2018b). According to a large study of patients with GTS who present for clinical care, about 58% have two psychiatric comorbidities (Freeman et al., 2000; Khalifa and von Knorring 2003; Hirschtritt et al., 2015). Based on the clinical spectrum of comorbidity, it has been suggested the GTS spectrum could be usefully divided into three different phenotypes: "pure GTS," consisting of motor and vocal tics only; "full-blown GTS," which includes the presence of coprophenomena, echophenomena, and paliphenomena; and "GTS-plus," with coexisting obsessive-compulsive disorder, attention-deficit and hyperactivity disorder, self-injurious behaviors, anxiety, affective disorders, or personality disorders (Packer, 1997). A number of studies using different analytic approaches (mainly principal

Table 2.2

Diagnostic criteria for neurodevelopmental tic disorders according to the Diagnostic and Statistical Manual for Mental Disorders, Fifth Edition, Text Revision (American Psychiatric Association, 2022).

Tic disorder	Number of motor tics	Number of vocal tics	Tics persisting for more than 1 year	Onset before 18 years of age
Gilles de la Tourette syndrome	2+	1+	Yes	Yes
Persistent motor tic disorder	1+	0	Yes	Yes
Persistent vocal tic disorder	0	1+	Yes	Yes
Provisional tic disorder	0 or 1+[a]	0 or 1+[a]	No	Yes

[a] Patients with provisional tic disorder present with at least one motor and/or vocal tic.

Table 2.3

European Society for the Study of Tourette Syndrome criteria for the clinical diagnosis of functional tics (clinically definite diagnosis of functional tics confirmed by the presence of all three major criteria; clinically probable diagnosis of functional tics confirmed by the presence of two major criteria and one minor criterion) (Pringsheim et al., 2023).

Criterion	Domain	Definition
Major criterion 1	Age at symptom onset	Age at first symptom onset of 12 years or older
Major criterion 2	Rapid evolution of symptoms	Rapid onset and evolution of symptoms (over hours to days)
Major criterion 3	Phenomenology	At least 4 of 9 phenomenologic features: - Multiple tic-like movements and/or vocalizations, with a larger number of complex than simple manifestations - The same tic-like behavior is variably reproduced - Motor tic-like behaviors include complex arm and hand movements (e.g., banging, hitting, throwing, rude gestures, etc.) and can be context-dependent or potentially leading to injury to self or others - Motor tic-like behaviors do not to follow the typical rostro-caudal progression in their first appearance - Vocal tic-like behaviors include several words and statements, and can be context-dependent or potentially offensive - Tic-like behaviors resemble popular cultural influences or patient's acquaintances - Patients experience a large variation in symptom frequency and intensity over the course of a single day, with symptom-free activities for several hours, and severe symptoms in specific contexts - Tic-like behaviors change rapidly, with the constant onset of new ones on a daily basis or every few days - The examining clinician observes an increase in tic-like behaviors during the patient's physical examination
Minor criterion 1	Comorbidity profile	Comorbid psychiatric symptoms and diagnoses, with a predominance of preceding anxiety and depressive symptoms, self-harm, suicidal ideation, preceding traumatic/stressful life events, and difficulties with social relationships
Minor criterion 2	Other functional neurologic symptoms or somatic symptom disorders	History or concurrent presence of other functional neurologic symptoms (e.g., functional seizures) or somatoform disorders (e.g., functional abdominal pain)

component factor analysis, hierarchical cluster analysis, and latent class analysis) have revealed different factors/clusters/classes of symptoms that tend to co-occur in the same patient. This work paved the way to the identification of multiple clinical phenotypes that are currently subsumed under the umbrella term of GTS (Cavanna and Rickards, 2013; Martino et al., 2013; Cravedi et al., 2018). Although these phenotypes vary depending on methodology and patient populations, most studies replicated selected findings and therefore confirmed the presence of the "pure GTS" phenotype. This phenotype is also referred to as "uncomplicated GTS" and includes the 10% of patients who do not present with comorbid psychiatric disorders (Freeman et al., 2000; Khalifa and von Knorring 2003; Cavanna and Rickards, 2013; Hirschtritt et al., 2015). As for the remaining patients, who represent the vast majority of individuals with GTS seen in specialist clinics, the most common behavioral comorbidities are tic-related obsessive-compulsive disorder and attention-deficit hyperactivity disorder, with an estimated 72% of individuals with GTS receiving a diagnosis of either disorder (Robertson, 2000; Hirschtritt et al., 2015). According to the developmental trajectory of behavioral comorbidities, attention-deficit hyperactivity symptoms tend to appear first, followed by obsessive-compulsive behaviors, anxiety, and affective symptoms (Hirschtritt et al., 2015). These observations are of great practical value, as behavioral comorbidities often have a greater detrimental effect on health-related quality of life than tics themselves (Bernard et al., 2009; Cavanna et al., 2013c, 2013d; Silvestri et al., 2016; Huisman-van Dijk et al., 2019; Isaacs et al., 2021). Patients with GTS plus comorbid conditions have been found to have an increased risk of mortality by natural or unnatural causes (such as suicide) when compared with patients with GTS only, although the contributing factors are yet to be fully delineated (Meier et al., 2017).

TIC-RELATED OBSESSIVE-COMPULSIVE BEHAVIORS

Tic-related obsessive-compulsive behaviors are reported by a considerable proportion (20%–60%) of patients with GTS (Freeman et al., 2000). Some of these patients fulfill validated diagnostic criteria for obsessive-compulsive disorder, whereas others present with subthreshold obsessive-compulsive behaviors (Cavanna, 2018a, 2018b). Despite the wide variability in estimates, these figures are consistently higher than those reported in otherwise healthy children and adolescents (0.5%–3.6%) (Mol Debes, 2013). Beyond the epidemiologic data, there is increasing evidence for genetic and pathophysiologic associations between GTS and obsessive-compulsive disorder (Pauls et al., 1986; Mathews and Grados, 2011). The link between the two conditions was formally acknowledged in the fifth edition of the DSM (DSM-5, 2013), which included tic-related obsessive-compulsive disorder as a specifier for the obsessive-compulsive disorder category (American Psychiatric Association, 2022).

Interestingly, tic-related obsessive-compulsive disorder (also referred to as "Tourettic obsessive-compulsive disorder") is characterized by phenomenologic features that are clinically different obsessive-compulsive disorder alone (Frankel et al., 1986; George et al., 1993; Mula et al., 2008; Eddy and Cavanna, 2014b; Hartmann and Millet, 2018; Szejko and Müller-Vahl, 2021; Katz et al., 2022). Patients with obsessive-compulsive disorder in the context of GTS commonly report checking and touching rituals, often driven by "just right" experiences, as well as obsessional counting (arithmomania), concerns with symmetry and evening-up behaviors (Cath et al., 2001a, 2001b; Worbe et al., 2010). By contrast, a relatively high number of patients with pure obsessive-compulsive disorder report obsessions concerning dirt and fears of contamination, resulting in compulsions that involve ordering, arranging, or washing and cleaning compulsions (Frankel et al., 1986; George et al., 1993; Mula et al., 2008; Eddy and Cavanna, 2014b; Szejko and Müller-Vahl, 2021; Katz et al., 2022). In general, compulsive behaviors in patients with pure obsessive-compulsive disorder are typically preceded by intrusive and disruptive thoughts that are perceived as anxiety-provoking and are qualitatively different from the premonitory urges that often drive tic expression (Miguel et al., 1997, 2000). Moreover, the obsessional thoughts associated with compulsions are subjectively perceived as intrusive or "egodystonic," whereas the premonitory urges associated with tics are more likely to be perceived as "egosyntonic" (Eddy and Cavanna, 2014b; Szejko and Müller-Vahl, 2021). By comparison, other repetitive behaviors of neurodevelopmental origin, such as complex motor stereotypies, are characteristically driven by reward-seeking processes related to specific situations (self-soothing function, "stimming," etc.). The spectrum of the subjective correlates of these ritualistic behaviors is graphically represented in Fig. 2.3.

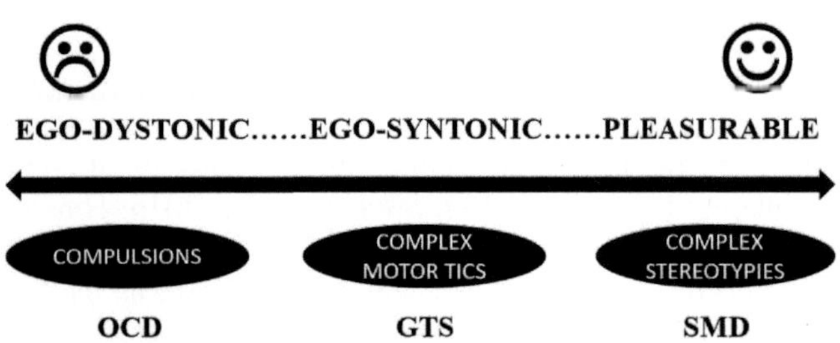

Fig. 2.3. Subjective correlates of different types of repetitive behaviors: compulsions (obsessive-compulsive disorder, OCD), complex motor tics (Gilles de la Tourette syndrome, GTS), and complex stereotypies (stereotypic movement disorder, SMD).

The differential diagnosis between compulsions and complex motor tics displayed by the same patient is not always easy, because of the many similarities and shared features. Like GTS, obsessive-compulsive disorder is characterized by onset in childhood and waxing and waning symptoms. However, tic-related obsessive-compulsive behaviors usually appear a few years after tics onset (Bloch and Leckman, 2009; Wanderer et al., 2012). Finally, the presence of comorbid obsessive-compulsive behaviors or disorder in patients with GTS is associated with a higher rate and an earlier onset of other comorbidities, including attention-deficit and hyperactivity disorder, autism spectrum disorder, anxiety, and affective symptoms (Wanderer et al., 2012).

OTHER BEHAVIORAL COMORBIDITIES

Attention-deficit and hyperactivity disorder is probably the most common behavioral comorbidity, particularly at young ages, affecting approximately 60% of patients with GTS (range 21%–90%), compared with 2%–12% of otherwise healthy youngsters (Freeman et al., 2000; Roessner et al., 2007a; Mol Debes, 2013; Cavanna, 2018a, 2018b). Symptoms characteristically include inattentiveness, distractibility, hyperactivity, and impulsivity, as well as features such as fidgetiness, that can easily be mistaken for tics. The symptoms of attention-deficit and hyperactivity disorder often precede the onset of tics (Bloch and Leckman, 2009). Their presence in young patients with GTS has been associated with male sex, earlier GTS diagnosis, and the development of further comorbidities, especially obsessive-compulsive disorder, anxiety, and affective symptoms (Freeman et al., 2000; Roessner et al., 2007a; Roessner et al., 2007b; Ganos and Martino, 2015). Moreover, the symptoms of attention-deficit and hyperactivity disorder may contribute to problems with executive function and poor academic performance, as well as anger control problems and other behavioral disturbances (Robertson, 2006a). Children with GTS and comorbid attention-deficit and hyperactivity disorder report an overall poorer social adaptation compared to those with uncomplicated GTS (Stephens and Sandor, 1999; Carter et al., 2000; Sukhodolsky et al., 2003). A significant problem in the investigation of cognitive functions in patients with tic disorders is the need to disentangle the differential impact of tics and associated comorbidities, especially attention deficit-hyperactivity disorder (Cavanna et al., 2009). Subtle alterations at the level of specific cognitive domains seem to be intrinsic to GTS. These might include executive dysfunction (mild deficits in attention and inhibition-related processes) (Eddy et al., 2012) and social cognition (unconventional responses to social information, rather than a failure to attribute mental states) (Eddy et al., 2010, 2011; Eddy and Cavanna, 2013b).

Neurodevelopmental conditions are notoriously characterized by high comorbidity rates (Canitano and Vivanti, 2007; Kim et al., 2023), because of shared genetic and pathophysiologic factors (State, 2010; Clarke et al., 2012; Fernandez et al., 2012). A significant proportion of patients with GTS, especially those with comorbid attention-deficit and hyperactivity disorder and/or obsessive-compulsive disorder, fulfill diagnostic criteria for autism spectrum disorder (up to 13%) (Burd et al., 2009). Moreover, patients with autism spectrum disorder often present with motor disorders: not only stereotypies, but also tics, as shown by both clinical reports (Sverd, 1991; Ringman and Jankovic, 2000) and systematic studies (Baron-Cohen et al., 1999; Canitano and Vivanti, 2007). Tic severity in the context of autism spectrum disorder is usually mild (Canitano and Vivanti, 2007).

Affective symptoms are commonly reported by patients with GTS, who are significantly more likely to experience depression than comparison subjects (Piedad and Cavanna, 2016). Prevalence figures of 13%–20% have been reported for the diagnosis of clinical depression across community and epidemiologic studies (Piedad and Cavanna, 2016; Gill and Kompoliti, 2020). The etiology of depression in patients with GTS is thought to be multifactorial, associated with neurobiologic alterations, social factors including stigma (Robertson, 2006b), comorbid conditions such as obsessive-compulsive disorder and attention-deficit and hyperactivity disorder, and adverse effects of medications used to treat GTS, especially antidopaminergic agents (Badenoch and Cavanna, 2020). It has been shown that affective symptoms, like other comorbidities of GTS, can develop or change in different ages: these findings highlight the need for regular follow-ups in specialist clinics (Hirschtritt et al., 2015).

Affective symptoms in the context of GTS often reported alongside anxiety, which can be more prevalent and severe (Lewin et al., 2011; Isaacs et al., 2023). Anxiety disorders have an estimated lifetime prevalence of 36% in patients with GTS (Hirschtritt et al., 2015) and are associated with obsessive-compulsive symptoms, as well as a higher risk for sleep disturbances (Coffey et al., 2000; Hibberd et al., 2020). Prevalence figures vary, with sleep problems encompassing insomnia, nightmares, night terrors, nightmares, and somnambulism affecting 12%–62% of patients (Mol Debes, 2013). Anger control problems (often referred to as explosive outbursts, rage attacks, or temper tantrums) are also frequently observed in GTS (Kano et al., 2008; Chen et al., 2013), with estimated prevalence figures ranging from 20% to 67% (Conte et al., 2020). These externalizing symptoms bear a significant impact on psychosocial functioning and are more common in patients with more severe tics and with other behavioral comorbidities, in particular attention-deficit and hyperactivity disorder (Stephens and Sandor, 1999; Sukhodolsky et al., 2003; Conte et al., 2020). Finally, initial reports suggested that patients with GTS might have an increased prevalence of personality disorders, most commonly cluster C personality disorders, schizotypal personality disorder, and narcissistic personality disorder (Robertson et al., 1997; Cavanna et al., 2007; Trillini and Müller-Vahl, 2015a, 2015b). However, these findings are yet to be consistently replicated in wider GTS populations.

CONCLUSION

In the absence of reliable biomarkers, clinical knowledge and experience about the multifaceted presentations of tics are of paramount importance for the diagnosis of tic disorders. Neurodevelopmental conditions like GTS pose further diagnostic challenges, in consideration of their frequent association with comorbid disorders. With the development of disease-specific health-related quality of life measures for children and adults with GTS (Cavanna et al., 2008, 2013e; Su et al., 2017), it has been possible to reliably assess the differential impact of both tics and tic-related behavioral symptoms on aspects of patients' well-being across the lifespan (Cavanna et al., 2013a; Evans et al., 2016). It has been consistently observed that behavioral comorbidities can exert greater influence on health-related quality of life than tics themselves, especially in children with milder tics and in adults (Bernard et al., 2009; Cavanna et al., 2013c, 2013d; Silvestri et al., 2016; Huisman-van Dijk et al., 2019; Isaacs et al., 2021). Moreover, comorbid conditions appear to affect both self-concept (self-perception, or what patients think about themselves) and self-esteem (how they feel about their self-concept) to a greater extent than tics persisting into adulthood (Silvestri et al., 2017, 2018; Weingarden et al., 2018).

A number of longstanding questions about tic disorders are still unanswered, and new questions, such as the ones posed by the recent rise of functional tics, have been formulated over the last few years (Cavanna et al., 2013b). Clinical phenomenology is still the prerequisite for a more in-depth understanding of GTS and other tic disorders. Only astute clinical observations have the potential to generate testable hypotheses that can lead to mechanistic advances and, ultimately, better treatment options.

References

American Psychiatric Association (2022). Diagnostic and statistical manual of mental disorders, fifth edition, text revision (DSM-5-TR). Arlington, VA: American Psychiatric Publishing.

Amorelli G, Martino D, Pringsheim T (2022). Rapid onset functional tic-like disorder outbreak: a challenging differential diagnosis in the COVID-19 pandemic. J Can Acad Child Adolesc Psychiatry 31: 144–151.

Andersen K, Jensen I, Okkels KB et al. (2023). Clarifying the differences between patients with organic tics and functional tic-like behaviors. Healthcare 11: 1481.

Badenoch J, Cavanna AE (2020). Pharmacotherapy for tics in adult patients with Tourette syndrome and other tic disorders. Neurol Sci 41: 1923–1926.

Baizabal-Carvallo JF, Alonso-Juarez M, Jankovic J (2022). Self-injurious behavior in Tourette syndrome. J Neurol 269: 2453–2459.

Baizabal-Carvallo JF, Alonso-Juarez M, Jankovic J (2023). Contrasting features between Tourette syndrome and secondary tic disorders. J Neural Transm 130: 931–936.

Baizabal-Carvallo JF, Fekete R (2015). Recognizing uncommon presentations of psychogenic (functional) movement disorders. Tremor Other Hyperkinet Mov 5: 279.

Baizabal-Carvallo JF, Jankovic J (2014). The clinical features of psychogenic movement disorders resembling tics. J Neurol Neurosurg Psychiatry 85: 573–575.

Baizabal-Carvallo JF, Jankovic J (2023). The clinical characterization of blocking tics in patients with Tourette syndrome. J Mov Disord 16: 163–167.

Barnea M, Benaroya-Milshtein N, Gilboa-Sechtman E et al. (2016). Subjective versus objective measures of tic severity in Tourette syndrome: the influence of environment. Psychiatry Res 242: 204–209.

Baron-Cohen S, Mortimore C, Moriarty J et al. (1999). The prevalence of Gilles de la Tourette's syndrome in children and adolescents with autism. J Child Psychol Psychiatry 40: 213–218.

Bernard BA, Stebbins GT, Siegel S et al. (2009). Determinants of quality of life in children with Gilles de la Tourette syndrome. Mov Disord 24: 1070–1073.

Bliss J (1980). Sensory experiences of Gilles de la Tourette syndrome. Arch Gen Psychiatry 37: 1343–1347.

Bloch MH, Leckman JF (2009). Clinical course of Tourette syndrome. J Psychosom Res 67: 497–501.

Bloch MH, Leckman JF, Zhu H et al. (2005). Caudate volumes in childhood predict symptom severity in adults with Tourette syndrome. Neurology 65: 1253–1258.

Bloch MH, Peterson BS, Scahill L et al. (2006). Adulthood outcome of tic and obsessive-compulsive symptom severity in children with Tourette syndrome. Arch Pediatr Adolesc Med 160: 65–69.

Bloch MH, Sukhodolsky DG, Dombrowski PA et al. (2011). Poor fine motor and visuospatial skills predict persistence of pediatric-onset obsessive-compulsive disorder into adulthood. J Child Psychol Psychiatry 52: 974–983.

Brandt V, Essing J, Jakubovski E et al. (2023). Premonitory urge and tic severity, comorbidities, and quality of life in chronic tic disorders. Mov Disord Clin Pract 10: 922–932.

Burd L, Li Q, Kerbeshian J et al. (2009). Tourette syndrome and comorbid pervasive developmental disorders. J Child Neurol 24: 170–175.

Buts S, Duncan M, Owen T et al. (2022). Paediatric tic-like presentations during the COVID-19 pandemic. Arch Dis Child 107: e17.

Calder-Sprackman S, Sutherland S, Doja A (2014). The portrayal of Tourette Syndrome in film and television. Can J Neurol Sci 41: 226–232.

Canitano R, Vivanti G (2007). Tics and Tourette syndrome in autism spectrum disorders. Autism 11: 19–28.

Carter AS, O'Donnell DA, Schultz RT et al. (2000). Social and emotional adjustment in children affected with Gilles de la Tourette's syndrome: associations with ADHD and family functioning. Attention Deficit Hyperactivity Disorder. J Child Psychol Psychiatry 41: 215–223.

Cath DC, Spinhoven P, Hoogduin CA et al. (2001). Repetitive behaviors in Tourette's syndrome and OCD with and without tics: what are the differences? Psychiatry Res 101: 171–185.

Cath DC, Spinhoven P, van Woerkom TC et al. (2001). Gilles de la Tourette's syndrome with and without obsessive-compulsive disorder compared with obsessive-compulsive disorder without tics: which symptoms discriminate? J Nerv Ment Dis 189: 219–228.

Cavanna A, Cavanna AE (2014). Christian VII of Denmark and Tourette syndrome: fact or fiction? Neurol Sci 35: 1611–1612.

Cavanna AE (2014). Improved criteria for the diagnosis of tic disorders in DSM-5. Future Neurol 9: 251–253.

Cavanna AE (2014). What is misophonia and how can we treat it? Expert Rev Neurother 14: 357–359.

Cavanna AE (2018). Gilles de la Tourette syndrome as a paradigmatic neuropsychiatric disorder. CNS Spectr 23: 213–218.

Cavanna AE (2018). The neuropsychiatry of Gilles de la Tourette syndrome: the état de l'art. Rev Neurol 174: 621–627.

Cavanna AE (2022). 'Le tic': a short story on psychogenic tics by Guy de Maupassant. Rev Neurol 178: 628–629.

Cavanna AE, Ali F, Leckman JF et al. (2010). Pathological laughter in Gilles de la Tourette syndrome: an unusual phonic tic. Mov Disord 25: 2233–2239.

Cavanna AE, Black KJ, Hallett M et al. (2017). Neurobiology of the premonitory urge in Tourette's syndrome: pathophysiology and treatment implications. J Neuropsychiatry Clin Neurosci 29: 95–104.

Cavanna AE, Damodaran L, Purpura G et al. (2022). Tourette syndrome with functional overlay: a case series. Arch Med Health Sci 10: 312–316.

Cavanna AE, David K, Bandera V et al. (2013). Health-related quality of life in Gilles de la Tourette syndrome: a decade of research. Behav Neurol 27: 83–93.

Cavanna AE, David K, Orth M et al. (2012). Predictors during childhood of future health-related quality of life in adults with Gilles de la Tourette syndrome. Eur J Paediatr Neurol 16: 605–612.

Cavanna AE, Eddy C, Rickards HE (2009). Cognitive functioning in Tourette syndrome. Discov Med 8: 191–195.

Cavanna AE, Kavanagh C, Robertson MM (2013). The future of research in Tourette syndrome. Behav Neurol 27: 139–142.

Cavanna AE, Luoni C, Selvini C et al. (2013). Disease-specific quality of life in young patients with Tourette syndrome. Pediatr Neurol 48: 111–114.

Cavanna AE, Luoni C, Selvini C et al. (2013). Parent and self-report health-related quality of life measures in young patients with Tourette syndrome. J Child Neurol 28: 1305–1308.

Cavanna AE, Luoni C, Selvini C et al. (2013). The Gilles de la Tourette Syndrome-Quality of Life Scale for Children and Adolescents (C&A-GTS-QOL): development and validation of the Italian version. Behav Neurol 27: 95–103.

Cavanna AE, Nani A (2013). Tourette syndrome and consciousness of action. Tremor Other Hyperkinet Mov 3: tre-03-181-4368-1.

Cavanna AE, Nani A, Trimble MR (2014). Emperor Claudius' neuropsychiatric presentation: from tics to behavioral symptoms. J Neuropsychiatry Clin Neurosci 26: E36–E37.

Cavanna AE, Pattumelli MG, Quarto T et al. (2010). The "imprisoned illness": motor tic disorder in Rainer Maria Rilke's Notebooks of Malte Laurids Brigge. Mov Disord 25: 1980–1982.

Cavanna AE, Purpura G, Nacinovich R (2022). Neurodevelopmental versus functional tics: the state of the art. Arch Med Health Sci 10: 239–246.

Cavanna AE, Purpura G, Riva A et al. (2023). Functional tics: expanding the phenotypes of functional movement disorders? Eur J Neurol 30: 3353–3356.

Cavanna AE, Purpura G, Riva A et al. (2023). Neurodevelopmental versus functional tics: a controlled study. J Neurol Sci 451120725.

Cavanna AE, Purpura G, Riva A et al. (2023). New-onset functional tics during the COVID-19 pandemic: clinical characteristics of 105 cases from a single centre. Eur J Neurol 30: 2411–2417.

Cavanna AE, Rickards H (2013). The psychopathological spectrum of Gilles de la Tourette syndrome. Neurosci Biobehav Rev 37: 1008–1015.

Cavanna AE, Robertson MM, Critchley HD (2007). Schizotypal personality traits in Gilles de la Tourette syndrome. Acta Neurol Scand 116: 385–391.

Cavanna AE, Schrag A, Morley D et al. (2008). The Gilles de la Tourette Syndrome – Quality of Life Scale (GTS-QOL): development and validation. Neurology 71: 1410–1416.

Cavanna AE, Seri S (2013). Tourette's syndrome. BMJ 347: f4964.

Cavanna AE, Seri S (2015). Misophonia: current perspectives. Neuropsychiatr Dis Treat 11: 2117–2123.

Cavanna AE, Seri S (2019). Georges Gilles de la Tourette and his legacy. Arch Med Health Sci 7: 303–308.

Chen K, Budman CL, Diego Herrera L et al. (2013). Prevalence and clinical correlates of explosive outbursts in Tourette syndrome. Psychiatry Res 205: 269–275.

Clarke RA, Lee S, Eapen V (2012). Pathogenetic model for Tourette syndrome delineates overlap with related neurodevelopmental disorders including autism. Transl Psychiatry 2: e158.

Claudio-Campos K, Stevens D, Koo SW et al. (2021). Is persistent motor or vocal tic disorder a milder form of Tourette syndrome? Mov Disord 36: 1899–1910.

Coffey BJ, Biederman J, Smoller JW et al. (2000). Anxiety disorders and tic severity in juveniles with Tourette's disorder. J Am Acad Child Adolesc Psychiatry 39: 562–568.

Conelea CA, Wellen B, Woods DW et al. (2018). Patterns and predictors of tic suppressibility in youth with tic disorders. Front Psychiatry 9: 188.

Conelea CA, Woods DW (2008). The influence of contextual factors on tic expression in Tourette's syndrome: a review. J Psychosom Res 65: 487–496.

Conte G, Valente F, Fioriello F et al. (2020). Rage attacks in Tourette syndrome and chronic tic disorder: a systematic review. Neurosci Biobehav Rev 119: 21–36.

Cox JH, Seri S, Cavanna AE (2018). Sensory aspects of Tourette syndrome. Neurosci Biobehav Rev 88: 170–176.

Cravedi E, Deniau E, Giannitelli M et al. (2018). Disentangling Tourette syndrome heterogeneity through hierarchical ascendant clustering. Dev Med Child Neurol 60: 942–950.

Demartini B, Ricciardi L, Parees I et al. (2015). A positive diagnosis of functional (psychogenic) tics. Eur J Neurol 22: 527–536.

Eapen V, Robertson MM (2015). Are there distinct subtypes in Tourette syndrome? Pure-Tourette syndrome versus Tourette syndrome-plus, and simple versus complex tics. Neuropsychiatr Dis Treat 11: 1431–1436.

Eddy CM, Cavanna AE (2013). "It's a curse!": coprolalia in Tourette syndrome. Eur J Neurol 20: 1467–1470.

Eddy CM, Cavanna AE (2013). Altered social cognition in Tourette syndrome: nature and implications. Behav Neurol 27: 15–22.

Eddy CM, Cavanna AE (2013). On being your own worst enemy: an investigation of socially inappropriate symptoms in Tourette syndrome. J Psychiatr Res 47: 1259–1263.

Eddy CM, Cavanna AE (2014). Premonitory urges in adults with complicated and uncomplicated Tourette syndrome. Behav Modif 38: 264–275.

Eddy CM, Cavanna AE (2014). Tourette syndrome and obsessive compulsive disorder: compulsivity along the continuum. J Obsessive Compuls Relat Disord 3: 363–371.

Eddy CM, Mitchell IJ, Beck SR et al. (2010). Impaired comprehension of nonliteral language in Tourette syndrome. Cogn Behav Neurol 23: 178–184.

Eddy CM, Mitchell IJ, Beck SR et al. (2011). Social reasoning in Tourette syndrome. Cogn Neuropsychiatry 16: 326–347.

Eddy CM, Rickards HE, Cavanna AE (2012). Executive functions in uncomplicated Tourette syndrome. Psychiatry Res 200: 46–48.

Essing J, Jakubovski E, Psathakis N et al. (2022). Premonitory urges reconsidered: urge location corresponds to tic location in patients with primary tic disorders. J Mov Disord 15: 43–52.

Evans J, Seri S, Cavanna AE (2016). The effects of Gilles de la Tourette syndrome and other chronic tic disorders on quality of life across the lifespan: a systematic review. Eur Child Adolesc Psychiatry 25: 939–948.

Ferenczi S (1921). Psychoanalytical observations on tic. Int J Psychoanal 2: 1–30.

Fernandez TV, Sanders SJ, Yurkiewicz IR et al. (2012). Rare copy number variants in Tourette syndrome disrupt genes in histaminergic pathways and overlap with autism. Biol Psychiatry 71: 392–402.

Frankel M, Cummings JL, Robertson MM et al. (1986). Obsessions and compulsions in Gilles de la Tourette's syndrome. Neurology 36: 378–382.

Freeman RD, Fast DK, Burd L et al. (2000). An international perspective on Tourette syndrome: selected findings from 3,500 individuals in 22 countries. Dev Med Child Neurol 42: 436–447.

Freeman RD, Zinner SH, Muller-Vahl KR et al. (2009). Coprophenomena in Tourette syndrome. Dev Med Child Neurol 51: 218–227.

Fremer C, Szejko N, Pisarenko A et al. (2023). What distinguishes patients with mass social media-induced illness presenting with Tourette-like behavior from those with Tourette syndrome? Results of a prospective cohort study. Eur Arch Psychiatry Clin Neurosci 274: 227–233.

Frey J, Black KJ, Malaty IA (2022). TikTok Tourette's: are we witnessing a rise in functional tic-like behavior driven by adolescent social media use? Psychol Res Behav Manag 15: 3575–3585.

Ganos C, Al-Fatly B, Fischer JF et al. (2022). A neural network for tics: insights from causal brain lesions and deep brain stimulation. Brain 145: 4385–4397.

Ganos C, Bongert J, Asmuss L et al. (2015). The somatotopy of tic inhibition: where and how much? Mov Disord 30: 1184–1189.

Ganos C, Edwards MJ, Müller-Vahl K (2016). "I swear it is Tourette's!": on functional coprolalia and other tic-like vocalizations. Psychiatry Res 246: 821–826.

Ganos C, Kahl U, Schunke O et al. (2012). Are premonitory urges a prerequisite of tic inhibition in Gilles de la Tourette syndrome? J Neurol Neurosurg Psychiatry 83: 975–978.

Ganos C, Martino D (2015). Tics and Tourette syndrome. Neurol Clin 33: 115–136.

Ganos C, Martino D, Espay AJ et al. (2019). Tics and functional tic-like movements: can we tell them apart? Neurology 93: 750–758.

Ganos C, Mencacci N, Gardiner A et al. (2014). Paroxysmal kinesigenic dyskinesia may be misdiagnosed in co-occurring Gilles de la Tourette syndrome. Mov Disord Clin Pract 1: 84–86.

Ganos C, Munchau A, Bhatia KP (2014). The semiology of tics, Tourette's, and their associations. Mov Disord Clin Pract 1: 145–153.
Ganos C, Neumann WJ, Müller-Vahl KR et al. (2021). The phenomenon of exquisite motor control in tic disorders and its pathophysiological implications. Mov Disord 36: 1308–1315.
Ganos C, Ogrzal T, Schnitzler A et al. (2012). The pathophysiology of echopraxia/echolalia: relevance to Gilles de la Tourette syndrome. Mov Disord 27: 1222–1229.
Garcia-Delgar B, Servera M, Coffey BJ et al. (2022). Tic disorders in children and adolescents: does the clinical presentation differ in males and females? A report by the EMTICS group. Eur Child Adolesc Psychiatry 31: 1539–1548.
George MS, Trimble MR, Ring HA et al. (1993). Obsessions in obsessive-compulsive disorder with and without Gilles de la Tourette's syndrome. Am J Psychiatry 150: 93–97.
Gill CE, Kompoliti K (2020). Clinical features of Tourette syndrome. Journal of Child Neurology 353: 166–174.
Gilles de la Tourette G (1885). Étude sur une affection nerveuse caractérisée par de l'incoordination motrice accompagnée d'écholalie et de coprolalie. Arch Neurol 919-42158-200.
Greenberg E, Petriceks AH, Stern TA (2021). Intrusive-destructive behaviors: novel behavioral presentations of patients with co-occurring Tourette syndrome, obsessive-compulsive disorder, and attention-deficit/hyperactivity disorder. Prim Care Companion CNS Disord 23: 20f02670.
Groth C, Mol Debes N, Rask CU et al. (2017). Course of Tourette syndrome and comorbidities in a large prospective clinical study. J Am Acad Child Adolesc Psychiatry 56: 304–312.
Hallett M (2001). Neurophysiology of tics. Adv Neurol 85: 237–244.
Han VX, Kozlowska K, Kothur K et al. (2022). Rapid onset functional tic-like behaviours in children and adolescents during COVID-19: clinical features, assessment and biopsychosocial treatment approach. J Paediatr Child Health 58: 1181–1187.
Hartmann A, Millet B (2018). Repetitive movements and behaviors in neurological and psychiatric practice: distinctions and similarities between Tourette disorder and obsessive-compulsive disorder. Rev Neurol 174: 199–202.
Hassan N, Cavanna AE (2012). The prognosis of Tourette syndrome: implications for clinical practice. Funct Neurol 27: 23–27.
Hibberd C, Charman T, Bhatoa RS et al. (2020). Sleep difficulties in children with Tourette syndrome and chronic tic disorders: a systematic review of characteristics and associated factors. Sleep 43: zsz308.
Hirschtritt ME, Lee PC, Pauls DL et al. (2015). Lifetime prevalence, age of risk, and genetic relationships of comorbid psychiatric disorders in Tourette syndrome. JAMA Psychiatry 72: 325–333.
Huisman-van Dijk HM, Matthijssen SJMA, Stockmann RTS et al. (2019). Effects of comorbidity on Tourette's tic severity and quality of life. Acta Neurol Scand 140: 390–398.
Hull M, Parnes M (2021). Tics and TikTok: functional tics spread through social media. Mov Disord Clin Pract 8: 1248–1252.
Hull M, Parnes M, Jankovic J (2021). Increased incidence of functional (psychogenic) movement disorders in children and adults amid the COVID-19 pandemic: a cross-sectional study. Neurol Clin Pract 11: e686–e690.
Isaacs D, Riordan H (2020). Sensory hypersensitivity in Tourette syndrome: a review. Brain Dev 42: 627–638.
Isaacs DA, Narapareddy A, Eckland MR et al. (2023). Dimensional assessment of depression and anxiety in a clinical sample of adults with chronic tic disorder. J Neuropsychiatry Clin Neurosci 35: 352–360.
Isaacs DA, Riordan HR, Claassen DO (2021). Clinical correlates of health-related quality of life in adults with chronic tic disorder. Front Psychiatry 12: 619854.
Isung J, Isomura K, Larsson H et al. (2021). Association of Tourette syndrome and chronic tic disorder with cervical spine disorders and related neurological complications. JAMA Neurol 78: 1205–1211.
Itard JM (1825). Mémoire sur quelques fonctions involontaires des appareils de la locomotion de la préhension et de la voix. Arch Gen Med 8: 385–407.
Iverson AM, Black KJ (2022). Why tic severity changes from then to now and from here to there. J Clin Med 11: 5930.
Jack RH, Joseph RM, Coupland CAC et al. (2023). Impact of the COVID-19 pandemic on incidence of tics in children and young people: a population-based cohort study. eClinicalMedicine 57: 101857.
Jafari F, Abbasi P, Rahmati M et al. (2022). Systematic review and meta-analysis of Tourette syndrome prevalence; 1986 to 2022. Pediatr Neurol 137: 6–16.
Janiri D, Petracca M, Moccia L et al. (2023). Functional movement disorders during COVID-19: psychological distress, affective temperament and emotional dysregulation. J Pers Med 13: 175.
Jankovic J (1997). Tourette syndrome. Phenomenology and classification of tics. Neurol Clin 15: 267–275.
Jankovic J, Fahn S (1986). The phenomenology of tics. Mov Disord 1: 17–26.
Jankovic J, Stone L (1991). Dystonic tics in patients with Tourette's syndrome. Mov Disord 6: 248–252.
Kaczyńska J, Janik P (2021). Blocking tics in Gilles de la Tourette syndrome. Front Neurol 12: 686785.
Kano Y, Ohta M, Nagai Y et al. (2008). Rage attacks and aggressive symptoms in Japanese adolescents with Tourette syndrome. CNS Spectr 13: 325–332.
Katz TC, Bui TH, Worhach J et al. (2022). Tourettic OCD: current understanding and treatment challenges of a unique endophenotype. Front Psychiatry 13: 929526.
Khalifa N, von Knorring AL (2003). Prevalence of tic disorders and Tourette syndrome in a Swedish school population. Dev Med Child Neurol 45: 315–319.
Kim YR, Song DY, Bong G et al. (2023). Clinical characteristics of comorbid tic disorders in autism spectrum disorder: exploratory analysis. Child Adolesc Psychiatry Ment Health 17: 71.
Knight T, Steeves T, Day L et al. (2012). Prevalence of tic disorders: a systematic review and meta-analysis. Pediatr Neurol 47: 77–90.
Kompoliti K (2016). Sources of disability in Tourette syndrome: children vs. adults. Tremor Other Hyperkinet Mov 5: 318.
Kompoliti K, Goetz CG (1998). Hyperkinetic movement disorders misdiagnosed as tics in Gilles de la Tourette syndrome. Mov Disord 13: 477–480.
Kurlan R, Daragjati C, Como PG et al. (1996). Non-obscene complex socially inappropriate behavior in Tourette's syndrome. J Neuropsychiatry Clin Neurosci 8: 311–317.
Kurlan R, McDermott MP, Deeley C et al. (2001). Prevalence of tics in schoolchildren and association with placement in special education. Neurology 57: 1383–1388.
Kurvits L, Mainka T, Cavanna AE et al. (2021). Aggression toward others misdiagnosed as primary tics. Mov Disord Clin Pract 8: 769–771.
Kushner HI (1998). Freud and the diagnosis of Gilles de la Tourette's illness. Hist Psychiatry 9: 1–25.
Kushner HI (1999). From Gilles de la Tourette's disease to Tourette syndrome: a history. CNS Spectr 4: 24–35.

Kwak C, Dat Vuong K, Jankovic J (2003). Premonitory sensory phenomenon in Tourette's syndrome. Mov Disord 18: 1530–1533.

Lang A (1991). Patient perception of tics and other movement disorders. Neurology 41: 223–228.

Leckman JF (2003). Phenomenology of tics and natural history of tic disorders. Brain Dev 25: S24–S28.

Leckman JF, Riddle MA, Hardin MT et al. (1989). The Yale Global Tic Severity Scale (YGTSS): initial testing of a clinician-rated scale of tic severity. J Am Acad Child Adolesc Psychiatry 28: 566–573.

Leckman JF, Walker DE, Cohen DJ (1993). Premonitory urges in Tourette's syndrome. Am J Psychiatry 150: 98–102.

Leckman JF, Zhang H, Vitale A et al. (1998). Course of tic severity in Tourette syndrome: the first two decades. Pediatrics 102: 14–19.

Lewin AB, Storch EA, Conelea CA et al. (2011). The roles of anxiety and depression in connecting tic severity and functional impairment. J Anxiety Disord 25: 164–168.

Lin H, Katsovich L, Ghebremichael M et al. (2007). Psychosocial stress predicts future symptom severities in children and adolescents with Tourette syndrome and/or obsessive-compulsive disorder. J Child Psychol Psychiatry 48: 157–166.

Mahler MS, Rangell L (1943). A psychosomatic study of maladie des tics (Gilles de la Tourette's syndrome). Psychiatr Q 17: 579–603.

Martindale JM, Mink JW (2022). The rise of functional tic-like behaviors: what do the COVID-19 pandemic and social media have to do with it? A narrative review. Front Pediatr 10863919.

Martino D, Cavanna AE, Robertson MM et al. (2012). Prevalence and phenomenology of eye tics in Gilles de la Tourette syndrome. J Neurol 259: 2137–2140.

Martino D, Ganos C, Pringsheim TM (2017). Tourette syndrome and chronic tic disorders: the clinical spectrum beyond tics. Int Rev Neurobiol 134: 1461–1490.

Martino D, Hedderly T, Murphy T et al. (2023). The spectrum of functional tic-like behaviours: data from an international registry. Eur J Neurol 30: 334–343.

Martino D, Madhusudan N, Zis P et al. (2013). An introduction to the clinical phenomenology of Tourette syndrome. Int Rev Neurobiol 112: 1–33.

Martino D, Pringsheim TM, Cavanna AE et al. (2017). Systematic review of severity scales and screening instruments for tics: critique and recommendations. Mov Disord 32: 467–473.

Mathews CA, Grados MA (2011). Familiality of Tourette syndrome, obsessive-compulsive disorder, and attention-deficit/hyperactivity disorder: heritability analysis in a large sib-pair sample. J Am Acad Child Adolesc Psychiatry 50: 46–54.

Mathews CA, Waller J, Glidden D et al. (2004). Self injurious behaviour in Tourette syndrome: correlates with impulsivity and impulse control. J Neurol Neurosurg Psychiatry 75: 1149–1155.

McGuire JF, Piacentini J, Storch EA et al. (2018). A multicenter examination and strategic revisions of the Yale Global Tic Severity Scale. Neurology 90: 1711–1719.

Meier SM, Dalsgaard S, Mortensen PB et al. (2017). Mortality risk in a nationwide cohort of individuals with tic disorders and with Tourette syndrome. Mov Disord 32: 605–609.

Meige H, Feindel E (1902). Les tics et leur traitement. Paris: Masson.

Miguel EC, Baer L, Coffey BJ et al. (1997). Phenomenological differences appearing with repetitive behaviours in obsessive-compulsive disorder and Gilles de la Tourette's syndrome. Br J Psychiatry 170: 140–145.

Miguel EC, do Rosario-Campos MC, Prado HS et al. (2000). Sensory phenomena in obsessive-compulsive disorder and Tourette's disorder. J Clin Psychiatry 61: 150–156.

Mito T, Okemoto H, Kawasaki A et al. (2023). Exacerbated Tourette syndrome diagnosed in adulthood after recurrent bilateral retinal detachment. Retin Cases Brief Rep 18: 488–490.

Mol Debes NM (2013). Co-morbid disorders in Tourette syndrome. Behav Neurol 27: 7–14.

Monaco F, Servo S, Cavanna AE (2009). Famous people with Gilles de la Tourette syndrome? J Psychosom Res 67: 485–490.

Mula M, Cavanna AE, Critchley H et al. (2008). Phenomenology of obsessive compulsive disorder in patients with temporal lobe epilepsy or Tourette syndrome. J Neuropsychiatry Clin Neurosci 20: 223–226.

Müller-Vahl KR, Pisarenko A, Jakubovski E et al. (2022). Stop that! It's not Tourette's but a new type of mass sociogenic illness. Brain 145: 476–480.

Müller-Vahl KR, Sambrani T, Jakubovski E (2019). Tic disorders revisited: introduction of the term "tic spectrum disorders.". Eur Child Adolesc Psychiatry 28: 1129–1135.

Murad A (2010). A neurological mystery from history: the case of Claudius Caesar. J Hist Neurosci 19: 221–227.

Neal M, Cavanna AE (2013). "Not just right experiences" in patients with Tourette syndrome: complex motor tics or compulsions? Psychiatry Res 210: 559–563.

Neal M, Cavanna AE (2013). Selective sound sensitivity syndrome (misophonia) in a patient with Tourette syndrome. J Neuropsychiatry Clin Neurosci 25: E01.

Nilles C, Pringsheim TM, Martino D (2022). The recent surge of functional movement disorders: social distress or greater awareness? Curr Opin Neurol 35: 485–493.

Olvera C, Stebbins GT, Goetz CG et al. (2021). TikTok Tics: a pandemic within a pandemic. Mov Disord Clin Pract 8: 1200–1205.

Packer LE (1997). Social and educational resources for patients with Tourette syndrome. Neurol Clin 15: 457–473.

Pappert EJ, Goetz CG, Louis ED et al. (2003). Objective assessments of longitudinal outcome in Gilles de la Tourette's syndrome. Neurology 61: 936–940.

Paszek J, Pollok B, Biermann-Ruben K et al. (2010). Is it a tic? Twenty seconds to make a diagnosis. Mov Disord 25: 1106–1108.

Patel N, Jankovic J, Hallett M (2014). Sensory aspects of movement disorders. Lancet Neurol 13: 100–112.

Pauls DL, Towbin KE, Leckman JF et al. (1986). Gilles de la Tourette's syndrome and obsessive-compulsive disorder. Evidence supporting a genetic relationship. Arch Gen Psychiatry 43: 1180–1182.

Paulus T, Bäumer T, Verrel J et al. (2021). Pandemic tic-like behaviors following social media consumption. Mov Disord 36: 2932–2935.

Pearce JM (1994). Doctor Samuel Johnson: "The great convulsionary" a victim of Gilles de la Tourette's syndrome. J R Soc Med 87: 396–399.

Peterson BS, Leckman JF (1998). The temporal dynamics of tics in Gilles de la Tourette syndrome. Biol Psychiatry 44: 1337–1348.

Piedad JCP, Cavanna AE (2016). Depression in Tourette syndrome: a controlled and comparison study. J Neurol Sci 364: 128–132.

Prado HS, Rosário MC, Lee J et al. (2008). Sensory phenomena in obsessive-compulsive disorder and tic disorders: a review of the literature. CNS Spectr 13: 425–432.

Pringsheim T, Ganos C, McGuire JF et al. (2021). Rapid onset functional tic-like behaviors in young females during the COVID-19 pandemic. Mov Disord 36: 2707–2713.

Pringsheim T, Ganos C, Nilles C et al. (2023). European Society for the Study of Tourette Syndrome 2022 criteria for clinical diagnosis of functional tic-like behaviours: international consensus from experts in tic disorders. Eur J Neurol 30: 902–910.

Pringsheim T, Martino D (2021). Rapid onset of functional tic-like behaviours in young adults during the COVID-19 pandemic. Eur J Neurol 28: 3805–3808.

Rajagopal S, Seri S, Cavanna AE (2013). Premonitory urges and sensorimotor processing in Tourette syndrome. Behav Neurol 27: 65–73.

Rickards H, Woolf I, Cavanna AE (2010). "Trousseau's disease": a description of the Gilles de la Tourette syndrome 12 years before 1885. Mov Disord 25: 2285–2289.

Rigas A, Mainka T, Pringsheim T et al. (2023). Distinguishing functional from primary tics: a study of expert video assessments. J Neurol Neurosurg Psychiatry 94: 751–756.

Ringman JM, Jankovic J (2000). Occurrence of tics in Asperger's syndrome and autistic disorder. J Child Neurol 15: 394–400.

Rizzo R, Gulisano M, Cali PV et al. (2012). Long term clinical course of Tourette syndrome. Brain Dev 34: 667–673.

Robertson MM (2000). Tourette syndrome, associated conditions and the complexities of treatment. Brain 123: 425–462.

Robertson MM (2006). Attention deficit hyperactivity disorder, tics and Tourette's syndrome: the relationship and treatment implications. A commentary. Eur Child Adolesc Psychiatry 15: 1–11.

Robertson MM (2006). Mood disorders and Gilles de la Tourette's syndrome: an update on prevalence, etiology, comorbidity, clinical associations, and implications. J Psychosom Res 61: 349–358.

Robertson MM, Banerjee S, Hiley PJ et al. (1997). Personality disorder and psychopathology in Tourette's syndrome: a controlled study. Br J Psychiatry 171: 283–286.

Robertson MM, Banerjee S, Kurlan R et al. (1999). The Tourette syndrome Diagnostic Confidence Index: development and clinical associations. Neurology 53: 2108–2112.

Robertson MM, Cavanna AE, Eapen V (2015). Gilles de la Tourette syndrome and disruptive behavior disorders: prevalence, associations, and explanation of the relationships. J Neuropsychiatry Clin Neurosci 27: 33–41.

Robertson MM, Eapen V (2014). Tourette's: syndrome, disorder or spectrum? Classificatory challenges and an appraisal of the DSM criteria. Asian J Psychiatr 11: 106–113.

Robertson MM, Eapen V, Cavanna AE (2009). The international prevalence, epidemiology, and clinical phenomenology of Tourette syndrome: a cross-cultural perspective. J Psychosom Res 67: 475–483.

Robinson S, Hedderly T, Conte G et al. (2018). Misophonia in children with tic disorders: a case series. J Dev Behav Pediatr 39: 516–522.

Roessner V, Becker A, Banaschewski T et al. (2007). Developmental psychopathology of children and adolescents with Tourette syndrome – impact of ADHD. Eur Child Adolesc Psychiatry 16: 24–35.

Roessner V, Becker A, Banaschewski T et al. (2007). Psychopathological profile in children with chronic tic disorder and co-existing ADHD: additive effects. J Abnorm Child Psychol 35: 79–85.

Sanger TD, Chen D, Fehlings DL et al. (2010). Definition and classification of hyperkinetic movements in childhood. Mov Disord 25: 1538–1549.

Scharf JM, Miller LL, Gauvin CA et al. (2015). Population prevalence of Tourette syndrome: a systematic review and meta-analysis. Mov Disord 30: 221–228.

Schubert L, Verrel J, Behm A et al. (2021). Inter-individual differences in urge-tic associations in Tourette syndrome. Cortex 143: 80–91.

Shapiro AK, Shapiro E (1968). Treatment of Gilles de la Tourette's Syndrome with haloperidol. Br J Psychiatry 114: 345–350.

Silvestri PR, Baglioni V, Cardona F et al. (2018). Self-concept and self-esteem in patients with chronic tic disorders: a systematic literature review. Eur J Paediatr Neurol 22: 749–756.

Silvestri PR, Chiarotti F, Baglioni V et al. (2016). Health-related quality of life in patients with Gilles de la Tourette syndrome at the transition between adolescence and adulthood. Neurol Sci 37: 1857–1860.

Silvestri PR, Chiarotti F, Baglioni V et al. (2017). A preliminary examination of self-concept in older adolescents and young adults with Gilles de la Tourette syndrome. Eur J Paediatr Neurol 21: 468–474.

Specht MW, Woods DW, Nicotra CM et al. (2013). Effects of tic suppression: ability to suppress, rebound, negative reinforcement, and habituation to the premonitory urge. Behav Res Ther 51: 24–30.

Stafford M, Cavanna AE (2020). Prevalence and clinical correlates of self-injurious behavior in Tourette syndrome. Neurosci Biobehav Rev 113: 299–307.

State MW (2010). The genetics of child psychiatric disorders: focus on autism and Tourette syndrome. Neuron 68: 254–269.

Stephens RJ, Sandor P (1999). Aggressive behaviour in children with Tourette syndrome and comorbid attention-deficit hyperactivity disorder and obsessive-compulsive disorder. Can J Psychiatry 44: 1036–1042.

Su MT, McFarlane F, Cavanna AE et al. (2017). The English version of the Gilles de la Tourette Syndrome-Quality of Life Scale for Children and Adolescents (C&A-GTS-QOL). J Child Neurol 32: 76–83.

Sukhodolsky DG, Scahill L, Zhang H et al. (2003). Disruptive behavior in children with Tourette's syndrome: association with ADHD comorbidity, tic severity, and functional impairment. J Am Acad Child Adolesc Psychiatry 42: 98–105.

Sverd J (1991). Tourette syndrome and autistic disorder: a significant relationship. Am J Med Genet 39: 173–179.

Szejko N, Müller-Vahl KR (2021). Challenges in the diagnosis and assessment in patients with Tourette syndrome and comorbid obsessive-compulsive disorder. Neuropsychiatr Dis Treat 17: 1253–1266.

Termine C, Grossi E, Anelli V et al. (2021). Possible tics diagnosed as stereotypies in patients with severe autism spectrum disorder: a video-based evaluation. Neurol Sci 42: 1559–1561.

Trau SP, Quehl L, Tsujimoto THM et al. (2022). Creating a patient-based diagnostic checklist for functional tics during the COVID-19 pandemic. Neurol Clin Pract 12: 365–376.

Trillini MO, Müller-Vahl KR (2015). Narcissistic vulnerability is a common cause for depression in patients with Gilles de la Tourette syndrome. Psychiatry Res 230: 695–703.

Trillini MO, Müller-Vahl KR (2015). Patients with Gilles de la Tourette syndrome have widespread personality differences. Psychiatry Res 228: 765–773.

Trousseau A (1873). Des diverses espèces de chorées. Clin Méd l'Hôtel Dieu Paris 2: 264–271.

Ubhi M, Achinivu K, Seri S et al. (2020). Motor stereotypies in adult patients with Tourette syndrome. Future Neurol 15: FNL42.

Voderholzer U, Muller N, Haag C et al. (1997). Periodic limb movements during sleep are a frequent finding in patients with Gilles de la Tourette's syndrome. J Neurol 244: 521–526.

Voss H (2012). The representation of movement disorders in fictional literature. J Neurol Neurosurg Psychiatry 83: 994–999.

Walusinski O, Féray JC (2020). The Marquise de Dampierre identified at last, the first described clinical case of Gilles de la Tourette syndrome. Rev Neurol 176: 754–762.

Wanderer S, Roessner V, Freeman R et al. (2012). Relationship of obsessive-compulsive disorder to age-related comorbidity in children and adolescents with Tourette syndrome. J Dev Behav Pediatr 33: 124–133.

Weingarden H, Scahill L, Hoeppner S et al. (2018). Self-esteem in adults with Tourette syndrome and chronic tic disorders: the roles of tic severity, treatment, and comorbidity. Compr Psychiatry 84: 95–100.

Worbe Y, Mallet L, Golmard JL et al. (2010). Repetitive behaviours in patients with Gilles de la Tourette syndrome: tics, compulsions, or both? PLoS One 5: e12959.

Yorston G, Hindley N (1998). Study of a nervous disorder characterized by motor incoordination with echolalia and coprolalia (The introduction and case studies of Gilles de la Tourette's 1885 paper). Hist Psychiatry 9: 97–101.

Handbook of Clinical Neurology, Vol. 215 (3rd series)
Tic Disorders and Tourette Syndrome
C. Ganos and Y. Worbe, Editors
https://doi.org/10.1016/B978-0-443-13554-5.00010-9

Chapter 3

Secondary tics: Etiology, presentation, and management

JOSÉ FIDEL BAIZABAL-CARVALLO[1,2,*] AND JOSEPH JANKOVIC[1]

[1]*Parkinson Disease Center and Movement Disorders Clinic, Department of Neurology, Baylor College of Medicine, Houston, TX, United States*

[2]*Department of Sciences and Engineering, University of Guanajuato, León, Mexico*

Abstract

Tics are involuntary or semivoluntary, abrupt, brief, nonrhythmic, recurrent movements or sounds. Tourette syndrome (TS) is the most common cause of tics; however, several other disorders have been associated with tics and tic-like movements. Etiologies of secondary tic syndromes and disorders (STS) include hereditary, drug-induced (including tardive), toxins, traumatic, cerebrovascular, infectious, parainfectious, autoimmune, functional disorders, and others. Age at onset after 18 years, lack of comorbid attention-deficit/hyperactivity disorder, obsessive-compulsive disorder, and no family history of tics are considered typical features of STS. Atypical neurologic manifestations such as additional movement disorders, seizures, decreased level of consciousness, and neurologic deficits, temporally related to some triggering event, such as brain trauma, stroke, or drug exposure, should raise suspicion of STS. These patients usually show a more restricted body distribution, less severe and less complex tics compared with patients with TS. Some tics, typically observed in TS, such as eye-blinking, facial grimacing, and complex motor and phonic tics that are preceded by a premonitory urge may not be present in STS. However, there is a substantial overlap of features between patients with TS and STS. Pharmacologic treatment of STS is similar to TS. Additionally, these patients may require specific treatment, such as anticonvulsants, antibiotics, immunotherapy, or drug-discontinuation. Therefore, prompt recognition and early treatment intervention of STS is imperative for a favorable outcome.

INTRODUCTION

Tics are among the most common movement disorders (MDs). They are typically characterized by involuntary or semivoluntary, abrupt, brief, nonrhythmic, recurrent movements or sounds, often preceded by a premonitory urge. Tourette syndrome (TS) is the most common cause of tics, but tics may occur in many other neurologic disorders or they may be triggered by drugs, infections, injury, and many other secondary causes (Jankovic, 2001). Secondary tic syndromes and disorders (STS) are sometimes referred to as "tourettism," as their presentation and phenomenology may overlap with the clinical features of TS. The *Diagnostic and Statistical Manual of Mental Disorders, Fifth Edition* (DSM-5) states that for diagnosis of TS, "symptoms should not be related to drugs or other medical condition," such as autism spectrum disorder (ASD) (American Psychiatric Association, 2013). If patients do not fulfill diagnostic criteria for TS, persistent (chronic) motor tic disorder, persistent (chronic) vocal tic disorder, or provisional tic disorder, clinicians should consider "other specified tic disorder" category (American Psychiatric Association, 2013). This review will focus on the latter category, STS and its many causes (Table 3.1). The distinguishing features of the most common STS are highlighted in Table 3.2 and Figure 3.1.

HEREDITARY DISORDERS

Pervasive developmental and chromosomal disorders

Pervasive developmental disorders (PDDs) are not uncommonly associated with tics (Sverd, 1991; Gulisano et al., 2020). In a study of 155 patients with tics and coexistent disorders, 8.3% had Asperger's syndrome, 2.5% mental retardation, 1.9% autism, and 0.6% Savant's syndrome (Ringman and Jankovic, 2000; Mejia and Jankovic, 2005). Patients with ASD have shown similar distribution of tics but lower severity and reduced awareness of tics compared with patients with TS

[*]Correspondence to: José Fidel Baizabal-Carvallo, MD, MSc, Ave León 428, Jardines del Moral, León, Guanajuato C.P. 37320, México. Tel: +52-(477)-718-3083. E-mail: baizabaljf@hotmail.com.

Table 3.1

Etiology of secondary tic syndromes and disorders.

Disorder category
Pervasive developmental and chromosomal disorders
Autism spectrum disorders
Down syndrome
Chromosome 15 duplication
Fragile X syndrome
Hereditary neurodegenerative
Huntington disease
Chorea-acanthocytosis
McLeod syndrome
Pantothenate kinase-associated neurodegeneration
Neuroferritinopathy
Frontotemporal dementia
Progressive supranuclear palsy
Poretti–Boltshauser syndrome
Drugs
Drug-induced tics
Tardive tics
Toxins/poisons
Carbon monoxide poisoning
Mercury
Physical trauma
Traumatic brain injury
Peripheral trauma
Cerebrovascular disease
Ischemic stroke
Arteriovenous malformations
Cardiopulmonary bypass
Other
Brain tumors
Basal ganglia calcification
Multiple sclerosis
Infectious encephalitis
Virus
Herpes simplex virus
Varicella zoster virus
Flavivirus
Measles, mumps, rubella
Bacteria
Mycoplasma pneumonia
Neurobrucellosis
Prions
Creutzfeldt–Jakob disease
Parainfectious disease
Sydenham disease
PANDAS/PANS[a]
Encephalitis lethargica[a]
Autoimmune encephalitis/disorders
Anti-NMDAR encephalitis
Anti VGKC/LGI1 encephalitis
Anti-GlyR encephalitis
Anti-dopamine-2 receptor encephalitis
Hashimoto's encephalitis
Functional (psychogenic) disorder
Functional tics
"TikTok" tics

PANDAS, Pediatric Autoimmune Neuropsychiatric Disorder Associated with Streptococcal infection; *PANS*, Pediatric Acute-Onset Neuropsychiatric Syndrome.

[a] Debated etiopathogenesis or nosology.

Kahl et al., 2015). Common genetic background has been suspected between PDD and TS, which would explain their common co-occurrence (Comings and Comings, 1991; Stern and Robertson, 1997). Besides tics, stereotypies are common in patients with PDD. These movements may respond to dopamine-blocking drugs, such as aripiprazole or risperidone; pentoxifylline may provide extra benefit (Rajapakse and Pringsheim, 2010).

Tourettism or TS may be occasionally observed in individuals with Down syndrome, either as full or mosaic trisomy of chromosome 21 (Karlinsky and Berg, 1987; Collacott and Ismail, 1988). A common pathophysiology between both disorders has been suggested to underly the co-occurrence of both conditions (Barabas et al., 1986). Motor and phonic tics have also been described in patients with fragile X syndrome. Patients are male with motor and phonic tics, but with a later onset of tics compared to patients with TS (Schneider et al., 2008). Of note, tics have also been reported in cases with GNB1 encephalopathy, an autosomal dominant disorder with common de novo variants that is characterized by severe developmental delay, seizures, and hyperkinetic MDs (Reyes et al., 2023).

A de novo chromosome inversion at 13q31.1 involving the Slit and Trk-like 1 (SLITRK1) gene and a deletion at 7q31.1–7q31.2 of 7.2 Mb that encompasses the IMMPL2 gene have also been causally associated with tics (Abelson et al., 2005; Patel et al., 2011), but these results have not been observed in other patients. Further, a deletion at the neuroligin 4 gene, deletions in the 16p13.11 chromosomal region, and 9p monosomy have also been reported to be causally associated with tics and other neuropsychiatric syndromes, as autism or obsessive-compulsive disorder (OCD) (Taylor et al., 1991; Lawson-Yuen et al., 2008; McGrath et al., 2014), but again, replications in larger cohorts are lacking. Overall, structural chromosomal variations and increases of copy number variants are reported to be detected in about 1% of patients with TS (Lin et al., 2022).

Hereditary neurodegenerative

Huntington disease (HD) is an autosomal dominant disorder caused by CAG expansion of the HTT gene in chromosome 4. Patients with HD present with a typical clinical triad consisting of behavioral changes, cognitive decline, and chorea. Some patients with HD have been described with motor and phonic tics, presenting with adult-onset tourettism (Jankovic and Ashizawa, 1995). Tics have been described in 2 out of 33 patients with juvenile HD described in the literature (Cui et al., 2017). Excessive eye-blinking resembling ocular tics with or without eye-lid fluttering or blepharoclonus, has been reported as a presenting symptom in juvenile HD (Xing et al., 2008; Medarametla et al., 2021). Tics in HD may coexist with obsessive-compulsive symptoms in the prediagnosis state, such as obsessive worrying or checking, which are estimated to occur in between 15% and 50% of cases (Beglinger et al., 2008). In summary, focal tics may be the presenting symptom in selected patients with HD, when other manifestations are

Table 3.2

Summary of features of most common causes of tics.

	TS	Drug-induced (tardive)	Posttraumatic	Sydenham disease	Autoimmune encephalitis	Functional (psychogenic)
Cause	Neurodevelopmental	Stimulants	Basal ganglia and other injury	Humoral or cellular autoimmunity	Antibody-mediated	Complex/unclear
Sex predominance	Male	Male	Male	Male	Variable	Female
Typical age at onset	< 18 years	Adults	Adults	< 15 years	Variable	Adults
Type of tics	Simple/complex motor and phonic tics	Simple/complex motor and phonic tics	Simple/complex motor and phonic tics	Vocalizations, simple motor tics may represent mild chorea	Simple motor tics	All types usually with other abnormal movements or bizarre behaviors
Comorbid OCD/ADHD	Common, 60%–70% of cases	Uncommon	OCD 25%	OCD in 16% of cases. OCS are common	Uncommon	Probably less than 20%
Therapy	DRB, TBZ, BoNT, DBS	Withdrawal of offending drug. Avoid other first and second-generation antipsychotics, except for CZP. Consider CNZ, TBZ	Like TS	VPA, CBZ, Cs, AB therapy	Immunotherapy	Psychotherapy/physiotherapy
Prognosis	Most patients improve during adulthood	Variable	Usually persistent	Most patients improve following acute phase	Most patients improve following acute phase	Usually poor if therapy started beyond 1 year

AB, antibiotic; *ADHD*, attention-deficit/hyperactivity disorder; *BoNT*, botulinum toxin; *CBZ*, carbamazepine; *CNZ*, clonazepam; *Cs*, corticosteroids; *CZP*, clozapine; *DBS*, deep brain stimulation; *DRB*, dopamine receptor blockers; *OCD*, obsessive-compulsive disorder; *OCS*, obsessive-compulsive symptoms; *TS*, Tourette syndrome; *VPA*, valproic acid.

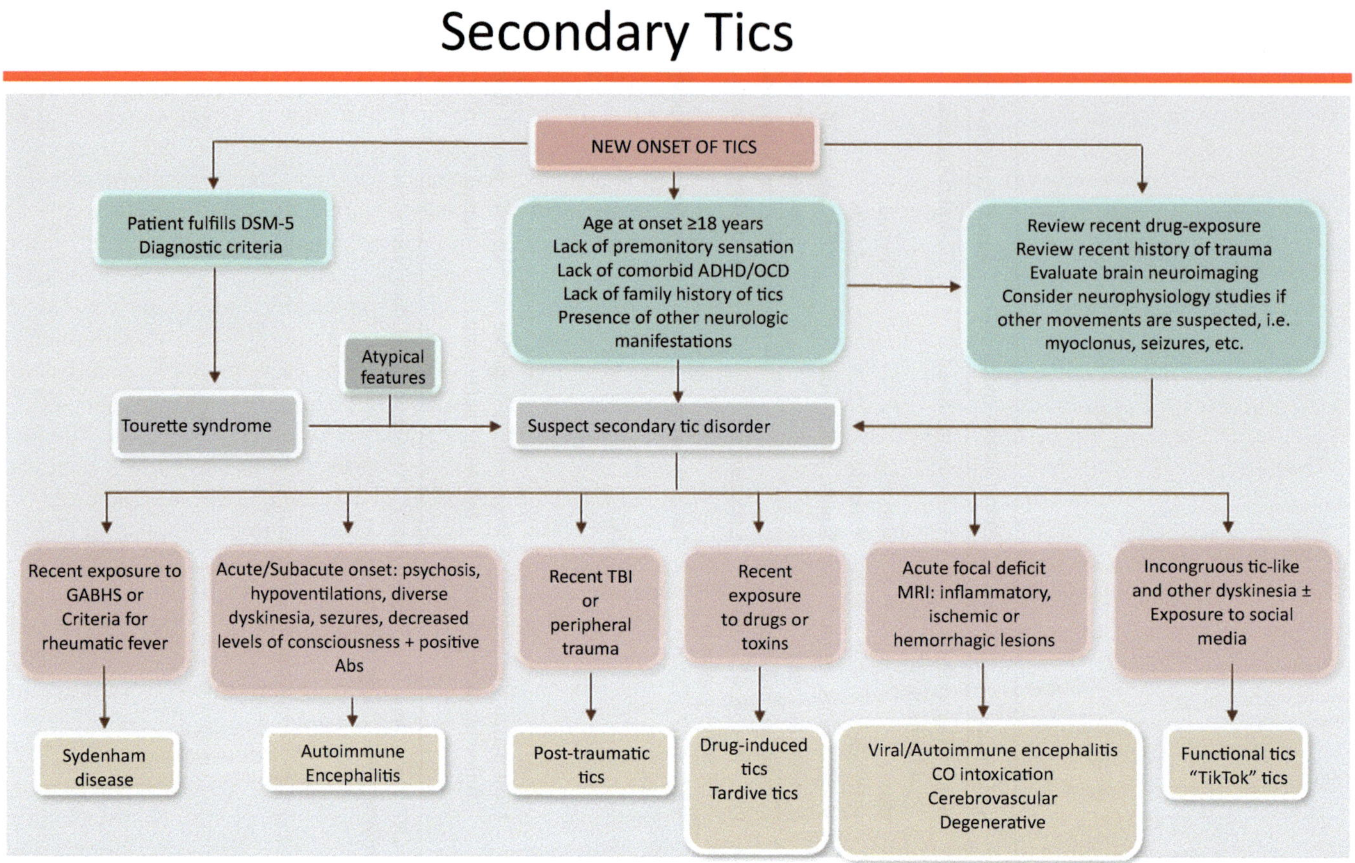

Fig. 3.1. Diagnostic algorithm for suspected secondary tic disorders. Reproduced with permission from Baizabal-Carvallo JF, Jankovic J (2024). *Handbook of clinical neurology.* Elsevier.

subclinical or lacking; they are described in juvenile and late-onset HD. A family history of HD should promptly raise suspicion that tics may be an early manifestation of this disease, although coincidental associations between HD and TS are possible (Kerbeshian et al., 1991). In patients with HD, tics may be replaced by frank chorea, and progressive cognitive and behavioral changes (Colosimo, 2015). However, the phenomenological distinction between certain tics and chorea in these patients may be blurred.

Neuroacanthocytosis syndromes, including chorea-acanthocytosis and McLeod syndrome, are a group of disorders characterized by cognitive decline, several abnormal movements, mostly chorea and oromandibular dystonia, with deformed erythrocytes (Hardie et al., 1991). Motor and phonic tics have been described in patients with neuroacanthocytosis syndromes, including facial tics and involuntary vocalizations, even coprolalia (Spitz et al., 1985; Danek et al., 2001; Dulski et al., 2016). Some of these patients may have comorbid OCD and ADHD (Saiki et al., 2004). In a study of 24 patients with neuroacanthocytosis from India, several phonic tics including grunting, loud throat clearing, sniffing, production of click consonants, and forceful exhalation were identified in 18 (75%) patients, whereas facial grimacing was observed in 8 (33%) cases (Neeraja et al., 2021). Tremor, parkinsonism, and myoclonus are not uncommon in patients with neuroacanthocytosis, while the "rubber man gait" with trunk flexion-extension during walking is considered a distinguishing feature in this disorder. These patients frequently develop self-injurious behavior, mostly confined to lip or cheek biting, whereas patients with TS will show different types of self-injurious behaviors comparatively less often (estimates are of the order of 17%) (Baizabal-Carvallo et al., 2022; Hull et al., 2023).

Panthotenate kinase-associated neurodegeneration (PKAN) is a rare autosomal recessive neurodegenerative disorder with brain iron accumulation (NBIA), secondary to mutations in the pantothenate kinase 2 gene. These patients frequently develop dystonia, parkinsonism, cognitive impairment, pigmentary retinopathy, and optic atrophy (Thomas et al., 2004). Selected patients with PKAN show a presentation that is indistinguishable from TS, including the presence of copropraxia (Molina da Costa et al., 2016; Rohani et al., 2018). These patients are usually young, and show a rapid progression and poor response to tics treatment; tic-like movements may be eventually replaced by dystonia, chorea, or parkinsonism, and cognitive decline may succeed (Nardocci et al., 1994; Scarano et al., 2002; Carod-Artal et al., 2004). The presence of bilateral globus pallidi interna hypo-intensity with a central hyperintensity on

T2-weighted MRI, the so-called "eye of the tiger" sign supports the diagnosis.

Neuroferritinopathy is an autosomal dominant disorder, considered within the spectrum of NBIA, caused by mutations in the ferritin light chain (FTL) gene (McNeill and Chinnery, 2012). Patients most frequently present with chorea, dystonia, and parkinsonism, although facial stereotypies and tics have been recognized in selected patients with appropriate response to tetrabenazine (Ondo et al., 2010). Progressive supranuclear palsy, a neurodegenerative tauopathy characterized by gaze palsy, parkinsonism, balance problems, and dementia, has been rarely associated with motor and phonic tics (Rojo et al., 1999).

Motor and vocal stereotypies (usually in a form of humming or other continuous sounds) or tic-like behaviors are frequently recognized in patients with frontotemporal dementia. Several mutations have been recognized in patients with fronto-temporal dementia (FTD), mostly microtubule-associated protein tau (MAPT), progranulin, or C9orf72 expansions (Baizabal-Carvallo and Jankovic, 2016). These movements have been reported to show an appropriate response to tetrabenazine (Ondo, 2012).

Few individuals with biallelic laminin α1 (LAMA1) mutations, the so-called Poretti–Boltshauser syndrome – a rare disorder characterized by cerebellar dysplasia, cerebellar cysts, myopia with or without retinal dystrophy (Elmas et al., 2020) – have also been described with tics, obsessive-compulsive traits, and anxiety (Vilboux et al., 2016). Neurons with knockdown LAMA1 genes show abnormal morphology and filopodia formation, whereas fibroblasts show multiple defects related to adhesion, migration morphology, and cell survival (Vilboux et al., 2016).

Although Wilson disease is often associated with a variety of hypokinetic and hyperkinetic MDs, there is no report of tics causally associated with Wilson disease.

DRUGS, TOXINS, POISONS

Medication-induced tics

Several medications have been associated with new-onset tics or tic exacerbation (Zesiewicz and Sullivan, 2011). Tics can be induced or paradoxically exacerbated by first- or second-generation antipsychotics. A clinical syndrome that is indistinguishable from TS characterizes tardive tourettism (Bharucha and Sethi, 1995). It usually follows a long exposure to dopamine receptor blockers, stimulants, antiepileptics, and other medications (Klawans et al., 1978; Stahl, 1980; Fountoulakis et al., 2011; Savitt and Jankovic, 2018). In a systematic review of the literature, 60 patients with antipsychotic-induced tics were identified (Kim et al., 2018). Tics occurred during antipsychotic treatment in 73.3% and after discontinuation in 26.6% of cases. It is not clear whether the tics that persisted represented a phenotype of tardive dyskinesia. About 30 (50%) cases were related to first-generation antipsychotics; the presence of phonic tics was statistically significant related to first-generation antipsychotic users. Patients are usually poorly responsive to medication changes and increasing the dose of antipsychotics may actually worsen or perpetuate tardive tourettism; therefore, neuroleptic withdrawal should be attempted; medications such as clonazepam, clozapine, or tetrabenazine have shown benefit in these patients (Klawans et al., 1978; Kuniyoshi et al., 1992; Reid, 2004). Other medications have been epidemiologically linked to tics. In an observational postmarketing study in France, involving cases registered in pharmacovigilance database (VigiBase) for more than three decades, methylphenidate, montelukast, and venlafaxine showed a statistically significant risk for tics, besides the antipsychotics aripiprazole and risperidone; however, it is unclear whether these cases represent exacerbation or new-onset tics (Touafchia et al., 2020). The Food and Drug Administration has approved two vesicular monoamine transporter 2 (VMAT2) inhibitors, valbenazine and deutetrabenazine, for the treatment of tardive dyskinesia, but none of the VMAT2 inhibitors have been approved for tics or TS, even though, in contrast to currently approved dopamine-receptor blocking drugs, VMAT2 inhibitors do not seem to cause tics or tardive phenomena (Savitt and Jankovic, 2018).

Patients with TS are frequently treated with dopamine receptor blockers and tardive phenomena have been identified in about 5.0% of these patients; however, classical tardive dyskinesia with the characteristic orolingual movements seems exceptional in patients with TS (Baizabal-Carvallo and Jankovic, 2021; Baizabal-Carvallo et al., 2023a). Comorbid tardive tourettism in patients with TS should be suspected when there is sudden exacerbation of tics, particularly if accompanied by other tardive syndromes, such as orofacial stereotypy or akathisia, in patients treated with dopamine receptor blockers. This scenario may also occur following withdrawal of these medications (Seeman et al., 1981; Mennesson et al., 1998).

Other medications, such as selective serotonin reuptake inhibitors (SSRIs), have been rarely reported in relation to tic emergence. Anecdotal reports of SSRI-induced tics, such as sertraline, have been reported in eight cases, mostly with adult-onset (Rua and Damásio, 2014; Arasu et al., 2022). Other sertraline-induced MDs such as akathisia or focal dystonia and trismus have been reported and should not be confused with tics (Wang et al., 2016). Most patients have a delayed onset of tics after few months of taking sertraline. Other SSRIs such as paroxetine and escitalopram have also been related to the emergence of tics (Rüth et al., 2000; Altindag et al., 2005). Discontinuation or switch to another SSRI has provided benefit in most of these patients.

Stimulants such as amphetamines, methylphenidate, or atomoxetine are frequently used to treat ADHD alone or comorbid with TS. Early case series in the decade of the 1970s and early 1980s linked the use of central nervous system (CNS) stimulants with tic onset or exacerbation in patients with TS (Fras and Karlavage, 1977; Pollack et al., 1977; Lowe et al., 1982). In a retrospective study including 39 patients with TS, stimulants increased tics in 11 of them (Erenberg et al., 1985). In another, short-term study, high-dose dextroamphetamine worsened tics (Castellanos et al., 1997). However, further randomized controlled trials did not confirm this association (Osland et al.,

2018). Although stimulants do not seem to increase tics in comparative studies, this phenomenon does occur in some individual cases (Erenberg, 2005). Cocaine is a drug that has potent stimulant effects in the nervous system with stimulation in the dopaminergic system. Cocaine has been associated with new-onset motor or phonic tics in adults or tic exacerbation in patients with TS (Pascual-Leone and Dhuna, 1990; Cardoso and Jankovic, 1993). This may occur in first- and chronic-time users (Deik et al., 2012).

Lithium has been associated with several MDs, mostly tremor and parkinsonism (Rissardo et al., 2022). In a literature review of 1100 cases treated with lithium, 6 individuals developed tics (Rissardo et al., 2022). Antiepileptics have also been associated with drug-induced tics. Among these medications, carbamazepine, oxcarbazepine, and eslicarbazepine have been associated with MDs (Robertson et al., 1993; Holtmann et al., 2000). In a study of 191 patients who developed MDs related to those medications, myoclonus was the most common in 33 cases, followed by dystonia in 23, and tics in 14 (Rissardo and Caprara, 2020). The dopaminergic effect of carbamazepine has been implicated in its potential for tic induction (Kurlan et al., 1989). Patients with carbamazepine-induced tics tend to show focal tics at nontoxic plasma levels of the medication, suggesting this is an idiosyncratic, rather than a dose-dependent reaction (Robertson et al., 1993).

Drug-induced tics should be suspected when they appear following the initiation of a new drug, particularly in adults. Although it is important to be vigilant about the possibility of drug-induced tics, clinicians must use their judgement before establishing that the tics in patients taking certain drugs are truly iatrogenic. The causative effect may not be established until the drug is gradually discontinued and the patient is observed for a period of time in a drug-free state.

Toxins and poisons

Intoxication with carbon monoxide (CO) is considered one of the most common causes of poisoning worldwide and it has been related with severe delayed neurologic sequelae (Hopkins et al., 2006). Gas CO is tasteless, odorless, and colorless; it is released into the environment by incomplete combustion of carbon-based substances (Hanley and Patel, 2023). It has 220% greater affinity to hemoglobin than oxygen, causing tissue hypoxia.

Tics have been reported in children and adults with CO intoxication, appearing 12 days following exposure to CO (Alioglu et al., 2004). They may be observed following a comatose phase and may present with several types of simple motor tics (Pulst et al., 1983; Tvedt et al., 1996). In selected cases, these tics may be transient and eventually disappear (Alioglu et al., 2004). However, coprolalia, fits of shouting, and "abnormal" vocal utterances have also been described in the context of tics related to CO intoxication (Pulst et al., 1983). Imaging findings typically show bilateral symmetric damage of basal ganglia in the acute phase that may be followed by atrophy or cavitation of the basal ganglia (Alioglu et al., 2004; Ko et al., 2004). The globus pallidus is the most commonly affected site; however, brain abnormalities are widespread and may include the caudate, putamen, white matter, cerebral cortex, or cerebellum (O'Donnell et al., 2000). In a review of 73 patients who suffered CO poisoning and underwent sequential MRI imaging, basal ganglia lesions occurred in the course of the disease in up to 88% of cases (Hopkins et al., 2006).

Patients with CO intoxication are believed to have pallidal and presynaptic dopaminergic dysfunction that may underlie motor abnormalities such as parkinsonism and may be involved in the pathogenesis of tics in selected cases (Chang et al., 2011). Supplemental oxygen is the cornerstone of therapy in patients with CO intoxication. However, patients with altered consciousness and cardiovascular depression should be emergently treated with hyperbaric oxygen (Hanley and Patel, 2023). Treatment with clonazepam or haloperidol has shown benefit in these patients (Alioglu et al., 2004; Ko et al., 2004).

High mercury level was associated with new-onset tics in a 5-year-old child, treated with "Watermelon Frost," a Chinese medicinal herb mouth spray (Li et al., 2000). Whether high mercury levels may trigger tics is unclear.

DISORDERS CAUSING IDENTIFIABLE STRUCTURAL BRAIN DAMAGE

Traumatic brain injury

Tics have been linked to diverse types of physical trauma, including traumatic brain injury (TBI) and peripheral trauma (Lenka and Jankovic, 2023). TBI is classified as mild, moderate, and severe, depending on the level of consciousness measured with the Glasgow coma scale, although this classification has been criticized for being simplistic (Hawryluk and Manley, 2015). Tics following TBI have been described in case reports and series. It is unclear, however, whether posttraumatic tics represent de novo tics or latent TS being triggered by trauma. In a series of 155 patients with tics, 2.5% occurred de novo or significantly worsened following head trauma, even without structural brain damage (Mejia and Jankovic, 2005). In a review of 25 patients (22 adults) with tics following TBI, 68% were male. Motor vehicle accidents were the most common cause of trauma (80%), followed by falls (8.0%) and being struck by an object (4.0%) (Ricketts et al., 2019).

The latency between TBI and tics has been reported to be between few hours or one day after TBI up to several months following head trauma (Krauss and Jankovic, 1997; Ranjan et al., 2011). Another study reported a latency between 2 weeks and 3 months, following TBI (Alegre et al., 1996). Patients with post-TBI tourettism present mainly with simple motor and phonic tics distributed in the face and neck; complex motor and phonic tics, coprolalia, and echolalia were identified in 20% of cases, similar to what it is observed in patients with TS (Ricketts et al., 2019). Moreover, obsessive-compulsive symptoms, a common comorbidity of TS, have been reported in about 20%–25% of patients with post-TBI tics (Krauss and Jankovic, 1997; Ricketts et al., 2019).

The frequency of comorbid ADHD is less clear. It is unclear whether patients with tics following penetrating brain injuries present with different phenomenology compared with patients with postconcussion tics. If the phenomenology or clinical course of the latter appear atypical, one should consider the possibility of a posttraumatic functional MD (see discussion below) (Ganos et al., 2016a, 2016b).

It is unclear, whether TBI may influence the course of tics in TS. About 16% of patients with post-TBI tics have been identified with a family history of tics, suggesting that some of these patients have underlying vulnerability to develop tics following traumatic events (Ricketts et al., 2019). Furthermore, there are reported cases where TS worsened following head trauma (Koller et al., 1989). However, about two-thirds of cases do not have any family history of tics or comorbid OCD/ADHD, making unclear if these individuals have a higher risk to develop tics (Ricketts et al., 2019).

Lesions in the basal ganglia and other brain structures have been identified in patients with post-TBI tics (Ranjan et al., 2011). Basal ganglia injuries, including caudate, putamen, or globus pallidus have been found to occur in 18% of cases without clear right/left laterality; other patients have variable unilateral or bilateral lesions in frontal, temporal, or parietal lobes (Ricketts et al., 2019). Volumetric assessment in 83 Vietnam War veterans who suffered penetrating brain injuries showed that volume loss in the left dorsolateral prefrontal cortex was associated with compulsive behavior, whereas volume loss in the basal ganglia was associated with development of tics, even though details on tic phenomenology are lacking (Fremont et al., 2022). Moreover, the role of allopregnanolone in the onset of tics following traumatic events has been recently highlighted. This neurosteroid increases rapidly in the brain following acute stressful events and is believed to have protective effects by providing mechanism of central nervous system coping and reducing allostatic load (Bortolato et al., 2022). However, studies in animal models have shown that increases in allopregnanolone may cause disinhibition and tic-like behaviors (Bortolato et al., 2022).

Peripheral trauma (not involving the CNS) has been associated with tics in 0.6% of patients with STS (Mejia and Jankovic, 2005; Lenka and Jankovic, 2023). Tics have been reported to develop between hours and 12 months following a peripheral injury (Factor and Molho, 1997). Tics developed in the anatomic site that suffered the injury, including the face, neck, and arm and were preceded by an inner urge (Factor and Molho 1997; Erer and Jankovic, 2008). These tics tend to wax and wane over time, are suppressible, and may remit spontaneously, thus suggesting the possibility of a functional component (Factor and Molho, 1997). Lack of comorbid ADHD or OCD has been noted in these patients (Erer and Jankovic, 2008). Abnormal sensory processing and maladaptative plasticity in the sensorimotor cortex are presumed to play a role in peripherally induced MDs (Lenka and Jankovic, 2023). A lack of an identifiable biomarker makes it difficult to establish a definite relationship between the peripheral trauma and tics; moreover, functional MDs, sometimes associated with complex regional pain syndrome, may also be seen in the context of peripheral trauma (Lenka and Jankovic, 2023).

Treatment of posttraumatic tics typically involves the use of VMAT2 inhibitors, benzodiazepines, or botulinum toxin (Lotia and Jankovic 2016; Billnitzer and Jankovic, 2020; Anandan and Jankovic, 2021). However, poor response has been reported in about a quarter of cases (Krauss and Jankovic, 1997; Poulopoulos and Hajjar, 2013).

Cerebrovascular disease

The onset of tics in some cases has been temporarily associated with subcortical ischemic damage involving the basal ganglia (Kwak and Jankovic, 2002; Gomis et al., 2008). Tics developed within 2– 3 weeks following the ischemic damage. One adult patient developed an isolated phonic tic (Gomis et al., 2008), while two children had isolated motor tics and hemidystonia (Kwak and Jankovic, 2002). Tics have been described in an 11-year-old girl with MELAS (Mitochondrial myopathy, Encephalopathy, Lactic Acidosis and Stroke-like episodes) and in a 16-year-old boy with a frontal hemorrhage secondary to an arteriovenous malformation (Yochelson and David, 2000; van Rossum and ten Houten, 2010). Diverse types of tics following cardiopulmonary bypass and hypothermia have been described in an adolescent male who showed progressive basal ganglia atrophy to that observed in TS (Singer et al., 1997). Tics following cerebrovascular events are considered rare and it is unclear if damage to brain parenchyma causes or just triggers tics in a vulnerable individual.

Other causes of structural brain lesions

Parenchymal brain tumors have been rarely associated with tics. There is a report of a child with an oligodendroglioma in the right temporal lobe extending to the ipsilateral basal ganglia, who developed stimulant-induced tics, OCD and ADHD has been reported (Luat et al., 2009). The role of basal ganglia lesions or drug exposure was unclear. Few patients with tourettism have been described in the context of multiple sclerosis (MS) (Mehanna and Jankovic, 2013). These patients may have motor and phonic tics and other features of tourettism that developed years after the onset of MS along with high burden of demyelinating brain lesions (Nociti et al., 2009). Whether they represent lesion-induced tics or a coincidental association between MS and TS is unclear. It has not been established that MS causes tics and each patient with MS who has coexistent tics must be evaluated individually in an attempt to postulate or refute a cause–effect relationship. Primary calcification syndromes (PCS) may present with variable cognitive, neuropsychiatric, or abnormal movements. Most of these syndromes are autosomal dominant and various genes have been implicated, including PDGFB, PDGFRB, SLC20A2, and XPR1, whereas MYORG and JAM2 are autosomal recessive (Xu et al., 2023). Three male patients between 40 and 64 years have been described with new-onset phonic tics, such as grunting, throat clearing associated with dystonia, parkinsonism, spasticity, stereotypies, etc. (Mulroy et al., 2021). Whether PCS have a selective predisposition for abnormal vocalization but not motor tics is unclear.

INFECTIOUS AND PARAINFECTIOUS DISORDERS

Viral infections

Viral infections affecting the brain parenchyma have been associated with tics in individual case reports or series. Herpes simplex virus (HSV) types 1 and 2 are double-stranded DNA neurotropic viruses that belong to the Herpesviridae family and a frequent cause of infectious encephalitis worldwide (Steiner and Benninger, 2013). Herpes simplex virus type 1 encephalitis most commonly affects the temporal lobes, but the basal ganglia may also be involved (Steiner and Benninger, 2013). Direct damage of brain parenchyma, associated with dyskinesia, has been also linked to the effects of anti-NMDAR antibodies. Simple and complex tics have been reported in children with HSV encephalitis, two weeks after recovering from this type of encephalitis (Turley, 1988; Northam and Singer, 1991). Tics have been observed during the acute course and following resolution of varicella zoster virus encephalitis (Dale et al., 2003). However, this type of encephalitis has not been associated with a specific humoral response, but abnormalities in basal ganglia may be observed in neuroimaging (Dale et al., 2003).

Encephalitis by flavivirus infections have been occasionally related to tics. Japanese encephalitis is a mosquito-borne infection caused by a single-stranded RNA virus (Ashraf et al., 2021). It is one of the most common forms of endemic encephalitis in humans, distributed mostly in Southeast Asia (Ashraf et al., 2021). MDs have been identified in up to 76% of patients with Japanese encephalitis (Kalita et al., 2003). Various hyperkinetic movements were identified in an early series of 68 patients affected by Japanese encephalitis virus; seven patients were found with motor tics, and chorea and athetosis at five years after the acute phase of the disease (Richter and Shimojyo, 1961). More recent series have identified parkinsonism and diverse types of focal dystonia as the most common MDs in patients suffering from Japanese encephalitis (Aryal et al., 2022). In addition, myoclonic jerks, orofacial dyskinesia, oculogyric crises, and ocular deviations have also been reported in some patients; the latter, however, may represent a form of oculogyric tics (Kalita et al., 2003; Baizabal-Carvallo and Jankovic, 2022). West Nile virus, another flavivirus-borne infection, can result in encephalitis in a small proportion of infected patients. The emergence of limb athetosis and facial tics was reported in a 5-year-old patient during the recovery phase of West Nile encephalitis (Francisco et al., 2006). Other MDs such as opsoclonus-myoclonus, tremor, parkinsonism, myoclonus, ataxia, and chorea have been identified in patients with West Nile encephalitis (Lenka et al., 2019). Bilateral thalamic, basal ganglia, and brainstem involvement is particularly common in patients with flavivirus encephalitis and are suspected to underlie the pathogenesis of MDs (Guth et al., 2014; Keng and Chang, 2018; Herring et al., 2019).

MDs have been identified as the presenting symptom in some patients with HIV infection including tics, dystonia, chorea, and hemiballism (Mirsattari et al., 1999). Tics have been described during the early or acute phase of HIV infection, but mostly during the chronic phase (Nath and Jankovic, 1989). Whether there is an intrinsic vulnerability of certain patients with HIV to develop tics is possible, as the majority of patients with HIV and CNS involvement do not develop tics (Antonelli et al., 2011). In HIV patients presenting with abnormal movements, direct neurotrophic effects of the virus or accompanying immunologic reactions have been implicated, whereas, in cases appearing years following initial infection, the effect of medications or opportunistic infections may play a role. A case of new-onset dystonic tics involving craniocervical muscles secondary to neurotoxoplasmosis, a well-identified opportunistic infection in HIV, has been described (Aquino et al., 2010).

Tics have been reported in children or adolescents during the acute or postinfectious phase of mumps, characterized by symptoms such as eye-blinking, throat clearing, and finger movements; in cases of measles leading to subacute sclerosing panencephalitis, where complex facial and shoulder tics may manifest as an initial symptom; and in conditions such as neurobrucellosis and *Mycoplasma pneumoniae* infections, where simple motor and phonic tics, including coughing, can occur (Badenoch et al., 2021). Several MDs have been identified in patients with COVID-19 infection. Among these movements, myoclonus, ataxia and tremor are the most commonly described (Brandão et al., 2021). Currently, there is no evidence that a neurotropic effect of COVID-19 causes tics.

Sydenham disease

Sydenham disease (SD) (also known as Sydenham chorea) is considered a parainfectious disorder, as it is a delayed clinical manifestation of Group A β-hemolytic streptococcal (GABHS) infections. SD is a major criterion for rheumatic fever in the modified Jones criteria (Cardoso, 2017). Indeed, SD can be observed in about a quarter of patients with rheumatic fever. The disorder is most commonly seen in children between 5 and 15 years old (Baizabal-Carvallo and Cardoso, 2020). Although chorea is the distinguishing feature, other motor and neuropsychiatric manifestations are frequently observed; therefore, the term "Sydenham disease" seems more appropriate. Motor and phonic tics have been reported in patients with SD (Mercadante et al., 1997) and in association with streptococcal infections (Baiardi et al., 2014). Vocalizations and other involuntary sounds have been detected in 7.8% of patients with SD; however, the lack of premonitory phenomena in those cases cast doubts whether they represent true tics (de Teixeira et al., 2009). In a study of 40 patients with dyskinesia and recent GABHS infections, chorea was the most common manifestation (*n*=20, 50%), followed by motor tics in 16 patients (40%); however, replication studies are lacking (Dale et al., 2004). Although the premonitory urge is a typical feature of tics in patients with TS, similar sensory phenomena have been reported in 10 of 13 (77%) of patients with SD and choreic movements (Rodopman-Arman et al., 2004).

Another overlapping feature between SD and TS is the frequent presence of neuropsychiatric comorbidities, including OCD, attention-deficit/hyperactivity disorder (ADHD), depression, anxiety, or conduct disorders (Dale et al., 2004). Obsessive-compulsive symptoms have been identified in up to 70% of individuals with SD and 16% fulfill diagnostic criteria for OCD (Asbahr et al., 1998). Moreover, there is evidence that patients with rheumatic fever (RF) with or without SD have a relatively high risk of developing OCD within the spectrum of TS (Mercadante et al., 2000; Hounie et al., 2004; Walker et al., 2005; de Alvarenga et al., 2009). Currently, there is no definite association between SD or RF and tics, as in many cases the involuntary movements seem to represent mild or fragments of chorea resembling tic-like movements.

The diagnosis of SD is based on documentation of a recent GABHS infection by clinical history, antibody testing (anti-streptolysin-O or anti-DNAse), or throat cultures. Clinicians should be aware for the possibility of carditis or cardiac valve involvement, which have been reported in up to 70.5% in a 7-year follow-up of 65 SD patients (Demiroren et al., 2007). The etiology of SD is not fully understood. Molecular mimicry with production of antibodies cross-reacting to GABHS and basal ganglia antigens, including D1 and D2 dopamine receptors, is presumed to play a role as an underlying pathogenic mechanism (Cunningham, 2019). However, these antibodies lack specificity, and a clear molecular target has not been identified (Baizabal-Carvallo and Cardoso, 2020).

A lively debate has arisen linking autoimmunity with tics, choreiform movements, obsessive-compulsive and other neuropsychiatric manifestations, but not carditis or arthritis in a model different from RF or SD. The term Pediatric Autoimmune Neuropsychiatric Disorder Associated with Streptococcal infection (PANDAS) was initially coined and then replaced by Pediatric Acute-Onset Neuropsychiatric Syndrome (PANS), as no consistent association was found with GABHS infections and other microorganisms such as *Mycoplasma* or influenza were considered to be implicated (Baizabal-Carvallo and Jankovic, 2018). Recently, the European Multicentre Tics in Children Studies (EMTICS) prospectively studied a large cohort of individuals; no evidence was found for the role of new GABHS infection with tic exacerbation or new-onset tics (Schrag et al., 2022). However, the study has been criticized owing to potential selection bias (Gupta et al., 2022). There is now a general consensus that the co-occurrence of GABHS and tic exacerbations occur by chance without direct etiopathogenic link (Martino et al., 2021). Therefore, assessing GABHS in patients with tic disorders may not be clinically meaningful and immunotherapy is not recommended.

Encephalitis lethargica

Encephalitis lethargica (EL) was first described in 1917 by Austrian neurologist Constantin von Economo (Hoffman and Vilensky, 2017). The disorder was first linked to the Spanish influenza pandemic, occurring between years 1917 and 1929; however, the etiology of the disorder has been debated, as cases have been described outside the pandemic. Moreover, an autoimmune cause has been suspected, as in modern reported cases of EL, an autoimmune cause has been identified (Di Vito et al., 2023). Three main clinical presentations have been described: (1) hypersomnolent-ophthalmoplegic form, (2) amyostatic-akinetic form, and (3) hyperkinetic form; all three preceding postencephalitic parkinsonism (PEP). However, the direct relationship between EL and PEP has been questioned (Hoffman and Vilensky, 2017). Several types of dyskinesia may present during the acute or during the chronic postencephalitic phase, including dystonia, catatonia, chorea, and tics (Vilensky et al., 2006). Diverse types of simple and complex tics have been described in patients with EL, such as oculogyric tics, phonic tics with compulsive shouting episodes (klazomania), and coprolalia (Badenoch et al., 2021). Tics in patients with PEP can occur following an initial good response with levodopa. This phenomenon is described in the book "Awakenings" by Oliver Sacks and portrayed in the movie with the same name, starring Robin Williams and Robert de Niro (Sacks, 1990; Hoffman and Vilensky, 2017).

AUTOIMMUNE ENCEPHALITIS

Anti-NMDAR encephalitis

Tics or tic-lie movements may also occur in patients with autoimmune encephalitis. Anti-*N*-methyl-D-aspartate receptor (anti-NMDAR) antibody-mediated encephalitis is the most common autoimmune MD (Florance et al., 2009). Patients with anti-NMDAR encephalitis may present with a wide variety of MDs including dystonia, chorea, tremor, stereotypies, orofacial dyskinesia, or myorhythmia, but tics are also rarely described in these patients (Baizabal-Carvallo et al., 2013). Patients may present with chorea, stereotypic movements, muscles jerks, or twitching, some of them resembling tics, but these movements may be difficult to characterize clinically as tics. In a case series of 11 children with anti-NMDAR encephalitis from Malaysia, two of them were observed with motor dystonic tic-like movements (Badenoch et al., 2021). Facial tics have been reported in a child from India, but without clear documentation, it is difficult to state whether the facial movement were tics or stereotypies (Datta et al., 2022). Loud "barky" phonic tics have been described in 11- and 4-year-old females with anti-NMDAR encephalitis, also displaying orofacial dyskinesia (Berkeley and Sohoni, 2010; Safadieh and Dabbagh, 2013). Owing to the variety and overlapping abnormal movements in patients with anti-NMDAR, it is difficult to establish whether reported movements are tics or other brief jerky phenomena. Prompt recognition of this disorder to initiate immunotherapy and search for an underlying teratoma or carcinoma are essential for a good prognosis (Nosadini et al., 2021). Improvement of hyperkinetic movements with tetrabenazine and other VMAT2 inhibitors has been described in patients with STS and other tic-like movements (Baizabal-Carvallo et al., 2013; Makhoul and Jankovic, 2023).

Other causes of autoimmune encephalitis

Tics have been identified in patients with other autoimmune disorders, including encephalitis syndromes associated with antivoltage-gated K channel/LGI1 or CASPR2 antibodies: a 13-year-old-male presented with severe cervical tics, whereas a 72-year-old female presented with orofacial tics and echolalia (Muñiz-Castrillo et al., 2020). A 72-year-old Hispanic woman with orofacial tics, echolalia, generalized hypertonia, dysmetria, and complex hallucinations associated with LGI1 antibodies has been described, but the phenomenology was not well characterized (Shin et al., 2013). Antibodies to glycine receptors (GlyR) usually present with progressive encephalitis with rigidity and myoclonus, and less commonly with epilepsy or atypical syndromes. A patient with anti-GlyR encephalitis, clonus, laryngeal dystonia, hemiballismus, and tics has been described (Swayne et al., 2018). New-onset tongue biting has been noted in a 13-year-old boy with tics and positive dopamine-2 receptor antibodies who improved with immunotherapy (Lai et al., 2022). Another 12-year-old boy was reported with Hashimoto's encephalitis (also known as steroid-responsive encephalopathy with autoimmune thyroiditis) and tics, but better characterization of the hyperkinetic disorder is needed, as they may represent myoclonus, which is a commonly recognized MD in these patients (Saygi et al., 2014).

FUNCTIONAL TIC DISORDERS

Functional (previously known as psychogenic) tics or tic-like movements have been detected in about 5% of patients with functional MDs. These patients may present with movements that highly resemble the tics of TS, including simple phonic tics such as sniffing. Coughing, moaning, gagging, groaning, repeating syllables "ha," "eh," and even complex vocalizations such as coprolalia (Baizabal-Carvallo and Jankovic, 2014; Ganos et al., 2016a, 2016b); however, they usually endorse a lack of transitory movement suppressibility, uncommon premonitory sensations, and frequent associated functional MDs (Baizabal-Carvallo and Jankovic, 2014). Of note, clinical differentiation between some cases with either functional or primary tics as in TS may be particularly difficult, and even more so, when a functional overlay is suspected on top of a primary tic disorder (Rigas et al., 2023). More recently, a syndrome consisting of rapid-onset and exaggerated tic-like movements or behaviors occurring after social media exposure to particular personalities and behaviors, so-called "Tik-Tok" tics, has been identified during the COVID-19 pandemic (Hull et al., 2021). These patients showed frequent coprolalia, aggressive behavior, object throwing, self-injurious behavior, and uttering long phrases and greater severity compared to patients with TS (Pringsheim et al., 2021; Zea Vera et al., 2022). A detailed discussion of functional tics is beyond the scope of this review and is provided in another chapter of this volume.

APPROACH AND MANAGEMENT

Overall STS seem substantially less common than TS. Secondary tics should be suspected when age at onset is later than 18 years of age, there are no typical comorbid OCD or ADHD, or family history of tics, but there are predisposing or triggering events, drugs or diseases, or abnormal brain neuroimaging. It should be noted that re-emergence of childhood-onset tics is the most common cause of new-onset tics during adulthood; therefore, a through clinical history is needed to rule out this possibility (Robakis, 2017). In one study of 43 adults with a mean age of 58.8 years at the first clinical assessment who presented with new-onset tics, 81.4% had in fact a history of tics before the age of 18 years (Jankovic et al., 2010). Adult patients with TS had lower rates of phonic tics and ADHD, but higher rates of substance abuse and mood disorders compared with children with TS (Jankovic et al., 2010). However, in another case series of 22 adults with tic onset after the age of 21 years, 13 patients did not have previous history of younger onset tics and in six cases, tics developed following an external trigger and were considered STS (Chouinard and Ford, 2000).

When STS is suspected, a review of clinical features for hereditary disorders, drug exposure, and recent history of head trauma with structural brain damage should be carried out. If there are clinical features that are not consistent with TS or functional tics, brain imaging or even cerebrospinal fluid (CSF) analysis (e.g., for anti-NDMAR antibodies) should be considered. This approach will help to detect inflammatory, autoimmune, traumatic, neoplastic, or ischemic lesions affecting basal ganglia or discrete lesions in the temporal lobes. Moreover, basal ganglia atrophy or iron accumulation (e.g., "eye of the tiger" sign) in some hereditary neurodegenerative disorders strongly support the diagnosis of neurodegenerative disorders such as HD or PKAN, to be confirmed by genetic tests. In case an autoimmune or parainfectious disorder is suspected, comprehensive blood and CSF antibody testing should be performed.

Patients with TS show a different clinical presentation compared to patients with STS. In a study comparing 156 patients with TS with 38 diagnosed with STS, 21 of them had functional tics. Patients with TS had greater frequency of tics in the craniocervical area and showed greater severity and complexity of tics when compared with patients with diverse STS (Baizabal-Carvallo et al., 2023b). When assessing diagnostic accuracy, phonic tics scored highest for sensitivity and specificity for the presence of TS; however, there is no definite clinical or diagnostic feature that reliably discriminate between the two types of tic disorders (Baizabal-Carvallo et al., 2023b).

Comorbid neuropsychiatric manifestations such as ADHD or OCD are described in about 60% of patients with TS (Hirschtritt et al., 2015). Interestingly, these disorders are not uncommon in patients with STS, with frequency roughly of 20% or 25%. Abnormalities in the cortical-basal ganglia-thalamic-cortex loop are deemed important in the pathogenesis of tics and comorbid neuropsychiatric disorders in TS and may be implicated in the pathogenesis of obsessive or compulsive symptoms (Lombroso and Scahill, 2008). It is possible that an acquired lesion strategically located in the limbic, associative, or motor cortical-basal ganglia-thalamic-cortex loop, particularly the anterior cingulate cortex and insula, may cause disruption in the dopaminergic and/or amma-aminobutyric acid (GABA) ergic neurotransmission leading to discrete disinhibition of these loops (Kalanithi et al., 2005; Johnson et al., 2023). Recently, the cerebellum has been highlighted as potentially important structure in the pathogenesis of tics (Caligiore et al., 2017). A common brain tic-generation network, which includes the insular cortex, cingulate gyrus, striatum, globus pallidus internus, thalamus, and cerebellum, has been identified by connectome analysis (Ganos et al., 2016a, 2016b). The same work identified the anterior striatum (putamen) as a highly specific area for tics (Ganos et al., 2016a, 2016b).

Pharmacologic treatment of STS follows the same principles as that of patients with TS. Antidopaminergic drugs, such as the VMAT2 inhibitors and dopamine receptor antagonists, clonazepam or botulinum toxin, have been used with variable success in patients with STS, and there is no clear evidence that they have lower efficacy than in TS, except for patients with functional tics where counseling and other behavioral approaches seem to be more appropriate. Identification of STS is important, as some patients may require etiology-specific treatments.

CONCLUSIONS

Several disorders, besides TS, may cause tics. In some cases, genetic or other predisposition, including TS, may play a role in the pathogenesis of these secondary tics. When the diagnosis of STS is suspected, a rapid evaluation is strongly recommended in order to initiate etiology-specific therapy.

FINANCIAL DISCLOSURE/CONFLICT OF INTEREST

None for all authors.

References

Abelson JF, Kwan KY, O'Roak BJ et al. (2005). Sequence variants in SLITRK1 are associated with Tourette's syndrome. Science 310: 317–320.

Alegre S, Chacón J, Redondo L et al. (1996). Tics postraumáticos [Post-traumatic tics]. Rev Neurol 24: 1280–1282.

Alioglu Z, Boz C, Sari A et al. (2004). Transient tic disorder following carbon monoxide poisoning. J Neuroradiol 31: 231–233.

Altindag A, Yanik M, Asoglu M (2005). The emergence of tics during escitalopram and sertraline treatment. Int Clin Psychopharmacol 20: 177–178.

American Psychiatric Association (2013). Diagnostic and statistical manual of mental disorders, 5th ed. Arlington, VA: Author.

Anandan C, Jankovic J (2021). Botulinum toxin in movement disorders: an update. Toxins (Basel) 13: 42.

Antonelli F, Borghi V, Galassi G et al. (2011). Can HIV infection cause a worsening of tics in Tourette patients? Neurol Sci 32: 191–192.

Aquino CC, Felício AC, Godeiro-Junior C et al. (2010). Tic disorder: an unusual presentation of neurotoxoplasmosis in a patient with AIDS. Case Rep Neurol 2: 145–149.

Arasu R, Badeshae S, Furlong Y et al. (2022). Sertraline-induced tics: a case report and narrative review. J Pharm Pract 36: 1528–1531.

Aryal R, Shrestha S, Homagain S et al. (2022). Clinical spectrum and management of dystonia in patients with Japanese encephalitis: a systematic review. Brain Behav 12: e2496.

Asbahr FR, Negrão AB, Gentil V et al. (1998). Obsessive-compulsive and related symptoms in children and adolescents with rheumatic fever with and without chorea: a prospective 6-month study. Am J Psychiatry 155: 1122–1124.

Ashraf U, Ding Z, Deng S et al. (2021). Pathogenicity and virulence of Japanese encephalitis virus: neuroinflammation and neuronal cell damage. Virulence 12: 968–980.

Badenoch J, Searle T, Watson I et al. (2021). Tics in patients with encephalitis. Neurol Sci 42: 1311–1323.

Baiardi S, Antelmi E, Filardi M et al. (2014). Remitting tics and narcolepsy overlap associated with streptococcal infection: a case report. Mov Disord Clin Pract 1: 374–376.

Baizabal-Carvallo JF, Alonso-Juarez M, Jankovic J (2022). Self-injurious behavior in Tourette syndrome. J Neurol 269: 2453–2459.

Baizabal-Carvallo JF, Alonso-Juarez M, Jankovic J (2023). Contrasting features between Tourette syndrome and secondary tic disorders. J Neural Transm (Vienna) 130: 1–6.

Baizabal-Carvallo JF, Alonso-Juarez M, Jankovic J (2023). Oromandibular tics associated with Tourette syndrome. J Neurol 270: 2591–2596.

Baizabal-Carvallo JF, Cardoso F (2020). Chorea in children: etiology, diagnostic approach and management. J Neural Transm (Vienna) 127: 1323–1342.

Baizabal-Carvallo JF, Jankovic J (2014). The clinical features of psychogenic movement disorders resembling tics. J Neurol Neurosurg Psychiatry 85: 573–575.

Baizabal-Carvallo JF, Jankovic J (2016). Parkinsonism, movement disorders and genetics in frontotemporal dementia. Nat Rev Neurol 12: 175–185.

Baizabal-Carvallo JF, Jankovic J (2018). Autoimmune and paraneoplastic movement disorders: an update. J Neurol Sci 385: 175–184.

Baizabal-Carvallo JF, Jankovic J (2021). Beyond tics: movement disorders in patients with Tourette syndrome. J Neural Transm (Vienna) 128: 1177–1183.

Baizabal-Carvallo JF, Jankovic J (2022). The clinical phenomenology and correlations of oculogyric tics. Acta Neurol Belg 122: 925–930.

Baizabal-Carvallo JF, Stocco A, Muscal E et al. (2013). The spectrum of movement disorders in children with anti-NMDA receptor encephalitis. Mov Disord 28: 543–547.

Barabas G, Wardell B, Sapiro M et al. (1986). Coincident Down's and Tourette syndromes: three case reports. J Child Neurol 1: 358–360.

Beglinger LJ, Paulsen JS, Watson DB et al. (2008). Obsessive and compulsive symptoms in prediagnosed Huntington's disease. J Clin Psychiatry 69: 1758–1765.

Berkeley RP, Sohoni A (2010). An 11-year-old female with altered mental status, speech changes, and abnormal jerking movements. Acad Emerg Med 17: 723–728.

Bharucha KJ, Sethi KD (1995). Tardive tourettism after exposure to neuroleptic therapy. Mov Disord 10: 791–793.

Billnitzer A, Jankovic J (2020). Current management of tics and Tourette syndrome: behavioral, pharmacologic, and surgical treatments. Neurotherapeutics 17: 1681–1693.

Bortolato M, Coffey BJ, Gabbay V et al. (2022). Allopregnanolone: the missing link to explain the effects of stress on tic exacerbation? J Neuroendocrinol 34: e13022.

Brandão PRP, Grippe TC, Pereira DA et al. (2021). New-onset movement disorders associated with COVID-19. Tremor Other Hyperkinet Mov (NY) 11: 26.

Caligiore D, Mannella F, Arbib MA et al. (2017). Dysfunctions of the basal ganglia-cerebellar-thalamo-cortical system produce motor tics in Tourette syndrome. PLoS Comput Biol 13: e1005395.

Cardoso F (2017). Autoimmune choreas. J Neurol Neurosurg Psychiatry 88: 412–417.

Cardoso FE, Jankovic J (1993). Cocaine-related movement disorders. Mov Disord 8: 175–178.

Carod-Artal FJ, Vargas AP, Marinho PB et al. (2004). Tourettism, hemiballism and juvenile Parkinsonism: expanding the clinical spectrum of the neurodegeneration associated to pantothenate kinase deficiency (Hallervorden Spatz syndrome). Rev Neurol 38: 327–331.

Castellanos FX, Giedd JN, Elia J et al. (1997). Controlled stimulant treatment of ADHD and comorbid Tourette's syndrome: effects of stimulant and dose. J Am Acad Child Adolesc Psychiatry 36: 589–596.

Chang CC, Chang WN, Lui CC et al. (2011). Clinical significance of the pallidoreticular pathway in patients with carbon monoxide intoxication. Brain 134: 3632–3646.

Chouinard S, Ford B (2000). Adult onset tic disorders. J Neurol Neurosurg Psychiatry 68: 738–743.

Collacott RA, Ismail IA (1988). Tourettism in a patient with Down's syndrome. J Ment Defic Res 32: 163–166.

Colosimo C (2015). A case of atypical adult-onset tic disorder. Neurologist 19: 99–100.

Comings DE, Comings BG (1991). Clinical and genetic relationships between autism-pervasive developmental disorder and Tourette syndrome: a study of 19 cases. Am J Med Genet 39 : 180–191.

Cui SS, Ren RJ, Wang Y et al. (2017). Tics as an initial manifestation of juvenile Huntington's disease: case report and literature review. BMC Neurol 17: 152.

Cunningham MW (2019). Molecular mimicry, autoimmunity, and infection: the cross-reactive antigens of group a streptococci and their sequelae. Microbiol Spectr 7: 10.

Dale RC, Church AJ, Heyman I (2003). Striatal encephalitis after varicella zoster infection complicated by Tourettism. Mov Disord 18: 1554–1556.

Dale RC, Heyman I, Surtees RA et al. (2004). Dyskinesias and associated psychiatric disorders following streptococcal infections. Arch Dis Child 89: 604–610.

Danek A, Tison F, Rubio J et al. (2001). The chorea of McLeod syndrome. Mov Disord 16: 882–889.

Datta AK, Ghosh PC, Bera M et al. (2022). A video-based discussion of movement disorders in paediatric anti NMDAR encephalitis: a case series from Eastern India. Eur J Paediatr Neurol 37: 40–45.

de Alvarenga PG, Floresi AC, Torres AR et al. (2009). Higher prevalence of obsessive-compulsive spectrum disorders in rheumatic fever. Gen Hosp Psychiatry 31: 178–180.

Deik A, Saunders-Pullman R, Luciano MS (2012). Substance of abuse and movement disorders: complex interactions and comorbidities. Curr Drug Abuse Rev 5: 243–253.

Demiroren K, Yavuz H, Cam L et al. (2007). Sydenham's chorea: a clinical follow-up of 65 patients. J Child Neurol 22: 550–554.

de Teixeira AL, Cardoso F, Maia DP et al. (2009). Frequency and significance of vocalizations in Sydenham's chorea. Parkinsonism Relat Disord 15: 62–63.

Di Vito A, Donato A, Bria J et al. (2023). Encephalitis lethargica. What is still wrong? Int J Immunopathol Pharmacol 37: 3946320231154997.

Dulski J, Sołtan W, Schinwelski M et al. (2016). Clinical variability of neuroacanthocytosis syndromes – a series of six patients with long follow-up. Clin Neurol Neurosurg 147: 78–83.

Elmas M, Gogus B, Solak M (2020). Understanding what you have found: a family with a mutation in the LAMA1 gene with literature review. Clin Med Insights Case Rep 13: 1179547620948666.

Erenberg G (2005). The relationship between Tourette syndrome, attention deficit hyperactivity disorder, and stimulant medication: a critical review. Semin Pediatr Neurol 12: 217–221.

Erenberg G, Cruse RP, Rothner AD (1985). Gilles de la Tourette's syndrome: effects of stimulant drugs. Neurology 35: 1346–1348.

Erer S, Jankovic J (2008). Adult onset tics after peripheral injury. Parkinsonism Relat Disord 14: 75–76.

Factor SA, Molho ES (1997). Adult-onset tics associated with peripheral injury. Mov Disord 12: 1052–1055.

Florance NR, Davis RL, Lam C et al. (2009). Anti-N-methyl-D-aspartate receptor (NMDAR) encephalitis in children and adolescents. Ann Neurol 66: 11–18.

Fountoulakis KN, Samara M, Siapera M et al. (2011). Tardive Tourette-like syndrome: a systematic review. Int Clin Psychopharmacol 26: 237–242.

Francisco AM, Glaser C, Frykman E et al. (2006). California pediatric West Nile virus case series. Pediatr Infect Dis J 25: 81–84.

Fras I, Karlavage J (1977). The use of methylphenidate and imipramine in Gilles de la Tourette's disease in children. Am J Psychiatry 134: 195–197.

Fremont R, Dworkin J, Manoochehri M et al. (2022). Damage to the dorsolateral prefrontal cortex is associated with repetitive compulsive behaviors in patients with penetrating brain injury. BMJ Neurol Open 4: e000229.

Ganos C, Edwards MJ, Bhatia KP (2016). Posttraumatic functional movement disorders. Handb Clin Neurol 139: 499–507.

Ganos C, Edwards MJ, Müller-Vahl K (2016). "I swear it is Tourette's!": on functional coprolalia and other tic-like vocalizations. Psychiatry Res 246: 821–826.

Gomis M, Puente V, Pont-Sunyer C et al. (2008). Adult onset simple phonic tic after caudate stroke. Mov Disord 23: 765–766.

Gulisano M, Barone R, Mosa MR et al. (2020). Incidence of autism spectrum disorder in youths affected by Gilles de la Tourette syndrome based on data from a large single Italian clinical cohort. Brain Sci 10: 812.

Gupta R, Agalliu D, Spalice A et al. (2022). Reader response: lack of association of Group A streptococcal infections and onset of tics: European Multicenter Tics in Children Study. Neurology 99: 445–446.

Guth JC, Futterer SA, Hijaz TA et al. (2014). Pearls & oysters: bilateral thalamic involvement in West Nile virus encephalitis. Neurology 83: e16–e17.

Hanley ME, Patel PH (2023). Carbon monoxide toxicityStatPearls [Internet]. Treasure Island, FL: StatPearls Publishing. https://www.ncbi.nlm.nih.gov/books/NBK430740/.

Hardie RJ, Pullon HW, Harding AE et al. (1991). Neuroacanthocytosis. A clinical, haematological and pathological study of 19 cases. Brain 114: 13–49.

Hawryluk GW, Manley GT (2015). Classification of traumatic brain injury: past, present, and future. Handb Clin Neurol 127: 15–21.
Herring R, Desai N, Parnes M et al. (2019). Pediatric West Nile virus-associated neuroinvasive disease: a review of the literature. Pediatr Neurol 92: 16–25.
Hirschtritt ME, Lee PC, Pauls DL et al. (2015). Lifetime prevalence, age of risk, and genetic relationships of comorbid psychiatric disorders in Tourette syndrome. JAMA Psychiatry 72: 325–333.
Hoffman LA, Vilensky JA (2017). Encephalitis lethargica: 100 years after the epidemic. Brain 140: 2246–2251.
Holtmann M, Korn-Merker E, Boenigk HE (2000). Carbamazepine-induced combined phonic and motor tic in a boy with Down's syndrome. Epileptic Disord 2: 39–40.
Hopkins RO, Fearing MA, Weaver LK et al. (2006). Basal ganglia lesions following carbon monoxide poisoning. Brain Inj 20: 273–281.
Hounie AG, Pauls DL, Mercadante MT et al. (2004). Obsessive-compulsive spectrum disorders in rheumatic fever with and without Sydenham's chorea. J Clin Psychiatry 65: 994–999.
Hull M, Parnes M, Jankovic J (2021). Increased incidence of functional (psychogenic) movement disorders in children and adults amid the COVID-19 pandemic: a cross-sectional study. Neurol Clin Pract 11: e686–e690.
Hull M, Parnes M, Jankovic J (2023). Botulinum neurotoxin injections in children with self-injurious behaviors. Toxins (Basel) 15: 236.
Jankovic J (2001). Tourette's syndrome. N Engl J Med 345: 1184–1192.
Jankovic J, Ashizawa T (1995). Tourettism associated with Huntington's disease. Mov Disord 10: 103–105.
Jankovic J, Gelineau-Kattner R, Davidson A (2010). Tourette's syndrome in adults. Mov Disord 25: 2171–2175.
Johnson KA, Worbe Y, Foote KD et al. (2023). Tourette syndrome: clinical features, pathophysiology, and treatment. Lancet Neurol 22: 147–158.
Kahl U, Schunke O, Schöttle D et al. (2015). Tic phenomenology and tic awareness in adults with autism. Mov Disord Clin Pract 2: 237–242.
Kalanithi PS, Zheng W, Kataoka Y et al. (2005). Altered parvalbumin-positive neuron distribution in basal ganglia of individuals with Tourette syndrome. Proc Natl Acad Sci USA 102: 13307–13312.
Kalita J, Misra UK, Pandey S et al. (2003). A comparison of clinical and radiological findings in adults and children with Japanese encephalitis. Arch Neurol 60: 1760–1764.
Karlinsky H, Berg JM (1987). Gilles de la Tourette's syndrome in Down's syndrome. Br J Psychiatry 151: 707.
Keng LT, Chang LY (2018). Japanese encephalitis. CMAJ 190: E657.
Kerbeshian J, Burd L, Leech C et al. (1991). Huntington disease and childhood-onset Tourette syndrome. Am J Med Genet 39: 1–3.
Kim DD, Barr AM, Chung Y et al. (2018). Antipsychotic-associated symptoms of Tourette syndrome: a systematic review. CNS Drugs 32: 917–938.
Klawans HL, Falk DK, Nausieda PA et al. (1978). Gilles de la Tourette syndrome after long-term chlorpromazine therapy. Neurology 28: 1064–1066.
Ko SB, Ahn TB, Kim JM et al. (2004). A case of adult onset tic disorder following carbon monoxide intoxication. Can J Neurol Sci 31: 268–270.
Koller WC, Wong GF, Lang A (1989). Posttraumatic movement disorders: a review. Mov Disord 4: 20–36.
Krauss JK, Jankovic J (1997). Tics secondary to craniocerebral trauma. Mov Disord 12: 776–782.
Kuniyoshi M, Inanaga K, Arikawa K et al. (1992). A case of tardive Tourette-like syndrome. Jpn J Psychiatry Neurol 46: 67–70.
Kurlan R, Kersun J, Behr J et al. (1989). Carbamazepine-induced tics. Clin Neuropharmacol 12: 298–302.
Kwak CH, Jankovic J (2002). Tourettism and dystonia after subcortical stroke. Mov Disord 17: 821–825.
Lai M, Li Y, Luo D et al. (2022). Dopamine-2 receptor antibody encephalitis presenting as pure tongue-biting in a tourette syndrome patient: a case report. BMC Psychiatry 22: 47.
Lawson-Yuen A, Saldivar JS, Sommer S et al. (2008). Familial deletion within NLGN4 associated with autism and Tourette syndrome. Eur J Hum Genet 16: 614–618.
Lenka A, Jankovic J (2023). Peripherally-induced movement disorders: an update. Tremor Other Hyperkinet Mov (NY) 13: 8.
Lenka A, Kamat A, Mittal SO (2019). Spectrum of movement disorders in patients with neuroinvasive West Nile virus infection. Mov Disord Clin Pract 6: 426–433.
Li AM, Chan MH, Leung TF et al. (2000). Mercury intoxication presenting with tics. Arch Dis Child 83: 174–175.
Lin WD, Tsai FJ, Chou IC (2022). Current understanding of the genetics of Tourette syndrome. Biomed J 45: 271–279.
Lombroso PJ, Scahill L (2008). Tourette syndrome and obsessive-compulsive disorder. Brain Dev 30: 231–237.
Lotia M, Jankovic J (2016). Botulinum toxin for the treatment of tremor and tics. Semin Neurol 36: 54–63.
Lowe TL, Cohen DJ, Detlor J et al. (1982). Stimulant medications precipitate Tourette's syndrome. JAMA 247: 1168–1169.
Luat AF, Behen ME, Juhász C et al. (2009). Secondary tics or tourettism associated with a brain tumor. Pediatr Neurol 41: 457–460.
Makhoul K, Jankovic J (2023). Real-world experience with VMAT2 inhibitors in Tourette syndrome. J Neurol 270: 4518–4522.
Martino D, Schrag A, Anastasiou Z et al. (2021). EMTICS Collaborative Group. Association of Group A streptococcus exposure and exacerbations of chronic tic disorders: a multinational prospective cohort study. Neurology 96: e1680–e1693.
McGrath LM, Yu D, Marshall C et al. (2014). Copy number variation in obsessive-compulsive disorder and Tourette syndrome: a cross-disorder study. J Am Acad Child Adolesc Psychiatry 53: 910–919.
McNeill A, Chinnery PF (2012). Neuroferritinopathy: update on clinical features and pathogenesis. Curr Drug Targets 13: 1200–1203.
Medarametla S, Boraiah N, Raju S (2021). A curious case of excessive winking. Mov Disord Clin Pract 8: 947–949.
Mehanna R, Jankovic J (2013). Movement disorders in multiple sclerosis and other demyelinating diseases. J Neurol Sci 328: 1–8.
Mejia NI, Jankovic J (2005). Secondary tics and tourettism. Braz J Psychiatry 27: 11–17.
Mennesson M, Klink BA, Fortin AH6th, (1998). Case study: worsening Tourette's disorder or withdrawal dystonia? J Am Acad Child Adolesc Psychiatry 37: 785–788.
Mercadante MT, Busatto GF, Lombroso PJ et al. (2000). The psychiatric symptoms of rheumatic fever. Am J Psychiatry 157: 2036–2038.
Mercadante MT, Campos MC, Marques-Dias MJ et al. (1997). Vocal tics in Sydenham's chorea. J Am Acad Child Adolesc Psychiatry 36: 305–306.
Mirsattari SM, Berry ME, Holden JK et al. (1999). Paroxysmal dyskinesias in patients with HIV infection. Neurology 52: 109–114.
Molina da Costa RQ, Marrocos RP, Araujo Leite MA et al. (2016). All that glitters are not gold, when motor and vocal tics in a child do

not match Tourette syndrome, a case report. Dement Neuropsychol 10: 251–253.

Mulroy E, Ilinca A, Gonzalez-Robles C et al. (2021). Throat-clearing vocalizations in primary brain calcification syndromes. Mov Disord Clin Pract 8: 627–630.

Muñiz-Castrillo S, Joubert B, Elsensohn MH et al. (2020). Anti-CASPR2 clinical phenotypes correlate with HLA and immunological features. J Neurol Neurosurg Psychiatry 91: 1076–1084.

Nardocci N, Rumi V, Combi ML et al. (1994). Complex tics, stereotypies, and compulsive behavior as clinical presentation of a juvenile progressive dystonia suggestive of Hallervorden-Spatz disease. Mov Disord 9: 369–371.

Nath A, Jankovic J (1989). Motor disorders in patients with human immunodeficiency virus infection. Prog AIDS Pathol 1: 159–166.

Neeraja K, Prasad S, Holla VV et al. (2021). The spectrum of movement disorders in neuroacanthocytosis syndromes: a video series. Mov Disord Clin Pract 8: 983–986.

Nociti V, Fasano A, Bentivoglio AR et al. (2009). Tourettism in multiple sclerosis: a case report. J Neurol Sci 287: 288–290.

Northam RS, Singer HS (1991). Postencephalitic acquired Tourette-like syndrome in a child. Neurology 41: 592–593.

Nosadini M, Thomas T, Eyre M et al. (2021). International consensus recommendations for the treatment of pediatric NMDAR antibody encephalitis. Neurol Neuroimmunol Neuroinflamm 8: e1052.

O'Donnell P, Buxton PJ, Pitkin A et al. (2000). The magnetic resonance imaging appearances of the brain in acute carbon monoxide poisoning. Clin Radiol 55: 273–280.

Ondo WG (2012). Tetrabenazine treatment for stereotypies and tics associated with dementia. J Neuropsychiatry Clin Neurosci 24: 208–214.

Ondo WG, Adam OR, Jankovic J et al. (2010). Dramatic response of facial stereotype/tic to tetrabenazine in the first reported cases of neuroferritinopathy in the United States. Mov Disord 25: 2470–2472.

Osland ST, Steeves TD, Pringsheim T (2018). Pharmacological treatment for attention deficit hyperactivity disorder (ADHD) in children with comorbid tic disorders. Cochrane Database Syst Rev 6: CD007990.

Pascual-Leone A, Dhuna A (1990). Cocaine-associated multifocal tics. Neurology 40: 999–1000.

Patel C, Cooper-Charles L, McMullan DJ et al. (2011). Translocation breakpoint at 7q31 associated with tics: further evidence for IMMP2L as a candidate gene for Tourette syndrome. Eur J Hum Genet 19: 634–639.

Pollack MA, Cohen NL, Friedhoff AJ (1977). Gilles de la Tourette's syndrome. Familial occurrence and precipitation by methylphenidate therapy. Arch Neurol 34: 630–632.

Poulopoulos M, Hajjar M (2013). Post-traumatic tics and tetrabenazine treatment: a blinded video assessment. BMJ Case Rep bcr2013201676.

Pringsheim T, Ganos C, McGuire JF et al. (2021). Rapid onset functional tic-like behaviors in young females during the COVID-19 pandemic. Mov Disord 36: 2707–2713.

Pulst SM, Walshe TM, Romero JA (1983). Carbon monoxide poisoning with features of Gilles de la Tourette's syndrome. Arch Neurol 40: 443–444.

Rajapakse T, Pringsheim T (2010). Pharmacotherapeutics of Tourette syndrome and stereotypies in autism. Semin Pediatr Neurol 17: 254–260.

Ranjan N, Nair KP, Romanoski C et al. (2011). Tics after traumatic brain injury. Brain Inj 25: 629–633.

Reid SD (2004). Neuroleptic-induced Tardive Tourette treated with clonazepam: a case report and literature review. Clin Neuropharmacol 27: 101–104.

Reyes NGD, Di Luca DG, McNiven V et al. (2023). Dystonia with myoclonus and vertical supranuclear gaze palsy associated with a rare GNB1 variant. Parkinsonism Relat Disord 106: 105239.

Richter RW, Shimojyo S (1961). Neurologic sequelae of Japanese B encephalitis. Neurology 11: 553–559.

Ricketts EJ, Wu MS, Leman T et al. (2019). A review of tics presenting subsequent to traumatic brain injury. Curr Dev Disord Rep 6: 145–158.

Rigas A, Mainka T, Pringsheim T et al. (2023). Distinguishing functional from primary tics: a study of expert video assessments. J Neurol Neurosurg Psychiatry 94: 751–756.

Ringman JM, Jankovic J (2000). Occurrence of tics in Asperger's syndrome and autistic disorder. J Child Neurol 15: 394–400.

Rissardo JP, Caprara ALF (2020). Carbamazepine-, oxcarbazepine-, eslicarbazepine-associated movement disorder: a literature review. Clin Neuropharmacol 43: 66–80.

Rissardo JP, Caprara ALF, Durante Í et al. (2022). Lithium-associated movement disorder: a literature review. Brain Circ 8: 76–86.

Robakis D (2017). How much do we know about adult-onset primary tics? Prevalence, epidemiology, and clinical features. Tremor Other Hyperkinet Mov (NY) 7: 441.

Robertson PL, Garofalo EA, Silverstein FS et al. (1993). Carbamazepine-induced tics. Epilepsia 34: 965–968.

Rodopman-Arman A, Yazgan Y, Berkem M et al. (2004). Are sensory phenomena present in Sydenham's Chorea? Evaluation of 13 cases. Neuropediatrics 35: 242–245.

Rohani M, Fasano A, Lang AE et al. (2018). Pantothenate kinase-associated neurodegeneration mimicking Tourette syndrome: a case report and review of the literature. Neurol Sci 39: 1797–1800.

Rojo A, Pernaute RS, Fontán A et al. (1999). Clinical genetics of familial progressive supranuclear palsy. Brain 122: 1233–1245.

Rua A, Damásio J (2014). Tics induced by sertraline: case report and literature review. Mov Disord Clin Pract 1: 243–244.

Rüth U, Mayer-Rosa J, Schlamp D et al. (2000). Tourette's syndrome and antidepressant therapy: exacerbation of nervous tics with paroxetine. Z Kinder Jugendpsychiatr Psychother 28: 105–108.

Sacks OW Postencephalitic syndromesStern GParkinson's disease. London: Chapman & Hall, pp. 415–428.

Safadieh L, Dabbagh O (2013). Anti-N-methyl-D-aspartate (NMDA) receptor encephalitis in a young Lebanese girl. J Child Neurol 28: 1222–1225.

Saiki S, Hirose G, Sakai K et al. (2004). Chorea-acanthocytosis associated with Tourettism. Mov Disord 19: 833–836.

Savitt D, Jankovic J (2018). Tardive syndromes. J Neurol Sci 389: 35–42.

Saygi S, Ozkale Y, Erol I (2014). Tic disorder probably associated with steroid responsive encephalopathy with autoimmune thyroiditis (SREAT). Indian J Pediatr 81: 1105–1107.

Scarano V, Pellecchia MT, Filla A et al. (2002). Hallervorden-Spatz syndrome resembling a typical Tourette syndrome. Mov Disord 17: 618–620.

Schneider SA, Robertson MM, Rizzo R et al. (2008). Fragile X syndrome associated with tic disorders. Mov Disord 23: 1108–1112.

Schrag AE, Martino D, Wang H et al. (2022). European Multicentre Tics in Children Study (EMTICS). Lack of association of Group A streptococcal infections and onset of tics: European Multicenter Tics in Children Study. Neurology 98: e1175–e1183.

Seeman MV, Patel J, Pyke J (1981). Tardive dyskinesia with Tourette-like syndrome. J Clin Psychiatry 42: 357–358.

Shin YW, Lee ST, Shin JW et al. (2013). VGKC-complex/LGI1-antibody encephalitis: clinical manifestations and response to immunotherapy. J Neuroimmunol 265: 75–81.

Singer HS, Dela Cruz PS, Abrams MT et al. (1997). A Tourette-like syndrome following cardiopulmonary bypass and hypothermia: MRI volumetric measurements. Mov Disord 12: 588–592.

Spitz MC, Jankovic J, Killian JM (1985). Familial tic disorder, parkinsonism, motor neuron disease, and acanthocytosis: a new syndrome. Neurology 35: 366–370.

Stahl SM (1980). Tardive Tourette syndrome in an autistic patient after long-term neuroleptic administration. Am J Psychiatry 137: 1267–1269.

Steiner I, Benninger F (2013). Update on herpes virus infections of the nervous system. Curr Neurol Neurosci Rep 13: 414.

Stern JS, Robertson MM (1997). Tics associated with autistic and pervasive developmental disorders. Neurol Clin 15: 345–355.

Sverd J (1991). Tourette syndrome and autistic disorder: a significant relationship. Am J Med Genet 39: 173–179.

Swayne A, Tjoa L, Broadley S et al. (2018). Antiglycine receptor antibody related disease: a case series and literature review. Eur J Neurol 25: 1290–1298.

Taylor LD, Krizman DB, Jankovic J et al. (1991). 9p monosomy in a patient with Gilles de la Tourette's syndrome. Neurology 41: 1513–1515.

Thomas M, Hayflick SJ, Jankovic J (2004). Clinical heterogeneity of neurodegeneration with brain iron accumulation (Hallervorden-Spatz syndrome) and pantothenate kinase-associated neurodegeneration. Mov Disord 19: 36–42.

Touafchia D, Montastruc F, Lapeyre-Mestre M et al. (2020). Drug-induced tics: an observational postmarketing study. Hum Psychopharmacol 35: e2734.

Turley JM (1988). Tourette-like disorder after herpes encephalitis. Am J Psychiatry 145: 1604–1605.

Tvedt B, Krogstad JM, Berstad J (1996). Hypoxic brain damage after carbon monoxide poisoning. Visual agnosia, reduced initiative and memory and delayed sequelae. Tidsskr Laegeforen 116: 3005–3008.

van Rossum IA, ten Houten R (2010). Unexceptional symptoms as expression of MELAS. Ned Tijdschr Geneeskd 154: A2168.

Vilboux T, Malicdan MC, Chang YM et al. (2016). Cystic cerebellar dysplasia and biallelic LAMA1 mutations: a lamininopathy associated with tics, obsessive compulsive traits and myopia due to cell adhesion and migration defects. J Med Genet 53: 318–329.

Vilensky JA, Goetz CG, Gilman S (2006). Movement disorders associated with encephalitis lethargica: a video compilation. Mov Disord 21: 1–8.

Walker KG, Lawrenson J, Wilmshurst JM (2005). Neuropsychiatric movement disorders following streptococcal infection. Dev Med Child Neurol 47: 771–775.

Wang LF, Huang JW, Shan SY et al. (2016). Possible sertraline-induced extrapyramidal adverse effects in an adolescent. Neuropsychiatr Dis Treat 12: 1127–1129.

Xing S, Chen L, Chen X et al. (2008). Excessive blinking as an initial manifestation of juvenile Huntington's disease. Neurol Sci 29: 275–277.

Xu X, Sun H, Luo J et al. (2023). The pathology of primary familial brain calcification: implications for treatment. Neurosci Bull 39: 659–674.

Yochelson MR, David RG (2000). New-onset tic disorder following acute hemorrhage of an arteriovenous malformation. J Child Neurol 15: 769–771.

Zea Vera A, Bruce A, Garris J et al. (2022). The phenomenology of tics and tic-like behavior in TikTok. Pediatr Neurol 130: 14–20.

Zesiewicz TA, Sullivan KL (2011). Drug-induced hyperkinetic movement disorders by nonneuroleptic agents. Handb Clin Neurol 100: 347–363.

Handbook of Clinical Neurology, Vol. 215 (3rd series)
Tic Disorders and Tourette Syndrome
C. Ganos and Y. Worbe, Editors
https://doi.org/10.1016/B978-0-443-13554-5.00022-5

Chapter 4

Acute tic exacerbations: Tic status, malignant Tourette syndrome, tic attacks, tic fits, and other misfits

BENEDETTA ANGELONI[1,2,3,*], ELIA ABI-JAOUDE[4,5], AND CHRISTOS GANOS[1,2,6]

[1]*Edmond J. Safra Program in Parkinson's Disease and the Morton and Gloria Shulman Movement Disorders Clinic, Toronto Western Hospital, Toronto, ON, Canada*

[2]*Division of Neurology, University of Toronto, Toronto, ON, Canada*

[3]*Department of Neurology, Catholic University of Sacred Heart, Rome, Italy*

[4]*Department of Psychiatry, Hospital for Sick Children, University of Toronto, Toronto, ON, Canada*

[5]*Department of Psychiatry, Toronto Western Hospital – University Health Network, University of Toronto, Toronto, ON, Canada*

[6]*Krembil Brain Institute, Toronto, ON, Canada*

Abstract

Acute tic exacerbations in Tourette syndrome (TS) and other tic disorders are influenced by biologic, psychologic, and social factors. Different terms have been used to describe acute tic exacerbations, including tic status, malignant TS, tic attacks, and tic fits. This review traces the evolution of our understanding of acute tic exacerbations and examines the emerging clinical constructs, highlighting the overlapping features and diagnostic ambiguities of such episodes. Common features include repetitive movements lasting from minutes to hours that are rarely suppressible, cause significant distress, functional impairment, and often require medical intervention. The complexity of these conditions and the potential co-occurrence of functional symptoms in some cases can complicate recognition of underlying triggers and appropriate management. Accurate differentiation from other hyperkinetic movement disorders demands a comprehensive assessment encompassing medical, pharmacologic, and psychiatric histories. By synthesizing current literature, this chapter aims to lay the groundwork for improving diagnostic precision and treatment, emphasizing the need for further research into the various mechanisms of acute tic exacerbations.

INTRODUCTION

Tics are brief, sudden, and involuntary movements or vocalizations that are stereotyped, repetitive, nonrhythmic, and often preceded by a premonitory urge (Ganos and Martino, 2015). Typically, tics manifest as exaggerated motor phenomena that resemble volitional actions but are out of context in a given situation. The premonitory urges that typically precede tics entail uncomfortable sensations characterized by feelings of tension or discomfort that are temporarily alleviated by the execution of tics. Moreover, tics may be temporarily suppressed through effortful volitional inhibition. During tic suppression, the intensity of premonitory urges typically increases to such an extent that tic execution feels inevitable (Eddy and Cavanna, 2014).

A notable feature of tics in idiopathic chronic tic disorders such as Tourette syndrome (TS) is that they typically follow a spontaneous "waxing and waning" course (Leckman, 2003). Interestingly, the variability in the manifestation of tics is influenced by a number of psychologic, physiologic, and environmental factors (Barber et al., 2024). For example, some studies have shown that exercise (Doja et al., 2018; Pringsheim et al., 2021), being engrossed in activities such as performing music (Bodeck et al., 2015), or receiving rewards for tic suppression (Woods and Himle, 2004; Iverson and Black, 2022), can all lead to a reduction of tics. Conversely, fatigue, stressful life events (Godar and Bortolato, 2017; Iverson and Black, 2022; Barber et al., 2024), and use of specific drugs (Lowe and Shayawitz, 1982) can exacerbate tics.

In some cases, tics can increase markedly in severity, even to the extent that urgent therapeutic measures may be required. Acute tic exacerbations have a range of different clinical presentations, which are often described with terms such as "tic bouts," "tic status," "malignant TS," and "tic attacks." Here, we examine these acute tic exacerbation phenomena through a review of available literature to highlight their clinical

*Correspondence to: Dr. Benedetta Angeloni, MD, Università Cattolica del Sacro Cuore, Rome, Italy; Krembil Research Institute, Toronto Western Hospital, 399 Bathurst Street, Toronto, ON, M5T 2S8, Canada Tel: +1 4376037285, E-mail: benedetta.angeloni@uhn.ca.

characteristics, including descriptions of unique and overlapping features. By integrating clinical and empirical insights, we aim to enhance diagnostic clarity and lay the foundation for standardized diagnostic labeling and treatment interventions.

TIC BOUTS

The recognition of phenomena characterized by acute tic worsening as distinct nosologic entities has been a subject of ongoing debate, primarily due to the absence of universally accepted diagnostic criteria and the complexities of underlying etiologies. This means that there is also an absence of consensus on the terminology used to describe these phenomena, with proposed terms that include "paroxysmal tic exacerbations," "tic status," "nonsuppressible tic episodes," "tic fits," and "tic attacks."

Historically, tic disorders have often been misclassified in medical literature and erroneously labeled as "intermittent chorea," epileptic manifestations, or hysteria, leading to inadequate treatment and a limited understanding of their pathophysiology (Rickards et al., 2010; Lavoie and Cavanna, 2022). Interestingly, Georges Gilles de la Tourette initially avoided the term "tic" and instead described the movements as "motor incoordination" (Lavoie and Cavanna, 2022). Following Jean-Martin Charcot's coinage of the term "Gilles de la Tourette's disease," Georges Guinon highlighted the relationship between tics, involuntary words, and obsessive thoughts, and ultimately proposed the term "convulsive tic disease" (Lavoie and Cavanna, 2022). By the early 20th century, Henry Meige and Édouard Feindel in their 1902 publication, "Les Tics et leur Traitement," introduced the concept of "tic attacks," characterized by sudden, intense bouts of tics. The authors describe those episodes as "(…) an alternating series of attacks and periods of respite. (…) The attacks vary with circumstances and environment." They observed that these episodes could be precipitated by factors such as stress or fatigue (Meige et al., 1907). However, their description of "tic attacks" more closely resembles what later became known as "tic bouts," which are clusters of tics occurring unpredictably, often following a premonitory urge, and usually brief and self-limiting (Leckman et al., 2006). Several authors have used the term "tic bouts" (or alternatively, "tic bursts") to describe short-term fluctuations in tic expression (Peterson and Leckman, 1998; Swain and Leckman, 2005; Leckman et al., 2014; Brandt et al., 2016). Tic bouts or bursts are, in fact, considered a typical clinical feature of tics in chronic tic disorders such as TS. Another term, "paroxysms of tics," is used in the original version of the Yale Global Tic Severity Scale as a separate rating item, where the duration is to be specified in seconds; however, there is no description of what is meant by the term (Leckman et al., 1989).

TIC STATUS

Stanley Fahn described "tic status" as a series of continuous clonic or dystonic tics lasting tens of minutes, with short periods of normal behavior between episodes (Kurlan, 2005). He likened this condition to status epilepticus, describing it as an extreme manifestation of severe tic disorders, but did not comment on etiology. Subsequent literature on the topic is scarce. Some case reports described acute tic exacerbations and linked them to various causes, such as drug withdrawal (Leckman et al., 1986; Max and Rasmussen, 1986) or stimulant drug prescription (Lowe and Shayawitz, 1982), but did not specifically mention "tic status." Sachdev et al. in 1996 did use the term "tic status" as defined by Fahn to describe two cases (Sachdev et al., 1996). The first case involved a 14-year-old girl who had acute tic worsening following an assault by a classmate. There were nonsuppressible jerky movements of her upper limbs and neck, facial grimaces, and eye-rolling, all consistent with her preexisting tics. These movements presented as a single episode with a duration of about 36 hours. She was treated with haloperidol (12.5 mg daily) and clonazepam (0.5 mg tid), which led to drastic symptom improvement. The second case was a 20-year-old woman who experienced an exacerbation of her tics after a dosage reduction of chronic haloperidol tic treatment. She presented with recurrent episodes of repetitive head shaking, which lasted for about 10–15 minutes each. These episodes were also consistent with her chronic tic repertoire, but were not suppressible, and she appeared preoccupied with tic recurrence. There were only a few seconds of tic-free intervals between each of the episodes. In her case, the treatment of choice consisted of increasing the dosage of haloperidol, which led to symptom amelioration. Although both cases describe acute tic worsening with nonsuppressible motor events consistent with the preexisting tic phenomenology, they differ in duration and triggering factors. Sachdev et al. proposed two key features that distinguish tic status from usual tic bouts: the nonsuppressibility of tic events during the episodes, and the severe interference of tics with other activities. They also suggested that, in order to diagnose tic status, the following criteria need to be met: "(1) tics occur in one or more groups of muscles in a stereotyped and repetitive manner; (2) the episode is sustained for at least several minutes and possibly for hours; (3) during the episode, the patient is unable to suppress the tics, and they intrude into normal activities, and (4) the episode is in clear consciousness and there is no suspicion of ictal phenomena."

Fifteen years later, in 2011, Kovacs et al. (2011) expanded the diagnostic criteria for "tic status," building upon the observations made by Sachdev. The authors reported the case of an 18-year-old man who presented with new-onset motor behaviors after withdrawing his tic medication (haloperidol, tiapride, clonazepam). The motor behaviors that occurred and were described as tics differed from his preexisting symptoms: the movements involved the limbs, the trunk, and the neck. At the time of admission to the emergency room, the patient was tachycardic, had excessive sweating and increased creatine kinase (CK) levels (up to 8000 U/L). Since the hyperkinetic symptoms were considered related to drug withdrawal, the patient was first treated with intravenous tiapride, haloperidol, and clonazepam, however, without any benefit. Continuous and severe motor tics caused pain, physical exhaustion, and cardiovascular demands. The clinical picture was considered life-threatening, and the patient was sedated and intubated. Based on this case description, the authors' experience, and previously reviewed literature (Leckman et al., 1986; Max and Rasmussen, 1986; Sachdev et al., 1996; Dehning et al., 2009), they proposed a new set of diagnostic criteria for "tic status," which encompassed positive or suggestive features and exclusion criteria. Their suggestive features include (1) preexistence of TS fulfilling the appropriate criteria of *Diagnostic and Statistical Manual of Mental Disorders*; (2) acute/subacute worsening in the hyperkinetic features of the preexisting TS usually preceded by acute infection or drug withdrawal/dosage reduction; (3) different tics from the preexisting tics (e.g., more severe, complex, and long-lasting); (4) the hyperkinetic movements dominate the clinical picture resulting in severe systemic effects (e.g., cardio-respiratory difficulties, autonomic signs, or rhabdomyolysis); (5) the duration of continuous and disabling motor symptoms may vary from minutes to days. Exclusion criteria were defined as the presence of neuroleptic malignant syndrome or altered consciousness, acute akathisia or other movement disorders (e.g., tardive or drug withdrawal dystonia or chorea), the presence of hyperkinetic movements due to obsessive-compulsive disorder (OCD) or attention-deficit/hyperactivity disorder (ADHD), and, lastly, fulfillment of the criteria of malignant TS (Cheung et al., 2007).

MALIGNANT TOURETTE SYNDROME

Malignant TS is a rare condition associated with severe tics or behavioral symptoms as part of commonly co-occurring comorbidities that are severe enough to lead to at least two emergency room visits or one hospitalization in a year (Cheung et al., 2007). This form often requires intensive medical management, including neuroleptic medications, deep brain stimulation (DBS), or other advanced interventions (Cheung et al., 2007). Historically, the condition has been narratively reported in scientific literature (Robertson et al., 1989; Krauss and Jankovic, 1996; Muroi et al., 2002; Dobbs and Berger, 2003; Hood et al., 2004; Lim et al., 2004; Fusco et al., 2006; Patterson et al., 2016) and was also described in the volume "Gilles de la Tourette Syndrome" by Arthur and Elaine Shapiro, which noted, "(…) A rare, potentially dangerous symptom which cannot be completely inhibited is hitting the eye with fingers, fists, or kitchen utensils, ultimately causing eye damage or blinding. Such symptoms constitute a medical emergency and should be treated vigorously with medication" (Shapiro et al., 1988). Malignant TS is, thereby, an umbrella term to encompass the more severe end of the chronic tic disorder spectrum, irrespective of the specific reasons that may lead to emergency room visits or hospitalization. In this sense, it remains unclear why "malignant TS" would be an exclusion criterion in the broad definition of "tic status" by Kovacs et al. (2011). For example, acute or subacute worsening of a tic disorder can be associated with complications from violent tic episodes, including self-injurious behavior and intense muscular contractions. In addition, it remains unclear whether the two presentations described by Sachdev et al. and the case of Kovacs et al. refer to the same type of phenomenon. For example, tic phenomenology in Sachdev's cases overlapped with preexisting tics, whereby Kovacs et al. propose criteria that require the presence of tic phenomena characteristically distinct from the typical tic repertoire.

TIC ATTACKS

In addition to the previous descriptions of acute tic worsening in the context of chronic tic disorders like TS, another term that has been introduced to tic literature is that of "tic attacks." In a conference abstract, Collicott et al. (2013) retrospectively screened for the presence of "clinically disabling bouts of nonsuppressible repetitive tics occurring in full consciousness" by reviewing the clinical records of 369 pediatric and adult tic cases. The authors identified 32 patients with such episodes, which they labeled as "tic attacks." Tic attacks lasted from 3 minutes to 3 hours. Provoking factors included periods of increased obsessive-compulsive symptomatology. They described that compared to the rest of the sample of patients with TS, the patients with tic attacks tended to be younger (mean age 14 vs 20 years) and had a higher prevalence of comorbid ADHD (56% vs 29%). They also highlighted that clinically, tic attacks had the potential to be mistaken for epileptic seizures.

Robinson and Hedderly (2016) used the same description as Collicott to characterize a case series involving 12 children and highlight the role of psychologic factors as modifiers or triggers of "tic attacks." The authors detail one of those cases, a 13-year-old boy with persistent leg tremor, alongside the emergence of intrusive thoughts of self-harm and violent behavior. He expressed social anxiety about leaving the house alone and avoided talking to unfamiliar people. These thoughts occurred daily and were linked with involuntary, brief limb movements, which the authors considered consistent with tics. Interestingly, the patient also exhibited twisting, extended, and exaggerated body movements, which were deemed inconsistent with tics and more aligned with a functional movement disorder. The patient described these movements as involuntary, and they prompted the family to seek emergency medical care multiple times. In anxiety-inducing situations, the patient would worry about experiencing a "tic attack" and the possible negative consequences. This anxiety led him to focus on his internal sensations, scanning his body for tic-related symptoms, and he reported physical signs of anxiety such as increased heart rate and sweating.

Similar clinical characteristics were observed across the described cohort: all children were reported to experience anxiety. Diagnoses included social anxiety disorder (n=6), OCD (n=4), specific phobias (n=1), and pica (n=1). Interestingly, as opposed to Collicott's cohort of patients, in this cohort, none of the children had a previous diagnosis of ADHD, and OCD was present in less than half of them. The attacks led to emergency department visits in five cases. Additionally, "tic attack" frequency varied considerably: half of the children reported episodes less than once a month, while three experienced weekly attacks and another three had daily episodes. Only one child reported multiple "tic attacks" within a single day.

Robinson and Hedderly hypothesized that the "tic attacks" they described are triggered by the misinterpretation of typical anxiety-related bodily sensations as being premonitory urges. In other words, for some people, the physical symptoms of anxiety are interpreted as premonitory urges, thus leading to an increase in tics and other physical symptoms, including functional movement disorders. These episodes may be interrupted by normal behavior (in contrast to epileptic seizures), but the patient is unable to regain control over their overall motor output during the episode (Sachdev et al., 1996). In this regard, some clinical presentations consistent with functional tic-like behaviors share several similarities with "tic attacks" (Martino et al., 2023).

TIC FITS

More recently, Au et al. described two cases of rhabdomyolysis secondary to severe tics in patients with Tourette syndrome, referring to these episodes as "tic fits" (Au et al., 2021). The first case involved a 12-year-old boy with ADHD, social anxiety, and post-traumatic stress disorder (PTSD), who was hospitalized three times due to escalating tic severity. It is unclear whether his repetitive movements, interpreted as motor tics in this chapter, characterized by continuous thrusting of the trunk and limbs, were consistent with his baseline. At initial presentation, his CK level was 6953 U/L. His medications included aripiprazole (2 mg daily), clonidine (0.1 mg nightly), guanfacine (4 mg daily), and methylphenidate (54 mg daily). Aripiprazole was discontinued to eliminate potential confounding effects on CK, and he was treated with intravenous (IV) fluids. Ten days later, he returned with similar symptoms, including forceful muscle stiffening unresponsive to oral lorazepam; CK was mildly elevated (~600 U/L). He received intramuscular haloperidol, increased clonidine, and new prescriptions for low-dose clonazepam and risperidone. CK levels normalized as his movements improved. Seven months later, he was re-hospitalized for another episode of rhabdomyolysis with severe repetitive movements involving the neck, trunk, and limbs. He was again successfully treated with fluids and a single dose of ziprasidone. The second case involved a 16-year-old boy with TS, ADHD, depression, and anxiety, who presented with an acute tic exacerbation marked by vocal tics, large-amplitude head and body jerks, violent thrashing, and self-injurious behaviors. CK was elevated to 2682 U/L. Symptoms were unresponsive to 4 mg IV lorazepam. Due to concerns over worsening rhabdomyolysis, he was treated with fluids and dexmedetomidine, though the latter was poorly tolerated due to bradycardia. He subsequently received a 1 g IV valproate load, followed by a maintenance dose of 500 mg bis in die (BID), based on prior evidence supporting valproate's benefit in refractory TS and its mood-stabilizing properties. His CK levels slowly decreased over time, and the patient was discharged.

Both patients experienced intense repetitive motor behaviors that were suggested to be tic exacerbations, resulting in elevated CK levels, requiring hospitalization and intravenous fluids. Neither case met criteria for neuroleptic malignant syndrome. Psychiatric comorbidities, such as anxiety, PTSD, and depression, were present in both individuals. Although the authors acknowledged the likely role of mood dysregulation in these "tic fits," they did not elaborate on potential triggers, provide a clear definition of the phenomenon, or propose a specific management approach, since their focus was primarily on urgent care and the treatment of potential complications associated with rhabdomyolysis.

FROM DIAGNOSTIC LABELS TO MANAGEMENT

In this chapter, we have reviewed reported cases of acute tic exacerbations, covering the various terms used to describe them, including tic status, malignant TS, tic attacks, and tic fits – see Table 4.1 for a full overview. Establishing a common nomenclature for acute tic exacerbations is crucial to avoid confusion and diagnostic mislabeling, which may lead to inappropriate treatment interventions. However, there are several overlapping features between the descriptions of tic status, malignant TS, tic attacks, and tic fits (see Table 4.1). Common characteristics in these descriptions include continuous and nonsuppressible repetitive movements. These may present as acute exacerbations of preexisting tics, the emergence of new tics, or the sudden onset of movements more consistent with functional neurologic symptoms as part of a novel stand-alone syndrome or overlaid on TS. These scenarios likely reflect different underlying pathophysiologic mechanisms, ranging from dysregulation within cortico-striato-thalamo-cortical circuits (Mink, 2003; Yael et al., 2015) to heightened interoception (Ganos et al., 2015; Rae et al., 2019), maladaptive metacognitive processing (Robinson and Hedderly, 2016), and altered sensory tolerance in functional or anxiety-driven presentations (Leckman et al., 2006; Robinson and Hedderly, 2016; Ganos et al., 2019), and thus require tailored therapeutic strategies. Importantly, none of these various terms have established validity – they represent descriptions of acute tic exacerbations based on small numbers of presenting cases.

Upon exclusion of other neurologic or paroxysmal movement disorders, management should begin with a comprehensive evaluation of medical, psychiatric, and pharmacologic history, including changes in medication and illicit drug use, as well as potential comorbidities such as anxiety disorders, OCD, ADHD, or self-injurious behaviors. It is crucial to understand the exact phenomenon at hand (e.g., acute worsening of tics versus functional movements) and the contributing factors that may be amenable to modification. Treatment may involve psychoeducation, cognitive-behavioral and metacognitive techniques targeting attentional focus, sensory tolerance, and ruminative thought patterns. Pharmacologic options may include short-term use of benzodiazepines, α-2 agonists, dopamine receptor-blocking agents, and, in rare and severe presentations the use of muscle relaxants or even deep sedation (e.g., propofol, midazolam), along with metabolic monitoring to prevent complications such as rhabdomyolysis, cervical myelopathy, vertebral artery dissection, corneal damage, or retinal detachment, all of which are examples of potential complications of severe tics. In select refractory cases,

Table 4.1

Summary of different terms used to describe acute tic exacerbations.

Term	Authors	No of patients	Age and sex	Clinical features	Comorbidities	Precipitating factors	Intervention and response	Description or proposed criteria
Tic status	Sachdev et al. (1996)	2	14 yo, F; 20 yo, F	Jerking movements of upper limbs, neck jerking and eye-rolling that occurred every few seconds and were not suppressible.	OCD, learning difficulties, ADHD, conduct problems.	Assault by classmate; reduction in antipsychotic dose.	Antipsychotic, resulting in symptom improvement.	Series of continuous clonic or dystonic tics lasting tens of minutes, with short periods of normal behavior between episodes (as per Fahn) (Bloch, 2006); nonsuppressible; severe interference in function; sustained for minutes to hours; conscious patient, no suspicion of ictal phenomena.
	Kovacs et al. (2011)	1	18 yo, M	Incessant movements with cardio-respiratory difficulties, autonomic signs, and increased creatine-kinase blood levels.	None reported.	Arbitrary medication withdrawal.	Intravenous tiapride, haloperidol, and clonazepam. Sedation with propofol and midazolam.	Preexisting TS: acute/subacute worsening, usually preceded by acute infection or drug withdrawal/dosage reduction; different tics from the preexisting tics (e.g., more severe, complex and long-lasting); severe systemic effects; duration of minutes to days; Exclusion: neuroleptic malignant syndrome, altered consciousness, acute akathisia or other movement disorders, hyperkinetic movements due to OCD or ADHD, malignant TS (MTS).
	Parmera et al. (2023)	1	29 yo, M	Hyperkinetic movements with muscle contractions in the trunk, neck, upper and lower limbs consistent with pattern of preexisting tics.	None reported.	DBS battery depletion.	Diazepam and battery replacement.	Nonrhythmic, continuous, generalized hyperkinesias associated with muscle contractions in the trunk, neck, and upper and lower limbs with preserved consciousness in a patient with DBS to treat severe TS.

(*Continued*)

Table 4.1 (Continued)

Term	Authors	No of patients	Age and sex	Clinical features	Comorbidities	Precipitating factors	Intervention and response	Description or proposed criteria
MTS	Cheung et al. (2007)	17	Mean age 20 y 4 m, 14 M, 3 F	Tic related injuries, at least 2 emergency room (ER) access and at least 1 hospitalization.	Self-injurious behavior, uncontrollable violence and temper, suicidal tendencies or attempts, OCD/ obsessive-compulsive behavior (OCB).	Not specified.		TS and its comorbid behavioral symptoms can result in potentially life-threatening outcomes due to the result of severe tics, extreme rage attacks, depression with suicidal ideation, and disabling self-injurious behavior.
	Motlagh et al. (2013)	3	Mean age 38 y 3 m, M	Tic related injuries in patients that underwent DBS implantation.	OCD.	Not specified.	DBS implantation.	Severe self-injurious tics.
	Zhang et al. (2019)	1	20 y, M	Involuntary blinking, shrugging, squatting; extreme rage attacks and self-injurious behavior (SIB) (left eye blindness); underwent DBS implantation.	ADHD, OCD.	No external trigger identified.	DBS implantation.	Rapidly evolving life-threatening symptoms, such as SIB and suicidal ideation.
	Monga et al. (2023)	1	20 yo, M	Tic related injuries and rhabdomyolysis in patient with MTS that underwent DBS implantation.	ADHD, OCD, OCB, and anxiety with depression.	Not specified.	Sertraline, clonazepam, melatonine, valproic acid, risperidone, dexmedetomidine infusion.	TS resistant to treatment, leading to significant complications.
	Krüger and Müller-Vahl (2015)	1	14 yo, M	Tic related injuries with extraction of four teeth.	ADHD, OCD.	Not reported.		Extreme SIB with severe tics and comorbid OCD, rage attacks, complex phonic tics, coprophenomena, mood disorder, suicidal ideation, and poor response to medication.
	Fasano and Galluccio (2018)	1	17 yo, M	Acute subdural and subarachnoid hemorrhages from head banging.	ADHD, celiac disease.	Not specified.	Decompressive craniectomy.	Life-threatening symptoms and TS.
Tic attacks	Collicott et al. (2013)	32	Mean age 14 y	Disabling bouts of continuous tics occurring in full consciousness.	56% hyperactivity disorder, OCD, oppositional defiant disorder (in children) and mood disorders (adults).	Reported provoking factors were periods of increased obsessive-compulsive symptomatology and preceding prolonged suppression of tics, distinct bouts of severe, continuous, nonsuppressible, and disabling tics lasting from 15 minutes to several hours.	Not specified.	Clinically disabling bouts of nonsuppressible repetitive tics occurring in full consciousness; 3 minutes to 3 hours.

	Robinson and Hedderly (2016)	12	Mean age 11 y 3 m, 9 M and 3 F	Distinct bouts of severe, continuous, nonsuppressible, and disabling tics lasting from 15 min to several hour.	100% anxiety, 50% social anxiety disorder, 33% OCD, 1 with specific phobias (agoraphobia, dark rooms, and heights), 1 with pica.	Triggered by the misinterpretation of typical anxiety-related bodily sensations as being premonitory urges.	Psychologic therapy.	Distinct bouts of severe, continuous, nonsuppressible, and disabling tics lasting from 15 min to several hours.
Tic fits	Au et al. (2021)	2	12 y and 16 y, M	Clinically disabling bouts of nonsuppressible repetitive tics occurring in full consciousness.	ADHD, OCD, depression, anxiety. Tics caused rhabdomyolysis and injuries from tics, required hospitalization.	Not provided.	Ziprasidone and fluids; lorazepam, valproic acid, deutetrabenazine.	Not provided.

The term “tic” is used here as per the original reports. *ADHD*, Attention-deficit/hyperactivity disorder; *DBS*, deep brain stimulation; *MTS*, malignant tourette syndrome; *OCD*, obsessive-compulsive disorder; *TS*, Tourette syndrome; *ER*, emergency room; *F*, female; *m*, months; *M*, male; *OCB*, obsessive-compulsive behavior; *SIB*, self-injurious behavior; *y*, years; *yo*, years old.

DBS may have value, and this requires consideration by a multidisciplinary team with specific expertise in the diagnosis and management of TS. Given the clinical overlap and potential for misdiagnosis, especially in acute settings where tics are a common reason for emergency department visits (Raucci et al., 2018), it is essential to adopt a nuanced and individualized approach that integrates phenomenology, pathophysiology, and psychosocial context to guide effective management.

CONCLUSIONS

The classification and management of acute tic exacerbations remain a significant clinical challenge, owing to the overlapping features and absence of standardized diagnostic criteria for phenomena such as acute worsening of tics, self-injurious behaviors in the context of TS, and functional tic-like behaviors. While these episodes may differ in duration, severity, and underlying triggers, they often share characteristics such as nonsuppressibility, functional impairment, and association with psychologic or psychopharmacologic factors. The importance of a comprehensive, individualized approach to diagnosis and management is critical. This should encompass careful differentiation of tic-related phenomena from other neurologic or psychiatric conditions, along with psychoeducation, behavioral therapies, and appropriate pharmacologic intervention. In rare and refractory cases, advanced neuromodulation treatments may be considered. Ultimately, the field would benefit from the development of consensus-based diagnostic frameworks and clearer terminology based on robust evidence, which can improve clinical communication, reduce misdiagnosis, and optimize treatment strategies. A unified understanding of these tic-related emergencies will not only enhance patient outcomes but also support future research into the pathophysiologic and psychosocial mechanisms driving these complex and distressing episodes.

References

Au KLK, Chiu S, Malaty IA (2021). Rhabdomyolysis secondary to severe tic fits. BMJ Case Rep 14: e239874.

Barber KE, Ding Q, Espil FM et al. (2024). Contextual triggers and tic severity across life periods: a retrospective analysis in adults with tic disorders. Child Psychiatry Hum Dev.

Bloch M (2006). Handbook of Tourette's syndrome and related tic and behavioral disorders: second edition. J Am Acad Child Adolesc Psychiatry 45: 494–495.

Bodeck S, Lappe C, Evers S (2015). Tic-reducing effects of music in patients with Tourette's syndrome: self-reported and objective analysis. J Neurol Sci 352: 41–47.

Brandt VC, Beck C, Sajin V et al. (2016). Temporal relationship between premonitory urges and tics in Gilles de La Tourette Syndrome. Cortex 77: 24–37.

Cheung M-YC, Shahed J, Jankovic J (2007). Malignant Tourette syndrome. Mov Disord 22: 1743–1750.

Collicott NJ, Stern JS, Williams D et al. (2013). Tic attacks in Tourette syndrome. J Neurol Neurosurg Psychiatry 84: e2.77–e2.

Dehning S, Matz J, Riedel M et al. (2009). Symptom exacerbation in Tourette syndrome due to bacterial reinfection. J Clin Psychiatry 70: 1606.

Dobbs M, Berger JR (2003). Cervical myelopathy secondary to violent tics of Tourette's syndrome. Neurology 60: 1862–1866.

Doja A, Bookwala A, Pohl D et al. (2018). Relationship between physical activity, tic severity and quality of life in children with Tourette syndrome. J Can Acad Child Adolesc Psychiatry 27: 222–227.

Eddy CM, Cavanna AE (2014). Premonitory urges in adults with complicated and uncomplicated Tourette syndrome. Behav Modif 38: 264–275.

Fasano A, Galluccio V (2018). Brain injury due to head banging in Tourette. Parkinsonism Relat Disord 49: 114–115.

Fusco C, Bertani G, Caricati G et al. (2006). Stress fracture of the peroneal bone secondary to a complex tic. Brain Dev 28: 52–54.

Ganos C, Garrido A, Navalpotro-Gómez I et al. (2015). Premonitory urge to tic in Tourette's is associated with interoceptive awareness. Mov Disord 30: 1198–1202.

Ganos C, Martino D (2015). Tics and Tourette syndrome. Neurol Clin 33: 115–136.

Ganos C, Martino D, Espay AJ et al. (2019). Tics and functional tic-like movements: can we tell them apart? Neurology 93: 750–758.

Godar SC, Bortolato M (2017). What makes you tic? Translational approaches to study the role of stress and contextual triggers in Tourette syndrome. Neurosci Biobehav Rev 76: 123–133.

Hood KK, Baptista-neto L, Beasley PJ et al. (2004). Case study: severe self-injurious behavior in comorbid Tourette's disorder and OCD. J Am Acad Child Adolesc Psychiatry 43: 1298–1303.

Iverson AM, Black KJ (2022). Why tic severity changes from then to now and from here to there. J Clin Med 11: 5930.

Kovacs N, Herold R, Janszky J et al. (2011). Tics status: a movement disorder emergency. J Neurol 258: 143–145.

Krauss JK, Jankovic J (1996). Severe motor tics causing cervical myelopathy in Tourette's syndrome. Mov Disord 11: 563–566.

Krüger D, Müller-Vahl KR (2015). Severe self-injurious behavior with teeth extraction in a boy with Tourette syndrome. Pediatr Neurol 52: e5.

Kurlan R (Ed.) (2005). Handbook of Tourette's syndrome and related tic and behavioral disorders, 2nd edn. In: Neurological disease and therapy, vol. 15. New York: Marcel Dekker.

Lavoie ME, Cavanna AE (2022). The neurobiology of the Gilles de La Tourette syndrome and chronic tics: Part A. Chantilly, VA: Elsevier Science & Technology. http://ebookcentral.proquest.com/lib/uhn-ebooks/detail.action?docID=7032748.

Leckman JF (2003). Phenomenology of tics and natural history of tic disorders. Brain Dev 25: S24–S28.

Leckman JF, Bloch MH, Scahill L et al. (2006). Tourette syndrome: the self under siege. J Child Neurol 21: 642–649.

Leckman JF, King RA, Bloch MH (2014). Clinical features of Tourette syndrome and tic disorders. J Obsessive Compuls Relat Disord 3: 372–379.

Leckman JF, Ort S, Caruso KA et al. (1986). Rebound phenomena in Tourette's syndrome after abrupt withdrawal of clonidine: behavioral, cardiovascular, and neurochemical effects. Arch Gen Psychiatry 43: 1168–1176.

Leckman JF, Riddle MA, Hardin MT et al. (1989). The Yale Global Tic Severity Scale: initial testing of a clinician-rated scale of tic severity. J Am Acad Child Adolesc Psychiatry 28: 566–573.

Lim S, Rezai KA, Abrams GW et al. (2004). Self-induced, bilateral retinal detachment in Tourette syndrome. Arch Ophthalmol 122: 930–931.

Lowe TL, Shayawitz B (1982). Stimulant medications precipitate Tourette's syndrome. JAMA 247: 1168–1169.

Martino D, Hedderly T, Murphy T et al. (2023). The spectrum of functional tic-like behaviours: data from an international registry. Eur J Neurol 30: 334–343.

Max JE, Rasmussen SA (1986). Clonidine in the treatment of Tourette's syndrome exacerbation due to haloperidol withdrawal. J Nerv Ment Dis 174: 243–246.

Meige H, Feindel E, Wilson SAK (1907). Tics and their treatment. New York: William Wood and Company.

Mink JW (2003). The basal ganglia and involuntary movements: impaired inhibition of competing motor patterns. Arch Neurol 60: 1365–1368.

Monga V, Madan R, Arora N (2023). Malignant Tourette's syndrome in an adult on deep brain stimulation presenting with rhabdomyolysis. Cureus 15: e44436.

Motlagh MG, Smith ME, Landeros-Weisenberger A et al. (2013). Lessons learned from open-label deep brain stimulation for Tourette syndrome: eight cases over 7 years. Tremor Other Hyperkinet Mov (NY) 3: tre-03-170-4428-1.

Muroi A, Matsumura A, Asakawa H et al. (2002). Myelopathy caused by tics in an adolescent, associated with T2 signal intensity changes of the spinal cord. Child's Nerv Syst 18: 191–194.

Parmera JB, Yamamoto JYS, Cury RG (2023). Tic status in Tourette syndrome due to depletion of the deep brain stimulation battery. JAMA Neurol 80: 320.

Patterson AL, Choudhri AF, Igarashi M et al. (2016). Severe neurological complications associated with Tourette syndrome. Pediatr Neurol 61: 99–106.

Peterson BS, Leckman JF (1998). The temporal dynamics of tics in Gilles de La Tourette Syndrome. Biol Psychiatry 44: 1337–1348.

Pringsheim T, Nosratmirshekarlou E, Doja A et al. (2021). Physical activity, sleep and neuropsychiatric symptom severity in children with Tourette syndrome. Eur Child Adolesc Psychiatry 30: 711–719.

Rae CL, Larsson DEO, Garfinkel SN et al. (2019). Dimensions of interoception predict premonitory urges and tic severity in Tourette syndrome. Psychiatry Res 271: 469–475.

Raucci U, Parisi P, Vanacore N et al. (2018). Acute hyperkinetic movement disorders in Italian paediatric emergency departments. Arch Dis Child 103: 790–794.

Rickards H, Woolf I, Cavanna AE (2010). 'Trousseau's Disease:' a description of the Gilles de La Tourette Syndrome 12 years before 1885. Mov Disord 25: 2285–2289.

Robertson MM, Trimble MR, Lees AJ (1989). Self-injurious behaviour and the Gilles de La Tourette syndrome: a clinical study and review of the literature. Psychol Med 19: 611–625.

Robinson S, Hedderly T (2016). Novel psychological formulation and treatment of 'Tic Attacks' in Tourette syndrome. Front Pediatr 4: 46.

Sachdev P, Chee KY, Wilson A (1996). Tics status. Aust N Z J Psychiatry 30: 392–396.

Shapiro AK, Shapiro ES, Young JG et al. (1988). Gilles de La Tourette Syndrome (1978). New York: Raven Press.

Swain JE, Leckman JF (2005). Tourette syndrome and tic disorders. Psychiatry (Edgmont) 2: 26–36.

Woods DW, Himle MB (2004). Creating tic suppression: comparing the effects of verbal instruction to differential reinforcement. J Appl Behav Anal 37: 417–420.

Yael D, Vinner E, Bar-Gad I (2015). Pathophysiology of tic disorders: pathophysiology of tic disorders. Mov Disord 30: 1171–1178.

Zhang C, Li H, Pan Y et al. (2019). Pallidal neurostimulation and capsulotomy for malignant Tourette's syndrome. Mov Disord Clin Pract 6: 393–395.

Handbook of Clinical Neurology, Vol. 215 (3rd series)
Tic Disorders and Tourette Syndrome
C. Ganos and Y. Worbe, Editors
https://doi.org/10.1016/B978-0-443-13554-5.00017-1

Chapter 5

Functional tics and tic-like behaviors

ABDULLAH A. ALRUWAITA[1,2], ANTHONY E. LANG[1,3,4], AND CHRISTOS GANOS[1,3,4,*]

[1]*Edmond J. Safra Program in Parkinson's Disease and the Morton and Gloria Shulman Movement Disorders Clinic, Toronto Western Hospital, Toronto, ON, Canada*

[2]*Neurology Department, Prince Sultan Military Medical City, Riyadh, Saudi Arabia*

[3]*Division of Neurology, University of Toronto, Toronto, ON, Canada*

[3]*Krembil Brain Institute, Toronto, ON, Canada*

Abstract

Functional tics and tic-like behaviors belong to the wide spectrum of functional movement disorders. Until the early 2010s, functional tic disorders were rather uncommon in the differential diagnosis of tics, and only few cases describing their features had been published. However, over the past 10 years there has been a steady increase in the frequency of these cases that peaked throughout the COVID-19 pandemic. On the one hand, the rise in functional tic cases created new challenges of diagnosis and treatment. At the same time, it also pushed the field forward to delineate helpful clinical clues, as well as to work toward specific consensus diagnostic criteria for functional tics and tic-like behaviors. Here, we first provide a historical summary on the debate between neurodevelopmental and functional tics. We then track relevant literature on functional tics and tic-like cases and discuss their salient features, such as an acute onset with severe symptoms and complex repetitive behaviors that typically occur in late adolescence or early adulthood, a large variability of symptoms including spontaneous symptom remissions and re-emergence, and a high prevalence of phonations and vocalizations with the common use of swearwords or variable sentences. In addition, it is common to see an overlap with additional functional neurologic symptoms, such as functional tremor or nonepileptic seizures. In diagnostically challenging cases, neurophysiologic evaluation, including surface electromyography and electroencephalography, may be useful, and markers such as the premotor potential (Bereitschaftspotential) and event-related desynchronization/synchronization may hold promise. Effective management of functional tics begins with an accurate diagnosis and often requires a multidisciplinary approach. Cognitive-behavioral therapy and the Comprehensive Behavioral Intervention for Tics may be particularly useful, alongside addressing comorbid psychiatric conditions. Currently there is an absence of standardized treatment protocols; individualized care plans tailored to each patient's specific needs are generally the most effective approach.

BACKGROUND

In 1885, Georges Gilles de la Tourette published his seminal case series on what is now recognized as Tourette syndrome (TS) (Gilles de la Tourette, 1885). He documented nine patients with motor and phonic tics, beginning with the case of the Marquise de Dampierre, a young French woman with repetitive movements and vocalizations, who was first described by Jean Itard many decades earlier (Itard, 1825). As in the example of the Marquise, Gilles de la Tourette provided a detailed account of tic phenomena in the cases he described and repeatedly noted their resistance to therapeutic interventions.

Gilles de la Tourette's publication came at an opportune time. This was a period rich with discussions on the nature of motor phenomena, including tics, and their interaction with volition. Some neurologists at the time viewed tics as an "illness of will," while others considered them to be a manifestation of hysteria (Kushner, 1998). Interestingly, both Gilles de la Tourette and Jean Martin Charcot postulated that although tics may be part of "hysteria," the syndrome of Gilles

*Correspondence to: Christos Ganos, MD, Wolf Chair in Neurodevelopmental Psychiatry, Associate Professor of Neurology, University of Toronto, Movement Disorders Clinic, Toronto Western Hospital, MCL7-402, 399 Bathurst Street, Toronto, ON M5T 2S8, Canada. Tel: (416) 603-5207, Fax: (416) 603-5004, Email: christos.ganos@uhn.ca.

de la Tourette is due to a hereditary (and degenerative) etiology, particularly when resistance to interventions was shown. The distinction between tic phenomena and "hysteria" was also discussed in Meige and Feindel's exquisite treatise on "Tics and their treatment" (Meige and Feindel, 1902). Specifically, they noted: "Our response to the question whether tics are hysterical in origin is a direct negative."

Following these early years of neuroscientific inquiry into the nature of tics, there was a constant debate as to the determinants that lead to tic manifestation and the role of "psychogenicity" in their emergence. Nevertheless, with pioneering works dealing with the clinical characteristics (Shapiro and Shapiro, 1968), the neurobiology (Balthasar, 1957), and the neuropharmacology of TS, coupled with the failure of psychoanalytic models in achieving sustained tic improvement, a significant shift occurred in the medical perception of tics. In fact, perhaps due to extreme caution not to err in diagnostic labeling, most discussions on tic syndromes as the result of a functional neurologic disorder were discontinued. Instead, an increasing array of signs and symptoms became categorized under TS. However, a more careful appraisal of certain – atypical – clinical signs, alongside a renewed interest in the neurobiology of functional neurologic syndromes, has reemphasized the century-old problem of distinguishing primary from functional tics. The notable rise in the prevalence of functional neurologic syndromes, particularly tic-like behaviors, during the COVID-19 pandemic further accelerated research in this complex clinical area. This chapter will showcase the findings of some of the key studies that examined the clinical characteristics of functional tics compared with primary tics and provide useful clinical handles for aiding clinicians in their distinction.

FEATURES OF FUNCTIONAL TICS

One of the first systematic studies conducted on the topic of functional tics and tic-like movements was conducted by Drs. Baizabal-Carvallo and Jankovic, titled "The clinical features of psychogenic movement disorders resembling tics" (Baizabal-Carvallo and Jankovic, 2014). In this study the authors reported nine patients with functional tics. The study utilized the Fahn–Williams clinical criteria for diagnosis, which rely on "movements inconsistent or incongruent with classic description of movement disorders associated with other psychogenic signs, multiple somatisations or obvious psychiatric disturbance." Although, it is difficult to apply these criteria in a movement disorder like tics, the authors felt confident in their diagnosis. Notably, none of the nine patients they reported had a family history of TS and symptom onset was in most cases in early adulthood, with a mean age of 34 years. Additionally, all patients exhibited other forms of functional movement disorders alongside tics, such as functional seizures, bizarre gait, or functional tremor. Unlike primary tic disorders, all patients lacked the classic suppressibility and premonitory urge associated with tics.

In the same year, Demartini et al. reported 11 patients with functional tics, highlighting further distinguishing features (Demartini et al., 2015). Indeed, a common characteristic among these patients was the abrupt onset of symptoms, with all patients displaying clear distractibility and suggestibility during neurologic examination (although it is important to point out that these two features are also typical of primary tics). None of the patients reported a waxing and waning course of symptoms or coprophenomena, coprolalia, or palilalia. Four out of the eleven patients experienced "blocking tics," that is sudden interruptions in ongoing physical activity or verbal communication due to co-occurring tics. Like the previous study, none of these patients had a family history of TS. Only two patients reported classic premonitory urges, and another two reported the ability to suppress their movements. Eight out of the eleven patients exhibited other functional symptoms, such as functional weakness, sensory changes, or seizures. Most patients experienced a stressful physical event preceding the occurrence of functional tics, such as physical trauma or illness requiring hospitalization (e.g., pneumonia). Three patients reported an acute psychologic stressor. An interesting observation was noted regarding the distribution of tics: unlike primary tics that typically start in the face and spread caudally, functional tics did not strictly follow this rostrocaudal somatotopic pattern.

In 2016, Ganos and colleagues reported 13 individuals with functional coprolalic and other tic-like complex vocalizations. Ten patients had an acute onset of symptoms, and seven experienced tic onset in adulthood. Coprolalia was predominant in seven patients, echolalia in five, and palilalia in 10 patients. Eleven patients exhibited jerky movements associated with vocalizations. Ten patients were able to temporarily suppress their vocalizations. Atypical courses of tics were evident in 11 patients, with major fluctuations and full remission in some cases. Most patients (10 out of 13) had been treated with typical anti-tic medications without clear improvement. None of these patients had a family history of TS. Six patients developed functional tics before the age of 18, emphasizing that functional tics can occur at any age. In contrast to previous studies on functional tics, nine patients in this cohort reported a premonitory urge, although it was atypical (e.g., "whole body pressure" or "a sudden energy pulse"). Another important distinguishing feature compared to TS coprolalia was the nature of vocalizations; in this study, the authors noted that vocalizations in functional coprolalia were longer, consisting of compound words or even short sentences of obscene content (Ganos et al., 2016). This contrasts with the common "curtailing" of obscenities in TS coprolalia (i.e., "fu-," "shi-"). Interestingly, none of the studies explicitly assessed the sense of agency in functional tics and tic-like movements, even though this may be used as a useful diagnostic hint for some patients (Lang, 1991). One study utilizing the premonitory urge for tics scale assessed the intensity and frequency of premonitory urges preceding tics in functional tics compared to primary tics. The study found similar premonitory urges with no significant differences between the two groups (Szejko et al., 2024).

Generally, comorbidity patterns may also provide useful clinical insights. Attention-deficit/hyperactivity disorder and obsessive-compulsive behaviors are frequently associated with primary tics and TS but were not commonly observed in patients with functional tics (Ganos et al., 2014, 2019). Conversely, both groups tend to experience anxiety and depression (Demartini et al., 2015; Hirschtritt et al., 2015).

COVID-19 AND FUNCTIONAL TICS

Since the onset of social restrictions during the COVID-19 pandemic, there was a notable worsening of tic symptoms in many patients with primary tics. In the early stages of the pandemic, a study involving 178 participants diagnosed with primary tics revealed that nearly half of them experienced a worsening in their tic symptoms. Factors presumed to contribute to this worsening included increased stress and health concerns, heightened anxiety about family and financial matters, reduced exercise, and the stigmatization of cough tics during that time (Mataix-Cols et al., 2020). However, since the COVID-19 lockdowns, clinicians noted a significant increase in tic consultations and referrals to their clinics (Pringsheim et al., 2021). They observed a profound rise among patients in their late teens, who were referred with sudden onset tic-like movements despite having no history of tics. The observations indicated that tics were more common among female patients in their teenage years, with the peak severity of tics occurring at onset. Similar to descriptions of functional tics before the pandemic (Ganos et al., 2019), they observed a variety of complex motor and phonic tics with frequent use of complex phrases and swearing, a lack of family history of tics, absence of rostrocaudal spreading, and frequent occurrences of self-injurious behaviors, including outward directed tic-like behaviors, such as throwing objects or even hitting and punching others. The majority of teenagers diagnosed with functional tics reported a relevant exposure to social media content creators who also presented with tic-like movements (Heyman et al., 2021; Kurvits et al., 2021; Paulus et al., 2021; Pringsheim et al., 2021; Buts et al., 2022; Han et al., 2022; Trau et al., 2022 Andersen et al., 2023; Cavanna et al., 2023; Martino et al., 2023; Okkels et al., 2023; Nilles et al., 2024b). This propelled some clinicians to consider the rise of cases with functional tics as part of a "mass sociogenic illness" prompting a lively scientific debate as to the role of social media in symptom onset, characteristics, and perpetuation (Ganos, 2021; Conelea et al., 2022). Table 5.1 provides a summary of most of the major studies we could identify that reported on patients with functional tics following the COVID-19 pandemic (see Table 5.1).

THE PROPOSED CRITERIA FOR DIAGNOSIS OF FUNCTIONAL TICS BY THE EUROPEAN SOCIETY FOR THE STUDY OF TOURETTE SYNDROME

As a result of the sudden increase of patients with functional tics and tic-like movements, international experts in the field of tics and TS collaborated to establish clinical diagnostic criteria for distinguishing functional tic-like behaviors (FTLBs) from primary tics (Pringsheim et al., 2023). The expert panel favored the term "functional tic-like behaviors" over "functional tics" to make a point that many of the clinical phenomena they observed differed from classic or typical tics in primary tic disorders and TS. Although the distinction lines may be blurred (e.g., what distinguishes a blink in a primary tic disorder from one at a functional tic-like disorder), the panel preferred the term FTLB to encompass the spectrum of all behavioral phenomena observed in this patient population, which include many complex motor and vocal behaviors that differ from the simple clonic tics seen in children and adolescents diagnosed with TS. The proposed consensus criteria discerned two possible diagnostic categories for FTLB: a clinically definite diagnosis and a clinically probable diagnosis. A definite diagnosis requires the presence of all three major criteria, which include age at symptom onset older than 12 years, rapid evolution of symptoms, and phenomenological markers underlining the variability and complexity of behaviors that typify functional tics or tic-like phenomena. A probable FTLB diagnosis necessitates two major criteria and at least one minor criterion, which includes the pattern of comorbidities typical for functional movement disorders, or indeed a history of other functional neurologic disorders (see Table 5.2). It is important to note that the majority of case series and registries have reported an association between FTLB, female sex, and a lack of response to anti-tic medications. However, these factors were not considered when developing these criteria as they were deemed to lack specificity. It is also crucial to recognize that these criteria are primarily based on the clinical expertise and data from published case series in specific regions of the world. They have not undergone independent prospective testing for sensitivity and specificity, which is considered a limitation. Indeed, only one study to date evaluated the specificity of the European Society for the Study of Tourette Syndrome (ESSTS) criteria in a tic disorders cohort and showed that the age at onset criterion (> 12 years) and the presence of at least two complex motor and one phonic tic-like behaviors at first visit had high specificity (94.2% and 96.2%, respectively) for the diagnosis of an FTLB (Nilles et al., 2024a). Importantly, it also remains unclear how to distinguish some functional tics and tic-like phenomena from primary tics, specifically in cases where they occur in the same patient (i.e., a patient with a primary tic disorder and functional overlay). Since there are no biomarkers available for diagnosing FTLB, physicians must rely on their clinical judgment informed not only by phenomenology, but also by the patient's medical history and physical examination. Indeed, a recent study that examined expert agreement on whether tic behaviors were due to a primary tic disorder, or a functional syndrome clearly showed that phenomenology alone does not suffice to reach diagnostic consensus (Rigas et al., 2023). Importantly, approximately 20% of clinician members of the International Parkinson and Movement Disorder Society who regularly see patients with tics confirmed that about one quarter of their

Table 5.1

Summary of the large studies that have reported patients with functional tics following the COVID-19 pandemic.

Study	Publication date	Number of subjects	Female (%)	Mean age of onset	Family history of tics (%)	Social media exposure to tics (%)	Tics phenomenology	Premonitory urge (%)	Suppressibility (%)	Comorbidities
(Pringsheim et al., 2021)	August 2021	20	95	13.9	NA	100	Complex arm movements, self-injurious behaviors (e.g., hitting self), wide range of odd words or phrases			Anxiety 75%, depression 55%, attention-deficit/hyperactivity disorder (ADHD) 25%, obsessive-compulsive disorder (OCD) 5%
(Buts et al., 2022)	November 2021	34	94	13.7	29	77	Injurious behaviors toward self or parents, head nodding, whistling, coprophenomena	62	59	Anxiety 68%, depression 24%, ADHD 26%, OCD 9%
(Paulus et al., 2021)	September 2021	13	37.5	15.3	8	100	Abrupt onset, complex slow movements involving extremities and trunk, worsening in the presence of others, goal-directed and complex vocalizations	83	40	OCD 23%, ADHD 8%
(Han et al., 2022)	February 2022	22	100	13.8	14	18	Abrupt onset, common movements of head nodding, shoulder shrugging, and facial twitching. Injurious behaviors both to self and others. Complex words and phrases	NA	NA	Anxiety/depression 95%, OCD 23%, ADHD 14%
(Trau et al., 2022)	October 2022	31	97	14	19	70	Acute, complex prolonged variable motor and vocal tics, injurious behaviors, and throwing objects, coprophenomenon and long phrases	77 (for premonitory urge and suppressibility)		Anxiety 90%, ADHD 68%, OCD 58%
(Okkels et al., 2023)	January 2023	28	96	14.4	25	96.4	50% abrupt onset, associated with complex movements, injurious behaviors to self and others, throwing objects, coprophenomena including long phrases. Some cases with simple motor movements (e.g., eye blinking) as well.	89	90	OCD 47%, ADHD 41%, anxiety 35%, depression 35%

(Prato et al., 2023)	January 2023	11	73	14	36	36	Abrupt onset in majority of patients, associated with complex motor and vocal tics, including coprophenomena	27	NA	NA
(Martino et al., 2023)	February 2023	294	87	15	12	92	Simple motor 96%, complex 94%, arms movements with injurious behaviors both directed to self or others 48%. Rare lower limb movements. Simple phonic tics 82%, complex phonic tics 81%, nonobscene inappropriate phrases 48%, context-dependent sentences 24%, coprophenomena 51%	47	57	Anxiety 66%, depression 22%, Autism spectrum disorder (ASD) 24%, Post-traumatic stress disorder (PTSD)6%
(Andersen et al., 2023)	May 2023	53	95	13.7	11	57	Complex motor tics, self-injurious behaviors, long phrases, and coprophenomena	NA	NA	Anxiety 26%, ADHD 19%, OCD 19%
(Cavanna et al., 2023)	August 2023	105	72	21	14	50	Abrupt onset 75%, complex motor 92%, and vocal tics 89%. Self-injurious behaviors 39%. Complex vocal 75%. Coprophenomena 49%	48	56	Anxiety 70%, affective disorder 40%, ADHD 18%, OCD 9%
(Nilles et al., 2024b)	January 2024	83	95	18	NA	NA	Simple motor tics 86%, complex self-injurious behaviors 42%, simple vocal tics 30%, and complex words 54%	NA	NA	Anxiety 65%, depression 42%, ADHD 35%, OCD 7%, ASD 4%

Table 5.2

European Society for the Study of Tourette Syndrome 2022 criteria for clinical diagnosis of FTLBs say (Adapted from Pringsheim et al., 2023).

Major criteria	1. *The age of symptom onset to be 12 years old or older.* 2. *Rapid evolution of symptoms, usually within hours to days.* Usually, family can recall the exact date or week when the symptoms start and progressed. 3. *Phenomenology* (at least four of the following nine proposed features): a. Multiple tic-like motor and/or vocal behaviors, mostly complex rather than simple tic-like movements. b. Inconsistence of the movements (variability rather than stereotypic movements). c. The motor FTLB, include complex hand-specific movements. For example, "banging chest/head, tapping, hitting others, sign language, throwing objects, offensive gestures, drop attacks, or freezing." These movements can be context-dependent, violent to the patient or others. d. Does not follow the typical rostrocaudal distribution of tics. e. Context-dependent and offensive vocal tics that might have many words and statements. f. Tics that resemble references or "popular cultural influencers" or other individuals, i.e., friends or family members. g. Significant variation in symptoms severity. Symptoms-free period in a single day and context-specific worsening in other time. h. New tic-like behaviors daily or every few days. i. Exacerbation of the FTLB during the physical examination encounter.
Minor criteria	1. *Comorbid psychiatric illness* (prior symptoms of anxiety and depression, behaviors of self-harm, suicidal ideation, difficulties with social relationships, or earlier traumatic events). 2. *Previous history or current other neurologic functional disorders* (functional neurologic symptoms (i.e., functional seizures, paralysis) or somatoform disorders (i.e., functional abdominal pain)).

FTLBs, functional tic-like behaviors.

patients might have functional tics overlaid to a primary tic syndrome (Ganos et al., 2021). The criteria that clinicians use to base their etiologic distinction between similar tic phenomena in a given patient remain difficult to confidently discern. In challenging cases, neurophysiologic investigations might also be of use.

NEUROPHYSIOLOGY ROLE IN FUNCTIONAL TICS

Several studies have also attempted to provide neurophysiologic markers as aids to differentiate primary or neurodevelopmental from functional tics. One line of research has focused on movement-related cortical potentials, and specifically the Bereitschaftspotential (BP) (also known as premotor potential). The BP is suggested to be a cortical correlate of volitional movements that appears about 1.5 seconds before the onset of an action. It is typically registered as a negative cortical potential, with an early bilateral component, and a late lateralized component that peaks over the contralateral precentral gyrus (Kornhuber and Deecke, 1965; Deecke et al., 1976). Numerous neurophysiologic studies have used the BP as a marker for volitional actions, tics, and functional jerky movements and have indeed provided useful distinctions. For instance, Obeso et al. observed that the majority of patients with primary tics lacked BP before tic movements (five out of the six study's subjects). One subject had an atypical late BP in comparison to the typical morphology of the BP found before voluntary actions (Obeso et al., 1981). However, Karp and colleagues found that some patients may or may have not have BP before their tics. They studied five individuals with tics. Two participants felt their tics were involuntary, and three believed it was a combination of voluntary and involuntary actions. All participants had BP before any self-initiated, or before imitation of their tics. Three out of five participants with primary tics exhibited no BP before their tics. Two had negative BP before their tics. The negativity was maximal centrally and contralateral to the movement. The BP latency, distribution, and amplitudes with tics in both patients were similar to the those obtained during voluntary movements. The first patient felt their tics were totally involuntary without an urge and without control over their presence, while the second participant felt their tics were triggered by internal urges and they were a response to inner tension (Karp et al., 1996). In a small further study that was published as a preliminary report, Duggal and colleagues examined three patients with primary motor tics and found that the BP was present in all of them. All three participants had premonitory urges, albeit with a shortened BP onset latency compared to normal BP in all cases (Duggal and Nizamie, 2002). Following these smaller sample studies, Van der Salm et al. assessed the presence of BP in jerky movement disorders among 48 participants: 29 patients with functional jerks, 14 with TS, and 5 with myoclonus, compared to 25 control subjects. The study revealed that 6 out of the 14 TS patients had a preceding BP before their motor tics, while 86% of the participants with functional jerks showed a preceding BP. The functional group had a significantly earlier onset of BP compared to the TS patients (p=0.020). None of the five participants with myoclonus had a BP before their involuntary movements. A unique finding was the absence of BP in most functional jerks patients before an intended action, which was highly specific to this group (Van der Salm et al., 2012). However, this finding needs considerable caution and replication as many different motor parameters may influence the presence of a BP (Jahanshahi and Hallett, 2003). In a recent study by Triggiani and colleagues, the Libet clock paradigm was used to compare patients with tics to healthy volunteers. The Libet paradigm measures the sense of the will to move (W time), which precedes the sense of movement (M time). This approach aimed to examine awareness and agency, as well as their alterations. The study also assessed the BP and event-related desynchronization (ERD)/event-related synchronization in the β frequency (13–30 Hz) within the same subjects, which is known to occur during the planning and execution phases of voluntary movements. The reported times for the intention to move and the tic movements did not differ significantly from those of healthy volunteers. However, in contrast to the earlier study by Obeso et al., five out of seven subjects with tics showed BP before the onset of their tics. This may be influenced by the use of the Libet-style paradigm, which required subjects to focus their attention on the movement. Interestingly, while most participants exhibited BP preceding their tics, only a minority (2 out of 7) showed ERD before their tic movements (Triggiani et al., 2023). Considering all these data collectively, it may be inferred that the presence of a late BP component only or an entirely absent BP before a motor tic when combined with the absence of ERD could be useful markers to distinguish tics from functional tic-like movements. However, these conclusions should be approached with caution due to the complexity of tic phenomenology and variations in clinical presentations, including the co-occurrence of psychiatric conditions such as obsessive-compulsive disorder. Additionally, neurophysiologic studies can contribute to characterizing phenomenology by observing muscle activation patterns through surface electromyography, offering a more sensitive approach to describing tic topography and assessing for variability (Vial et al., 2019).

TREATMENTS AND OUTCOMES

Similar to other conditions involving functional neurologic disorders, there is no evidence to support the usage of pharmacologic treatments to control FTLBs; most case series do not report substantial or sustained evidence of the effectiveness of pharmacologic interventions (Ganos et al., 2016). It is important to note that individuals with primary tics and TS may also present with coexisting functional tics. The ability to differentiate between primary symptoms and functional tics is crucial when addressing the most debilitating aspects of the condition, as each entity requires a distinct treatment approach (Cavanna et al., 2024).

Early diagnosis and treatment of comorbid psychiatric disorders has been found to correlate with improved outcomes. For example, the diagnosis and treatment of mood disorders, such as anxiety and depression, has proven to be beneficial in the treatment of patients and their capacity to participate in further treatment programs (Pehlivantürk and Unal, 2002). A multidisciplinary approach, including cognitive and behavioral therapy, has shown considerable effectiveness. Interventions like Comprehensive Behavioral Intervention for Tics (CBIT) and trigger awareness have yielded positive outcomes. For instance, in a case series involving 20 adolescents with functional tics who underwent cognitive-behavioral therapy (CBT) and received selective serotonin reuptake inhibitors targeting anxiety and depression, 15 of them returned for follow-up at 6 months and demonstrated a significant reduction in tic severity scales (Howlett et al., 2022). Another prospective study further supported these findings, showing good outcomes for patients with functional tics over a 12-month period when they were treated with CBT for anxiety and/or depression (Nilles et al., 2024b). Both studies suggested that better outcomes were achieved in participants under the age of 18.

In general, initiating treatment with a clear diagnosis is a crucial aspect of understanding and accepting the condition. Educating patients about the nature of functional neurologic disorders is equally important. Identifying triggers and focusing on their reduction and avoidance, such as limiting exposure to social media and videos with tic-like behaviors content that can provoke tic-like manifestations, may also play an important role (Paulus et al., 2021). Additionally, reinforcing normal motor and speech functions through physical, occupational, and speech therapy may be helpful and should be determined on a case-by-case basis. While CBIT was initially developed to treat patients with primary tics, some recent studies have reported similar outcomes in patients with functional tics (Martino et al., 2023; Maxwell et al., 2023).

SUMMARY

Functional neurologic disorders, including functional tics and tic-like behaviors, constitute a complex entity with various clinical presentations, often causing significant disability to patients and profound distress to their families. These conditions are frequently subject to misdiagnosis (Ganos et al., 2019). Large-scale validations of the recent ESSTS criteria and further refinement will be necessary. In addition, the development and application of biomarkers, for example through cortical neurophysiology, will be valuable in differentiating between primary and functional tics, and also exploring potential pathophysiologic mechanisms underlying their occurrence. Additionally, larger-scale intervention studies are needed to provide much needed data for the development of evidence-based tailored treatments.

References

Andersen K, Jensen I, Okkels KB et al. (2023). Clarifying the differences between patients with organic tics and functional tic-like behaviors. Healthcare (Basel) 11: 1481.

Baizabal-Carvallo JF, Jankovic J (2014). The clinical features of psychogenic movement disorders resembling tics. J Neurol Neurosurg Psychiatry 85: 573–575.

Balthasar K (1957). Über das anatomische substrat der generalisierten tic-krankheit: entwicklungshemmung des corpus striatum. Arch Psychiatr Nervenkrankh 195: 531–549.

Buts S, Duncan M, Owen T et al. (2022). Paediatric tic-like presentations during the COVID-19 pandemic. Arch Dis Child 107: e17.

Cavanna AE, Purpura G, Riva A et al. (2023). New-onset functional tics during the COVID-19 pandemic: clinical characteristics of 105 cases from a single centre. Eur J Neurol 30: 2411–2417.

Cavanna AE, Spini L, Ferrari S et al. (2024). Functional tic-like behaviors: from the COVID-19 pandemic to the post-pandemic era. Healthcare (Basel) 12: 1106.

Conelea CA, Bervoets J, Bethan Davies E et al. (2022). A call for caution: "stop that" sentiments threaten tic research, healthcare and advocacy'. Brain 145: e18–e20.

Deecke L, Grözinger B, Kornhuber HH (1976). Voluntary finger movement in man: cerebral potentials and theory. Biol Cybern 23: 99–119.

Demartini B, Ricciardi L, Parees I et al. (2015). A positive diagnosis of functional (psychogenic) tics. Eur J Neurol 22: 527–e36.

Duggal HS, Nizamie SH (2002). Bereitschaftspotential in tic disorders: a preliminary observation. Neurol India 50: 487–489.

Ganos C (2021). Tics and tic-like phenomena – old questions on a grand new scale: invited editorial on TikTok and tics. Mov Disord Clin Pract 8: 1198–1199.

Ganos C, Edwards MJ, Müller-Vahl K (2016). "I swear it is Tourette's!": on functional coprolalia and other tic-like vocalizations. Psychiatry Res 246: 821–826.

Ganos C, Martino D, Espay AJ et al. (2019). Tics and functional tic-like movements: can we tell them apart? Neurology 93: 750–758.

Ganos C, Münchau A, Bhatia KP (2014). The semiology of tics, Tourette's, and their associations. Mov Disord Clin Pract 1: 145–153.

Ganos C, Sarva H, Kurvits L et al. (2021). Clinical practice patterns in tic disorders among movement disorder society members. Tremor Other Hyperkin Mov 11: 43.

Gilles de la Tourette G (1885). Étude sur une affection nerveuse caractérisée par de l'incoordination motrice accompagnée d'écholalie et de coprolalie. Arch Neurol (Paris). 9: 19–42, 158–200.

Han VX, Kozlowska K, Kothur K et al. (2022). Rapid onset functional tic-like behaviours in children and adolescents during COVID-19: clinical features, assessment and biopsychosocial treatment approach. J Paediatr Child Health 58: 1181–1187.

Heyman I, Liang H, Hedderly T (2021). COVID-19 related increase in childhood tics and tic-like attacks. Arch Dis Child 106: 420–421.

Hirschtritt ME, Lee PC, Pauls DL et al. (2015). Lifetime prevalence, age of risk, and genetic relationships of comorbid psychiatric disorders in Tourette syndrome. JAMA Psychiatry 72: 325–333.

Howlett M, Martino D, Nilles C et al. (2022). Prognosis of rapid onset functional tic-like behaviors: prospective follow-up over 6 months. Brain Behav 12: e2606.

Itard JM (1825). Mémoire sur quelques fonctions involontaires des appareils de la locomotion, de la préhension et de la voix. Arch Gén Méd 8: 385–407.

Jahanshahi M, Hallett M (2003). The Bereitschaftspotential: what does it measure and where does it come from? In: Jahanshahi M, Hallett M. (Eds.). The Bereitschaftspotential. Boston, MA: Springer, pp. 1–7.

Karp BI, Porter S, Toro C et al. (1996). Simple motor tics may be preceded by a premotor potential. J Neurol Neurosurg Psychiatry 61: 103–106.

Kornhuber HH, Deecke L (1965). Hirnpotentialänderungen bei willkürbewegungen und passiven bewegungen des menschen: Bereitschaftspotential und reafferente potentiale. Pflügers Arch 284: 1–17.

Kurvits L, Mainka T, Cavanna AE et al. (2021). Aggression toward others misdiagnosed as primary tics. Mov Disord Clin Pract 8: 769–771.

Kushner HI (1998). Freud and the diagnosis of Gilles de la Tourette's illness. Hist Psychiatry 9: 1–25.

Lang A (1991). Patient perception of tics and other movement disorders. Neurology 41: 223–228.

Martino D, Hedderly T, Murphy T et al. (2023). The spectrum of functional tic-like behaviours: data from an international registry. Eur J Neurol 30: 334–343.

Mataix-Cols D, Ringberg H, Fernández de la Cruz L (2020). Perceived worsening of tics in adult patients with Tourette syndrome after the COVID-19 outbreak'. Mov Disord Clin Pract 7: 725–726.

Maxwell A, Zouki JJ, Eapen V (2023). Integrated cognitive behavioral intervention for functional tics (I-CBiT): case reports and treatment formulation. Front Pediatr 11: 1265123.

Meige H, Feindel E (1902). Les tics et leur traitement. Paris: Masson.

Nilles C, Martino D, Pringsheim T (2024a). Testing the specificity of phenomenological criteria for functional tic-like behaviours in youth with Tourette syndrome. Eur J Neurol 31: e16262.

Nilles C, Szejko N, Martino D et al. (2024b). Prospective follow-up study of youth and adults with onset of functional tic-like behaviours during the COVID-19 pandemic. Eur J Neurol 31e16051.

Obeso JA, Rothwell JC, Marsden CD (1981). Simple tics in Gilles de la Tourette's syndrome are not prefaced by a normal premovement EEG potential. J Neurol Neurosurg Psychiatry 44: 735–738.

Okkels KB, Skov L, Klansø S et al. (2023). Increased number of functional tics seen in Danish adolescents during the COVID-19 pandemic. Neuropediatrics 54: 113–119.

Paulus T, Bäumer T, Verrel J et al. (2021). Pandemic tic-like behaviors following social media consumption. Mov Disord 36: 2932–2935.

Pehlivantürk B, Unal F (2002). Conversion disorder in children and adolescents: a 4-year follow-up study. J Psychosom Res 52: 187–191.

Prato A, Saia F, Milana MC et al. (2023). Functional tic-like behaviours during the COVID-19 pandemic: follow-up over 12 months. Frontiers in Pediatrics 10: 1003825.

Pringsheim T, Ganos C, McGuire JF et al. (2021). Rapid onset functional tic-like behaviors in young females during the COVID-19 pandemic. Mov Disord 36: 2707–2713.

Pringsheim T, Ganos C, Nilles C et al. (2023). European Society for the Study of Tourette Syndrome 2022 criteria for clinical diagnosis of functional tic-like behaviours: international consensus from experts in tic disorders. Eur J Neurol 30: 902–910.

Rigas A, Mainka T, Pringsheim T et al. (2023). Distinguishing functional from primary tics: a study of expert video assessments. J Neurol Neurosurg Psychiatry 94: 751–756.

Shapiro AK, Shapiro E (1968). Treatment of Gilles de la Tourette's syndrome with haloperidol. Br J Psychiatry 114: 345–350.

Szejko N, Fletcher J, Martino D et al. (2024). Premonitory urge in patients with tics and functional tic-like behaviors. Mov Disord Clin Pract 11: 276–281.

Trau SP, Quehl L, Tsujimoto THM et al. (2022). Creating a patient-based diagnostic checklist for functional tics during the COVID-19 pandemic. Neurol Clin Pract 12: 365–376.

Triggiani AI, Scheman K, Pirio Richardson S et al. (2023). Physiological and introspective antecedents of tics and movements in adults with tic disorders. Clin Neurophysiol 151: 143–150.

Van der Salm SM, Tijssen MA, Koelman JH et al. (2012). The bereitschaftspotential in jerky movement disorders. J Neurol Neurosurg Psychiatry 83: 1162–1167.

Vial F, Attaripour S, Hallett M (2019). Differentiating tics from functional (psychogenic) movements with electrophysiological tools. Clin Neurophysiol Pract 4: 143–147.

Handbook of Clinical Neurology, Vol. 215 (3rd series)
Tic Disorders and Tourette Syndrome
C. Ganos and Y. Worbe, Editors
https://doi.org/10.1016/B978-0-443-13554-5.00001-8

Chapter 6

Repetitive and restricted behaviors, habits, body-focused repetitive behaviors, and motor stereotypies

HARVEY S. SINGER[1,2,*], SAHAR MAHATE[1], AND CHRISTINA MORRIS-BERRY[1,2]

[1]*Department of Neurology, Johns Hopkins University School of Medicine, Baltimore, MD, United States*

[2]*Department of Neurology, Kennedy Krieger Institute, Baltimore, MD, United States*

Abstract

Movement disorder definitions that include the clinical characteristics "sudden, rapid, repetitive, brief, and involuntary" are not limited to motor tic disorders. Hence, it is essential that caregivers be aware of other relatively common conditions that could be confused with the diagnosis of tics. Included in this differential are entities such as repetitive and restricted behaviors (RRBs), habits, body-focused repetitive behaviors (BFRBs), and motor stereotypies. The importance of separation of these individual conditions is further emphasized by the recognition that they can be coexisting problems in individuals with tic disorders, and thus therapeutic approaches may differ. For example, some individuals diagnosed with Tourette syndrome have co-occurring complex motor stereotypies or recurring skin picking, nail biting, and hair pulling. This chapter reviews the definition of a movement disorder and the evolving entity labeled as RRBs. It also contains a more in-depth review of our knowledge regarding the clinical features, etiology, pathophysiology, and therapeutic approaches for the more specific diagnoses including habits, BFRBs, and stereotypies.

The goal of this chapter is to provide the reader with information required to identify and characterize several selected repetitive movement disorder conditions that potentially could be confused with a tic disorder. To achieve this aim, this report begins with definitions of two broad terms: movement disorders and the slightly narrowed but still confusing designation of "repetitive and restricted behaviors" (RRBs). The text then reviews three assigned repetitive behavioral conditions: habits, body-focused repetitive behaviors (BFRBs), and complex motor stereotypies (CMS). Within each of these latter topics, descriptions review the clinical characteristics, pathophysiologic features, and therapeutic options. As will become clear, there is much to be learned about each of these three repetitive behavioral diagnoses.

MOVEMENT DISORDER DEFINITION

A movement disorder is typically defined as a broad neurologic condition that involves either an impaired voluntary movement, an abnormality of posture, the presence of an abnormal involuntary movement, or a normal movement occurring at inappropriate or unintended times. Movement abnormalities should not be due to weakness, spasticity, or rigidity. Most childhood disorders are hyperkinetic, abnormal, involuntary, repetitive movements that include tics, stereotypies, dystonia, chorea, athetosis, myoclonus, startle, and tremors. The placement of a specific movement abnormality within a defined category (e.g., stereotypy) requires knowledge of the movement's clinical characteristics, observation of the movement, preferably in-person, and a comprehensive physical examination. The term "extrapyramidal disorder" should be abandoned since it lacks specificity, and most movement disorders cannot be accurately attributed to a particular anatomic abnormality.

REPETITIVE AND RESTRICTED BEHAVIORS

The term repetitive and restricted behaviors is used within the *Diagnostic and Statistical Manual of Psychiatric Disorders, Fifth Edition* (DSM-V) to define the condition Autistic Spectrum Disorders (American Psychiatric Association, 2013). In this context, RRBs include a wide variety of "restricted, repetitive patterns of behavior, interests, or

*Correspondence to: Harvey S. Singer, MD, Kennedy Krieger Institute, Department of Neurology, 707 North Broadway, Room 400N, Baltimore, MD 21205, United States. Tel: +1-410-245-2076, E-mail: hsinger@jhmi.edu.

activities." Others have expanded this definition to include the diagnostic terms cognitively inflexible and socially inappropriate (Faja and Nelson Darling, 2019; Brierley et al., 2021; Comparan-Meza et al., 2021). Researchers interested in improving the specificity of the RRB classification have also suggested dividing the disorder into "lower" and "higher" levels (Turner, 1999; Jiujias et al., 2017; Kuzminskaite et al., 2020; Comparan-Meza et al., 2021). The former includes body movements (stereotypies, self-injurious behaviors, tics, object manipulation, dyskinetic movements) and the latter, an adherence to rituals, routines, or concerns for an object. In contrast, others have suggested subdivisions including "repetitive sensory motor" and "insistence on sameness" groupings (Bishop et al., 2013) as well as a greater emphasis on sensory behaviors (hyperresponsiveness, hyporesponsiveness, and sensory seeking) (Kirby et al., 2017; Faja and Nelson Darling, 2019). The DSM-V suggests that unusual sensory responses and repetitive/stereotypic behaviors are distinct, but each are subsumed under the restricted and repetitive symptom grouping for the diagnosis of ASD (Kirby et al., 2017). In summary, recognizing that the term RRB lacks specificity, it is not recommended as a substitute for a specific diagnosis.

SPECIFIC REPETITIVE BEHAVIORS

Habits

In general, the term habit is defined as a behavior that is performed quickly, semiautomatically, triggered by stimuli or contexts, and is not causing a problem (Robbins and Costa, 2017; Du et al., 2022; Watson et al., 2022). For example, in the Merriam-Webster Dictionary, habits are defined as (1) "a settled tendency or usual manner of behavior" or (2) "an acquired mode of behavior that has become nearly or completely involuntary." Others have noted habit is frequently used to describe behaviors that have "an inexorable pull over us," either positive (healthy exercise) or negative (drug use) (Watson et al., 2022). Common examples range from *lifestyle habits* such as going to the gym or driving to work following the same route, to *movement habits* that include repetitive actions (i.e., foot tapping). Both types may increase during periods of anxiety, stress, fatigue, or boredom. Regarding movement habits, other than fulfilling the broad definition and having a normal neurologic examination, there are no specific confirmatory criteria.

Biologically, the factors determining whether a behavior becomes a goal-directed or habitual action remain unclear. The dual-process model hypothesis suggests that the control of action is determined by stimulus–response and response–outcome (Adams, 1982; Dickinson and Mulatero, 1989; Robbins and Costa, 2017). In the *stimulus–response hypothesis* it is proposed that via the process of repetition and learning, environmental stimuli can lead to spontaneous, automatically elicited responses. Response–outcome recognizes that the resulting outcome reinforces links between the stimulus and response. For example, in animal models, initial food reward behaviors (e.g., pressing a lever earns a food reward) may begin as goal-directed but with training can develop into an established habit that is difficult to eradicate (Gasbarri et al., 2014; Dickinson and Pérez, 2018; Balleine and Dezfouli, 2019). Determining that the lever pressing has become a habit is based upon a "reward devaluation measurement." In other words, a movement is called a habit, if the frequency of lever pressing persists after the food reward has been devalued by its removal or switch to a noxious food. Unfortunately, despite an initial enthusiasm regarding the dual-process model, a variety of concerns have arisen regarding the interpretation of observations, outcome devaluation insensitivity, and the potential production of a narrow definition for habits (Robbins and Costa, 2017; de Wit et al., 2018; Du et al., 2022; Watson et al., 2022). In addition, the application of reward-devaluation and contingency degradation testing, used to confirm the definition of habits in animals, is not usually applicable to humans (Robbins and Costa, 2017; Du et al., 2022; Watson et al., 2022).

Despite the definitional vagueness regarding habitual behaviors, movement habits, and dual-process hypotheses, investigators have neuroanatomically dichotomized movements into goal-directed (voluntary) and habitual (nongoal-directed) with the aim of better understanding their underlying developmental mechanisms and pathophysiology. Goal-directed movements are defined as consciously performed, controlled, flexible actions that adapt quickly, and are undertaken to achieve a desired purposeful outcome (Balleine and Dickinson, 1998). In contrast, habitual behaviors are automatically performed, slow to change, repetitive, stimulus-responsive actions, driven by reinforcement, and having no link to current or future goals (Colwill and Rescorla, 1990; Graybiel, 2008). The goal-directed cortical-striatal circuit originates in the ventral-medial prefrontal cortex, and projects to the caudate (Fig. 6.1). The habitual circuit commences in the premotor/supplementary motor cortex and projects to the putamen. Posterior putamen involvement in habitual learning is supported by functional magnetic resonance imaging (MRI) research (Tanaka et al., 2008; Tricomi et al., 2009) and cortico-striatal connectivity studies using diffusion tensor imaging (Lehéricy et al., 2004; Draganski et al., 2008; de Wit et al., 2012).

In animal models, lesion studies have identified that both the dorsal and ventral striatum modulate behavior during reward-related tasks (Nishizawa et al., 2012). Both regions also compensate for the loss of functionality in goal-directed and habitual tasks (Burton et al., 2014). Other studies demonstrate that amphetamine sensitization enhances the progression from goal-directed to habitual responses (Nelson and Killcross, 2013), and damage to dopaminergic nigrostriatal inputs prevents habit learning, but not goal-directed learning (Faure et al., 2005).

Body-focused repetitive behaviors

BFRBs include activities such as hair pulling, skin picking, nail biting, lip/cheek biting, teeth grinding, and similar recurrent actions, potentially causing both physical and psychologic

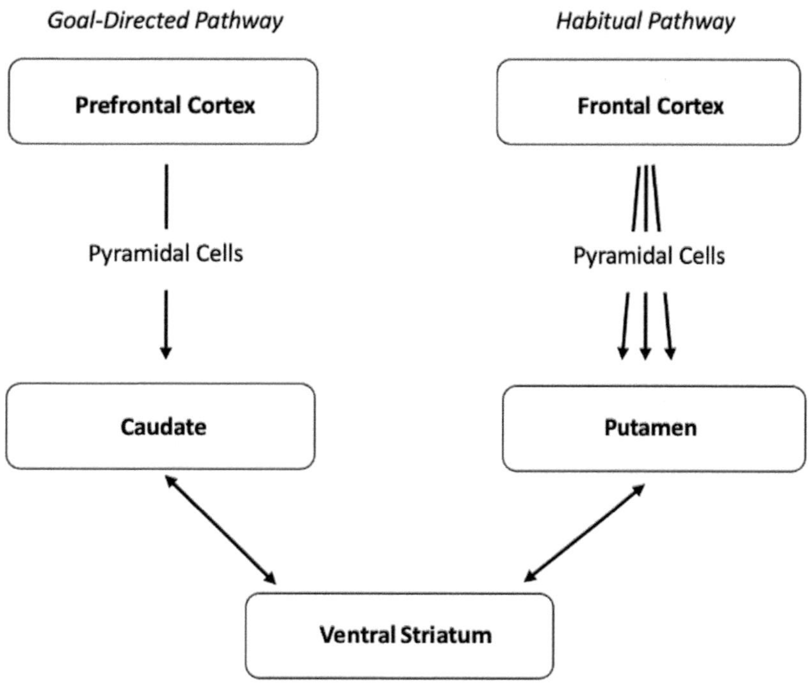

Fig. 6.1. Comparison of goal-directed and habitual circuits in the brain.

distress. Despite the common use of this term, BFRBs is not officially recognized in DSM-V, and it does not have an International Classification of Diseases code. Several subtypes of BFRBs, previously labeled an impulse control abnormality, have been recategorized as an obsessive-compulsive and related disorder (American Psychiatric Association, 2013). The latter include trichotillomania (i.e., recurrent hair pulling resulting in hair loss), excoriation disorder (repetitive skin picking), and some severe forms of onychophagia (nail biting). All three possess continuous repetition of a specific behavior, dysfunctional repetitive attempts to reduce the behavior, and a significant clinical impairment (Sani et al., 2019).

Using standard diagnostic criteria, the prevalence rate ranges from 1.4% to 5.4% for excoriation disorder, and 0.5% to 2% for trichotillomania (Sani et al., 2019). Despite these published reports, in a study of 4335 college students examining the self-reported prevalence of BFRBs, 60% reported occasional behaviors and 12% pathologic conditions (i.e., occurred five or more times a day, caused physical damage and significant distress) (Houghton et al., 2018). BFRBs are exacerbated by anxiety, stress, and boredom and are often associated with sleep disturbances (Clark et al., 2022; Najera, 2022). Clinically, some BFRBs have been subdivided into two groups based on their premonitory sensation: one group having an automatic, non-sensory-related movement and the other a preceding uncomfortable sensation or urge, analogous to a tic disorder (Badenoch et al., 2020). Also similar to chronic tic disorders is the finding that both excoriation disorder and trichotillomania have significant relationships with conditions including attention-deficit hyperactivity disorder, obsessive-compulsive disorder, anxiety, and depression (Lochner et al., 2017; Solley and Turner, 2018; Grant et al., 2020; Najera, 2022).

The underlying pathophysiology of BFRBs is unknown. Limited MRI studies comparing patients with trichotillomania to controls have identified variable results. One study in adults suggested significant volumetric reductions in the left putamen and right amygdala, and localized shape deformities in bilateral nucleus accumbens, bilateral amygdala, right caudate, and putamen (Isobe et al., 2018). The defined regions have a role in affect regulation, inhibitory control, and habit generation (Stein et al., 2006). In contrast, functional imaging studies in children with hair pulling suggested changes in the left temporal lobe, dorsal posterior cingulate gyrus, and putamen (Lee et al., 2010). Structural imaging studies in individuals with excoriation disorder have also been inconsistent (Harries et al., 2017; Schienle et al., 2018).

Prior reports have emphasized that the medical profession is currently not well prepared to identify, intervene, or treat individuals with BFRBs (Houghton et al., 2018; Henkel et al., 2019). Therapeutically, limited randomized studies have suggested that cognitive behavioral therapy with habit reversal therapy can be beneficial (Lochner et al., 2017; Farhat et al., 2020). Pharmacologically, literature searches have suggested that the selective serotonin reuptake inhibitors (SSRIs) and clomipramine are considered first-line in treating trichotillomania with other treatments including lamotrigine, olanzapine, *N*-acetylcysteine, inositol, and naltrexone (Henkel et al., 2019; Sani et al., 2019; Baczynski and Sharma, 2020; Hossain et al., 2022).

Stereotypies

Historically, the word stereotypy appears to have evolved from the French word "stereotype," an 18th century term used to describe a process of making copies of texts or pictures using an inked copper plate (Edwards et al., 2012). The word "stereotyped" is commonly used to describe something that is repeated in the same fashion. When used to define a movement disorder category it continues to evolve. For example, several cited definitions have included: (1) "Stereotypies are repetitive, predictable, coordinated movements that resemble a fragment of a normal action and are continually repeated without any purpose" (Josephs et al., 2008); (2) "involuntary, rhythmic, repetitive, fixed (fashion, form, amplitude, and location) movements that appear purposeful, but are purposeless (serve no obvious adaptive function or purpose), and stop with distraction" (Singer, 2011); (3) "a non-goal-directed movement pattern that is repeated continuously for a period of time in the same form and on multiple occasions, and which is typically distractible" (Edwards et al., 2012); and (4) "involuntary, coordinated, patterned, repetitive, rhythmic, purposeless but seemingly purposeful or ritualistic movement, posture or utterance" (Jankovic et al., 2021). Excluded movements are those that arise from the physiologic effect of a substance, delusional beliefs or obsessive thoughts, mannerisms, dyskinesias, and actions typically performed in situations of boredom (drumming fingers and tapping feet). Unfortunately, the DSM-V, as currently written, fails to provide a clinically meaningful definition, (see Table 6.1). In summary, the precise definition of the movement disorder stereotypy continues to evolve (Bhatoa et al., 2021; Edwards et al., 2012; Singer, 2011; Singer et al., 2022). Nevertheless, essential diagnostic features include *repetitive* (recurrent), *rhythmic* ("moving in a pronounced rhythm" at regular intervals), *fixed* (recur in similar patterns), prolonged (lasting for multiple seconds, minutes, or even hours), *purposeless* (are not a pathologic movement and activity has no identified purpose, i.e., nongoal-directed), and is *distractible* (stops with distraction).

1. Distinguishing stereotypies from other movement disorders
   a. *Mannerisms/habits:* These movements tend to be unique to the individual, do not occur in clusters, are brief, and less complex.
   b. *Tics*: Stereotypies are often misdiagnosed as tics, and both may occur in the same individual. See Table 6.2 for a list of factors distinguishing stereotypies from tics.
   c. *Compulsions*: Compulsions are behaviors performed repetitively in a fixed fashion. They are driven by obsessive thoughts or rigidly applied rules, must be done a certain number of times, on both sides of the body, and typically achieve a certain feeling before one is able to stop.
2. Subdivision into Primary and Secondary Variants
   Stereotypies should initially be classified into two etiologic subgroups, that is, "primary" and "secondary" (Singer, 2009). The primary form indicates an otherwise normally developing child, where the movement is not associated with another condition. The secondary stereotypy category includes those with (1) developmental problems (autism (Loh et al., 2007; Goldman and Greene, 2013)), developmental delay, sensory deprivation (deafness and visual impairment (Fazzi et al., 1999), Rett syndrome (Temudo et al., 2007; Dy et al., 2017)), and Lesch–Nyhan disease (Harris, 2018); or (2) conditions potentially associated with genetic syndromes (Moss et al., 2009); autoimmune encephalitis (Varley et al., 2018), metabolic disorders (Mostile et al., 2019), and degenerative conditions. An additional potential cause of secondary stereotypies, especially in adults, is a drug-induced movement disorder (Stacy et al., 1993; Jankovic et al., 2021). Further information regarding the diagnosis of primary complex motor stereotypies (pCMS; clinical characteristics, comorbidities, outcome, pathophysiology, and treatment) are discussed in the following sections.

Table 6.1

Diagnostic and Statistical Manual of Mental Disorders, Fifth Edition criteria for stereotypic movement disorder

Criterion A	Repetitive, seemingly driven, apparently purposeless
Criterion B	Interferes with pursuits and may result in self-injury
Criterion C	Onset in early developmental period (by 3 years, often before 1 year)
Criterion D	Must not be better explained by another neurodevelopmental or mental disorders[a], or physiologic effects of substance or neurologic condition

[a] Obsessive-compulsive disorder, trichotillomania, skin-picking, or body-focused repetitive behaviors (American Psychiatric Association, 2013).

Subdivision into three types of stereotypies

Motor stereotypies have been divided by some investigators (Mahone et al., 2004) into three phenomenological movement types:

1. *Simple*: Indicates the presence of a wide variety of repetitive behaviors (e.g., thumb sucking, nail/lip biting, body rocking, shaking, or tapping of an extremity, and hair twirling). These movements are estimated to occur in about 20% of healthy children, are often labeled as "habits" or "repetitive behaviors," and most regress (Sallustro and Atwell, 1978; Foster, 1998; MacDonald et al., 2007; Lotia et al., 2018). One of the most common forms of simple stereotypies in children and adults, reported to be present in 7% of the general population, is the leg stereotypy syndrome (Lotia et al., 2018). This condition consists of repetitive, continuous movement present almost exclusively in the legs while the patient is seated. Whether body rocking should be considered as a separate entity remains controversial (Hansen et al., 1990; Rafaeli-Mor et al., 1999). Studies in college students have also reported frequent repetitive movements (touching one's face; playing with hair, pens, or jewelry; shaking a leg; tapping

Table 6.2

Distinguishing stereotypies and tics.

	Stereotypies	Tics
Age at onset	Typically, < 3 years	Typically, 4–8 years, must be < 18 years
Pattern	Rhythmic, repetitive, fixed, identical, and patterned	Varies, waxes and wanes, brief
Movements	Arms/hands flapping and waving, body rocking/head nodding	Blink, grimace, shrug, head jerks, neck stretch, twist
Vocalizations	Occasional moan or hum with movement	Grunt, bark, hoot, holler, moan, groan, throat clear
Rhythm	Rhythmic	Rapid, sudden, random
Duration	Intermittent, prolonged (seconds to minutes, or longer)	Intermittent, brief, abrupt
Precipitant	Excitement, stress, boredom	Excitement, stress, anxiety, fatigue
Distractibility	Stops abruptly	Continues, reduces with involvement in other activities
Suppressibility	Ability to suppress increases with aging	Brief (often leads to an increased premonitory urge)
Premonitory urge	No	Yes
Family history	Positive, 25%	Frequently positive, > 50% have a positive family history
Treatment	Behavioral therapy	Behavioral and pharmacologic therapy

Source: 391–395.Adapted from Singer HS, Mink JW, Gilbert DL, Jankovic J (2022) Movement disorders in childhood, 3rd ed. Elsevier;, Mahone EM, Bridges D, Prahme C, Singer HS (2004) Repetitive arm and hand movements (complex motor stereotypies) in children. J Pediatr, 145: 391–395.

fingers; scratching one's head; etc.), but most were not time-consuming or disruptive (Niehaus et al., 2000; Houghton et al., 2018).

2. *Head nodding*: Periodic head nodding, either side-to-side, up-and-down, or shoulder-to-shoulder, has a frequency of 1–2 per second and occurs without associated nystagmus or torticollis. This primary entity should be differentiated from individuals with similar head movements due to specific secondary etiologies (e.g., rhombencephalosynapsis, Sandifer's syndrome, spasmus nutans, bobble-head doll syndrome, and developmental brain malformations (Tully et al., 2013; Reddy et al., 2014; Dean et al., 2019)). In a study following eight children with typical development and head nodding, three stopped entirely, suggesting that outcome may differ in this category as compared to children with CMS (Harris et al., 2008).
3. *CMS*: Includes bilateral hand shaking, flapping or waving, opening and closing one's hands, finger wiggling, arm flapping, and wrist rotations. Other frequent accompanying activities include mouth opening, grimacing, pacing, rocking, and occasionally an accompanying sound. Stereotypies are typically exacerbated by excitement, when engrossed in activities, stress, boredom, and fatigue (Mahone et al., 2004; Harris et al., 2008; Singer, 2011). Recognizing that CMS can occur as either a primary or secondary disorder, the remainder of this chapter focuses on the expanding topic of pCMS.

Primary complex motor stereotypies

Clinical characteristics

Motor stereotypies in typically developing children usually begin in the first three years of life (Mahone et al., 2004; Harris et al., 2008; Freeman et al., 2010; Singer, 2011). Movements occur in bursts, multiple times a day, with durations ranging from seconds to minutes or hours. Each child tends to have their own repertoire with additional associated motor features evolving over time and, in some, movement associated sounds. The most common movements include arm/hand flapping or waving, rotating/flexing/extending the wrists, fluttering fingers in front of the face, and opening and closing the hands. Activities are associated with periods of engrossment, excitement, stress, fatigue, or boredom. Children lack a premonitory urge, and either report they are "unaware of the movements," "enjoy doing the activity," or "are mentally visualizing an activity, TV program, or event." The latter has been labeled by some as "intense imagery movements" (IIM), and a few investigators have suggested that it be considered a separate entity (Robinson et al., 2014). Nevertheless, recognizing that many subjects had pCMS for years, before developing their visual imagery associations, the authors do not support the concept that IIM is a different condition.

Comorbid conditions

In children with pCMS evaluated between ages 7 and 12 years and 9 and 19 years, 30% and 63%, respectively, had attention-deficit hyperactivity disorder; 18% and 22% had a tic disorder; and 9% and 35% had obsessive-compulsive disorder (Harris et al., 2008; Oakley et al., 2015). In the older group, 73% had anxiety (Oakley et al., 2015). As expected, a case-controlled study in children with pCMS, showed intact neuropsychologic profiles (Mahone et al., 2014). In contrast, using the Physical and Neurological Assessment of Subtle Signs, about one-third of children in the CMS group showed evidence of atypical motor coordination and slower overall motor speed (Mahone et al., 2014). Preliminary results from an ongoing quality-of-life study in teenagers with pCMS suggests an association between stereotypy severity, as measured by the Stereotypy Severity Scale (Miller et al., 2006), and quality of life. While children with autism spectrum disorder also frequently have stereotypies, these are secondary stereotypies and do not fit under the category of pCMS.

Outcome

Although many pediatricians suggest to parents that pCMS will resolve in childhood, existing data support that of the majority of pCMS cases persist (Mahone et al., 2004; Harris et al., 2008; Oakley et al., 2015). In several studies of children aged 6–12 years, only 3%–5% of subjects had a resolution of movements (Tan et al., 1997; Harris et al., 2008; Freeman et al., 2010). In a follow-up study in 9–19-year-olds with pCMS, although many reported a reduction in frequency and duration, only 2% were stereotypy-free (Oakley et al., 2015). Studies are currently in progress to assess the outcome in adulthood. Anecdotally, many adults with a history of pCMS report that when stressed or anxious, they will often seek a private area and do their predictable movements.

Etiology

The precise etiology of pCMS is unknown. A genetic etiology has been suggested by data showing that an affected individual has a 20%–40% likelihood of having an affected family member (Harris et al., 2008; Freeman et al., 2010). In a whole exome sequencing study of 94 individuals with pCMS and their unaffected parents, two children had independent de novo nonsense variants in the KDM5B gene (Fernandez et al., 2023). The latter is a member of the histone lysine demethylase family with links to developmental disorders (Lebrun et al., 2018). Additionally, a chromosome microarray analysis in 42 patients with motor stereotypies has identified 24 different microdeletions (Soliani et al., 2023).

Pathophysiology

The underlying pathophysiologic mechanism for CMS in children is unknown, with hypotheses ranging from an arousal to a motor control disorder. In a study of children with pCMS, no movement-related cortical potentials were identified before the appearance of a CMS, indicating that premotor areas are likely not involved in the preparation of these complex movements and suggesting that stereotypies are initiated by mechanisms different from voluntary movements (Houdayer et al., 2014). Support for an underlying neurobiologic mechanism is expanding with abnormalities within cortical-basal ganglia-thalamo-cortical (CBGTC) circuits receiving special interest, more specifically the habitual/goal-directed circuits (see Fig. 6.1). For example, a resting state functional connectivity MRI study comparing children with pCMS to typically developing children, showed that functional connectivity between prefrontal cortical and striatal regions, considered to be essential for developing goal-directed behaviors, was reduced in children with pCMS (Augustine et al., 2021). In contrast, functional connectivity between motor/premotor and striatal regions, critical for the development and regulation of habitual behaviors, did not differ between groups. These data have suggested that an imbalance between the development of goal-directed and habitual pathways has a clinical role in pCMS. Animal studies further support the imbalance hypothesis between goal-directed and habitual circuits. In a rodent model, lesioning the dorsomedial striatum, a region that facilitates goal-directed actions, increases habit-like responses (Corbit and Janak, 2010). In contrast, lesioning the dorsolateral striatum, a region associated with habitual behaviors, changed habit reward-seeking behaviors into goal-directed actions without altering goal-directed behaviors (Yin et al., 2004; Gremel and Costa, 2013). Additional studies have demonstrated that both the dorsal and ventral striatum modulate behavior in goal-directed tasks (Nishizawa et al., 2012) and both compensate for the loss of functionality in goal-directed and habitual tasks (Burton et al., 2014). Other potential sites have included the CBGTC reactive/proactive inhibitory circuits (Aron, 2011) and the cerebellum (Dean et al., 2022).

Treatment

Initial steps in caring for patients with pCMS include (1) the assessment of stereotypies, their severity, and resulting impairment; (2) determining the presence of coexisting conditions and their impact on functioning; and (3) educating the patient, family, and school about the diagnosis. Clearly, knowledge about what to expect can assist in making affected children and their families feel more secure. Additional therapy should be considered if the stereotypic movements are causing comments from the child's peers or family, physical problems, or disruptions in the academic setting (Bhatoa et al., 2021).

Behavioral therapy Multiple behavioral interventions have been utilized to treat repetitive behaviors in autistic children (Boyd et al., 2012; Giles et al., 2012). In children aged 8–14 years with pCMS, the combination of awareness training and differential reinforcement of other behaviors, was beneficial in reducing motor stereotypies (Miller et al., 2006). A study evaluating the efficacy of a home-based, parent-administered instructional digital versatile disk was beneficial in reducing movements, but there were many dropouts (Specht et al., 2017). A third study in children with pCMS, ages 5–7 years, using a combination of home-based, parent-administered behavioral therapy plus telephone support provided by a therapist, showed the best improvement and had few dropouts (Singer et al., 2018). A current behavioral trial in progress is designed to quantify the variables which trigger CMS activity and then implement an assessment-driven innovative, game-based, habit-reversal treatment program (Edelstein et al., 2024).

Pharmacotherapy There are no reported formal pharmacologic trials for the reduction of stereotypies in children with pCMS. A historical review of parent reports in this population has suggested that several medications (methylphenidate, atomoxetine, amphetamine mixed salts, lisdexamfetamine, dexmethylphenidate, guanfacine, fluoxetine, lamotrigine, topiramate, and escitalopram) are not effective in reducing pCMS (Mahone et al., 2014). In autistic and intellectually delayed childhood populations, many with self-injurious behaviors, atypical antipsychotics (risperidone and aripiprazole) have been shown to be effective in reducing repetitive behaviors (Rajapakse and Pringsheim, 2010; Hirsch and Pringsheim, 2016). Enthusiasm, however, regarding the use of these medications is diminished by concerns regarding study methodology and the presence of significant side-effects, including tardive dyskinesia. Studies (Carrasco et al., 2012) have questioned reported beneficial effects associated with the use of serotonin receptor inhibitors (clomipramine and tianeptine) (Lewis et al., 1995; Niederhofer et al., 2003). Low-dose buspirone in the autistic population has reduced stereotypies (Chugani et al., 2016). Further therapeutic trials are indicated.

References

Adams CD (1982). Variations in the sensitivity of instrumental responding to reinforcer devaluation. Q J Exp Psychol Sect B 34: 77–98.

American Psychiatric Association (APA) (2013). Diagnostic and statistical manual of mental disorders, 5th ed. Arlington, VA: APA.

Aron AR (2011). From reactive to proactive and selective control: developing a richer model for stopping inappropriate responses. Biol Psychiatry 69: e55–e68.

Augustine F, Nebel MB, Mostofsky SH et al. (2021). Aberrant prefrontal cortical-striatal functional connectivity in children with primary complex motor stereotypies. Cortex 142: 272–282.

Baczynski C, Sharma V (2020). Pharmacotherapy for trichotillomania in adults. Expert Opin Pharmacother 21: 1455–1466.

Badenoch J, Searle T, Watson I et al. (2020). Sensory symptoms in body-focused repetitive behaviors, restless legs syndrome, and Tourette syndrome: an overlap? Neurosci Biobehav Rev 119: 320–332.

Balleine BW, Dezfouli A (2019). Hierarchical action control: adaptive collaboration between actions and habits. Front Psychol 10: 2735.

Balleine BW, Dickinson A (1998). Goal-directed instrumental action: contingency and incentive learning and their cortical substrates. Neuropharmacology 37: 407–419.

Bhatoa RS, Malik O, Robinson S et al. (2021). Clinical management of complex motor stereotypies. Arch Dis Child 107: 861–865.

Bishop SL, Hus V, Duncan A et al. (2013). Subcategories of restricted and repetitive behaviors in children with autism spectrum disorders. J Autism Dev Disord 43: 1287–1297.

Boyd BA, McDonough SG, Bodfish JW (2012). Evidence-based behavioral interventions for repetitive behaviors in autism. J Autism Dev Disord 42: 1236–1248.

Brierley NJ, McDonnell CG, Parks KMA et al. (2021). Factor structure of repetitive behaviors across autism spectrum disorder and attention-deficit/hyperactivity disorder. J Autism Dev Disord 51: 3391–3400.

Burton AC, Bissonette GB, Lichtenberg NT et al. (2014). Ventral striatum lesions enhance stimulus and response encoding in dorsal striatum. Biol Psychiatry 75: 132–139.

Carrasco M, Volkmar FR, Bloch MH (2012). Pharmacologic treatment of repetitive behaviors in autism spectrum disorders: evidence of publication bias. Pediatrics 129: e1301–e1310.

Chugani DC, Chugani HT, Wiznitzer M et al. (2016). Efficacy of low-dose buspirone for restricted and repetitive behavior in young children with autism spectrum disorder: a randomized trial. J Pediatr 170: 45–53.e1–4.

Clark SDB, Lahoud AA, Gladstone TR et al. (2022). Sleep disturbances in pediatric body-focused repetitive behaviors: a preliminary investigation. Child Psychiatry Hum Dev 54: 1396–1403.

Colwill RM, Rescorla RA (1990). Effect of reinforcer devaluation on discriminative control of instrumental behavior. J Exp Psychol Anim Behav Process 16: 40–47.

Comparan-Meza M, Vargas de la Cruz I, Jauregui-Huerta F et al. (2021). Biopsychological correlates of repetitive and restricted behaviors in autism spectrum disorders. Brain Behav 11: e2341.

Corbit LH, Janak PH (2010). Posterior dorsomedial striatum is critical for both selective instrumental and Pavlovian reward learning. Eur J Neurosci 31: 1312–1321.

Dean SL, Huisman T, Poretti A et al. (2019). Figure of eight stereotypies in a young girl with a prenatal cerebellar injury. In: Movement disorders clinical practice, Vol. 6, pp. 488–490. United States.

Dean SL, Tochen L, Augustine F et al. (2022). The role of the cerebellum in repetitive behavior across species: childhood stereotypies and deer mice. Cerebellum 21: 440–451.

Dickinson A, Mulatero CW (1989). Reinforcer specificity of the suppression of instrumental performance on a non-contingent schedule. Behav Process 19: 167–180.

Dickinson A, Pérez OD (2018). Actions and habits: psychological issues in dual-system theory. In: Goal-directed decision making: computations and neural circuits, pp. 1–25. San Diego, CA: Elsevier/Academic Press.

Draganski B, Kherif F, Klöppel S et al. (2008). Evidence for segregated and integrative connectivity patterns in the human Basal Ganglia. J Neurosci 28: 7143–7152.

Du Y, Krakauer JW, Haith AM (2022). The relationship between habits and motor skills in humans. Trends Cogn Sci 26: 371–387.

Dy ME, Waugh JL, Sharma N et al. (2017). Defining hand stereotypies in Rett syndrome: a movement disorders perspective. Pediatr Neurol 75: 91–95.

Edelstein M, Pogue E, Singer H (2024). Development and validation of an assessment-driven behavioral intervention for primary complex motor stereotypies in young children. Behavior Modification 485–6: 477–501.

Edwards MJ, Lang AE, Bhatia KP (2012). Stereotypies: a critical appraisal and suggestion of a clinically useful definition. Mov Disord 27: 179–185.

Faja S, Nelson Darling L (2019). Variation in restricted and repetitive behaviors and interests relates to inhibitory control and shifting in children with autism spectrum disorder. Autism 23: 1262–1272.

Farhat LC, Olfson E, Nasir M et al. (2020). Pharmacological and behavioral treatment for trichotillomania: an updated systematic review with meta-analysis. Depress Anxiety 37: 715–727.

Faure A, Haberland U, Condé F et al. (2005). Lesion to the nigrostriatal dopamine system disrupts stimulus-response habit formation. J Neurosci 25: 2771–2780.

Fazzi E, Lanners J, Danova S et al. (1999). Stereotyped behaviours in blind children. Brain Dev 21: 522–528.

Fernandez TV, Williams ZP, Kline T et al. (2023). Primary complex motor stereotypies are associated with de novo damaging DNA coding mutations that identify KDM5B as a risk gene. PLoS One 18: e0291978.

Foster LG (1998). Nervous habits and stereotyped behaviors in preschool children. J Am Acad Child Adolesc Psychiatry 37: 711–717.

Freeman RD, Soltanifar A, Baer S (2010). Stereotypic movement disorder: easily missed. Dev Med Child Neurol 52: 733–738.

Gasbarri A, Pompili A, Packard MG et al. (2014). Habit learning and memory in mammals: behavioral and neural characteristics. Neurobiol Learn Mem 114: 198–208.

Giles AF, St Peter CC, Pence ST et al. (2012). Preference for blocking or response redirection during stereotypy treatment. Res Dev Disabil 33: 1691–1700.

Goldman S, Greene P (2013). Stereotypies in autism: a video demonstration of their clinical variability. Front Integr Neurosci 6: 121.

Grant JE, Dougherty DD, Chamberlain SR (2020). Prevalence, gender correlates, and co-morbidity of trichotillomania. Psychiatry Res 288: 112948.

Graybiel AM (2008). Habits, rituals, and the evaluative brain. Annu Rev Neurosci 31: 359–387.

Gremel CM, Costa RM (2013). Orbitofrontal and striatal circuits dynamically encode the shift between goal-directed and habitual actions. Nat Commun 4: 2264.

Hansen DJ, Tishelman AC, Hawkins RP et al. (1990). Habits with potential as disorders. Prevalence, severity, and other characteristics among college students. Behav Modif 14: 66–80.

Harries MD, Chamberlain SR, Redden SA et al. (2017). A structural MRI study of excoriation (skin-picking) disorder and its relationship to clinical severity. Psychiatry Res Neuroimaging 269: 26–30.

Harris JC (2018). Lesch-Nyhan syndrome and its variants: examining the behavioral and neurocognitive phenotype. Curr Opin Psychiatry 31: 96–102.

Harris KM, Mahone EM, Singer HS (2008). Nonautistic motor stereotypies: clinical features and longitudinal follow-up. Pediatr Neurol 38: 267–272.

Henkel ED, Jaquez SD, Diaz LZ (2019). Pediatric trichotillomania: review of management. Pediatric Dermatology 36: 803–807.

Hirsch LE, Pringsheim T (2016). Aripiprazole for autism spectrum disorders (ASD). Cochrane Database Syst Rev 2016: Cd009043.

Hossain R, Leung-Yee J, Sinyor M (2022). Body-focused repetitive disorders. CMAJ 194: E1381.

Houdayer E, Walthall J, Belluscio BA et al. (2014). Absent movement-related cortical potentials in children with primary motor stereotypies. Mov Disord 29: 1134–1140.

Houghton DC, Alexander JR, Bauer CC et al. (2018). Body-focused repetitive behaviors: more prevalent than once thought? Psychiatry Res 270: 389–393.

Isobe M, Redden SA, Keuthen NJ et al. (2018). Striatal abnormalities in trichotillomania: a multi-site MRI analysis. Neuroimage Clin 17: 893–898.

Jankovic J, Hallett M, Okun MS et al. (2021). Principles and practice of movement disorders. Elsevier Health Sciences.

Jiujias M, Kelley E, Hall L (2017). Restricted, repetitive behaviors in autism spectrum disorder and obsessive-compulsive disorder: a comparative review. Child Psychiatry Hum Dev 48: 944–959.

Josephs KA, Whitwell JL, Jack CR Jr (2008). Anatomic correlates of stereotypies in frontotemporal lobar degeneration. Neurobiol Aging 29: 1859–1863.

Kirby AV, Boyd BA, Williams KL et al. (2017). Sensory and repetitive behaviors among children with autism spectrum disorder at home. Autism 21: 142–154.

Kuzminskaite E, Begeer S, Hoekstra RA, Grove R (2020). Short report: Social communication difficulties and restricted repetitive behaviors as predictors of anxiety in adults with autism spectrum disorder. Autism 247: 1917–1923.

Lebrun N, Mehler-Jacob C, Poirier K et al. (2018). Novel KDM5B splice variants identified in patients with developmental disorders: functional consequences. Gene 679: 305–313.

Lee JA, Kim CK, Jahng GH et al. (2010). A pilot study of brain activation in children with trichotillomania during a visual-tactile symptom provocation task: a functional magnetic resonance imaging study. Prog Neuropsychopharmacol Biol Psychiatry 34: 1250–1258.

Lehéricy S, Ducros M, Krainik A et al. (2004). 3-D diffusion tensor axonal tracking shows distinct SMA and pre-SMA projections to the human striatum. Cereb Cortex 14: 1302–1309.

Lewis MH, Bodfish JW, Powell SB et al. (1995). Clomipramine treatment for stereotypy and related repetitive movement disorders associated with mental retardation. Am J Ment Retard 100: 299–312.

Lochner C, Roos A, Stein DJ (2017). Excoriation (skin-picking) disorder: a systematic review of treatment options. Neuropsychiatr Dis Treat 13: 1867–1872.

Loh A, Soman T, Brian J et al. (2007). Stereotyped motor behaviors associated with autism in high-risk infants: a pilot videotape analysis of a sibling sample. J Autism Dev Disord 37: 25–36.

Lotia M, York MK, Strutt AM et al. (2018). Leg stereotypy syndrome: phenomenology and prevalence. J Neurol Neurosurg Psychiatry 89: 692–695.

MacDonald R, Green G, Mansfield R et al. (2007). Stereotypy in young children with autism and typically developing children. Res Dev Disabil 28: 266–277.

Mahone EM, Bridges D, Prahme C et al. (2004). Repetitive arm and hand movements (complex motor stereotypies) in children. J Pediatr 145: 391–395.

Mahone EM, Ryan M, Ferenc L et al. (2014). Neuropsychological function in children with primary complex motor stereotypies. Dev Med Child Neurol 56: 1001–1008.

Miller JM, Singer HS, Bridges DD et al. (2006). Behavioral therapy for treatment of stereotypic movements in nonautistic children. J Child Neurol 21: 119–125.

Moss J, Oliver C, Arron K, Burbidge C et al. (2009). The prevalence and phenomenology of repetitive behavior in genetic syndromes. J Autism Dev Disord 39: 572–588.

Mostile G, Barone R, Nicoletti A et al. (2019). Hyperkinetic movement disorders in congenital disorders of glycosylation. Eur J Neurol 26: 1226–1234.

Najera DB (2022). Body-focused repetitive behaviors: beyond bad habits. JAAPA 35: 28–33.

Nelson AJ, Killcross S (2013). Accelerated habit formation following amphetamine exposure is reversed by D1, but enhanced by D2, receptor antagonists. Front Neurosci 7: 76.

Niederhofer H, Staffen W, Mair A (2003). Tianeptine: a novel strategy of psychopharmacological treatment of children with autistic disorder. Hum Psychopharmacol 18: 389–393.

Niehaus DJ, Emsley RA, Brink P et al. (2000). Stereotypies: prevalence and association with compulsive and impulsive symptoms in college students. Psychopathology 33: 31–35.

Nishizawa K, Fukabori R, Okada K et al. (2012). Striatal indirect pathway contributes to selection accuracy of learned motor actions. J Neurosci 32: 13421–13432.

Oakley C, Mahone EM, Morris-Berry C et al. (2015). Primary complex motor stereotypies in older children and adolescents: clinical features and longitudinal follow-up. Pediatr Neurol 52: 398–403.e1.

Rafaeli-Mor N, Foster L, Berkson G (1999). Self-reported body-rocking and other habits in college students. Am J Ment Retard 104: 1–10.

Rajapakse T, Pringsheim T (2010). Pharmacotherapeutics of Tourette syndrome and stereotypies in autism. Semin Pediatr Neurol 17: 254–260.

Reddy OJ, Gafoor JA, Suresh B et al. (2014). Bobble head doll syndrome: a rare case report. J Pediatr Neurosci 2: 175–177.

Robbins TW, Costa RM (2017). Habits. Curr Biol 27: R1200–R1206.

Robinson S, Woods M, Cardona F et al. (2014). Intense imagery movements: a common and distinct paediatric subgroup of motor stereotypies. Dev Med Child Neurol 56: 1212–1218.

Sallustro F, Atwell CW (1978). Body rocking, head banging, and head rolling in normal children. J Pediatr 93: 704–708.

Sani G, Gualtieri I, Paolini M et al. (2019). Drug treatment of trichotillomania (hair-pulling disorder), excoriation (skin-picking) disorder, and nail-biting (onychophagia). Curr Neuropharmacol 17: 775–786.

Schienle A, Potthoff J, Wabnegger A (2018). Voxel-based morphometry analysis of structural brain scans in skin-picking disorder. Compr Psychiatry 84: 82–86.

Singer HS (2009). Motor stereotypies. Semin Pediatr Neurol 16: 77–81.

Singer HS (2011). Stereotypic movement disorders. Handb Clin Neurol 100: 631–639.

Singer HS, Mink JW, Gilbert DL et al. (2022). Movement disorders in childhood, 3rd ed. Elsevier.

Singer HS, Rajendran S, Waranch HR et al. (2018). Home-based, therapist-assisted, therapy for young children with primary complex motor stereotypies. Pediatr Neurol 85: 51–57.

Soliani L, Alcalá San Martín A, Balsells S et al. (2023). Chromosome microarray analysis for the investigation of deletions in pediatric movement disorders: a systematic review of the literature. Mov Disord Clin Pract 10: 547–557.

Solley K, Turner C (2018). Prevalence and correlates of clinically significant body-focused repetitive behaviors in a non-clinical sample. Compr Psychiatry 86: 9–18.

Specht MW, Mahone EM, Kline T et al. (2017). Efficacy of parent-delivered behavioral therapy for primary complex motor stereotypies. Dev Med Child Neurol 59: 168–173.

Stacy M, Cardoso F, Jankovic J (1993). Tardive stereotypy and other movement disorders in tardive dyskinesias. Neurology 43: 937–941.

Stein DJ, Chamberlain SR, Fineberg N (2006). An A-B-C model of habit disorders: hair-pulling, skin-picking, and other stereotypic conditions. CNS Spectr 11: 824–827.

Tan A, Salgado M, Fahn S (1997). The characterization and outcome of stereotypical movements in nonautistic children. Mov Disord 12: 47–52.

Tanaka SC, Balleine BW, O'Doherty JP (2008). Calculating consequences: brain systems that encode the causal effects of actions. J Neurosci 28: 6750–6755.

Temudo T, Oliveira P, Santos M et al. (2007). Stereotypies in Rett syndrome: analysis of 83 patients with and without detected MECP2 mutations. Neurology 68: 1183–1187.

Tricomi E, Balleine BW, O'Doherty JP (2009). A specific role for posterior dorsolateral striatum in human habit learning. Eur J Neurosci 29: 2225–2232.

Tully HM, Dempsey JC, Ishak GE et al. (2013). Persistent figure-eight and side-to-side head shaking is a marker for rhombencephalosynapsis. Mov Disord 28: 2019–2023.

Turner M (1999). Annotation: repetitive behaviour in autism: a review of psychological research. J Child Psychol Psychiatry 40: 839–849.

Varley J, Taylor J, Irani SR (2018). Autoantibody-mediated diseases of the CNS: structure, dysfunction and therapy. Neuropharmacology 132: 71–82.

Watson P, O'Callaghan C, Perkes I et al. (2022). Making habits measurable beyond what they are not: a focus on associative dual-process models. Neurosci Biobehav Rev 142: 104869.

de Wit S, Kindt M, Knot SL et al. (2018). Shifting the balance between goals and habits: five failures in experimental habit induction. J Exp Psychol Gen 147: 1043–1065.

de Wit S, Watson P, Harsay HA et al. (2012). Corticostriatal connectivity underlies individual differences in the balance between habitual and goal-directed action control. J Neurosci 32: 12066–12075.

Yin HH, Knowlton BJ, Balleine BW (2004). Lesions of dorsolateral striatum preserve outcome expectancy but disrupt habit formation in instrumental learning. Eur J Neurosci 19: 181–189.

Section II

Comorbidities in primary tic disorders

Handbook of Clinical Neurology, Vol. 215 (3rd series)
Tic Disorders and Tourette Syndrome
C. Ganos and Y. Worbe, Editors
https://doi.org/10.1016/B978-0-443-13554-5.00024-9

Chapter 7

Obsessive-compulsive disorder in primary tic disorders and Gilles de la Tourette syndrome

YULIA WORBE[1,2,*], EDOUARD COURTIN[3,4], AND PIERRE BURBAUD[3,4]

[1]*Department of Clinical Neurophysiology, Sorbonne University, APHP-Sorbonne University – Hospital Saint-Antoine, Paris, France*

[2]*Paris Brain Institute (ICM), Sorbonne University, Inserm U1127, CNRS UMR7225, UM75, Movement Investigation and Therapeutics Team, Paris, France*

[3]*Department of Clinical Neurophysiology, Pellegin Hospital, Bordeaux, France*

[4]*Institute for Neurodegenerative Disorders, CNRS – UMR 5293, Bordeaux, France*

Abstract

This chapter provides a review of obsessive-compulsive disorder (OCD), as it manifests in primary tic disorders and Gilles de la Tourette syndrome (GTS). We examine the clinical features of OCD in GTS, focusing on how these symptoms diverge from typical OCD presentations, often blending with or mimicking tics. Key pathophysiologic mechanisms are explored, with an emphasis on the neurobiologic substrates that underlie this unique OCD subtype and distinctions from tics. Finally, the chapter discusses therapeutic approaches specifically adapted for this population, highlighting both established and emerging treatments. This chapter aims to deepen understanding of the complex interplay between OCD and GTS, offering insights into more effective clinical management and therapeutic strategies for affected individuals.

PREVALENCE AND NATURAL HISTORY OF ASSOCIATION OF TICS WITH OBSESSIVE-COMPULSIVE DISORDER

The prevalence of obsessive-compulsive disorder (OCD) in individuals with Gilles de la Tourette syndrome (GTS) varies widely, with estimates ranging from 10% to 35% (Ferrao et al., 2022). Conversely, the presence of tics in OCD patients also varies significantly, with studies showing that 7% of OCD patients meet full criteria for GTS, while up to 53% exhibit at least some tics. This variability may be partly due to difficulties in differentiating repetitive behaviors in GTS, as well as differences in diagnostic criteria across studies (Eddy and Cavanna, 2014). The complex relationship between OCD and tics in GTS has spurred interest in understanding the unique clinical, neurobiologic, and genetic factors underlying this comorbidity.

Tics and GTS typically develop in childhood, usually before the age of 12 (Sarchioto et al., 2024). OCD, on the other hand, can emerge at various stages in life and is generally classified as either early-onset (EO-OCD) or late-onset OCD (LO-OCD). EO-OCD patients display distinct etiologic and phenotypic traits when compared to those with LO-OCD. Studies have shown that EO-OCD is more prevalent among males (Torresan et al., 2013; Girone et al., 2024) and demonstrates higher genetic heritability (Bolton et al., 2007; Narayanaswamy et al., 2012). Additionally, the early OCD type has a higher rate of comorbidity with other neurodevelopmental disorders, such as tics and GTS (De Mathis et al., 2008; Janowitz and Leckman, 2009; Taylor, 2011). This subtype is also associated with poorer treatment responses (Ravi Kishore et al., 2004; Van Roessel et al., 2023) and generally higher overall severity of OCD symptoms (Taylor, 2011; Anholt et al., 2014).

With regard to natural history of tics and EO-OCD, a Danish longitudinal study (Groth et al., 2018) of 227 children with GTS, followed over six years, found that tic severity generally declined over time, with only 23% of patients exhibiting moderate to severe tics at follow-up. However, 63% of the cohort continued to show associated psychiatric comorbidities, with OCD and attention deficit hyperactivity disorder (ADHD) being the most common. This suggests that while tics often decrease with age, OCD symptoms may persist or even worsen, as indicated by a smaller yearly decline in OCD severity compared to tics (Pringsheim et al., 2017). Also, in a study of 167 children with chronic tic disorder at Denmark's National

*Correspondence to: Pr. Yulia Worbe, MD, PhD, Department of Clinical Neurophysiology, Sorbonne University – Hospital Saint Antoine, Paris, France, E-mail: yulia.worbe@icm-institute.org.

Tourette Clinic (Jensen and Mol Debes, 2024), 152 patients with tics were analyzed for comorbidities, including OCD. Children with both tics and OCD showed significantly higher simple phonic tics scores than those with tics only, indicating that OCD as a comorbidity may intensify specific tic symptoms.

CLINICAL CHARACTERISTICS OF GILLES DE LA TOURETTE SYNDROME ASSOCIATED WITH OBSESSIVE-COMPULSIVE DISORDER

OCD associated with GTS differs from OCD without tics in several key characteristics, particularly in the nature of obsessions and compulsions.

For instance, Leckman et al. (1998) found that tic-related OCD is more frequently associated with intrusive thoughts that involve aggressive, religious, and sexual themes. Compulsions in individuals with GTS-related OCD are also distinct, tending to include checking, counting, ordering, touching, and hoarding behaviors, all more common than in cases of OCD without tics. Banaschewski et al. (2003) analyzed data from a global database of 4833 individuals with GTS and found that co-occurring obsessive-compulsive behaviors were frequently linked to impulsive and aggressive behavior, as well as depression and anxiety.

Second, OCD symptoms associated with GTS frequently present unique clinical features shaped by the interaction of compulsions, tics, and premonitory urges. Unlike typical OCD, where compulsions are primarily anxiety-driven responses to obsessions (such as contamination fears or intrusive thoughts), OCD in GTS often involves sensory-driven experiences rather than anxiety, making it difficult to distinguish between tics and compulsions (Ferrao et al., 2022). In GTS, these sensory-motor phenomena can precede both tics and compulsions, creating specific diagnostic challenges (Eddy and Cavanna, 2014; Szejko and Müller-Vahl, 2021). Figure 7.1 illustrates the overlapping and distinct characteristics of tics and compulsions in this context.

Worbe et al. (2010) explored this spectrum of behaviors through structured interviews with 166 GTS patients, categorizing repetitive behaviors into three types: "tic-like" (24% of patients), which included actions such as touching and symmetry-seeking; "OCD-like" (20%), featuring classic OCD compulsions like washing and checking; and a "mixed group" (13%), exhibiting a blend of tic-like and OCD-like behaviors. An additional 6% of patients could not be classified within these categories, further suggesting that repetitive behaviors in GTS may span a continuum from tics to compulsions, with intermediate expressions that share characteristics of both. Aligned with this concept, additional terms have been introduced to capture intermediate forms of repetitive behaviors in GTS associated with OCD. For example, "compulsive tics" describe repetitive actions in patients with GTS that blend characteristics of both compulsions and complex tics, emphasizing their dual nature (Palumbo and Kurlan, 2007). Another term, "cognitive tics" or "mental tics," is often used synonymously to describe intrusive, obsession-like thoughts that resemble obsessions but function more like tics in their repetitive, automatic nature (O'Connor, 2005).

A distinctive feature of OCD associated with GTS involves "just-right" and "not-just-right" experiences. Unlike typical OCD, where behaviors are often driven by harm-avoidance, just-right and not-just-right in GTS are motivated by sensations of incompleteness.

Leckman et al. (1998) first introduced the concept of "just-right" sensations in GTS, which is now widely accepted as a prevalent obsessive-compulsive behavior in this population, where actions are

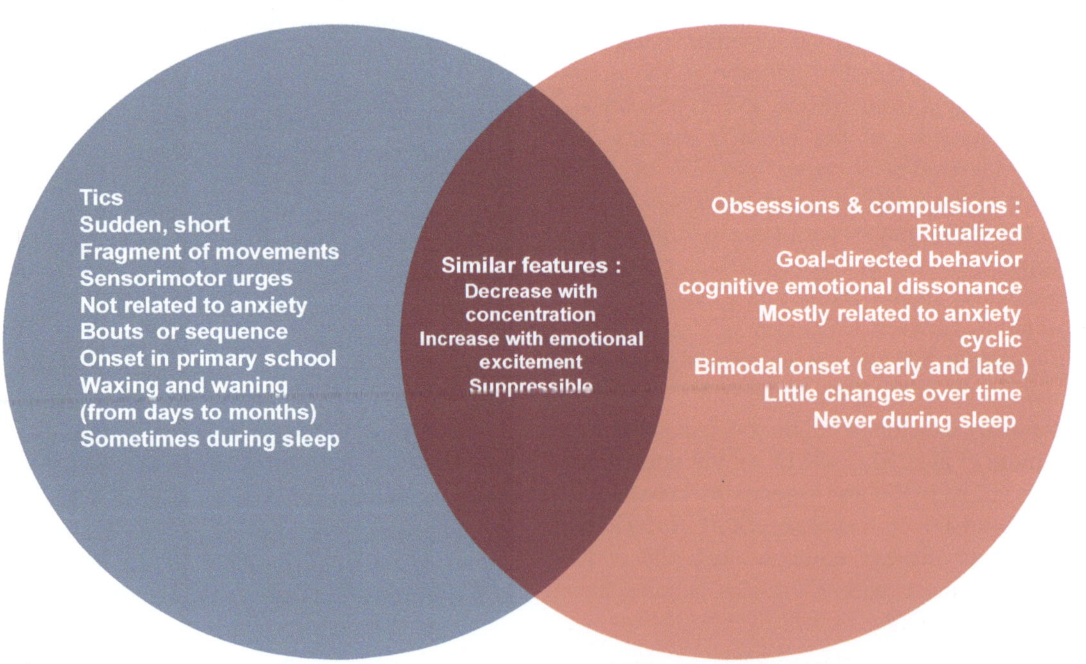

Fig. 7.1. Common and distinct features of tics and compulsions. Adapted from Szejko N, Müller-Vahl KR (2021) Challenges in the diagnosis and assessment in patients with Tourette syndrome and comorbid obsessive-compulsive disorder. Neuropsychiatr Dis Treat 17: 1253–1266.

performed in a specific manner to achieve internal relief (Leckman et al., 1998; Neal and Cavanna, 2013). In 2013, Neal and Cavanna further expanded GTS research by introducing not-just-right, a term previously used in OCD studies since 1992.

Using the "Not-Just-Right Experiences Questionnaire-Revised" (NJRE-QR), Neal and Cavanna (2013) assessed NJREs in 71 adults with GTS, finding that 80% reported at least one NJRE. Notably, individuals with comorbid OCD experienced significantly more not-just-right than those with tics alone. The strongest correlation was found between NJRE-QR scores and compulsivity measures, suggesting NJREs may be more closely associated with OCD than with tics alone. A study of 111 GTS patients investigated whether not-just-right experiences are more closely linked to tics and premonitory urges or to obsessive-compulsive behaviors. Not-just-right showed a stronger correlation with OCD symptoms than with tics or premonitory urges, as measured by the Premonitory Urge for Tics Scale (Brandt et al., 2023). This aligns with other studies reporting high not-just-right prevalence in pure OCD patients (Coles et al., 2003, 2005; Nissen and Mikkelsen, 2018).

BODY-FOCUSED REPETITIVE BEHAVIORS AND THEIR RELATIONSHIP WITH OBSESSIVE-COMPULSIVE DISORDER AND TICS

In addition to OCD, individuals with GTS frequently display body-focused repetitive behaviors (BFRBs) like nail-biting (onychophagia), hair-pulling (trichotillomania), and skin-picking (excoriation disorder) (see also Chapter 6). These behaviors add complexity to the clinical presentation of GTS, as they share similarities with both compulsions and tics. Typically, BFRBs are driven by a strong urge to relieve tension, providing temporary relief and engaging the brain's reward system, which connects them to addictive behaviors (Badenoch et al., 2020). Much like tics, BFRBs are preceded by a rising urge that momentarily alleviates discomfort, making them challenging to control.

Several similarities and distinctions can be identified between GTS, OCD, and BFRBs.

Both BFRBs and tics are often preceded by a sensation of urge that prompts the behavior. Certain situations can intensify symptoms due to the involvement of a reward system – specifically, negative reinforcement – which sustains these behaviors. Relief typically follows the execution of these acts, and both anatomic and functional evidence suggests that the cortico-striato-thalamo-cortical loops are implicated in the symptomatology of these disorders. BFRBs are often associated with pleasure and relief following a decrease in inner tension, similar to the sensations experienced after tics execution. These shared features have led to the suggestion that GTS and BFRBs may belong to a related spectrum.

A study involving 811 GTS patients reported that 3.8% met the criteria for trichotillomania and 13% for excoriation disorder, with increased BFRB rates linked to greater tic severity and co-occurring OCD (Greenberg et al., 2018). Notably, trichotillomania and excoriation disorder are observed at higher rates among GTS patients with comorbid OCD, which supports the theory that BFRBs could represent an extended spectrum of GTS-related psychopathology. Studies by Rozenman et al. (2016) suggest that hair-pulling disorder may have closer associations with GTS or GTS-OCD than with OCD alone, indicating a potential genetic connection within the GTS spectrum.

The intricate interactions among OCD, BFRBs, and tics can lead to substantial psychosocial impairment, as BFRBs often cause visible physical harm (e.g., hair loss, skin damage) that may provoke feelings of shame and reinforce a cycle of compulsive behavior. This shame can heighten psychologic distress and contribute to a repetitive behavior cycle, which is particularly burdensome for individuals with GTS. Greenberg et al. also observed that in cases of co-occurring OCD and GTS, additional comorbidities, such as ADHD, were more common, further complicating treatment approaches and emphasizing the need for comprehensive evaluations in GTS-OCD cases.

PATHOPHYSIOLOGIC MECHANISMS OF OBSESSIVE-COMPULSIVE DISORDER AND TICS

Several studies have addressed neurobiologic distinctions between OCD subtypes based on age of onset, suggesting that EO-OCD, often associated with tics and GTS, and LO-OCD may involve distinct pathophysiologic mechanisms.

Structural neuroimaging studies pointed out differences in OCD dependent on the age of onset. Thus, Boedhoe et al. (2017) conducted a large-scale structural neuroimaging study to clarify inconsistencies in previous research on OCD, which were often limited by small sample sizes and clinical heterogeneity. By analyzing T1-weighted magnetic resonance imaging (MRI) scans from 1830 OCD patients and 1759 healthy controls across multiple international sites, the authors applied standardized methods to perform both meta- and mega-analyses of subcortical brain volumes. The study revealed age-dependent neuroanatomic differences: adult patients with OCD showed smaller hippocampal volumes and larger pallidal volumes compared to controls, with these differences being more pronounced in medicated individuals. In contrast, unmedicated pediatric patients exhibited significantly larger thalamic volumes relative to healthy peers. These findings were not influenced by demographic or technical variables such as age or MRI field strength. Overall, the results suggest distinct patterns of subcortical abnormalities in pediatric and adult OCD, supporting the hypothesis of neurodevelopmental alterations underlying the disorder.

Park et al. (2023) reported increased cortical gyrification in EO-OCD patients, particularly in the frontoparietal and cingulate regions, compared to those with LO-OCD. Complementing these findings, Busatto et al. (2001) observed reduced cerebral blood flow in the left anterior cingulate cortex (ACC) and right thalamus among EO-OCD patients. Together, these results suggest that EO-OCD may constitute a distinct neurodevelopmental subtype, characterized by unique neural pathways that shape its clinical course. This distinction highlights the need for age-specific diagnostic and therapeutic strategies tailored to the neurobiologic features of EO-OCD and LO-OCD presentations (Burchi and Pallanti, 2019).

The relationship between GTS and EO-OCD is often conceptualized along a spectrum; however, the precise neurobiologic underpinnings remain unclear. Diffusion tensor imaging studies investigating white matter microstructure have revealed alterations that differentiate GTS from OCD. In a recent study involving drug-naïve children with GTS, GTS +OCD, and pure OCD, compared to healthy controls, increased fractional anisotropy (FA) was observed in specific white matter (WM) tracts – including the corticospinal tract, anterior thalamic radiations, and corpus callosum – in the GTS and GTS +OCD groups. These FA increases were negatively correlated with tic severity. Conversely, children with pure OCD exhibited widespread reductions in FA, suggesting divergent patterns of WM development. These findings support the classification of GTS+OCD as a subtype within the GTS spectrum, while reinforcing the view that pure OCD follows a distinct neurobiologic trajectory.

The neurobiologic etiology of BFRBs has been clarified through neuroimaging studies on skin-picking disorder (SPD) and trichotillomania (hair-pulling). Structural imaging has revealed reduced cortical thickness in the supramarginal gyrus, inferior parietal lobule, and temporal gyrus in BFRBs groups compared to controls and these changes have been associated with greater symptom severity, underlining the role of these regions in sensory integration and emotional regulation (Harries et al., 2017). Notably, no statistically significant differences were observed in the structural volumes of the basal ganglia (caudate, putamen, and nucleus accumbens) between affected individuals and controls. However, within the SPD group, greater impulsivity was positively correlated with increased cortical thickness in the left insula, while skin-picking severity showed a negative correlation with cortical thickness in the left supramarginal gyrus, as well as in a region encompassing the right inferior parietal, right temporal, and right supramarginal gyrus. These findings suggest that while SPD shares certain features with other obsessive-compulsive and related disorders, it also presents distinct neurobiologic characteristics, highlighting both commonalities and differences in symptom expression across the spectrum.

Grant et al. (2022) performed functional magnetic resonance imaging (fMRI) study examined reward processing in individuals with trichotillomania and SPD. Compared to healthy controls, participants with trichotillomania and SPD showed hyperactivation in the inferior frontal gyrus and increased brain activity during loss anticipation in several regions, including the fusiform gyrus and cerebellum. Reward and loss processing abnormalities have also been observed in patients with tics and those with comorbid tics and OCD (Palminteri et al., 2009; Worbe et al., 2011). Collectively, these findings suggest that repetitive behaviors in may potentially share some overlapping pathophysiologic mechanisms. This assumption is also in line with some recent neurophysiologic studies directly comparing GTS, OCD, and BFRB using event-related potentials (ERPs) to assess the timing ("when") and localization ("where") of brain activity. Participants with GTS, OCD, and BFRBs were compared to controls during a visual oddball task. While early ERP components (P200 and N200) showed no significant group differences, the anterior P300 component – associated with later cognitive processing – was reduced in both GTS and OCD groups but preserved in BFRB participants (Desfossés-Vallée et al., 2024). Source localization analyses revealed distinct brain activation patterns, with the OCD group showing unique involvement of the lingual and middle occipital gyri. These findings suggest shared deficits in late-stage cognitive processing in GTS and OCD, alongside differential neural activation patterns, supporting both overlap and divergence in the underlying neurobiology of these conditions.

Finally, genetic investigations highlight a significant overlap between GTS and OCD. A recent genome-wide association study identified 21 single nucleotide polymorphisms (SNPs) in the LINC01122 gene on chromosome 2p16.1 associated with both conditions, suggesting a shared genetic basis exclusive to GTS-OCD, distinct from other psychiatric disorders (Yang et al., 2021). Additional genes – CADM2, LY6G6F, MEGT1, and APOM – were also implicated, potentially informing novel treatment approaches. However, more functional analyses are needed to confirm the roles of these genes in predisposition to GTS and OCD and to evaluate their potential as therapeutic targets (Khoodoruth et al., 2023).

Despite growing evidence of both shared and distinct neurobiologic mechanisms across OCD, GTS, and their comorbid presentations, significant gaps remain. Many studies rely on small, heterogeneous samples, and few have focused specifically on well-defined subgroups such as GTS+OCD or EO-OCD. The lack of longitudinal, multimodal research limits our ability to understand developmental trajectories and the clinical significance of observed brain and genetic differences.

Future research should prioritize stratified designs that account for age of onset, comorbidity profiles, and symptom dimensions. Integrating neuroimaging, neurophysiologic, and genetic data within these frameworks will be critical for identifying reliable biomarkers and developing targeted, mechanism-based interventions for these complex and overlapping disorders.

PHARMACOLOGIC TREATMENTS IN PATIENTS WITH TICS AND OBSESSIVE-COMPULSIVE DISORDER

Several recent meta-analyses have evaluated the efficacy of pharmacologic and behavioral interventions in pediatric OCD, with varying conclusions and limited data specific to individuals with comorbid tic disorders.

Maneeton et al. (2020) conducted a systematic review focused on fluoxetine and found that it significantly reduced OCD symptom severity compared to placebo across several standardized scales (Children's Yale-Brown Obsessive Compulsive Scale [CY-BOCS], national institute of mental health – obsessive-compulsive scale (NIMH-OC), clinical global impression – severity (CGI-S)). Despite small sample sizes, fluoxetine was also found to be well tolerated and comparable to placebo in terms of acceptability. Uhre et al. (2020) and Locher et al. (2017) reviewed randomized controlled trials of cognitive-behavioral therapy (CBT) and concluded that CBT was more effective than no intervention and comparable to selective serotonin reuptake inhibitors (SSRIs) in reducing OCD severity. However, the certainty of evidence in both cases was rated as low to very low due to methodological limitations. Öst et al. (2016) performed a broad meta-analysis comparing CBT, SSRIs, and combination therapy. Their results showed that CBT and CBT+SSRIs produced higher response and remission rates than SSRI alone, and that combination therapy did not outperform CBT alone, regardless of initial severity.

Although OCD and GTS are closely related and often co-occur, no clinical trials to date have specifically targeted the treatment of OCD symptoms in individuals with GTS. Most available insights stem from a limited number of studies assessing anti-OCD treatments in patients with chronic tic disorders. Importantly, many clinical trials evaluating pharmacologic or behavioral treatments for OCD have systematically excluded patients with tics, resulting in a persistent lack of evidence regarding treatment efficacy in this subgroup. This significantly limits the generalizability of existing findings to individuals with comorbid tics, despite some evidence suggesting that treatment response may differ in this population.

To address this gap, a number of studies – including some from earlier decades – have explored the unique clinical presentation and treatment response of patients with comorbid OCD and GTS or primary tics, providing preliminary insights into whether standard interventions may require adaptation for this subgroup.

A recent individual participant data meta-analysis (Cohen et al., 2025), pooling data from several randomized controlled trials, confirmed that SSRIs are the first-line pharmacologic treatment in pediatric OCD. The analysis included over 600 participants, representing more than 80% of all individuals ever enrolled in double-blind, placebo-controlled SSRI trials for pediatric OCD. While the overall treatment effect was modest, the analysis showed that baseline symptom severity was associated with a reduced likelihood of response. This finding raises important questions about the effectiveness of SSRIs in specific subgroups, including those with comorbid tics. As such, more targeted research is needed to determine how children and adolescents with both OCD and tics respond to SSRI treatment, and whether alternative strategies may be warranted in this group.

Geller et al. (2004) conducted a 16-week, double-blind, placebo-controlled study on OCD patients responding to paroxetine. Results showed a significant reduction in CY-BOCS scores from an average of 26 to 13 by the endpoint of the initial phase, with 71% of patients meeting response criteria. However, those with comorbid conditions, including tic disorders, ADHD, and oppositional defiant disorder, had lower response rates (39%–56%) and higher relapse rates following the withdrawal phase, indicating that comorbidities may impact treatment efficacy.

The Pediatric OCD Treatment Study (POTS) (2004) further investigated the effects of tic comorbidity on OCD treatment outcomes. Among the 112 OCD subjects aged 7–17, approximately 15% had GTS or a chronic motor tic disorder. This study found that combined treatment with CBT and sertraline was more effective than sertraline alone, which was, in turn, more effective than a placebo. By the study's end, CY-BOCS scores did not differ significantly between the OCD groups with (16%) and without tics (17%). However, there was a notable interaction between tic disorders and treatment type, where sertraline showed significant benefits over placebo only in the absence of tics. This suggests that tics may reduce the impact of SSRIs on OCD symptoms in patients with both OCD and tic disorders.

The first study to examine an augmentative approach was conducted on 34 adult OCD patients with and without chronic tic disorders, involving a 4-week, double-blind trial comparing fluvoxamine plus placebo against fluvoxamine plus haloperidol (McDougle et al., 1994). Patients receiving fluvoxamine with haloperidol showed improved OCD symptoms (measured by Y-BOCS scores) and a reduction in tics, while those with only fluvoxamine showed less improvement. However, in OCD patients without tics, haloperidol offered only a slight benefit, and other small studies in this area have yielded similar trends without achieving statistical significance, likely due to limited sample sizes.

In terms of additional pharmacologic options, some clinical experience suggests that sulpiride, a benzamide antipsychotic, may benefit patients with a combination of OCD, tics, and anxious-depressive symptoms (George et al., 1993). In a double-blind study, sulpiride was found effective in reducing tics, particularly in individuals with comorbid compulsive-anxious symptoms. Its efficacy in alleviating depressive symptoms was also noted, with minimal side-effects at lower doses. Given the broad positive clinical experience with sulpiride, including its use in minors over several decades, it is considered a valuable option for treating patients with co-occurring tics and compulsive-anxious symptoms.

Also, in the prospective open-label study investigated the efficacy and safety of aripiprazole in 44 drug-free adults with GTS, focusing on its effects on tics, comorbidities, premonitory urges (PU), and quality of life (QoL) (Gerasch et al., 2016). Eighteen patients elected to undergo treatment and completed follow-up assessments after 4–6 weeks. The findings showed that aripiprazole significantly reduced tic severity and improved obsessive-compulsive symptoms, with a trend toward improvement in depression, anxiety, and ADHD. However, it had no effect on premonitory urges. Interestingly, patients' decisions to initiate treatment were not influenced by tic severity or QoL, but rather by the type of comorbidity – those with OCD were more likely, and those with ADHD less likely, to opt for treatment. Sleep disturbances were the most common side-effect. Overall, the study suggests that aripiprazole may provide broad therapeutic benefits for patients with GTS, potentially due to its adaptive pharmacologic action on dopaminergic and other neurotransmitter systems. It also highlights, for the first time, that treatment decisions in GTS may be shaped more by comorbid conditions than by tic severity alone.

New pharmacologic options are emerging for the treatment of tics, with ecopipam – an innovative dopamine D1 receptor antagonist – showing promising results in recent clinical trials. A recent systematic review evaluated the efficacy and safety of ecopipam in individuals with GTS, synthesizing findings from six publications based on three completed trials (including two randomized controlled trials and one open-label extension), encompassing a total of 251 participants (Panda et al., 2025). Ecopipam demonstrated a statistically significant reduction in tic severity, with a pooled mean decrease of 3.0 points in the Yale Global Tic Severity Scale (YGTSS) total tic score compared to placebo. Additional improvements were observed in motor and phonic tic sub-scores, as well as in clinical global impression – Tourette syndrome (CGI-TS), highlighting its potential as a targeted therapeutic option for core tic symptoms.

However, when it comes to comorbid symptoms commonly associated with GTS, such as OCD, depression, and ADHD, the review found no significant differences between ecopipam and placebo groups. This suggests that while ecopipam may effectively address tics, it does not appear to impact OCD symptoms in patients with GTS – a notable consideration given the high prevalence of comorbidity in this population.

Ecopipam showed a favorable tolerability profile, with no major adverse effects reported. Despite these encouraging findings, the authors of the review emphasized the limited number and small scale of existing studies, many of which had short follow-up durations. Consequently, there is a clear need for larger, well-powered, and longer-term clinical trials to fully establish ecopipam's efficacy, safety, and its potential role – particularly in the management of GTS with comorbid OCD (Table 7.1).

Taken together, the available evidence highlights both the progress and persistent gaps in the treatment of pediatric OCD, particularly in patients with comorbid tic disorders such as GTS. While SSRIs and CBT remain first-line treatments for OCD, findings from both meta-analyses and clinical trials suggest that these interventions may be less effective in the GTS+OCD subgroup, with lower response rates and greater relapse risk – especially following SSRI monotherapy. Augmentation strategies using antipsychotics like haloperidol, sulpiride, or aripiprazole have shown promise in addressing both tics and obsessive-compulsive symptoms, particularly in patients with multiple comorbidities, including ADHD, anxiety, and depression. However, these insights are largely drawn from small-scale or open-label studies, and robust, controlled trials specifically designed for individuals with GTS+OCD are still lacking.

The recent development of novel treatments such as ecopipam, which demonstrates efficacy for core tic symptoms without clear benefit for OCD or associated comorbidities, further illustrates the complexity of managing this dual-diagnosis population. Importantly, emerging data suggest that treatment decisions in patients with GTS may be shaped more by the nature of comorbid conditions than by tic severity alone, underscoring the need for more individualized, patient-centered treatment strategies.

Future research must prioritize stratified trial designs and mechanistic studies that account for the distinct neurobiologic and clinical profiles of patients with comorbid GTS and OCD.

Table 7.1

Summary of treatment approaches for OCD, GTS, and GTS+OCD: efficacy, evidence level, and comorbidity considerations.

Treatment modality	OCD	GTS	GTS+OCD	Comments
SSRIs (e.g., fluoxetine, sertraline, paroxetine)	First-line; moderate efficacy (CY-BOCS reduction)	Not indicated for tics	Reduced efficacy in presence of tics	Higher relapse rates in comorbid cases (Geller et al., 2004)
CBT (ERP-based)	First-line; strong evidence	May help with comorbid anxiety/OCD	Effective when combined with SSRIs	Often underused in tic populations due to behavioral complexity
Combined CBT+SSRI	Superior to monotherapy	Not studied for tics	Variable efficacy in GTS+OCD	POTS study suggests tics may reduce SSRI effectiveness
Antipsychotic augmentation (e.g., haloperidol, aripiprazole, sulpiride)	Used in SSRI-refractory OCD	Effective for tics (especially D2 antagonists)	Dual benefit for tics and OCD in selected cases	Evidence mostly from small-scale or open-label studies
Aripiprazole	Limited stand-alone effect on OCD	Effective for tics and comorbid symptoms	Some benefits for OCD, ADHD, anxiety	Well-tolerated; flexible dopaminergic profile (Gerasch et al., 2016)
Sulpiride	Not first-line	Some efficacy in GTS+anxiety/OCD	May help in cases with anxious-depressive profile	Positive clinical experience, especially in minors
Ecopipam (D1 antagonist)	Not indicated	Reduces tic severity	No effect on OCD symptoms	Promising for tics only (Panda et al., 2025)

CBT, Cognitive behavioral therapy; *CY-BOCS*, Children's Yale-Brown Obsessive Compulsive Scale; *ERP*, exposure and response prevention; *GTS*, Gilles de la Tourette syndrome; *OCD*, obsessive-compulsive disorder; *POTS*, Pediatric OCD Treatment Study; *SSRI*, selective serotonin reuptake inhibitor.

Only through such targeted efforts can we develop truly effective, evidence-based interventions tailored to this high-need population.

DEEP BRAIN STIMULATION IN PATIENTS WITH TICS AND OBSESSIVE-COMPULSIVE DISORDER

As previously discussed, GTS and OCD, which have proven to be highly comorbid pathologies, are thought to involve different levels of dysregulation within the cortico-basal ganglia-thalamo-cortical network (Stein et al., 2019; Johnson et al., 2023). Despite frequent co-occurrence, suggesting some degree of shared circuit dysfunction, neuroimaging and connectomic studies have demonstrated distinct (dys)connectivity profiles in each condition (Bhikram et al., 2020; Hollunder et al., 2024). Within this partially overlapping pathophysiologic framework, pharmacologic approaches may differ, but in severe and treatment-resistant cases, both conditions may ultimately lead to invasive neuromodulation strategies.

Indeed, deep brain stimulation (DBS) has become an increasingly refined and promising intervention for both diseases (Lipsman et al., 2013; Visser-Vandewalle and Kuhn, 2013). Interestingly, according to the International Tourette DBS Database and Registry (Tourette Deep Brain Stimulation Registry, n.d.), over 50% of patients undergoing DBS for GTS present with comorbid OCD. Similarly, the baseline values in the Multinational Tourette Syndrome DBS cohort reported that 64.2% of patients met criteria for comorbid OCD (Martinez-Ramirez et al., 2018).

DBS of the anterior limb of the internal capsule (ALIC) has been approved under a Humanitarian Device Exemption by the food and drug administration (FDA) since 2009 for treatment-resistant OCD, as an alternative to anterior capsulotomy. It also received CE mark approval in Europe, although this was not renewed in 2022, and clinical implementation and reimbursement have varied across countries. Common DBS targets for OCD include the ALIC, the anteromedial subthalamic nucleus (STN), the bed nucleus of the stria terminalis, and the superolateral medial forebrain bundle, among others (for review, see Mar-Barrutia et al., 2021). Recent advances in the field of connectomic DBS have further refined our understanding of the optimal networks and tract(s) to stimulate for therapeutic benefit in OCD (Li et al., 2020; Baldermann et al., 2021; Meyer et al., 2024; Coenen et al., 2025; Horn et al., 2025).

In contrast, DBS for GTS remains categorized as investigational, due to more heterogeneous clinical outcomes and limited high-quality evidence. Over nine targets have been reported in the literature (Visser-Vandewalle and Kuhn, 2013), though clinical practice tends to favor two structures: the internal segment of the globus pallidus (in its motor [posterolateroventral] or limbic [anteromedial] region) and the centromedian nucleus of the thalamus. These regions are key regulators of motor and limbic circuitry and are also implicated in habit formation and inhibition – core dysfunctions in GTS. The most used targets include centromedian nucleus – ventral oral internal (CM-Voi), centromedian-parafascicular complex (CM-Pf), postrior ventro-lateral GPI (pvGPi), and anteromedial GPI (amGPi).

While DBS target selection might prioritize tic suppression in GTS with comorbid OCD, significant improvements in obsessive-compulsive symptoms have been observed in patients treated with DBS for Tourette syndrome. Multiple randomized controlled trials (Ackermans et al., 2011; Welter et al., 2017; Müller-Vahl et al., 2021) and small open-label studies have evaluated this effect across different targets, with generally encouraging albeit heterogeneous results. A 2021 meta-analysis (Wehmeyer et al., 2021) assessing target-specific outcomes in GTS DBS reported significant reductions in Y-BOCS scores with stimulation of CM-Voi (43.2%), posteroventral globus pallidus internus (GPi) (pvGPi, 55.2%), and amGPi (87%). Only CM-Pf stimulation failed to reach statistical significance, though it still demonstrated a median 44.4% Y-BOCS reduction, likely underpowered due to the small sample size. Pairwise comparisons suggested superior OCD symptom reduction with pallidal over thalamic targets. While these findings support pallidal stimulation as promising for GTS+OCD, caution is warranted due to heterogeneity and potential confounds in patient selection and study design. More recently, another meta-analysis encompassing both invasive and noninvasive neurostimulation in GTS (transcranial magnetic stimulation (TMS) and DBS) confirmed significant reductions in both tic severity (YGTSS) and OCD symptoms (Y-BOCS) across studies (Aloufi et al., 2025). These results reinforce the clinical observation that GTS-DBS targets can modulate OCD-relevant circuits, suggesting network convergence beyond classical OCD targets (Horn et al., 2025).

Recent work by Johnson et al. supports the concept of circuit-specific neuromodulation (Johnson et al., 2021). In a cohort of 35 patients with Tourette syndrome receiving bilateral GPi DBS, the authors employed computational models of fiber tract activation to identify stimulation patterns associated with therapeutic response. At the latest follow-up, YGTSS scores improved by an average of 35.6 points, while Y-BOCS scores improved by 7.5 points. Tic improvement was best predicted by activation of the associative pallido-subthalamic pathway, the ansa lenticularis, and internal capsule projections to the prefrontal cortex. In contrast, improvement in comorbid obsessive-compulsive behavior was more strongly associated with stimulation of the sensorimotor pallido-subthalamic pathway and internal capsule fibers projecting to prefrontal, premotor, and motor cortices. Although some overlap existed between tic- and OCD-related pathways, they remain functionally and anatomically distinct. These findings suggest that although the GPi may represent a viable single anatomic target, optimal outcomes in dual-diagnosis patients may require subregion-specific stimulation or even multilead strategies to engage distinct therapeutic networks.

Other exploratory targets have been investigated in GTS, with some also quantifying concurrent changes in OCD symptoms. The compact anatomy of the STN, situated at the intersection of motor, associative, and limbic loops, could offer a promising but complex target in GTS-OCD comorbidity. In a small open-label trial of anterior STN DBS in GTS, involving 10 patients (three with comorbid OCD), mean Y-BOCS improvement reached 52%, with concomitant 50% tic reduction, though one did not show meaningful OCD symptom relief. Recently, Hollunder et al. explored the subthalamo-frontal pathways implicated in DBS outcomes across multiple disorders, including OCD and GTS (Hollunder et al., 2024). In this large connectomic study, effective STN stimulation in GTS was linked primarily to connectivity with M1 and the supplementary motor area, while in OCD, therapeutic effects were associated with projections to the ventromedial prefrontal cortex, ACC, dorsolateral prefrontal cortex, and orbitofrontal cortex. Notably, 14 patients with GTS were included, three of whom had comorbid OCD. The study reported a mean ± SD YGTSS improvement of 62% ± 18% in the Pisa/Milan cohort and 59% ± 24% in the Shanghai cohort, which suggests that STN-DBS can achieve substantial tic reduction in treatment-refractory GTS. OCD symptom outcomes were not detailed in this subgroup. Patients with OCD treated with STN-DBS experienced a mean Y-BOCS improvement of 45% ± 29% in the London cohort and 44% ± 32% in the Grenoble cohort. Within the compact anatomy of the STN, it may be possible to engage both networks through refined stimulation strategies and advanced programming.

Still, no single target has yet proven universally effective for both symptom domains, and the absence of clearly identified overlap in optimal stimulation networks suggests a need for alternative approaches in dual-diagnosis patients. Dual-target strategies have thus emerged as a logical extension. Najera et al. (2023) described two patients with lifelong, treatment-refractory GTS and OCD who received dual-lead DBS, one in the ventral capsule/ventral striatum and one in the pvGPi (Najera et al., 2023). Both patients achieved improvements in Y-BOCS (−40% and −56%) and YGTSS (−36% and −80%), respectively. In contrast, attempts to rescue OCD symptoms in GTS patients via secondary implantation of OCD-specific leads have yielded more inconsistent results (Servello et al., 2009).

New targeting strategies informed by stereoelectroencephalography (SEEG) recordings are currently emerging in the field of complex movement disorders (Sanger et al., 2018) and neuropsychiatric conditions (MacLean et al., 2021; Sheth et al., 2022, 2023), following a framework inspired by epilepsy monitoring units. This approach enables temporally and spatially precise mapping of symptom-relevant circuits, allowing for tailored DBS targeting based on individual neurophysiologic signatures. Future DBS paradigms for GTS+OCD may benefit from such methodologies to achieve more selective and effective modulation of overlapping motor and limbic networks.

In sum, while DBS remains an evolving therapy for GTS and OCD, growing evidence suggests that – when carefully selected and precisely targeted – it can offer substantial dual benefits in this challenging population. Ongoing innovations in connectomic mapping, individualized targeting, and circuit-level understanding will likely refine and expand both DBS and less-invasive neuromodulation approaches for these indications in the coming years.

References

Ackermans L, Duits A, van der Linden C et al. (2011). Double-blind clinical trial of thalamic stimulation in patients with Tourette syndrome. Brain 134: 832–844.

Aloufi AK, Zahhar JA, Bader MW et al. (2025). Tourette syndrome and brain stimulation therapy: a systematic review and meta-analysis of current evidence. Front Psychiatry 16: 1478503.

Anholt GE, Aderka IM, van Balkom AJ et al. (2014). Severity and course of OCD in early-onset cases compared to late-onset cases. Psychiatry Res 219: 363–368.

Badenoch J, Searle T, Watson I et al. (2020). Sensory symptoms in body-focused repetitive behaviors, restless legs syndrome, and Tourette syndrome: an overlap? Neurosci Biobehav Rev 119: 320–332.

Baldermann JC, Schüller T, Kohl S et al. (2021). Connectomic deep brain stimulation for obsessive-compulsive disorder. Biol Psychiatry 90: 678–688.

Banaschewski T, Woerner W, Rothenberger A et al. (2003). Obsessive-compulsive disorders in children and adolescents. Eur Child Adolesc Psychiatry 12: 262–272.

Bhikram T, Arnold P, Crawley A et al. (2020). The functional connectivity profile of tics and obsessive-compulsive symptoms in Tourette Syndrome. J Psychiatr Res 123: 128–135.

Boedhoe P, Schmaal L, Abe Y et al. (2017). Distinct subcortical volume alterations in pediatric and adult OCD: a worldwide meta- and mega-analysis. Am J Psychiatry 174: 60–69.

Bolton D, Rijsdijk F, O'Connor TG et al. (2007). Genetic heritability in early-onset obsessive-compulsive disorder and comorbid neurodevelopmental disorders. Am J Psychiatry 164: 898–904.

Brandt V, Otte J-H, Fremer C et al. (2023). Non-just-right experiences are more closely related to OCD than tics in Tourette patients. Sci Rep 13: 19627.

Burchi E, Pallanti S (2019). Diagnostic issues in early-onset obsessive-compulsive disorder and their treatment implications. Curr Neuropharmacol 17: 672–680.

Busatto G, Buchpiguel C, Zamignani DR (2001). Regional cerebral blood flow abnormalities in early-onset obsessive-compulsive disorder: an exploratory SPECT study. J Am Acad Child Adolesc Psychiatry 40: 347–354.

Coenen VA, Polosan M, Schläpfer TE et al. (2025). Deconstructing a common pathway concept for deep brain stimulation in the case of obsessive-compulsive disorder. Mol Psychiatry, 1–12.

Cohen SE, de Boer A, Storosum BWC et al. (2025). Systematic review and meta-analysis of individual participant data: randomized, placebo-controlled trials of selective serotonin reuptake inhibitors for pediatric obsessive-compulsive disorder. J Am Acad Child Adolesc Psychiatry 64: 775–785.

Coles ME, Frost RO, Heimberg RG et al. (2003). "Not just right experiences": perfectionism, obsessive-compulsive features, and general psychopathology. Behav Res Ther 41: 681–700.

Coles ME, Frost RO, Heimberg RG et al. (2005). Not just right experiences: an examination of perfectionism in obsessive-compulsive disorder and eating disorders. J Anxiety Disord 19: 684–698.

Desfossés-Vallée S, Leclerc JB, Blanchet P et al. (2024). Comparing the 'When' and the 'Where' of electrocortical activity in patients with Tourette syndrome, body-focused repetitive behaviors, and obsessive compulsive disorder. J Clin Med 13: 2489.

Eddy CM, Cavanna AE (2014). On the overlap between obsessive-compulsive disorder and Gilles de la Tourette syndrome: neuropsychiatric aspects. J Psychosom Res 76: 205–210.

Ferrao YA, de Alvarenga PG, Hounie AG et al. (2022). The phenomenology of obsessive-compulsive symptoms in Tourette syndrome. In: Martino D., Leckman J.F. (Eds.). Tourette syndrome, 2nd ed. Oxford University Press.

Geller DA, Wagner KD, Emslie G et al. (2004). Paroxetine treatment in children and adolescents with obsessive-compulsive disorder: a randomized, multicenter, double-blind, placebo-controlled trial. J Am Acad Child Adolesc Psychiatry 43: 1387–1396.

George MS, Trimble MR, Robertson MM (1993). Fluvoxamine and sulpiride in comorbid obsessive–compulsive disorder and Gilles de la Tourette syndrome. Hum Psychopharmacol Hum Exp 8: 327–334.

Gerasch S, Kanaan AS, Jakubovski E et al. (2016). Aripiprazole improves associated comorbid conditions in addition to tics in adult patients with Gilles de la Tourette Syndrome. Front Neurosci 10: 416.

Girone N, Benatti B, Vismara M et al. (2024). Early-onset obsessive-compulsive disorder: sociodemographic and clinical characterization of a large outpatient cohort. J Psychiatr Res 172: 1–8.

Grant JE, Peris TS, Ricketts EJ et al. (2022). Reward processing in trichotillomania and skin picking disorder. Brain Imaging Behav 16: 547–556.

Greenberg E, Tung ES, Gauvin C et al. (2018). Prevalence and predictors of hair pulling disorder and excoriation disorder in Tourette syndrome. Eur Child Adolesc Psychiatry 27: 569–579.

Groth C, Debes NM, Rask CU (2018). Long-term prognosis of tics and comorbid conditions in a cohort of children with Gilles de la Tourette syndrome: a six-year follow-up. Eur Child Adolesc Psychiatry 27: 1211–1220.

Harries MD, Chamberlain SR, Redden SA et al. (2017). A structural MRI study of excoriation (skin-picking) disorder and its relationship to clinical severity. Psychiatry Res Neuroimaging 269: 26–30.

Hollunder B, Ostrem JL, Sahin IA et al. (2024). Mapping dysfunctional circuits in the frontal cortex using deep brain stimulation. Nat Neurosci, 1–14.

Horn A, Li N, Meyer GM et al. (2025). Deep brain stimulation response circuits in obsessive compulsive disorder. Biol Psychiatry S0006-3223(25)01096-0.

Janowitz D, Leckman JF (2009). Comparative symptomatology of obsessive-compulsive disorder in Tourette syndrome patients. CNS Neurosci Ther 15: 263–270.

Jensen I, Mol Debes NMM (2024). Correlation of comorbidities and variability of tics in children with chronic tic disorder. Neuropediatrics 55: 382–388.

Johnson KA, Duffley G, Foltynie T et al. (2021). Basal ganglia pathways associated with therapeutic pallidal deep brain stimulation for Tourette syndrome. Biol Psychiatry Cogn Neurosci Neuroimaging 6: 961–972.

Johnson KA, Worbe Y, Foote KD et al. (2023). Tourette syndrome: clinical features, pathophysiology, and treatment. Lancet Neurol 22: 147–158.

Khoodoruth MAS, Ahammad F, Khan YS et al. (2023). The shared genetic risk factors between Tourette syndrome and obsessive-compulsive disorder. Front Neurol 14: 1283572.

Leckman JF, Zhang H, Vitale A et al. (1998). Course of tic severity in Tourette syndrome: the first two decades. Pediatrics 102: 14–19.

Li N, Baldermann JC, Kibleur A et al. (2020). A unified connectomic target for deep brain stimulation in obsessive-compulsive disorder. Nat Commun 11: 3364.

Lipsman N, Giacobbe P, Lozano AM (2013). Chapter 19 – Deep brain stimulation in obsessive–compulsive disorder neurocircuitry and clinical experience. Handb Clin Neurol 116: 245–250.

Locher C, Koechlin H, Zion SR et al. (2017). Efficacy and safety of selective serotonin reuptake inhibitors, serotonin-norepinephrine reuptake inhibitors, and placebo for common psychiatric disorders among children and adolescents: a systematic review and meta-analysis. JAMA Psychiatry 74: 1011–1020.

MacLean JA, Ferman D, Chu JK et al. (2021). Transient complete resolution of Tourette syndrome symptoms following personalized depth electrode placement. Brain Sci 11: 1559.

Maneeton N, Maneeton B, Karawekpanyawong N et al. (2020). Fluoxetine in acute treatment of children and adolescents with obsessive-compulsive disorder: a systematic review and meta-analysis. Nord J Psychiatry 74: 461–469.

Mar-Barrutia L, Real E, Segalás C et al. (2021). Deep brain stimulation for obsessive-compulsive disorder: a systematic review of world-wide experience after 20 years. World J Psychiatry 11: 659–680.

Martinez-Ramirez D, Jimenez-Shahed J, Leckman JF et al. (2018). Efficacy and safety of deep brain stimulation in Tourette syndrome: the International Tourette Syndrome Deep Brain Stimulation Public Database and Registry. JAMA Neurol 75: 353.

De Mathis MA, do Rosario MC, Diniz JB et al. (2008). Obsessive-compulsive disorder and tics in Tourette syndrome patients: a comparative analysis. CNS Spectr 13: 379–386.

McDougle CJ, Goodman WK, Leckman JF et al. (1994). Haloperidol addition in fluvoxamine-refractory obsessive-compulsive disorder. A double-blind, placebo-controlled study in patients with and without tics. Arch Gen Psychiatry 51: 302–308.

Meyer GM, Hollunder B, Li N et al. (2024). Deep brain stimulation for obsessive-compulsive disorder: optimal stimulation sites. Biol Psychiatry 96: 101–113.

Müller-Vahl KR, Szejko N, Saryyeva A et al. (2021). Randomized double-blind sham-controlled trial of thalamic versus GPi stimulation in patients with severe medically refractory Gilles de la Tourette syndrome. Brain Stimul 14: 662–675.

Najera RA, Provenza N, Dang H et al. (2023). Dual-target deep brain stimulation for obsessive-compulsive disorder and Tourette syndrome. Biol Psychiatry 93: e53–e55.

Narayanaswamy JC, Viswanath B, Cherian AV et al. (2012). Heritability and clinical relevance of obsessive-compulsive disorder in Tourette syndrome: a family study. J Clin Psychiatry 73: 869–875.

Neal M, Cavanna AE (2013). Not just right experiences in patients with Tourette syndrome: a questionnaire study. J Neurol Neurosurg Psychiatry 84: e1.

Nissen JB, Mikkelsen E (2018). Not just right experiences and perfectionism in adolescents with obsessive-compulsive disorder. J Obsessive Compuls Relat Disord 17: 82–87.

O'Connor KP (2005). Cognitive-behavioral management of tic disorders. J Cogn Psychother 19: 215–230.

Öst LG, Riise EN, Wergeland GJ et al. (2016). Cognitive behavioral and pharmacological treatments of OCD in children: a systematic review and meta-analysis. J Anxiety Disord 43: 58–69.

Palminteri S, Lebreton M, Worbe Y et al. (2009). Pharmacological modulation of subliminal learning in Parkinson's and Tourette's syndromes. Proc Natl Acad Sci USA 106: 19179–19184.

Palumbo D, Kurlan R (2007). Complex obsessive-compulsive and impulsive symptoms in Tourette syndrome. Adv Neurol 99: 104–110.

Panda PK, Panda P, Dawman L et al. (2025). Safety and efficacy of ecopipam in patients with Tourette syndrome: a systematic review and meta-analysis. CNS Drugs 39: 127–142.

Park I, Ha M, Kim T et al. (2023). Cortical gyrification differences between early- and late-onset obsessive–compulsive disorder: neurobiological evidence for neurodevelopmentally distinct subtypes. Psychol Med 53: 5976–5985.

Pediatric OCD Treatment Study (POTS) Team(2004). Cognitive-behavior therapy, sertraline, and their combination for children and adolescents with obsessive-compulsive disorder: the Pediatric OCD Treatment Study (POTS) randomized controlled trial. JAMA 292: 1969–1976.

Pringsheim T, Davenport W, Lang A (2017). Comorbid psychiatric disorders and tic severity in children with Tourette syndrome: an epidemiological analysis of the Canadian pediatric population. J Child Neurol 32: 650–658.

Ravi Kishore V, Sinha VK, Shrivastava A et al. (2004). Comparative response rates to treatment in early versus late-onset OCD patients. J Child Adolesc Psychopharmacol 14: 485–493.

Van Roessel I, de Haan E, Hendriks M et al. (2023). Predictors of treatment response in childhood-onset OCD: the impact of tics and Tourette syndrome. J Am Acad Child Adolesc Psychiatry 62: 32–43.

Rozenman M, Peris TS, Gonzalez A et al. (2016). Clinical characteristics of pediatric trichotillomania: comparisons with obsessive-compulsive and tic disorders. Child Psychiatry Hum Dev 47: 124–132.

Sanger TD, Liker M, Arguelles E et al. (2018). Pediatric deep brain stimulation using awake recording and stimulation for target selection in an inpatient neuromodulation monitoring unit. Brain Sci 8: 135.

Sarchioto M, Frey J, Ganos C et al. (2024). Diagnostic criteria for primary tic disorders: time for reappraisal. Mov Disord 39: 1276–1283.

Servello D, Sassi M, Brambilla A et al. (2009). De novo and rescue DBS leads for refractory Tourette syndrome patients with severe comorbid OCD: a multiple case report. J Neurol 256: 1533.

Sheth SA, Bijanki KR, Metzger B et al. (2022). Deep brain stimulation for depression informed by intracranial recordings. Biol Psychiatry 92: 246–251.

Sheth SA, Shofty B, Allawala A et al. (2023). Stereo-EEG-guided network modulation for psychiatric disorders: surgical considerations. Brain Stimul 16: 1792–1798.

Stein DJ, Costa DLC, Lochner C et al. (2019). Obsessive–compulsive disorder. Nat Rev Dis Primers 5: 52.

Szejko N, Müller-Vahl KR (2021). Challenges in the diagnosis and assessment in patients with Tourette syndrome and comorbid obsessive-compulsive disorder. Neuropsychiatr Dis Treat 17: 1253–1266.

Taylor S (2011). Symptom dimensions and severity in obsessive-compulsive disorder: examining early versus late onset. J Affect Disord 130: 374–382.

Torresan RC, Ramos-Cerqueira AT, de Mathis MA et al. (2013). Early-onset OCD: clinical characteristics and implications. Psychiatry Res 209: 682–687.

Tourette Deep Brain Stimulation Registry (n.d.). https://tourettedeepbrainstimulationregistry.ese.ufhealth.org.

Uhre CF, Uhre VF, Lønfeldt NN et al. (2020). Systematic review and meta-analysis: cognitive-behavioral therapy for obsessive-compulsive disorder in children and adolescents. J Am Acad Child Adolesc Psychiatry 59: 64–77.

Visser-Vandewalle V, Kuhn J (2013). Chapter 20: Deep brain stimulation for Tourette syndrome. Handb Clin Neurol 116: 251–258.

Wehmeyer L, Schüller T, Kiess J et al. (2021). Target-specific effects of deep brain stimulation for tourette syndrome: a systematic review and meta-analysis. Front Neurol 12: 769275.

Welter M-L, Houeto J-L, Thobois S et al. (2017). Anterior pallidal deep brain stimulation for Tourette's syndrome: a randomised, double-blind, controlled trial. Lancet Neurol 16: 610–619.

Worbe Y, Mallet L, Golmard JL et al. (2010). Repetitive behaviours in patients with Gilles de la Tourette Syndrome: tics, compulsions, or both? PLoS One 5: e12959.

Worbe Y, Palminteri S, Hartmann A et al. (2011). Reinforcement learning and Gilles de la Tourette syndrome: dissociation of clinical phenotypes and pharmacological treatments. Arch Gen Psychiatry 68: 1257–1266.

Yang Z, Wu H, Lee PH et al. (2021). Investigating shared genetic basis across tourette syndrome and comorbid neurodevelopmental disorders along the impulsivity-compulsivity spectrum. Biol Psychiatry 90: 317–327.

Handbook of Clinical Neurology, Vol. 215 (3rd series)
Tic Disorders and Tourette Syndrome
C. Ganos and Y. Worbe, Editors
https://doi.org/10.1016/B978-0-443-13554-5.00003-1

Chapter 8

Attention-deficit/hyperactivity disorder in individuals with primary tic disorders and Tourette syndrome

VALERIE BRANDT[1], NADER A. PERROUD[2], AND KERSTIN JESSICA PLESSEN[3,*]

[1]*School of Psychology, Centre for Innovation in Mental Health, University of Southampton, Southampton, United Kingdom*

[2]*Division of Psychiatric Specialties, Department of Psychiatry, University Hospitals of Geneva, Geneva, Switzerland*

[3]*Division of Child and Adolescent Psychiatry, Department of Psychiatry, Lausanne University Hospital (CHUV) and University of Lausanne, Lausanne, Switzerland*

Abstract

The co-occurrence of tic disorders and ADHD (attention-deficit/hyperactivity disorder) is common; however, the significant overlap in their symptoms and underlying mechanisms frequently poses questions for diagnosis and treatment. The interaction between tic disorders and ADHD involves intricate genetic and neurobiologic factors, particularly related to inhibitory processes and cortico-striato-thalamic pathways in the brain. While both conditions present shared and specific challenges, studies suggest that ADHD symptoms often pose a greater burden on patients than tics. Therefore addressing ADHD symptoms alongside tics is essential for achieving optimal outcomes and those may differ with respect to the developmental stage and/or age of the individual. Psychoeducation, as well as behavioral and psychopharmacologic treatment strategies, have evolved during the last decade, demonstrating that managing both conditions concurrently can be done safely and effectively. This approach involves comprehensive evaluations and prioritizing treatments that target both tic symptoms and those of ADHD. By focusing on therapies that address these overlapping issues, individuals may experience improved overall outcomes.

TICS AND ATTENTION-DEFICIT/ HYPERACTIVITY DISORDER

Phenomenology

Individuals with tic disorders suffer frequently from comorbidities, which are often the reason for seeking help, because their symptoms may result in more impairment of everyday function than the tics themselves (Storch et al., 2007; Bernard et al., 2009; Brandt et al., 2023a). The co-occurrence of comorbidities with tics is less common in tic disorders other than Tourette syndrome (TS), such as in transient/provisional tics or in chronic motor and chronic vocal tics, as well as in populations that do not seek treatment for their tics (Martino et al., 2017). Approximately 90% of patients with TS seen in clinical populations across ages present with additional obsessive-compulsive disorder (OCD), attention-deficit/hyperactivity disorder (ADHD), major depression, or anxiety (Hirschtritt et al., 2015). The co-occurrence of TS and ADHD presents in particular during childhood, but may also persist later in life, posing a number of challenges for evaluation and treatment, which will be discussed in this chapter. Approximately 60% of individuals with TS fulfill criteria of ADHD at one point in their life, whereas epidemiologic studies estimate that up to 20% of individuals with ADHD also have chronic tics (Rothenberger et al., 2007; Centers for Disease C, Prevention, 2009). The point prevalence of tic disorders in ADHD appears to be closer to 10%, with TS at around 1% (Steinhausen et al., 2006; Poh et al., 2018; Mohammadi et al., 2021; Seo et al., 2022). The vast majority of adolescents and adults with a chronic tic disorder also have one or more lifetime psychiatric disorders, with between 10% and 80% having comorbid ADHD depending on the study population (Freeman et al., 2000; Gorman et al., 2010; Hirschtritt et al., 2015; Huisman-van Dijk et al., 2016; Sambrani et al., 2016; Ricketts et al., 2022a). This association, however, is not specific to tics. Lifetime mental disorders lead, in general, to an increased vulnerability for developing other mental disorders, irrespective of the disorder (McGrath et al., 2020), as well as an increased risk for a large range of somatic

*Correspondence to: Kerstin Jessica Plessen, Division of Child and Adolescent Psychiatry, Department of Psychiatry, Lausanne University Hospital (CHUV); Av. d'Echallens 9, 1004 Lausanne, Switzerland. Tel: +41 (0)21 3141960, E-mail: Kerstin.plessen@chuv.ch.

illnesses (Brandt et al., 2023b). For instance, in a large cross-sectional international clinical cohort study of 6805 individuals with TS, 61% of children and 39% of adults were also diagnosed with ADHD (Freeman and Tourette Syndrome International Database, 2007). In a Swedish population cohort of 3590 children and adults diagnosed with TS or a chronic tic disorders in specialist care, 52.2% were also diagnosed with ADHD (Perez-Vigil et al., 2018).

ADHD is a neurodevelopmental disorder characterized by persistent and pervasive inattention and/or hyperactivity and impulsivity that exceed normal levels of these behaviors for a specific age group (American Psychiatric Association, 2013). With a prevalence of approximately 6% in children and 2.5% in adults worldwide, ADHD is a common childhood onset disorder (Faraone et al., 2021; Coghill et al., 2023). Individuals with ADHD present with a typical clinical phenomenology that varies based on the principal symptom complex: first, the predominantly inattentive presentation, characterized by attention difficulties, problems organizing and completing tasks, or following instructions; second, the predominantly hyperactive/impulsive presentation, characterized by disruptive behaviors such as interrupting others, an inability to sit still, fidgeting, or restlessness; third, the combined presentation, where both inattentive and hyperactive/impulsive symptoms are present. ADHD is more common in boys than girls during childhood (2.5:1) (Mowlem et al., 2019), whereas the gender gap tends to diminish, with the ratio being roughly 1:1 in adults (Friedrichs et al., 2012).

Clinicians may struggle to disentangle symptoms and impact of tics and ADHD because both disorders are characterized by increased motor output, although the onset and course typically differ between tics and ADHD. Tic disorders typically emerge around the age of 6 years, peak in severity around the age of 10–12, and then typically diminish (Sambrani et al., 2016). In contrast, symptoms of ADHD tend to occur 2–3 years earlier, thus typically preceding the onset of tics and may persist across the life span with a change in symptom presentation in adolescence and adulthood, where symptoms of hyperactivity and impulsivity tend to abate (Dopfner et al., 2015). Thus symptoms of ADHD typically occur first during childhood, followed by tics, whereas comorbidities such as OCD, anxiety, or depression typically appear later, often just before or during adolescence or in early adulthood (Hirschtritt et al., 2015). Children with ADHD who develop tics later in the course are often diagnosed earlier with a tic disorder compared with other children with tics (Freeman and Tourette Syndrome International Database, 2007). They are treated earlier and are more likely to show additional comorbidities, such as oppositional defiant disorder, learning disabilities, and self-injurious behavior (Freeman and Tourette Syndrome International Database, 2007; Pringsheim, 2017). The majority of children with mild tics improve in adolescence, and the remission rate of tics (approximately 50%) (Spencer et al., 1999) is likely higher than for ADHD (approximately 20%–30%) (Karam et al., 2015; Black et al., 2021). The remission rate for tics does not appear to be affected by comorbid ADHD or vice versa (Spencer et al., 1999).

In contrast, comorbid ADHD does have a significant impact over and above tics on several psychosocial outcomes. Individuals with TS and ADHD frequently attain lower educational and occupational levels, while those outcomes are less clear in individuals with TS only (Kuriyan et al., 2013; Perez-Vigil et al., 2018; Ricketts et al., 2022b). Comorbid ADHD is often more impairing in children and youths than tics, (Gorman et al., 2010; Bitsko et al., 2014) and is associated with poor academic attainment (Kadesjo and Gillberg, 2000; Freeman and Tourette Syndrome International Database, 2007; Perez-Vigil et al., 2018), decreased quality of life (Storch et al., 2007; Bernard et al., 2009; Brandt et al., 2023a), aggressive or delinquent behavior (Budman et al., 2000; Sukhodolsky et al., 2003; Freeman and Tourette Syndrome International Database, 2007), self-injurious behavior, deficits in social skills, mood disorders (Freeman and Tourette Syndrome International Database, 2007), an increased risk for insomnia (Isomura et al., 2022), as well as worse cognitive performance on some IQ tests (Wanderer et al., 2021). Adults with tics and comorbid ADHD experience higher rates of mood disorders, anxiety, and OCD (Haddad et al., 2009). It is thus important to detect children with ADHD early, before gaps in their knowledge at school widen and lead to a sudden risk of decline in grades when the gaps become too large to compensate for. Therefore, children with tics should always be screened for ADHD, and appropriate measures should be discussed with parents and schools. This also means that children with tics and ADHD should have access to ADHD treatment when they present with both sets of symptoms (see details for treatment below). Depending on the severity of the symptoms and the impairment experienced, treatment for both disorders starts in general with a psychoeducational approach directed toward the affected individual and their family, as well as their academic or work environment (NICE Guideline, 2019; Andren et al., 2022). The indication for behavioral therapy and/or medication is then evaluated in a second step. In line with the fact that ADHD symptoms often are more impairing than tics in individuals with TS, whereas tic disorders typically have less impact on outcomes in children with ADHD (Spencer et al., 2001).

In late adolescence and adulthood, due to the low number of studies at present, it remains difficult to assess whether individuals with both disorders (ADHD+TS) have worse quality of life and lower functioning levels than those suffering from only one of the two disorders. Some studies suggest that chronic tic disorders in adults are more negatively influenced by comorbid depression and anxiety than by ADHD, whereas in children, quality of life is mainly impacted by comorbid ADHD (Evans et al., 2016). A number of studies also suggest that having comorbid ADHD in adolescence or adulthood is associated with sleep disturbances (Keenan et al., 2021), lower educational level (Perez-Vigil et al., 2018), poorer psychosocial outcomes (Gorman et al., 2010), lower quality of life (Ricketts et al., 2022a), and further psychiatric comorbidities (Freeman and Tourette Syndrome International Database, 2007). Apart from an increased risk of psychiatric comorbidities and low academic attainment, ADHD and chronic tic disorders are also associated with an increased risk for a number of physical disorders, such as heart conditions and epilepsy (Brander et al., 2019; Reed et al., 2023).

Sleep problems are well recognized in children, adolescents, and adults with ADHD, with data clearly confirming the presence of nighttime sleep disturbances in children with ADHD, which is not attributable to comorbidities (Virring et al., 2016). A clinical review focusing on polysomnographic findings in TS identified several common objective sleep disturbances, including decreased sleep efficiency and increased periodic limb movements (Jimenez-Jimenez et al., 2020). Moreover, a meta-analysis (Keenan et al., 2021) comparing objective measures of sleep (i.e., polysomnography and actigraphy) of individuals of all ages with persistent tic disorders and ADHD (and the comorbidity of the two) found that individuals with tic disorders only as well as with tic-disorders and co-occurring ADHD (a group with comorbidity) had more pronounced problems during nighttime sleep compared to those with ADHD only and controls. The group of tic disorders and ADHD also had increased sleep time. Thus this is an important and often underestimated domain of symptoms with several studies pinpointing problems with a crucial impact on the everyday life of individuals with TS and ADHD.

SPECIFIC CLINICAL FEATURES OF ADHD IN INDIVIDUALS WITH TOURETTE SYNDROME

The clinical presentation of ADHD may differ slightly in the presence of tics and TS. Clinically, it has been stated a difference in the dimensional symptoms of ADHD in that individuals with both ADHD and TS may experience more severe maladaptive behavior, which could have their origin in the impulsivity and hyperactivity symptoms, especially in adults (Haddad et al., 2009). However, in a large database of patients with TS (across all ages, but predominantly minors, with several comorbidities), the type of ADHD did not differ (Freeman and Tourette Syndrome International Database, 2007). Moreover, several studies have shown that impaired inhibitory performance and increased neural activity in TS appear to manifest predominantly in relation to ADHD symptoms (Openneer et al., 2021) and the individual clinical presentation may thus depend on additional other factors that are less well-identified. As the presence of tics can further exacerbate symptoms of impulsivity in patients with TS and vice versa, in the extreme this combination may make these individuals more prone to developing symptoms of self-injurious behavior linked to their tics (Freeman and Tourette Syndrome International Database, 2007).

Moreover, a growing body of evidence indicates that children and adolescents with ADHD often experience challenges in social cognition (Gardner and Gerdes, 2015), albeit to a lesser degree than adults with ADHD (Morellini et al., 2022). These social impairments associated with the motor manifestation of tics, particularly during childhood when individuals may feel less socially accepted than their peers, may have a cumulative impact and are well-documented (Malli et al., 2016). Therefore it is crucial to recognize the potential for heightened deficits in social cognition among children with ADHD and comorbid TS.

NEUROPSYCHOLOGY

Several brain functions may be altered in individuals with TS and comorbid ADHD, which may have a specific or overlapping impact on various neurocognitive processes and neuropsychologic functions, probably also as a result of shared heritability (Yang et al., 2021). Although problems of attention including difficulties in sustaining attention, shifting focus, or filtering out distractions, are a typical presentation of children, adolescents, and adults with ADHD, tics may likewise disturb their span of attention simply by the interruption of a flow of movement, or of an act of concentration in progress (Wadman et al., 2016). It is well known that tics typically disappear in situations that require a high degree of attention and concentration (Caurin et al., 2014), but tics may also present distractions and provoke problems of attention.

Impaired inhibitory control is a hallmark feature of ADHD, with impulsivity being one of the most important characteristics of the disorder. The question of reduced inhibitory control in individuals with tic disorders requires a perspective that also considers the developmental age of the person. From earlier studies, it has been speculated that young individuals with TS show an important capacity for compensation and tic suppression and even may present with a comparable (Mancini et al., 2018) or superior ability of inhibitory control compared with controls (Jackson et al., 2007; Maigaard et al., 2019; Schuller et al., 2021). However, the frequent comorbidity with ADHD, as well as the well-documented presence of rage attacks (Conte et al., 2020), suggests that inhibitory control may be less developed in individuals with tic disorders. Along the same lines, a recent review and meta-analysis proposed that individuals with a pure tic disorder may exhibit a deficit in inhibitory control in selected tasks (Morand-Beaulieu et al., 2017). It may thus be relevant to look further into the different types of inhibitory control, such as, reactive inhibitory control in adults with TS may be reduced, whereas proactive inhibitory motor control may be intact (Indrajeet et al., 2022). (This subject is focused on more in depth in Chapter 21).

In terms of the larger concept of executive functions, an abundance of literature documents the impact of ADHD on these functions, encompassing a range of higher-order cognitive processes, including planning, organization, problem-solving, and self-regulation. This leads to the well-known difficulties of individuals with ADHD in managing time, setting goals, and regulating emotions. To date only a few studies have tried to disentangle the influence of comorbid ADHD regarding executive functions in individuals with TS.

One of these features within the wider domain of executive functions is emotional regulation (ER). ER is a facet of self-regulatory control central to typical and neuro-diverse development and is particularly important for the psychosocial "outcome" in individuals with ADHD (Bodalski et al., 2019).

However, the question of whether the influence of this trait in chronic tic disorders solely depends on comorbidity with ADHD, or if children with TS without this comorbidity also have difficulties regulating their emotions has been scarcely addressed in the literature on adults with TS (Drury et al., 2016) and in youths with TS (Quast et al., 2020). This is, however, an important clinical question and its integration into the framework of a diagnostic evaluation may allow a more individualized treatment approach. One of the main reasons for the absence of evidence is the difficulty in performing valid measurements of this complex behavioral component. External cues, such as the psychosocial context and the relationships with other people surrounding the child play an impo. Ecologically valid measurements should thus create a setting that includes those elements. A recent study assessed participants' ER ability, as well as parent–child interactions, in the context of a complex puzzle task and coded the observed behavior with the "Tangram Emotion Coding Manual" (Hagstrom et al., 2019). Children with TS did not differ from controls in their ER ability (Hagstrom et al., 2021). However, children with TS and co-occurring ADHD and those with "ADHD only" regulated their emotions less well than those with TS only or controls. Moreover, parents of children with ADHD displayed more tension during the experimental task. Interestingly, neither tic severity nor premonitory urges were associated with ER; however, better ER ability was linked to less severe symptoms of ADHD.

The impact of ADHD, however, is highly variable across individuals with TS and depends highly on the developmental age. Additionally, other comorbid conditions can interact and influence each other, further complicating the neurocognitive function of those children. A comprehensive evaluation by a healthcare professional or specialist may help to provide a better understanding of the specific neurocognitive challenges faced by someone with comorbid ADHD and TS or chronic tic disorder.

PATHOPHYSIOLOGY AND NEUROBIOLOGY

It is currently assumed that the cortico-striato-thalamo-cortical (CSTC) loops are implicated in the emergence of involuntary motor behaviors. The CSTC loops are involved in motor planning and motor execution, and subtle imbalances in frontal and striatal areas can lead to unwanted or context-inappropriate motor output, such as tics and impulsive behavior in ADHD (El Malhany et al., 2015). The balance between selecting context-appropriate actions and suppressing context-inappropriate actions is maintained by a complex balance between the "direct pathway" and the "indirect pathway" in the basal ganglia (Rothenberger and Heinrich, 2022), connecting different structures within the basal ganglia, which are part of the CSTC loop. An imbalance between the direct and the indirect pathway is assumed to be associated with an increased selection and output of context-inappropriate actions, such as tics or impulsive actions (Sonuga-Barke, 2003; Kalanithi et al., 2005; Kataoka et al., 2010; Colzato et al., 2022).

While some studies have suggested that the caudate might be smaller in children and adults with TD with or without comorbid ADHD (Peterson et al., 2003; Bloch et al., 2005), a recent study, including TS and ADHD patients as separate groups, concluded that the shape and volume of basal ganglia nuclei did not differ between children (aged 8–12) with ADHD, or TS, and healthy controls (Forde et al., 2017). On a neurobiologic level, deficits in executive functioning in ADHD have been associated with the prefrontal cortex, whereas tic disorders are thought to be mainly associated with alterations in the basal ganglia, both being part of the CSTS loop (Chapter 16). Furthermore, several neurotransmitter abnormalities are involved in the etiology of tic disorders and ADHD, such as noradrenergic, glutamatergic, and GABAergic abnormalities. It has been hypothesized that dopamine in the striatum is down-regulated in ADHD, whereas the opposite has been suggested for tic disorders (Hienert et al., 2018; Rothenberger and Heinrich, 2022). In addition, as the proposed dopaminergic abnormalities are located in similar neuroanatomic regions in both disorders, it is not surprising to find a high co-occurrence of the two conditions (Rothenberger and Heinrich, 2022).

A recent review proposed that the neurobiology of ADHD and TS may be additive rather than interactional, that is, each disorder presents structural, functional, and neurophysiologic alterations, whereas the co-occurrence does not induce additional changes (Rothenberger and Heinrich, 2022). Moreover, there are some alterations common to ADHD and TS but these seem to apply to other neurodevelopmental disorders as well, such as the systems associated with stress and immune response (Rothenberger and Heinrich, 2022).

THE CHALLENGE OF MAPPING THE COMORBIDITY IN THE BRAIN

Findings derived from studies of brain mapping have helped to improve our understanding of the pathophysiology of ADHD and TS, and the neurobiologic determinants of the clinical course of these disorders. However, the consensus on the main deviations is difficult to confirm in meta-analyses (Samea et al., 2019). The evidence for both disorders rests as of today still largely on developmental interpretations of cross-sectional data, with some exceptions (Shaw et al., 2006; Norman et al., 2021) and should therefore be regarded as preliminary. If individuals with ADHD and TS share a common genetic vulnerability, even only in some subtypes of TS and ADHD, several morphometric similarities should be present with either of the two conditions. The definition of their clinical coexistence in imaging studies, however, is scarce as of today. Even though a large percentage of patients in clinical samples of individuals with TS have coexisting psychiatric disorders, (Plessen et al., 2007; Ganos et al., 2013; Church and Schlaggar, 2014) such as ADHD and OCD, analyses have typically controlled for their potential confounding effects rather than possessing the size (and power) to address the question as a specific hypothesis. Most studies considered the presence of comorbid disorders by testing their confounding influence in statistical models. Further, analyses limited to all individuals with "pure" TS and a control group ensured that the primary findings did not result from the presence of comorbid disorders. Other studies have included contrasting groups (e.g., ADHD only) or separate groups composed of participants with comorbid disorders (e.g., TS +ADHD) to map specific anatomic deviations in a particular subgroup. Although the samples are clinically well defined, and the comorbidity is well screened for, this approach often produced small groups with low statistical power. In one of the largest studies in children and adults with TS, cortical areas have been found to be larger in children with TS and comorbid ADHD (and those with comorbid OCD); however, the difference was at a trend level only, even in this considerable sample (Peterson et al., 2001). In conclusion, evidence from neuroimaging studies with both individuals with TS alone and individuals with TS and comorbid ADHD suggests that comorbid ADHD does not significantly alter the primary findings in samples of individuals with TS, thus providing support for the hypothesis of a shared genetic vulnerability, as expressed in deviations of brain morphology in these coexisting symptoms. However, this evidence is derived in the context of small study populations and in the general absence of longitudinal studies. Moreover, the advent of MRI with higher field strength, as well as the use of multimodal imaging studies, paired with genetic analyses will contribute to better mapping of neurobiologic similarities and differences in comorbid disorders in the future.

Assessment of Tourette syndrome and attention-deficit/hyperactivity disorder

Children with tics and ADHD should be assessed according to recommendations by the European Study of Tourette Syndrome Society guidelines (NICE Guideline, 2019; Szejko et al., 2022), and the NICE guidelines for ADHD (NICE Guideline, 2019). In children with a suspected tic disorder and ADHD, a family history should be taken, along with information from the parents. Chronic tic disorders, OCD, and ADHD are not only strongly genetically related but also highly heritable. This is important, particularly in children, because patients with tics tend to suppress their tics in certain contexts and it can therefore be difficult to gain an understanding of the severity of the symptoms in their everyday lives from a clinical setting. Parents can be encouraged to record videos of "high-ticcing" phases, such as when children watch TV or when tired at the end of day (Szejko et al., 2022).

Symptoms of ADHD may be more difficult to record but it is also important to gain observer views and multiple reports. Parents are encouraged to bring reports from school if teachers also observe issues with hyperactivity, impulsivity, or inattention. This is crucial to assess whether the symptoms are relevant, which means are impairing their everyday function, in different situations and contexts, as specified by the diagnostic criteria (American Psychiatric Association, 2013; NICE Guideline, 2019).

In children with multiple disorders, it is particularly important to assess which symptoms are most debilitating. The first occurrence of symptoms and their presentation (ADHD or TS) requires documentation. It is also relevant to co-construct with the family how symptoms and functional impairment have developed over time, as well as to determine which environmental triggers and stressors aggravate symptoms, such as test-taking in schools or other significant events in the life of the child. It should also be determined whether and how the context at school may be accommodated to the specific neuropsychiatric symptoms, such as allocating more time on a test, allowing the student to leave the classroom and go for a quick walk if they have difficulties siting still, or accommodating a desk that hold little distraction for children with ADHD, etc. Psychosocial functioning should be assessed, meaning how symptoms affect the social life of the patient, employment, family issues, or provoke conflicts. In children and adolescents with tics, it is particularly important to define the functional impairment of the young person and to determine together whether it is the child or the parent who wishes to treat the symptoms.

Finally, it is important to assess other prior symptoms and evaluations, past and present medication, potential drug use, and any neurologic and psychiatric antecedents. A full developmental and psychiatric history is required to explore whether tics and/or ADHD were present during childhood and have persisted throughout life (NICE Guideline, 2019; Australian Government NHaMRC, 2023). Even though most patients with suspected ADHD and tics will be minors, an increasing number of adults are being diagnosed with ADHD. School reports can be particularly helpful, as they can sometimes give an insight into how others perceived the patient as a child. In adults, other observers, such as partners, can be included in the assessment process to gain a better picture of the severity and frequency of the reported symptoms. If possible and wished by the patient, parents of adult patients can also be involved in the diagnostic process. Just as with a diagnosis of TS, a diagnosis of ADHD requires the workup by a mental health specialist or by another appropriately qualified healthcare professional trained in the diagnosis and treatment of ADHD, because it is crucial to confirm that ADHD is not better explained by other psychiatric diagnoses. The diagnosis of ADHD is based on DSM-5 criteria assessed during a clinical interview and no formal diagnostic test exists.

Assessments should ideally include standardized procedures.

Attention-deficit/hyperactivity disorder questionnaires and diagnostic interviews

Standardized interviews are helpful tools to inquire symptoms such as present and past inattention, impulsivity, and hyperactivity occurring or starting in childhood. Furthermore, different sources should be consulted, such as parents and teachers, to make sure the symptoms occur across contexts. It is important to assess the level of impairment (Murphy and Adler, 2004), also in relation to tics and ADHD. Rating scales may be useful tools when screening for ADHD symptoms, even though a formal diagnosis requires a thorough and direct evaluation by a clinician in view of the diagnostic criteria.

Conners' ADHD Rating Scale (CAARS) is a widely used tool with good psychometric properties and good content validity for use in childhood. It has several advantages, including multiple parallel versions that can be used to triangulate symptoms from different raters' perspectives, by providing self-report scales (CAARS-S/Conners 3-SR) and observer-report scales (CAARS-O (observer)/Conners 3-P (parent)/Conners 3-T (teacher)), available in both long and short versions, and it has a children's version (Conners 3) and an adult's version (CAARS) (Conners et al., 1999). It also assesses impulsivity, inattention, and hyperactivity respectively in subscales and provides separate norms for different age groups and gender.

While the Conners scale assesses current ADHD symptoms, the short version of the Wender Utah Rating Scale (Taylor et al., 2011) used in adults assesses ADHD symptoms in childhood retrospectively and has good psychometric properties (Ward et al., 1993). It is useful in determining whether symptoms were already present in childhood. This questionnaire is thus often used in adults in combination with the Adult ADHD Self-Report Scale-V1.1 Symptoms Checklist and its DSM.5 version in order to assess whether or not ADHD may be present (Kessler et al., 2005; Ustun et al., 2017). As mentioned previously, these are screening tools and ADHD diagnosis should not be done solely based on these questionnaires.

The gold standard for a semistructured diagnostic interview for children and adolescents is the Kiddie Schedule for Affective Disorders and Schizophrenia-Present and Lifetime (K-SADS PL) (Kaufman et al., 2016), which covers a wide range of diagnostic entities and allows for precise mapping of psychopathologic phenomena in that age range, which is key for mapping comorbidities in children with TS (Szejko et al., 2022). In adults, the structured Diagnostic Interview for Adult ADHD (DIVA-5) should ideally be used to aid formal diagnosis of ADHD in the presence of psychiatric comorbidities such as tic disorders (Zamani et al., 2021).

Treatment

As mentioned earlier, treatment starts with psychoeducation and assessing the specific needs of the patient, their coping skills, and the constellation of present challenges within their family and broader environment. Parents and children should be educated on the neurobiologic background of the symptoms and the developmental aspects. Heritability can be explained to help parents understand that tics are a disorder to which some families are more vulnerable than others and that they can run in families without being caused necessarily by a specific event or stressor. Exemples of other family-m The usual course of tics, with their patterns of waxing and waning, sometimes makes it difficult for parents to monitor the symptoms in a longer perspective, however, in most cases, tics abate with age, and parents should be aware of this often benign course (see Chapter 2).

These aspects of psychoeducation help explain to parents, and sometimes teachers, that children have limited control over their tics, as well as their impulsive or hyperactive behaviors, without being "misbehaved." The waxing and waning nature of tics should be addressed, as well as their association with potential and individual contexts (e.g., the relaxation when coming home from school that often provokes an increase in tics). The joint identification of stressors together with family members helps educating the family and raising awareness concerning patterns of tic emergence where they, themselves, may play a role as maintaining factors.

Behavioral therapy

TS cannot be cured, but tic frequency and severity can be influenced and treated to a certain degree by behavioral adaptation and feedback loops. If treatment is necessary and requested, the recommended first-line treatment is behavioral therapy (Andren et al., 2022). The main behavioral treatments for chronic tic disorders are habit reversal training and exposure with response prevention (see Chapter 22 for details). It is sometimes considered whether patients with ADHD should take stimulants before receiving behavioral training to increase attention span and the ability to suppress tics. However, it has been shown that patients with tics and ADHD, with and without treatment, may subjectively feel like they are not well able to suppress their tics but objectively do not differ with respect to this question from individuals with tic disorder without comorbid ADHD (Brandt et al., 2017; Conelea et al., 2018). A meta-analysis suggested smaller therapeutic effects for children with tics and comorbid ADHD (McGuire et al., 2014), whereas the largest study conducted using HRT in patients with tic disorders showed that children with comorbid ADHD also benefitted from HRT for their tics (Piacentini et al., 2010).

Moreover, the recommended first-line treatment for children and adolescents with ADHD is also behavioral therapy (NICE Guideline, 2019). Despite little research on cognitive behavioural therapy (CBT) in comorbid conditions, clinical guidelines on how to adapt behavioral treatment in children with tics and comorbidities have recently been developed (Sanderson et al., 2022). These guidelines recommend adapting therapeutic sessions for children with comorbid ADHD according to their symptoms. This can be accommodated by making sessions shorter, including breaks, and allowing fidget toys. Using consistent reward schedules, praise, and immediate feedback may yield better effects because children with ADHD have been shown to react particularly well to rewards. They may require more support with their practice between sessions than children without ADHD (Sanderson et al., 2022).

Medical treatment

It is recommended to treat impairing ADHD symptoms in individuals with tic disorders with the appropriate ADHD medication (Pringsheim et al., 2019a). The recommended first-line psychopharmacologic treatment for ADHD is stimulants (Cortese et al., 2018). It was previously assumed that stimulants (either methylphenidate (MPH)- or amphetamine-based medications) might worsen tics due to their dopaminergic potential or even lead to a first manifestation of tics. However, this notion was based on several observational studies from the 1970s, which reported the onset of tics after treatment with stimulants (Golden, 1974, 1977; Pollack et al., 1977; Bremness and Sverd, 1979). Although these observations were based on few cases, one influential report in the 80s, including 15 children with ADHD, probably led to the introduction of the warning by the FDA (Lowe et al., 1982). However, data about the onset or worsening of tics are difficult to interpret in light of the fact that the typical onset for ADHD tends to be several years before the onset of tics and TS and that tics tend to wax and wane in severity (Leckman, 2002). The onset or worsening of tics, thus, may often represent the natural course of a percentage of individuals with ADHD, particularly when the time period of this onset is many months to years after first stimulant intake.

Later, well-designed studies have not shown a tic-exacerbation or onset in patients treated with stimulants (Gadow et al., 1999; Tourette's Syndrome Study Group, 2002; Roessner et al., 2006; Gadow et al., 2007; Bloch et al., 2009; Man et al., 2023). A randomized controlled trial showed that MPH and clonidine decreased ADHD symptoms, as well as tic severity on average. A subset of patients reported a worsening in tics but the percentage was comparable across all groups, including the placebo group (Tourette's Syndrome Study Group, 2002). The causal association between ADHD and tic disorders is difficult to study under experimental conditions because drug studies including control groups usually have a relatively limited time-frame for observation and follow-up. A prospective long-term study in a decent sample of male children with ADHD (n=128) over a 4-year observation period showed that the influence of stimulant medication on the development of tic disorders was no different in those who had ever been exposed to stimulants and those who received the medication during the observational period (Spencer et al., 1999). Moreover, a review of high-quality studies and a meta-analysis concluded that the risk for a new tic onset as a consequence of stimulant use was not elevated (Roessner et al., 2006; Cohen et al., 2015). Finally, a systematic recent review concluded that stimulants can even help relieve tics in young patients with comorbid tics and ADHD; a limiting factor is that all studies were conducted in children, and no data-driven recommendations can be made for adults with comorbid tics and ADHD (Pringsheim et al., 2019b).

Treatment with atomoxetine (ATX) is considered less effective than stimulants (Cortese et al., 2018), but safe and also recommended for patients with comorbid ADHD and TS (Cheng et al., 2007; Poncin et al., 2007), because it did not increase tic severity in the only existing controlled trial and its posthoc analysis that has addressed this question (Allen et al., 2005; Spencer et al., 2008). However, one important study (Spencer et al., 2008) has been criticized for its large attrition rate and a potential selection bias (Gilbert, 2006). A meta-analysis concluded that the effect size for the treatment of ADHD with ATX was marginally lower in those with tic disorder than those with ADHD alone (Bloch et al., 2009). Furthermore, several case studies reported a correlation between the treatment with ATX and the exacerbation of tics in patients with both disorders that was reversible when the medication was stopped (Lee et al., 2004; Ledbetter, 2005; Parraga et al., 2007, 2008). A single study even reported a new onset of tics after a single dose of ATX (Sears and Patel, 2008).

Alpha-agonists, such as guanfacine and clonidine, can be a useful alternative to stimulants in children who do not tolerate stimulants well and are considered in several recommendations as the first choice targeting tics in a person with ADHD (DGKJP, 2018; NICE Guideline, 2019). Both medications can significantly improve symptoms of ADHD. Guanfacine appears to reduce tics (Weisman et al., 2013), although a small randomized controlled trial (RCT) in children with tics did not show a significant effect on tic severity, but ADHD symptoms reduced significantly (Murphy et al., 2017). Clonidine might have a larger effect on tics in children with ADHD and tics than those with tics only (Weisman et al., 2013; Roessner et al., 2022).

Subsequently, several reviews (Erenberg, 2005; Robertson, 2006; Osland et al., 2011, 2018) and meta-analyses (Leckman, 2002) focused on randomized, double-blind, controlled trials of any pharmacologic treatment for ADHD in children with both ADHD and tic disorder (Kaufman et al., 2016). These reviews explored the effect of medication compared to a placebo intervention in home and school settings both on ADHD and on tics. The reviews confirmed that medication for ADHD in general is effective for symptoms of ADHD, regardless of the presence of additional tics. Moreover, tics do not appear to worsen with MPH, ATX, or alpha-agonists. However, these systematic reviews still are largely based on relatively few observations, with the largest review including less than 500 individuals.

A Cochrane review of eight studies concluded that in comorbid TS and ADHD, MPH, clonidine, desipramine, dextroamphetamine, guanfacine, and ATX reduced ADHD symptoms, while guanfacine, desipramine, MPH, clonidine, and the combination of MPH and clonidine improved tics significantly in children (Osland et al., 2018).

If psychoeducation and environmental modifications have not been able to reduce significantly the impairments related to ADHD in adults, pharmacologic treatment should be considered (Guidelines N, 2018). Stimulants, such as MPH or lisdexamfetamine, are also the first-line pharmacologic treatment for adults with ADHD (Cortese, 2023) with or without comorbid conditions including tic disorders. The dose titration of stimulants should nevertheless be slower with more frequent monitoring in the presence of comorbid tic disorders (OHNF Trust, 2021). If an adult taking stimulants develops tics, one should be sure that tics are truly related to this medication as they naturally wax and wane. If a sudden exacerbation of tics occurs, and if tics outweigh the benefits of ADHD treatment, stimulant medication may be reduced or stopped. ATX should then be offered as a second option in most of the cases (OHNF Trust, 2021). Guanfacine and clonidine should not be offered to adults with ADHD unless advice from a tertiary ADHD service has been received (NICE Guideline, 2019).

Neurostimulation

Few studies of noninvasive neurostimulation as a therapeutic tool in TS exist (Munchau et al., 2002; Orth et al., 2005; Kwon et al., 2011; Le et al., 2013). Overall, it appears that transcranial magnetic stimulation and transcranial direct current stimulation affect premonitory urges more than tic severity (Steuber and McGuire, 2024). Tic severity only significantly reduced in patients with comorbid OCD or ADHD (Chae et al., 2004; Le et al., 2013; Bloch et al., 2016) (the topic has been recently reviewed (Grados et al., 2018) and meta-analyzed (Steuber and McGuire, 2024)).

CONCLUSIONS

To summarize, tic disorders and ADHD are common comorbidities with overlapping neurobiologic backgrounds. Their co-occurrence is often associated with a higher social and psychopathologic impairment, where ADHD may be more debilitating than tics and can impact attainment, as well as peer and family relationships. It is, therefore, of high importance to perform a detailed evaluation of co-occurring symptoms in children with tics, not only to exclude ADHD but also to determine whether other comorbidities are present. As described in this chapter, the co-occurrence of the two disorders requires complex treatment considerations with a comprehensive treatment approach that addresses symptoms of both conditions. If present, treatment of symptoms of ADHD is recommended and treating ADHD can also improve tics. Moreover, coexisting TS does not necessarily change the treatment recommendations for ADHD. Importantly, ADHD is not a counterindication for prescribing stimulants or for conducting behavioral therapy, such as habit reversal training or exposure with response prevention. The underlying genetic and neurobiologic causes for the high co-occurrence of the two disorders still await more studies including a higher number of participants.

References

Allen AJ, Kurlan RM, Gilbert DL et al. (2005). Atomoxetine treatment in children and adolescents with ADHD and comorbid tic disorders. Neurology 65: 1941–1949.

American Psychiatric Association (2013). American Psychiatric Association: Diagnostic and Statistical Manual of Mental Disorders, Fifth Edition. Arlington, VA: American Psychiatric Association.

Andren P, Jakubovski E, Murphy TL et al. (2022). European clinical guidelines for Tourette syndrome and other tic disorders-version 2.0. Part II: psychological interventions. Eur Child Adolesc Psychiatry 31: 403–423.

Australian Government NHaMRC (2023). Australian Evidence-Based Clinical Practice Guideline For Attention Deficit Hyperactivity Disorder (ADHD). https://adhdguideline.aadpa.com.au.

Bernard BA, Stebbins GT, Siegel S et al. (2009). Determinants of quality of life in children with Gilles de la Tourette syndrome. Mov Disord 24: 1070–1073.

Bitsko RH, Holbrook JR, Visser SN et al. (2014). A national profile of Tourette syndrome, 2011-2012. J Dev Behav Pediatr 35: 317–322.

Black KJ, Kim S, Yang NY et al. (2021). Course of tic disorders over the lifespan. Curr Dev Disord Rep 8: 121–132.

Bloch MH, Leckman JF, Zhu H et al. (2005). Caudate volumes in childhood predict symptom severity in adults with Tourette syndrome. Neurology 65: 1253–1258.

Bloch MH, Panza KE, Landeros-Weisenberger A et al. (2009). Meta-analysis: treatment of attention-deficit/hyperactivity disorder in children with comorbid tic disorders. J Am Acad Child Adolesc Psychiatry 48: 884–893.

Bloch Y, Arad S, Levkovitz Y (2016). Deep TMS add-on treatment for intractable Tourette syndrome: a feasibility study. World J Biol Psychiatry 17: 557–561.

Bodalski EA, Knouse LE, Kovalev D (2019). Adult ADHD, emotion dysregulation, and functional outcomes: examining the role of emotion regulation strategies. J Psychopathol Behav Assess 41: 81–92.

Brander G, Isomura K, Chang Z et al. (2019). Association of Tourette syndrome and chronic tic disorder with metabolic and cardiovascular disorders. JAMA Neurol 76: 454–461.

Brandt V, Essing J, Jakubovski E et al. (2023). Premonitory urge and tic severity, comorbidities, and quality of life in chronic tic disorders. Mov Disord Clin Pract 10: 922–932.

Brandt VC, Herrmann K, Kerner Auch Koerner J et al. (2017). Perceived and real tic suppression ability and its relation to impulsivity. Mov Disord 32: 1795–1796.

Brandt V, Zhang Y, Carr H et al. (2023). First evidence of a general disease ("d") factor, a common factor underlying physical and mental illness. World Psychiatry 22: 335–337.

Bremness AB, Sverd J (1979). Methylphenidate-induced Tourette syndrome: case report. Am J Psychiatry 136: 1334–1335.

Budman CL, Bruun RD, Park KS et al. (2000). Explosive outbursts in children with Tourette's disorder. J Am Acad Child Adolesc Psychiatry 39: 1270–1276.

Caurin B, Serrano M, Fernandez-Alvarez E et al. (2014). Environmental circumstances influencing tic expression in children. Eur J Paediatr Neurol 18: 157–162.

Centers for Disease C, Prevention. (2009). Prevalence of diagnosed Tourette syndrome in persons aged 6-17 years – United States, 2007. MMWR Morb Mortal Wkly Rep 58: 581–585.

Chae JH, Nahas Z, Wassermann E et al. (2004). A pilot safety study of repetitive transcranial magnetic stimulation (rTMS) in Tourette's syndrome. Cogn Behav Neurol 17: 109–117.

Cheng JY, Chen RY, Ko JS et al. (2007). Efficacy and safety of atomoxetine for attention-deficit/hyperactivity disorder in children and adolescents-meta-analysis and meta-regression analysis. Psychopharmacology (Berl) 194: 197–209.

Church JA, Schlaggar BL (2014). Pediatric Tourette syndrome: insights from recent neuroimaging studies. J Obsessive Compuls Relat Disord 3: 386–393.

Coghill D, Banaschewski T, Cortese S et al. (2023). The management of ADHD in children and adolescents: bringing evidence to the clinic: perspective from the European ADHD Guidelines Group (EAGG). Eur Child Adolesc Psychiatry 32: 1337–1361.

Cohen SC, Mulqueen JM, Ferracioli-Oda E et al. (2015). Meta-analysis: risk of tics associated with psychostimulant use in randomized, placebo-controlled trials. J Am Acad Child Adolesc Psychiatry 54: 728–736.

Colzato LS, Hommel B, Zhang W et al. (2022). The metacontrol hypothesis as diagnostic framework of OCD and ADHD: a dimensional approach based on shared neurobiological vulnerability. Neurosci Biobehav Rev 137: 104677.

Conelea CA, Wellen B, Woods DW et al. (2018). Patterns and predictors of tic suppressibility in youth with tic disorders. Front Psychiatry 9: 188.

Conners CK, Erhardt D, Sparrow E (1999). CAARS adult ADHD rating scales: a technical manual. North Tonawanda, NY: Multi Health Systems Inc.

Conte G, Valente F, Fioriello F et al. (2020). Rage attacks in Tourette syndrome and chronic tic disorder: a systematic review. Neurosci Biobehav Rev 119: 21–36.

Cortese S (2023). Evidence-based prescribing of medications for ADHD: where are we in 2023? Expert Opin Pharmacother 24: 425–434.

Cortese S, Adamo N, Del Giovane C et al. (2018). Comparative efficacy and tolerability of medications for attention-deficit hyperactivity disorder in children, adolescents, and adults: a systematic review and network meta-analysis. Lancet Psychiatry 5: 727–738.

DGKJP D (2018). Deutsche Gesellschaft für Sozialpädiatrie und Jugendmedizin e.V. Kurzfassung der interdisziplinären evidenz- und konsensbasierten (S3) Leitlinie, Aufmerksamkeitsdefizit- / Hyperaktivitätsstörung (ADHS) im Kindes-, Jugend- und Erwachsenenalter.

Dopfner M, Hautmann C, Gortz-Dorten A et al. (2015). Long-term course of ADHD symptoms from childhood to early adulthood in a community sample. Eur Child Adolesc Psychiatry 24: 665–673.

Drury H, Wilkinson V, Robertson MM et al. (2016). Self-reported emotion regulation in adults with Tourette's syndrome. Psychiatry Res 245: 157–163.

Erenberg G (2005). The relationship between tourette syndrome, attention deficit hyperactivity disorder, and stimulant medication: a critical review. Semin Pediatr Neurol 12: 217–221.

Evans J, Seri S, Cavanna AE (2016). The effects of Gilles de la Tourette syndrome and other chronic tic disorders on quality of life across the lifespan: a systematic review. Eur Child Adolesc Psychiatry 25: 939–948.

Faraone SV, Banaschewski T, Coghill D et al. (2021). The World Federation of ADHD International Consensus Statement: 208 evidence-based conclusions about the disorder. Neurosci Biobehav Rev 128: 789–818.

Forde NJ, Zwiers MP, Naaijen J et al. (2017). Basal ganglia structure in Tourette's disorder and/or attention-deficit/hyperactivity disorder. Mov Disord 32: 601–604.

Freeman RD, Fast DK, Burd L et al. (2000). An international perspective on Tourette syndrome: selected findings from 3,500 individuals in 22 countries. Dev Med Child Neurol 42: 436–447.

Freeman RD, Tourette Syndrome International Database C (2007). Tic disorders and ADHD: answers from a world-wide clinical dataset on Tourette syndrome. Eur Child Adolesc Psychiatry 16: 15–23.

Friedrichs B, Igl W, Larsson H et al. (2012). Coexisting psychiatric problems and stressful life events in adults with symptoms of ADHD – a large Swedish population-based study of twins. J Atten Disord 16: 13–22.

Gadow KD, Sverd J, Sprafkin J et al. (1999). Long-term methylphenidate therapy in children with comorbid attention-deficit hyperactivity disorder and chronic multiple tic disorder. Arch Gen Psychiatry 56: 330–336.

Gadow KD, Sverd J, Nolan EE et al. (2007). Immediate-release methylphenidate for ADHD in children with comorbid chronic multiple tic disorder. J Am Acad Child Adolesc Psychiatry 46: 840–848.

Ganos C, Roessner V, Munchau A (2013). The functional anatomy of Gilles de la Tourette syndrome. Neurosci Biobehav Rev 37: 1050–1062.

Gardner DM, Gerdes AC (2015). A review of peer relationships and friendships in youth with ADHD. J Atten Disord 19: 844–855.

Gilbert D (2006). Treatment of children and adolescents with tics and Tourette syndrome. J Child Neurol 21: 690–700.

Golden GS (1974). Gilles de la Tourette's syndrome following methylphenidate administration. Dev Med Child Neurol 16: 76–78.

Golden GS (1977). The effect of central nervous system stimulants on Tourette syndrome. Ann Neurol 2: 69–70.

Gorman DA, Thompson N, Plessen KJ et al. (2010). Psychosocial outcome and psychiatric comorbidity in older adolescents with Tourette syndrome: controlled study. Br J Psychiatry 197: 36–44.

Grados M, Huselid R, Duque-Serrano L (2018). Transcranial magnetic stimulation in Tourette syndrome: a historical perspective, its current use and the influence of comorbidities in treatment response. Brain Sci 8: 129.

Haddad ADM, Umoh G, Bhatia V et al. (2009). Adults with Tourette's syndrome with and without attention deficit hyperactivity disorder. Acta Psychiat Scand 120: 299–307.

Hagstrom J, Spang KS, Christiansen BM et al. (2019). The puzzle of emotion regulation: development and evaluation of the tangram emotion coding manual for children. Front Psychiatry 10: 723.

Hagstrom J, Spang KS, Vangkilde S et al. (2021). An observational study of emotion regulation in children with Tourette syndrome. J Child Psychol Psychiatry 62: 790–797.

Hienert M, Gryglewski G, Stamenkovic M et al. (2018). Striatal dopaminergic alterations in Tourette's syndrome: a meta-analysis based on 16 PET and SPECT neuroimaging studies. Transl Psychiatry 8: 143.

Hirschtritt ME, Lee PC, Pauls DL et al. (2015). Lifetime prevalence, age of risk, and genetic relationships of comorbid psychiatric disorders in Tourette syndrome. JAMA Psychiatry 72: 325–333.

Huisman-van Dijk HM, Schoot R, Rijkeboer MM et al. (2016). The relationship between tics, OC, ADHD and autism symptoms: a cross- disorder symptom analysis in Gilles de la Tourette syndrome patients and family-members. Psychiatry Res 237: 138–146.

Indrajeet I, Atkinson-Clement C, Worbe Y et al. (2022). Compromised reactive but intact proactive inhibitory motor control in Tourette disorder. Sci Rep 12: 2193.

Isomura K, Sidorchuk A, Sevilla-Cermeno L et al. (2022). Insomnia in Tourette syndrome and chronic tic disorder. Mov Disord 37: 392–400.

Jackson GM, Mueller SC, Hambleton K et al. (2007). Enhanced cognitive control in Tourette syndrome during task uncertainty. Exp Brain Res 182: 357–364.

Jimenez-Jimenez FJ, Alonso-Navarro H, Garcia-Martin E et al. (2020). Sleep disorders in tourette syndrome. Sleep Med Rev 53: 101335.

Kadesjo B, Gillberg C (2000). Tourette's disorder: epidemiology and comorbidity in primary school children. J Am Acad Child Adolesc Psychiatry 39: 548–555.

Kalanithi PS, Zheng W, Kataoka Y et al. (2005). Altered parvalbumin-positive neuron distribution in basal ganglia of individuals with Tourette syndrome. Proc Natl Acad Sci USA 102: 13307–13312.

Karam RG, Breda V, Picon FA et al. (2015). Persistence and remission of ADHD during adulthood: a 7-year clinical follow-up study. Psychol Med 45: 2045–2056.

Kataoka Y, Kalanithi PS, Grantz H et al. (2010). Decreased number of parvalbumin and cholinergic interneurons in the striatum of individuals with Tourette syndrome. J Comp Neurol 518: 277–291.

Kaufman J, Birmaher B, Axelson D et al. (2016). K-SADS-PL DSM-5. Pittsburgh: Western Psychiatric Institute and Clinic.

Keenan L, Sherlock C, Bramham J et al. (2021). Overlapping sleep disturbances in persistent tic disorders and attention-deficit hyperactivity disorder: a systematic review and meta-analysis of polysomnographic findings. Neurosci Biobehav Rev 126: 194–212.

Kessler RC, Adler L, Ames M et al. (2005). The World Health Organization Adult ADHD Self-Report Scale (ASRS): a short screening scale for use in the general population. Psychol Med 35: 245–256.

Kuriyan AB, Pelham WE Jr, Molina BS et al. (2013). Young adult educational and vocational outcomes of children diagnosed with ADHD. J Abnorm Child Psychol 41: 27–41.

Kwon HJ, Lim WS, Lim MH et al. (2011). 1-Hz low frequency repetitive transcranial magnetic stimulation in children with Tourette's syndrome. Neurosci Lett 492: 1–4.

Le K, Liu L, Sun M et al. (2013). Transcranial magnetic stimulation at 1 Hertz improves clinical symptoms in children with Tourette syndrome for at least 6 months. J Clin Neurosci 20: 257–262.

Leckman JF (2002). Tourette's syndrome. Lancet 360: 1577–1586.

Ledbetter M (2005). Atomoxetine use associated with onset of a motor tic. J Child Adolesc Psychopharmacol 15: 331–333.

Lee TS, Lee TD, Lombroso PJ et al. (2004). Atomoxetine and tics in ADHD. J Am Acad Child Adolesc Psychiatry 43: 1068–1069.

Lowe TL, Cohen DJ, Detlor J et al. (1982). Stimulant medications precipitate Tourette's syndrome. JAMA 247: 1729–1731.

Maigaard K, Nejad AB, Andersen KW et al. (2019). A superior ability to suppress fast inappropriate responses in children with Tourette syndrome is further improved by prospect of reward. Neuropsychologia 131: 342–352.

El Malhany N, Gulisano M, Rizzo R et al. (2015). Tourette syndrome and comorbid ADHD: causes and consequences. Eur J Pediatr 174: 279–288.

Malli MA, Forrester-Jones R, Murphy G (2016). Stigma in youth with Tourette's syndrome: a systematic review and synthesis. Eur Child Adolesc Psychiatry 25: 127–139.

Man KKC, Hage A, Banaschewski T et al. (2023). Long-term safety of methylphenidate in children and adolescents with ADHD: 2-year outcomes of the Attention Deficit Hyperactivity Disorder Drugs Use Chronic Effects (ADDUCE) study. Lancet Psychiatry 10: 323–333.

Mancini C, Cardona F, Baglioni V et al. (2018). Inhibition is impaired in children with obsessive-compulsive symptoms but not in those with tics. Mov Disord 33: 950–959.

Martino D, Ganos C, Pringsheim TM (2017). Tourette syndrome and chronic tic disorders: the clinical spectrum beyond tics. Int Rev Neurobiol 134: 1461–1490.

McGrath JJ, Lim CCW, Plana-Ripoll O et al. (2020). Comorbidity within mental disorders: a comprehensive analysis based on 145 990 survey respondents from 27 countries. Epidemiol Psychiatr Sci 29: e153.

McGuire JF, Piacentini J, Brennan EA et al. (2014). A meta-analysis of behavior therapy for Tourette Syndrome. J Psychiatr Res 50: 106–112.

Mohammadi MR, Zarafshan H, Khaleghi A et al. (2021). Prevalence of ADHD and its comorbidities in a population-based sample. J Atten Disord 25: 1058–1067.

Morand-Beaulieu S, Grot S, Lavoie J et al. (2017). The puzzling question of inhibitory control in Tourette syndrome: a meta-analysis. Neurosci Biobehav Rev 80: 240–262.

Morellini L, Ceroni M, Rossi S et al. (2022). Social cognition in adult ADHD: a systematic review. Front Psychol 13: 940445.

Mowlem FD, Rosenqvist MA, Martin J et al. (2019). Sex differences in predicting ADHD clinical diagnosis and pharmacological treatment. Eur Child Adolesc Psychiatry 28: 481–489.

Munchau A, Bloem BR, Thilo KV et al. (2002). Repetitive transcranial magnetic stimulation for Tourette syndrome. Neurology 59: 1789–1791.

Murphy KR, Adler LA (2004). Assessing attention-deficit/hyperactivity disorder in adults: focus on rating scales. J Clin Psychiatry 65: 12–17.

Murphy TK, Fernandez TV, Coffey BJ et al. (2017). Extended-release guanfacine does not show a large effect on tic severity in children with chronic tic disorders. J Child Adolesc Psychopharmacol 27: 762–770.

NICE Guideline (2019). Attention deficit hyperactivity disorder: diagnosis and management. https://www.nice.org.uk/guidance/ng87/chapter/Recommendations#managing-adhd.

Norman LJ, Sudre G, Bouyssi-Kobar M et al. (2021). A longitudinal study of resting-state connectivity and response to psychostimulant treatment in ADHD. Am J Psychiatry 178: 744–751.

OHNF Trust (2021). Methylphenidate, atomoxetine, lisdexamfetamine and dexamfetamine for Attention Deficit Hyperactivity Disorder (ADHD) in adults. Shared Care Protocol. Guideliner Number 796FM.

Openneer TJC, van der Meer D, Marsman JC et al. (2021). Impaired response inhibition during a stop-signal task in children with Tourette syndrome is related to ADHD symptoms: a functional magnetic resonance imaging study. World J Biol Psychiatry 22: 350–361.

Orth M, Kirby R, Richardson MP et al. (2005). Subthreshold rTMS over pre-motor cortex has no effect on tics in patients with Gilles de la Tourette syndrome. Clin Neurophysiol 116: 764–768.

Osland ST, Steeves TD, Pringsheim T (2011). Pharmacological treatment for attention deficit hyperactivity disorder (ADHD) in children with comorbid tic disorders. Cochrane Database Syst Rev 6: CD007990.

Osland ST, Steeves TD, Pringsheim T (2018). Pharmacological treatment for attention deficit hyperactivity disorder (ADHD) in children with comorbid tic disorders. Cochrane Database Syst Rev 6: CD007990.

Parraga HC, Parraga MI, Harris DK (2007). Tic exacerbation and precipitation during atomoxetine treatment in two children with attention-deficit hyperactivity disorder. Int J Psychiatry Med 37: 415–424.

Parraga HC, Parraga KL, Harris DK et al. (2008). Abdominal tics during atomoxetine treatment in a child with ADHD: evaluation and differential diagnosis. CNS Spectr 13: E1.

Perez-Vigil A, Fernandez de la Cruz L, Brander G et al. (2018). Association of Tourette syndrome and chronic tic disorders with objective indicators of educational attainment: a population-based sibling comparison study. JAMA Neurol 75: 1098–1105.

Peterson BS, Staib L, Scahill L et al. (2001). Regional brain and ventricular volumes in Tourette syndrome. Arch Gen Psychiatry 58: 427–440.

Peterson BS, Thomas P, Kane MJ et al. (2003). Basal ganglia volumes in patients with Gilles de la Tourette syndrome. Arch Gen Psychiatry 60: 415–424.

Piacentini J, Woods DW, Scahill L et al. (2010). Behavior therapy for children with Tourette disorder: a randomized controlled trial. JAMA 303: 1929–1937.

Plessen KJ, Royal JM, Peterson BS (2007). Neuroimaging of tic disorders with co-existing attention-deficit/hyperactivity disorder. Eur Child Adolesc Psychiatry 16: 60–70.

Poh W, Payne JM, Gulenc A et al. (2018). Chronic tic disorders in children with ADHD. Arch Dis Child 103: 847–852.

Pollack MA, Cohen NL, Friedhoff AJ (1977). Gilles de la Tourette's syndrome. Familial occurrence and precipitation by methylphenidate therapy. Arch Neurol 34: 630–632.

Poncin Y, Sukhodolsky DG, McGuire J et al. (2007). Drug and non-drug treatments of children with ADHD and tic disorders. Eur Child Adolesc Psychiatry 16: 78–88.

Pringsheim T (2017). Tic severity and treatment in children: the effect of comorbid attention deficit hyperactivity disorder and obsessive compulsive behaviors. Child Psychiatry Hum Dev 48: 960–966.

Pringsheim T, Okun MS, Muller-Vahl K et al. (2019). Practice guideline recommendations summary: treatment of tics in people with Tourette syndrome and chronic tic disorders. Neurology 92: 896–906.

Pringsheim T, Holler-Managan Y, Okun MS et al. (2019). Comprehensive systematic review summary: treatment of tics in people with Tourette syndrome and chronic tic disorders. Neurology 92: 907–915.

Quast LF, Rosenthal LD, Cushman GK et al. (2020). Relations between tic severity, emotion regulation, and social outcomes in youth with Tourette syndrome. Child Psychiatry Hum Dev 51: 366–376.

Reed C, Cortese S, Larsson H et al. (2023). Longitudinal associations between physical health conditions in childhood and attention-deficit/hyperactivity disorder symptoms at age 17 years. J Am Acad Child Adolesc Psychiatry 63: 245–254.

Ricketts EJ, Montalbano GE, Burgess HJ et al. (2022). Sleep and chronotype in adults with persistent tic disorders. J Clin Psychol 78: 1516–1539.

Ricketts EJ, Wolicki SB, Danielson ML et al. (2022). Academic, interpersonal, recreational, and family impairment in children with Tourette syndrome and attention-deficit/hyperactivity disorder. Child Psychiatry Hum Dev 53: 3–15.

Robertson MM (2006). Attention deficit hyperactivity disorder, tics and Tourette's syndrome: the relationship and treatment implications. A commentary. Eur Child Adolesc Psychiatry 15: 1–11.

Roessner V, Robatzek M, Knapp G et al. (2006). First-onset tics in patients with attention-deficit-hyperactivity disorder: impact of stimulants. Dev Med Child Neurol 48: 616–621.

Roessner V, Eichele H, Stern JS et al. (2022). European clinical guidelines for Tourette syndrome and other tic disorders-version 2.0. Part III: pharmacological treatment. Eur Child Adolesc Psychiatry 31: 425–441.

Rothenberger A, Heinrich H (2022). Co-occurrence of tic disorders and attention-deficit/hyperactivity disorder-does it reflect a common neurobiological background? Biomedicines 10: 2950.

Rothenberger A, Roessner V, Banaschewski T et al. (2007). Co-existence of tic disorders and attention-deficit/hyperactivity disorder-recent advances in understanding and treatment. Eur Child Adolesc Psychiatry 16: 1–4.

Sambrani T, Jakubovski E, Muller-Vahl KR (2016). New insights into clinical characteristics of Gilles de la Tourette syndrome: findings in 1032 patients from a Single German Center. Front Neurosci 10: 415.

Samea F, Soluki S, Nejati V et al. (2019). Brain alterations in children/adolescents with ADHD revisited: a neuroimaging meta-analysis of 96 structural and functional studies. Neurosci Biobehav Rev 100: 1–8.

Sanderson C, Verdellen C, Debes N et al. (2022). Addressing co-occurring conditions in behavioural therapy for tic disorders: a review and guideline. Eur Child Adolesc Psychiatry 33: 2111–2127.

Schuller CB, Wagner BJ, Schuller T et al. (2021). Temporal discounting in adolescents and adults with Tourette syndrome. PLoS One 16: e0253620.

Sears J, Patel NC (2008). Development of tics in a thirteen-year-old male following atomoxetine use. CNS Spectr 13: 301–303.

Seo JC, Jon DI, Shim SH et al. (2022). Prevalence and comorbidities of attention deficit hyperactivity disorder among adults and children/adolescents in Korea. Clin Psychopharmacol Neurosci 20: 126–134.

Shaw P, Lerch J, Greenstein D et al. (2006). Longitudinal mapping of cortical thickness and clinical outcome in children and adolescents with attention-deficit/hyperactivity disorder. Arch Gen Psychiatry 63: 540–549.

Sonuga-Barke EJ (2003). The dual pathway model of AD/HD: an elaboration of neuro-developmental characteristics. Neurosci Biobehav Rev 27: 593–604.

Spencer T, Biederman M, Coffey B et al. (1999). The 4-year course of tic disorders in boys with attention-deficit/hyperactivity disorder. Arch Gen Psychiatry 56: 842–847.

Spencer TJ, Biederman J, Faraone S et al. (2001). Impact of tic disorders on ADHD outcome across the life cycle: findings from a large group of adults with and without ADHD. Am J Psychiatry 158: 611–617.

Spencer TJ, Sallee FR, Gilbert DL et al. (2008). Atomoxetine treatment of ADHD in children with comorbid Tourette syndrome. J Atten Disord 11: 470–481.

Steinhausen HC, Novik TS, Baldursson G et al. (2006). Co-existing psychiatric problems in ADHD in the ADORE cohort. Eur Child Adolesc Psychiatry 15: I25–I29.

Steuber ER, McGuire JF (2024). A meta-analysis of transcranial magnetic stimulation in Tourette syndrome. J Psychiatr Res 173: 34–40.

Storch EA, Merlo LJ, Lack C et al. (2007). Quality of life in youth with Tourette's syndrome and chronic tic disorder. J Clin Child Adolesc Psychol 36: 217–227.

Sukhodolsky DG, Scahill L, Zhang H et al. (2003). Disruptive behavior in children with Tourette's syndrome: association with ADHD comorbidity, tic severity, and functional impairment. J Am Acad Child Adolesc Psychiatry 42: 98–105.

Szejko N, Robinson S, Hartmann A et al. (2022). European clinical guidelines for Tourette syndrome and other tic disorders-version 2.0. Part I: assessment. Eur Child Adolesc Psychiatry 31: 383–402.

Taylor A, Deb S, Unwin G (2011). Scales for the identification of adults with attention deficit hyperactivity disorder (ADHD): a systematic review. Res Dev Disabil 32: 924–938.

Tourette's Syndrome Study Group.(2002). Treatment of ADHD in children with tics: a randomized controlled trial. Neurology 58: 527–536.

Ustun B, Adler LA, Rudin C et al. (2017). The World Health Organization Adult Attention-Deficit/Hyperactivity Disorder Self-Report Screening Scale for DSM-5. JAMA Psychiatry 74: 520–527.

Virring A, Lambek R, Thomsen PH et al. (2016). Disturbed sleep in attention-deficit hyperactivity disorder (ADHD) is not a question of psychiatric comorbidity or ADHD presentation. J Sleep Res 25: 333–340.

Wadman R, Glazebrook C, Beer C et al. (2016). Difficulties experienced by young people with Tourette syndrome in secondary school: a mixed methods description of self, parent and staff perspectives. BMC Psychiatry 16: 14.

Wanderer S, Roessner V, Strobel A et al. (2021). WISC-IV performance of children with chronic tic disorder, obsessive-compulsive disorder and attention-deficit/hyperactivity disorder: results from a German clinical study. Child Adolesc Psychiatry Ment Health 15: 44.

Ward MF, Wender PH, Reimherr FW (1993). The Wender Utah Rating Scale: an aid in the retrospective diagnosis of childhood attention deficit hyperactivity disorder. Am J Psychiatry 150: 885–890.

Weisman H, Qureshi IA, Leckman JF et al. (2013). Systematic review: pharmacological treatment of tic disorders – efficacy of antipsychotic and alpha-2 adrenergic agonist agents. Neurosci Biobehav Rev 37: 1162–1171.

Yang Z, Wu H, Lee PH et al. (2021). Investigating shared genetic basis across Tourette syndrome and comorbid neurodevelopmental disorders along the impulsivity-compulsivity spectrum. Biol Psychiatry 90: 317–327.

Zamani L, Shahrivar Z, Alaghband-Rad J et al. (2021). Reliability, criterion and concurrent validity of the Farsi translation of DIVA-5: a semi-structured diagnostic interview for adults With ADHD. J Atten Disord 25: 1666–1675.

Handbook of Clinical Neurology, Vol. 215 (3rd series)
Tic Disorders and Tourette Syndrome
C. Ganos and Y. Worbe, Editors
https://doi.org/10.1016/B978-0-443-13554-5.00014-6

Chapter 9

Explosive outbursts in Tourette syndrome and chronic tic disorders

ERICA GREENBERG[1,*], FLORENTINA LAZAROAIE[2], AND CATHY BUDMAN[2]

[1]*Department of Psychiatry, Massachusetts General Hospital, Harvard Medical School, Boston, MA, United States*

[2]*Department of Psychiatry, Zucker School of Medicine at Hofstra/Northwell Health, Hempstead, NY, United States*

Abstract

Explosive outbursts, often referred to as "rage attacks," affect a significant number of individuals with Tourette syndrome (TS) and chronic tic disorders (CTDs), and are associated with increased disorder-related morbidity and psychosocial impairment. Tic severity, psychiatric comorbidity, psychosocial and other factors appear relevant. This chapter provides an overview of the phenomenology, possible neurobiologic and environmental underpinnings, and evaluation and treatment of explosive outbursts in TS. Existing literature on explosive outbursts, rage attacks, impulsive anger and aggression, and episodic dyscontrol in TS/CTD was reviewed and summarized. Explosive outbursts are common transdiagnostic symptoms and occur with increased frequency in individuals with TS/CTD. The etiology of these symptoms in TS/CTD is still unknown, appears multifactorial, and may reflect factors that are intrinsic to tic disorders, associated with their common neuropsychiatric comorbidities, and/or extrinsic influences. At present, evidence-based treatments for explosive outbursts in TS/CTD are limited. There remains an urgent need to advance the investigation of these disabling symptoms in TS/CTD. Use of consistent terminology and standardized assessment tools for studying and treating explosive outbursts in TS and CTDs are much needed.

BACKGROUND

Tourette syndrome (TS) is a developmental neuropsychiatric condition characterized by the presence of at least two motor and at least one vocal tic, not necessarily concurrently. Tics must have been present for at least one year, and not due to an underlying medical or substance-related cause. Persistent (chronic) motor or vocal tic disorder (CTD) is characterized by the presence of either motor or vocal tics that have been present for at least one year, and is believed to be part of the same clinical spectrum as TS (Scharf et al., 2012; Müller-Vahl et al., 2019; Yu et al., 2019).

It is well established that there is a high prevalence of co-occurring psychiatric conditions in those with TS and CTD (Hirschtritt et al., 2015). Almost 90% of those with TS have co-occurring disorders, with obsessive-compulsive disorder (OCD) and attention-deficit/hyperactivity disorder (ADHD) having the greatest prevalence, about 50% each (Roessner et al., 2007; Eddy et al., 2011; Hirschtritt et al., 2015). A variety of other conditions (e.g., affective disorders, anxiety, body-focused repetitive behaviors, autism spectrum symptoms, learning disorders) also occur at increased rates in TS/CTD, ranging from approximately 10% to 30% (Burd et al., 2005; Canitano and Vivanti, 2007; Hirschtritt et al., 2015; Groth et al., 2017; Rizzo et al., 2017; Greenberg et al., 2018). Additional disruptive behavioral disorders, including oppositional defiant disorder (ODD), conduct disorder (CD), and other impulse control disorders, are highly represented in TS/CTD populations (Frank et al., 2011; Cavanna and Termine, 2012; Hirschtritt et al., 2015; Groth et al., 2017; Rizzo et al., 2017), and individuals with TS/CTD often have difficulty inhibiting reward-seeking and/or impulsive behaviors (Conte et al., 2020).

Impulsive aggression (IA) and problems regulating anger are common transdiagnostic symptoms in those with psychiatric disorders (Carlson et al., 2009; Carlson et al., 2023), and have been reported at increased rates worldwide among populations with TS/CTD (Wand et al., 1993; Budman et al., 2000; Freeman et al., 2000; Kano et al., 2008; Frank et al., 2011). Explosive outbursts, a form of severe, recurrent impulsive aggressive episodes, have been reported to occur significantly more frequently in individuals with TS/CTD compared to a control population (Müller-Vahl et al., 2020). When present, they are often considered the "most challenging and disruptive symptom" in TS/CTD (Dooley et al., 1999; Atkinson-Clement et al., 2020).

*Correspondence to: Erica Greenberg, MD, Massachusetts General Hospital, Department of Psychiatry, 185 Cambridge Street, Suite 2000, Boston, MA 02114, United States. Tel: +1-617-643-2780, Fax: +1-534-429-4390, E-mail: EGREENBERG@partners.org.

This chapter focuses on explosive outbursts (also known as "rage attacks") in TS/CTD. The authors will review current findings relevant to the phenomenology, epidemiology, neurobiology of explosive episodes, and review the association between psychiatric co-occurring conditions and environmental influences on explosive outbursts in TS/CTD. Contemporary and potential future approaches for evaluation and treatment of explosive outbursts in TS/CTD will be discussed.

TRANSDIAGNOSTIC DEFINITIONS AND CONSIDERATIONS

As we proceed to consider explosive outbursts in TS/CTD, it is helpful to review commonly used, clinically relevant, transdiagnostic terms that relate to anger and mood regulation.

Anger is a universal, often adaptive emotion, that is thought to arise from a combination of unpleasant circumstances in conjunction with a complex network of feelings, thoughts and memories, and expressive reactions (Berkowitz, 1990). *Rage* is characterized by excessive and contextually inappropriate anger.

Aggression implies destructive or punitive behavior directed toward other persons or objects (Spielberger et al., 1983). It can be conceptualized as the "behavioral experience of anger" (Blair, 2012) "aimed at the injury of a target" (Berkowitz, 1990). Though anger is not necessary (nor sufficient) for aggression, anger can lead to aggressive behavior, particularly when the intensity of the anger emotion overrides regulatory control mechanisms (Novaco, 2020). Maladaptive aggression occurs in response to minimal or absent provocation, and is commonly abrupt, inappropriately intense, and often excessive in duration (Bambauer and Connor, 2005; Connor et al., 2019). Aggression is commonly divided into "proactive" (e.g., predatory, nonimpulsive) and "reactive" aggression, whereby the behavior is in response to a perceived slight or provocation (Crick and Dodge, 1996; Vitiello and Stoff, 1997; Hubbard et al., 2010; Connor et al., 2019). In general, the presence of proactive aggression has been linked to higher psychopathic traits (Coccaro et al., 2014; Conte et al., 2020), whereas reactive aggression is associated with predicting future internalizing disorders (i.e., anxiety, depression), externalizing conditions, and substance use (Carlson et al., 2023).

Emotion regulation is characterized by the ability to modulate the experience and expression of emotions, and adapt those responses to contextual factors (Gross, 1998). Emotion regulation requires the ability to simultaneously direct one's attention, modify situation-specific cognitive appraisals, and modulate physiologic arousal and behavior (Gross, 2013). When emotion regulation is effective, it helps facilitate positive and inhibit negative emotional states (Fernandez et al., 2016). In contrast, *emotion dysregulation* is characterized by "too much emotion that is expressed too quickly and for too long relative to the triggering event" (Blader et al., 2021). Though emotion dysregulation is commonly used to describe anger dysregulation, the term can apply to any emotion (e.g., fear, sadness, happiness) (Blader et al., 2021). Difficulty with emotion regulation has been observed in a number of neuropsychiatric conditions, including TS/CTD (Compas et al., 2017), and is thought to reflect shared, transdiagnostic underlying neuropathology (Sheppard et al., 1999; Yael et al., 2015).

Frustration is an affective experience that occurs in the setting of "blocked-goal attainment" (Leibenluft et al., 2003). Frustration tolerance is an outcome of normal emotion regulation and is characterized by the developmentally appropriate capacity to handle goal-blocking unexpected events in a flexible manner, including the ability to delay gratification, manage obstacles, or manage unpleasant emotions (Blair, 2010). *Distress intolerance* refers to difficulty coping with uncomfortable emotional states (such as frustration), which may lead to overwhelmed feelings, and escape and avoidance behaviors (Linehan, 2014). Impaired distress tolerance is frequently associated with emotion dysregulation difficulties (Lewin et al., 2011).

Irritability is frequently described in conjunction with anger and emotional dysregulation. While a consensus on the definition of irritability is lacking, it commonly refers to a lower threshold to experience anger in response to perceived frustrations or provocations (Leibenluft and Stoddard, 2013), and a "mood of easy annoyance and touchiness characterized by anger and temper outbursts" (Stringaris, 2011). Irritability can have both tonic (i.e., sustained) and phasic (i.e., intermittent) presentations (McLaughlin et al., 2012; Avenevoli et al., 2015), whereby phasic irritability often refers to explosive outbursts or distinct episodes of irritability (Copeland et al., 2015; Carlson et al., 2023).

Explosive outbursts are common transdiagnostic pathologic behaviors that have been described as abrupt, extreme displays of emotion (most commonly of excessive anger or rage) that appear inappropriate to the triggering event. These occur within several different psychiatric diagnoses, including disruptive behavior disorders, such as ADHD, ODD, CD, and intermittent explosive disorder (IED) (American Psychiatric Association, 2022). Explosive outbursts can also occur in mood disorders, including major depressive disorder, bipolar affective disorder, disrupted mood dysregulation disorder, and in certain personality disorders (Fava et al., 1991; Fernandez and Johnson, 2016; Mesbah et al., 2021). Increased rates of aggressive behaviors and rage attacks have also been reported in those with OCD, posttraumatic stress disorder anxiety disorders, and autism spectrum disorder (ASD) (Painuly et al., 2011; Krebs and Johansen, 2013; Zhang et al., 2019; Palm et al., 2021; Carlson et al., 2023). Such symptoms may also manifest in neurologic conditions, such as Huntington disease, Wilson's disease, traumatic brain injury, encephalitis, and dementia (Lennox and Lennox, 2002; Butler and Zeman, 2005; Roffman et al., 2006; Lane et al., 2011).

Vulnerabilities for experiencing these explosive episodes vary and can include different state and trait risk factors. State-associated factors include current emotions (e.g., fear, anger) and/or sensory overstimulation, whereas trait-associated factors may include chronic impulsivity, poor distress tolerance, cognitive rigidity, and irritable temperament (Copeland et al., 2015; Carlson et al., 2023). Environmental precipitants (e.g., medication side-effects, poor sleep hygiene) and interpersonal triggers (e.g., family conflict, bully victimization) also associate with explosive behaviors (Zinner et al., 2012; Steinberg et al., 2013; Horesh et al., 2018; Gruhn and Compas, 2020; Müller-Vahl et al., 2020; Carlson et al., 2023).

EXPLOSIVE OUTBURSTS AND TOURETTE SYNDROME/CHRONIC TIC DISORDERS

General features

Explosive outbursts in TS are characterized by sudden, excessive, seemingly uncontrollable, developmentally and contextually inappropriate, episodic displays of anger and/or aggression, typically in response to trivial provocation or frustration (Budman et al., 1998; Budman et al., 2003; Müller-Vahl et al., 2019). These outbursts may include verbal and/or physical attacks of other people and/or property (Budman et al., 1998; Budman et al., 2003; Kano et al., 2008). These explosive outbursts usually demonstrate the following characteristics, physiologic antecedents (such as fatigue, hunger, and sensory overstimulation) and psychologic antecedents (including feelings of frustration, not getting one's way, being told one is wrong about something or being told "no," being teased or embarrassed, or experiencing an unexpected change in plans) (Budman et al., 2003). The overwhelming majority of these outbursts occur at home and are usually targeted at the child's mother or caregiver, or at inanimate objects. The outburst itself often feels uncontrollable and is often accompanied by a sense of relief once it is over. Expressions of regret or guilt, while not always immediately forthcoming, are usually expressed by the majority of youth with TS/CTD following such symptoms (Budman et al., 2003; Kano et al., 2008).

In the TS/CTD population, explosive outbursts are typically impulsive and reactive in character, though they are distinguishable from simple impulsivity (Müller-Vahl et al., 2020). Explosive outbursts are also distinguishable from the cluster of behaviors and symptoms associated with premeditated (proactive) aggression, and callous, unemotional traits as observed in CD and certain personality disorders (Budman and Feirman, 2001; Connor et al., 2010; Coccaro et al., 2014). Explosive outbursts and rage symptoms in TS/CTD are often associated with signs of difficulty regulating anger beginning in early childhood, and can often continue into adolescence and adulthood (Frank et al., 2011; Wright et al., 2012).

Prevalence

Explosive outbursts occur relatively frequently in the TS/CTD population and have been reported by 20%–70% of individuals with TS/CTD worldwide (Wand et al., 1993; Freeman et al., 2000; Budman et al., 2003; Kano et al., 2008; Mol Debes et al., 2008; Ghanizadeh and Mosallaei, 2009; Sambrani et al., 2016; Conte et al., 2020). Although these symptoms have been described in both biologic males and females of all ages, they appear to occur more commonly in male youth and usually arise during early childhood (Chen et al., 2013). Explosive outbursts in TS/CTD are associated with higher rates of family stress and conflict, worsened academic and occupational functioning, impaired social relationships, higher rates of psychiatric hospitalization, and higher rates of suicidal thoughts and behaviors (Coffey and Park, 1997; Dooley et al., 1999; Kadesjö and Gillberg, 2000; Budman et al., 2003; Frank et al., 2011; Robertson et al., 2015; Storch et al., 2015; Kumar et al., 2016; Sukhodolsky et al., 2017; Carlson et al., 2023).

A recent systematic review of 32 studies conducted between January 2008 and December 2019 found the prevalence for rage attacks and related constructs (including irritability, emotional dysregulation, aggressive behaviors, IA, etc.) to be 20%–67% (Conte et al., 2020). Two of the larger cross-sectional studies in youth with TS/CTD found that explosive anger and IA occurred in 51% and 58% of youth, respectively (Mol Debes et al., 2008; Sambrani et al., 2016). A relatively large study that evaluated individuals with TS/CTD from the community and compared them with individuals with TS recruited from specialty clinics found that 20% experienced explosive outbursts across both samples (Chen et al., 2013). Separate from explosive outbursts, higher rates of anger, hostility, and nonsecure attachment styles compared to healthy controls have been observed in youth with TS/CTD (Dehning et al., 2014). In one clinical study, 45% of adolescents with TS/CTD endorsed clinically significant aggressive symptoms (Kano et al., 2008). Even when evaluating individuals with "uncomplicated TS" (i.e., those without comorbidities), 40% scored higher on specific anger and hostility subscales compared to age-matched healthy controls (Dehning et al., 2014).

Interestingly, when self-report assessments were utilized in some studies, youth with TS did not endorse greater anger or aggression compared to their counterparts (Cavanna et al., 2015; Benaroya-Milshtein et al., 2019). However, the proxy-reported measures in the study by Cavanna et al. did demonstrate significantly higher anger and externalizing behavior (Cavanna et al., 2015), thus highlighting the impact of heterogeneous methodology when assessing these symptoms.

Ultimately, the actual prevalence of rage attacks in TS/CTD is unknown, at least in part due to the use of multiple different methodologies that have been applied to study what may or may not pertain to the same phenomenon. Explosive outbursts have been characterized in both clinical and community TS samples using a variety of different terms such as "episodic dyscontrol," "anger attacks," "rage attacks," "repetitive anger-generated episodes," and "impulsive aggression." Research in this area has also been stymied by a conspicuous lack of consistency in the measures used to assess explosive outbursts qualitatively and quantitatively. Many studies employed scales that selectively repurposed items from other existing measures of different types of aggression, and/or used scales that were validated in non-TS clinical populations (Carlson et al., 2022).

Measuring explosive outbursts

Few scales are specific for assessing explosive outbursts/rage attacks. One of the more recent scales is the Emotional Outbursts Index (EMO-I), a transdiagnostic screening tool that measures phasic irritability in youth, and assesses severity, frequency, and duration of explosive outbursts (Carlson et al., 2022). Other transdiagnostic scales are the affective reactivity index, a validated seven-question self-report transdiagnostic questionnaire assessing chronic irritability in youth via threshold, frequency, and duration for angry reactions/feelings/behaviors (Stringaris, 2011) and the IA diary, a reliable, validated 15-question IA measurement tool specifically designed for youth aged 6–12 with ADHD (Ceresoli-Borroni et al., 2019).

Early studies of explosive outbursts in TS attempted to characterize its phenomenology using a TS-specific rage attack scale developed from modified IED diagnostic criteria (Budman et al., 1998; Budman et al., 2003). The original Rage Attack Questionnaire was first devised by Budman et al. in attempt to better characterize episodic rage youth with TS (Budman et al., 2003), and more recently has been further revised and validated (Rage Attack Questionnaire – Revised) in a large mixed psychiatric and healthy control sample (Müller-Vahl et al., 2020; Palm et al., 2021). The Rage Outbursts and Anger Rating Scale (ROARS), a three-item clinician-rated scale that assesses frequency, intensity, and duration of anger attacks in the past week was initially devised for evaluating and monitoring rage attacks in both youth and adults with TS/CTD, and more recently has been applied transdiagnostically; the ROARS has demonstrated acceptable internal consistency and validity (Budman et al., 2008; Storch et al., 2012; Wu and Storch, 2015). Consensus on the use of a standard scale would considerably benefit the research and treatment of explosive outbursts in TS/CTD.

Relationship with tic severity

Exploration of the relationship between explosive outbursts in TS/CTD and tic severity is ongoing, with evidence to date generally supporting a positive correlation, although varying criteria, constructs, and assessment tools for these symptoms limit meaningful comparisons among existing studies. A 2020 systematic review of rage attacks in TS/CTD reported that among the 32 clinical studies reviewed, most supported a direct correlation between tic severity and rage attacks (Conte et al., 2020). A study of 220 child and adult subjects from the community in Costa Rica and TS specialty clinics in the United States who were participating in a large genetic study found evidence that the presence of explosive outbursts was associated with both greater severity of tics and an earlier age of tic onset (Chen et al., 2013). A study of 29 Japanese adolescents with TS and rage attacks who were recruited from a specialty clinic also found a positive association between tic severity and rage attacks (Kano et al., 2008). A positive correlation between tic severity and number of diagnosed impulse control disorders (where IED was the most prevalent subtype) has also been reported (Frank et al., 2011). Though not specific to explosive outbursts, other investigators noted a correlation between tic severity and externalizing behaviors including aggression (Pollak et al., 2009), and a positive relationship between (vocal) tic severity and irritability (Cox and Cavanna, 2015).

That said, given the association between increased tic severity and increased burden of co-occurring conditions/ increased overall disease burden (Thériault et al., 2014; Hirschtritt et al., 2015; Martino et al., 2017), the direction and degree of causality between tic severity and explosive outbursts in TS remain elusive. Furthermore, under-recognized and undertreated psychiatric comorbidity in TS/CTD may exacerbate tic severity and/or increase predisposition toward explosive episodes (Himle et al., 2014). The interrelationship between tic severity and psychiatric comorbidity, as well as the differential contributions of specific comorbidities (and/or of pharmacologic treatments) to tic severity and explosive outbursts over the course of development is ongoing and complex.

Clinical correlates

Transdiagnostically, the psychiatric disorders with the greatest effect sizes on explosive episodes appear to be ADHD, borderline personality disorder, and bipolar affective disorder (Palm et al., 2021). ADHD is a developmental neuropsychiatric syndrome with high prevalence in TS/CTD (54%), and is characterized by a persistent pattern of difficulty regulating attention, hyperactivity, and/or impulsivity such that it interferes with one's life. Explosive outbursts and emotion dysregulation are common symptoms in ADHD, though the severity of the outbursts are typically less than those observed in the context of a mood disorder (Mick et al., 2005). ADHD is also associated with "deficient emotion self-regulation" (Spencer et al., 2011; Surman et al., 2011), a concept characterized by quickness to frustration or anger, frequent anger or resentment, emotional over-reactivity, "loss of temper," and low distress tolerance (Barkley, 1997; Surman et al., 2011; Faraone et al., 2019).

Some studies that have attempted to explore the role of psychiatric comorbidity in explosive outbursts and TS have suggested that ADHD may be an independent and primary cause (Mol Debes et al., 2008; Pollak et al., 2009). Other investigations also implicate ADHD (and disruptive behavior disorders in general) to explain TS-related emotion dysregulation, irritability, and aggressive behaviors (Ghanizadeh and Mosallaei, 2009; Drury et al., 2012; Cox and Cavanna, 2015). In a large clinical study comparing those with TS and ADHD to those with TS only, explosive outbursts occurred more often in subjects with TS and ADHD, though both groups showed higher rates of rage episodes compared to control groups (Müller-Vahl et al., 2020).

A clinical investigation of 150 youth comparing those with TS only, ADHD only, TS and ADHD, and neurotypical controls found that deficiencies in emotional regulation in the youth with TS and rage attacks were more strongly associated with co-occurring ADHD than with the presence of tics or tic severity (Hagstrøm et al., 2021). The investigators actually reported that youth with TS-only did not differ in their emotional regulation abilities from neurotypical controls. Rather, the study subjects with ADHD alone and those with TS and ADHD had greater challenges with emotion regulation, and less severe ADHD symptoms were associated with stronger emotion regulation skills (Hagstrøm et al., 2021).

A 2003 study of 207 youth with TS/CTD found that youth with TS-only did not differ from unaffected controls in measures related to aggression, but that those with ADHD-only, and those with TS and ADHD rated similarly to one another and significantly higher for disruptive behaviors than the study subjects with TS without ADHD (Sukhodolsky et al., 2003).

Another recent study examining 140 youth with either TS, ADHD, TS and ADHD, or neurotypical controls found that youth in the ADHD groups (i.e., ADHD alone, TS and ADHD) performed worse on the Go-No Go task (a measure of inhibitory control and possible proxy for explosive episodes) compared to neurotypical controls (Morand-Beaulieu et al., 2022).

A smaller study of 50 youth with TS and aggressive outbursts found that the probability of aggression in TS was increased by co-occurring ADHD (and OCD), and that the severity of ADHD symptoms was a statistically significant predictor of aggression (Benaroya-Milshtein et al., 2019). This particular study found no independent differences in the rates of aggressive symptoms in youth with TS and controls, and no correlation between tic severity and the presence of aggressive behaviors.

A meta-analysis examining inhibitory control in patients with TS and common psychiatric comorbidities found that while there appeared to be a small to medium effect of inhibitory deficits in TS alone, the effect was significantly larger when there was comorbid ADHD (Morand-Beaulieu et al., 2017).

Alternatively, other studies have found a strong association between TS/CTD and explosive outbursts/aggressive phenomena independent from ADHD diagnosis. A clinical case series by Cavanna et al. reported no statistically significant difference in rates or types of aggressive symptoms when comparing subjects with TS and ADHD to those with TS without ADHD (Cavanna et al., 2015). An earlier clinical study found that youth with TS had a greater risk for externalizing behaviors independent of ADHD symptoms (Pollak et al., 2009). Another clinical study indicated that 44% of the subjects with TS and rage attacks did not have co-occurring ADHD (Sambrani et al., 2016), and a larger case-controlled study by Müller-Vahl et al. (2019) showed only "moderate correlation" between ADHD and rage attacks in TS.

Although Atkinson-Clement et al. (2020) found that the presence of explosive outbursts in TS was frequently associated with ADHD, they demonstrated that there were ADHD-independent structural brain changes associated with explosive episodes in TS. As such, certain researchers have proposed that explosive outbursts in TS/CTD may be uniquely and independently related to TS/CTD itself, and even may represent a discrete comorbidity (Sambrani et al., 2016; Müller-Vahl et al., 2019; Conte et al., 2020).

Some evidence suggests that under-recognized and/or untreated OCD may also play a role in explosive outbursts in TS. A higher incidence of explosive outbursts was reported in a study of 88 youth with TS and comorbid OCD (Storch et al., 2012). In this study, rage severity was associated with functional impairment beyond the influence of obsessive-compulsive symptom severity, and was most often explained by the impact of family accommodation (Storch et al., 2012).

A study of Japanese youth noted that aggression and impulsiveness were significantly more frequent in youth with TS and co-occurring OCD than in those with TS-only (Kano et al., 2010). Thériault et al. (2014) reported that in youth with TS and ODD (and explosive outbursts), higher scores in the ODD-irritability subgroup correlated more strongly with obsessive-compulsive behaviors. That said, in contrast, Pollak et al. did not find an association between OCD and "externalizing behaviors" in youth with TS (Pollak et al., 2009).

ASD has also been suggested as a contributing factor of explosive episodes in TS/CTD. In a 2013 study by Pringsheim and Hammer, youth with TS/CTD and co-occurring ADHD and ODD were more likely to have rage episodes when there were co-occurring ASD symptoms compared to when there were not (Pringsheim and Hammer, 2013).

Lastly, a case series study on 12 TS youth with misophonia found that outside of tic exacerbations, explosive outbursts were the most frequent response to misophonic stimuli (Robinson et al., 2018). Future studies should evaluate the role of misophonia on rage attacks in a TS/CTD population (Storch et al., 2023).

Psychosocial and environmental considerations

In recent years, there has been increased interest in the role that family accommodation may play in explosive outbursts (Storch et al., 2017). In a study of 75 youth and their respective parent(s), family accommodation was associated with increased anxiety and depressive symptoms, higher externalizing, rule-breaking, aggressive behaviors, and social problems, and greater tic-related functional impairment over and above the effect of tic severity, anxiety, depression, and externalizing problems (Storch et al., 2017). Nadeau and Hieneman (2018) also reported increased explosive episodes in youth with TS directly related to a pattern of conflict avoidance, lack of parental limit setting, and family accommodation of OCD and tic symptoms.

Additional psychosocial factors have been associated with increased rates of explosive behaviors in TS/CTD as well. A history of being bullied is associated with increased explosive outbursts and internalizing symptoms, greater tic frequency, complexity and severity, and lower quality of life (Zinner et al., 2012). Additional environmental risk factors for explosive episodes in TS/CTD include mismatches in parent–child temperament, a history of trauma, and severe parental conflict or family-related events (Steinberg et al., 2013; Horesh et al., 2018; Gruhn and Compas, 2020).

There appears to be a bidirectional model between stressful life events and explosive episodes in youth with TS/CTD. Horesh et al. found that youth with TS are more vulnerable to stressful life events, and in turn, stressful life events are associated with increased explosive episodes (whereas positive major life events were protective) (Horesh et al., 2018).

It is important to remember that given the elevated heritability of TS and its co-occurring conditions, multiple family members may struggle with similar symptomatology leading to high levels of family expressed emotion and overstimulation. That said, it has been demonstrated that when parents learn to regulate their emotions, it can have a positive impact on the youth's regulation (Lobo and Lunkenheimer, 2020).

The course of explosive outbursts for any particular individual with TS/CTD likely depends on the combination of factors most responsible for driving the episodes. For example, rage attacks that are more associated with underlying disinhibition may demonstrate improvement as the affected individual undergoes brain maturation. In contrast, if an untreated mood disorder and/or a history of/ongoing trauma are contributing to the explosive episodes, these symptoms may be less likely to resolve without intervention.

NEUROBIOLOGY

TS/CTD and its related comorbidities have long been associated with dysregulated fronto-cortico-striatal-thalamo-cortical circuitry (CSTC), which is responsible for regulation of movements, thoughts, attention, and mood, and is implicated in disorders of compulsivity (e.g., OCD) and impulsivity (e.g., ADHD). When disrupted, one commonly sees signs of dysregulated habit formation, dysregulated reward processing and inhibitory control, and trouble with regulating movement, cognition, and affect (Haber and Knutson 2010; Worbe et al., 2010; Ito and Doya, 2011; Jahanshahi et al., 2015; Worbe et al., 2015). The combination and overlapping features of the pathways associated with anger, aggression, impulsivity, and disinhibition may predispose and amplify risk for experiencing recurrent explosive episodes.

The neurocircuitry of aggression is believed to involve processes related to arousal, motivation, impulse control, and sensory perception (Flanigan and Russo, 2019). It has been proposed that reactive aggression (in TS/CTD) may arise secondary to impulse control deficits and inappropriate emotional reactivity to perceived social threats (Rosell and Siever, 2015; Flanigan and Russo, 2019). Studies in aggression have demonstrated a correlation with reduced top-down recruitment of the inferior frontal gyrus, a region that works with prefrontal cortex to support cognitive and emotional control (Wager and Smith, 2003; Buhle et al., 2013). A recent study by Atkinson-Clement additionally demonstrated less efficient top-down control in those with TS and explosive outbursts, and impairments related to emotion regulation, aggressive behaviors, and impulse control (Atkinson-Clement et al., 2020).

Historically, neuroimaging studies have focused on disrupted dopamine modulation in the striatum and its relationship with aggression and reward processing (Anstrom et al., 2009; Ramírez-Bermudez et al., 2010; Augustine and Singer, 2019). A study by Schlüter et al. (2013) in healthy volunteers demonstrated that men with low dopamine synthesis and transmission capacity in the striatum and midbrain were more vulnerable to impulsive aggressive responses when provoked. A dysregulated serotonergic system with abnormal prefrontal cortex (PFC)-amygdala connectivity may also be implicated in explosive outbursts and rage (Witte et al., 2009; Rylands et al., 2012).

Studies of reactive aggression have also demonstrated involvement of limbic prefrontal regions (i.e., orbitofrontal cortex and anterior cingulate cortex), which help to integrate affective, motor, sensory, and cognitive processes (Neuner et al., 2010; Rosell and Siever, 2015; Tinaz et al., 2015; Conte et al., 2020), hypofunctioning of the medial prefrontal cortex, and hyperfunctioning of the amygdala (Coccaro et al., 2014; Hwang et al., 2015; Flanigan and Russo, 2019).

An event-related functional MRI (fMRI) study in adults with TS demonstrated increased amygdala activation in response to angry, fearful, and neutral facial expressions (Neuner et al., 2010). The authors suggested that this finding was likely associated with deficient frontal inhibition of the limbic system, or disruptions in the limbic structures associated with the CSTC network (Neuner et al., 2010). Kalsi et al. (2019) also demonstrated that youth with TS process emotions, particularly anger, differently from neurotypical controls, and that in some youth with TS, emotion regulation plays a role in modulating the observed cognitive and motor deficits.

In a recent study comparing nine patients with TS (without ADHD) and explosive episodes to 21 patients with TS (without ADHD) and without explosive episodes, Atkinson-Clement found structural changes in the right supplementary motor area, the right hippocampus, and the left orbitofrontal cortex. They reported lower connectivity within the sensorimotor cortico-basal network, and altered connectivity among the orbitofrontal cortex, amygdala, and hippocampus (Atkinson-Clement et al., 2020).

In sum, current consensus supports treatment strategies that may improve behavioral inhibition and enhance "top-down" regulation of impulses and cognitions as a way to benefit individuals with TS who are struggling with explosive outbursts (Flanigan and Russo, 2019).

EVALUATION AND TREATMENT

Youth who present for clinical evaluation of tics should be screened at their initial visit and subsequently thereafter at regular intervals for the presence of explosive outbursts, particularly during the critical developmental window (~ages 5–12 years) whereby many of the common co-occurring conditions in TS will have emerged (Budman et al., 2003; Chen et al., 2013; Hirschtritt et al., 2015). Screening youth with established diagnoses of TS/CTD for explosive outbursts and rage symptoms during adolescence is also important, given that other psychiatric comorbidities associated with explosive outbursts (e.g., substance and alcohol misuse, eating disorders) typically arise during this developmental phase (Lichter, 2011; Hirschtritt et al., 2015; Mataix-Cols et al., 2022).

When explosive outbursts are the presenting problem in a youth already diagnosed with TS/CTD, it is necessary to probe further about the onset and history of these symptoms, their frequency and intensity, potential association with pharmacologic changes, and potential environmental triggers and psychosocial stressors, in addition to assessing for current tic and common psychiatric comorbidity symptoms.

It should be noted that parental and/or clinician preoccupation with addressing tic symptoms can at times eclipse the need for prioritized treatment of the psychiatric comorbidities that may be more directly impacting tic severity, explosive outbursts, and reduced quality of life (Cavanna and Rickards, 2013; Rizzo et al., 2014; Eapen et al., 2016; Rizzo et al., 2017; Huisman-van Dijk et al., 2019).

A comprehensive review of current and prior pharmacologic treatments is critical, since many psychotropic medications alone or in combination may cause activation and/or aggression (Luft et al., 2018; Virtanen et al., 2023). Similarly, it is important to critically review current and prior nonpharmacologic treatments to ascertain whether these are evidence-based and appropriate for the particular conditions (e.g., comprehensive behavioral intervention for tics (CBIT), cognitive behavioral therapy/exposure response prevention for OCD).

Treatment strategies for explosive outbursts in TS/CTD require careful systematic identification and prioritization of target symptoms and conditions.

When explosive outbursts and TS/CTD occur in context of an unrecognized and/or untreated primary mood disorder such as major depression or bipolar affective disorder, mood stabilization should be top priority, particularly when enduring periods of aggression and/or suicidal or self-injurious behaviors are a concern. However, when mood symptoms develop as a secondary complication of other unrecognized and untreated psychiatric comorbidities, such as anxiety disorders, OCD, and/or ADHD, it may be prudent to first target the particular psychiatric condition that is most disturbing for that individual with TS/CTD and explosive outbursts. For example, one may ask if the explosive outbursts occur only when exposed to specific anxiety-provoking stimuli? If so, one should focus first on treating the underlying anxiety.

If there is concurrent ADHD, how well are those symptoms controlled? Untreated or undertreated ADHD poses hazards for an array of problems in overall psychosocial functioning: academic underachievement, poor peer and family relationships, accidents and physical injuries, alcohol, and substance misuse, etc. Fear about potentially worsening tics often inhibit clinicians from using the appropriate evidence-based pharmacotherapy for treatment of ADHD despite extensive literature that documents the safety and efficacy of using psychostimulants in youth with comorbid ADHD and tics (Tourette's Syndrome Study Group, 2002; Cohen et al., 2015).

Tic-related OCD symptoms are also extremely important to identify and may not become apparent or disclosed during an initial evaluation. Of note, tic-related OCD symptoms may not be as responsive to usual evidence-based OCD pharmacotherapy (selective serotonin reuptake inhibitors, SSRIs) and may preferentially require high-quality cognitive behavioral therapy (March et al., 2006).

Other important questions to ask include whether there a history of ongoing trauma, physical, sexual, or emotional abuse? Bullying and/or social estrangement? Family conflict, loss, or disruption?

Effective treatment strategies for targeting explosive outbursts in youth with TS/CTD commonly involve a combination of psychoeducation, behavioral treatments, and pharmacologic interventions. Psychoeducation, parent, and teacher training are particularly important given the frequent confusion about which TS/CTD-associated behaviors are intentional and which are more intimately related to symptoms associated with underlying TS/CTD and/or its comorbidities (Ashurova et al., 2021).

There is excellent evidence for behavioral therapy strategies to reduce explosive episodes and aggressive behavior in youth with and without TS/CTD (Sukhodolsky et al., 2003; Scahill et al., 2006; Sukhodolsky et al., 2009; Ashurova et al., 2021). Some of these behavioral treatments target the explosive outbursts directly, and some target these symptoms indirectly by focusing on improving overall emotion regulation. Some treatments are directed toward parents, and others are directed toward the individual with the explosive outbursts.

Regarding parent-directed therapies, Scahill et al. (2006) demonstrated that parent management training in youth with co-occurring tic disorders was comparably effective in reducing disruptive behaviors to that in youth without co-occurring tic disorders. Specht et al. (2018) reported benefit in reducing aggressive behaviors in youth with tic disorders using "brief transdiagnostic parent training," a six-session treatment that strives to help parents shape their child's disruptive behaviors into more prosocial ones.

Sukhodolsky et al. (2009) conducted a clinical trial aimed at reducing aggressive and disruptive behaviors in youth with TS called "Anger Control Training" (ACT). The study enrolled youth with TS and "moderate" disruptive behaviors on the Disruptive Behavior Rating Scale, and provided 10 one-hour long sessions consisting of a combination of anger management strategies, cognitive restructuring, and behavioral interventions. Fifty percent of participants in the ACT group demonstrated reduced aggressive behaviors versus 11% in the treatment as usual group, and this was sustained three months later (Sukhodolsky et al., 2009).

A recent behavioral treatment study in youth with co-occurring CTD and ADHD utilized cognitive reappraisal approaches to reduce emotion dysregulation and found that cognitive reappraisal negated the differences in emotion regulation seen between youth with and without ADHD (Hagstrøm et al., 2020).

There is some evidence that CBIT treatment may indirectly reduce disruptive behaviors in individuals with TS/CTD (Zimmerman-Brenner et al., 2022). Zimmerman-Brenner et al. compared CBIT to an educational intervention for tics. The group that received CBIT demonstrated reduced anxiety, behavioral problems, and aggressive behaviors compared to the education group (Zimmerman-Brenner et al., 2022). The corollary of using CBIT, a tic-specific treatment that may indirectly help disruptive behaviors, is utilizing a transdiagnostic treatment approach. Given the association between tics, comorbidity severity and emotion dysregulation, Ruhrman et al. proposed developing a transdiagnostic treatment that targets emotion dysregulation and distress tolerance (using dialectical behavioral therapy techniques) in individuals with TS/CTD (Ruhrman et al., 2023).

A recent small pilot study of a novel nonpharmacologic intervention for youth with TS/CTD and explosive outbursts, called "Mightier," employed a biofeedback-based video game, which utilizes heart rate control to teach emotion regulation. Preliminary results among the three of four participating subjects revealed a reduction in rage attack severity (Tae et al., 2023).

Despite the efficacy of these nonpharmacologic interventions, the widespread availability of evidence-based behavioral therapies is still lacking (Ashurova et al., 2021). Hence, many cases of explosive outbursts in TS/CTD are ultimately empirically treated using psychotropic medications.

There is a paucity of well-designed pharmacologic treatment trials for explosive outbursts/rage attacks in TS/CTD. Medication studies are predominantly retrospective or open-label, and have involved small numbers of clinically heterogenous TS/CTD subjects (Budman et al., 2008; McCracken et al., 2008; Ashurova et al., 2021). As such, much of the proposed pharmacologic suggestions are extrapolated from transdiagnostic clinical trials on generalized aggression, often occurring in the context of ADHD, ODD, or CD diagnoses.

Pharmacologic classes of medication that have historically been used to target explosive outbursts and aggression transdiagnostically include atypical and typical antipsychotics (i.e., dopamine modulators), α-agonists, stimulants, antiepileptic mood stabilizers, and lithium. Antipsychotic agents tend to be reserved for more severe and/or treatment refractory presentations, given their greater potential side-effect burden. Common adverse effects of typical antipsychotics, for example, include sedation, weight gain, dulling, tiredness, and extrapyramidal symptoms (e.g., acute dystonic reaction, parkinsonism, akathisia, and tardive dyskinesia). While generally less likely to cause extrapyramidal symptoms than typical antipsychotics, atypical antipsychotics are more likely to cause metabolic side-effects and weight gain.

Canadian guidelines published in 2015 for disruptive and aggressive behavior in youth with ADHD, ODD, or CD (independent from TS/CTD diagnosis) gave a strong recommendation in favor of use of stimulants, and a conditional recommendation in favor of use of atomoxetine and α-agonists (Gorman et al., 2015). Gorman et al. (2015) assigned risperidone, a conditional recommendation for its use in aggressive/disruptive behavior – conditional because of its elevated burden of side-effects. Vaudreuil et al. (2021) recently published a comprehensive review on the treatment of aggression in youth, and concluded that risperidone has the best evidence available for reactive aggression. Divalproex also has some evidence for its use as an adjunctive medication to stimulants for explosive outbursts (Blader et al., 2021). Other antiepileptics, including carbamazepine and oxcarbazepine, have either mixed or limited evidence in ameliorating aggression or explosive outbursts (Budman, 2014; Vaudreuil et al., 2021). Lithium does have some evidence in decreasing severe aggression and CD in placebo-controlled studies (Pine and Cohen, 1999).

Based on current guidelines and existing literature, treatment with methylphenidate should be an early consideration for individuals with comorbid ADHD and TS/CTD who present with explosive episodes. Stimulants, particularly methylphenidate, have been shown to reduce disruptive behavior in individuals with co-occurring TS/CTD and ADHD (Connor et al., 2002; Srour et al., 2008; Gadow and Nolan, 2011; Osland et al., 2018). Methylphenidate may even confer additional benefits in reducing oppositional behaviors and aggressive symptoms in TS and ADHD when there is co-occurring ODD as well (Gadow et al., 2008). α-Agonists have similarly demonstrated benefit in reducing irritability and aggression in those with co-occurring TS/CTD and ADHD (Pisano and Masi, 2019).

Given the preponderance of evidence for disrupted dopaminergic systems in TS/CTD (Augustine and Singer, 2019), atypical antipsychotics, such as risperidone, aripiprazole, and olanzapine, have been trialed to target co-occurring rage, aggression, and tics in those with TS/CTD (Conte et al., 2020). Budman et al. (2008) investigated aripiprazole in youth with TS and explosive outbursts and reported significant reductions of both tics and explosive outbursts; however, side-effects with aripiprazole were considerable and led to a 22% discontinuation rate. McCracken et al. (2008) completed a small open-label trial of olanzapine for treatment of disruptive behaviors in youth with TS, which similarly showed a reduction in both tics and disruptive behaviors, but was accompanied by excessive weight gain (i.e., an average of 4 kg over six weeks) limiting its tolerability.

A more recent case series of adolescents with TS and treatment-resistant aggressive behaviors that utilized lurasidone as an add-on to risperidone or aripiprazole reported improved reduction of aggression with limited side-effects (Colizzi et al., 2021).

Given that disturbances of serotonergic systems are also implicated in aggression, the use of SSRIs has also been explored for treatment of explosive outbursts in TS/CTD. An open-label study used paroxetine to treat 45 youth with TS and explosive outbursts and demonstrated reduced outbursts in approximately 75%. However, almost 10% developed worsened symptoms, and one developed hypomania (Bruun and Budman, 1998). In general, serotonergic agents can be extremely helpful for treatment of anxiety and/or OCD when presenting with explosive outbursts and are generally safe and well tolerated (Strawn et al., 2018). However, one should remain vigilant, as the use of SSRIs in individuals with bipolar spectrum disorders (or predisposition to bipolar disorder) can lead to destabilization and an increase in aggressive behaviors (Virtanen et al., 2023).

Novel pharmacologic approaches were explored in small samples of clinically referred adults with TS/CTD and explosive outbursts. In a trial using cannabis, self-reported outcomes showed a reduction of rage outbursts (Abi-Jaoude et al., 2023). A recent study on *Lactobacillus plantarum* PS128, a probiotic that modulates neurotransmitters, demonstrated some (weak) evidence for reducing oppositional/defiant behaviors (Liu et al., 2019) and is currently being trialed in 80 Taiwanese youth with TS and ASD (Wu et al., 2021).

Trials examining the use of deep brain stimulation on self-destructive and aggressive behaviors in TS/CTD are generally lacking. A small open-label case series utilizing deep brain stimulation (DBS) (targeting nucleus accumbens) in six patients (three with TS) with severe self-destructive and aggressive behaviors demonstrated symptom reduction by almost 75% (Harat et al., 2021).

FUTURE DIRECTIONS

Aristotle stated, "Anybody can become angry, that is easy; but to be angry with the right person, and to the right degree, and at the right time, and for the right purpose, and in the right way, that is not within everybody's power, that is not easy."

The above quote speaks to the complex series of neurologic and psychologic actions that need to be successfully anticipated, established, coordinated, and executed to optimize healthy regulation of anger and aggressive impulses. As discussed in this chapter, there are numerous mitigating influences, risk factors, and neuropsychiatric as well as environmental pathologies that can disrupt this process, alone and in combination. Factors that seem to yield the greatest support include tic severity, presence of ADHD and other conditions with impulsive/compulsion dysregulation, overall disease burden, psychosocial and family related factors, and underlying tendency toward poor emotion regulation, increased irritability, and poor distress tolerance (Atkinson-Clement et al., 2020; Budman, 2014; Conte et al., 2020; Müller-Vahl et al., 2020).

Investigating the neurobiologic as well as psychosocial correlates and influences associated with explosive outbursts in TS/CTD over the life cycle remains an important challenge. Standardized terminology, and the use of appropriately validated screening tools and rating scales in both clinical and research settings for this very disabling symptom in TS/CTD are critical to facilitating future collaborative research endeavors. Larger, prospective, longitudinal studies are also necessary to better study this transdiagnostic symptom. As the complex neurobiologic and environmental templates and risk factors for explosive outbursts in TS become better clarified, more meaningful treatment approaches and clinical trials can be developed.

References

Abi-Jaoude E, Bhikram T, Parveen F et al. (2023). A double-blind, randomized, controlled crossover trial of *Cannabis* in adults with Tourette Syndrome. Cannabis Cannabinoid Res 8: 835–845.

American Psychiatric Association (2022). Neurodevelopmental disorders. In: Diagnostic and statistical manual of mental disorders, 5th ed., text rev.

Anstrom KK, Miczek KA, Budygin E et al. (2009). Increased phasic dopamine signaling in the mesolimbic pathway during social defeat in rats. Neuroscience 161: 3–12.

Ashurova M, Budman C, Coffey BJ (2021). Ticked off: anger outbursts and aggressive symptoms in Tourette Disorder. Child Adolesc Psychiatr Clin N Am 30: 361–373.

Atkinson-Clement C, Sofia F, Fernandez-Egea E et al. (2020). Structural and functional abnormalities within sensori-motor and limbic networks underpin intermittent explosive symptoms in Tourette disorder. J Psychiatr Res 125: 1–6.

Augustine F, Singer HS (2019). Merging the pathophysiology and pharmacotherapy of tics. Tremor Other Hyperkinet Mov (N Y) 8: 595.

Avenevoli S, Blader J, Leibenluft E (2015). Irritability in youth: an update. J Am Acad Child Adolesc Psychiatry 54: 881–883.

Bambauer K, Connor D (2005). Characteristics of aggression in clinically referred children. CNS Spectr 10: 709–718.

Barkley RA (1997). Behavioral inhibition, sustained attention, and executive functions: constructing a unifying theory of ADHD. Psychol Bull 121: 65–94.

Benaroya-Milshtein N, Shmuel-Baruch S, Apter A et al. (2019). Aggressive symptoms in children with TIC disorders. Eur Child Adolesc Psychiatry 29: 617–624.

Berkowitz L (1990). On the formation and regulation of anger and aggression: a cognitive-neoassociationistic analysis. Am Psychol 45: 494–503.

Blader J, Pliszka S, Kafantaris V et al. (2021). Stepped treatment for attention-deficit/hyperactivity disorder and aggressive behavior: a randomized, controlled trial of adjunctive risperidone, divalproex sodium, or placebo after stimulant medication optimization. J Am Acad Child Adolesc Psychiatry 60: 236–251.

Blair R (2010). Psychopathy, frustration, and reactive aggression: the role of ventromedial prefrontal cortex. Br J Psychol Gen Sect 101: 383–399.

Blair R (2012). Considering anger from a cognitive neuroscience perspective. Wiley Interdiscip Rev Cogn Sci 3: 65–74.

Bruun RD, Budman CL (1998). Paroxetine treatment of episodic rages associated with Tourette's disorder. J Clin Psychiatry 59: 581–584.

Budman C, Coffey BJ, Shechter R et al. (2008). Aripiprazole in children and adolescents with Tourette disorder with and without explosive outbursts. J Child Adolesc Psychopharmacol 18: 509–515.

Budman C, Feirman L (2001). The relationship of Tourette's syndrome with its psychiatric comorbidities: is there an overlap? Psychiatr Ann 31: 541–548.

Budman C, Rockmore L, Stokes J et al. (2003). Clinical phenomenology of episodic rage in children with Tourette syndrome. J Psychosom Res 55: 59–65.

Budman CL (2014). The role of atypical antipsychotics for treatment of Tourette's syndrome: an overview. Drugs 74: 1177–1193.

Budman CL, Bruun RD, Park KS et al. (1998). Rage attacks in children and adolescents with Tourette's disorder. J Clin Psychiatry 59: 576–580.

Budman CL, Bruun RD, Park KS et al. (2000). Explosive outbursts in children with Tourette's disorder. J Am Acad Child Adolesc Psychiatry 39: 1270–1276.

Buhle JT, Silvers JA, Wager TD et al. (2013). Cognitive reappraisal of emotion: a meta-analysis of human neuroimaging studies. Cereb Cortex 24: 2981–2990.
Burd L, Freeman RD, Klug MG et al. (2005). Tourette Syndrome and learning disabilities. BMC Pediatr 5: 34Published 2005 Sep 1.
Butler C, Zeman A (2005). Neurological syndromes which can be mistaken for psychiatric conditions. J Neurol Neurosurg Psychiatry 76 Suppl 1: i31–38.
Canitano R, Vivanti G (2007). Tics and Tourette syndrome in autism spectrum disorders. Autism 11: 19–28.
Carlson GA, Potegal M, Margulies D et al. (2009). Rages—what are they and who has them? J Child Adolesc Psychopharmacol 19: 281–288.
Carlson GA, Silver J, Klein DN (2022). Psychometric properties of the Emotional Outburst Inventory (EMO-I): rating what children do when they are irritable. J Clin Psychiatry 83: 21m14015.
Carlson GA, Singh MK, Amaya-Jackson L et al. (2023). Narrative review: impairing emotional outbursts: what they are and what we should do about them. J Am Acad Child Adolesc Psychiatry 62: 135–150.
Cavanna AE, Selvini C, Luoni C et al. (2015). Measuring anger expression in young patients with Tourette syndrome. Child Health Care 44: 264–276.
Cavanna AE, Rickards H (2013). The psychopathological spectrum of Gilles de la Tourette syndrome. Neurosci Biobehav Rev 37: 1008–1015.
Cavanna AE, Termine C (2012). Tourette syndrome. Adv Exp Med Biol 724: 375–383.
Ceresoli-Borroni G, Liranso T, Brittain ST et al. (2019). A novel assessment tool for impulsive aggression in children with attention-deficit/hyperactivity disorder. J Child Adolesc Psychopharmacol 29: 592–598.
Chen K, Budman C, Diego Herrera L et al. (2013). Prevalence and clinical correlates of explosive outbursts in Tourette syndrome. Psychiatry Res 205: 269–275.
Coccaro E, Lee R, McCloskey M (2014). Relationship between psychopathy, aggression, anger, impulsivity, and intermittent explosive disorder. Aggress Behav 40: 526–536.
Coffey B, Park S (1997). Behavioral and emotional aspects of Tourette syndrome. Neurol Clin 15: 277–289.
Cohen SC, Mulqueen JM, Ferracioli-Oda E et al. (2015). Meta-analysis: risk of tics associated with psychostimulant use in randomized, placebo-controlled trials. J Am Acad Child Adolesc Psychiatry 54: 728–736.
Colizzi M, Bortoletto R, Zoccante L (2021). The effectiveness of lurasidone add-on for residual aggressive behavior and obsessive symptoms in antipsychotic-treated children and adolescents with Tourette syndrome: preliminary evidence from a case series. Child (Basel) 8: 121.
Compas B, Jaser S, Bettis A et al. (2017). Coping, emotion regulation, and psychopathology in childhood and adolescence: a meta-analysis and narrative review. Psychol Bull 143: 939–991.
Connor D, Chartier K, Preen E et al. (2010). Impulsive aggression in attention-deficit/hyperactivity disorder: symptom severity, co-morbidity, and attention-deficit/hyperactivity disorder subtype. J Child Adolesc Psychopharmacol 20: 119–126.
Connor DF, Glatt SJ, Lopez ID et al. (2002). Psychopharmacology and aggression. I: a meta-analysis of stimulant effects on overt/covert aggression-related behaviors in ADHD. J Am Acad Child Adolesc Psychiatry 41: 253–261.
Connor DF, Newcorn JH, Saylor KE et al. (2019). Maladaptive aggression: with a focus on impulsive aggression in children and adolescents. J Child Adolesc Psychopharmacol 29: 576–591.
Conte G, Valente F, Fioriello F et al. (2020). Rage attacks in Tourette Syndrome and chronic tic disorder: a systematic review. Neurosci Biobehav Rev 119: 21–36.
Copeland W, Brotman M, Costello E (2015). Normative irritability in youth: developmental findings from the Great Smoky Mountains Study. J Am Acad Child Adolesc Psychiatry 54: 635–642.
Cox J, Cavanna A (2015). Irritability symptoms in Gilles de la Tourette syndrome. J Neuropsychiatry Clin Neurosci 27: 42–47.
Crick N, Dodge A (1996). Social information-processing mechanisms in reactive and proactive aggression. Child Dev 67: 993–1002.
Dehning S, Burger MB, Krause D et al. (2014). Tourette syndrome is associated with insecure attachment and higher aggression. Int J Neurosci 125: 521–525.
Dooley J, Brna P, Gordon K (1999). Parent perceptions of symptom severity in Tourette's syndrome. Arch Dis Child 81: 440–441.
Drury S, Theall K, Gleason M et al. (2012). Telomere length and early severe social deprivation: linking early adversity and cellular aging. Mol Psychiatry 17: 719–727.
Eapen V, Cavanna AE, Robertson MM (2016). Comorbidities, social impact, and quality of life in Tourette syndrome. Front Psychiatry 7: 97.
Eddy CM, Rizzo R, Gulisano M et al. (2011). Quality of life in young people with Tourette syndrome: a controlled study. J Neurol 258: 291–301.
Faraone SV, Rostain AL, Blader J et al. (2019). Practitioner review: emotional dysregulation in attention-deficit/hyperactivity disorder – implications for clinical recognition and intervention. J Child Psychol Psychiatry 60: 133–150.
Fava M, Rosenbaum J, McCarthy M et al. (1991). Anger attacks in depressed outpatients and their response to fluoxetine. Psychopharmacol Bull 27: 275–279.
Fernandez E, Johnson L (2016). Anger in psychological disorders: prevalence, presentation, etiology and prognostic implications. Clin Psychol Rev 46: 124–135.
Fernandez K, Jazaieri H, Gross J (2016). Emotion regulation: a transdiagnostic perspective on a new RDoC domain. Cogn Ther Res 40: 426–440.
Flanigan ME, Russo SJ (2019). Recent advances in the study of aggression. Neuropsychopharmacology 44: 241–244.
Frank M, Piedad J, Rickards H et al. (2011). The role of impulse control disorders in Tourette syndrome: an exploratory study. J Neurol Sci 310: 276–278.
Freeman RD, Fast DK, Burd L et al. (2000). An international perspective on Tourette syndrome: selected findings from 3500 individuals in 22 countries. Dev Med Child Neurol 42: 436–447.
Gadow KD, Devincent C, Schneider J (2008). Predictors of psychiatric symptoms in children with an autism spectrum disorder. J Autism Dev Disord 38: 1710–1720.
Gadow KD, Nolan EE (2011). Methylphenidate and comorbid anxiety disorder in children with both chronic multiple tic disorder and ADHD. J Atten Disord 15: 246–256.
Ghanizadeh A, Mosallaei S (2009). Psychiatric disorders and behavioral problems in children and adolescents with Tourette syndrome. Brain Dev 31: 15–19.
Gorman DA, Gardner DM, Murphy AL et al. (2015). Canadian guidelines on pharmacotherapy for disruptive and aggressive behaviour in children and adolescents with attention-deficit

hyperactivity disorder, oppositional defiant disorder, or conduct disorder. Can J Psychiatry 60: 62–76.

Greenberg E, Tung ES, Gauvin C et al. (2018). Prevalence and predictors of hair pulling disorder and excoriation disorder in Tourette syndrome. Eur Child Adolesc Psychiatry 27: 569–579.

Gross JJ (1998). The emerging field of emotion regulation: an integrative review. Rev Gen Psychol 2: 271–299.

Gross JJ (2013). Emotion regulation: taking stock and moving forward. Emotion 13: 359–365.

Groth C, Mol Debes N, Rask CU et al. (2017). Course of Tourette syndrome and comorbidities in a large prospective clinical study. J Am Acad Child Adolesc Psychiatry 56: 304–312.

Gruhn M, Compas B (2020). Effects of maltreatment on coping and emotion regulation in childhood and adolescence: a meta-analytic review. Child Abuse Negl 103: 104446.

Haber SN, Knutson B (2010). The reward circuit: linking primate anatomy and human imaging. Neuropsychopharmacology 35: 4–26.

Hagstrøm J, Maigaard K, Pagsberg AK et al. (2020). Reappraisal is an effective emotion regulation strategy in children with Tourette syndrome and ADHD. J Behav Ther Exp Psychiatry 68: 101541.

Hagstrøm J, Spang KS, Vangkilde S et al. (2021). An observational study of emotion regulation in children with Tourette syndrome. J Child Psychol Psychiatry 62: 790–797.

Harat M, Kiec M, Rudaś M et al. (2021). Treating aggression and self-destructive behaviors by stimulating the nucleus accumbens: a case series. Front Neurol 12: 706166.

Himle M, Capriotti M, Hayes L (2014). Variables associated with tic exacerbation in children with chronic tic disorders. Behav Modif 38: 163–183.

Hirschtritt ME, Lee PC, Pauls DL et al. (2015). Tourette Syndrome Association International Consortium for Genetics. Lifetime prevalence, age of risk, and genetic relationships of comorbid psychiatric disorders in Tourette syndrome. JAMA Psychiatry 72: 325–333.

Horesh N, Shmuel-Baruch S, Farbstein D et al. (2018). Major and minor life events, personality and psychopathology in children with Tourette syndrome. Psychiatry Res 260: 1–9.

Hubbard J, McAuliffe D, Morrow M et al. (2010). Reactive and proactive aggression in childhood and adolescence: precursors, outcomes, processes, experiences, and measurement. J Pers 78: 95–118.

Huisman-van Dijk HM, Matthijssen SJMA, Stockmann RTS et al. (2019). Effects of comorbidity on Tourette's tic severity and quality of life. Acta Neurol Scand 140: 390–398.

Hwang S, White SF, Nolan ZT et al. (2015). Executive attention control and emotional responding in attention-deficit/hyperactivity disorder – a functional MRI study. Neuroimage Clin 9: 545–554.

Ito M, Doya K (2011). Multiple representations and algorithms for reinforcement learning in the cortico-basal ganglia circuit. Curr Opin Neurobiol 21: 368–373.

Jahanshahi M, Obeso I, Rothwell JC et al. (2015). A fronto-striato-subthalamic-pallidal network for goal-directed and habitual inhibition. Nat Rev Neurosci 16: 719–732.

Kadesjö B, Gillberg C (2000). Tourette's disorder: epidemiology and comorbidity in primary school children. J Am Acad Child Adolesc Psychiatry 39: 548–555.

Kalsi N, Tambelli R, Altavilla D et al. (2019). Neurophysiological correlate of emotional regulation in cognitive and motor deficits in Tourette's syndrome. World J Biol Psychiatry 20: 647–661.

Kano Y, Ohta M, Nagai Y et al. (2008). Rage attacks and aggressive symptoms in Japanese adolescents with Tourette syndrome. CNS Spectr 13: 325–332.

Kano Y, Ohta M, Nagai Y et al. (2010). Association between Tourette syndrome and comorbidities in Japan. Brain Dev 32: 201–207.

Krebs T, Johansen P (2013). Psychedelics and mental health: a population study. PLoS One 8: e63972.

Kumar A, Trescher W, Byler D (2016). Tourette syndrome and comorbid neuropsychiatric conditions. Curr Dev Disord Rep 3: 217–221.

Lane S, Kjome K, Moeller F (2011). Neuropsychiatry of aggression. Neurol Clin 29: 49–64, vii.

Leibenluft E, Charney DS, Towbin KE et al. (2003). Defining clinical phenotypes of juvenile mania. Am J Psychiatry 160: 430–437.

Leibenluft E, Stoddard J (2013). The developmental psychopathology of irritability. Dev Psychopathol 25: 1473–1487.

Lennox BR, Lennox GG (2002). Mind and movement: the neuropsychiatry of movement disorders. J Neurol Neurosurg Psychiatry 72 Suppl 1: I28–I31.

Lewin A, Peris T, Lindsey Bergman R et al. (2011). The role of treatment expectancy in youth receiving exposure-based CBT for obsessive compulsive disorder. Behav Res Ther 49: 536–543.

Lichter DG (2011). Substance use disorder in adult Tourette syndrome: frequency, profile, and comorbidities. Mind Brain 2.

Linehan M (2014). DBT? Skills training manual. Guilford Publications.

Liu L, Chen W, Vitoratou S et al. (2019). Is emotional lability distinct from "Angry/Irritable Mood," "Negative Affect," or other subdimensions of oppositional defiant disorder in children with ADHD? J Atten Disord 23: 859–868.

Lobo F, Lunkenheimer E (2020). Understanding the parent-child coregulation patterns shaping child self-regulation. Dev Psychol 56: 1121–1134.

Luft MJ, Lamy M, DelBello MP et al. (2018). Antidepressant-induced activation in children and adolescents: risk, recognition and management. Curr Probl Pediatr Adolesc Health Care 48: 50–62.

March JS, Franklin ME, Leonard H et al. (2006). Tics moderate treatment outcome with sertraline but not cognitive-behavior therapy in pediatric obsessive-compulsive disorder. Biol Psychiatry 61: 344–347.

Martino D, Ganos C, Pringsheim TM (2017). Tourette syndrome and chronic tic disorders: the clinical spectrum beyond tics. Int Rev Neurobiol 134: 1461–1490.

Mataix-Cols D, Andersson E, Aspvall K et al. (2022). Operational definitions of treatment response and remission in obsessive-compulsive disorder capture meaningful improvements in everyday life. Psychother Psychosom 91: 424–430.

McCracken JT, Suddath R, Chang S et al. (2008). Effectiveness and tolerability of open label olanzapine in children and adolescents with Tourette syndrome. J Child Adolesc Psychopharmacol 18: 501–508.

McLaughlin K, Green J, Hwang I et al. (2012). Intermittent explosive disorder in the National Comorbidity Survey Replication Adolescent Supplement. Arch Gen Psychiatry 69: 1131–1139.

Mesbah R, de Bles N, Ruis-Ottenheim N et al. (2021). Anger and cluster B personality traits and the conversion from unipolar depression to bipolar disorder. Depress Anxiety 38: 671–681.

Mick E, Spencer T, Wozniak J et al. (2005). Heterogeneity of irritability in attention-deficit/hyperactivity disorder subjects with and without mood disorders. Biol Psychiatry 58: 576–582.

Mol Debes N, Hjalgrim H, Skov L et al. (2008). Validation of the presence of comorbidities in a Danish clinical cohort of children with Tourette syndrome. J Child Neurol 23: 1017–1027.

Morand-Beaulieu S, Grot S, Lavoie J et al. (2017). The puzzling question of inhibitory control in Tourette syndrome: a meta-analysis. Neurosci Biobehav Rev 80: 240–262.

Morand-Beaulieu S, Smith S, Ibrahim K (2022). Electrophysiological signatures of inhibitory control in children with Tourette syndrome and attention-deficit/hyperactivity disorder. Cortex 147: 157–168.

Müller-Vahl K, Sambrani T, Jakubovski E (2019). Tic disorders revisited: introduction of the term "tic spectrum disorders. Eur Child Adolesc Psychiatry1129–1135.

Müller-Vahl KR, Kayser L, Pisarenko A et al. (2020). The Rage Attack Questionnaire-Revised (RAQ-R): assessing rage attacks in adults with Tourette syndrome. Front Psychiatry 10: 956Published 2020 Jan 28.

Nadeau JM, Hieneman M (2018). Managing avoidance and accommodation of tics and related behaviors. In: The clinician's guide to treatment and management of youth with Tourette syndrome and tic disorders. Academic Press, pp.177–200.

Neuner I, Kellermann T, Stöcker T et al. (2010). Amygdala hypersensitivity in response to emotional faces in Tourette's patients. World J Biol Psychiatry 11: 858–872.

Novaco R (2020). Anger. Encyclopedia of personality and individual differences. Springer, pp. 166–171.

Osland ST, Steeves TD, Pringsheim T (2018). Pharmacological treatment for attention deficit hyperactivity disorder (ADHD). in children with comorbid tic disorders. Cochrane Database Syst Rev 6: CD007990.

Painuly N, Grover S, Mattoo S et al. (2011). Anger attacks in obsessive compulsive disorder. Ind Psychiatry J 20: 115–119.

Palm L, Haas M, Pisarenko A et al. (2021). Validation of the Rage Attack Questionnaire-Revised (RAQ-R) in a mixed psychiatric population. Front Psychiatry 12: 724802.

Pine DS, Cohen E (1999). Therapeutics of aggression in children. Paediatr Drugs 1: 183–196.

Pisano S, Masi G (2019). Recommendations for the pharmacological management of irritability and aggression in conduct disorder patients. Expert Opin Pharmacother 21: 5–7.

Pollak Y, Benarroch F, Kanengisser L et al. (2009). Tourette syndrome-associated psychopathology: roles of comorbid attention-deficit hyperactivity disorder and obsessive-compulsive disorder. J Dev Behav Pediatr 30: 413–419.

Pringsheim T, Hammer T (2013). Social behavior and comorbidity in children with tics. Pediatr Neurol 49: 406–410.

Ramírez-Bermudez J, Perez-Neri I, Montes S et al. (2010). Imbalance between nitric oxide and dopamine may underly aggression in acute neurological patients. Neurochem Res 35: 1659–1665.

Rizzo R, Gulisano M, Martino D et al. (2017). Gilles de la Tourette syndrome, depression, depressive illness, and correlates in a child and adolescent population. J Child Adolesc Psychopharmacol 27: 243–249.

Rizzo R, Gulisano M, Pellico A et al. (2014). Tourette syndrome and comorbid conditions: a spectrum of different severities and complexities. J Child Neurol 29: 1383–1389.

Robertson M, Cavanna A, Eapen V (2015). Gilles de la Tourette syndrome and disruptive behavior disorders: prevalence, associations, and explanation of the relationships. J Neuropsychiatry Clin Neurosci 27: 33–41.

Robinson S, Hedderly T, Conte G et al. (2018). Misophonia in children with tic disorders: a case series. J Dev Behav Pediatr 39: 516–522.

Roessner V, Becker A, Banascheski T et al. (2007). Developmental psychopathology of children and adolescents with Tourette syndrome – impact of ADHD. Eur Child Adolesc Psychiatry 16: 24–35.

Roffman J, Eisenberg T, Stern T (2006). Neuropsychiatric phenomena associated with movement disorders. Prim Care Companion J Clin Psychiatry 8: 39–42.

Rosell DR, Siever LJ (2015). The neurobiology of aggression and violence. CNS Spectr 20: 254–279.

Ruhrman D, Mikulincer M, Apter A (2023). Emotion regulation and tic disorders in children. Eur Child Adolesc Psychiatry 32: 893–902.

Rylands AJ, Hinz R, Jones M et al. (2012). Pre- and postsynaptic serotonergic differences in males with extreme levels of impulsive aggression without callous unemotional traits: a positron emission tomography study using (11)C-DASB and (11)C-MDL100907. Biol Psychiatry 72: 1004–1011.

Sambrani T, Jakubovski E, Müller-Vahl K (2016). New insights into clinical characteristics of Gilles de la Tourette Syndrome: findings in 1032 patients from a single German Center. Front Neurosci 10: 415.

Scahill L, Sukhodolsky DG, Bearss K et al. (2006). Randomized trial of parent management training in children with tic disorders and disruptive behavior. J Child Neurol 21: 650–656.

Scharf JM, Miller LL, Mathews CA et al. (2012). Prevalence of Tourette syndrome and chronic tics in the population-based Avon longitudinal study of parents and children cohort. J Am Acad Child Adolesc Psychiatry 51: 192–201.e5.

Schlüter T, Winz O, Henkel K et al. (2013). The impact of dopamine on aggression: an [18F]-FDOPA PET Study in healthy males. J Neurosci 33: 16889–16896.

Sheppard D, Bradshaw J, Purcell R et al. (1999). Tourette's and comorbid syndromes obsessive compulsive and attention deficit hyperactivity disorder. A common etiology? Clin Psychol Rev 19: 531–552.

Specht M, Edwards KR, Perry-Parrish C et al. (2018). Brief transdiagnostic parent training. In: The clinician's guide to treatment and management of youth with Tourette syndrome and tic disorders. Elsevier, pp. 225–253.

Spencer TJ, Faraone SV, Surman CB et al. (2011). Toward defining deficient emotional self-regulation in children with attention-deficit/hyperactivity disorder using the Child Behavior Checklist: a controlled study. Postgrad Med 123: 50–59.

Spielberger CD, Gorsuch RL, Lushene R et al. (1983). Manual for the State-Trait Anxiety Inventory. Palo Alto, CA: Consulting Psychologists Press.

Srour M, Lespérance P, Richer F et al. (2008). Psychopharmacology of tic disorders. J Can Acad Child Adolesc Psychiatry 17: 150–159.

Steinberg T, Shmuel-Baruch S, Horesh N et al. (2013). Life events and Tourette syndrome. Compr Psychiatry 54: 467–473.

Storch EA, Guzick AG, D'Souza J et al. (2023). Family accommodation in children and adolescents with misophonia. Behav Ther 55: 595–604.

Storch EA, Hanks C, Mink J et al. (2015). Suicidal thoughts and behaviors in children and adolescents with chronic tic disorders. Depress Anxiety 32: 744–753.

Storch EA, Johnco C, McGuire JF et al. (2017). An initial study of family accommodation in children and adolescents with chronic tic disorders. Eur Child Adolesc Psychiatry 26: 99–109.

Storch EA, Jones AM, Lack CW et al. (2012). Rage attacks in pediatric obsessive-compulsive disorder: phenomenology and clinical correlates. J Am Acad Child Adolesc Psychiatry 51: 582–592.

Strawn JR, Geracioti L, Rajdev N et al. (2018). Pharmacotherapy for generalized anxiety disorder in adult and pediatric patients: an evidence-based treatment review. Expert Opin Pharmacother 19: 1057–1070.

Stringaris A (2011). Irritability in children and adolescents: a challenge for DSM-5. Eur Child Adolesc Psychiatry 20: 61–66.

Sukhodolsky D, Woods D, Piacentini J et al. (2017). Moderators and predictors of response to behavior therapy for tics in Tourette syndrome. Neurology 88: 1029–1036.

Sukhodolsky DG, Scahill L, Zhang H et al. (2003). Disruptive behavior in children with Tourette's syndrome: association with ADHD comorbidity, tic severity, and functional impairment. J Am Acad Child Adolesc Psychiatry 42: 98–105.

Sukhodolsky DG, Vitulano LA, Carroll DH et al. (2009). Randomized trial of anger control training for adolescents with Tourette's syndrome and disruptive behavior. J Am Acad Child Adolesc Psychiatry 48: 413–421.

Surman CB, Biederman J, Spencer T et al. (2011). Deficient emotional self-regulation and adult attention deficit hyperactivity disorder: a family risk analysis. Am J Psychiatr 168: 617–623.

Tae M., Myers S., Kahn J., et al. (2023) Biofeedback-based video game device to treat rage attacks in youth with Tourette syndrome. In: Abstract from 75th American Academy of Neurology annual conference.

Thériault M, Lespérance P, Achim A et al. (2014). Odd irritability is associated with obsessive–compulsive behavior and not ADHD in chronic tic disorders. Psychiatry Res 220: 447–452.

Tinaz S, Malone P, Hallett M et al. (2015). Role of the right dorsal anterior insula in the urge to tic in Tourette syndrome. Mov Disord 30: 1190–1197.

Tourette's Syndrome Study Group(2002). Treatment of ADHD in children with tics: a randomized controlled trial. Neurology 58: 527–536.

Vaudreuil C, Farrell A, Wozniak J (2021). Psychopharmacology of treating explosive behavior. Child Adolesc Psychiatr Clin N Am 30: 537–560.

Virtanen S, Lagerberg T, Takami Lageborn C et al. (2023). Antidepressant use and risk of manic episodes in children and adolescents with unipolar depression. JAMA Psychiatry 81: 25–33.

Vitiello B, Stoff D (1997). Subtypes of aggression and their relevance to child psychiatry. J Am Acad Child Adolesc Psychiatry 36: 307–315.

Wager TD, Smith EE (2003). Neuroimaging studies of working memory: a meta-analysis. Cogn Affect Behav Neurosci 3: 255–274.

Wand R, Matazow G, Shady G et al. (1993). Tourette syndrome: associated symptoms and most disabling features. Neurosci Biobehav Rev 17: 271–275.

Witte AV, Flöel A, Stein P et al. (2009). Aggression is related to frontal serotonin-1A receptor distribution as revealed by PET in healthy subjects. Hum Brain Mapp 30: 2558–2570.

Worbe Y, Gerardin E, Hartmann A et al. (2010). Distinct structural changes underpin clinical phenotypes in patients with Gilles de la Tourette syndrome. Brain 133: 3649–3660.

Worbe Y, Marrakchi-Kacem L, Lecomte S et al. (2015). Altered structural connectivity of cortico-striato-pallido-thalamic networks in Gilles de la Tourette syndrome. Brain 138: 472–482.

Wright A, Thomas K, Hopwood C et al. (2012). The hierarchical structure of DSM-5 pathological personality traits. J Abnorm Psychol 121: 951–957.

Wu CC, Wong LC, Hsu CJ et al. (2021). Randomized controlled trial of probiotic PS128 in children with Tourette syndrome. Nutrients 13: 3698. 21.

Wu M, Storch E (2015). Rage outbursts in adult obsessive-compulsive disorder. Ann Clin Psychiatry 27: 304–305.

Yael D, Vinner E, Bar-Gad I (2015). Pathophysiology of tic disorders. Mov Disord 30: 1171–1178.

Yu D, Sul JH, Tsetsos F et al. (2019). Interrogating the genetic determinants of Tourette's syndrome and other tic disorders through genome-wide association studies. Am J Psychiatry 176: 217–227.

Zhang YY, Gong HF, Zhang XL et al. (2019). Incidence and clinical correlates of anger attacks in Chinese patients with obsessive-compulsive disorder. J Zhejiang Univ Sci B 20: 363–370.

Zimmerman-Brenner S, Pilowsky-Peleg T, Rachamim L et al. (2022). Group behavioral interventions for tics and comorbid symptoms in children with chronic tic disorders. Eur Child Adolesc Psychiatry 31: 637–648.

Zinner SH, Conelea CA, Glew GM et al. (2012). Peer victimization in youth with Tourette syndrome and other chronic tic disorders. Child Psychiatry Hum Dev 43: 124–136.

Handbook of Clinical Neurology, Vol. 215 (3rd series)
Tic Disorders and Tourette Syndrome
C. Ganos and Y. Worbe, Editors
https://doi.org/10.1016/B978-0-443-13554-5.00006-7

Chapter 10

Sleep in tic disorders and Tourette syndrome

GLORIA P. MINGOLLA[1], ILARIA A. DI VICO[2], AND ELENA ANTELMI[2,3,*]

[1]*Department of Neurosciences, Biomedicine and Movement Sciences, University of Verona, Verona, Italy*

[2]*Neurology Unit, Parkinson Disease and Movement Disorders Division, AOUI, Verona, Italy*

[3]*DIMI, Department of Engineering and Medicine of Innovation, University of Verona, Verona, Italy*

Abstract

Sleep disturbances are highly prevalent in patients with tic disorders and Tourette syndrome, with up to 80% of patients experiencing sleep difficulties. Insomnia is notably present, and pediatric patients often experience parasomnias, including sleepwalking and night terrors. Disruptions in circadian rhythms are also reported, with a pronounced evening chronotype. Concurrently, patients may also suffer from sleep-related movement disorders, and tics are reported to often persist during sleep, further contributing to significant sleep fragmentation. The presence of comorbid attention-deficit hyperactivity disorder may further exacerbate sleep disturbances. Current genetic and neurobiologic evidence suggests shared pathophysiologic mechanisms, particularly involving the dopaminergic system, that underlie both tic manifestations and sleep disruptions. While pharmacologic and nonpharmacologic interventions offer potential therapeutic avenues, their efficacy in a combined tic-sleep disorder context requires further investigation.

INTRODUCTION

In his seminal work, in 1884, the French Neurologist Gilles de la Tourette described nine patients who presented with multiple tics, coprolalia, echolalia, and echopraxia, marking the first systematic description of Tourette syndrome (TS). He also reported the case of one patient who exhibited violent behaviors in a state of confusion upon sudden awakening from profound sleep, pinpointing to the possible co-occurrence of sleep disorders in TS patients (Gilles de la Tourette, 1885; Lajonchere et al., 1996; Jiménez-Jiménez et al., 2022).

Over the years, subsequent research has confirmed the frequent coexistence of sleep disorders with TS (Jiménez-Jiménez et al., 2020, 2022). Indeed, sleep disorders may pose a significant concern, as they may affect up to 80% of individuals with chronic tic disorders (CTDs) and TS (Jiménez-Jiménez et al., 2020). Older age and female sex increase the risk for sleep disturbances, with females exhibiting heightened vulnerability to disrupted sleep and reduced sleep sufficiency compared to their male counterparts within the TS spectrum (Sambrani et al., 2016; Groth et al., 2017). The interplay between sleep disruption and clinical factors commonly associated with TS is intricate. Stimulant medications prescribed for attention-deficit hyperactivity disorder (ADHD), as well as tic-specific medications, may disrupt sleep patterns (Ghosh et al., 2014). Conversely, certain medications, such as clonidine, have shown potential in enhancing sleep quality in neurodevelopmental disorders, although their efficacy in TS necessitates further exploration (Bruni et al., 2019). Comorbid ADHD or anxiety may also often exacerbate sleep disturbances, highlighting the need for comprehensive assessments and tailored interventions (Martino et al., 2017).

Insomnia rates are higher in individuals with TS compared to the general population (Isomura et al., 2022), and parasomnias, including sleepwalking and night terrors, are also frequently reported, especially among children (Ghosh et al., 2014). Additionally, patients with CTD often experience sleep-related movement disorders, such as restless legs syndrome (RLS), periodic limb movement disorder (PLMD), and nocturnal bruxism (Jiménez-Jiménez et al., 2020, 2022). Moreover, tics have been reported to occur during sleep in most TS patients, significantly impacting sleep structure and efficiency (Glaze et al., 1983; Jankovic and Rohaidy, 1987; Drake et al., 1992; Cohrs et al., 2001; Ricketts et al., 2022b).

Measurements from wrist-worn devices have shown increased sleep onset latency, reduced sleep efficiency, and heightened sleep fragmentation in patients with TS compared to

*Correspondence to: Elena Antelmi, MD, PhD, Neurology Unit, Parkinson Disease and Movement Disorders Division, DIMI, Department of Engineering and Medicine of Innovation, University of Verona, Piazzale L.A. Scuro 10, 37134, Verona, Italy. E-mail: elenaantelmi@gmail.com.

controls (Ricketts et al., 2022b). Polysomnographic (PSG) studies further corroborated these findings by demonstrating increased night-time awakenings, increased movements during sleep, and diminished sleep duration and efficiency (Jiménez-Jiménez et al., 2020).

The investigation of sleep microstructure in TS has provided inconsistent findings. Some studies suggest a reduction in slow-wave sleep (SWS; nonrapid eye movement (NREM) sleep N3) in TS, contrasting with others reporting an opposite pattern (Mendelson et al., 1980; Glaze et al., 1983; Cohrs et al., 2001). Additionally, discrepancies in rapid eye movement (REM) sleep have been noted. Observations include both decreased and increased REM sleep (Drake et al., 1992; Kirov et al., 2017). Methodological differences, including the use of psychotropic medications, psychiatric comorbidities, and differences in the age of study participants, can partly account for the variability in findings.

It is noteworthy that tics tend to persist across all sleep stages, albeit with reduced frequency, intensity, and complexity (Jiménez-Jiménez et al., 2020). Some studies have reported a correlation between tic severity and sleep disruption, postulating that heightened tic severity might disrupt the settling process crucial for optimal sleep onset and maintenance and, vice versa, loss of sleep could lead to increased severity of tics. Moreover, sleep disturbances can also significantly impact psychiatric comorbidities. Indeed, poor sleep is linked to emotional deregulation and impulse modulation difficulties, and it can itself manifest as hyperactivity, inattention, and temper outbursts, often resembling ADHD (Stephens et al., 2013). Moreover, the role of sleep loss-induced stress and anxiety in waking tic exacerbations should also be examined to understand the complex interplay between comorbid symptomatology, sleep, and tics (Ricketts et al., 2022b).

Most studies available to date have primarily focused on investigating sleep-related disturbances within the context of TS rather than isolated chronic motor or phonic tic disorder (CTD). This differential prevalence of sleep issues in TS may be attributed to its pronounced tic severity, early tic onset, familial predisposition, and psychiatric comorbidities compared to isolated CTD (Claudio-Campos et al., 2021). Additionally, dopaminergic pathways, central to both movement and arousal regulation, are suggested to be more severely affected as tic severity increases (Baym et al., 2008), potentially elucidating the heightened susceptibility to sleep disruption in TS patients.

This bidirectional relationship underscores the plausible involvement of neural processes rendering TS patients susceptible to both tics and sleep disruption (Kirov et al., 2014; Isomura et al., 2022). Understanding this common ground for TS and sleep patterns within TS could significantly advance our comprehension of the underlying pathophysiology of TS, potentially paving the way for targeted interventions addressing both disorders.

In the following sections, we will dissect the intersection of sleep disturbances with CTD/TS, examining how these conditions may exacerbate or be influenced by changes in sleep architecture and quality. We will begin by discussing the various sleep disorders commonly associated with CTD/TS, following the third edition of the International Classification of Sleep Disorders (American Academy of Sleep Medicine, 2014), which categorizes sleep disorders into several major groups: insomnia, circadian rhythm sleep-wake disorders (CRSWDs), central disorders of hypersomnolence, parasomnias, sleep-related movement disorders, and sleep-related breathing disorders. We will then examine possible mechanisms underlying this relationship, with a concluding discussion on the available pharmacologic and nonpharmacologic treatment options.

SLEEP DISTURBANCES IN TIC DISORDERS AND TOURETTE SYNDROME

Sleep is a fundamental physiologic process that plays a critical role in cognitive processes, emotional regulation, and overall physical health (Goldstein and Walker, 2014; Zielinski et al., 2016). It is characterized by a biphasic cycle, alternating between NREM and REM sleep. NREM sleep is further segmented into three distinct stages: N1, N2, and N3 (SWS), with each subsequent stage transitioning to a deeper level of sleep. NREM and REM sleep serve complementary functions in the sleep process, including cellular repair, synaptic regulation, brain plasticity, and memory consolidation (Vyazovskiy and Delogu, 2014; Miletínová and Bušková, 2021).

The architecture of sleep undergoes significant transformations across the human lifespan, reflecting developmental and aging processes (Garbarino, 2020). Infants exhibit a sleep pattern dominated by REM sleep, crucial for brain development. As individuals transition into adulthood, a stable pattern emerges, with NREM sleep constituting approximately 75% of sleep. The elderly exhibit a significant decrease in the N3 stage of NREM sleep and alterations in sleep quality and duration.

Disturbances in sleep can profoundly affect the quality of life of individuals and significantly impact cognitive, emotional, and physiologic domains (O'Brien, 2009; Goldstein and Walker, 2014). When such disturbances manifest in the context of neurologic disorders, such as TS, they become integral components of a complex interplay involving neurobiologic mechanisms, clinical symptoms, and treatments.

INSOMNIA

Insomnia is a sleep disorder characterized by difficulty initiating or maintaining sleep, that occurs despite sufficient opportunities and suitable conditions for sleep (Sateia, 2014). Insomnia can manifest as difficulty falling asleep, frequent awakenings during the night, early morning awakenings, or nonrestorative sleep. These symptoms lead to a variety of consequences, including decreased daytime functioning, impaired cognitive performance, mood disturbances, and increased risk of accidents (Léger et al., 2002, 2006; Riemann, 2007; Daley et al., 2009; Ustinov et al., 2010). Behavioral issues such as oppositional behaviors and hyperactivity are often observed in childhood insomnia (Gregory et al., 2004; Calhoun et al., 2017). Approximately 10% of adults experience insomnia (Morin and Jarrin, 2022), while in general pediatric populations, its prevalence is around 20% (Camhi et al., 2000; Calhoun et al., 2014; Honaker and Meltzer, 2014).

Insomnia has been reported to be more frequent in patients with CTD/TS than in the general population. A recent large population cohort study (Isomura et al., 2022) conducted in Sweden evaluated the prevalence of insomnia in individuals with CTD/TS and found that patients affected with CTD/TS were almost seven times more likely to suffer from insomnia (32.2%) than those without (13.7%). The study also revealed that ADHD and the use of ADHD medication further increased the risk of insomnia in individuals with CTD/TS. This finding was consistent with a previous study (Ghosh et al., 2014), which reported a higher prevalence of both sleep initiation and maintenance difficulties among TS patients with comorbid ADHD, compared to those with TS alone.

Both subjective and objective outcomes have confirmed the higher prevalence of insomnia in individuals with CTD/TS. Standardized questionnaires based on parental reports have revealed that pediatric patients with CTD/TS encountered greater difficulties initiating and maintaining sleep, which were positively correlated with tic severity, relative to typically developing children (Romano et al., 2004). Additionally, these patients exhibited lower sleep duration, an increased prevalence of hypnagogic jerks, higher levels of anxiety or fear when falling asleep, and a greater need for environmental aids in their bedroom to facilitate sleep when compared to their typically developing counterparts, with these symptoms more commonly reported among patients not undergoing medication treatment (Modafferi et al., 2016). Notably, TS patients with more severe symptoms were more likely to exhibit higher tendencies toward bedtime resistance, nocturnal awakenings, and anxiety related to sleep (Mi et al., 2022). Similarly, tic severity in TS children showed a weak positive correlation with sleep quality measures based on parental reports (Pringsheim et al., 2021).

These subjective outcomes collectively align with the objective findings obtained through PSG and actigraphy assessments. Ranging from children to adults, CTD/TS patients exhibited significantly more disrupted sleep compared to controls. Specifically, patients showed decreased sleep efficiency, prolonged sleep onset and SWS latencies, and increased wakefulness, with further disruption marked by higher movement frequencies, as compared to healthy controls (Cohrs et al., 2001; Kostanecka-Endress et al., 2003; Kirov et al., 2007, 2017). These sleep impairments were found to correlate with the severity of daytime tic-related symptoms (Cohrs et al., 2001). The sleep-stage distribution also varied in CTD/TS patients, showing increased wake after sleep onset, reduced SWS percentage, and more microarousals during REM sleep (Cohrs et al., 2001; Kostanecka-Endress et al., 2003; Kirov et al., 2007). Inconsistencies were observed regarding the percentage of REM sleep, as some research indicated an increase in CTD/TS patients (Kirov et al., 2007, 2017), while other investigations reported no significant difference relative to controls (Cohrs et al., 2001; Kostanecka-Endress et al., 2003). Finally, a recent study using actigraphy revealed that TS patients exhibited significantly prolonged sleep onset latency, increased sleep disruption, and a marked decrease in overall sleep efficiency when compared to healthy controls (Ricketts et al., 2022a).

Hyperarousal, characterized by increased cognitive and physiologic activation, is a common underlying mechanism in both CTD/TS and insomnia. In CTD/TS, hyperarousal leads to increased internal and external sensory responsiveness, intensifying tic symptoms (Cohen and Leckman, 1992; Woods et al., 2005). In insomnia, cognitive hyperarousal, manifested as intrusive thoughts and worry, combined with physiologic changes like increased heart rate and body temperature, contribute to sleep initiation and maintenance difficulties (Bonnet and Arand, 2010; Riemann et al., 2010).

CIRCADIAN RHYTHM SLEEP-WAKE DISORDERS

CRSWDs delineate a class of sleep disorders in which individuals experience persistent or recurring patterns of sleep disruption, primarily resulting from a desynchronization between the internal circadian rhythm and the external environment or social obligations (American Academy of Sleep Medicine, 2014).

Ghosh et al. (2014) described the occurrence of CRSWDs in 10% (3 out of 31) of patients with TS alone, and in 4% (2 out of 48) of patients with both TS and ADHD. Two more recent studies investigated the association between chronotype and CTD/TS in adults (Ricketts et al., 2022a, 2022b). Using actigraphy, a stronger preference for evening activities and schedules for late bedtime (evening circadian preference or eveningness) was observed in TS patients compared to healthy controls, whereas the endogenous circadian phase (as assessed by Dim Light Melatonin Onset) did not differ between the two groups (Ricketts et al., 2022a). Interestingly, higher tic severity in TS patients correlated with an increased eveningness preference, as indicated by a lower score in the Morningness-Eveningness Questionnaire (Ricketts et al., 2022b). With increasing age, a shift toward morningness was observed across both TS and CTD groups (Ricketts et al., 2022b), consistent with general population trends (Druiven et al., 2021).

Potential shared mechanisms between the dopaminergic system and the suprachiasmatic nucleus (SCN) (Mendoza and Challet, 2014; Kant et al., 2021), which regulates the circadian function (Moore et al., 2002), may underlie the tic-circadian rhythms association. In rat models of TS, dopamine expression within the striatum – suggested to be dysregulated in TS (Hienert et al., 2018) – also shows a circadian rhythm (Castaneda et al., 2004), which in turn is modulated by the SCN via intermediary pathways (Mendoza and Challet, 2014). Given the interplay between dopamine and the SCN, any dysregulation in the dopaminergic system associated with TS could potentially impact both circadian function and clock gene expression (Imbesi et al., 2009; Mesgar et al., 2022).

CENTRAL DISORDERS OF HYPERSOMNOLENCE

Hypersomnia disorders represent a group of sleep-related conditions wherein individuals experience persistent excessive daytime sleepiness (EDS) and extended sleep episodes. Central disorders of hypersomnolence include narcolepsy type 1, characterized by EDS with sleep onset in REM sleep periods (SOREMPs), the presence of cataplexy and reduced levels of hypocretin in the cerebrospinal fluid; narcolepsy type 2, marked by EDS and SOREMPs but without cataplexy; idiopathic hypersomnia, which is characterized by sleep inertia, EDS, and prolonged nocturnal sleep; and Kleine–Levin syndrome, a rare disorder distinguished by recurrent episodes of excessive sleepiness and altered behavior or cognition. Other specified hypersomnolence disorders encompass conditions with hypersomnia-like symptoms that do not precisely match the criteria for central disorders (American Academy of Sleep Medicine, 2014).

In a study conducted by Ghosh et al. (2014), narcolepsy was observed in a modest subset of patients diagnosed with either TS alone (6%) or TS comorbid with ADHD (4%). Thompson et al. (2004) found that among 70 TS patients, one with ADHD comorbidity was identified to carry a mutation in the orexin-2/hypocretin-2 receptor gene, pointing to a possible bridge with hypocretin deficiency which is the biologic hallmark of narcolepsy type 1. A comprehensive cohort study (Lee et al., 2017) reported only a marginally elevated incidence of hypersomnia in newly diagnosed TS patients (0.3%) compared to controls (0.1%). Similarly, Ghosh et al. (2014) observed that both TS and TS+ADHD groups exhibited a relatively low prevalence (ranging from 2% to 6%) of conditions such as primary hypersomnia, primary hypersomnia with concurrent sleep apnea, and hypersomnia due to medication.

Using validated sleep questionnaires, several studies revealed a recurrent trend of hypersomnolence-related symptoms in CTD/TS patients, particularly pronounced in those with comorbid ADHD. This latter group reported significantly higher scores in both "daytime drowsiness" (33.3%) and "irresistible daytime sleep" (23%) items compared to controls (3%), while such difference was not observable in the TS-only group (10.7%) (Allen et al., 1992). Another study found comparable rates of EDS across TS-only (52%) and TS+ADHD (52%) cohorts (Ghosh et al., 2014). Modafferi et al. (2016) further highlighted an increased frequency of "daytime somnolence" (19.4%) and "falling asleep at school" (5.6%) in patients with CTD/TS compared to healthy controls (5.6% and 0.4%, respectively), especially in those exhibiting greater tic severity. Lastly, a recent case-control study conducted on 271 children with CTD/TS and 271 non-CTD/TS children found that CTD/TS patients were 1.72 times more likely to have daytime sleepiness compared to controls (Mi et al., 2022).

Data, however, are still insufficient to confidently corroborate an established relationship between CTD/TS and hypersomnolence, and this is further complicated by the common coexistence of tics with ADHD. ADHD indeed is one of the most frequent misdiagnoses in patients with narcolepsy, and pediatric patients with narcolepsy might have comorbid hyperkinetic movement disorders as well (Plazzi et al., 2011). However, medical literature suggests some pathophysiologic links between hypersomnia and CTD/TS. Orexin, also known as hypocretin, plays a pivotal role in regulating arousal, wakefulness, and the sleep-wake cycle. The signature biomarker of narcolepsy, specifically narcolepsy type 1, is the deficiency of this neurotransmitter (Chemelli et al., 1999; Scammell, 2015; Branch et al., 2016). Given the existing evidence pointing to an association between narcolepsy and ADHD (Modestino and Winchester, 2013; Lecendreux et al., 2015) due to overlapping neurobiologic dysregulations of dopamine and hypocretin/orexin signaling pathways (Kim et al., 2020), it might be expected to observe increased symptoms of hypersomnolence in TS patients codiagnosed with ADHD.

Moreover, the histaminergic system, another crucial component in maintaining wakefulness and arousal, might also be involved in TS with the identification of a mutation of the histidine decarboxylase in a single family (Ercan-Sencicek et al., 2010), a finding further confirmed in a larger European cohort (Karagiannidis et al., 2013).

PARASOMNIAS

Parasomnias refer to distressing experiential or unwanted behavioral phenomena that occur during sleep initiation, throughout any phase of NREM or REM sleep, or during sleep–wake transitions. These events include motor activities or disruptive behaviors such as sleepwalking and sleep terrors (NREM parasomnias), as well as nightmares and REM sleep behavior disorder (RBD) (REM parasomnias) (American Academy of Sleep Medicine, 2014).

A recent case-control study (Mi et al., 2022) assessed sleep patterns and disturbances using the Children Sleep Habits Questionnaire in children with TS and controls. The study reported that children with CTD/TS (mean age 8.47 years) had a significantly higher prevalence of parasomnias (24%) compared to those without CTD/TS (7.7%). Notably, the severity and specific type of TS, as well as the presence of comorbid ADHD, were associated with an increased risk of these sleep disturbances.

Sleepwalking, also known as somnambulism, is characterized by ambulation or other complex activities typically performed during NREM sleep stages. A recurrent pattern of somnambulism has been highlighted in TS patients across varied age groups, although most of the available studies are primarily dated and reliant on self-report measures (Barabas et al., 1983; Comings and Comings, 1987; Wand et al., 1993; Młodzikowska-Albrecht et al., 2010). These studies consistently demonstrated a significantly higher occurrence of sleepwalking episodes in TS patients, ranging from 17% to 41.5%, compared to healthy controls or other clinical groups with learning disability or seizure disorders (1.7%–12.8%).

Furthermore, somnambulism appeared to be more prevalent in individuals with TS who had comorbid ADHD (Allen et al., 1992), as well as among younger TS patients (Wand et al., 1993). This latter finding is consistent with the evidence coming from the general population in which sleepwalking predominantly occurs in children rather than adults (Petit et al., 2015).

Sleep terrors consist of abrupt and intense screaming episodes that occur during SWS. Similar to sleepwalking, individuals experiencing night terrors usually have no recollection of the event once they awaken. Studies have shown higher rates of sleep terrors in TS patients, with percentages ranging from 15.8% (Barabas et al., 1984) to 48.9% (Comings and Comings, 1987), significantly higher compared to those observed in individuals with learning disabilities (1.7%) or seizure disorders (3.8%) and the control group (21.2%), respectively. Modafferi et al. (2016) did not observe a significant difference in the prevalence of sleep terrors between children with CTD/TS and healthy controls. However, within the same study, the authors reported a higher frequency of "waking up screaming in the night" episodes in the CTD/TS group relative to the control group (13.9% vs 4.9%).

Nightmares are intense, distressing dreams occurring during REM sleep. They involve intense emotions, often including fear or anxiety, and a sense of impending threat. Nightmares typically lead to full or partial awakening, with the individual recalling vivid details of the dream upon awakening. Dream disturbances were significantly more prevalent in children with CTD/TS in comparison to control groups (Allen et al., 1992; Modafferi et al., 2016). The presence of comorbid ADHD further exacerbated the occurrence of this phenomenon in TS patients (Allen et al., 1992).

RBD is a sleep condition characterized by the absence of muscle atonia during REM sleep, resulting in potentially aggressive or violent dream enactment. Although less common in children compared to adults, RBD in childhood can occur, particularly in cases of narcolepsy, specific neurodevelopmental disorders, or the use of medications that enhance muscle tone during REM sleep (Nevsimalova et al., 2007; Lloyd et al., 2012). To date, a single-case study by Trajanovic et al. (2004) reported the presence of RBD in a seven-year-old child with TS.

The pathophysiologic mechanisms underlying parasomnias in CTD/TS are currently unknown. However, it can be hypothesized that the presence of movements combined with alterations in neurotransmitter systems, as observed in CTD/TS patients, may contribute to the development of parasomnias. Indeed, persistent tics can occur during both REM and NREM sleep (Jankovic and Rohaidy, 1987; Cohrs et al., 2001), disrupting sleep continuity and potentially leading to parasomnias, which are frequently triggered by external or internal events that disrupt sleep continuity (Howell, 2012).

SLEEP-RELATED MOVEMENT DISORDERS

Sleep-related movement disorders are neurologic disorders that can be categorized into several distinct subtypes, including RLS, PLMD, sleep-related leg cramps, sleep-related bruxism, and sleep-related rhythmic movement disorder. These disorders share the common feature of involuntary movements during sleep, leading to disruption of normal sleep architecture and subsequent daytime impairment (American Academy of Sleep Medicine, 2014).

RLS is a neurologic disorder characterized by an irresistible urge to move the legs, usually accompanied by uncomfortable sensations. There are sporadic case-reports of co-occurrence of TS and RLS (Lespérance et al., 2004), but a study on larger cohort did not confirm this association, that is, Walters et al. (2003) found a 1.7% frequency of TS among 118 adult patients diagnosed with RLS, which was comparable to the expected frequency in the general population. However, a genetic study (Rivière et al., 2009) reported a significant association between specific genetic markers in the BTBD9 gene (related to both RLS and periodic limb movements (PLMS) risk) and TS, particularly in individuals without concurrent obsessive-compulsive disorder and regardless of ADHD comorbidity. Of note, both tics and RLS share a compulsion to move, which is triggered by interoceptive input (Antelmi et al., 2022).

PLMD is characterized by repetitive, stereotyped limb movements during sleep, typically involving the lower extremities. These movements are usually rhythmic, lasting from 0.5 to 5 seconds, and are associated with brief arousals or microarousals, resulting in sleep fragmentation. Parental reports indicated a higher prevalence of movements during sleep in children with CTD/TS compared to controls, particularly in those untreated or with comorbid ADHD (Ghosh et al., 2014; Modafferi et al., 2016). In a PSG study (Voderholzer et al., 1997), a significant increase in PLMS, both with and without arousal, was observed in a majority (5 out of 7) of adult TS patients compared to controls, with a subset also displaying reduced sleep efficiency. Moreover, Stephens et al. (2013) found a higher frequency of PLMS in TS patients with comorbid ADHD compared to those with TS only. Other PSG findings on TS revealed increased motor activity during both REM and NREM sleep (Cohrs et al., 2001; Stephens et al., 2013), as well as higher number of short motor-related arousals (Kostanecka-Endress et al., 2003; Kirov et al., 2007).

Sleep-related bruxism is a condition defined by the involuntary clenching or grinding of teeth during sleep. Two studies (Młodzikowska-Albrecht et al., 2010; Modafferi et al., 2016) reported a significantly higher prevalence of bruxism in children and adolescents with CTD/TS, ranging from 19.4% to 35%, compared to those observed in age-matched controls (4.1% to 15%, respectively).

However, above all, the most disturbing movement during sleep in this population is the persistence of tic movements during sleep (Table 10.1). A recent investigation (Ricketts et al., 2022b) highlighted the common occurrence and impact of tics during sleep, revealing that around 21% of CTD/TS patients experienced tics during sleep, which was significantly associated with sleep disturbances in 56.1% of them. Similarly, an earlier study (Jankovic and Rohaidy, 1987) found the presence of motor or vocal tics during sleep in 32% of TS patients who

Table 10.1

Studies on tic persistence during sleep and main polysomnographic outcomes.

Study	Subjects	Method(s)	Tics/movements during sleep (% of sample)[a]	Sleep stage of tic occurrence	Sleep-related parameters
Glaze et al. (1983)	TS=14 HC=11	PSG and video monitoring	Motor tics (NS)	All stages	↑ WASO ↓ REM sleep % ↑ NREM sleep %
Jankovic and Rohaidy (1987)	TS=34	Interview and PSG	Motor (68%) and vocal (12%) tics	All stages	↓ REM sleep %
Silvestri et al. (1990)	TS=4	PSG	Movements (NS)	All stages, prominent in stage 1 and 2	NS/NA
Drake et al. (1992)	TS=20 HC=7	EEG	Tics (NS)	NS/NA	↓ SE ↑ WASO ↑ Stage 1% ↓ Stage 2% ↓ SWS % ↓ REM sleep % ↑ REM latency (min)
Silvestri et al. (1995)	TS=9	PSG	Movements and vocalizations (NS)	All stages, prominent in stage 1 and 2	↓ TST ↓ SE ↑ WASO ↑ REM latency (min) ↓ SWS latency (min)
Cohrs et al. (2001)	TS=25 HC=11	PSG	Tics (100%)	All stages	↓ SE ↑ SOL (min) ↑ WASO ↑ Stage 1% ↓ SWS %
Stephens et al. (2013)	TS=20 TS+ADHD=21 ADHD=33 HC=16	PSG	Movements in ADHD and TS+ADHD groups (NS)	REM and NREM sleep	↑PLMI in TS+ADHD
Modafferi et al. (2016)	TD=36	Sleep Disturbance Questionnaire	Movements (22%)	NA	NA
Ricketts et al. (2022b)	CTD/TS=125	Online survey	Tics (21%)	NA	NA

[a]Not all studies specified the difference between tics and general movements during sleep; thus, the reported data are reflective of each study's individual assessment and terminology. ↑ indicates an increase; ↓ indicates a decrease; *ADHD*, attention-deficit hyperactivity disorder; *CTD*, chronic tic disorder; *EEG*, electroencephalography; *HC*, healthy controls; *NA*, not assessed; *NREM*, nonrapid eye movement; *NS*, not specified; *PLMI*, periodic leg movement index; *PSG*, polysomnography; *REM*, rapid eye movement; *SE*, sleep efficiency; *SOL*, sleep onset latency; *SWS*, slow-wave sleep; *TD*, tic disorders; *TS*, Tourette syndrome; *TST*, total sleep time; *WASO*, wake after sleep onset.

reported sleep-related complaints. PSG data from the same study revealed motor tics during all sleep stages in 23 of 34 TS patients (68%), while Cohrs et al. (2001) observed tics in all examined patients (100%), with higher tic index and intensity during REM compared to NREM sleep, despite a consistent tic complexity across both stages. Nevertheless, while informative, available studies in the literature do not consistently distinguish between tics and other movements during sleep (Table 10.1). The lack of standardized criteria for tic identification in PSG recordings complicates the interpretation of such movements as true tic manifestations, especially since motor activities during sleep can span a spectrum from physiologic body movements during sleep to sleep-related movement disorders.

As aforementioned, there is documented evidence supporting a genetic component underlying RLS and PLMS in patients with CTD/TS. Moreover, iron deficiency has been suggested as a common mechanism in RLS, TS, and ADHD (Cortese et al., 2008). Studies have indicated significantly lower ferritin and serum iron levels in TS patients (Peterson et al., 1994; Gorman et al., 2006), which could affect dopamine metabolism and neurotransmitter function (Erikson et al., 2000, 2001; Bianco et al., 2008). This hypothesis aligns with consistent reports of dopaminergic system dysfunctions in all three disorders (Arnsten, 2006; Cervenka et al., 2006; Hienert et al., 2018), although this association has not been extensively validated in subsequent larger-scale studies, indicating a need for further research in this field.

SLEEP-RELATED BREATHING DISORDERS

Sleep-related breathing disorders are a group of conditions characterized by abnormal respiratory patterns during sleep. Studies addressing these disorders in CTD/TS are limited, and current evidence are controversial.

A recent study reported a significantly higher frequency of general sleep disordered breathing in CTD/TS children compared to non-CTD/TS children based on parental reports (Mi et al., 2022). Among the most prevalent sleep breathing disorders, obstructive sleep apnea (OSA) is defined by recurrent episodes of upper airway obstruction during sleep that cause interrupted breathing patterns and sleep fragmentation (Sankri-Tarbichi, 2012; Franklin and Lindberg, 2015). In their study, Sverd and Montero (1993) observed the occurrence of OSA and sudden infant death in children from families affected by TS, suggesting that carriers of TS genes may have an increased risk of experiencing life-threatening apnea during infancy. Contrastingly, more recent investigations have reported no significant difference between CTD/TS and controls in sleep apnea prevalence (Ghosh et al., 2014; Modafferi et al., 2016; Lee et al., 2017). It is worth noting, however, that none of these studies relied on PSG, the gold standard for sleep apnea diagnosis. Only one earlier study confirmed the presence of sleep apnea in approximately 23.5% of TS patients evaluated with PSG (Jankovic and Rohaidy, 1987). Future research should prioritize the use of PSG to yield more definitive and standardized, comparable data. This could elucidate whether CTD/TS directly impacts respiratory function during sleep and, if so, how this translates into significant clinical risk for these patients.

PATHOPHYSIOLOGIC MECHANISMS

The pathogenesis of sleep disturbances in TS is intricate, involving a complex interplay of neurobiologic mechanisms. The frequent coexistence of sleep disorders and CTD/TS, some phenomenological commonalities and overlapping neurotransmitter systems involved suggest shared mechanisms among these disorders (Ricketts et al., 2022b).

Current research emphasizes the importance of striatal microcircuits and excessive postsynaptic dopamine in tic generation in TS (Hienert et al., 2018; Maia and Conceição, 2018). Dopamine is essential in regulating movement, reward-related behavior, and habit formation (Arias-Carrión and Pŏppel, 2007). Critically, it also influences circadian rhythms. Indeed, striatal dopamine exhibits a rhythmic pattern, with a peak in the morning and a nadir in the late evening and night, reaching peak levels during periods of heightened activity (Ji et al., 2016). This rhythm is modulated by the daily variations in the function of the dopamine transporter and the transcription of monoamine oxidase A, both of which are under the control of the SCN, that is the circadian master clock (Hampp et al., 2008; Ferris et al., 2014; Khaldy et al., 2002). The SCN, the principal circadian pacemaker in mammals, receives direct dopamine input from the midbrain, enabling this neurotransmitter to influence the effects of light on the circadian cycle through the release of molecules such as melatonin and cortisol. Evidence points to a direct dopamine input originating in the midbrain and reaching the SCN (Grippo et al., 2017). Notably, dopamine has been found to play a crucial role in regulating the expression of clock genes within the SNC and in the striatum, with D2/D3 receptor agonists and D1 receptor agonists demonstrating distinct effects on clock genes (Imbesi et al., 2009). This highlights the potential role of dopaminergic dysfunction in influencing circadian disruption and disorders that exhibit atypical circadian patterns such as RLS or circadian rhythm disorders (Antelmi et al., 2022). In TS, dopaminergic dysregulation is considered a key pathophysiologic feature (Maia and Conceição, 2018). Clinically, individuals with TS often exhibit a greater evening chronotype, indicating a delay in their rest-activity preferences (Ricketts et al., 2022b). While the exact role of circadian disruption in CTD/TS remains under investigation, studies suggest that the timing of peak tic occurrence might influence circadian rhythm entrainment (Ricketts et al., 2022b).

Importantly, one animal study shed light on the mechanisms underlying tic manifestation during sleep in tic disorders. In a rat model of CTD/TS, infusion of the gamma-aminobutyric acid type A ($GABA_A$) antagonist bicuculline into the motor regions of the striatum induced tic-like behaviors (Harduf et al., 2021). Tic frequency and intensity significantly decreased during sleep compared to wakefulness, hinting that neural mechanisms during sleep might override the typical striatal effects associated with tics. Moreover, during wakefulness, each tic occurrence was associated with a single local field potential (LFP) striatal spike. However, during the transition from wake to sleep and during sleep itself, this association weakened, and although LFP spikes persisted, their association with tic expression disappeared, with a concurrent reduction in maximal striatal neural activity. These findings suggest that while LFP spikes are strongly linked to tic expression during wakefulness, their ongoing activity during sleep does not correspond to tic expression. Thus, the reduction in maximal striatal neural activity during sleep may play a crucial role in blocking or significantly reducing tic expression, offering promise for a potential mechanism for tic inhibition/modulation. Interestingly, dopaminergic levels are also decreased during night sleep (Eban-Rothschild et al., 2016; Dong et al., 2019), further emphasizing the intricate role of dopamine in tic expression and inhibition.

Additionally, iron deficiency has emerged as a common factor relevant to sleep, and potentially also tic pathophysiology and psychiatric comorbidities (Cortese et al., 2008; Kanaan et al., 2023). The iron deficiency metabolic theory posits that brain iron deficiency (BID) contributes to neurochemical alterations, including increased presynaptic glutamatergic function and dopamine dysregulation, which may underlie sensory-motor dysfunction seen in disorders like TS or alterations in sleep structure. Notably, iron supplementation has been suggested in one study to offer some benefits in managing tic severity, providing a possible therapeutic avenue in TS and related RLS (Ghosh and Burkman, 2017). TS patients with high frequency of tics, have shown lower ferritin levels, indicating a potential link between iron deficiency and tic severity (Ghosh and Burkman, 2017). Moreover, BID plays a central role in the pathophysiology of RLS and related symptoms (Allen and Earley, 2007). BID is associated with amplified neuropathic pain, hyperarousal, and insomnia, all of which significantly affect sleep quality (Ferré et al., 2018). It is also known that iron deficiency alters dopamine release and function, and affects both presynaptic and postsynaptic receptors, thereby contributing to central sensory-motor dysfunction observed in RLS (Silvestri and Ipsiroglu, 2022). Therapeutic interventions involving oral and intravenous iron have shown promise in ameliorating various aspects of sleep and behavior in individuals with RLS and related disorders (Allen et al., 2018). In TS, where sleep disruptions are prevalent, addressing iron deficiency through appropriate supplementation could hold potential for alleviating both tic severity and associated sleep disturbances. However, this approach has not been tested experimentally.

A connection between RLS and TS is further reinforced by shared genetic factors, like variants in the BTBD9 gene associated with both disorders (Rivière et al., 2009). These shared genes shed light on the involvement of the histaminergic system, a neurotransmitter crucial for arousal control. Histamine, released by histaminergic neurons, plays a role in arousal regulation and originates from the tuberomammillary nucleus in the hypothalamus. Mutations in the HDC gene, which encodes the enzyme L-histidine decarboxylase, vital for histamine production, were discovered in a family with TS, suggesting a possible involvement of the histaminergic system in tic generation (Ercan-Sencicek et al., 2010). However, these findings have not been replicated further. Additionally, altered dopamine D2/D3 receptor binding in the basal ganglia in TS patients and those with HDC mutations suggest a potential interplay between the histaminergic and dopaminergic systems in TS (Baldan et al., 2014). Also, in terms of phenomenology, RLS and TS exhibit commonalities in urges and the nature of movements. Individuals with RLS often experience a compelling urge to move their legs or other body parts due to uncomfortable sensations, which intensify during rest, mirroring the inner tension felt by individuals with TS. These urges in both conditions lead to movement, which in RLS alleviates discomfort temporarily and in TS manifests as motor or vocal tics, alleviating as well the "urge to perform" sensation.

Anatomically, the cortico-basal ganglia-thalamo-cortical circuitry is strongly implicated in tic generation. While extensive research on neural mechanisms during wakefulness in TS is available, our understanding of neural activity during sleep in individuals with TS remains limited (Ricketts et al., 2022b). Previous studies have predominantly focused on wakefulness using functional magnetic resonance imaging or positron emission tomography (PET) imaging, revealing altered neural activity in motor areas, including the premotor and prefrontal cortex, crucial for motor planning, execution, and inhibition (Polyanska et al., 2017).

Investigations utilizing [15O] H_2O PET imaging during tic release in TS patients during both wakefulness and N2 sleep revealed insights into the neural dynamics associated with TS. These studies confirmed the involvement of various brain regions, particularly strong activation of the insula and bilateral cerebellum (Lerner et al., 2007). The cerebellum and insula influence structures such as the thalamus, putamen, motor cortex, and supplementary motor area, forming a network implicated in tic generation and sensory abnormalities associated with tic experience. The insula's role in integrating external stimuli and internal states suggests its involvement in the affective aspects of TS. Moreover, robust cerebellar activation indicates its potential role in tic pathophysiology, possibly related to learned stereotyped movements.

Notably, abnormal cerebellar and insula activity have been reported in sleep disorders and in sleep-related movement disorders (Del Rosso and Hoque, 2014; Wang et al., 2022; Benarroch, 2023), suggesting an overlap with TS. Shared neural markers, such as increased connectivity between the amygdala and premotor cortex and sensorimotor cortex, have been proposed to suggest an overlap between the neurologic underpinnings of TS and insomnia (Ricketts et al., 2022b).

Importantly, hyperarousal, a state of heightened cognitive and physiologic activation, serves as a shared underlying mechanism in both CTD/TS and insomnia. Within CTD/TS, hyperarousal typically manifests as heightened attention to internal sensations, increased sensitivity to sensory input, and enhanced responsiveness to external stimuli (Cohen and Leckman, 1992). This elevated state of arousal is closely tied to the exacerbation of tic symptoms (Woods et al., 2005). In parallel, insomnia is characterized by intrusive thoughts and persistent worry (cognitive hyperarousal), coupled with elevated heart rate, metabolic rate, and body temperature (physiologic hyperarousal) (Riemann et al., 2010). This state of hyperarousal is believed to underlie the challenge of initiating and maintaining sleep experienced in insomnia (Bonnet and Arand, 2010).

TREATMENT

Sleep disturbances are important challenges in the management of CTD/TS. The bidirectional relationship between tics and sleep issues – where sleep disruption can intensify tics and vice versa – underscores the need for effective treatment modalities that address both concerns concurrently. It is paramount to evaluate the sleep habits, behaviors, and comorbidities of children and adults with CTD/TS, as these concomitant disorders can further exacerbate sleep disruption, leading to cognitive, executive function, emotional regulation, and other consequences.

To date, there are no established guidelines for addressing sleep complaints in children with CTD/TS. In general, treatment strategies for tics in CTD/TS encompass both pharmacologic and nonpharmacologic interventions (Table 10.2). Pharmacologic therapy in TS becomes the focus when tics cause significant emotional, physical, or social impairment. Pharmacologic interventions commonly used to manage tics in TS may also have implications for improving sleep disturbances in affected individuals. Atypical antipsychotics, recognized for their antidopaminergic activity, are frequently employed to address both the tics associated with TS and accompanying sleep complaints. For example, risperidone administration resulted in improved sleep latency and efficiency in a documented case involving a 12-year-old patient (Arana-Lechuga et al., 2008). Similarly, aripiprazole has shown promise in managing tics and associated sleep issues (Cavanna et al., 2012). However, it is noteworthy that a prospective uncontrolled open-label study with 44 adults suggested that aripiprazole may induce sleep disorders as a side-effect (Gerasch et al., 2016).

Furthermore, while not directly studied for sleep complaints in children with TS, other pharmacologic agents commonly used to manage tics may also exert effects on sleep disturbances. These include medications such as the antinoradrenergic drug clonidine, clonazepam, selective serotonin reuptake inhibitors (e.g., fluvoxamine), and dopamine, as well as noradrenaline agonists (e.g., methylphenidate and atomoxetine) (Blaty and DelRosso, 2022). These medications, often prescribed to address comorbid conditions like ADHD, may indirectly impact sleep regulatory mechanisms and produce various alterations in sleep patterns. The specific effects on sleep can vary depending on factors such as the medication type, the patient's psychopathologic situation, and other comorbidities. Therefore, the choice of medication should be made carefully, considering the overall well-being of the patient, including their sleep quality.

Iron supplementation has been found to mitigate tic severity in an uncontrolled trial on 57 patients (Ghosh and Burkman, 2017). As mentioned above, the connection between iron levels and TS pathophysiology is intriguing and also extends to children with ADHD and RLS, although more research is needed to understand this relationship fully. Caution is warranted before establishing a definitive conclusion regarding the efficacy of iron supplementation in managing tic severity in CTD/TS.

Other possibilities include nonpharmacologic strategies, which should be regarded as the first-line treatment for most sleep disorders. Cognitive behavioral therapy for insomnia (CBT-I), while primarily developed for chronic insomnia, may offer synergistic benefits for TS patients experiencing concurrent sleep issues; however, to date, there have been no studies investigating the effectiveness of CBT-I in these patients. CBT-I comprises a set of strategies, ranging from sleep restriction and stimulus control to cognitive restructuring and relaxation techniques, tailored to break the cycle of sleep-related anxieties and maladaptive behaviors that perpetuate insomnia (Morin, 2004). Moreover, CBT-I not only is effective in treating insomnia, but also promotes sleep stability by establishing a more consolidated sleep architecture, which in turn reduces nocturnal movements and the onset of parasomnias (Song et al., 2020; O'Regan et al., 2021). In the context of TS, cognitive behavioral therapy (CBT) has been shown to be effective (O'Connor et al., 2009; Flessner, 2011; Shou et al., 2022). Within the CBT framework, comprehensive behavioral intervention for tics (CBIT) is recommended as the first-line treatment by the American Academy of Neurology (Pringsheim et al., 2019). CBIT combines habit reversal training with other behavioral strategies, providing patients with tools to counteract the premonitory urges that precede tics (Woods et al., 2008; Hwang et al., 2012). The theoretical alignment of CBT-I and CBIT suggests that behavioral and cognitive restructuring, pivotal in both approaches, may offer a bridge between addressing sleep disturbances and tic management. Furthermore, inherent relaxation techniques of CBT-I could serve as a valuable tool against the high level of stress and muscle tension known to precede and exacerbate tics (Hoogduin et al., 1997; O'Connor, 2002).

Mindfulness-based techniques have shown promise in treating sleep disturbances by helping individuals become more attuned to their bodily sensations and thought patterns (Black et al., 2015; Rash et al., 2019). A pilot study on a modified mindfulness-based stress reduction approach for CTD/TS showed effectiveness, with 58.8% of patients experiencing significant reductions in tic severity, which correlated with increased self-awareness of those factors influencing tic expression (Reese et al., 2015).

Light therapy is recognized as an effective nonpharmacologic intervention in CRSWDs for its potential to modulate circadian rhythms and influence melatonin secretion (Gooley, 2008). The light exposure is typically administered through

Table 10.2

Comparative overview of pharmacologic and nonpharmacologic interventions for tics and sleep disorders.

	Type of intervention	Description	Advantages	Limitations	References
Pharmacologic interventions	Antipsychotics (e.g., risperidone, aripiprazole)	Target dopamine receptors to manage tics and concurrent sleep complaints	Improvements in sleep latency and efficiency while managing tic severity	May induce sleep disorders as side-effects	Arana-Lechuga et al. (2008), Cavanna et al. (2012), Gerasch et al. (2016)
	Iron supplementation	Addresses iron deficiency, potentially linked to tic severity in TS	Reduces tic severity, indirectly improving sleep quality	Mechanism and optimal dosage unclear; requires careful monitoring to prevent iron overload	Ghosh and Burkman (2017)
	Adrenergic agents (e.g., clonidine)	Modulate noradrenergic activity to reduce tics, with potential secondary benefits for sleep	First-line treatment for TS with moderate efficacy for tic reduction; sedative properties can improve sleep	Side-effects include sedation, dizziness, and hypotension	Roessner et al. (2011), Malaty and Akbar (2014), Blaty and DelRosso (2022)
	Benzodiazepines (e.g., clonazepam)	Medications with sedative effects, traditionally used for TD	Sleep-promoting effect; effective for occasional tic control, particularly with concurrent anxiety	Risk of tolerance	Gonce and Barbeau (1977), Roessner et al. (2011), Raggi et al. (2023)
	Psychostimulants (e.g., methylphenidate)	Used for tic management and to address co-occurring ADHD	Beneficial when comorbid conditions exacerbate tics	Higher doses may temporarily worsen tics; limited synergistic effect on sleep issues	Castellanos et al. (1997), Kurlan and Tourette's Syndrome Study Group (2002)
Nonpharmacologic interventions	Cognitive behavioral therapy for insomnia	A set of strategies targeting dysfunctional sleep-related beliefs and behaviors	Improves sleep architecture and reduces nocturnal movements and parasomnias	Limited studies specifically investigating effectiveness in TD	Morin (2004), Song et al. (2020), O'Regan et al. (2021), Shou et al. (2022)
	Comprehensive behavioral intervention for tics	Employs structured behavioral techniques to reduce tic severity by addressing the underlying premonitory urges	Recommended as a first-line intervention for TD; may improve associated sleep issues	Focuses primarily on tic management; direct effects on sleep not extensively studied	Woods et al. (2008), Hwang et al. (2012), Pringsheim et al. (2019)
	Mindfulness-based techniques	Strategies that increase self-awareness of bodily sensations and thought patterns	Effective in reducing tic severity and improving sleep quality	More research needed to elucidate the synergistic benefits in managing sleep issues in TD	Black et al. (2015), Reese et al. (2015), Rash et al. (2019)
	Light therapy	Exposure to specific light wavelengths and intensities to regulate circadian rhythms and melatonin secretion	Aligns circadian rhythms to environmental cues and reduces tic severity and associated symptoms in TD	Effectiveness may depend on individual response and adherence to therapy; limited studies in TD	Gooley (2008), Coles and Strauss (2015), Wahl et al. (2019), Ricketts et al. (2022a)
	Repetitive transcranial magnetic stimulation and transcranial direct current stimulation	Neuromodulatory interventions that influence neuronal activity through magnetic fields or electric currents	Effective in improving sleep quality in various sleep disorders; evidence of tic severity reduction in TD	TD symptom relief is inconsistent; efficacy for sleep disturbances in TD not well established	Mrakic-Sposta et al. (2008), Jiang et al. (2013), Le et al. (2013), Lin et al. (2015), Lai et al. (2017), Zhou et al. (2020), Kahl et al. (2021)

ADHD, attention deficit hyperactivity disorder; *TD*, tic disorder; *TS*, Tourette syndrome.

specially designed light boxes or lamps. Notably, the wavelength, intensity, and duration of light exposure are crucial parameters that determine its therapeutic impact. Blue-wavelength light, typically between 460 and 480 nm, has demonstrated significant efficacy in shifting the endogenous circadian phase to align with the external environment (Gooley, 2008; Wahl et al., 2019). In a recent pilot study (Ricketts et al., 2022a) involving 14 TS patients with eveningness chronotype, morning light therapy led to significant phase shifts in melatonin onset, reduced tic severity, and decreased levels of daytime sleepiness and anxiety. Furthermore, Coles and Strauss (2015) documented a 52-year-old CTD patient who experienced a significant decrease in tic frequency after using a 10,000-lux light box. Interestingly, her tic severity seemed to depend on regular exposure to both the light box and natural sunlight.

Finally, noninvasive neuromodulatory interventions, such as repetitive transcranial magnetic stimulation (rTMS) and transcranial direct current stimulation (tDCS), may be valuable therapeutic options for addressing sleep disturbances in CTS/TS, although not investigated yet. rTMS utilizes rapidly changing magnetic fields to induce electric currents in targeted brain regions, potentially "resetting" aberrant neural circuits; tDCS, on the other hand, delivers a continuous low electric current, which can either facilitate or inhibit neuronal firing based on the current polarity, allowing for the selective enhancement or suppression of specific neural pathways. These interventions have demonstrated potential in improving sleep quality and alleviating sleep-related symptoms in multiple sleep disturbances, such as insomnia, RLS, hypersomnia, and narcolepsy by targeting prefrontal, premotor, and parietal areas (Jiang et al., 2013; Lin et al., 2015; Galbiati et al., 2016; Lai et al., 2017; Sheng et al., 2018; Sonmez et al., 2020; Zhou et al., 2020). However, when applied to alleviate CTD/TS motor symptoms, especially targeting the supplementary motor area, results vary, with some studies suggesting tic severity reduction (Kwon et al., 2011; Le et al., 2013; Carvalho et al., 2015; Kahl et al., 2021), whereas others reporting limited benefits (Mrakic-Sposta et al., 2008; Landeros-Weisenberger et al., 2015; Bloch et al., 2016).

In light of the current evidence, it becomes clear that a multifaceted approach may be crucial in addressing the complex interplay between CTD/TS and sleep disturbances. While the discussed interventions have each shown potential benefits in either the context of sleep or tic management, their collective efficacy in a comprehensive treatment regimen for TS needs further investigation.

CONCLUSION

The bidirectional relationship between tics and sleep disturbances presents a multifaceted clinical challenge. Sleep disturbances can have significant negative effects on quality of life, cognitive functioning, and emotional regulation. Poor sleep has been linked to an increased risk of anxiety and depression and could exacerbate existing emotional and behavioral problems in CTD/TS patients (Sukhodolsky et al., 2003; Leisman and Sheldon, 2022). Furthermore, the presence of tics during sleep can lead to sleep fragmentation, reducing the overall sleep quality and duration. This, in turn, can amplify tic severity and frequency during the daytime, creating a vicious cycle. Adding complexity to this scenario are the psychiatric comorbidities often associated with CTD/TS, particularly ADHD. This condition, with its inherent potential to disrupt sleep architecture, can amplify sleep issues in CTD/TS cohorts.

However, despite the high prevalence and impact of sleep disturbances in this population, there is limited understanding of the underlying mechanisms and effective treatments. Several studies have identified potential common mechanisms underlying these disorders. For instance, alterations in the levels of neurotransmitters such as dopamine, histamine, and BID have been linked to both CTD/TS and sleep disturbances. This "hyperarousal"/"hyperexcitability" model has been proposed for both CTD/TS and sleep-related disorder such as RLS and insomnia and likely reflects specific changes in the neural mechanism that governs both sensorimotor system and sleep-wake patterns.

Sleep disturbances may exacerbate tic severity by increasing fatigue and reducing the threshold for tic expression. Both pharmacologic and nonpharmacologic treatments offer promise in addressing sleep in CTD/TS however, their combined efficacy within a comprehensive treatment approach remains to be robustly validated. Moreover, the persistence of tics during sleep underscores the need for a more integrated clinical perspective. Longitudinal studies are paramount to delineate the trajectory of sleep disturbances and tic severity over time, offering insights for tailored treatment strategies.

References

Allen RP, Earley CJ (2007). The role of iron in restless legs syndrome. Mov Disord 22: S440–S448.

Allen RP, Picchietti DL, Auerbach M et al. (2018). Evidence-based and consensus clinical practice guidelines for the iron treatment of restless legs syndrome/Willis-Ekbom disease in adults and children: an IRLSSG task force report. Sleep Med 41: 27–44.

Allen RP, Singer HS, Brown JE et al. (1992). Sleep disorders in Tourette syndrome: a primary or unrelated problem? Pediatr Neurol 8: 275–280.

American Academy of Sleep Medicine (2014). International classification of sleep disorders: diagnostic and coding manual 3rd ed. Darien, IL: American Academy of Sleep Medicine.

Antelmi E, Rocchi L, Latorre A et al. (2022). Restless legs syndrome: known knowns and known unknowns. Brain Sci 12: 118.

Arana-Lechuga Y, Sanchez-Escandon O, de Santiago-Trevino N et al. (2008). Risperidone treatment of sleep disturbances in Tourette's syndrome. J Neuropsychiatry Clin Neurosci 20: 375–376.

Arias-Carrión O, Pöppel E (2007). Dopamine, learning, and reward-seeking behavior. Acta Neurobiol Exp (Wars) 67: 481–488.

Arnsten AF (2006). Fundamentals of attention-deficit/hyperactivity disorder: circuits and pathways. J Clin Psychiatry 67: 7–12.

Baldan LC, Williams KA, Gallezot J-D et al. (2014). Histidine decarboxylase deficiency causes Tourette syndrome: parallel findings in humans and mice. Neuron 81: 77–90.

Barabas G, Ferrari M, Matthews WS (1983). Childhood migraine and somnambulism. Neurology 33: 948–949.

Barabas G, Matthews WS, Ferrari M (1984). Disorders of arousal in Gilles de la Tourette's syndrome. Neurology 34: 815–817.

Baym CL, Corbett BA, Wright SB et al. (2008). Neural correlates of tic severity and cognitive control in children with Tourette syndrome. Brain 131: 165–179.

Benarroch E (2023). What is the involvement of the cerebellum during sleep? Neurology 100: 572–577.

Bianco LE, Wiesinger JA, Earley CJ et al. (2008). Iron deficiency alters dopamine uptake and response to L-DOPA injection in Sprague–Dawley rats. J Neurochem 106: 205–215.

Black DS, O'Reilly GA, Olmstead R et al. (2015). Mindfulness meditation and improvement in sleep quality and daytime impairment among older adults with sleep disturbances: a randomized clinical trial. JAMA Intern Med 175: 494–501.

Blaty JL, DelRosso LM (2022). Tourette disorder and sleep. Biomed J 45: 240–249.

Bloch Y, Arad S, Levkovitz Y (2016). Deep TMS add-on treatment for intractable Tourette syndrome: a feasibility study. World J Biol Psychiatry 17: 557–561.

Bonnet MH, Arand DL (2010). Hyperarousal and insomnia: state of the science. Sleep Med Rev 14: 9–15.

Branch AF, Navidi W, Tabuchi S et al. (2016). Progressive loss of the orexin neurons and wakefulness. Sleep 39: 369–377.

Bruni O, Angriman M, Melegari MG et al. (2019). Pharmacotherapeutic management of sleep disorders in children with neurodevelopmental disorders. Expert Opin Pharmacother 20: 2257–2271.

Calhoun SL, Fernandez-Mendoza J, Vgontzas AN et al. (2014). Prevalence of insomnia symptoms in a general population sample of young children and preadolescents: gender effects. Sleep Med 15: 91–95.

Calhoun SL, Fernandez-Mendoza J, Vgontzas AN et al. (2017). Behavioral profiles associated with objective sleep duration in young children with insomnia symptoms. J Abnorm Child Psychol 45: 337–344.

Camhi SL, Morgan WJ, Pernisco N et al. (2000). Factors affecting sleep disturbances in children and adolescents. Sleep Med 1: 117–123.

Carvalho S, Gonçalves ÓF, Soares JM et al. (2015). Sustained effects of a neural-based intervention in a refractory case of Tourette syndrome. Brain Stimul 8: 657–659.

Castaneda TR, de Prado BM, Prieto D et al. (2004). Circadian rhythms of dopamine, glutamate and GABA in the striatum and nucleus accumbens of the awake rat: modulation by light. J Pineal Res 36: 177–185.

Castellanos FX, Giedd JN, Elia J et al. (1997). Controlled stimulant treatment of ADHD and comorbid Tourette's syndrome: effects of stimulant and dose. J Am Acad Child Adolesc Psychiatry 36: 589–596.

Cavanna AE, Selvini C, Termine C et al. (2012). Tolerability profile of aripiprazole in patients with Tourette syndrome. J Psychopharmacol 26: 891–895.

Cervenka S, Pålhagen SE, Comley RA et al. (2006). Support for dopaminergic hypoactivity in restless legs syndrome: a PET study on D2-receptor binding. Brain 129: 2017–2028.

Chemelli RM, Willie JT, Sinton CM et al. (1999). Narcolepsy in orexin knockout mice: molecular genetics of sleep regulation. Cell 98: 437–451.

Claudio-Campos K, Stevens D, Koo SW et al. (2021). Is persistent motor or vocal tic disorder a milder form of Tourette syndrome? Mov Disord 36: 1899–1910.

Cohen AJ, Leckman JF (1992). Sensory phenomena associated with Gilles de la Tourette's syndrome. J Clin Psychiatry 53: 319–323.

Cohrs S, Rasch T, Altmeyer S et al. (2001). Decreased sleep quality and increased sleep related movements in patients with Tourette's syndrome. J Neurol Neurosurg Psychiatry 70: 192–197.

Coles ME, Strauss GP (2015). Shedding light on tics. Psychiatry Res 225: 743.

Comings DE, Comings BG (1987). A controlled study of Tourette syndrome. VI. Early development, sleep problems, allergies, and handedness. Am J Hum Genet 41: 822–838.

Cortese S, Lecendreux M, Dalla Bernardina B et al. (2008). Attention-deficit/hyperactivity disorder, Tourette's syndrome, and restless legs syndrome: the iron hypothesis. Med Hypotheses 70: 1128–1132.

Daley M, Morin CM, LeBlanc M et al. (2009). Insomnia and its relationship to health-care utilization, work absenteeism, productivity and accidents. Sleep Med 10: 427–438.

Del Rosso LM, Hoque R (2014). The cerebellum and sleep. Neurol Clin 32: 893–900.

Dong H, Wang J, Yang YF et al. (2019). Dorsal striatum dopamine levels fluctuate across the sleep-wake cycle and respond to salient stimuli in mice. Front Neurosci 13: 242.

Drake ME Jr, , Hietter SA, Bogner JE et al. (1992). Cassette EEG sleep recordings in Gilles de la Tourette syndrome. Clin Electroencephalogr 23: 142–146.

Druiven SJM, Riese H, Kamphuis J et al. (2021). Chronotype changes with age; seven-year follow-up from the Netherlands study of depression and anxiety cohort. J Affect Disord 295: 1118–1121.

Eban-Rothschild A, Rothschild G, Giardino WJ et al. (2016). VTA dopaminergic neurons regulate ethologically relevant sleep-wake behaviors. Nat Neurosci 19: 1356–1366.

Ercan-Sencicek AG, Stillman AA, Ghosh AK et al. (2010). L-histidine decarboxylase and Tourette's syndrome. N Engl J Med 362: 1901–1908.

Erikson KM, Jones BC, Beard JL (2000). Iron deficiency alters dopamine transporter functioning in rat striatum. J Nutr 130: 2831–2837.

Erikson KM, Jones BC, Hess EJ et al. (2001). Iron deficiency decreases dopamine D1 and D2 receptors in rat brain. Pharmacol Biochem Behav 69: 409–418.

Ferré S, Quiroz C, Guitart X et al. (2018). Pivotal role of adenosine neurotransmission in restless legs syndrome. Front Neurosci 11: 722.

Ferris MJ, España RA, Locke JL et al. (2014). Dopamine transporters govern diurnal variation in extracellular dopamine tone. Proc Natl Acad Sci USA 111: E2751–E2759.

Flessner CA (2011). Cognitive-behavioral therapy for childhood repetitive behavior disorders: tic disorders and trichotillomania. Child Adolesc Psychiatr Clin N Am 20: 319–328.

Franklin KA, Lindberg E (2015). Obstructive sleep apnea is a common disorder in the population – a review on the epidemiology of sleep apnea. J Thorac Dis 7: 1311–1322.

Galbiati A, Abutalebi J, Iannaccone S et al. (2016). The effects of transcranial direct current stimulation (tDCS) on idiopathic hypersomnia: a pilot study. Arch Ital Biol 154: 1–5.

Garbarino S (2020). Sleep disorders across the lifespan: a different perspective. Int J Environ Res Public Health 17: 9025.

Gerasch S, Kanaan AS, Jakubovski E et al. (2016). Aripiprazole improves associated comorbid conditions in addition to tics in adult patients with Gilles de la Tourette Syndrome. Front Neurosci 10: 416.

Ghosh D, Burkman E (2017). Relationship of serum ferritin level and tic severity in children with Tourette syndrome. Childs Nerv Syst 33: 1373–1378.

Ghosh D, Rajan PV, Das D et al. (2014). Sleep disorders in children with Tourette syndrome. Pediatr Neurol 51: 31–35.

Gilles de la Tourette G (1995). Étude sur une affection nerveuse caractérisée par de l'incoordination motrice accompagnée d'écholalie et de coprolalie. Arch Neurol 9: 19–42.

Glaze DG, Frost JD Jr, , Jankovi J (1983). Sleep in Gilles de la Tourette's syndrome: disorder of arousal. Neurology 33: 586–592.

Goldstein AN, Walker MP (2014). The role of sleep in emotional brain function. Annu Rev Clin Psychol 10: 679–708.

Gonce M, Barbeau A (1977). Seven cases of Gilles de la Tourette's syndrome: partial relief with clonazepam: a pilot study. Can J Neurol Sci 4: 279–283.

Gooley JJ (2008). Treatment of circadian rhythm sleep disorders with light. Ann Acad Med Singap 37: 669–676.

Gorman DA, Zhu H, Anderson GM et al. (2006). Ferritin levels and their association with regional brain volumes in Tourette's syndrome. Am J Psychiatry 163: 1264–1272.

Gregory AM, Eley TC, O'Connor TG et al. (2004). Etiologies of associations between childhood sleep and behavioral problems in a large twin sample. J Am Acad Child Adolesc Psychiatry 43: 744–751.

Grippo RM, Purohit AM, Zhang Q et al. (2017). Direct midbrain dopamine input to the suprachiasmatic nucleus accelerates circadian entrainment. Curr Biol 27: 2465–2475.e3.

Groth C, Debes NM, Rask CU et al. (2017). Course of Tourette syndrome and comorbidities in a large prospective clinical study. J Am Acad Child Adolesc Psychiatry 56: 304–312.

Hampp G, Ripperger JA, Houben T et al. (2008). Regulation of monoamine oxidase A by circadian-clock components implies clock influence on mood. Curr Biol 18: 678–683.

Harduf EV, Matzner A, Belelovsky K et al. (2021). Dissociation of tic generation from tic expression during the sleep-wake cycle. iScience 24: 102380.

Hienert M, Gryglewski G, Stamenkovic M et al. (2018). Striatal dopaminergic alterations in Tourette's syndrome: a meta-analysis based on 16 PET and SPECT neuroimaging studies. Transl Psychiatry 8: 143.

Honaker SM, Meltzer LJ (2014). Bedtime problems and night wakings in young children: an update of the evidence. Paediatr Respir Rev 15: 333–339.

Hoogduin K, Verdellen C, Cath D (1997). Exposure and response prevention in the treatment of Gilles de la Tourette's syndrome: four case studies. Clin Psychol Psychother 4: 125–135.

Howell MJ (2012). Parasomnias: an updated review. Neurotherapeutics 9: 753–775.

Hwang GC, Tillberg CS, Scahill L (2012). Habit reversal training for children with Tourette syndrome: update and review. J Child Adolesc Psychiatr Nurs 25: 178–183.

Imbesi M, Yildiz S, Arslan AD et al. (2009). Dopamine receptor-mediated regulation of neuronal "clock" gene expression. Neuroscience 158: 537–544.

Isomura K, Sidorchuk A, Sevilla-Cermeño L et al. (2022). Insomnia in Tourette syndrome and chronic tic disorder. Mov Disord 37: 392–400.

Jankovic J, Rohaidy H (1987). Motor, behavioral and pharmacologic findings in Tourette's syndrome. Can J Neurol Sci 14: 541–546.

Ji X, Wu M, Zhang X (2016). Dopamine level in the stratium exhibits circadian rhythms in the rat model of Tourette syndrome. Int J Clin Exp Med 9: 9015–9020.

Jiang CG, Zhang T, Yue FG et al. (2013). Efficacy of repetitive transcranial magnetic stimulation in the treatment of patients with chronic primary insomnia. Cell Biochem Biophys 67: 169–173.

Jiménez-Jiménez FJ, Alonso-Navarro H, Garcia-Martin E et al. (2020). Sleep disorders in Tourette syndrome. Sleep Med Rev 53: 101335.

Jiménez-Jiménez FJ, Alonso-Navarro H, García-Martín E et al. (2022). Sleep disorders and sleep problems in patients with Tourette syndrome and other tic disorders: current perspectives. Nat Sci Sleep 14: 1313–1331.

Kahl CK, Kirton A, Pringsheim T et al. (2021). Bilateral transcranial magnetic stimulation of the supplementary motor area in children with Tourette syndrome. Dev Med Child Neurol 63: 808–815.

Kanaan AS, Yu D, Metere R et al. (2023). Convergent imaging-transcriptomic evidence for disturbed iron homeostasis in Gilles de la Tourette syndrome. Neurobiol Dis 185: 106252.

Kant R, Meena MK, Pathania M (2021). Dopamine: a modulator of circadian rhythms/biological clock. Int J Adv Med 8: 316–324.

Karagiannidis I, Dehning S, Sandor P et al. (2013). Support of the histaminergic hypothesis in Tourette syndrome: association of the histamine decarboxylase gene in a large sample of families. J Med Genet 50: 760–764.

Khaldy H, León J, Escames G et al. (2002). Circadian rhythms of dopamine and dihydroxyphenyl acetic acid in the mouse striatum: effects of pinealectomy and of melatonin treatment. Neuroendocrinology 75: 201–208.

Kim J, Lee GH, Sung SM et al. (2020). Prevalence of attention deficit hyperactivity disorder symptoms in narcolepsy: a systematic review. Sleep Med 65: 84–88.

Kirov R, Becker A, Rothenberger A (2014). Sleep in Tourette syndrome. Curr Dev Disord Rep 1: 252–259.

Kirov R, Brand S, Banaschewski T et al. (2007). Sleep patterns in children with attention-deficit/hyperactivity disorder, tic disorder, and comorbidity. J Child Psychol Psychiatry 48: 561–570.

Kirov R, Brand S, Banaschewski T et al. (2017). Opposite impact of rem sleep on neurobehavioral functioning in children with common psychiatric disorders compared to typically developing children. Front Psychol 7: 2059.

Kostanecka-Endress T, Banaschewski T, Kinkelbur J et al. (2003). Disturbed sleep in children with Tourette syndrome: a polysomnographic study. J Psychosom Res 55: 23–29.

Kurlan R Tourette's Syndrome Study Group(2002). Treatment of ADHD in children with tics: a randomized controlled trial. Neurology 58: 527–536.

Kwon HJ, Lim WS, Lim MH et al. (2011). 1-Hz low frequency repetitive transcranial magnetic stimulation in children with Tourette's syndrome. Neurosci Lett 492: 1–4.

Lai JB, Han MM, Xu Y et al. (2017). Effective treatment of narcolepsy-like symptoms with high-frequency repetitive transcranial magnetic stimulation: a case report. Medicine 96: e8645.

Lajonchere C, Nortz M, Finger S (1996). Gilles de la Tourette and the discovery of Tourette syndrome. Includes a translation of his 1884 article. Arch Neurol 53: 567–574.

Landeros-Weisenberger A, Mantovani A, Motlagh MG et al. (2015). Randomized sham controlled double-blind trial of repetitive

transcranial magnetic stimulation for adults with severe Tourette syndrome. Brain Stimul 8: 574–581.
Le K, Liu L, Sun M et al. (2013). Transcranial magnetic stimulation at 1 Hertz improves clinical symptoms in children with Tourette syndrome for at least 6 months. J Clin Neurosci 20: 257–262.
Lecendreux M, Lavault S, Lopez R et al. (2015). Attention-deficit/ hyperactivity disorder (ADHD) symptoms in pediatric narcolepsy: a cross-sectional study. Sleep 38: 1285–1295.
Lee WT, Huang HL, Wong LC et al. (2017). Tourette Syndrome as an independent risk factor for subsequent sleep disorders in children: a nationwide population-based case-control study. Sleep 40.
Léger D, Guilleminault C, Bader G et al. (2002). Medical and socio-professional impact of insomnia. Sleep 25: 625–629.
Léger D, Massuel MA, Metlaine A et al. (2006). Professional correlates of insomnia. Sleep 29: 171–178.
Leisman G, Sheldon D (2022). Tics and emotions. Brain Sci 12: 242.
Lerner A, Bagic A, Boudreau EA et al. (2007). Neuroimaging of neuronal circuits involved in tic generation in patients with Tourette syndrome. Neurology 68: 1979–1987.
Lespérance P, Djerroud N, Diaz Anzaldua A et al. (2004). Restless legs in Tourette syndrome. Mov Disord 19: 1084–1087.
Lin YC, Feng Y, Zhan SQ et al. (2015). Repetitive transcranial magnetic stimulation for the treatment of restless legs syndrome. Chin Med J 128: 1728–1731.
Lloyd R, Tippmann-Peikert M, Slocumb N et al. (2012). Characteristics of REM sleep behavior disorder in childhood. J Clin Sleep Med 8: 127–131.
Maia TV, Conceição VA (2018). Dopaminergic disturbances in Tourette syndrome: an integrative account. Biol Psychiatry 84: 332–344.
Malaty IA, Akbar U (2014). Updates in medical and surgical therapies for Tourette syndrome. Curr Neurol Neurosci Rep 14: 458.
Martino D, Ganos C, Pringsheim TM (2017). Tourette syndrome and chronic tic disorders: the clinical spectrum beyond tics. Int Rev Neurobiol 134: 1461–1490.
Mendelson WB, Caine ED, Goyer P et al. (1980). Sleep in Gilles de la Tourette syndrome. Biol Psychiatry 15: 339–343.
Mendoza J, Challet E (2014). Circadian insights into dopamine mechanisms. Neuroscience 282: 230–242.
Mesgar S, Jameie SB, Aliaghaei A et al. (2022). Dopamine D1 receptor-mediated regulation of Per1, Per2, CLOCK, and BMAL1 expression in the suprachiasmatic nucleus in adult male rats. J Mol Neurosci 72: 618–625.
Mi Y, Zhao R, Sun X et al. (2022). Sleep disturbances and sleep patterns in children with tic disorder: a case-control study. Front Pediatr 10: 911343.
Miletínová E, Bušková J (2021). Functions of sleep. Physiol Res 70: 177–182.
Młodzikowska-Albrecht J, Żarowski M, Steinborn B (2010). Sleep disorders in Tourette syndrome in children and adolescents. Neurol Dziecięca 11: 10–14.
Modafferi S, Stornelli M, Chiarotti F et al. (2016). Sleep, anxiety and psychiatric symptoms in children with Tourette syndrome and tic disorders. Eur J Paediatr Neurol 20: 696–703.
Modestino EJ, Winchester J (2013). A retrospective survey of childhood ADHD symptomatology among adult narcoleptics. J Atten Disord 17: 574–582.
Moore RY, Speh JC, Leak RK (2002). Suprachiasmatic nucleus organization. Cell Tissue Res 309: 89–98.
Morin CM (2004). Cognitive-behavioral approaches to the treatment of insomnia. J Clin Psychiatry 65: 33–40.
Morin CM, Jarrin DC (2022). Epidemiology of insomnia: prevalence, course, risk factors, and public health burden. Sleep Med Clin 17: 173–191.
Mrakic-Sposta S, Marceglia S, Mameli F et al. (2008). Transcranial direct current stimulation in two patients with Tourette syndrome. Mov Disord 23: 2259–2261.
Nevsimalova S, Prihodova I, Kemlink D et al. (2007). REM behavior disorder (RBD) can be one of the first symptoms of childhood narcolepsy. Sleep Med 8: 784–786.
O'Brien LM (2009). The neurocognitive effects of sleep disruption in children and adolescents. Child Adolesc Psychiatr Clin N Am 18: 813–823.
O'Connor K (2002). A cognitive-behavioral/psychophysiological model of tic disorders. Behav Res Ther 40: 1113–1142.
O'Connor KP, Laverdure A, Taillon A et al. (2009). Cognitive behavioral management of Tourette's syndrome and chronic tic disorder in medicated and unmedicated samples. Behav Res Ther 47: 1090–1095.
O'Regan D, Nesbitt A, Biabani N et al. (2021). A novel group cognitive behavioral therapy approach to adult non-rapid eye movement parasomnias. Front Psychiatry 12: 679272.
Peterson BS, Gore JC, Riddle MA et al. (1994). Abnormal magnetic resonance imaging T2 relaxation time asymmetries in Tourette's syndrome. Psychiatry Res 55: 205–221.
Petit D, Pennestri MH, Paquet J et al. (2015). Childhood sleepwalking and sleep terrors: a longitudinal study of prevalence and familial aggregation. JAMA Pediatr 169: 653–658.
Plazzi G, Pizza F, Palaia V et al. (2011). Complex movement disorders at disease onset in childhood narcolepsy with cataplexy. Brain 134: 3477–3489.
Polyanska L, Critchley HD, Rae CL (2017). Centrality of prefrontal and motor preparation cortices to Tourette syndrome revealed by meta-analysis of task-based neuroimaging studies. NeuroImage Clin 16: 257–267.
Pringsheim T, Nosratmirshekarlou E, Doja A et al. (2021). Physical activity, sleep and neuropsychiatric symptom severity in children with Tourette syndrome. Eur Child Adolesc Psychiatry 30: 711–719.
Pringsheim T, Okun MS, Müller-Vahl K et al. (2019). Practice guideline recommendations summary: treatment of tics in people with Tourette syndrome and chronic tic disorders. Neurology 92: 896–906.
Raggi A, Mogavero MP, DelRosso LM et al. (2023). Clonazepam for the management of sleep disorders. Neurol Sci 44: 115–128.
Rash JA, Kavanagh VAJ, Garland SN (2019). A meta-analysis of mindfulness-based therapies for insomnia and sleep disturbance: moving towards processes of change. Sleep Med Clin 14: 209–233.
Reese HE, Vallejo Z, Rasmussen J et al. (2015). Mindfulness-based stress reduction for Tourette syndrome and chronic tic disorder: a pilot study. J Psychosom Res 78: 293–298.
Ricketts EJ, Burgess HJ, Montalbano GE et al. (2022). Morning light therapy in adults with Tourette's disorder. J Neurol 269: 399–410.
Ricketts EJ, Montalbano GE, Burgess HJ et al. (2022). Sleep and chronotype in adults with persistent tic disorders. J Clin Psychol 78: 1516–1539.
Riemann D (2007). Insomnia and comorbid psychiatric disorders. Sleep Med 8: S15–S20.
Riemann D, Spiegelhalder K, Feige B et al. (2010). The hyperarousal model of insomnia: a review of the concept and its evidence. Sleep Med Rev 14: 19–31.

Rivière JB, Xiong L, Levchenko A et al. (2009). Association of intronic variants of the BTBD9 gene with Tourette syndrome. Arch Neurol 66: 1267–1272.

Roessner V, Plessen KJ, Rothenberger A et al. (2011). European clinical guidelines for Tourette syndrome and other tic disorders. Part II: pharmacological treatment. Eur Child Adolesc Psychiatry 20: 173–196.

Romano A, Cundari G, Bruni O et al. (2004). Disturbi da tic e disturbi dell'arousal. Valutazione clinica di 49 bambini e adolescenti [Tic disorders and arousal dysfunction: clinical evaluation of 49 children and adolescents]. Minerva Pediatr 56: 327–334.

Sambrani T, Jakubovski E, Müller-Vahl KR (2016). New insights into clinical characteristics of Gilles de la Tourette syndrome: findings in 1032 patients from a single German center. Front Neurosci 10: 415.

Sankri-Tarbichi AG (2012). Obstructive sleep apnea-hypopnea syndrome: etiology and diagnosis. Avicenna J Med 2: 3–8.

Sateia MJ (2014). International classification of sleep disorders – third edition: highlights and modifications. Chest 146: 1387–1394.

Scammell TE (2015). Narcolepsy. N Engl J Med 373: 2654–2662.

Sheng J, Xie C, Fan DQ et al. (2018). High definition-transcranial direct current stimulation changes older adults' subjective sleep and corresponding resting-state functional connectivity. Int J Psychophysiol 129: 1–8.

Shou S, Li Y, Fan G et al. (2022). The efficacy of cognitive behavioral therapy for tic disorder: a meta-analysis and a literature review. Front Psychol 13: 851250.

Silvestri R, De Domenico P, Di Rosa AE et al. (1990). The effect of nocturnal physiological sleep on various movement disorders. Mov Disord 5: 8–14.

Silvestri R, Ipsiroglu OS (2022). Behavioral sleep medicine – the need for harmonization of clinical best practice outcome measures in children and adolescents with intellectual or developmental disabilities and restless sleep. Front Psychiatry 13: 1003019.

Silvestri R, Raffaele M, De Domenico P et al. (1995). Sleep features in Tourette's syndrome, neuroacanthocytosis, and Huntington's chorea. Neurophysiol Clin 25: 66–77.

Song ML, Park KM, Motamedi GK et al. (2020). Cognitive behavioral therapy for insomnia in restless legs syndrome patients. Sleep Med 74: 227–234.

Sonmez AI, Kucuker MU, Lewis CP et al. (2020). Improvement in hypersomnia with high frequency repetitive transcranial magnetic stimulation in depressed adolescents: preliminary evidence from an open-label study. Prog Neuropsychopharmacol Biol Psychiatry 97: 109763.

Stephens RJ, Chung SA, Jovanovic D et al. (2013). Relationship between polysomnographic sleep architecture and behavior in medication-free children with TS, ADHD, TS and ADHD, and controls. J Dev Behav Pediatr 34: 688–696.

Sukhodolsky DG, Scahill L, Zhang H et al. (2003). Disruptive behavior in children with Tourette's syndrome: association with ADHD comorbidity, tic severity, and functional impairment. J Am Acad Child Adolesc Psychiatry 42: 98–105.

Sverd J, Montero G (1993). Is Tourette syndrome a cause of sudden infant death syndrome and childhood obstructive sleep apnea? Am J Med Genet 46: 494–496.

Thompson MD, Comings DE, Abu-Ghazalah R et al. (2004). Variants of the orexin2/hcrt2 receptor gene in patients with excessive daytime sleepiness and other conditions. Am J Med Genet B Neuropsychiatr Genet 129B: 69–75.

Trajanovic NN, Voloh I, Shapiro CM et al. (2004). REM sleep behaviour disorder in a child with Tourette's syndrome. Can J Neurol Sci 31: 572–575.

Ustinov Y, Lichstein KL, Wal GS et al. (2010). Association between report of insomnia and daytime functioning. Sleep Med 11: 65–68.

Voderholzer U, Müller N, Haag C et al. (1997). Periodic limb movements during sleep are a frequent finding in patients with Gilles de la Tourette's syndrome. J Neurol 244: 521–526.

Vyazovskiy VV, Delogu A (2014). NREM and REM sleep: complementary roles in recovery after wakefulness. Neuroscientist 20: 203–219.

Wahl S, Engelhardt M, Schaupp P et al. (2019). The inner clock-Blue light sets the human rhythm. J Biophotonics 12: e201900102.

Walters AS, LeBrocq C, Passi V et al. (2003). A preliminary look at the percentage of patients with Restless Legs Syndrome who also have Parkinson Disease, Essential Tremor or Tourette Syndrome in a single practice. J Sleep Res 12: 343–345.

Wand RR, Matazow GS, Shady GA et al. (1993). Tourette syndrome: associated symptoms and most disabling features. Neurosci Biobehav Rev 17: 271–275.

Wang Y, Li M, Li W et al. (2022). Is the insula linked to sleep? A systematic review and narrative synthesis. Heliyon 8: e11406.

Woods DW, Piacentini J, Chang S et al. (2008). Managing Tourette syndrome: a behavioral intervention for children and adults therapist guide. Oxford University Press.

Woods DW, Piacentini J, Himle MB et al. (2005). Premonitory Urge for Tics Scale (PUTS): initial psychometric results and examination of the premonitory urge phenomenon in youths with Tic disorders. J Dev Behav Pediatr 26: 397–403.

Zhou Q, Yu C, Yu H et al. (2020). The effects of repeated transcranial direct current stimulation on sleep quality and depression symptoms in patients with major depression and insomnia. Sleep Med 70: 17–26.

Zielinski MR, McKenna JT, McCarley RW (2016). Functions and mechanisms of sleep. AIMS Neurosci 3: 67–104.

Handbook of Clinical Neurology, Vol. 215 (3rd series)
Tic Disorders and Tourette Syndrome
C. Ganos and Y. Worbe, Editors
https://doi.org/10.1016/B978-0-443-13554-5.00027-4

Chapter 11

Tics in autism spectrum and in intellectual disability

TAMMY HEDDERLY[1,2,*], AND OSMAN MALIK[1,3]

[1]*Tic and Neurodevelopmental Movements Service (TANDeM), Evelina London Children's Hospital, Guy's and St Thomas' NHS Foundation Trust, London, United Kingdom*

[2]*School of Life Sciences and Medicine, King's College London, London, United Kingdom*

[3]*South London and Maudsley NHS Foundation Mental Health Trust, London, United Kingdom*

Abstract

Tourette syndrome (TS) is a neuropsychiatric and a neurodevelopmental disorder often presenting alongside other neurodevelopmental conditions such as attention deficit/hyperactivity disorder (ADHD), autism spectrum disorder (ASD), learning difficulties, and dyspraxia. TS shares genetic and neurobiologic features with ASD and ADHD, suggesting common underlying mechanisms. Recent diagnostic frameworks such as the DSM-5 (Diagnostic and Statistical Manual of Mental Disorders, Fifth Edition, APA, 2013) allow for multiple co-occurring diagnoses, reflecting the complex overlap among neurodevelopmental disorders. Whilst there are several studies stating high rates of TS in ASD and vice versa, the data about the nature of tics in those with ASD is sparse; nevertheless, it shows that they are probably not different in nature to tics present in TS. Tics in individuals with ASD can be diagnostically challenging due to overlapping motor behaviors like stereotypies. Intellectual disability (ID) further complicates tic identification, given communication difficulties and co-occurring behaviors. A pragmatic, needs-focused clinical approach is recommended, emphasizing prioritisation of presenting problems over an accumulation of diagnostic labels. This strategy facilitates targeted interventions where needed. In children with ID, differentiating between tics and stereotypies is important to avoid inappropriate management.

Tourette syndrome (TS) is described as a *neuropsychiatric* condition in which motor and phonic manifestations are often accompanied by co-occurring psychiatric and neurodevelopmental conditions; In the DSM-5 (Diagnostic and Statistical Manual of Mental Disorders, Fifth Edition, American Psychiatric Association, 2013), TS belongs to the "neurodevelopmental disorders" group. Given TS is considered a neurodevelopmental disorder, it naturally co-occurs with other neurodevelopmental conditions such as attention deficit hyperactivity disorder (ADHD), learning difficulties, dyspraxia, and autism spectrum disorder (ASD). In the current classifications such as DSM-5, individuals with a diagnosis of ASD can be diagnosed with TS, if they have co-occurring tics that meets the criteria for a diagnosis of TS. In the previous mental health classification systems, there were some arbitrary rules such as having only one main diagnosis, for example, if someone had a diagnosis of ASD, then an additional diagnosis such as that of ADHD was not allowed. Previous classification systems were criticized and replaced with newer diagnostic systems,where clinicians are able to give multiple diagnoses, for instance TS, ASD, ADHD, dyslexia, obsessive-compulsive disorder (OCD), phobia, and anxiety disorder to the same young person; one does not preclude the other. A clinician is not required to choose one diagnosis over another when multiple conditions are present (DSM-5, APA 2013, 2022). On a psychologic level we question the usefulness of giving many labels, and it can be debated whether this is sometimes enabling or disabling for the individual. For clinical purposes, there are some advantages to having one main predominant diagnosis, as long as there is awareness that others can be important secondary diagnoses. Multiple labels can dilute the main impairing symptom and become confusing and we recommend a pragmatic approach of leading with the label which best describes the actual presenting problem or impairment.

TS shares similar multifactorial, complex genetic predispositions and perinatal risk factors with the other neurodevelopmental disorders (Cukier et al., 2014). Some of the proposed brain pathophysiologic mechanisms overlap in the different conditions (Kern et al., 2015) which describe similar brain

*Correspondence to: Tammy Hedderly, MBBS, BSc, FRCPCH, Dept of Children's Neuroscience, 2nd Floor Becket House, St Thomas' Hospital, Westminster Bridge Road, London, SE1 7EH, United Kingdom. Tel: 0207 188 3997, E-mail: t.hedderly@nhs.net.

connectivity alterations in ASD, ADHD, and TS patients (Worbe, 2015). Individuals with TS often have family members with TS, but they also have higher rates of family members experiencing ADHD symptoms, autism, and developmental coordination difficulties without TS. Genetic overlap between TS and ASD has been described by several groups (Fernandez et al., 2012 and Carias and Wevrick, 2019). These overlaps strengthen our understanding of the shared mechanisms between TS and other neurodevelopmental conditions at many levels, that is, genetic, neurobiologic, and structural questionning the idea of these being simply comorbid conditions.

Cravedi et al. (2017) examined 36 studies addressing TS and neurodevelopmental conditions, and their primary findings were as follows: (1) neurodevelopmental comorbidities in TS are the rule, rather than the exception; (2) ADHD is the most frequent; and (3) there is a continuum from simple TS only or TS + ADHD or/and learning disorder and more complex phenotypes (TS + ASD) (Table 11.1).

In this chapter, we will discuss tics in ASD and LD. The relationship between tics and ADHD is discussed in Chapter 8.

AUTISM SPECTRUM DISORDER

ASDs are a diverse group of conditions related to differences in brain development (World Health Organization – https://www.who.int/news-room/fact-sheets/detail/autism-spectrum-disorders). ASD is characterized by difficulties in social interaction and communication. Other characteristics include atypical patterns of activities (with focus on attention to detail and repetitiveness in play or action) and behavioral rigidity such as difficulty with transition from one activity to another and problems with change of routine. Unusual reactions to sensory stimuli are common, such as extreme noise sensitivity or dislike of touch of certain textures. Around 30%–50% of individuals with ASD also have a co-occurring learning disability (LD) and between a quarter to a third have epilepsy. ASD manifests in individuals in multiple different ways and is best consideredas being a spectrum, with some individuals being markedly impaired with little to no language, self-injury, and functional impairment whilst others showing marked islets of ability in specific areas, and showing advanced language skills. One of the challenges in studying ASD and the coexisting personality features is the marked heterogeneity and variation within this spectrum as well as the variation in diagnosis cutoffs used for an identifiable disorder for the person seeking the label.

Table 11.1

Estimated prevalence of neurodevelopmental disorders in Tourette syndrome (TS).

Condition	Prevalence (range) in TS (%)
Autism spectrum disorder	3–20
Attention deficit/hyperactivity disorder	40–60
Dyspraxia	20
Specific learning disability (LD, dyslexia/dyscalculia)	16–23
LD	22.7

The diagnostic system has been continually refined over the years, with standardization of assessment tools; nevertheless, a diagnosis of autism is often a subjective observation of a particular social communication style. Diagnosis will vary from clinician to clinician, within different cultures, countries, and even within local clinical services. There are few currently available biomarkers to aid classification or to help standardize language or diagnostic labels around the various clinical presentations. Lord (2012) attempted to verify the traditional clinical subtyping of autism (such as atypical autism, childhood autism, pervasive developmental disorder, and Asperger syndrome) across 12 specialist centers in the United States, which were employing commonly available standard diagnostic assessments. The authors found that the use of these subtype diagnoses was idiosyncratic within the different centers, and had little transferable validity across the whole cohort. This finding influenced the removal of these clinical subtypes within the Diagnostic and Statistical Manual of Mental Disorders in its latest edition (DSM-5, American Psychiatric Association 2013, American Psychiatric Association, 2022). By contrast, new nosologic classifications across large-scale databases successfully identified two independent symptom domains (rather than the previous so-called triad of impairments) as the best description of the behavioral phenotype of ASD, with some convergent support from genetic analysis (Snow et al., 2009; Frazier et al., 2012; Veatch et al., 2014). The two core domains include persistent deficits in social communication and social interaction, together with restricted, repetitive patterns of behavior, interests, or activities.

The DSM-5 not only removed the subgroups of pervasive developmental disorders (renamed autism spectrum disorder, ASD) but explicitly encouraged the use of "qualifiers" to describe coexisting medical, neurodevelopmental, and mental health disorders; the International Classifications of Disease, 11th revision (ICD-11) is a globally recognized classification system for diseases, which used a similar approach. These changes recognized that comorbidity in ASD was being increasingly reported (Simonoff et al., 2008) and therefore enabled co-diagnosis and improved access for the treatments of common coexisting disorders with ASD, for example ADHD, anxiety, and OCD. The hope was that these changes would improve the quality of care for young people experiencing difficulties from ASD, rather than clinicians describing the problems as "all being due to the autism itself."

There is a current move toward discussions around being neurodiverse, and some clinicians prefer to use the term autistic or autism spectrum condition (ASC) as opposed to applying the word "disorder." There is controversy in this area, as young people with significant impairments can feel that their own problems are being undermined to some extent, by not using the term "disorder" and readers should be familiar with the general arguments for and against diagnostic labeling and the use of neuroaffirmative language. It is recommended that clinicians hold discussions within the wider multidisciplinary forums and also directly with the presenting adult, child, young person and their family . This will allow for canvassing of the wider views and an appreciation of the benefit or not of using the label ASD as well as allowing for discussions around some of the complexities of the use of labels.

PREVALENCE OF TICS/TS IN ASD AND ASD IN TS

Within the context of the changing landscape of ASD, studies nevertheless highlight that almost all neurodevelopmental disorders frequently co-exist with TS and viewing this in a circular way, the prevalence of TS is higher in most neurodevelopmental disorders, indicating that having a neurodevelopmental diversity or disorder is a risk factor for having tics. A discussion needs to be held around whether having tics within a child already diagnosed with autism, autism spectrum condition (ASC), or ASD should be accepted as an expected and common feature, rather than needing an additional diagnosis of TS. It may be meaningful to add a Tourette diagnosis where tics are the prominent or impairing symptom, especially when an intervention is required. A TS diagnosis could be problematic in someone with ASD who is experiencing some mild multiple but nonimpairing chronic motor and phonic tics, even though strictly speaking they meet the DSM-5 diagnostic criteria for a TS label. It can lead to a redirection of care to focus on the motor symptoms when the social difficulties could predominate. Kahl et al. (2015) looked at tics of 21 individuals with ASD and compared these to tics in 16 individuals with TS, and they found that tics in ASD were indistinguishable from tics in TS and were similarly distributed, but less severe. They did find that tic awareness was limited in ASD group, again highlighting a potential harm that could come from additional focus on these motor symptoms.

Individuals with ASD have been reported to show a six times higher chance of having features of TS compared with the general population (Baron-Cohen et al., 1999). The reported TS rate in ASD population varies from 2.6% to 11% as noted in the following studies. In a study from Italy (Canitano and Vivanti, 2007) with a clinical cohort of 105 young people with ASD, 11% were diagnosed with TS (based on DSM-IV criteria, Yale global tic severity scale YGTSS). In England, Baron-Cohen's group (1999) had a clinical cohort of 458 young people with ASD and found TS in 6.2% (based on NHIS, YGTSS, DSM-III–IV criteria).

Looking from the alternative perspective, studies looking for comorbid ASD in individuals with a TS diagnosis highlight a prevalence of ASD from 2.9% to 20% in TS (Canitano and Vivanti, 2007). Kadesjö and Gillberg (2000) examined 435 school-age children and found a 0.15% prevalence of TS. Among the 10 children diagnosed with TS, only 1 had a comorbid ASD (20%).

Another larger epidemiologic study conducted in Sweden (Cravedi et al., 2017) on a non-clinical population of 4479 children measured a 0.6% frequency of TS. In the TS group, the rate of comorbid ADHD, evaluated using the DSM-IV criteria, was 68%; the rate of ASD was 20% (16% Asperger, 4% PDD); the rate of dyslexia was 16%; and the rate of developmental coordination disorder was 20%. This study also included a broader phenotype and older classification terms such as Asperger's. In the previous version of diagnostic classifications systems (both DSM and ICD), autism was divided into subtypes and atypical forms which did not include the full range of ASD features.

These studies indicate a wide variation in ASD prevalence in TS, but if we look in more detail, we can highlight some limitations in this research. As discussed above, the definitions of ASD vary widely and the diagnostic assessments have also changed over time. Some studies include a broader description of ASD or just traits of ASD, which makes it appear as if the prevalence is very high within the TS group. What these studies are actually probably reporting is a high prevalence of *Traits of ASD*.

Eapen and colleagues conducted a study (Eapen et al., 2019), looking at autistic traits, as measured by the social communication questionnaire (SCQ) in individuals with TS ($n = 44$), comparing them to those with ASD ($n = 26$) and a general population cohort ($n = 133$); they found that the TS sample had significantly higher SCQ scores compared to the general populations but lower scores than the ASD sample. The group differences in the mean SCQ scores between the TS and ASD sample were significant except in the domain of restricted repetitive behaviors. Authors concluded that ASD traits occur commonly in the TS population with a significant overlap in certain clinical features, especially the case for complex movements or repetitive behaviors, which they believed may represent a shared phenotype which is subclinical, or a phenocopy where some clinical symptoms mimic each other or a comorbidity.

In our clinical experience we would promote the use of a dimensional system approach in a clinical setting. This is a system which allows for identification of the important psychosocial and environmental problems in a young person's life, whilst addressing this in the context of the choice of the most useful clinical label to apply. This system helps to then target interventions to support these specific presenting difficulties; for example, a young person with labels of ASD, and ADHD with minor tics, may fulfill the theoretical criteria for a diagnosis of TS but the tics may not be bothering them at all. In a child who presents with motor tics, but where bullying, social communication, and friendship issues are more significant problems, we would then recommend considering the diagnosis of ASD to be the primary one in this case. In this situation, if autism is then diagnosed, it can be helpful to discuss the tics as being expected within this spectrum. We discuss the shared mechanisms and neurogenetic underpinnings within TS/ASD. We would go on to explore the specific reported difficulties and plan interventions to target these areas. Conversely, a young person with a diagnosis of ASD for example, who has settled socially in college life, is functioning well in the family but who reports being "tormented by tics," could benefit from an additional TS label. This would allow for wider access to tic therapies and targeted tic psychoeducation. By using this axis approach, young people could be given more than one primary diagnostic label but these would be ranked using the primary problems as the area of main focus, rather than leaving a long list of labels, of uncertain significance with little weight on impairment.

Case example 1

AT is an 11-year-old shy quiet boy who presented with tics that had become bothersome, impacting on his ability to stay in the classroom at school, due to increasing anxiety. He was described to have bouts of tic attacks. His classmates were bullying him.

Interventions for tics such as comprehensive behavioral intervention for tics (CBIT) and clonidine medication were only partially effective in alleviating the tic-like attacks and wider social difficulties. Following an assessment of his developmental, cognitive, and psychologic profile, combined with observations in school, he received a diagnosis of ASD. He suffered high levels of sensory processing difficulties, anxiety, and comorbid dyslexia. Interventions included literacy support, emotional well-being counseling, and a friendship mentor at school. These led to an improvement in his general well-being as well as a marked reduction of the tic attacks. In his case, we formulated that the tic-like attacks had been surface markers for the ASD and the autism-related problems with marked social distress. Putting in place further modifications to reduce sensory overload, and supporting the academic difficulties made a huge difference to the tics, and the tic attacks resolved over a six-month period. Tic therapies and tic-related medication have very little effect for clear reasons.

People with ASD can present with symptoms also commonly reported in TS, examples being compulsive behaviors, sensory hypersensitivity, echolalia (Ganos et al., 2012) and misophonia (Robinson et al., 2018; Balachandar et al., 2025).

There can be an overlap between features of OCD and ASD as well with OCD being very common in young people with TS. A meta-analysis by van Steensel showed that 17.4% of young people with ASD also met the diagnostic criteria for having OCD. Conversely, looking at a cohort of individuals with OCD, Martin et al. found that 25% may also meet the diagnostic criteria for ASD. This highlights the fact that when clinicians are seeing young people in the TS clinic without a formal diagnosis of ASD but with obsessive compulsive behaviours, they may have undiagnosed ASD or many may have traits of ASD (broader phenotype). As mentioned above, the diagnostic criteria for ASD have changed over the years and the diagnostic scope is broad now. This can be a challenge for both patients and clinicians but leads to an important question of whether we need to label and diagnose all possible conditions, many of which may simply be differences than disorders.

In our clinical experience working with such children who have been diagnosed with ASD, there can be certain tics that are especially bothersome. This is due, in part, to attentional bias, or difficulty managing sensory urges and/or having high baseline anxiety. There is anecdotal observation that ASD can, therefore, be a risk factor for tic attacks or functional tic-like behaviors (FTLBs). Recent studies indicate that ASD is highly prevalent in people with a functional neurologic disorder (González-Herrero et al., 2022). We and others have also found a high prevalence of ASD in young people with FTLBs (Buts et al., 2021; Martino et al., 2022). The diagnosis of FTLBs often requires the clinician working with this group of children to have specialist knowledge of both TS and ASD.

Case example 2

An adolescent female AM, aged 15 years, presented with functional tic-like movements. She had a past history of minor tics which had settled several years before. There was history of recent trauma predating the onset of the current tic-like movements, which had been diagnosed as a "functional neurologic disorder." There was a family history of tics, a clear history of early onset of mild motor and phonic tics, aged 5 years, and then almost complete remission of these minor tics, aged around 13 years, in keeping with TS. The family reported that over the last 6 months it had become impossible to go on a car journey together or on a family holiday as AM was shouting and screaming at her brother, becoming severely distressed whenever he was speaking, singing, or eating next to her. Following an assessment, AM was showing misophonia to her brother's voice, and she showed severe aversive emotional arousal with irritation, which led to an increase in functional tic-like movements. We explored the associated difficulties she had with recognizing emotions (alexithymia) and performed a full MDT (multi disciplinary team) comprehensive assessment revealing that AM also reached diagnostic cutoff for ASD, which had not been recognized in the past. This formulation led to signposting to ASD strategies around routine, managing change, exposure work for the sensory sensitives together with some emotional literacy training. The misophonia settled and the functional symptoms subsided over the next 4 months.

It is not known yet empirically whether sitting with the physical tic urge during ERP is harder for children with ASD, even though anecdotally it is reported. Interoception is the ability to pick up internal bodily sensations, and this may be different and deficient in individuals with ASD, which has also been shown at the structural-and-functional level with hypo-connectivity of the insula and less gray matter volume in the insula (Radeloff et al., 2014) in autistic individuals. At the same time, Ganos et al. (2015) found that patients with TS have overall less interoceptive awareness when compared to controls, although within TS subjects high levels of interoceptive awareness likely reflects a self-attentive capacity to perceive urges. The impact and awareness of interoception have not been studied empirically in those with both ASD and TS. This remains an important area to be explored, given that tic severity can be linked with urge perception and the fact that current tic treatments rely heavily on management of tic through awareness of the premonitory urges.

Another area of challenge for individuals with ASD with co-occurring tics, is the impact of emotions and emotional processing. It is well known that anxiety and other emotions (anger/excitement) can worsen tics. Often the management of tics includes managing emotional difficulties and anxiety. This work can result in a reduction in tic intensity, frequency, and impact. However, individuals with ASD have difficulties with the recognition of emotion (alexithymia) and vary in emotional processing (Philip et al., 2010). This makes it harder to recognize and address emotional factors contributing to any tic worsening, nevertheless, highlights an important area that clinicians should try to uncover in the TS patients with ASD. It is by including an exploration of emotions and environmental factors, that clinicians working with young people can improve their care with targeted interventions. Indeed, one further case example below highlights the complexity that presents in children with ASD and tics. In our clinical experience (at a tertiary Tourette service in London, United Kingdom), children with comorbid ASD often present with the most impairing symptoms, presenting some additional challenges for optimal intervention and a need to especially support the mental health. An audit conducted in our service indicated that 48% of all the parents attending the tic psychoeducation group had a child with TS + ASD (all parents attending the initial assessment are offered this group). Children with comorbid ASD form a higher proportion of our clinical case load based on the expected prevalence of ASD in TS (3%–20%). The case below captures an example of some complexity and need.

Case example 3 (with treatment details)

AW, a 15-year-old male, presented to clinic with severe tiredness and worsening tics. He had a prior diagnosis of TS, ASD, and OCD. He had a history of chronic fatigue syndrome (from 8 years of age), with early separation anxiety seen soon after primary school entry aged 5 years. On assessment, AW had florid motor and phonic tics, which included explosive speech, coprolalia, facial grimacing, tapping/hitting self, and stamping. Some tics were reported to cause physical harm (e.g., falling up the stairs), with him experiencing extended bouts of tics and anxiety-related "tic-like attacks" (Robinson & Hedderly, 2016) that were reported to cause muscular aches/pains. AW had difficulties with impulse control that included rage where items may be broken (e.g., plates, cups), episodes where he would pick up knives and run toward people whilst stating that he will harm them (e.g., brother), and stripping his clothes off and running around naked. AW had been home educated since being aged 10 years, due to problems reintegrating at school following a period of illness with chronic fatigue symptoms and frequent absconding. AW was diagnosed with FTLB and anxiety. He was referred for psychologic therapy (Table 11.2).

Table 11.2

AW treatment plan.

Sessions 1 and 2: Psychoeducation and goal setting
• Understanding functional neurologic disorder, tics, and anxiety • External attention experiments (e.g., finger counting and the five senses game) • Discussions around possible advantages/disadvantages of symptoms (e.g., being left alone, sleeping less, missing college, not playing rugby) • Treatment goals (attending college, going to music gigs)
Sessions 3 and 4: Role of attention (self/others) and formulation
• Behavioral analysis/monitoring of tic-like attacks/recording antecedent, behavior, and consequences • CBT (Cognitive behaviour therapy) formulation of tic-like attacks, with consideration of attentional focus, triggering thoughts (e.g., I am going to have one) and safety behaviors (e.g., parents putting cushion under head, standing at bottom of stairs) • External attention practice (e.g., daily listening to sounds)
Sessions 5–8: Behavioral experiments and thought challenging
• Role of attention on symptoms (e.g., performing daily tasks whilst focusing attention on symptoms/away from symptoms such as by humming/singing). • Shifting attentional focus (e.g., walking up/downstairs whilst focusing externally by thinking about the purpose of going up/downstairs) • Graded exposure (e.g., going into town, music gigs, college interview, catching bus)
Sessions 9 and 10: Review and relapse prevention
• Review of goals and treatment gains • Plan for the future

Following treatment, the motor and phonic tics had settled and were no longer interfering with day-to-day functioning. Functional tic-like attacks had stopped and an improvement in his impulse control difficulties was reported. AW exhibited increased insight into his emotions and a better understanding of anxiety and tic triggers, with him proactively managing these situations by anticipating his attention to what was going on around him and using cognitive behavioral (CBIT) techniques. Parents reported reduced concerns and increased confidence in him going out independently. He had enrolled in college and had started to engage in a range of social and recreational activities. On the clinician-rated CGAS (clinical global assessment scale), pretreatment score = 31 (Serious Problems) and post-treatment score = 78 (Doing All right).

INTELLECTUAL DISABILITY/LEARNING DISABILITY

Intellectual disability (ID) involves problems with general mental abilities that affect functioning in two areas: (1) intellectual functioning (such as learning, problem-solving, judgment) and (2) adaptive functioning (activities of daily life such as communication and independent living). Additionally, the intellectual and adaptive deficit begin early in the developmental period (American Psychiatric Association, 2013). ID affects about 1% of the population, and of those approximately 85% have mild ID. Males are more likely than females to be diagnosed with ID (American Psychiatric Association, 2013). This is further divided into mild, moderate, severe, and profound based on IQ (intelliegnce qoutient). The term learning disability (LD) refers to weaknesses in certain academic skills such as reading, writing, and math being the main ones, although this term is used mostly in the educational/academic context but some clinicians use these interchangeably.

In children with ID and LD, tics are common (as we will see in the prevalence below); however, it can be difficult to identify tics because many children with a ID have co-occurring motor stereotypies and stereotyped behaviors. Although clinicians can often differentiate common stereotypies such as hand-flapping or rocking from tics some more complex motor stereotypies may be inadvertently interpreted as motor tics. Conversely, some tics may be missed in children with prominent motor stereotypies and motor restlessness. Lastly some children have both stereotypies and tics; for example, a study of children with excessive motor stereotypies (diagnosed as stereotypic movement disorder) showed that 23% of them also had chronic tics which presents clinical challenge in diagnosis.

We have also seen many children with vocalizations that have been described as tics, when they are consistent with more pleasurable phonic stereotypies. It is not always necessary to distinguish between the two movements, but it is important if either of the two is problematic and requires an intervention, as management approaches between stereotypies and tics vary.

In pediatric special education LD classes, the prevalence of individuals with tic disorders is reported to be around 20%–45% (Eapen et al., 2013).

Burd et al. (2005) identified phenotype differences between TS patients with LD and TS patients without LD. They used the Tourette Syndrome International Consortium (TIC) database, with data from 5450 subjects. The overall prevalence of LD in this sample of TS patients was 22.7%. Subjects with TS + LD had an increased proportion of males compared to TS − LD ($p < 0.001$) and had an age of onset of TS before 8 years of age ($p = 0.030$). The average age first seen for those with TS + LD was 12.5 (s.d. = 7.5), while it was over three years later (mean = 15.7, s.d. = 11.6) for those with TS − LD. Seventy-four percent of the cases of TS + LD were diagnosed by a psychiatrist, and 19% were diagnosed by a neurologist ($p < 0.001$).

In another evaluation of 151 children with LD on the *TIC* database, it was noted that 24% had ASD, 33% had pre- and perinatal problems, and 25% had seizures (in contrast to 5% children on the database without LD having seizures). This indicates that children with TS and comorbid LD present a complex group. In another study (William et al., 2013), the rate of seizures in sample of 347 patients with TS was 4.8% in those without LD, and 18.2% in those with a LD.

There a number of studies looking at cognitive and neuropsychologic functioning in TS such as having lower IQ scores than the general population (by less than one standard deviation) (Debes et al., 2011). The authors of a study postulated that early onset of tics and the presence of comorbidities might cause specific deficits on cognitive performance. Other studies have reported similar results such as the impact of comorbid ADHD in TS on lowering of full-scale IQ (Schuerholz et al., 1996). We would advise the reader to look at the detail in the chapter on neuropsychology for comprehensive discussion in this area.

Children with ID/LD may struggle to talk about or describe the problems they experience from tics, or they might not be able to tell parents that they are not as bothered by the tics, as they appear. We have seen many children who have seemingly extreme tics, but are fine with them and happily refuse medications and other interventions; it is therefore important to establish a good rapport with a child with ID/LD and facilitate communication to understand how much are they affected by the tics. Children with ID/LD who have Tourette-related phenomenon such as coprolalia may sometimes be seen as deliberately using the inappropriate words. It is important to differentiate between behavior and coprophenomenon. Giving them consequences or highlighting the coprophenomenon only make these more entrenched due to the associated suggestibility phenomena. The case example below highlights this issue.

Case example 4

SN aged 13 years, presented with swearing in school, which had led to a recent school expulsion. He suffered from a range of motor and vocal tics for many years. SN also had a diagnosis of moderate ID, TS, and a deletion on chromosome 9. He was attending a special needs school and was facing difficulties with the threat of expulsion. The swearing was considered "bad behaviour," which was not initially recognized at school as being coprolalia or as being tic-related. As as SN showed no embarrassment or remorse, due to the lack of social awarness related to a co-exisiting ASD diagnosis. We presented the formulation to the parents and the school staff, that this was best managed as coprophenomena. Various tic-related phenomena (such as coprolalia, copropraxia, echopraxia, compulsive touching, or tapping) are misconstrued as deliberate behaviors by some professionals and even by parents. This under-recognition of coprophenomena in this group can result in poor management of tic-related phenomenon, with unnecessary punishment/repercussion for children with ID. This can impact negatively on their self-esteem and mental health. In SN's case, this coprolalia was becoming entrenched, as he was being told off frequently and punished for it at school and at home. The school followed a strict plan of ignoring the coprolalia and allowing him a safe space to release his tics. SN's mother found the swear words to be extremely upsetting; especially at night and when at church. This resulted in him being told off, and when he was attending church with his mother, it became very embarrassing for her when he swore. Thes added pressure on SN to not swear, led to anxiety and worsening of coprolalia. The focus of our therapeutic work was with his parents, family, and school to remind the family of focused non suggestibility, diverting of attention away from the episodes and general stress reduction. The outcome of the intervention led to a significant reduction in coprolalia.

SUMMARY

Tics are common in individuals with neurodevelopmental disorders, ASD and ID. This group of young often present with complex challenges and they have high needs. For people experience combination of Tics, ASD and ID, the management will be benefit if the clinicians have a wider understanding of co-occuring conditions. Assessments and formulations will often need this wider view of the developmental differences in order to target therapy or advice. The clinical focus should be on the primary presenting problem even when there are multiple diagnoses. More research is needed to fully understand the interplay between certain phenomenon such as tic urges in those with ASD, ERP (exposure and response prevention) treatment response, and interoceptive awareness in those with both ASD and tics.☆

References

American Psychiatric Association (2013). Diagnostic and statistical manual of mental disorders, 5th ed. American Psychiatric Publishing.

American Psychiatric Association (2022). Diagnostic and statistical manual of mental disorders (DSM). 5th ed., text revised. Arlington, VA: American Psychiatric Publishing.

Balachandar A, Grippe TC, Lim WK et al. (2025). Misophonia in tic disorders and their neuropsychiatric associations. Mov Disord Clin Pract 12: 858–859.

Baron-Cohen S, Scahill VL, Izaguirre J et al. (1999). The prevalence of Gilles de la Tourette syndrome in children and adolescents with autism: a large scale study. Psychol Med 29: 1151–1159.

Burd L, Freeman RD, Klug MG et al. (2005). Tourette syndrome and learning disabilities. BMC Pediatr 5: 34.

Buts S, Duncan M, Owen T et al. (2021). Paediatric tic-like presentations during the COVID-19 pandemic. Arch Dis Child 107: e17.

Canitano R, Vivanti G (2007). Tics and Tourette syndrome in autism spectrum disorders. Autism 11: 19–28.

Carias KV, Wevrick R (2019). Clinical and genetic analysis of children with a dual diagnosis of Tourette syndrome and autism spectrum disorder. J Psychiatr Res 111: 145–153.

Cravedi E, Deniau E, Giannitelli M et al. (2017). Tourette syndrome and other neurodevelopmental disorders: a comprehensive review. Child Adolesc Psychiatry Ment Health 11: 59.

Cukier HN, Dueker ND, Slifer SH et al. (2014). Exome sequencing of extended families with autism reveals genes shared across neurodevelopmental and neuropsychiatric disorders. Mol Autism 5: 1.

Debes NM, Lange T, Jessen TL et al. (2011). Performance on Wechsler intelligence scales in children with Tourette syndrome. Eur J Paediatr Neurol 15: 146–154.

Eapen V, Črnčec R, McPherson S et al. (2013). Tic disorders and learning disability: clinical characteristics, cognitive performance and comorbidity. Australas J Spec Educ 37: 162–172.

Eapen V, McPherson S, Karlov L et al. (2019). Social communication deficits and restricted repetitive behavior symptoms in Tourette syndrome. Neuropsychiatr Dis Treat 15: 2151–2160.

Fernandez TV, Sanders SJ, Yurkiewicz IR et al. (2012). Rare copy number variants in Tourette syndrome disrupt genes in histaminergic pathways and overlap with autism. Biol Psychiatry 71: 392–402.

Frazier TW, Youngstrom EA, Speer L et al. (2012). Validation of proposed DSM-5 criteria for autism spectrum disorder. J Am Acad Child Adolesc Psychiatry 51: 28–40.e3.

Ganos C, Garrido A, Navalpotro-Gómez I et al. (2015). Premonitory urge to tic in Tourette's is associated with interoceptive awareness. Mov Disord 30: 1198–1202.

Ganos C, Ogrzal T, Schnitzler A et al. (2012). The pathophysiology of echopraxia/echolalia: relevance to Gilles De La Tourette syndrome. Mov Disord 27: 1222–1229.

González-Herrero B, Morgante F, Pagonabarraga J et al. (2022). Autism spectrum disorder may be highly prevalent in people with functional neurological disorders. J Clin Med 12: 299.

Kadesjö B, Gillberg C (2000). Tourette's disorder: epidemiology and comorbidity in primary school children. J Am Acad Child Adolesc Psychiatry 39: 548–555.

Kahl U, Schunke O, Schöttle D et al. (2015). Tic phenomenology and tic awareness in adults with autism. Mov Disord Clin Pract 2: 237–242.

Kern JK, Geier DA, King PG et al. (2015). Shared brain connectivity issues, symptoms, and comorbidities in autism spectrum disorder, attention deficit/hyperactivity disorder, and Tourette syndrome. Brain Connect 5: 321–335.

Lord C (2012). A multisite study of the clinical diagnosis of different autism spectrum disorders. Arch Gen Psychiatry 69: 306.

Martino D, Hedderly T, Murphy T et al. (2022). The spectrum of functional tic-like behaviours: data from an international registry. Eur J Neurol 30: 334–343.

Philip RCM, Whalley HC, Stanfield AC et al. (2010). Deficits in facial, body movement and vocal emotional processing in autism spectrum disorders. Psychol Med 40: 1919–1929.

Radeloff D, Ciaramidaro A, Siniatchkin M et al. (2014). Structural alterations of the social brain: a comparison between schizophrenia and autism. PLoS One 9: e106539.

Robinson S, Hedderly T (2016). Novel psychological formulation and treatment of 'Tic Attacks' in Tourette syndrome. Front Pediatr 4: 46.

Robinson S, Hedderly T, Conte G et al. (2018). Misophonia in children with tic disorders. J Dev Behav Pediatr 39: 516–522.

Schuerholz L, Baumgartner T, Singer H, Reiss A, Denckla M (1996). Neuropsychological status of children with Tourette's syndrome with and without attention deficit hyperactivity disorder. Neurology 46: 958–965

Simonoff E, Pickles A, Charman T, Chandler S, Loucas T, Baird G (2008). Psychiatric disorders in children with autism spectrum disorders: prevalence, comorbidity, and associated factors in a population-derived sample. J Am Acad Child Adolesc Psychiatry 47: 921–929

Snow AV, Lecavalier L, Houts C (2009). The structure of the Autism Diagnostic Interview-Revised: diagnostic and phenotypic implications. J Child Psychol Psychiatry 50: 734–742.

Veatch OJ, Veenstra-VanderWeele J, Potter M et al. (2014). Genetically meaningful phenotypic subgroups in autism spectrum disorders. Genes Brain Behav 13: 276–285.

William D, Stern J, Grabecki K et al. (2013). Epilepsy in Tourette Syndrome. J Neurol Neurosurg Psychiatry 84: e1

Worbe Y (2015). Neuroimaging signature of neuropsychiatric disorders. Curr Opin Neurol 28: 358–364.

☆Covering the association between tics and neurodevelopmental disorders, including autism spectrum disorder, intellectual disability/learning disabilities.

Further reading

Huisman-van Dijk HM, van de Schoot R, Rijkeboer MM et al. (2016). The relationship between tics, OC, ADHD and autism symptoms: a cross-disorder symptom analysis in Gilles de la Tourette syndrome patients and family-members. Psychiatry Res 237: 138–146.

Khalifa NA, Von Knorring A-L (2006). Psychopathology in a Swedish population of school children with tic disorders. J Am Acad Child Adolesc Psychiatry 45: 1346–1353.

Martin AF, Jassi A, Cullen AE et al. (2020). Co-occurring obsessive–compulsive disorder and autism spectrum disorder in young people: prevalence, clinical characteristics and outcomes. Eur Child Adolesc Psychiatry 29: 1603–1611.

van Steensel FJA, Bögels SM, Perrin S (2011). Anxiety disorders in children and adolescents with autistic spectrum disorders: A meta-analysis. Clin Child Fam Psychol Rev 14: 302–317.

Handbook of Clinical Neurology, Vol. 215 (3rd series)
Tic Disorders and Tourette Syndrome
C. Ganos and Y. Worbe, Editors
https://doi.org/10.1016/B978-0-443-13554-5.00005-5

Chapter 12

Sexuality and sexual health in primary tic disorders

AGNES SANTER[1,2,*], TILL AMELUNG[3,4], TINA MAINKA[5], AND CHRISTOS GANOS[6,7,8]

[1]*Department of Neurology, Medical University of Vienna, Vienna, Austria*

[2]*Department of Neurology, Klinik Ottakring, Vienna, Austria*

[3]*Institute of Sexology and Sexual Medicine, Charité - University Medicine Berlin, Corporate Member of Free University of Berlin and Humboldt University of Berlin, Berlin, Germany*

[4]*Department of Psychology, Medical School Berlin, Berlin, Germany*

[5]*Department of Neurology with Experimental Neurology, Charité - University Medicine Berlin, Berlin, Germany*

[6]*Edmond J. Safra Program in Parkinson's Disease and the Morton and Gloria Shulman Movement Disorders Clinic, Toronto Western Hospital, Toronto, ON, Canada*

[7]*Division of Neurology, University of Toronto, Toronto, ON, Canada*

[8]*Krembil Brain Institute, Toronto, ON, Canada*

Abstract

Sexuality and sexual health are essential for the overall well-being of all individuals. In neurology, research traditionally focused on disorders of sexual function associated with syndromes that present with specific lesions of the spinal cord and brain. In neuropsychiatric tic disorders, however, early research addressed sexuality and sexual behaviors in a wider context and beyond the strict diagnostic categories of sexual dysfunction, including gender identity and sexual orientation. Although these studies provided important insights in the behavioral structure of adults with primary tic disorders and Tourette syndrome, many of these early findings were not replicated subsequently and should, therefore, be critically reconsidered. This chapter provides a comprehensive catalog of existing data on sexuality and sexual health, including gender identity, sexual orientation, sexual dysfunction, adverse sexual experiences, compulsive sexual behaviors, and paraphilic disorders in adults with primary tic disorders. It also discusses the role of comorbidities, particularly attention-deficit/hyperactivity disorder, as well as the impact of neuropsychiatric medication. Additionally, it catalogs, for the first time, different therapeutic approaches that may improve sexual health in adults with primary tic disorders. Finally, knowledge gaps and methodological limitations are also addressed. Ultimately, this chapter's goal is to carefully present existing data, map underexplored needs, and highlight ways to respectfully address them both in clinical practice and through future research.

INTRODUCTION TO SEXUALITY AND SEXUAL HEALTH IN PEOPLE WITH PRIMARY TIC DISORDERS

Sexual health is defined by the World Health Organization (WHO) as "a state of physical, emotional, mental, and social well-being related to sexuality" (World Health Organization, 2006). This definition acknowledges many different aspects of human sexuality and sexual health that also include sexual wellbeing, sexual justice, and sexual pleasure (Mitchell et al., 2021). In neurology, most research in the field of human sexuality focused on the impact of specific disorders on sexual function (Rees et al., 2007). However in tic disorders, perhaps due to the rather unusual position they traditionally occupied in neurology, several early studies addressed aspects of human sexuality beyond sexual function, including gender identity, sexual orientation, and sexual practice (Moldofsky et al., 1974; Nee et al., 1980; Comings and Comings, 1982; Robertson et al., 1988; Sandyk, 1988; Comings, 1994). Although these studies offered some important insights into the structure of human sexual behavior in adults with tic disorders, it is important to

*Correspondence to: Agnes Santer, Department of Neurology, Klinik Ottakring, Montleartstraße 37, 1160 Vienna. Austria. Tel: +43-1-49150-2001. E-mail: agnes.santer@gesundheitsverbund.at.

note that many medical concepts of human sexuality have drastically changed since then, including research methodology. Moreover, knowledge about tic disorders and Tourette syndrome (TS) has also increased and, therefore, the adult populations that were included in early studies may strongly differ from those who present to clinics today. Given the relevance and increasing scientific interest in exploring human sexual behavior with a positive regard to healthy sexual expression, this chapter attempts to critically review the existing literature on sexuality and sexual health in primary tic disorders. The chapter's goal is to carefully present existing knowledge, map underexplored needs, and highlight ways to respectfully address them both in clinical practice and through future research.

SEXUAL DYSFUNCTION IN PRIMARY TIC DISORDERS

In the International Statistical Classification of Diseases and Related Health Problems (ICD)-11, sexual dysfunctions are defined as "syndromes that comprise the various ways in which adult people may have difficulty experiencing personally satisfying, non-coercive sexual activities" (WHO, 2019/2021). These include hypoactive sexual desire dysfunction, sexual arousal dysfunctions, orgasmic dysfunctions, and ejaculatory dysfunctions (WHO, 2019/2021). Clinically significant distress must be present to receive an ICD-11 diagnosis.

Several case reports of adults with primary tic disorders, and specifically TS, have documented the presence of sexual dysfunction, typically as a secondary consequence of medication (Duncan and Bateman, 1993; Meston et al., 1997; Bobes et al., 2003; Scahill et al., 2003; Coskun and Zoroglu, 2009; Roessner et al., 2011; Park et al., 2012; Camporeale et al., 2013; Ganos and Martino, 2015; Chen et al., 2017; Pizzol et al., 2019) or tic-related injury (Krauss and Jankovic, 1996). In the example of antipsychotics, drugs such as haloperidol can lead to hyperprolactinemia, which can result in side-effects such as galactorrhea, gynecomastia, and sexual dysfunction (Bobes et al., 2003; Scahill et al., 2003; Dossenbach et al., 2006). Selective serotonin reuptake inhibitors (SSRIs), such as paroxetine and escitalopram, may also typically lead to different sexual dysfunctions related to reduced sexual desire, low arousal, and orgasmic and ejaculatory dysfunction. Table 12.1 summarizes some of the typical medication-related side-effects relevant to sexual function. In addition, thalamic deep brain stimulation led to sexual dysfunction in a 21-year-old person with TS. Approximately 3 months after the stimulation was started, he noted a reduction in his libido. Sexual dysfunction persisted for the entire period of neurostimulation and improved only when the neurostimulator was removed due to infection of the device (Zhang et al., 2016).

To date only one questionnaire-based study provided some insights into sexual desire in a sample of 358 people with TS (mean age 28 years). In this 1994 study, participants were asked whether they experienced an "aversion to being touched" and an "aversion to sex." More than half of the female participants with TS and comorbid attention-deficit/hyperactivity disorder (ADHD) reported being averse to being touched (53.1%) and sex (52.2%). This number dropped to 16.3% and 15.2%, when examining men and women with TS but without comorbid ADHD. Notably, there was also a gender difference when healthy controls were asked the same questions: 7.1% of male healthy controls reported an aversion to being touched and none of male healthy controls reported an aversion to sex, whereas 11.4% and 15.9% of female healthy controls said that they had an aversion to being touched and an aversion to sex, respectively (Comings, 1994). Despite this interesting observation, the lack of specificity in the questions makes it difficult to distinguish between a primary sexual arousal disorder, specific phobias, or also disorders related to sexual pain. Indeed, more research is needed to carefully explore primary sexual dysfunction and separate it from secondary problems, for example treatment-related side-effects.

There may be a relationship between sexual arousal and tic frequency, as sexual arousal is associated with increases in sympathetic tone at all stages of the sexual response cycle (Meston and Bradford, 2007). Given that tics may increase under conditions of autonomic arousal (Nagai et al., 2009), it is plausible that tic frequency also increases during sexual arousal.

HYPERSEXUALITY AND COMPULSIVE SEXUAL BEHAVIOR IN PRIMARY TIC DISORDERS

The concepts and definitions of hypersexuality and sexual compulsivity have changed over time making it challenging to interpret data of earlier medical reports from today's perspective. In the Diagnostic and Statistical Manual of Mental Disorders(DSM)-IV, the diagnosis of "sexual disorders not otherwise specified" allowed for a diagnosis that included hypersexual behavior (Krueger, 2016). However, the diagnosis and definition were ultimately excluded by the American Psychiatric Association in the DSM-V (American Psychiatric Association, 2013). The ICD-10 included the category "excessive sexual drive," which was also excluded from the ICD-11 (Gola et al., 2020). The ICD-11 instead covers "compulsive sexual behavior disorder" (CSBD) as an impulse control disorder "characterized by a persistent pattern of failure to control intense, repetitive sexual impulses or urges resulting in repetitive sexual behavior." This pattern may manifest in repetitive sexual behavior leading to neglecting health and personal care or other interests, activities and responsibilities, unsuccessful efforts to control or significantly reduce such behavior, persistence despite adverse consequences (e.g., marital conflict, financial or legal consequences, negative impact on health), or absence of satisfaction from it (WHO, 2019/2021). Importantly, accompanying distress and impairment in functioning must not be solely related to moral judgments and disapproval about sexual impulses, urges, or behaviors. While the diagnostic entity of CSBD excludes the aforementioned behaviors in the presence of another mental or neurologic disorder, compulsive sexual behavior (CSB) in primary tic

Table 12.1

Medications commonly used for the treatment of tics and comorbid conditions and possible sexual health-related side-effects.

Medication	Possible side-effects on sexuality and sexual health	Suggested mechanisms of action
Medication for tics		
α-Agonists	Decrease in erectile function and libido; sexual dysfunction, inhibition of subjective and physiologic sexual arousal in women (Clonidine) (Duncan and Bateman, 1993; Meston et al., 1997); one case of inappropriate touching in a male with schizophrenia (Clonidine) (Yadav et al., 2021).	Central α2 agonism; increases synaptic norepinephrine levels via its action on postsynaptic α2 receptors. Increased synaptic norepinephrine levels compete with dopamine reuptake, increasing synaptic dopamine levels, which stimulate central dopaminergic projections to the nucleus accumbens, resulting in an increase in disinhibited and impulsive behavior (Yadav et al., 2021)
Antipsychotics	Haloperidol: decreased libido (30.8%), erectile dysfunction (30.8%), and ejaculatory disorder (27.7%) in males (n = 78); decreased libido (23.8%), arousal disorder (9.4%), and vaginal dryness (8.1%) in females (n=51) (Bobes et al., 2003). Risperidone: 37.8% of male participants (n=149) reported decreased libido, 32.1% erectile dysfunction, and 32.6% ejaculatory disorder. 40.5% of female patients (n=83) reported a decrease in libido, 19.0% arousal disorders, 15% vaginal dryness (Bobes et al., 2003). Gynecomastia, galactorrhea, and priapism in male patients, amenorrhea and galactorrhea in females with schizophrenia (Park et al., 2012). Aripiprazole: case of aripiprazole-induced priapism in a patient with paranoid schizophrenia (Trivedi et al., 2016). Aripiprazole-associated hypersexuality in two female patients with schizophrenia (Cheon et al., 2013). Ziprasidone: male and female sexual dysfunction and gynecomastia occurred in less than 1/1000 patients (Cutler, 2003); case of spontaneous orgasm with ziprasidone in a bipolar patient (Boora et al., 2010).	Haloperidol, risperidone, tiapride, pimozide, metoclopramide: dopamine receptor blockers, prolactin-elevating antipsychotics Ziprasidone, and aripiprazole: prolactin-sparing agents
Topiramate	Sexual dysfunction, libido disorder, anorgasmia, priapism, erectile dysfunction (males), anorgasmia, increased sexual arousal (females) (Carpenter et al., 2002; Chen et al., 2017; McElroy et al., 2000).	Unknown
Cannabis-based medications	Erectile dysfunction (Pizzol et al., 2019). Alterations in sexual experience (Wiebe and Just, 2019). Dronabinol or cannabis intoxication may be therapeutic for sexual dysfunction or sexual arousal disorders (Salerian, 2004; Androvicova et al., 2017).	Increases the activation of the right nucleus accumbens toward erotic stimuli but only in users whose prolactin levels do not rise in response to the intoxication (Androvicova et al., 2017)
Medication for ADHD		
Methylphenidate	Cases of priapism and hypersexual behavior, also effective treatment of hypersexual behavior with methylphenidate (Coskun and Zoroglu, 2009).	Pharmacologic stimulation of dopaminergic neurotransmission enhances sexual drive (Schmid et al., 2015)
Atomoxetine	Decreased libido, ejaculation disorder, and erectile dysfunction (males) (Camporeale et al., 2013).	
Medication for obsessive-compulsive disorder, anxiety disorder, and depression		
Selective serotonin reuptake inhibitors (SSRIs)	High risk of sexual dysfunction; also post-SSRI sexual dysfunction: decreased libido, genital anesthesia, pleasureless or weak orgasm, erectile dysfunction, delayed ejaculation, loss of lubrication in women, and anorgasmia (Rothmore, 2020).	Downregulation and desensitization of 5-HT1A receptors (involved in the regulation of sexual motivation), serotonergic neurotoxicity (peripheral damage) (Rothmore, 2020)

This table provides an overview of the sexual health-related side-effects of medications used to treat primary tic disorders and primary tic disorders with comorbid attention-deficit/hyperactivity disorder (ADHD). The potential side-effects of drugs used in our clinic, as well as treatment recommendations from the "Practice Guideline Summary: Treatment of Tics" (Pringsheim et al., 2019), were reviewed for their impact on sexual health and sexuality. We selected the evidence through a systematic search on PubMed with an additional manual search for relevant studies.

disorders may qualify for the diagnosis "secondary impulse control syndrome" according to the ICD-11.

Comings and Comings have investigated the presence of increased sexual behaviors in people with TS in several studies. In their 1987 study of 246 people with TS (mean age 17 years, 80.1% males and 19.1% females; Comings and Comings, 1987a), they specifically asked about precocious interest in sex ("Did your parents feel that you had a precocious (very early) interest in sex?") and sexual drive ("Do you feel that you think about sex more than others your age?"). The study found that 19.6% of enquired participants responded positively to both questions (Comings and Comings, 1987b). A subsequent 1994 study (also see section above) included questions about "greater than average sex drive," and found a prevalence of up to 28.7%. Notably, 7% of the control group also responded positively (Comings, 1994). The definition of hypersexuality in these two studies (Comings and Comings, 1987a; Comings, 1994) leaves room for internalized stigma in the absence of absolute measurements. Indeed, as aforementioned, should moral judgement alone lead to distress, a diagnosis of CSBD cannot be made according to ICD-11 (WHO, 2019/2021). In addition, possible medication-related side-effects might have also contributed to these results, but were not specifically examined in these studies (Comings and Comings, 1987b; Comings, 1994). Beyond the Comings' studies, case reports and case series documented behaviors consistent with sexual compulsivity in TS (Sandyk, 1988; McDougle et al., 1990; Yero et al., 2006; Mclaughlin et al., 2013; Teive et al., 2016).

In contrast to the Comings' data, a more recent study by Kurvits et al. screened 62 adults with primary tic disorders using the validated Hypersexual Behavior Inventory (HBI) (Klein et al., 2014; Kurvits et al., 2023). In this study, there was no evidence of increased sexual compulsivity in primary tic disorders as compared to existing data of comparable nonclinical populations (12.9% of total participants compared to the range in previous community-based studies of 5.7%–15.8%). There was also no correlation between HBI scores and sex or antipsychotic medication (Kurvits et al., 2023). Of note, all participants who met the criteria for clinically significant CSB were male. The prevalence of CSB was even lower compared to other male subgroups, such as male university students (27.9%) (Kingston et al., 2018; Kurvits et al., 2023). The observations of the Kurvits et al.'s study align with Frank et al., who also reported a much lower prevalence of CSB of 6.5% in their cohort of 31 people (20 males and 11 females) with TS (Frank et al., 2011).

Another observation of the 1994 Comings' study was that higher scores of obsessive-compulsive symptoms were associated with affirmative answers to questionnaires about sexual symptoms, including CSB (Comings, 1994). This finding was also supported by several case reports which described CSB in people with TS and obsessive-compulsive disorder (OCD) (Comings and Comings, 1982; McDougle et al., 1990; Krauss and Jankovic, 1996; Nunes and Telles, 2019). Generally, the prevalence of CSB in individuals with OCD has been shown to be comparable to the reported prevalence of sexual compulsivity in the general population (Fuss et al., 2019).

PARAPHILIAS IN INDIVIDUALS WITH PRIMARY TIC DISORDERS

Paraphilic disorders are defined in the ICD-11 as conditions "characterized by persistent and intense patterns of atypical sexual arousal, manifested through sexual thoughts, fantasies, urges, or behaviors, the focus of which involves others whose age or status renders them unwilling or unable to consent and on which the person has acted or by which he or she is markedly distressed" (WHO, 2019/2021). It should be noted that distinguishing between paraphilic disorders and CSB in past research is challenging, because the applicability of the WHO ICD-11 definition of paraphilic disorders is difficult to assess retrospectively and a differentiation of single paraphilic acts from repeated sexual urges or stable patterns of sexual arousal is not always possible. This notwithstanding, several reports described paraphilic behaviors in individuals with tic disorders and TS. For example, Moldofsky et al. described "sexual behavioral problems" like exhibitionism, voyeurism, and sadism in their 1974 cohort of 15 people with TS (Moldofsky et al., 1974). Nee et al. reported exhibitionism in a cohort of 50 cases (age and gender not explicitly reported) "inappropriate sexual activity" including exhibitionism in 16 of 50 cases (32%), and Robertson et al. reported exhibitionism and frotteurism in 9 (10%) of their cohort of 90 participants (70% male, age not reported) with TS; however, the authors did not use valid or reliable diagnostic criteria in their observations and again it remains unclear, whether people in their cohort acted on paraphilic urges or showed repetitive paraphilic behavior (Nee et al., 1980; Robertson et al., 1988). Comings reported exhibitionistic urges, sadistic tendencies, and pedophilic interests in their cohort of 358 people with TS. In this study, exhibitionism was present in 12.5% and exhibitionistic urges present in 14.2% of participants, respectively, but again not using valid or reliable diagnostic criteria (Comings, 1994). Several cases have also mentioned exhibitionistic, voyeuristic, frotteuristic, and sadistic urges and actions in people with TS and chronic tic disorders (Comings and Comings, 1982; Sandyk, 1988; McDougle et al., 1990; Comings and Comings, 1993). For example, Comings and Comings described the case of a male with TS and exhibitionistic and pedophilic urges and reported successful treatment with haloperidol (Comings and Comings, 1982).

The already aforementioned study by Kurvits et al. also examined paraphilic interests in their adult sample of 62 people with primary tic disorders (Kurvits et al., 2023). Questions about potential paraphilic interests derived from the ICD-11 definitions and were used to explore sexual arousal patterns focused on nonconsenting others and associated with significant distress or risk of harm. They found that 19.4% (n=12) of the participants reported clinically significant paraphilic interests and about 33% of them responded affirmatively to feeling distressed by these interests. About 8% (1 participant) reported having acted out on their paraphilic interests. Additionally, the study found a significant association with a previous diagnosis of ADHD. Voyeuristic (12.9%), sadistic (9.7%), frotteuristic (4.8%), exhibitionistic (3.2%), and pedophilic interests (1.6%) were among the paraphilic interests reported by the participants. Importantly, the prevalence of these interests is comparable to the prevalence of paraphilic interests in the general population (Bártová et al., 2021). Among the participants showing paraphilic interests, eight identified as men, two as women, and two as transgender. Kurvits et al. also did not find any association between the presence of paraphilic interests, CSB, and the use of antipsychotic medication (Kurvits et al., 2023).

An association between exhibitionistic behavior and a diagnosis of TS with comorbid ADHD was shown in Comings' study cohort (Comings, 1994). In addition, an influence of substance abuse on the sexual behavior score was also evident (Comings, 1994). Notably, the prevalence of ADHD is higher in individuals with hypersexuality and paraphilic disorders compared to the general population (Korchia et al., 2022). However, there is no clear evidence to suggest that hypersexuality and paraphilic disorders are more frequent in individuals with ADHD (Soldati et al., 2020).

Comings analyzed their 1994 study cohort for gender differences and found no difference in the frequency of exhibitionism between male and female study participants, and both male and female participants exhibited similar frequencies of recurrent thoughts about sex and precocious interest in sex. However, the study reported a higher prevalence of pedophilia in male individuals with TS and comorbid ADHD (9.5%) compared to females with TS (2%). Sadism was more frequent in male participants with TS and comorbid ADHD (10.9% for males versus 4.1% for females), while females showed a higher prevalence for masochism (12.5% versus 8.7% for males).

EXPERIENCES OF SEXUAL ASSAULT VICTIMIZATION AND CRIMINAL CONVICTIONS OF INDIVIDUALS WITH PRIMARY TIC DISORDERS

Comings systematically investigated the prevalence of sexual molestation experiences among 459 TS individuals in 1994 and found it to be as high as 25.7% (Comings, 1994). In a recent study, Mataix-Cols et al. conducted a comprehensive investigation using health and administrative registers in Sweden including 13,819,284 individuals between 1973 and 2013. This study identified all individuals living in Sweden during the specified time frame using nationwide health and administrative registers. The risk of violent assault and criminal convictions among people with TS was estimated using Cox proportional hazards regression models and was compared with the general population and unaffected full siblings. The study showed that individuals with TS and chronic tic disorders were significantly more prone to sexual assault victimization than the general population and their unaffected siblings. However, the numbers differed considerably from the high prevalence in the Comings' study cohort, as Mataix-Cols only found a prevalence of 0.8% for sexual assault victimization in their study cohort. There was a twofold increased risk of sexual and nonsexual violent assault victimization, compared to the general population. People with comorbid ADHD or substance abuse had an increased risk of sexual assault victimization (Mataix-Cols et al., 2022).

GENDER DIFFERENCES, GENDER IDENTITY, AND SEXUAL ORIENTATION OF PEOPLE WITH PRIMARY TIC DISORDERS

Comings explored sexual orientation and gender identity among study participants with TS and comorbid ADHD in 1994. Among the cohort, 4.8% of participants identified as homosexual, and an equal percentage identified as bisexual. For individuals with TS without comorbid ADHD, 3.1% identified as homo- or bisexual. These numbers are not significantly different from the overall prevalence of individuals identifying as nonheterosexual, which is estimated to be under 5% of the general population (Bailey et al., 2016). Thirteen percent of female participants with TS and comorbid ADHD, and 5.3% of male participants with TS and comorbid ADHD reported feelings of being born to the wrong sex. Additionally, when asked about cross-dressing, participants with TS and comorbid ADHD had a positive response rate of 9.2%, which was significantly higher than that of relatives without TS and controls (Comings, 1994). Research on the topic is limited, so it is important to approach it with caution.

There has been research on the relationship between androgen and other steroid hormone levels and the development of TS. Alexander and Peterson tested whether prenatal masculinization of the brain increases the risk of tic disorders in postnatal life and measured gender and gender role behavior in people with tic disorders and controls. They showed that the behavioral profiles of people with tic disorders are consistent with those of children who have demonstrably elevated prenatal androgen levels and concluded that an altered androgen-dependent process of sexual differentiation during prenatal life may contribute to the development of tic-related disorders (Alexander and Peterson, 2004). However, large scale studies on the topic are scarce. Therefore, such results should be interpreted with caution.

THERAPEUTIC APPROACHES TO IMPROVE SEXUAL HEALTH

Sexual health and function may be impacted in many different ways and therefore, requires a combination of therapeutic approaches for effective management. Savoy et al. recommend that physicians take a proactive approach to discuss sexual health with their patients. To organize sexual health histories, models such as the "five Ps" (asking patients about their partners, practices, protection from sexually transmitted diseases, past history of sexually transmitted diseases, and pregnancy plans) can be used. Common concerns, such as erectile dysfunction and arousal disorders, should be openly addressed (Savoy et al., 2020). Generally, it should be considered whether there is a connection between sexual health problems and current medication for individuals with chronic tic disorders or TS. If necessary, changes or adaptations to ongoing therapies for tics or their neuropsychiatric associations may be required. In such cases, sexual therapy or medication in addition may be appropriate. Generally, phosphodiesterase-5 inhibitors such as sildenafil, tadalafil, or vardenafil are effective treatment options for erectile dysfunction. However, their effectiveness in drug induced conditions may be limited. For drug-induced sexual dysfunctions caused by SSRIs or antipsychotics, addition of Bupropion 150–300 mg daily may yield improvements (Taylor et al., 2013). Also, switching an existing antipsychotic treatment causing sexual dysfunction to aripiprazole may be beneficial (Park et al., 2012). Pharmacologic interventions may prove of limited effectiveness when insufficient sexual stimulation is part of the cause of sexual dysfunction, for example, through impaired sexual communication, or sexual insecurity. These problems may be addressed more adequately in sexual therapy. Beginning with the classic sex therapy by Masters and Johnson in the 1970s (Masters and Johnson, 1970), several different approaches have been developed, which often diverge from more traditional approaches, in terms of frequency of delivery, single therapist versus therapeutic team, or the applied theoretical framework. Psychologic interventions have been found effective in improving sexual function (d=0.58, 95% confidence interval (CI): 0.40, 0.77; n=33) and sexual satisfaction (d=0.47, 95% CI: 0.27, 0.68; n=32). Data providing differential evidence on the efficacy of different therapeutic approaches for the different types of sexual dysfunction are missing (Frühauf et al., 2013).

The guidelines for treating CSBD and paraphilias may be applied to treat individuals with comorbid chronic tic disorders. Guidelines for the pharmacologic treatment of CSBD and paraphilic disorders comprise the use of opioid antagonists, high-dosage SSRIs, androgen receptor antagonists, and Gonadotropin-releasing hormone agonists. Involvement of mental health professionals, as well as endocrinologists, in the case of testosterone-lowering medications is recommended (Thibaut et al., 2020; Turner et al., 2022). The evidence base for a reduction of unwanted sexual behaviors of any of these substances lacks well-controlled studies, especially for modern substances. The evidence on the effectiveness of psychotherapeutic interventions to reduce unwanted sexual behavior in convicted sexual offenders is contradictory. Meta-analyses have yielded both positive and negative effects of psychotherapeutic interventions on recidivism (see, for example, Dennis et al., 2012; Grønnerød et al., 2015 for negative effects, and Schmucker and Lösel, 2015 for positive effects). Reasons for this confusion lie in the lack of well-controlled studies and the diagnostic heterogeneity of the populations studied.

Current knowledge about a specific therapy of sexual health disorders in tic disorders and TS remains scarce. Haloperidol has been used in addressing CSBD in individuals with chronic tic disorders and TS. McDougle et al. demonstrated that haloperidol, administered at a daily dose of 1.5 mg for six weeks, not only significantly reduced tics but also improved associated "sexually inappropriate behavior," mainly consisting of CSB, in their case of a 40-year-old man with TS (McDougle et al., 1990). Comings and Comings recommend several treatment options for "inappropriate sexual behavior." These include haloperidol, fluoxetine, clomipramine hydrochloride, and clonidine hydrochloride. They argue that such therapies have the potential to effectively address sexually inappropriate behavior, with positive outcomes often observed following treatment administration (Comings and Comings, 1993). Comings also reported that the treatment of TS patients in his cohort with serotonergic antidepressants improved symptoms of comorbid CSB (Comings, 1994).

LIMITATIONS OF PAST AND CURRENT KNOWLEDGE ABOUT SEXUALITY AND SEXUAL HEALTH IN INDIVIDUALS WITH PRIMARY TIC DISORDERS AND FUTURE CONSIDERATIONS

Sexuality and sexual health in individuals with primary tic disorders is under-researched. Outdated diagnostic definitions with broad descriptions in earlier literature and shifting societal norms contribute to potential biases. There is limited literature on gender differences and therapy options for potential problems with sexuality and sexual health in individuals with primary tic disorders. Longitudinal studies of individuals could help uncover patterns and trends that may be overseen in cross-sectional studies. Diversity, including elements like age, gender identity, sexual orientation, cultural background, and socioeconomic status, along with additional sociocultural factors like stigma, self-esteem, body image, and social interactions, may affect sexuality and the sexual health of people with primary tic disorders and require further exploration.

References

Alexander GM, Peterson BS (2004). Testing the prenatal hormone hypothesis of tic-related disorders: gender identity and gender role behavior. Dev Psychopathol 16: 407–420.

American Psychiatric Association (2013). Diagnostic and statistical manual of mental disorders: DSM-5, 5th ed. Washington, DC: American Psychiatric Association.

Androvicova R, Horacek J, Tintera J et al. (2017). Individual prolactin reactivity modulates response of nucleus accumbens to erotic stimuli during acute cannabis intoxication: an fMRI pilot study. Psychopharmacology 234: 1933–1943.

Bailey JM, Vasey PL, Diamond LM et al. (2016). Sexual orientation, controversy, and science. Psychol Sci Public Interest 17: 45–101.

Bártová K, Androvičová R, Krejčová L et al. (2021). The prevalence of paraphilic interests in the Czech population: preference, arousal, the use of pornography, fantasy, and behavior. J Sex Res 58: 86–96.

Bobes J, Garc A-Portilla MP, Rejas J et al. (2003). Frequency of sexual dysfunction and other reproductive side-effects in patients with schizophrenia treated with risperidone, olanzapine, quetiapine, or haloperidol: the results of the EIRE study. J Sex Marital Ther 29: 125–147.

Boora K, Chiappone K, Dubovsky S et al. (2010). Ziprasidone-induced spontaneous orgasm. J Psychopharmacol (Oxford, Engl) 24: 947–948.

Camporeale A, Day KA, Ruff D et al. (2013). Profile of sexual and genitourinary treatment-emergent adverse events associated with atomoxetine treatment: a pooled analysis. Drug Saf 36: 663–671.

Carpenter LL, Leon Z, Yasmin S et al. (2002). Do obese depressed patients respond to topiramate? A retrospective chart review. J Affect Disord 69: 251–255.

Chen LW-H, Chen MY-S, Chen K-Y et al. (2017). Topiramate-associated sexual dysfunction: a systematic review. Epilepsy Behav 73: 10–17.

Cheon E, Koo B-H, Seo SS et al. (2013). Two cases of hypersexuality probably associated with aripiprazole. Psychiatry Investig 10: 200–202.

Comings DE (1994). Role of genetic factors in human sexual behavior based on studies of Tourette syndrome and ADHD probands and their relatives. Am J Med Genet 54: 227–241.

Comings DE, Comings BG (1982). A case of familial exhibitionism in Tourette's syndrome successfully treated with haloperidol. Am J Psychiatry 139: 913–915.

Comings DE, Comings BG (1987). A controlled study of Tourette syndrome. I. Attention-deficit disorder, learning disorders, and school problems. Am J Hum Genet 41: 701–741.

Comings DE, Comings BG (1987). A controlled study of Tourette syndrome. IV. Obsessions, compulsions, and schizoid behaviors. Am J Hum Genet 41: 782–803.

Comings DE, Comings BG (1993). Sexual abuse or Tourette syndrome? Soc Work 38: 347–350.

Coskun M, Zoroglu S (2009). A report of two cases of sexual side effects with OROS methylphenidate. J Child Adolesc Psychopharmacol 19: 477–479.

Cutler AJ (2003). Sexual dysfunction and antipsychotic treatment. Psychoneuroendocrinology 28: 69–82.

Dennis JA, Khan O, Ferriter M et al. (2012). Psychological interventions for adults who have sexually offended or are at risk of offending. Cochrane Database Syst Rev 12: CD007507.

Dossenbach M, Dyachkova Y, Pirildar S et al. (2006). Effects of atypical and typical antipsychotic treatments on sexual function in patients with schizophrenia: 12-month results from the Intercontinental Schizophrenia Outpatient Health Outcomes (IC-SOHO) study. Eur Psychiatry 21: 251–258.

Duncan L, Bateman DN (1993). Sexual function in women. Do antihypertensive drugs have an impact? Drug Saf 8: 225–234.

Frank MC, Piedad J, Rickards H et al. (2011). The role of impulse control disorders in Tourette syndrome: an exploratory study. J Neurol Sci 310: 276–278.

Frühauf S, Gerger H, Schmidt HM et al. (2013). Efficacy of psychological interventions for sexual dysfunction: a systematic review and meta-analysis. Arch Sex Behav 42: 915–933.

Fuss J, Briken P, Stein DJ et al. (2019). Compulsive sexual behavior disorder in obsessive-compulsive disorder: prevalence and associated comorbidity. J Behav Addict 8: 242–248.

Ganos C, Martino D (2015). Tics and tourette syndrome. Neurol Clin 33: 115–136.

Gola M, Lewczuk K, Potenza MN et al. (2020). What should be included in the criteria for compulsive sexual behavior disorder? J Behav Addict 11: 160–165.

Grønnerød C, Grønnerød JS, Grøndahl P (2015). Psychological treatment of sexual offenders against children: a meta-analytic review of treatment outcome studies. Trauma Violence Abuse 16: 280–290.

Kingston DA, Walters GD, Olver ME et al. (2018). Understanding the latent structure of hypersexuality: a taxometric investigation. Arch Sex Behav 47: 2207–2221.

Klein, V., Rettenberger, M., Boom, K.D., Briken, P. (2014). A validation study of the German version of the Hypersexual Behavior Inventory (HBI)]. Psychother Psychosom Med Psychol 64: 136-140.

Korchia T, Boyer L, Deneuville M et al. (2022). ADHD prevalence in patients with hypersexuality and paraphilic disorders: a systematic review and meta-analysis. Eur Arch Psychiatry Clin Neurosci 272: 1413–1420.

Krauss JK, Jankovic J (1996). Severe motor tics causing cervical myelopathy in Tourette's syndrome. Mov Disord 11: 563–566.

Krueger RB (2016). Diagnosis of hypersexual or compulsive sexual behavior can be made using ICD-10 and DSM-5 despite rejection of this diagnosis by the American Psychiatric Association. Addiction (Abingdon, England) 111: 2110–2111.

Kurvits L, Tozdan S, Mainka T et al. (2023). Compulsive sexual behavior and paraphilic interests in adults with chronic tic disorders and Tourette syndrome: a survey-based study. Int J Impot Res 37: 233–237.

Masters WH, Johnson VE (1970). Human sexual inadequacy. Boston: Little, Brown.

Mataix-Cols D, Virtanen S, Sidorchuk A et al. (2022). Association of Tourette syndrome and chronic tic disorder with violent assault and criminal convictions. JAMA Neurol 79: 459–467.

McDougle CJ, Southwick SM, Rohrbaugh RM (1990). Tourette's disorder and associated complex behaviors: a case report. Yale J Biol Med 63: 209–214.

McElroy SL, Suppes T, Keck PE et al. (2000). Open-label adjunctive topiramate in the treatment of bipolar disorders. Biol Psychiatry 47: 1025–1033.

Mclaughlin T, Oscar-Berman M, Simpatico T et al. (2013). Hypothesizing repetitive paraphilia behavior of a medication refractive Tourette's syndrome patient having rapid clinical attenuation with KB220Z-nutrigenomic amino-acid therapy (NAAT). J Behav Addict 2: 117–124.

Meston C, Bradford A (2007). Autonomic nervous system influences: the role of the sympathetic nervous system in female sexual arousal. The psychophysiology of sex. Indiana University Press.

Meston CM, Gorzalka BB, Wright JM (1997). Inhibition of subjective and physiological sexual arousal in women by clonidine. Psychosom Med 59: 399–407.

Mitchell KR, Lewis R, O'Sullivan LF et al. (2021). What is sexual wellbeing and why does it matter for public health? Lancet Public Health 6: e608–e613.

Moldofsky H, Tullis C, Lamon R (1974). Multiple tic syndrome (Giles de la Tourette's syndrome). J Nerv Ment Dis 159: 282–292.

Nagai Y, Cavanna A, Critchley HD (2009). Influence of sympathetic autonomic arousal on tics: implications for a therapeutic behavioral intervention for Tourette syndrome. J Psychosom Res 67: 599–605.

Nee LE, Caine ED, Polinsky RJ et al. (1980). Gilles de la Tourette syndrome: clinical and family study of 50 cases. Ann Neurol 7: 41–49.

Nunes FT, Telles LEB (2019). Tourette's disorder and sexual offenses: psychiatric-forensic considerations. Rev Bras Psiquiatr 41: 364–365.

Park YW, Kim Y, Lee JH (2012). Antipsychotic-induced sexual dysfunction and its management. World J Mens Health 30: 153–159.

Pizzol D, Demurtas J, Stubbs B et al. (2019). Relationship between cannabis use and erectile dysfunction: a systematic review and meta-analysis. Am J Mens Health 13: 1557988319892464.

Pringsheim T, Okun MS, Müller-Vahl K et al. (2019). Practice guideline recommendations summary: treatment of tics in people with Tourette syndrome and chronic tic disorders. Neurology 92: 896–906.

Rees PM, Fowler CJ, Maas CP (2007). Sexual function in men and women with neurological disorders. Lancet (London, England) 369: 512–525.

Robertson MM, Trimble MR, Lees AJ (1988). The psychopathology of the Gilles de la Tourette syndrome. A phenomenological analysis. Br J Psychiatry 152: 383–390.

Roessner V, Plessen KJ, Rothenberger A et al. (2011). European clinical guidelines for Tourette syndrome and other tic disorders. Part II: pharmacological treatment. Eur Child Adolesc Psychiatry 20: 173–196.

Rothmore J (2020). Antidepressant-induced sexual dysfunction. Med J Aust 212: 329–334.

Salerian AJ (2004). Successful treatment of sexual dysfunction with dronabinol: a case report. J Clin Psychiatry 65: 1146–1147.

Sandyk R (1988). A case of Tourette's syndrome with midbrain involvement. Int J Neurosci 43: 171–175.

Savoy M, O'Gurek D, Brown-James A (2020). Sexual health history: techniques and tips. Am Fam Physician 101: 286–293.

Scahill L, Leckman JF, Schultz RT et al. (2003). A placebo-controlled trial of risperidone in Tourette syndrome. Neurology 60: 1130–1135.

Schmid Y, Hysek CM, Preller KH et al. (2015). Effects of methylphenidate and MDMA on appraisal of erotic stimuli and intimate relationships. Eur Neuropsychopharmacol 25: 17–25.

Schmucker M, Lösel F (2015). The effects of sexual offender treatment on recidivism: an international meta-analysis of sound quality evaluations. J Exp Criminol 11: 597–630.

Soldati L, Bianchi-Demicheli F, Schockaert P et al. (2020). Sexual function, sexual dysfunctions, and ADHD: a systematic literature review. J Sex Med 17: 1653–1664.

Taylor MJ, Rudkin L, Bullemor-Day P et al. (2013). Strategies for managing sexual dysfunction induced by antidepressant medication. Cochrane Database Syst Rev 2013: CD003382.

Teive HAG, Moro A, Moscovich M et al. (2016). Increased sexual arousal in patients with movement disorders. Arq Neuropsiquiatr 74: 303–306.

Thibaut F, Cosyns P, Fedoroff JP et al. (2020). The World Federation of Societies of Biological Psychiatry (WFSBP) 2020 guidelines for the pharmacological treatment of paraphilic disorders. World J Biol Psychiatry 21: 412–490.

Trivedi SK, Mangot AG, Sinha S (2016). Aripiprazole-induced priapism. Ind Psychiatry J 25: 119–121.

Turner D, Briken P, Grubbs J et al. (2022). The World Federation of Societies of Biological Psychiatry guidelines on the assessment and pharmacological treatment of compulsive sexual behaviour disorder. Dialogues Clin Neurosci 24: 10–69.

WHO (2019/2021) International classification of diseases, eleventh revision (ICD-11).

Wiebe E, Just A (2019). How cannabis alters sexual experience: a survey of men and women. J Sex Med 16: 1758–1762.

World Health Organization (2006). Defining sexual health: report of a technical consultation on sexual health, 28–31 January 2002, Geneva.

Yadav S, Takács A, Pantelis C et al. (2021). Clonidine induced sexual disinhibition in a patient with treatment-resistant schizophrenia: a case report. Aust N Z J Psychiatry 55: 522.

Yero SA, McKinney T, Petrides G et al. (2006). Successful use of electroconvulsive therapy in 2 cases of persistent sexual arousal syndrome and bipolar disorder. J ECT 22: 274–275.

Zhang X-H, Li J-Y, Zhang Y-Q et al. (2016). Deep brain stimulation of the globus pallidus internus in patients with intractable Tourette syndrome: a 1-year follow-up study. Chin Med J 129: 1022–1027.

Section III

Etiology and pathophysiology

Handbook of Clinical Neurology, Vol. 215 (3rd series)
Tic Disorders and Tourette Syndrome
C. Ganos and Y. Worbe, Editors
https://doi.org/10.1016/B978-0-443-13554-5.00011-0

Chapter 13

Epidemiology of Tourette syndrome

DONALD L. GILBERT[1,2,*]

[1]*Division of Neurology, Cincinnati Children's Hospital Medical Center, University of Cincinnati College of Medicine, Cincinnati, OH, United States*

[2]*Department of Pediatrics, University of Cincinnati Collecge of Medicine, Cincinnati, OH, United States*

Abstract

The objective of this chapter is to provide a clinically useful review of epidemiologic studies of Tourette syndrome (TS). The viewpoint of this chapter is that there are significant challenges to validity in epidemiologic studies of TS and co-occurring conditions but that, nonetheless, understanding this literature has value for clinicians, educators, and healthcare policymakers. The topics covered in the chapter, following background provided in the introduction, are as follows: (1) prevalence estimates, (2) risk factors, and (3) co-occurring conditions.

INTRODUCTION: CAN NEUROLOGISTS USE EPIDEMIOLOGIC STUDIES OF TOURETTE SYNDROME TO TAKE BETTER CARE OF PATIENTS?

The objective of this chapter is to provide a clinically useful review of epidemiologic studies of Tourette syndrome (TS). The viewpoint of this chapter is that there are significant challenges to validity in epidemiologic studies of TS and co-occurring conditions but that, nonetheless, understanding this literature has value for clinicians, educators, and healthcare policymakers. The topics covered in this chapter, following background provided in this introduction, are as follows: (1) prevalence estimates, (2) risk factors, and (3) co-occurring conditions.

Epidemiology, the study of how often and why diseases or health states occur in groups of people, depends on disease and health state definitions and accurate ascertainment of relevant states and risk factors in the population. All estimates regarding demographics, comorbidities, and risk factors related to TS depend to some degree on the precision of clinicians' TS diagnoses. These estimates are also influenced by decisions by individuals and families to seek a diagnosis. In the case of TS, a consensus in high-resource countries is that the prevalence of TS in children and adolescents is approximately 5 per 1000, or 1 in 200 (Hartmann et al., 2021). However, behind this estimate lie many methodological challenges. Thus, an important framework for this chapter is understanding threats to validity of epidemiologic studies of TS.

TS is a childhood-onset neurodevelopmental condition, which means that often the decision to seek a diagnosis depends on the views or concerns of the parent or other caregiver. The cardinal TS symptom is chronic, waxing and waning, migratory, motor and phonic tics (Leckman, 2022). Tics occur commonly, especially in children. Tics may be subtle. Some tics, such as throat clearing or eye blinking, may be overlooked or misattributed to other causes, such as allergies. A discussion of the epidemiology of mild, generally nonimpairing tics for which no medical attention is sought will for the most part be considered outside the scope of this chapter. However, the key point here is that many epidemiologic studies likely fail to ascertain mild TS cases.

Many parents and children with tics as well as adults with tics do seek medical attention due to problems or concerns related to the tics themselves, due to commonly associated symptoms, or just to make sure there is nothing else seriously harmful occurring. Persons with chronic tics of sufficient severity to generate medical care (primary care visits, specialty referrals, treatments) are the focus of this chapter.

PREVALENCE ESTIMATES OF TOURETTE SYNDROME

This section will be organized as follows: (1) a brief review of threats to validity in TS epidemiology research, (2) tic phenomenology as it relates to epidemiologic ascertainment,

*Correspondence to: Donald L. Gilbert, MD, MS, Professor of Pediatrics and Neurology, Cincinnati Children's Hospital Medical Center and University of Cincinnati College of Medicine, 3333 Burnet Avenue, ML # 2015 Neurology, Cincinnati, OH, United States. Tel: +1-513-636-4222, Fax: +1-513-636-3980, E-mail: donald.gilbert@cchmc.org.

(3) TS diagnostic criteria, and (4) estimates of prevalence. The prevalence section will discuss children versus adults, males versus females, white versus non-white individuals, and high- versus low-resource countries. The conclusion of this section is that TS prevalence is sufficiently high that knowledge of the diagnosis is important for clinicians, healthcare policymakers, and educators.

Threats to validity in Tourette syndrome epidemiology research

In 2007, the prevalence of TS along with 11 other common neurologic conditions was estimated based on the published English medical literature (Hirtz et al., 2007). At that time, the most common five neurologic conditions in children, listed in the decreasing order of prevalence for all ages, were migraine, epilepsies, autism spectrum disorder (ASD), TS, and cerebral palsy (CP). Of these five conditions, the authors judged that only TS had no Class I epidemiologic studies, and only TS was characterized as having "data inadequate for firm estimate" of prevalence.

To provide context for some of the challenges faced in the epidemiologic studies of TS to be reviewed in this chapter, it is helpful to consider this question: Why, among these five common conditions, would TS be the only diagnosis considered to have no firm prevalence estimate as recently as the year 2007? Or, more generally, what features of TS might make prevalence estimation difficult?

Hirtz and colleagues designated studies as Class I (highest value) if they had all of the following: (1) Time Frame: time period includes years from 1990 or later; (2) Case-finding and sample size: evaluation of all eligible population members (or adequate-sized random sample) or search of all relevant referral sources, with likelihood of identifying substantial majority of targeted cases; (3) Case definition: clear definition, consistent with generally accepted clinical/laboratory criteria; (4) Sources of diagnosis: specialist or fully validated source (including self-report) with known positive predictive values. The authors concluded that TS studies were limited by the following problems: (1) Sample size: relatively small samples resulting in wide estimated confidence intervals; (2) Case-finding: varying methods of case ascertainment. Further, they noted that variable symptom frequency and intensity might result in under-recognition by patients themselves, families, and physicians (Hirtz et al., 2007).

It is informative to consider why the other four diagnoses achieved more valid estimates despite facing some of these same problems. Table 13.1 shows how TS compares to the other four diagnoses with Class I and Class II studies. In addition, the following considerations are pertinent:

CP and epilepsy: Similar to TS, CP and epilepsy are umbrella terms, meaning that they categorize a group of patients with heterogeneous etiologies. Also similarly, the diagnosis in some cases involves clinical-decision classification. However, several symptom-based and healthcare-related clinical factors might underlie the differences in the number, size, and quality classification of studies. Unlike TS, CP and epilepsy often involve high rates of morbidity and mortality. Diagnosis commonly involves utilization of diagnostic neuroimaging and EEG. Patients with both CP and epilepsy incur high lifetime healthcare costs. These features likely provide both incentive for and validation of epidemiologic research in CP and epilepsy that are lacking in TS.

Migraine: Similar to TS, many individuals with mild, infrequent migraines may never seek medical attention. However, when severe, migraines more often result in healthcare utilization in the urgent care center and emergency department. Although, similar to TS, there is not a medical diagnostic test to rule in migraine, unlike TS, many migraineurs may undergo brain imaging to rule out other medical conditions. Further, the emergence of expensive and effective 5-$HT_{1B/1D}$ receptor agonists in the 1990s may have driven some of this epidemiologic research. To date, no such specific drug has emerged for TS.

ASD: Like TS, ASD is not diagnosed with the aid of any objective medical diagnostic tests and has undergone evolution of diagnostic criteria, performed using an observational rating scale (Kaufman, 2022). Similarly, like both TS and migraine, the large number of children with ASD and mild symptoms likely results in underdiagnosis or prevalence estimates heavily dependent on healthcare-seeking behaviors and clinician or study methodology. However, compared to TS, ASD has higher impairment and need for school-based and medical/behavioral services among both moderately and severely affected children. This may have contributed to the larger number of studies.

With this context, a comprehensive survey or quantitative analysis of all epidemiologic estimates likely would have limited value for clinicians due to many studies suffering from imprecision and bias. Therefore, the present chapter will provide a selective overview of the more rigorous studies of epidemiology of TS.

Tic phenomenology

Tics are patterned, repetitive movements that look the same or sound the same each time they are performed (Nilles et al., 2023). They can be performed automatically, with limited ongoing awareness, or in association with premonitory sensations or urges (Langelage et al., 2022).

Individuals with TS may perform or "emit" their tics automatically with limited ongoing awareness or in association with or in response to premonitory sensations or urges (Essing et al., 2022; Langelage et al., 2022). Persons with TS can often suppress or postpone tics, even in childhood (Kim et al., 2019). Tics diminish during cued and performed tasks within their same bodily distribution and tend to diminish during periods of engaged mental and physical focus, such as during musical performing (Bodeck et al., 2015). Tics have a consistent appearance or sound each time they are performed (Nilles et al., 2023).

The differential diagnosis of tics in children is broad, including non-tic behaviors such as fidgeting, compulsions, stereotypies, functional tic-like behaviors, myoclonus, epileptic

Table 13.1

Comparison of epidemiologic studies of five most common neurologic conditions presenting in children.

	Symptom factors		Clinical factors			No. of Epidemiologic studies		Median sample size		Total sample size	
	Timing of primary symptom	Typical frequency of primary symptom	Primary symptom commonly observed by physician in clinic	Biologic test useful to verify diagnosis	Commonly requires healthcare services in emergency department or hospital	Class I	Class II	Class I	Class II	Class I	Class II
Migraine	Episodic	Less than daily	No	No	Yes	7	9	3471	4192	27,168	78506
Epilepsy	Episodic	Less than daily	No	Yes	Yes	10	11	86,851	89,656	873,784	349,8833
Autism spectrum disorder	Static	Constant	Yes	No	No	7	7	17,792	289,456	923,555	508,1605
Tourette syndrome	Episodic	Daily, hourly	Yes	No	No	0	4		4490		37,451
Cerebral palsy	Static	Constant	Yes	Yes	Yes	9	8	683,66	92,947	2,866,934	86,7571

Source: 326–337.From Hirtz D, Thurman DJ, Gwinn-Hardy K, et al. (2007) How common are the "common" neurologic disorders? Neurology 68: 326–337.

automatisms, and secondary tic disorders. Some studies report high rates of misdiagnosis (Jin et al., 2005), while still others suggest a short latency to correct diagnosis (Shilon et al., 2008). Some surveys of movement disorder specialists suggest that in clinical practice medical diagnostic testing may be important to rule out secondary tic disorders (Ganos et al., 2021), while other reviews by pediatric neurologists state that a skilled clinician can usually make the diagnosis without any laboratory or imaging studies, and that secondary tic disorders are extremely uncommon (Zinner and Mink, 2010). To the extent that clinician or researcher training in recognition of tic phenomenology varies within and across studies, the diagnosis of TS might also vary, and estimates of prevalence would vary along with it. In high-resource countries, general neurology graduate medical education certification should be sufficient to allow clinicians to readily distinguish tics, in most cases, from other movements such as stereotypies, myoclonus, or functional movements.

Diagnostic criteria for Tourette syndrome

TS is a clinical diagnosis for which various diagnostic classification frameworks have been developed over time. Diagnostic criteria include the symptoms, duration, and age of onset (see Table 13.2). Generally, the required features are consistent across classifications: multiple motor tics and at least one vocal tic, with duration of tics of at least one year, and with childhood onset. One inconstant feature that influenced prevalence estimates was the requirement in the Diagnostic and Statistical Manual of Mental Disorders-IV (DSM-IV) that symptoms not only be present but also cause impairment. The effect of this criterion was to reduce the prevalence of TS (Cubo et al., 2011). Considering whether tics cause impairment has relevance for clinical decision making (Garris et al., 2021) but the addition of impairment in the DSM-IV caused strong objections by geneticists and epidemiologists (Comings, 1995a). The impairment criterion was subsequently removed in the DSM-IV-TR and the DSM-5. The upper age limit for onset criteria varies somewhat, but nearly always the onset of tics is in the first half of that age range (Nilles et al., 2023). So, while application of the impairment criterion would be expected to lower the estimated prevalence of TS, shifts in the upper limit of age of onset would exert only a small effect.

Prevalence estimates of Tourette syndrome

The previously discussed study of common neurologic conditions by Hirtz and colleagues, published in 2007, yielded an estimate of 3.5 TS cases per 1000 individuals aged 7–17 years (Hirtz et al., 2007). A number of epidemiologic studies and meta-analyses have provided updated estimates. One of the most recent is a formal meta-analysis by Jafari et al. (2022). This study yielded a somewhat higher estimate of 5 per 1000 for all ages and 7 per 1000 for children and adolescents. However, this meta-analysis identified significant heterogeneity across studies. Although it is reassuring that Hirtz and Jafari produced similar estimates, understanding how these estimates were generated remains important.

Jafari and colleagues applied strict criteria for inclusion and ultimately analyzed data from 30 published studies with a combined sampling frame of approximately 1.1 million persons. Their quality review utilized the Joanna Briggs Institute checklist, which contains nine factors: “sample frame, participants, sample size, study subjects and setting described in detail, data analysis, valid methods for identifying conditions, measuring the situation, statistical analysis, and an adequate response rate” (Munn et al., 2015). Out of nine possible quality points, the median quality score for these 30 studies was 5 and the average score was 5.9 (Jafari et al., 2022).

Table 13.3 shows the prevalence estimates from the six studies that specifically distinguished TS from other tic disorders and that had the highest quality scores of 7–9. Where studies presented age strata, these are separated, and the table is sorted by age of the cohorts.

The statistical heterogeneity in TS prevalence estimates identified by the authors is apparent even when viewing only these studies with the highest quality scores. Within the European studies, the lowest estimates came from the study utilizing medical claims data. The most important lesson from this table is that the higher estimates of prevalence occur in more labor-intensive school-based studies in which a two-step process, screening followed by direct interview by trained personnel, was used. This method probably captures children whose parents do not seek medical care for tic and supports the decades-old observation that many children with TS are not brought to physicians, likely for a variety of reasons (Kurlan, 1994). Other studies with school-based observers have also identified higher proportions of children with tics, although such studies can be biased by low-participation rates (Kurlan et al., 2001; Snider et al., 2002). An additional important factor from a recent systematic review of studies worldwide of prevalence of neurodevelopmental disorders is the training of professionals in diagnosis of children and adolescents (Frances et al., 2022). Thus, research methodology, cultural factors, and medical practitioner education should always be considered when interpreting epidemiologic studies of TS.

Age, sex, and geography

Two consistent findings across studies of multiple methodologies are (1) higher prevalence estimates in school-aged children and adolescents compared to the youngest children and to adults, and (2) higher prevalence in males than females (Scahill et al., 2009). This age result supports findings from retrospective self-report studies where adolescents and adults report their tic onset in earlier childhood, peak tic frequency in mid-childhood/preadolescence, and reduction in late adolescence (Leckman et al., 1998). However, this should also be placed in context of studies showing the persistence of tics into adulthood is not always reported or recognized by individuals with TS (Pappert et al., 2003). That is, lower prevalence estimates in adults could result from factors such as clinic ascertainment,

Table 13.2

Classifications for Tourette syndrome (TS; Robertson and Eapen, 2014).

Classification	Diagnostic criteria	Age of onset upper limit	Duration	Impairment or distress requirement
DSM-I (1952)	Included various symptoms potentially related to TS, such as echolalia, interjectional speech, habit spasm, spasm nutans, and tic.	Not specified	Not specified	Not specified
DSM-II (1968)	Tic included under special symptoms not classified elsewhere.	Not specified	Not specified	Not specified
DSM-III (1980)	Recurrent, involuntary, repetitive, rapid motor movements affecting multiple muscle groups and multiple vocal tics; ability to suppress movements voluntarily for some time; variations in symptom intensity.	2–15 years	More than 1 year	Not specified
DSM-III-TR (1987)	Multiple motor and one or more vocal tics; tics occur many times a day; anatomic location, number, frequency, complexity, and severity of tics change over time.	Before 21 years	More than 1 year	Not specified
DSM-IV (1994)	Both multiple motor and one or more vocal tics present, the tics occur many times a day, disturbance causes marked distress or significant impairment in functioning.	Before 18 years	More than 1 year, no tic-free period of more than three consecutive months	Yes
DSM-IV-TR (2000)	Both multiple motor and one or more vocal tics present, the tics occur many times a day, disturbance not due to direct physiologic effects of a substance or a general medical condition.	Before 18 years	More than 1 year, no tic-free period of more than three consecutive months	Not specified
DSM-5 (2013)	Both multiple motor and one or more vocal tics present, the tics may wax and wane in frequency, disturbance not attributable to the physiologic effects of a substance or other medical condition.	Before 18 years	More than 1 year	Not specified
Tourette Syndrome Study Group criteria (1993)	Both multiple motor and one or more vocal tics present; tics occur many times a day; anatomic location, number, frequency, complexity, type, or severity of tics changes over time; involuntary movements and noises cannot be explained by other medical conditions.	Before 21 years	More than 1 year	Not specified
ICD-10 criteria (WHO, 1992/3) F95.2	Multiple motor tics and one or more vocal tics present, the frequency of tics must be many times a day, nearly every day.	Before 18 years	More than 1 year, with no period of remission during that year lasting longer than 2 months	Not specified

Table 13.3

Prevalence estimates of Tourette syndrome.

Reference	Publication year	Country	Region	Age	Quality score	Prevalence estimates per 1000 Total	Male	Female	Male:female ratio	Detail
Schlander	2011	Germany	Europe	0–6	8	0.07	0.08	0.04	2.0	Based on outpatient medical claims data
Giraldo	2013	Colombia	America	9 ± 1.5	8	34.00	55.00	12.70	4.3	School children screening, followed by researcher interview
Schlander	2011	Germany	Europe	7–12	8	0.40	0.60	0.19	3.2	Based on outpatient medical claims data
Khalifa	2003	Sweden	Europe	7–15	7	6.00	9.00	1.00	9.0	Parent questionnaire of school children, followed by clinician interview
Cubo	2011	Spain	Europe	6–17	7	16.10	22.30	6.80	3.3	Screening in school, then clinician interview
Jin	2005	China	Asia	7– 17	7	4.30	7.40	0.70	10.6	Screening in school, then clinician interview
Yang	2016	Canada	America	12–17	9	3.33	6.03	0.48	12.6	Based on outpatient medical claims data
Schlander	2011	Germany	Europe	13–18	8	0.38	0.57	0.18	3.2	Based on outpatient medical claims data
Schlander	2011	Germany	Europe	19–29	8	0.09	0.17	0.01	17.0	Based on outpatient medical claims data
Schlander	2011	Germany	Europe	30–49	8	0.06	0.08	0.04	2.0	Based on outpatient medical claims data
Schlander	2011	Germany	Europe	> 50	8	0.03	0.03	0.03	1.0	Based on outpatient medical claims data
Yang	2016	Canada	America	> 17	9	0.66	0.89	0.44	2.0	Based on data from Canadian Community Health Survey

Estimates confirmed from primary sources (Khalifa and Von Knorring, 2003; Jin et al., 2005; Cubo et al., 2011; Schlander et al., 2011; Giraldo et al., 2013; Yang et al., 2016b).

Source: 6-16.Data from Jafari F, Abbasi P, Rahmati M, et al. (2022) Systematic review and meta-analysis of Tourette syndrome prevalence; 1986 to 2022. Pediatr Neurol 137: 6-16.

patterns of healthcare utilization, lower levels of recognition decades ago, or habituation to symptoms. As in children, prevalence estimates in adults may be reduced when methodologies do not involve direct observation.

Nearly all studies of TS show a male predominance (see male to female ratios in Table 13.3). This sex ratio is less marked in adult cohorts (Yang et al., 2016b). Although this sex ratio seems less marked in the oldest and youngest cohorts, this should be interpreted with caution due to the smaller case numbers and thus less precise estimates in some age strata.

Another key observation supported by data in Table 13.3 is that vast areas of the world have no high-quality epidemiologic studies of TS or of other neurodevelopmental disorders (Frances et al., 2022). Thus, we have limited or no reliable information about how common TS is in Africa, South Asia, and Central and South America.

Ancestry

High-resource countries in North America and Europe in which many epidemiologic studies of TS have occurred have majority-white populations. In these countries, minority demographic groups will therefore have smaller numbers and less precise prevalence estimates. Societal and institutional factors disproportionately reduce access to healthcare. Nonetheless, a consistently reported finding is that TS is more common in whites – individuals with European ancestry. In North America, data from a home survey conducted by the Centers for Disease Control and Prevention supported that TS diagnosis is more common in whites (estimated prevalence 3.9/1000). Prevalence estimates were approximately 60% lower in Hispanics and Blacks (Scahill et al., 2009). The impression that TS is less common among blacks within and outside of Africa has been reported elsewhere as well (Robertson et al., 2009). Although clearly more research needs to be done in South America, Asia, and Africa, it is useful to examine some existing data in order to understand challenges moving forward.

Estimates in lower resource countries

Africa

Epidemiologic studies from Africa are important from a public health standpoint but also for what they reveal about challenges to global neurology healthcare (Wilmshurst et al., 2023) and therefore research in TS. There are no large community-based or direct interview-based studies estimating overall prevalence of tic disorders in African countries. Some important work addressing the prevalence of neurologic and developmental disorders, including tic disorders, in children with HIV/AIDS in Sub-Saharan Africa has been conducted in Uganda. In a formal study conducted in rural (Masaka) and urban (Kampala) clinics in Uganda (Mpango et al., 2019), prevalence was determined among 677 children aged 5–11 years and 393 adolescents aged 12–17 years diagnosed with HIV/AIDS. Methodology involved the locally adapted Child and Adolescent Symptom Inventory-5 administered to the child's guardian. Diagnosis was based on a cutoff, essentially parents or other caregivers reporting that tics are present "often" or "very often." Impairment was not used for the tic disorder diagnosis and the presence or absence of TS was not determined separately. In this study, the prevalence of tics was 2.1% in children and 10.9% in adolescents. Further, tics were reported in 3.9% of males and 6.7% of females. There are multiple difficulties generalizing from this study, as the authors acknowledge. First, there is the problem of this adapted scale in this population not being validated for tics. Introducing clinician or parent surveys or rating scales into another language and culture with limited opportunity for validation and working with clinicians with differing medical training likely influence findings. Second, while obtaining neurodevelopmental data from youth with HIV/AIDS is important, generalizability of the findings to individuals not infected with HIV/AIDs, or more broadly to Ugandans, East Africans, or Sub-Saharan Africans groups cannot be assured. Finally, there are unexpected findings that need further study.

The first unexpected finding is the reversed sex ratio (female preponderance), which the authors did not explain. The second is that the study reported the prevalence of tics differed among Mugandan (largest ethnic group) versus non-Mugandan participants. They attributed this possibly to both genetic and socioeconomic factors but did not provide further rationale. To understand why unexpected or unexplained findings might have occurred, important insights can be gained from a fascinating study of the knowledge, attitudes, and experiences of 148 healthcare professionals (including physicians, psychologists, nurses, therapists, and social workers) in Uganda by Rodin et al. (2021). A slight majority (52%) reported having seen a patient with a tic disorder but misdiagnoses such as epilepsy were common. The majority of respondents (79%) considered making a tic disorder diagnosis to be important. As expected (Nilles et al., 2023), the most common motor tics recognized were simple: eye blinking (78%) and facial grimacing (64%). Surprisingly, however, the most common vocal tics recognized were shouting (43%) and saying words (38%). Simple tics like sniffing (34%) and throat clearing (33%), which are more common in all other surveys (Nilles et al., 2023), here were less commonly recognized. Importantly, a minority (only 32%) felt comfortable diagnosing a tic disorder in a young person. Barriers to diagnosis included provider self-reported lack of knowledge (76%), and families not being concerned about tics (60%) or feeling too guilty and not reporting tics to professionals (28%).

China

A rigorous meta-analysis of 13 published studies (11 in Chinese) from 1983 to 2010 involving surveys of 269,571 participants concluded that in general, quality of studies is not high. For example, in 61.5% of studies it was not clear if data collection methods were standardized. Prevalence overall of tic disorders was 6.1% and of TS 0.3%, or 3 per 1000 (Yang et al., 2016a). As discussed above, estimates from studies not employing direct interviews are likely to be lower, as is the case here. No other Asian countries had epidemiologic studies with rigorous methodology (Jafari et al., 2022). That the summary estimate of TS prevalence was 3 per 1000, however, is in reasonable accord with the recent meta-analysis estimate of 5 per 1000 (Jafari et al., 2022).

RISK FACTORS FOR TOURETTE SYNDROME

Similar to studies of prevalence, studies of risk factors in TS merit close scrutiny based on methodology. The organization of this section will be to review findings from three large cohort studies. These were selected due to their more rigorous methodologies, assessments of potential confounding, and thorough comparisons of their findings to other published reports. Factors investigated in these reports are mainly pre/perinatal and socioeconomic. Interested readers are encouraged to obtain the primary reference papers, available open access.

There are a large number of studies of possible infectious or inflammatory risk factors for TS (Gilbert et al., 2018; Gilbert, 2019). The most prevalent etiologic question has been whether Group A β-hemolytic streptococcal infections might induce, in susceptible children, an inflammatory response that caused a distinct form of TS (Kurlan, 2004; Gilbert and Kurlan, 2009). This hypothesized distinct form of TS was labeled Pediatric Autoimmune Neuropsychiatric Disorder Associated with Streptococcal infections (Swedo et al., 1998). Intensive, longitudinal studies failed to support this hypothesis (Kurlan et al., 2008; Leckman et al., 2011; Schrag et al., 2022). Further discussion of the large body of literature surrounding this topic lies outside the scope of this chapter.

The conclusion of this risk factors section is that some prenatal, perinatal, and socioeconomic factors may modestly influence the prevalence of TS, but there are no highly predictive risk factors, even in combination with genetics, that are useful for clinicians, healthcare policymakers, or educators at this time.

The Danish National Birth Cohort

The Danish National Birth Cohort study includes data from 100,000 pregnant residents, enrolled from 1996 to 2002 (Olsen et al., 2001). This registry contains detailed medical diagnostic information as well as quantitative information on lifestyles. Based on this medical registry data, with mean follow-up of 13 years, the incidence of TS or chronic tic (CT) disorder is estimated to be approximately 0.4 per 1000 per year. If the majority of diagnoses are confirmed during a 10-year span from age 3 to 13 and persist for 10 years, one could estimate the prevalence as 4 per 1000 by age 13, approximately consistent with prevalence estimates discussed above. The authors identified as a major finding a significantly increased risk identified based on maternal smoking. Heavy smoking was associated with an approximately 65% higher risk for TS/CT. The authors did not find that this was attributable to maternal psychiatric history but acknowledged that psychiatric confounders could not be completely ruled out (Browne et al., 2016).

The Swedish Cohort

The Swedish Cohort study consists of all singletons live births from January 1, 1973 to December 31, 2003, involving ascertainment from multiple archival and modern sources, linkage through registers, tracking of emigration and travel data, and resulting in a sample of 3,026,861 individuals of whom 5597 were diagnosed with TS/chronic tics. Their estimated prevalence is 2.6 per 1000 by age 40, lower than prevalence estimates discussed above. A primary source of medical information is the Swedish National Patient Register, which is based on inpatient hospital admissions. Importantly, outpatient specialist care as a data source was only available since 2001, 2 years before the end of the study cohort. Identified risk factors, after incorporating a sibling comparison analysis (which controlled for shared possible familial confounders), included breech presentation, C section delivery, impaired fetal growth, and preterm birth. Maternal smoking was not a risk factor. The authors note that their cohort is weighted toward more severe cases, in part related to the incomplete outpatient coverage (Brander et al., 2018). It may also be the case that during this time period, 1973 to 2003, recognition of tics and TS by patients, families, and doctors was much lower. Similarly, the ability to evaluate associations with psychiatric illnesses may have varied during those decades.

The Avon Longitudinal Cohort

An important series of studies addressing risk factors for TS (Scharf et al., 2012; Miller et al., 2014; Ben-Shlomo et al., 2016; Abdulkadir et al., 2023) was generated by the Avon Longitudinal Study of Parents and Children (ALSPAC). The primary dataset is mother-completed questionnaires from 14,541 pregnant residents in Avon, United Kingdom, enrolled with expected delivery dates from April 1, 1991 to December 31, 1992 (Golding et al., 2001). The authors utilized several methods to mitigate risks of misdiagnosis, including defining TS with three levels of certainty – Narrow (n=23), Intermediate (n=50), and Broad (n=217). Age 13-year point-prevalence estimates for TS for these categories ranged from 3 per 1000 (Narrow) to 32 per 1000 (Broad) (Scharf et al., 2012). Using surrogate factors for socioeconomic status such as car ownership and reported financial difficulties, the authors identified modest increases in risk related to lower socioeconomic status (SES) (Miller et al., 2014). They also queried maternal reports of psychiatric diagnosis, finding most prominently that chronic maternal anxiety increased the odds of TS (Ben-Shlomo et al., 2016).

Although this study is smaller, investigators have leveraged concurrent participation in genome-wide association studies (GWAS; Abdulkadir et al., 2019, 2023). Genotyping data available on 9915 children in the ALSPAC cohort was available. Researchers assigned polygenic risk scores based on alleles and effects sizes reported in a GWAS of TS (Yu et al., 2019). The key finding emerging from this effort is that genetic risks explain very little: between 0.1% and 0.7% of variance of tics within the ALSPAC sample. The authors found that genetic and pregnancy-related risk factors, the latter mediated by maternal anxiety/depression, contribute independently to the risk of development of tics (Abdulkadir et al., 2023).

CO-OCCURRING NEURODEVELOPMENTAL AND PSYCHIATRIC DIAGNOSES IN TOURETTE SYNDROME

Essentially all studies of TS have found strong associations with attention-deficit/hyperactivity disorder (ADHD), disruptive behaviors, and obsessive-compulsive disorder (OCD; Leckman, 2022). The association was sufficiently strong that even in the early days of neurodevelopmental and psychiatric genetics, some prominent investigators hypothesized that TS was polygenic (Comings, 1995b; Comings et al., 1996) and a spectrum disorder of altered basal ganglia function (Kurlan, 1994; Palumbo et al., 1997). The challenges discussed above in estimating prevalence of TS also apply to estimates of commonly co-occurring neurodevelopmental and psychiatric diagnoses. Problems of referral bias, imprecision of small samples, and variability of diagnostic processes have induced multiple groups, taking various approaches, to arrive at a more accurate understanding of the relationship of TS to co-occurring diagnoses. The organization of this section is a summary and comparison of findings from two large studies with very different methodologies. However, despite the variation of methods, clinicians can conclude that there is a strong, evidence-based consensus that individuals with TS are substantially more likely to have neurodevelopmental and psychiatric disorders than are individuals in the general population. The implication of this section for clinicians and educators is that identification of TS or chronic tics should raise suspicions for presence of a co-occurring diagnosis, especially ADHD or OCD. Clinicians who regularly see individuals with TS should have screening, testing, or referring for evaluation of these other diagnoses as a standard practice (Pringsheim et al., 2019; Szejko et al., 2022). Educators who observe that a student has tics should also be vigilant for co-occurring problems with learning, behavior, and mood (Claussen et al., 2018).

Multisite, international database

In the 1990s, Dr. Roger Freeman established an international database, which ultimately involved clinicians and investigators from 22 countries. The methodology was that Dr. Freeman sent a one-page data entry form to each participating clinician who "agreed to record the results of their usual assessments." No attempt was made to establish identical assessment procedures or to determine the interrater reliability of diagnostic categories, nor was there any on-site validation of recruitment methods. Basically, Dr. Freeman trusted the clinicians to do a reasonable job, to recruit consecutive or representative individuals from their clinics, and to record the data in his data entry forms reliably. He/his team then entered the form data into the database and performed all analyses. In the paper, the authors acknowledged high variability of prevalence of co-occurring diagnoses, even among sites that recruited larger numbers of participants (Freeman et al., 2000).

Genetic consortium phenotypic study

Participants for this study were collected from 1374 individuals with TS participating in genetic studies from 1992 to 2008 in the United States, Canada, Great Britain, and the Netherlands. Thus, biases may involve willingness to participate in a genetic study as well as having sought care at a specialty clinic for TS and related diagnoses. Investigator training, rigor for making all diagnoses, and overall standardization were much higher in this study. A psychiatrist, neurologist, or psychologist trained in all assessments performed structured clinical interviews on all participants. All final diagnoses were assigned using a best-estimate process involving at least two raters (Hirschtritt et al., 2015).

As seen in Table 13.4, the findings of both studies are reasonably similar. Some differences may be attributable to

Table 13.4

Prevalence of co-occurring conditions in Tourette syndrome.

	Freeman et al. (2000)	Hirschtritt et al. (2015)
Cohort features		
Sample size	3500	1374
Tic onset (mean age in years)	6.5	6
Male to female ratio	4.3	2.7
Diagnoses		
TS only	12%	14%
Attention-deficit/hyperactivity disorder	50%	54%
Obsessive-compulsive disorder spectrum	59%	66%
Disruptive behavior disorders	15%	30%
Mood disorders	20%	30%
Anxiety disorders	18%	36%

recruitment, diagnostic practices, and sex ratios in the two cohorts. In both samples, more than 85% of individuals with TS had at least one co-occurring diagnosis.

CONCLUSIONS

This chapter reviewed epidemiologic studies of TS, emphasizing the topics of prevalence estimation, risk factor analysis, and co-occurring neurodevelopmental and psychiatric diagnoses. High-quality rigorous studies have provided a range of estimates of prevalence, with a rough approximation of 1 in 200 children. Nearly all studies show a male predominance, particularly during childhood, with greater prevalence in mid-childhood and adolescence than in early childhood or adulthood. These estimates come largely from higher resourced, predominantly white countries. Studies employing observers in schools yield higher childhood estimates.

We also surveyed literature on risk factors. Studies of TS risk factors primarily identify, at best, contributions with effects sizes too small to influence clinical practice or healthcare policy.

Finally, we surveyed literature showing marked associations of TS with ADHD, OCD, and other psychiatric conditions. As is the case for many human studies of clinically diagnosed entities, there are substantial methodological challenges. Despite these, it is clear that TS has sufficient prevalence that clinicians should be ready to accurately diagnose it. Educators should also be aware of the very high association with disorders that can affect behavior and learning in the classroom. Of course, educating educators has typically required substantial resources. Early work on a scale that could be administered by teachers to assess prevalence, without need for rater training, shows some promise (Adams et al., 2023).

References

Abdulkadir M, Mathews CA, Scharf JM et al. (2019). Polygenic risk scores derived from a Tourette syndrome genome-wide association study predict presence of tics in the Avon Longitudinal Study of Parents and Children Cohort. Biol Psychiatry 85: 298–304.

Abdulkadir M, Tischfield JA, Heiman GA et al. (2023). Polygenic and environmental determinants of tics in the Avon Longitudinal Study of Parents and Children. Am J Med Genet B Neuropsychiatr Genet 192: 73–84.

Adams HR, Augustine EF, Bonifacio K et al. (2023). Evaluation of new instruments for screening and diagnosis of tics and tic disorders in a well-characterized sample of youth with tics and recruited controls. Evid Based Pract Child Adolesc Ment Health 9: 216–230.

Ben-Shlomo Y, Scharf JM, Miller LL et al. (2016). Parental mood during pregnancy and post-natally is associated with offspring risk of Tourette syndrome or chronic tics: prospective data from the Avon Longitudinal Study of Parents and Children (ALSPAC). Eur Child Adolesc Psychiatry 25: 373–381.

Bodeck S, Lappe C, Evers S (2015). Tic-reducing effects of music in patients with Tourette's syndrome: self-reported and objective analysis. J Neurol Sci 352: 41–47.

Brander G, Rydell M, Kuja-Halkola R et al. (2018). Perinatal risk factors in Tourette's and chronic tic disorders: a total population sibling comparison study. Mol Psychiatry 23: 1189–1197.

Browne HA, Modabbernia A, Buxbaum JD et al. (2016). Prenatal maternal smoking and increased risk for Tourette syndrome and chronic tic disorders. J Am Acad Child Adolesc Psychiatry 55: 784–791.

Claussen AH, Bitsko RH, Holbrook JR et al. (2018). Impact of Tourette syndrome on school measures in a nationally representative sample. J Dev Behav Pediatr 39: 335–342.

Comings DE (1995). The role of genetic factors in conduct disorder based on studies of Tourette syndrome and attention-deficit hyperactivity disorder probands and their relatives. J Dev Behav Pediatr 16: 142–157.

Comings DE (1995). DSM-IV criteria for Tourette's. J Am Acad Child Adolesc Psychiatry 34: 401–402.

Comings DE, Wu S, Chiu C et al. (1996). Polygenic inheritance of Tourette syndrome, stuttering, attention deficit hyperactivity, conduct, and oppositional defiant disorder: the additive and subtractive effect of the three dopaminergic genes-DRD2, D beta H, and DAT1. Am J Med Genet 67: 264–288.

Cubo E, Gabriel Y, Galan JM et al. (2011). Prevalence of tics in schoolchildren in central Spain: a population-based study. Pediatr Neurol 45: 100–108.

Essing J, Jakubovski E, Psathakis N et al. (2022). Premonitory urges reconsidered: urge location corresponds to tic location in patients with primary tic disorders. J Mov Disord 15: 43–52.

Frances L, Quintero J, Fernandez A et al. (2022). Current state of knowledge on the prevalence of neurodevelopmental disorders in childhood according to the DSM-5: a systematic review in accordance with the PRISMA criteria. Child Adolesc Psychiatry Ment Health 16: 27.

Freeman RD, Fast DK, Burd L et al. (2000). An international perspective on Tourette syndrome: selected findings from 3,500 individuals in 22 countries. Dev Med Child Neurol 42: 436–447.

Ganos C, Sarva H, Kurvits L et al. (2021). Clinical practice patterns in tic disorders among Movement Disorder Society members. Tremor Other Hyperkinet Mov (NY) 11: 43.

Garris JF, Huddleston DA, Jackson HS et al. (2021). Implementation of the Mini-Child Tourette Syndrome Impairment Scale: relationships to symptom severity and treatment decisions. J Child Neurol 36: 288–295.

Gilbert DL (2019). Inflammation in tic disorders and obsessive-compulsive disorder: are PANS and PANDAS a path forward? J Child Neurol 34: 598–611.

Gilbert DL, Kurlan R (2009). PANDAS: horse or zebra? Neurology 73: 1252–1253.

Gilbert DL, Mink JW, Singer HS (2018). A pediatric neurology perspective on pediatric autoimmune neuropsychiatric disorder associated with streptococcal infection and pediatric acute-onset neuropsychiatric syndrome. J Pediatr 199: 243–251.

Giraldo BO, David M, Sanchez Y et al. (2013). Prevalence of tics among 6- to 12-year-old schoolchildren in the Itagui Municipality, Colombia, in 2010. J Child Neurol 28: 1406–1411.

Golding J, Pembrey M, Jones R et al. (2001). ALSPAC – the Avon Longitudinal Study of Parents and Children. I. Study methodology. Paediatr Perinat Epidemiol 15: 74–87.

Hartmann A, Szejko N, Mol Debes N et al. (2021). Is Tourette syndrome a rare condition? F1000Res 10: 434.

Hirschtritt ME, Lee PC, Pauls DL et al. (2015). Lifetime prevalence, age of risk, and genetic relationships of comorbid psychiatric disorders in Tourette syndrome. JAMA Psychiatry 72: 325–333.

Hirtz D, Thurman DJ, Gwinn-Hardy K et al. (2007). How common are the "common" neurologic disorders? Neurology 68: 326–337.

Jafari F, Abbasi P, Rahmati M et al. (2022). Systematic review and meta-analysis of Tourette syndrome prevalence; 1986 to 2022. Pediatr Neurol 137: 6–16.

Jin R, Zheng RY, Huang WW et al. (2005). Epidemiological survey of Tourette syndrome in children and adolescents in Wenzhou of P.R. China. Eur J Epidemiol 20: 925–927.

Kaufman NK (2022). Rethinking "gold standards" and "best practices" in the assessment of autism. Appl Neuropsychol Child 11: 529–540.

Khalifa N, Von Knorring AL (2003). Prevalence of tic disorders and Tourette syndrome in a Swedish school population. Dev Med Child Neurol 45: 315–319.

Kim S, Greene DJ, Robichaux-Viehoever A et al. (2019). Tic suppression in children with recent-onset tics predicts 1-year tic outcome. J Child Neurol 34: 757–764.

Kurlan R (1994). Hypothesis II: Tourette's syndrome is part of a clinical spectrum that includes normal brain development. Arch Neurol 51: 1145–1150.

Kurlan R (2004). The PANDAS hypothesis: losing its bite? Mov Disord 19: 371–374.

Kurlan R, Johnson D, Kaplan EL et al. (2008). Streptococcal infection and exacerbations of childhood tics and obsessive-compulsive symptoms: a prospective blinded cohort study. Pediatrics 121: 1188–1197.

Kurlan R, Mcdermott MP, Deeley C et al. (2001). Prevalence of tics in schoolchildren and association with placement in special education. Neurology 57: 1383–1388.

Langelage J, Verrel J, Friedrich J et al. (2022). Urge-tic associations in children and adolescents with Tourette syndrome. Sci Rep 12: 16008.

Leckman JF (2022). The neurobiology of Gilles de la Tourette syndrome and chronic tics. International Review of Movement Disorders 3: 69–101.

Leckman JF, King RA, Gilbert DL et al. (2011). Streptococcal upper respiratory tract infections and exacerbations of tic and obsessive-compulsive symptoms: a prospective longitudinal study. J Am Acad Child Adolesc Psychiatry 50: 108–118.e3.

Leckman JF, Zhang H, Vitale A et al. (1998). Course of tic severity in Tourette syndrome: the first two decades. Pediatrics 102: 14–19.

Miller LL, Scharf JM, Mathews CA et al. (2014). Tourette syndrome and chronic tic disorder are associated with lower socio-economic status: findings from the Avon Longitudinal Study of Parents and Children cohort. Dev Med Child Neurol 56: 157–163.

Mpango RS, Rukundo GZ, Muyingo SK et al. (2019). Prevalence, correlates for early neurological disorders and association with functioning among children and adolescents with HIV/AIDS in Uganda. BMC Psychiatry 19: 34.

Munn Z, Moola S, Lisy K et al. (2015). Methodological guidance for systematic reviews of observational epidemiological studies reporting prevalence and cumulative incidence data. Int J Evid Based Healthc 13: 147–153.

Nilles C, Martino D, Fletcher J et al. (2023). Have we forgotten what tics are? A re-exploration of tic phenomenology in youth with primary tics. Mov Disord Clin Pract 10: 764–773.

Olsen J, Melbye M, Olsen SF et al. (2001). The Danish National Birth Cohort – its background, structure and aim. Scand J Public Health 29: 300–307.

Palumbo D, Maughan A, Kurlan R (1997). Hypothesis III. Tourette syndrome is only one of several causes of a developmental basal ganglia syndrome. Arch Neurol 54: 475–483.

Pappert EJ, Goetz CG, Louis ED et al. (2003). Objective assessments of longitudinal outcome in Gilles de la Tourette's syndrome. Neurology 61: 936–940.

Pringsheim T, Okun MS, Muller-Vahl K et al. (2019). Practice guideline recommendations summary: treatment of tics in people with Tourette syndrome and chronic tic disorders. Neurology 92: 896–906.

Robertson MM, Eapen V (2014). Tourette's: syndrome, disorder or spectrum? Classificatory challenges and an appraisal of the DSM criteria. Asian J Psychiatry 11: 106–113.

Robertson MM, Eapen V, Cavanna AE (2009). The international prevalence, epidemiology, and clinical phenomenology of Tourette syndrome: a cross-cultural perspective. J Psychosom Res 67: 475–483.

Rodin A, Fleetwood-Meade K, Gilmour J et al. (2021). Why don't children in Uganda have tics? A mixed-methods study of beliefs, knowledge, and attitudes of health professionals. Child Adolesc Ment Health 26: 47–53.

Scahill L, Bitsko R, Visser S et al. (2009). Prevalence of diagnosed Tourette syndrome in persons aged 6-17 years – United States, 2007. Morbidity Mortal Wkly Rep 58: 581–585.

Scharf JM, Miller LL, Mathews CA et al. (2012). Prevalence of Tourette syndrome and chronic tics in the population-based Avon longitudinal study of parents and children cohort. J Am Acad Child Adolesc Psychiatry 51: 192–201.e5.

Schlander M, Schwarz O, Rothenberger A et al. (2011). Tic disorders: administrative prevalence and co-occurrence with attention-deficit/hyperactivity disorder in a German community sample. Eur Psychiatry 26: 370–374.

Schrag AE, Martino D, Wang H et al. (2022). Lack of association of Group A streptococcal infections and onset of tics: European Multicenter Tics in Children Study. Neurology 98: e1175–e1183.

Shilon Y, Pollak Y, Benarroch F et al. (2008). Factors influencing diagnosis delay in children with Tourette syndrome. Eur J Paediatr Neurol 12: 398–400.

Snider LA, Seligman LD, Ketchen BR et al. (2002). Tics and problem behaviors in schoolchildren: prevalence, characterization, and associations. Pediatrics 110: 331–336.

Swedo SE, Leonard HL, Garvey M et al. (1998). Pediatric autoimmune neuropsychiatric disorders associated with streptococcal infections: clinical description of the first 50 cases. Am J Psychiatry 155: 264–271.

Szejko N, Robinson S, Hartmann A et al. (2022). European clinical guidelines for Tourette syndrome and other tic disorders-version 2.0. Part I: assessment. Eur Child Adolesc Psychiatry 31: 383–402.

Wilmshurst JM, Catsman-Berrevoets C, Gilbert DL et al. (2023). Access to pediatric neurology training and services worldwide: a survey by the International Child Neurology Association. Neurology 101: 798–808.

Yang C, Zhang L, Zhu P et al. (2016). The prevalence of tic disorders for children in China: a systematic review and meta-analysis. Medicine (Baltimore) 95: e4354.

Yang J, Hirsch L, Martino D et al. (2016). The prevalence of diagnosed tourette syndrome in Canada: a national population-based study. Mov Disord 31: 1658–1663.

Yu D, Sul JH, Tsetsos F et al. (2019). Interrogating the genetic determinants of Tourette's syndrome and other tic disorders through genome-wide association studies. Am J Psychiatry 176: 217–227.

Zinner SH, Mink JW (2010). Movement disorders I: tics and stereotypies. Pediatr Rev 31: 223–233.

Handbook of Clinical Neurology, Vol. 215 (3rd series)
Tic Disorders and Tourette Syndrome
C. Ganos and Y. Worbe, Editors
https://doi.org/10.1016/B978-0-443-13554-5.00020-1

Chapter 14

Genetics of Tourette syndrome and tic disorders

PRITESH JAIN, PERISTERA PASCHOU*, AND APOSTOLIA TOPALOUDI

Department of Biological Sciences, Purdue University, West Lafayette, IN, United States

Abstract

Tourette syndrome (TS) is a childhood-onset neurodevelopmental disorder characterized by multiple motor tics and at least one vocal tic lasting for more than a year. It is a complex, polygenic, and multifactorial disorder. In this chapter, we briefly discuss the early genetic studies on TS, and then provide a detailed overview of current efforts to uncover common and rare variants contributing to TS susceptibility. We also summarize pathway analyses which combine information from multiple loci to identify converging cellular processes and pathways associated with TS pathogenesis. Since TS is highly comorbid with other psychiatric disorders, we highlight the cross-disorder studies exploring the shared genetic basis of TS and these disorders, as well as the phenome-wide association study which revealed associations of multiple phenotypes with TS genetic risk. We provide an overview of the efforts to investigate the contribution of environmental factors and gene–environment interactions on TS development. Finally, we discuss some future directions that we believe the genetic research on TS will follow.

INTRODUCTION

Tourette syndrome (TS) is a complex childhood-onset neurodevelopmental disorder characterized by multiple motor tics and vocal tics (Robertson et al., 2015, 2017). The prevalence of TS is estimated in the range of 0.6%–1% in school-aged children (Robertson et al., 2009; Scharf et al., 2015), and studies have shown its genetic architecture is complex (Robertson et al., 2017; Yang et al., 2021). Same as in other common neuropsychiatric disorders, a combination of genetic variants, both common and rare, and their interaction with each other and the environment contribute to TS risk. The proportion of the contribution of genetic factors to the phenotypic variability of complex traits is measured by the heritability (h^2) estimate, and TS is a highly heritable disorder (Davis et al., 2013) with the heritability estimated from genome-wide association studies (GWAS) ranging from 0.21 (Yu et al., 2019) to 0.58 (Davis et al., 2013). TS is highly comorbid with other neuropsychiatric disorders; about 50% of the patients are comorbid with obsessive-compulsive disorder (OCD), attention-deficit/hyperactivity disorder (ADHD, up to 54%), autism spectrum disorders (ASDs, up to 20%), major depressive disorder (MDD, up to 26%), and anxiety disorders (AXDs, up to 36%) (Hirschtritt et al., 2015; Robertson et al., 2015, 2017).

TYPES OF GENETIC VARIATION CONTRIBUTING RISK TO TOURETTE SYNDROME

There are mainly two types of variation in human DNA: single nucleotide polymorphisms (SNPs) (sometimes also referred to as single nucleotide variants – SNVs) and structural variants (Frazer et al., 2009). Both types have been found to underlie TS genetic risk. SNPs are variations in the DNA sequence where a single nucleotide is altered. Other types of variations that are larger than point mutations and involve segments of the DNA sequence are considered structural variants, including insertion–deletions, block substitutions, inversions of DNA sequences, and copy number variants (CNVs) (Frazer et al., 2009). Both types of genetic variation based on the observed frequency of the minor allele (MAF) in the studied population are often classified as rare (MAF < 1%) or common (MAF ≥1% or 5% in some studies). Both rare and common SNVs and structural variants have been associated with TS risk as described in detail below.

*Correspondence to: Peristera Paschou, PhD, Department of Biological Sciences, Purdue University, 915 Mitch Daniels Blvd, Room 1-225, West Lafayette, IN 47907, United States. Tel: +1-765-494-4407, E-mail: ppaschou@purdue.edu.

LINKAGE ANALYSIS FOR TOURETTE SYNDROME

Linkage analysis represents a type of approach that was used in the early days of TS genetic research. Such analyses investigated large, multigenerational pedigrees with the hypothesis that TS cases present within these pedigrees were due to few high-penetrant risk variants (Kurlan et al., 1986; Curtis et al., 1992; Eapen et al., 1993). The majority of these studies identified variants with nominal degree of significance and failed to replicate across families (O'Rourke et al., 2009; Paschou, 2013).

One notable exception was a study by Ercan-Sencicek et al., which analyzed data from a two-generation family with nine TS-affected members including an affected father, an unaffected mother, and eight affected children (Ercan-Sencicek et al., 2010). Through linkage analysis, they identified a shared region on chromosome 15, around the histidine decarboxylase (HDC) gene. This region contained a heterozygous G-to-A transition at nucleotide position 951 in exon 9 of the HDC gene, resulting in a premature termination codon (W317X). Following up on this study, Karagiannidis et al. analyzed tagging SNPs from 520 TS trio families and also found significant overtransmission of SNPs in the HDC gene, thus providing additional support for implication of this region (Karagiannidis et al., 2013).

COMMON VARIANTS UNDERLYING TOURETTE SYNDROME RISK

In the early 2000s, with the advancement of genotyping approaches, the first GWAS were published, and soon became the mainstream approach in genetic analyses for complex traits. Rather than testing targeted regions as in candidate gene studies, GWAS test millions of variants that are spread across the genome. Following multiple candidate gene studies, and linkage analysis scans that yielded inconsistent results, the first GWAS for TS was published in 2013 (Scharf et al., 2013). In this study, Scharf and colleagues analyzed 1285 TS cases of European ancestry and 4964 ancestry-matched controls as the primary GWAS, and performed a secondary analysis with additional 211 TS cases and 285 controls from Latin America, resulting in a total sample of 1496 TS cases and 5249 controls (Scharf et al., 2013). In both primary and secondary analyses, no variant reached the genome-wide statistical significance level (5×10^{-8}), which corresponds to p=$0.05/10^6$, where 10^6 controls for the independent regions in Europeans (Marees et al., 2018). In the primary analysis, the variant with the lowest p-values (p=1.85×10^{-6}) was located on an intron of COL27A1 on chromosome 9q32. It was also the most significant locus in the secondary analysis (p=3.61×10^{-7}). COL27A1 is a fibrillar collagen that is mainly expressed in cartilage (Pace et al., 2003). It is also expressed in the cerebellum during many stages of human development; however, its role in neural development is not fully understood (Pace et al., 2003). Since the variant did not reach the genome-wide significance level and it was not supported by an independent replication sample, COL27A1 cannot be considered a TS risk gene at this time.

Six years later, Yu et al. (2019) performed the second TS GWAS and meta-analysis with an increased sample of 4819 TS cases and 9488 controls of European ancestry (Yu et al., 2019). This effort revealed the first genome-wide significant SNP for TS (rs2504235, p=2.1×10^{-8}), which is located within an intron of FMS-like tyrosine kinase 3 (FLT3). However, this variant was not replicated in an independent population-based Icelandic sample of 706 cases and 6068 controls. This could reflect population differences and heterogeneity in ascertainment of cases.

In 2023, the largest TS GWAS meta-analysis to date was published by Tsetsos et al. (2024). They analyzed a new dataset (TS EUROTRAIN) of 1314 TS cases of European ancestry and 4077 controls and performed a meta-analysis with the TS GWAS2 data, resulting in a total sample size of 6133 cases and 13565 controls. The increased sample size provided power to detect a novel genome-wide significant locus on chromosome 5q15, upstream of the NR2F1 gene. The top variant, rs2453763, reached the genome-wide significance threshold in the TS EUROTRAIN dataset (p=2.62×10^{-8}), as well as in the meta-analysis (p=4.05×10^{-8}). NR2F1 is a nuclear receptor that is a regulator of transcription. It belongs to the steroid/thyroid hormone nuclear receptor superfamily, which is involved in a wide range of roles, including cell differentiation, cancer progression, and central and peripheral neurogenesis (Manikandan et al., 2018).

Since the first TS GWAS, the continuously increased sample size of subsequent studies revealed the importance of increasing statistical power in order to dissect the genetic architecture of TS. As this chapter is being submitted for publication, an updated TS GWAS of 13,000 cases and 500,000 controls is currently underway made possible through collaboration of multiple international groups. Although such efforts to collect and analyze large samples will contribute substantially in detecting novel TS risk loci, we should not overlook the fact that the majority of participants are of European ancestry. It is thus crucial for future efforts to recruit study participants from populations that are under-represented in genetics research.

POLYGENIC RISK SCORES IN TOURETTE SYNDROME

While GWAS for TS implicate several risk loci, no single one of them is considered a definitive risk variant for TS at this time. Nevertheless, multiple studies showed that genetic risk for TS can be detected by combining the effect of multiple alleles across the genome (Davis et al., 2013; Abdulkadir et al., 2019; Yu et al., 2019; Tsetsos et al., 2024). It is well known that independently, common variants account for a small amount of the heritability of complex traits, including TS (Gibson, 2012; Yu et al., 2019; Tsetsos et al., 2024). Thus, by creating the polygenic risk score (PRS), researchers can estimate the person's genetic liability to a trait. Typically, it is calculated as the weighted sum of the risk alleles an individual carries, and the weights are calculated by the effect sizes the alleles have in a GWAS of the same phenotype (Choi et al., 2020).

Yu et al., using the TS GWAS2 summary statistics, calculated the TS PRS in the independent population-based Icelandic dataset, and they observed it was predictive of TS case status (p=5.3×10^{-9}) and tic spectrum disorders status (p=4.2×10^{-4}) (Yu et al., 2019). Within the analyzed data, TS GWAS1 included many cases from families with multiple affected siblings (multiplex TS families), so they carried out additional PRS analyses. They observed that multiplex TS cases carried higher PRS values than simplex TS cases (p=0.027), and they also demonstrated a significant positive correlation between TS PRS and lifetime, worst-ever tic severity.

Additionally, Abdulkadir and colleagues used the TS GWAS2 summary to calculate TS PRS in the Avon Longitudinal Study of Parents and Children participants cohort. They observed that TS PRS significantly predicted the presence but not the chronicity of tics in the studied cohort (Abdulkadir et al., 2019).

RARE VARIATION IN TOURETTE SYNDROME

While the genetic studies focusing on common genetic variation in TS revealed multiple risk loci, they can only account for a modest proportion of disease heritability (Manolio et al., 2009). Thus, except for the GWAS studies, efforts to recover parts of the missing heritability for TS through the analysis of rare variants have been made. The recent advancement and cost reduction of genotyping and sequencing technologies made the analysis of rare variants in larger cohorts possible.

In genetic studies, rare variants are considered the ones with MAF < 1%, and similar to common ones can be SNVs or structural variants. Despite their rarity, these variants usually have a larger contribution to disease susceptibility than the common ones. A special category of rare mutations is the de novo variants, which occur during the formation of the gametes or postzygotically and thus are not inherited by the parents to the child. It is estimated that de novo SNVs occur about 44–82 times throughout an individual's genome and only one or two of them are in coding regions, while small de novo indels arise on average 2.9–9 times (Acuna-Hidalgo et al., 2016). Efforts to detect and test the association of rare de novo variants are done through the analysis of whole exome (WES) and whole genome sequencing (WGS) data collected from parent–child trios, including the affected offspring and the unaffected child.

SINGLE NUCLEOTIDE AND INDEL VARIANTS

SNVs and small insertions or deletions (indels) are two types of rare markers that are widely studied when analyzing sequencing data. Willsey et al. investigated rare de novo SNVs and indels using WES data from 511 TS trios, and by applying a Transmission and De Novo Association framework, they detected a high confidence TS risk gene (WWC1) and three probable TS risk genes (CELSR3, NIPBL, FN1) (Willsey et al., 2017). They also observed an enrichment of de novo damaging variants in cases compared to controls (Willsey et al., 2017). One year later, with the sequencing of 291 additional TS trios, resulting in a total sample size of 802 trios, Wang and colleagues performed a follow-up analysis (Wang et al., 2018). They detected a new high-confidence TS risk gene, CELSR3, and observed increased de novo mutations in genes involved in cell polarity. From their total sample, around 10% of the cases carried rare damaging de novo variants, and they estimated that 483 genes contribute to TS risk through de novo variants. Additionally, compound heterozygous mutation on CELSR3 was detected in an independent sample of 15 TS trios of Chinese families (Zhao et al., 2020); however, due to the small sample size, they did not detect a significant enrichment of de novo variants in TS cases compared to controls. A smaller study analyzing WES data from 97 TS trios identified de novo variants in ASH1L and KIF26B genes, which were further supported by rare variant transmission disequilibrium testing (Liu et al., 2020). In addition, they did functional experiments with Ash1l$^{+/-}$ mouse line, which showed tic-like and compulsive behaviors. Ryan and colleagues performed an identity-by-descent analysis using WGS data from 19 individuals from an extended pedigree from a Costa Rican family (17 TS-affected and two controls) (Ryan et al., 2022). Among the implicated TS loci, they observed a rare deleterious missense variation in the RAPGEF1 gene and two ultra-rare deleterious intronic variants in the ERBB4 and IKZF2 genes. All three genes play a role in neurodevelopment. Bae and colleagues identified a missense mutation in the ARHGEF6 gene when analyzed WGS human brain data from 19 TS donors (Bae et al., 2022). Lastly, in a study of 17 multiplex TS families, performing a co-segregation analysis and variant prioritization using WGS data, Fichna and colleagues detected 37 rare and possible pathogenic variants shared by the cases within a family and three ultra-rare variants in two families (Fichna et al., 2023). These variants are located in multiple genes and the majority of them on introns; thus, additional studies are needed to help understand their role.

So far, the studies focusing on rare variants using sequencing data revealed multiple high-confidence and probable TS risk genes; thus, generating larger datasets will help to replicate and clarify the risk loci with a goal to guide the research for new treatment targets.

COPY NUMBER VARIATION

CNVs are large deletions or duplications on the genome that usually cover several megabases and thus could span multiple genes. The detection of CNVs is possible using microarray data (Iafrate et al., 2004; Sebat et al., 2004; Redon et al., 2006), but recently CNV calling can also be done through sequencing data. Genome-wide scans for structural variants on different neuropsychiatric disorders revealed large CNVs that impact the function of multiple genes (Levy et al., 2012; Guyatt et al., 2018).

The first CNV study of TS was performed by Sundaram and colleagues in 2010 (Sundaram et al., 2010). They analyzed DNA microarray data from 111 TS patients and 73 controls of

European ancestry and identified associations with five rare exonic CNVs, some of which had been previously implicated in other neuropsychiatric disorders, such as neurexin 1 (NRXN1) and catenin α-3 (CTNNA3). However, they did not detect a significant difference in rare CNV burden between cases and controls, which could be attributed to the small sample size. Two years later, Fernandez et al. conducted a larger CNV analysis including 460 TS cases and 1131 controls (Fernandez et al., 2012). They also did not find a significant increase in the number of rare CNVs in cases versus controls, but they detected rare TS CNVs overlapping regions previously identified in ASD. The pathway analysis showed significant enrichment of CNVs overlapping genes involved in histaminergic and GABAergic pathways in TS cases. Similarly, with the previous two studies, McGrath and colleagues in 2014, did not detect a significant difference in CNV burden when they analyzed 1086 TS cases and 1789 controls (McGrath et al., 2014).

The first study that detected a significant increase of large (> 500 kb) CNVs in TS cases compared to controls was published by Nag et al. (2013). They analyzed 179 TS cases and 234 controls from two Latin American populations, and they also identified CNV duplications in COL8A1 and deletions in NRXN1 genes among TS patients but not in controls. NRXN1 encodes a neurexin and CNVs covering that gene were previously detected in TS and other neurodevelopmental traits. COL8A1 encodes a collagen subunit, a protein group that studies indicate to have a role in neural development.

The largest study using DNA microarray data to date was performed by Huang and colleagues. They performed a genome-wide rare CNV scan on 2434 TS cases and 4093 controls and identified an increased rare CNV burden of large CNVs (> 1 Mb) in TS cases compared to controls (Huang et al., 2017). Additionally, they also found two genome-wide significant CNV loci associated with TS: NRXN1 deletions and CNTN6 duplications. CNTN6 encodes a CNS-expressed contactin involved in nervous system development.

Wang and colleagues analyzing WES data from 802 TS parent–proband trios found a statistically significant excess of rare de novo CNVs in TS versus unaffected controls (rate ratio=2.2; $p=2.5\times10^{-3}$), which was also replicated by case and controls microarray data. Additional analyses showed an overrepresentation of de novo deletions (rate ratio=3.8; $p=0.02$), while de novo duplications were not significant (rate ratio=1.9; $p=0.15$) (Wang et al., 2018). In another study, Bae and colleagues analyzed WGS data from 131 human brains, 19 of which were from donors with TS, to detect somatic mutations (Bae et al., 2022). In the TS samples, they detected two duplications on chromosome 3 in two different brains present in all bulk samples, suggesting that these occurred during development. Genes in duplicated regions included RP11-553D4.2, SPICE1, and WDR52 in one sample and FHIT1 and U3 in the other.

In a recent study, Mahjani and colleagues analyzed data from a population-based cohort study in Sweden including 993 individuals with OCD, 217 individuals with chronic tic disorders (CTD), and 91 diagnosed with both OCD and CTD (Mahjani et al., 2022). They observed 19 potentially damaging CNV (pdCNV) events in 18 CTD individuals; three of them were within known genomic disorders (16p13.11 deletion, 15q25.2 deletion, and 17q12 duplication). The 16p13.11 deletion is a locus associated with multiple neurodevelopmental disorders including AXDs, ASD, epilepsy, and learning difficulties. Similarly, 15q25.2 was previously reported in individuals with ASD. They also detected nine pdCNV events in eight individuals diagnosed with both OCD and CTD.

All these studies highlight the importance of further exploring the contribution of rare inherited and de novo CNVs to TS etiology as well as their overlap with other neurodevelopmental studies. Thus, currently, a larger CNV scan utilizing microarray data from about 5500 cases and 11,000 controls is underway, including samples contributed by multiple international cohorts.

PATHWAY ANALYSIS IN TOURETTE SYNDROME

Genomic analysis of both rare and common variants of TS has also helped understand the biologic pathways and mechanisms involved in the etiology of the disorder. Pathway analysis allows studies to combine information from multiple genes and variants to identify converging cellular processes and pathways associated with TS pathogenesis even if the individual variants themselves do not reach genome-wide significance. Fig. 14.1 shows the network of the enriched gene ontology (GO) terms using the genes implicated in TS by large-scale studies.

Some of these processes have also been found and discussed in previous studies. Fernandez et al. (2012) conducted a pathway analysis on individuals with TS to examine genes that overlap with high-confidence CNVs. They discovered a notable enrichment of genes related to gamma-aminobutyric acid (GABA) receptor signaling and the life cycle of GABA type A receptors. However, it is important to note that this effect was primarily driven by a single large de novo CNV that overlapped with multiple GABA receptor genes. Other studies investigating rare variants in TS have also implicated genes such as NRXN1 (Sundaram et al., 2010; Nag et al., 2013; Huang et al., 2017), which has been shown to interact with GABA receptors and regulate synaptic transmission (Pak et al., 2015; Luo et al., 2021); ASH1L, which has been demonstrated to influence GABAergic inhibition in the prefrontal cortex (Qin et al., 2021); and also ALDH2, DLD, and ALDH1B1 genes which could influence oxidoreductase activity in the brain (Fichna et al., 2023).

Histaminergic activity has also been implicated in the development of TS. In a study by Ercan-Sencicek et al. (2010), a specific genetic mutation in the HDC gene was identified, resulting in a truncated protein with key segments of the active domain missing. Studies in mice with a similar version of the mutated protein indicated that heterozygous HDC knockouts resulted in approximately 50% reduction in brain histamine levels (Ohtsu et al., 2001). HDC is an enzyme crucial in histamine biosynthesis, suggesting that histaminergic

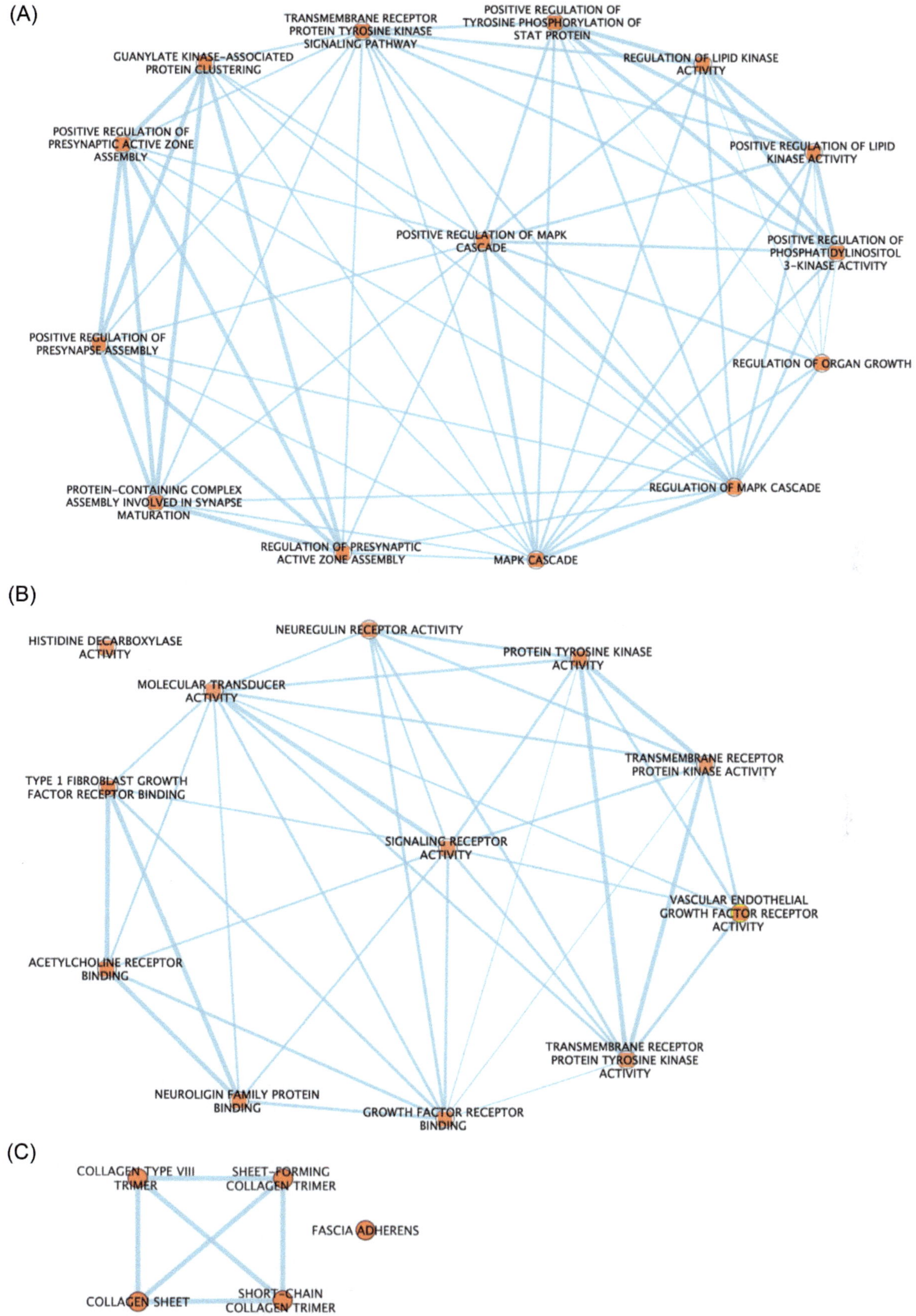

Fig. 14.1. Network plot of significantly enriched gene ontology (GO) terms using genes implicated in Tourette syndrome by large-scale genetic studies. The enrichment analysis was performed using the online tool g:Profiler. The statistically significance threshold was set at $p < 0.05$ after false discovery rate correction. The plots were generated with EnrichmentMap in Cytoscape. (**A**) Biologic processes GO terms, (**B**) molecular function GO terms, and (**C**) cellular component GO terms.

neurotransmission may play a role in the pathobiology of TS. Histamine signaling in the central nervous system involves four G protein-coupled receptors, located both presynaptically (H3, H4) and postsynaptically (H1–H3). H2 and H3 receptors are particularly abundant in the striatum and cortex, brain regions implicated in TS (Haas et al., 2008). Notably, in vitro studies of mouse brains have demonstrated that histamine modulates various aspects of functional connectivity within the striatum. During heightened attention and wakefulness, histaminergic activity affects the flow of excitatory inputs to the striatum and the intrastriatal processing of those inputs. Consequently, this results in feedforward inhibition and suppressed excitatory drive, influencing striatal functioning (Ellender et al., 2011). HDC knockout mice exhibit increased grooming or other motor stereotypies and upregulated H3 receptor expression in the striatum, leading to their proposition as an animal model for TS (Baldan et al., 2014; Xu et al., 2015; Pittenger, 2017, 2020).

A genomics-based approach to pathway analysis have also helped identify biologic processes involved in the neurobiology of TS. Tsetsos et al. (2021) conducted gene set analysis on a large TS GWAS dataset and identified three significant sets of genes associated with the TS phenotype. These gene sets included ligand-gated ion channel signaling (LICS), lymphocytic genes, and cell adhesion and trans-synaptic signaling (CATS) genes. The LICS gene set consists of genes implicated in ion channel signaling, particularly through γ-aminobutyric acid and glutamate. Some of these genes have previously been linked to neuropsychiatric conditions. For instance, HCN1, involved in pacemaker currents, has associations with schizophrenia and ASDs (Okbay et al., 2016; Autism Spectrum Disorders Working Group of The Psychiatric Genomics Consortium, 2017; Pardiñas et al., 2018). GRIK4, which encodes a glutamate-gated ionic channel, has been associated with mathematical ability, educational attainment, and ADHD (Yang et al., 2013; Lee et al., 2018). The association with the lymphocytic gene set was driven by the FLT3 gene, suggesting a potential role for neuroimmunity in TS etiology. A transcriptome-wide association study (TWAS) of TS indicated that an increased expression of FLT3 in the dorsolateral prefrontal cortex is significantly associated with the disorder and also showed that there is global dysregulation of FLT3 across several brain regions and probabilistic causal fine-mapping of the TWAS signal prioritizes FLT3 as a likely causal gene (Liao et al., 2022). The FLT3 variant rs1933437 is an eQTL in the brain cortex and cerebellum (Carithers et al., 2015) and has also been implicated in the age at the onset of menarche (Kichaev et al., 2019). FLT3 has also been associated with peripheral neuropathic pain, indicating its involvement in neuroimmune interactions and potential role in neuroinflammation (Tsetsos et al., 2021). The CATS gene set, associated with TS, encompasses various cell adhesion molecules, with CDH26 showing the strongest signals. CDH26 is a cadherin involved in regulating leukocyte migration, adhesion, and activation (Caldwell et al., 2017). Cell adhesion molecules have consistently been implicated in brain-related phenotypes, and their association with TS has been well documented, including neurexins, contactins, neuroligins, and their associated proteins (Verkerk et al., 2003; Clarke et al., 2012; Sun et al., 2016; Huang et al., 2017).

In the cross-disorder study of TS and related disorders, Yang et al. (2021) investigated gene and tissue enrichment to gain further insights into the underlying mechanisms. Their findings supported the involvement of the basal ganglia in TS, ADHD, ASD, and OCD, consistent with previous independent studies (Calderoni et al., 2014; Shaw et al., 2014; Maciel et al., 2016; Caligiore et al., 2017; Yang et al., 2021). Additionally, significant enrichment of cross-disorder gene expression was observed in the pituitary and hypothalamus, indicating the potential involvement of the hypothalamus-pituitary-adrenal axis and stress responses aligning with previous clinical studies that have implicated this system in various childhood-onset psychiatric traits, including TS and ADHD (van Goozen et al., 2000; Lopez-Duran et al., 2009; Hankin et al., 2010; Martino et al., 2013; Reichl et al., 2016).

The study by Mufford et al. (2019) used summary statistics from the TS GWAS2 (Yu et al., 2019) as well as the ENIGMA GWAS of subcortical volumes (30,717 individuals) to understand the relation between TS and subcortical brain structures. They employed SNP effect concordance analysis and linkage disequilibrium score regression (LDSR) to examine genetic pleiotropy and direction of effect. The analysis revealed significant pleiotropy between TS and the volumes of the putamen and caudate, independent of the direction of effect. Additionally, there was significant concordance between TS and lower thalamic volume. The LDSR analysis further supported the association between TS and thalamic volume. Tsetsos et al. (2024) used a phenome-wide association approach to identify associations between TS PRS and subcortical brain volumes. The study found significant association between TS risk and right putamen volume as well as bilateral thalamus volume.

EPIGENETIC ANALYSIS AND GENE–ENVIRONMENT INTERACTION

While a substantial portion of the variation in TS status can be attributed to genetic factors, environmental factors and gene–environment interactions also play a significant role in the development of the disorder. One of these studies by Abelson et al. (2005) identified a nucleotide variant (var321) in the 3′-UTR of the *SLITRK1* gene, leading to its stronger repression by miR-189. This rare variant has also been investigated in several TS patient studies (Abelson et al., 2005). A study investigated six TS patients and three controls, revealing that the microRNA miR-429 was under-expressed in TS patients. This under-expression of miR-429 could potentially impact the regulation of neurodifferentiation and synaptic transmission, contributing to TS pathology (Rizzo et al., 2015). In the first epigenome-wide association study of tic disorders, DNA methylation patterns were compared between 188 cases and 1490 controls from the Netherlands Twin Register. Although no specific methylation sites reached genome-wide significance, permutation testing of the top 57 sites ($p < 10^{-4}$) suggested that these sites were not randomly occurring ($p < 0.05$). Furthermore, 8 of the top 57 sites mapped to genes that had been previously associated with various neuropsychiatric disorders such as OCD, ASD, bipolar disorder, schizophrenia, and intellectual disability (Zilhão et al., 2015).

Gene–environment interaction studies between TS GWAS SNPs and adverse childhood experiences have also shown positive results. In a study by Abdulkadir et al. (2021), they investigated genetic loci previously implicated in TS and their interaction with pre- and perinatal adversity and their association with tic severity. They found a significant interaction of rs6539267, a top SNP from a TS GWAS ($F = 6.80$, p=0.001) with the cumulative pre- and perinatal adversity score. Another study by Yang et al. (2022) aimed to assess the association between adverse childhood experiences and tic severity and if TS genetic risk modifies this association. They determined that TS genetic risk moderated the relationship between ACE score and tic impairment, but not tic severity, particularly for individuals with higher TS polygenic risk.

Although the studies are limited, there have been multiple findings about the contribution of epigenetic factors and gene–environment interaction to TS risk and severity. It has also been proposed that epigenetic mechanisms observed in other neuropsychiatric disorders may also contribute to TS. The exploration of epigenetic associations with TS remains an ongoing and open area of research.

CROSS-DISORDER ANALYSIS

TS often occurs with other neuropsychiatric disorders, particularly those within the impulsivity-compulsivity spectrum such as ADHD (54% of TS cases), ASD (20% of TS cases), and OCD (50% of TS cases) and pure TS cases accounting for only 10% of all TS cases (Hirschtritt et al., 2015; Karagiannidis et al., 2016; Cravedi et al., 2017). The high rates of comorbidity among TS patients support the idea of shared neural circuitry alterations and genetic susceptibility across these disorders (Mathews and Grados, 2011; Hirschtritt et al., 2015; Huisman-van Dijk et al., 2016; Karagiannidis et al., 2016).

Cross-disorder analyses have been performed to understand the genetic overlap of TS with other psychiatric disorders, revealing broad genetic correlations (Abdulkadir et al., 2018; Brainstorm Consortium et al., 2018; Cross-Disorder Group of the Psychiatric Genomics Consortium, 2013, 2019; Taylor et al., 2019). Rare CNVs associated with TS risk have also been implicated in neurodevelopmental disorders like ASD, schizophrenia, and epilepsy. Examples include deletions in 1q21, NRXN1, 16p13.11, and duplications in 22q11 (Sundaram et al., 2010; Fernandez et al., 2012; McGrath et al., 2014) found an enrichment of TS-associated CNVs in genes also associated with ASD, with cell adhesion molecules like neurexins and neuroligins being notable examples of genes repeatedly implicated in the etiology of TS and other neurodevelopmental phenotypes (Clarke et al., 2012).

OCD has long been recognized as the neuropsychiatric disorder most strongly associated with TS (Davis et al., 2013; Yu et al., 2015). Davis et al. (2013) observed a significant shared heritability between the two disorders, although their overall genetic architecture differed. Yu et al. (2015), using polygenic risk scoring, identified a significant polygenic component for predicting OCD. However, when including OCD cases with TS in the dataset used to create the polygenic model, the ability to predict OCD status in the independent dataset decreased. This suggests that OCD with and without tics may have distinct genetic architectures.

In a comprehensive cross-disorder study encompassing 23 neurologic and psychiatric disorders, a large-scale GWAS investigation revealed that a significant portion of TS polygenic heritability is shared with OCD, ADHD, and migraine (Brainstorm Consortium et al., 2018). While the shared heritability between TS and OCD and between TS and ADHD has been previously established (Hirschtritt et al., 2015), the newly discovered shared genetic relationship between TS and migraine was not entirely unexpected. It has been observed that TS and migraine co-occur more frequently than expected in comparison to control rates (Barabas et al., 1984). This association was also validated in the latest TS GWAS (Tsetsos et al., 2024), which showed significant genetic correlation between TS and migraine. Fig. 14.2 shows in detail the genetic correlation analysis of TS with 16 psychiatric and

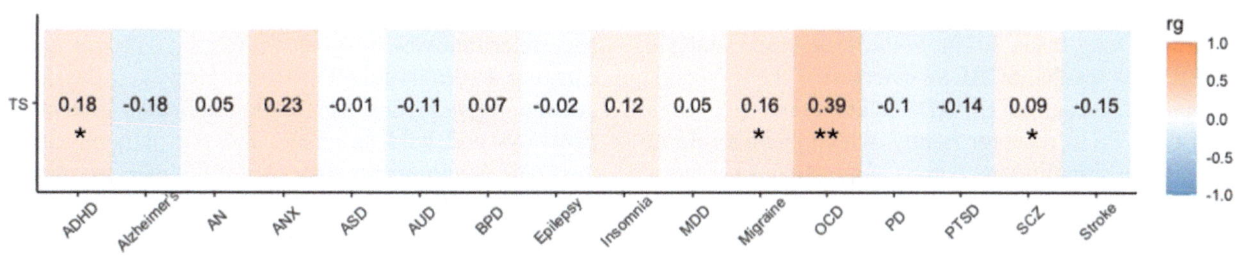

Fig. 14.2. Genetic correlations with TS. The plot shows the correlations between the most recent TS genome-wide association study (Tsetsos et al. 2024) and 16 psychiatric and neurologic traits. The number in each square denotes the correlation *rg*. Asterisk (*) shows the correlations with nominal $p < 0.05$. Double asterisks (**) denote the correlations that were identified as statistically significant after Benjamini–Hochberg false discovery rate correction (α=0.05). *ADHD*, attention-deficit/hyperactivity disorder; *AN*, anorexia nervosa; *ANX*, anxiety disorders; *ASD*, autism spectrum disorder; *AUD*, alcohol use disorder; *BPD*, bipolar disorder; *MDD*, major depressive disorder; *OCD*, obsessive-compulsive disorder; *PD*, Parkinson disease; *PTSD*, posttraumatic stress disorder; *SCZ*, schizophrenia; *TS*, Tourette syndrome.

neurologic traits. Interestingly, a cross-disorder meta-analysis examining the top loci from GWAS of TS and ADHD (Tsetsos et al., 2016) identified TBC1D7 (tuberous sclerosis protein complex gene) as the leading signal, which has also been associated with migraine (Brainstorm Consortium et al., 2018).

The Psychiatric Genomics Consortium conducted a large-scale GWAS meta-analysis of eight common psychiatric disorders, including TS, ADHD, ASD, OCD, anorexia nervosa, bipolar disorder, MDD, and schizophrenia (Cross-Disorder Group of the Psychiatric Genomics Consortium, 2019). Through exploratory factor analysis, they identified three factors, with early-onset disorders (ADHD, ASD, and TS) falling into one factor along with MDD, which is not typically early onset. TS showed a weak correlation with compulsive disorders (OCD and anorexia nervosa) in another factor. Pairwise analysis did not find a significant correlation between anorexia nervosa and TS. Three years later, a larger study exploring the genetic architecture of 11 psychiatric disorders was published (Grotzinger et al., 2022). In this study, TS was also part of two factors, the first "compulsive disorders" included TS, anorexia nervosa, and OCD. TS was also part of the neurodevelopmental disorders factor, primarily made up of ASD and ADHD, but also including posttraumatic stress disorder, MDD, and problematic alcohol use. Additionally, the authors performed cross-disorder meta-analyses within each factor to identify shared genetic risk variants of the traits in each cluster. They found one risk locus shared across disorders in the compulsive disorders factor, and nine shared loci for the neurodevelopmental disorders factor; two of them were not significant in the individual GWAS included in the analysis.

Focusing on the disorders along the impulsivity-compulsivity spectrum, Yang et al. (2021) performed a cross-disorder GWAS analysis including TS, ADHD, ASD, and OCD. Factor analysis revealed and validated the strong correlation of genetic factors across TS, ADHD, and ASD and reproduced the high genetic correlation of TS and OCD that appeared to be separate from the TS-ADHD-ASD factor. The results of the meta-analysis between TS, ADHD, and ASD, highlighted 16 genomic risk regions, 13 of which exhibited high pleiotropy across all analyzed disorders. The top significant genomic risk locus associated with TS-ADHD-ASD was found in LINC00461 on chromosome 5, which is highly expressed in the brain and visual cortex and has also been previously linked to tumorigenesis (Dong et al., 2019). In the gene-based analysis of the meta-analysis results, SORCS3 emerged as the top result with pleiotropic effects across TS, ADHD, and ASD. It encodes a member of the VPS10 receptor family, which plays a role in intracellular protein signaling in neurons and regulates neuronal viability through multiple pathways (Willnow et al., 2008). It is highly expressed in brain tissues (GTEx Consortium et al., 2017) and has also been implicated in neurologic diseases, including ADHD, ASD, Alzheimer disease, and MDD (Small et al., 2005; Small and Petsko, 2015; Hyde et al., 2016; Demontis et al., 2019; Grove et al., 2019). Furthermore, its interaction with postsynaptic proteins, such as PICK1, suggests its involvement in regulating glutamate receptor function (Breiderhoff et al., 2013; Christiansen et al., 2017). The glutamate pathway has long been hypothesized to be involved in abnormalities associated with ADHD, ASD, and TS, making it a potential therapeutic target for these disorders (Chakrabarty et al., 2005; Maltezos et al., 2014).

Jain et al. (2023) used a phenome-wide approach to compare the associations of the genetic risk of TS, ADHD, ASD, and OCD with different phenotypes. The study identified that the disorders shared associations with multiple health and behavioral phenotypes such as neuroticism, body pain, and indices of multiple deprivation. Interestingly, ADHD and ASD had similar direction of effect as TS for almost all phenotypes, but OCD had an opposite direction of effect compared to TS (for all phenotypes except mental health traits). These results suggest that although these disorders are highly comorbid with TS and have strong genetic overlap, there are still considerable differences.

CONCLUSION AND FUTURE DIRECTIONS

Since the early days of genetic research on TS, numerous studies implementing different approaches implicated multiple genetic factors that could underlie risk for TS. Both family- and population-based studies demonstrated the high heritability of TS, pointing to the important role of genetics in TS pathogenesis. GWAS exploring the common variants involved in TS, although with relatively small sample size, pointed to risk loci and PRS constructed with those variants, are gaining increasing predicting power of TS risk. Additionally, analyses of rare variants revealed many loci associated with TS risk. So far, rare CNV analyses identified genes affected by structural changes associated with TS, some of which were also involved in other psychiatric disorders. Most recently, studies analyzing whole genome and whole exome sequencing data detected various de novo mutations in TS cases. Finally, the constantly increasing sample sizes in TS GWAS analyses point to the aggregate effect of common variants on the genetic risk of TS, which could help in developing diagnostics and prediction markers and tools for the disorder.

Similarly to other psychiatric disorders, TS is a complex polygenic trait; thus, multiple mutations in many genes, as well as the environment, and gene–environment interactions contribute to its pathogenesis This makes it challenging to quantify and determine the risk of developing TS for an individual carrying a certain genetic variant. However, the discoveries of genetic studies thus far have implicated potential biologic pathways and help us understand the pathophysiology of the disorder. Such knowledge could be used to develop novel treatments. Additionally, with TS being highly comorbid with other psychiatric traits, cross-disorder studies help us move beyond diagnostic categories and individualized risk prediction. As population-based biobanks with detailed electronic health records for thousands of individuals are emerging, they offer opportunities for exploring further the relationship of TS risk with other phenotypes. This is an exciting time for TS genetic research, as studies with increased sample sizes are underway and we are expecting many additional risk loci to be revealed. The larger sample sizes together with the recruitment of individuals of non-European ancestries and the increasing sequencing data, will be catalytic to gain further insight into TS biology and move toward informing clinical care.

References

Abdulkadir M, Londono D, Gordon D et al. (2018). Investigation of previously implicated genetic variants in chronic tic disorders: a transmission disequilibrium test approach. Eur Arch Psychiatry Clin Neurosci 268: 301–316.

Abdulkadir M, Mathews CA, Scharf JM et al. (2019). Polygenic risk scores derived from a Tourette syndrome genome-wide association study predict presence of tics in the Avon Longitudinal Study of parents and children cohort. Biol Psychiatry 85: 298–304.

Abdulkadir M, Yu D, Osiecki L et al. (2021). Investigation of gene-environment interactions in relation to tic severity. J Neural Transm 128: 1757–1765.

Abelson JF, Kwan KY, O'Roak BJ et al. (2005). Sequence variants in SLITRK1 are associated with Tourette's syndrome. Science 310: 317–320.

Acuna-Hidalgo R, Veltman JA, Hoischen A (2016). New insights into the generation and role of de novo mutations in health and disease. Genome Biol 17: 241.

Autism Spectrum Disorders Working Group of The Psychiatric Genomics Consortium(2017). Meta-analysis of GWAS of over 16,000 individuals with autism spectrum disorder highlights a novel locus at 10q24.32 and a significant overlap with schizophrenia. Mol Autism 8: 21.

Bae T, Fasching L, Wang Y et al. (2022). Analysis of somatic mutations in 131 human brains reveals aging-associated hypermutability. Science 377: 511–517.

Baldan LC, Williams KA, Gallezot JD et al. (2014). Histidine decarboxylase deficiency causes Tourette syndrome: parallel findings in humans and mice. Neuron 81: 77–90.

Barabas G, Matthews WS, Ferrari M (1984). Tourette's syndrome and migraine. Arch Neurol 41: 871–872.

Brainstorm Consortium(2018). Analysis of shared heritability in common disorders of the brain. Science 360: eaap8757.

Breiderhoff T, Christiansen GB, Pallesen LT et al. (2013). Sortilin-related receptor SORCS3 is a postsynaptic modulator of synaptic depression and fear extinction. PLoS One 8: e75006.

Calderoni S, Bellani M, Hardan AY et al. (2014). Basal ganglia and restricted and repetitive behaviours in autism spectrum disorders: current status and future perspectives. Epidemiol Psychiatr Sci 23: 235–238.

Caldwell JM, Collins MH, Kemme KA et al. (2017). Cadherin 26 is an alpha integrin-binding epithelial receptor regulated during allergic inflammation. Mucosal Immunol 10: 1190–1201.

Caligiore D, Mannella F, Arbib MA et al. (2017). Dysfunctions of the basal ganglia-cerebellar-thalamo-cortical system produce motor tics in Tourette syndrome. PLoS Comput Biol 13: e1005395.

Carithers LJ, Ardlie K, Barcus M et al. (2015). A novel approach to high-quality postmortem tissue procurement: the GTEx project. Biopreserv Biobank 13: 311–319.

Chakrabarty K, Bhattacharyya S, Christopher R et al. (2005). Glutamatergic dysfunction in OCD. Neuropsychopharmacology 30: 1735–1740.

Choi SW, Mak TS-H, O'Reilly PF (2020). Tutorial: a guide to performing polygenic risk score analyses. Nat Protoc 15: 2759–2772.

Christiansen GB, Andersen KH, Riis S et al. (2017). The sorting receptor SorCS3 is a stronger regulator of glutamate receptor functions compared to GABAergic mechanisms in the hippocampus. Hippocampus 27: 235–248.

Clarke RA, Lee S, Eapen V (2012). Pathogenetic model for Tourette syndrome delineates overlap with related neurodevelopmental disorders including autism. Transl Psychiatry 2: e158.

Cravedi E, Deniau E, Giannitelli M et al. (2017). Tourette syndrome and other neurodevelopmental disorders: a comprehensive review. Child Adolesc Psychiatry Ment Health 11: 59.

Cross-Disorder Group of the Psychiatric Genomics Consortium (2013). Identification of risk loci with shared effects on five major psychiatric disorders: a genome-wide analysis. Lancet 381: 1371–1379.

Cross-Disorder Group of the Psychiatric Genomics Consortium (2019). Genomic relationships, novel loci, and pleiotropic mechanisms across eight psychiatric disorders. Cell 179: 1469–1482.e11.

Curtis D, Robertson MM, Gurling HM (1992). Autosomal dominant gene transmission in a large kindred with Gilles de la Tourette syndrome. Br J Psychiatry 160: 845–849.

Davis LK, Yu D, Keenan CL et al. (2013). Partitioning the heritability of Tourette syndrome and obsessive compulsive disorder reveals differences in genetic architecture. PLoS Genet 9: e1003864.

Demontis D, Walters RK, Martin J et al. (2019). Discovery of the first genome-wide significant risk loci for attention deficit/hyperactivity disorder. Nat Genet 51: 63–75.

Dong L, Qian J, Chen F et al. (2019). LINC00461 promotes cell migration and invasion in breast cancer through miR-30a-5p/ integrin β3 axis. J Cell Biochem 120: 4851–4862.

Eapen V, Pauls DL, Robertson MM (1993). Evidence for autosomal dominant transmission in Tourette's syndrome. United Kingdom cohort study. Br J Psychiatry 162: 593–596.

Ellender TJ, Huerta-Ocampo I, Deisseroth K et al. (2011). Differential modulation of excitatory and inhibitory striatal synaptic transmission by histamine. J Neurosci 31: 15340–15351.

Ercan-Sencicek AG, Stillman AA, Ghosh AK et al. (2010). L-histidine decarboxylase and Tourette's syndrome. N Engl J Med 362: 1901–1908.

Fernandez TV, Sanders SJ, Yurkiewicz IR et al. (2012). Rare copy number variants in Tourette syndrome disrupt genes in histaminergic pathways and overlap with autism. Biol Psychiatry 71: 392–402.

Fichna JP, Borczyk M, Piechota M et al. (2023). Genomic variants and inferred biological processes in multiplex families with Tourette syndrome. J Psychiatry Neurosci 48: E179–E189.

Frazer KA, Murray SS, Schork NJ et al. (2009). Human genetic variation and its contribution to complex traits. Nat Rev Genet 10: 241–251.

Gibson G (2012). Rare and common variants: twenty arguments. Nat Rev Genet 13: 135–145.

van Goozen SH, Matthys W, Cohen-Kettenis PT et al. (2000). Hypothalamic-pituitary-adrenal axis and autonomic nervous system activity in disruptive children and matched controls. J Am Acad Child Adolesc Psychiatry 39: 1438–1445.

Grotzinger AD, Mallard TT, Akingbuwa WA et al. (2022). Genetic architecture of 11 major psychiatric disorders at biobehavioral, functional genomic and molecular genetic levels of analysis. Nat Genet 54: 548–559.

Grove J, Ripke S, Als TD et al. (2019). Identification of common genetic risk variants for autism spectrum disorder. Nat Genet 51: 431–444.

GTEx Consortium(2017). Genetic effects on gene expression across human tissues. Nature 550: 204–213.

Guyatt AL, Stergiakouli E, Martin J et al. (2018). Association of copy number variation across the genome with neuropsychiatric traits in the general population. Am J Med Genet B Neuropsychiatr Genet 177: 489–502.

Haas HL, Sergeeva OA, Selbach O (2008). Histamine in the nervous system. Physiol Rev 88: 1183–1241.

Hankin BL, Badanes LS, Abela JRZ et al. (2010). Hypothalamic-pituitary-adrenal axis dysregulation in dysphoric children and adolescents: cortisol reactivity to psychosocial stress from pre-school through middle adolescence. Biol Psychiatry 68: 484–490.

Hirschtritt ME, Lee PC, Pauls DL et al. (2015). Lifetime prevalence, age of risk, and genetic relationships of comorbid psychiatric disorders in Tourette syndrome. JAMA Psychiatry 72: 325–333.

Huang AY, Yu D, Davis LK et al. (2017). Rare copy number variants in NRXN1 and CNTN6 increase risk for Tourette syndrome. Neuron 94: 1101–1111.e7.

Huisman-van Dijk HM, Schoot R, Rijkeboer MM et al. (2016). The relationship between tics, OC, ADHD and autism symptoms: a cross-disorder symptom analysis in Gilles de la Tourette syndrome patients and family-members. Psychiatry Res 237: 138–146.

Hyde CL, Nagle MW, Tian C et al. (2016). Identification of 15 genetic loci associated with risk of major depression in individuals of European descent. Nat Genet 48: 1031–1036.

Iafrate AJ, Feuk L, Rivera MN et al. (2004). Detection of large-scale variation in the human genome. Nat Genet 36: 949–951.

Jain P, Miller-Fleming T, Topaloudi A et al. (2023). Polygenic risk score-based phenome-wide association study identifies novel associations for Tourette syndrome. Transl Psychiatry 13: 69.

Karagiannidis I, Dehning S, Sandor P et al. (2013). Support of the histaminergic hypothesis in Tourette syndrome: association of the histamine decarboxylase gene in a large sample of families. J Med Genet 50: 760–764.

Karagiannidis I, Tsetsos F, Padmanabhuni SS et al. (2016). The genetics of Gilles de la Tourette syndrome: a common aetiological basis with comorbid disorders? Curr Behav Neurosci Rep 3: 218–231.

Kichaev G, Bhatia G, Loh PR et al. (2019). Leveraging polygenic functional enrichment to improve GWAS power. Am J Hum Genet 104: 65–75.

Kurlan R, Behr J, Medved L et al. (1986). Familial Tourette's syndrome: report of a large pedigree and potential for linkage analysis. Neurology 36: 772–776.

Lee JJ, Wedow R, Okbay A et al. (2018). Gene discovery and polygenic prediction from a genome-wide association study of educational attainment in 1.1 million individuals. Nat Genet 50: 1112–1121.

Levy RJ, Xu B, Gogos JA et al. Copy number variation and psychiatric disease risk. In: Feuk L (Ed.). Genomic structural variants: methods and protocols. New York, NY: Springer New York, pp. 97–113.

Liao C, Vuokila V, Catoire H et al. (2022). Transcriptome-wide association study reveals increased neuronal FLT3 expression is associated with Tourette's syndrome. Commun Biol 5: 289.

Liu S, Tian M, He F et al. (2020). Mutations in ASH1L confer susceptibility to Tourette syndrome. Mol Psychiatry 25: 476–490.

Lopez-Duran NL, Kovacs M, George CJ (2009). Hypothalamic-pituitary-adrenal axis dysregulation in depressed children and adolescents: a meta-analysis. Psychoneuroendocrinology 34: 1272–1283.

Luo F, Sclip A, Merrill S et al. (2021). Neurexins regulate presynaptic GABAB-receptors at central synapses. Nat Commun 12: 2380.

Maciel ROH, Ferreira GA, Akemy B et al. (2016). Executive dysfunction, obsessive-compulsive symptoms, and attention deficit and hyperactivity disorder in systemic lupus erythematosus: evidence for basal ganglia dysfunction? J Neurol Sci 360: 94–97.

Mahjani B, Birnbaum R, Buxbaum Grice A et al. (2022). Phenotypic impact of rare potentially damaging copy number variation in obsessive-compulsive disorder and chronic tic disorders. Genes 13: 1796.

Maltezos S, Horder J, Coghlan S et al. (2014). Glutamate/glutamine and neuronal integrity in adults with ADHD: a proton MRS study. Transl Psychiatry 4: e373.

Manikandan M, Abuelreich S, Elsafadi M et al. (2018). NR2F1 mediated down-regulation of osteoblast differentiation was rescued by bone morphogenetic protein-2 (BMP-2) in human MSC. Differentiation 104: 36–41.

Manolio TA, Collins FS, Cox NJ et al. (2009). Finding the missing heritability of complex diseases. Nature 461: 747–753.

Marees AT, de Kluiver H, Stringer S et al. (2018). A tutorial on conducting genome-wide association studies: quality control and statistical analysis. Int J Methods Psychiatr Res 27: e1608.

Martino D, Macerollo A, Leckman JF (2013). Neuroendocrine aspects of Tourette syndrome. Int Rev Neurobiol 112: 239–279.

Mathews CA, Grados MA (2011). Familiality of Tourette syndrome, obsessive-compulsive disorder, and attention-deficit/hyperactivity disorder: heritability analysis in a large sib-pair sample. J Am Acad Child Adolesc Psychiatry 50: 46–54.

McGrath LM, Yu D, Marshall C et al. (2014). Copy number variation in obsessive-compulsive disorder and Tourette syndrome: a cross-disorder study. J Am Acad Child Adolesc Psychiatry 53: 910–919.

Mufford M, Cheung J, Jahanshad N et al. (2019). Concordance of genetic variation that increases risk for Tourette syndrome and that influences its underlying neurocircuitry. Transl Psychiatry 9: 120.

Nag A, Bochukova EG, Kremeyer B et al. (2013). CNV analysis in Tourette syndrome implicates large genomic rearrangements in COL8A1 and NRXN1. PLoS One 8: e59061.

Ohtsu H, Tanaka S, Terui T et al. (2001). Mice lacking histidine decarboxylase exhibit abnormal mast cells. FEBS Lett 502: 53–56.

Okbay A, Beauchamp JP, Fontana MA et al. (2016). Genome-wide association study identifies 74 loci associated with educational attainment. Nature 533: 539–542.

O'Rourke JA, Scharf JM, Yu D et al. (2009). The genetics of Tourette syndrome: a review. J Psychosom Res 67: 533–545.

Pace JM, Corrado M, Missero C et al. (2003). Identification, characterization and expression analysis of a new fibrillar collagen gene, COL27A1. Matrix Biol 22: 3–14.

Pak C, Danko T, Zhang Y et al. (2015). Human neuropsychiatric disease modeling using conditional deletion reveals synaptic transmission defects caused by heterozygous mutations in NRXN1. Cell Stem Cell 17: 316–328.

Pardiñas AF, Holmans P, Pocklington AJ et al. (2018). Common schizophrenia alleles are enriched in mutation-intolerant genes and in regions under strong background selection. Nat Genet 50: 381–389.

Paschou P (2013). The genetic basis of Gilles de la Tourette syndrome. Neurosci Biobehav Rev 37: 1026–1039.

Pittenger C (2017). The neurobiology of tic disorders and obsessive-compulsive disorder: animal and human studies. Oxford University Press.

Pittenger C (2020). The histidine decarboxylase model of tic pathophysiology: a new focus on the histamine H3 receptor. Br J Pharmacol 177: 570–579.

Qin L, Williams JB, Tan T et al. (2021). Deficiency of autism risk factor ASH1L in prefrontal cortex induces epigenetic aberrations and seizures. Nat Commun 12: 6589.

Redon R, Ishikawa S, Fitch KR et al. (2006). Global variation in copy number in the human genome. Nature 444: 444–454.

Reichl C, Heyer A, Brunner R et al. (2016). Hypothalamic-pituitary-adrenal axis, childhood adversity and adolescent nonsuicidal self-injury. Psychoneuroendocrinology 74: 203–211.

Rizzo R, Ragusa M, Barbagallo C et al. (2015). Circulating miRNAs profiles in Tourette syndrome: molecular data and clinical implications. Mol Brain 8: 44.

Robertson MM, Eapen V, Singer HS et al. (2017). Gilles de la Tourette syndrome. Nat Rev Dis Prim 3: 16097.

Robertson MM, Cavanna AE, Eapen V (2015). Gilles de la Tourette syndrome and disruptive behavior disorders: prevalence, associations, and explanation of the relationships. J Neuropsychiatry Clin Neurosci 27: 33–41.

Robertson MM, Eapen V, Cavanna AE (2009). The international prevalence, epidemiology, and clinical phenomenology of Tourette syndrome: a cross-cultural perspective. J Psychosom Res 67: 475–483.

Ryan N, Ormond C, Chang YC et al. (2022). Identity-by-descent analysis of a large Tourette's syndrome pedigree from Costa Rica implicates genes involved in neuronal development and signal transduction. Mol Psychiatry 27: 5020–5027.

Scharf JM, Miller LL, Gauvin CA et al. (2013). Genome-wide association study of Tourette's syndrome. Mol Psychiatry 18: 721–728.

Scharf JM, Yu D, Mathews CA et al. (2015). Population prevalence of Tourette syndrome: a systematic review and meta-analysis. Mov Disord 30: 221–228.

Sebat J, Lakshmi B, Troge J et al. (2004). Large-scale copy number polymorphism in the human genome. Science 305: 525–528.

Shaw P, De Rossi P, Watson B et al. (2014). Mapping the development of the basal ganglia in children with attention-deficit/hyperactivity disorder. J Am Acad Child Adolesc Psychiatry 53: 780–789.e11.

Small SA, Kent K, Pierce A et al. (2005). Model-guided microarray implicates the retromer complex in Alzheimer's disease. Ann Neurol 58: 909–919.

Small SA, Petsko GA (2015). Retromer in Alzheimer disease, Parkinson disease and other neurological disorders. Nat Rev Neurosci 16: 126–132.

Sun N, Tischfield JA, King RA et al. (2016). Functional evaluations of genes disrupted in patients with Tourette's disorder. Front Psychiatry 7: 11.

Sundaram SK, Huq AM, Wilson BJ et al. (2010). Tourette syndrome is associated with recurrent exonic copy number variants. Neurology 74: 1583–1590.

Taylor MJ, Martin J, Lu Y et al. (2019). Association of genetic risk factors for psychiatric disorders and traits of these disorders in a Swedish Population twin sample. JAMA Psychiatry 76: 280–289.

Tsetsos F, Padmanabhuni SS, Alexander J et al. (2016). Meta-analysis of Tourette syndrome and attention deficit hyperactivity disorder provides support for a shared genetic basis. Front Neurosci 10: 340.

Tsetsos F, Yu D, Sul JH et al. (2021). Synaptic processes and immune-related pathways implicated in Tourette syndrome. Transl Psychiatry 11: 56.

Tsetsos F, Topaloudi A, Jain P et al. (2024). Genome-wide association study points to novel locus for Gilles de la Tourette Syndrome. Biol Psychiatry 96: 114–124.

Verkerk AJMH, Mathews CA, Joosse M et al. (2003). CNTNAP2 is disrupted in a family with Gilles de la Tourette syndrome and obsessive compulsive disorder. Genomics 82: 1–9.

Wang S, Mandell JD, Kumar Y et al. (2018). De novo sequence and copy number variants are strongly associated with Tourette disorder and implicate cell polarity in pathogenesis. Cell Rep 24: 3441–3454.e12.

Willnow TE, Petersen CM, Nykjaer A (2008). VPS10P-domain receptors – regulators of neuronal viability and function. Nat Rev Neurosci 9: 899–909.

Willsey AJ, Fernandez TV, Yu D et al. (2017). De novo coding variants are strongly associated with Tourette disorder. Neuron 94: 486–499.e9.

Xu M, Li L, Ohtsu H et al. (2015). Histidine decarboxylase knockout mice, a genetic model of Tourette syndrome, show repetitive grooming after induced fear. Neurosci Lett 595: 50–53.

Yang K, Essa A, Noriega D et al. (2022). Relationship between adverse childhood experiences and symptom severity in adult men with Tourette syndrome. J Psychiatr Res 155: 252–259.

Yang L, Neale BM, Liu L et al. (2013). Polygenic transmission and complex neuro developmental network for attention deficit hyperactivity disorder: genome-wide association study of both common and rare variants. Am J Med Genet B Neuropsychiatr Genet 162B: 419–430.

Yang Z, Wu H, Lee PH et al. (2021). Investigating shared genetic basis across Tourette syndrome and comorbid neurodevelopmental disorders along the impulsivity-compulsivity spectrum. Biol Psychiatry 90: 317–327.

Yu D, Mathews CA, Scharf JM et al. (2015). Cross-disorder genome-wide analyses suggest a complex genetic relationship between Tourette's syndrome and OCD. Am J Psychiatry 172: 82–93.

Yu D, Sul JH, Tsetsos F et al. (2019). Interrogating the genetic determinants of Tourette's syndrome and other tic disorders through genome-wide association studies. Am J Psychiatry 176: 217–227.

Zhao X, Wang S, Hao J et al. (2020). A whole-exome sequencing study of Tourette disorder in a Chinese population. DNA Cell Biol 39: 63–68.

Zilhão NR, Padmanabhuni SS, Pagliaroli L et al. (2015). Epigenome-wide association study of tic disorders. Twin Res Hum Genet 18: 699–709.

Handbook of Clinical Neurology, Vol. 215 (3rd series)
Tic Disorders and Tourette Syndrome
C. Ganos and Y. Worbe, Editors
https://doi.org/10.1016/B978-0-443-13554-5.00016-X

Chapter 15

Neuroimmunology of tics and Gilles de la Tourette Syndrome

GABRIEL AMORELLI[1,2] AND DAVIDE MARTINO[3,4,*]

[1]*Division of Neurology, Department of Medicine, University of Ottawa, Ottawa, ON, Canada*

[2]*Ottawa Hospital Research Institute, Ottawa Hospital, Ottawa, ON, Canada*

[3]*Department of Clinical Neurosciences, University of Calgary, Calgary, AB, Canada*

[4]*Hotchkiss Brain Institute, University of Calgary, Calgary, AB, Canada*

Abstract

In the past decades, the connection between the immune system and Tourette syndrome (TS) has been explored as a potential pathophysiologic key aspect. This idea is supported by studies describing the effect of immune cells and inflammatory mediators in neurodevelopment by acting in synaptic arrangement, pruning, and maturation of neural circuitries. Supporting evidence for that correlation is varied, including genome-wide association studies, transcriptome analysis, animal models, and clinical populational-based studies. Moreover, the influence of environmental aspects typically involved in immunologic response and the association of TS with allergies and autoimmune diseases adds to the immune-based neurodevelopment theory. Overall, the research data looking at this association points toward a proinflammatory state, even though there is still uncertainty about the exact role and implication of this immunologic profile in TS, especially due to the lack of robust evidence regarding immunomodulatory treatments. Furthermore, the discrepancy of results in studies shed doubt on the immunologic hypothesis. In this chapter, we summarize the scientific knowledge regarding immunology in TS, discussing it from a historical perspective to more recent studies.

INTRODUCTION

Tourette syndrome (TS) and other persistent tic disorders are neurodevelopmental conditions characterized by motor and vocal tics with typical onset in childhood. Even though it was originally described in 1885 (Lajonchere, 1996), the pathophysiology of TS is still not fully elucidated. In recent decades, there has been a growing scientific interest around the association between immune-inflammatory processes and the development of neurodivergent traits and conditions, including TS and tic disorders.

During central nervous system (CNS) development, the immune system participates in molecular and cellular processes related to synaptogenesis, synaptic pruning, and maturation of brain connections using immune mediators and cells (Zengeler and Lukens, 2021). Microglia play a crucial role in modulating immune processes within the CNS by releasing inflammatory cytokines and promoting phagocytosis of cells and apoptotic debris. Moreover, this cell is entangled in brain development by promoting other processes, including glial cell differentiation and angiogenesis (Thion et al., 2018; Zengeler and Lukens, 2021). Integrating this conceptualization of neuroimmune interplay during brain development with the recent discoveries in TS-related genetic and environmental risk factors opens a vast territory to explore different aspects of the pathophysiology of tic disorders.

A broad introduction to immune function is needed to clarify the contribution of this neuroimmune interplay. First and foremost, the immune system is divided into innate and adaptive sections. The innate system is the first responder to a potential threat. It uses toll-like receptors (TLRs) to identify pathogen-associated molecular patterns and generate an inflammatory response (Takeda and Akira, 2015). Cells of the innate system include dendritic cells (DCs), phagocytes, and natural killer (NK). This initial phase is mediated by immune effector molecules (cytokines, chemokines) initiating multiple aspects of the immunologic response, including the activation of the adaptive system.

The adaptive immune system orchestrates a specific response to the identified pathogen and creates immunologic memory, expediting response after a second antigenic exposure. Adaptive

*Correspondence to: Davide Martino, MD, PhD, Cumming School of Medicine, University of Calgary, Health Sciences Centre, 3330 Hospital Drive NW, Calgary, AB T2N 4N1, Canada. Tel: +1-403-210–8726, Fax: +1-403-210-9149, E-mail: davide.martino@ucalgary.ca.

system cells comprise B and T lymphocytes. Cytotoxic T lymphocytes (Tc) can promote a direct cytolytic response against pathogens, whereas T-helper (Th) lymphocytes generate a coordinated immune response against an offending agent. A third subgroup, regulatory T cells (Treg), suppress the immune response. Finally, B lymphocytes generate the immunologic response mediated by antibodies (humoral) and create immunologic memory for a rapid reaction in case of another exposure to that pathogen. This broad description accounts for the primary and classical immune response.

The following sections will summarize the current knowledge on the role of immunity in TS pathogenesis, including immunogenetics, extra-genetic environmental factors, immunophenotypes in TS, and associations with autoimmune comorbidities. The relationship between infections and TS will be explored, as well as potential immune-based therapies.

IMMUNOGENETICS OF TOURETTE SYNDROME

Genomics

TS has a well-known heritability described in up to 30% of patients, which is most likely supported by a polygenic contribution. Several of the genes implicated in this polygenic risk are known to regulate immune mechanisms. Genome-wide association studies (GWAS) showed a direct correlation between TS and genes associated with allergy (Thion et al., 2018) as well as histamine-receptor signaling pathways (Fernandez et al., 2012). GWAS-related evidence in histamine metabolism includes a rare functional mutation in the *HDC* gene (Ercan-Sencicek et al., 2010), which encodes L-histidine decarboxylase, a rate-limiting enzyme in histamine synthesis. Another gene associated with TS is *FTL3*, which encodes FMS-like tyrosine kinase 3, a protein expressed in progenitor hematopoietic cells that participates in the differentiation of inflammatory cells (Kazi and Rönnstrand, 2019). Polymorphisms in the *FTL3* gene were described in a large GWAS study of tic disorders (Yu et al., 2019) and further observed in a second GWAS study with TS patients and ancestry-matched controls (Tsetsos et al., 2021). Another genetic susceptibility risk is the *ASH1L* gene, which negatively regulates cytokine pathways and inflammatory responses (Liu et al., 2020); in a study in a Chinese population, a significant over-transmission of two *ASH1L* gene single nucleotide polymorphisms (SNPs) was found to be associated with the diagnosis of TS (Liu et al., 2022).

A similar polygenic contribution to immune dysregulation is classically detected in systemic immune-mediated disorders, with differentially expressed gene products interacting within complex mechanistic pathways. In line with this, a study looking at densely affected TS families found that not a single high-risk polygenic variation, but a high load of common risk alleles to TS could underlie high heritability (Halvorsen et al., 2021). How these common variants interact to determine the increased risk for TS remains to be elucidated.

Finally, other studies have associated other genetic variants with the immune system in tic disorders. One research found a SNP in the tumor necrosis factor (*TNF*) gene (rs1800629) in tic patients (Keszler et al., 2014). This gene codes for the TNF-α cytokine, a proinflammatory mediator involved in the Th1 response and associated with several inflammatory and autoimmune diseases such as rheumatoid arthritis, inflammatory bowel disease, psoriasis, Graves' disease, and ankylosing spondylitis (Qidwai and Khan, 2011). In a different study, *TNF* rs361525, another SNP, was unrelated to TS (Liu et al., 2011). Similarly, a polymorphism in the gene encoding the interleukin (IL)-1 receptor antagonist molecule (*IL1RN*) has been associated with TS in one study (Chou et al., 2010), whereas other reports did not find a significant relationship (He et al., 2015). Studies examining possible associations between human leukocyte antigen (HLA) alleles and TS have also yielded conflicting results. One study in a Korean cohort found higher frequencies of HLA A11 and A26 and lower frequencies of HLA A24 and B13 in TS, although not formally correcting for multiple comparisons (Min et al., 1991). However, other studies have failed to reveal any association between HLA haplotypes and TS (Caine et al., 1985; Schoenian et al., 2003).

Transcriptomics

Studies analyzing the gene expression profile in postmortem brain tissue of TS patients and peripheral blood samples showed significant differences in the expression of immune-related products compared to controls. One microarray study showed upregulated expression of the inflammatory mediators IL-2 and IL-1 receptors in the putamen (Hong et al., 2004). The IL cytokine family is involved in T-cell and B-cell proliferation and differentiation (Akdis et al., 2011). Moreover, an RNA-sequencing transcriptome study showed several gene expression differences in striatal tissue, with the most upregulated module detected in genes encoding for key immune effector molecules like TNF-α, IL-6, and IL-12 (Lennington et al., 2016). Another transcriptome study reported upregulation of the genes coding for IL-2 and monocyte chemotactic factor 1 (Morer et al., 2010), which are directly involved in monocyte recruitment and infiltration into the brain, suggesting enhanced inflammatory responses (Yadav et al., 2010). Global dysregulation of gene expression was detected across different brain regions, predominantly in the dorsolateral prefrontal cortex, together with increased expression of *FTL3* in lymphoblastoid cell lines in the TS group (Liao et al., 2022). Moreover, a phenome-wide association study was conducted, which showed the involvement of *FTL3* in several immune pathways and markers, including monocyte count (Liao et al., 2022). More recently, a study looking at the molecular landscape in TS described an enriched *FTL3* expression in TS patients' cerebellum and thalamus.

With respect to studies exploring the peripheral blood transcriptome, one study on untreated TS patients detected a downregulation of killer cell lectin-like receptor subfamily D, member 1, involved in NK cells' function (Tian et al., 2011).

The same gene was upregulated in a second study that explored the peripheral blood transcriptome in treated TS patients (Lit et al., 2007), supporting a potential association between mechanisms of action of tic-suppressing medications and NK cell function. Furthermore, the transcriptomic profile of TS is likely to extend to a larger number of mechanistic pathways: a recent study of shared immune-related pathways and gene expression across transcriptomic studies of autism spectrum disorder and TS found 23 shared transcriptomic pathways involved with the regulation of innate immune response (Alshammery et al., 2022).

Epigenetics

Epigenetics investigates the mechanisms and the effects produced by environmental factors that affect gene expression. A preliminary observation reported altered methylation status of the dopaminergic D2 receptor in TS patients (Müller-Vahl et al., 2017), but this area remains heavily underexplored in tic disorders. Given the relevance of intrauterine and postnatal environmental epigenetic effects during brain development and their recognized importance for a mechanistic approach to precision medicine for neurodevelopmental disorders (Woolfenden et al., 2022), the exploration of epigenetic effects is a research priority area also for TS and tic disorders.

EXTRA-GENETIC AND ENVIRONMENTAL PREDISPOSING FACTORS

Prenatal period

During CNS development, both genetic and environmental aspects are determinant. The potential relevance of maternal tobacco smoking exposure is explained in part by its association with an immune-hormonal-metabolic disarrangement impacting the functions of the placenta (Sabra et al., 2017). Therefore, neurodevelopmental diseases have been explored in their association with maternal exposure during pregnancy. However, some of the evidence is conflicting. For example, some studies directly correlate TS diagnosis with maternal smoking exposure (Cubo et al., 2014; Browne et al., 2016), whereas others did not confirm this association (Motlagh et al., 2010; Mathews et al., 2014; Leivonen et al., 2015; Abdulkadir et al., 2016; Brander et al., 2018). On the other hand, maternal tobacco exposure was found to predict higher tic severity and obsessive-compulsive disorder (OCD) comorbidity (Mathews et al., 2006).

Two studies, one assessing the associations of maternal infections during pregnancy in general (Cubo et al., 2014) and one specifically exploring maternal influenza viral infection (Mathews et al., 2014), did not detect a significant association with tic disorders in the offspring. In a more recent, large sibling-controlled study, an association between OCD and/or TS and maternal or early childhood infections was not confirmed by sibling analyses, suggesting that familial factors might explain the predisposition to both neurodevelopmental traits and liability to infections (Zhang et al., 2022).

More robust evidence found a correlation between other maternal preexisting conditions and TS. Maternal autoimmune diseases are reported to be associated with TS diagnosis in the offspring and increased disease severity (Murphy et al., 2010; Dalsgaard et al., 2015; Mataix-Cols et al., 2018; Jones et al., 2021). A systematic review corroborated this correlation (Han et al., 2021). It is still undetermined if a specific autoimmune disease is connected to TS or if it may be seen in different diseases based on the inflammatory pattern. This notion is supported by transcriptome studies looking at genetic alterations in TS patients and mothers with autoimmune disease, which reported similar patterns in inflammatory cascades (Lennington et al., 2016; Jones et al., 2021).

Finally, a recent study looked at the association between pregnancy risk factors and TS-related genetic polymorphisms. This study found that the association of genetic polymorphisms linked to TS risk and environmental prenatal factors such as maternal depression, anxiety, and substance use translates into an increased likelihood of TS, and that may account for the diagnosis variability in the general population (Abdulkadir et al., 2022).

Delivery/perinatal period

Several factors related to delivery were implicated in tic disorders pathophysiology, including cesarean delivery. This is supported by a Swedish population study (Brander et al., 2018) and a smaller nested case-control study (Cubo et al., 2014) showing an increased risk of TS/tic disorders in the children born through a C-section. Different aspects may explain this association through an immunologic perspective, including the lack of exposure to the birth canal microbiome and lower stress hormones exposure (Cho and Norman, 2013). Moreover, the increased risk of fetal hypoxia and exposure to exogenous oxytocin and anesthetics may implicate altered immune system function and neurodevelopment (Cubo et al., 2014). Furthermore, an association between cesarean delivery and chronic immune diseases has been reported, but the related causative mechanisms are still speculative (Sevelsted et al., 2015).

The cited Swedish population study described an increased risk of TS in premature neonatal babies and impaired fetal growth (Brander et al., 2018). Preterm birth can be a consequence of prenatal infection and inflammation, and these have been connected with immune system immaturity (Melville and Moss, 2013). How perinatal factors participate in TS pathophysiology remains to be determined, especially if their influence is not primarily driven by maternal hereditary risk factors other than the delivery itself.

Postnatal period

The postnatal period and its environmental aspects are vital to the immune system's maturation. These factors have been looked at with particular interest in TS, especially infections and allergies. Group A Streptococcal (GAS) infection is commonly associated with TS. A case-control study reported that children with TS and OCD were more likely to have a previous GAS infection than controls and that risk increased in children with multiple GAS infections (Mell et al., 2005). Another study in the United States, but this time with 742 children, confirmed the increased risk of TS, tics, or OCD after a GAS infection compared to controls (Leslie et al., 2008). Moreover, a nationwide registry in Denmark showed an increased risk of developing mental health disorders with throat infections, particularly for TS and OCD in children with GAS throat infections (Orlovska et al., 2017). A similar indication was found in a specialty clinic, where 38% of the patients with TS/tic disorders had elevated antistreptolysin O (ASO) compared to controls with high titers in only 2% (Cardona and Orefici, 2001). On the other hand, a UK population study failed to demonstrate the association between TS and GAS infection (Schrag et al., 2009). A study comparing TS and other epilepsy or sleep disorders showed increased ASO titers and frequency of GAS throat infection in the tic disorders group (Martino et al., 2011). A prospective research with 25 subjects demonstrated the occurrence of more drastic fluctuations in tic symptomatology in TS subjects with higher ASO titers (Murphy et al., 2004). The extensive data regarding TS and GAS infection was recently reviewed in a meta-analysis, which confirmed a positive correlation between elevated ASO titers and TS, at the same time highlighting methodological limitations of several of the reviewed studies, and the authors describe that their inclusion criteria have excluded other research with negative results that may have altered the meta-analysis results and interpretation (Lamothe et al., 2021). Finally, a prospective European study of 715 children with chronic tic disorders failed to demonstrate an association between GAS exposure and tic exacerbations (Martino et al., 2021).

Other infections, aside from GAS, have been implicated in TS pathophysiology. Non-CNS *Enterovirus* infection was associated with tic disorders in a population study in Taiwan (Tsai et al., 2016). Another research, with 32 TS patients, found *Chlamydia trachomatis* elevated IgG titers compared to controls (Krause et al., 2009). *Toxoplasma gondii* infection was associated with TS in two studies, one which showed high titers of antibodies compared to controls (Akaltun et al., 2018) and the second with a nonstatistically significant trend toward *T. gondii* antibodies elevation in TS patients (Krause et al., 2009). Similarly, a direct correlation was found between *Mycoplasma pneumoniae* antibodies and tic disorders, and higher titers were associated with increased severity (Schnell et al., 2021). It is unclear whether a causative relationship exists between infections, including GAS, and tic disorders, or if there is a primary predisposition for infections in these subjects caused by immunologic dysfunction.

Psychosocial stress is another environmental factor implicated in TS and other tic disorders. The effects of psychosocial stress on the immune system are thought to be partially derived from the hyperactivation of the hypothalamus-pituitary-adrenal axis, a crucial endocrinologic system in the inflammatory response controlling cortisol and other peptidic hormones secretion, yet the full extent of this association in TS is to be elucidated (Martino et al., 2013). This environmental variable was tested in a study which found an increased tics severity with the combination of GAS infection and psychosocial stressors (Lin et al., 2010). Furthermore, a prospective clinic-based survey also demonstrated that increased psychosocial stress and depression were independent predictors for tics severity (Lin et al., 2007).

IMMUNOPHENOTYPES

Systemic immune activation

Studies exploring systemic immunophenotypes in TS patients provided conflicting results. These stem from differences in study design, clinical heterogeneity, symptom fluctuations, and effects of tic-suppressing treatments. However, a systemic proinflammatory profile is likely important in TS individuals, as demonstrated and replicated by several studies (see Table 15.1 for a summary of the most relevant findings).

In children, studies looking at the immune cells population in TS and OCD subjects compared to controls showed a decreased number of Treg cells in moderate and severe TS patients (Kawikova et al., 2007). Nonetheless, a second study in a pediatric cohort found a higher number of Treg cells in TS subjects than in the controls (Yildirim et al., 2020). Further studies in the pediatric population with tic disorders found decreased CD3+ and CD4+ T cells, reduced CD4+/CD8+ ratio, and increased NK cells and TNF-α (Li et al., 2015). The reduction of CD4+ T cells was more significant in the ASO-positive population, which also demonstrated lower IL-6 and IL-8 levels (Li et al., 2015). Recently, a meta-analysis reported disproportionate changes in Th cells in children with TS (Li et al., 2015).

Systemic immune studies in adults were heterogeneous and inconsistent. A pilot study with 20 TS subjects found a higher number of CD69+CD22+ B cells and CD95+CD4+ T cells compared to age and sex-matched controls (Möller et al., 2008). CD69+ is a marker of lymphocyte activation (Cibrián and Sánchez-Madrid, 2017), whereas CD95+ marks lymphocyte apoptosis (Peter et al., 2007), indicating increased peripheral immune activity. DCs in tic disorders compared to healthy controls were not different in the circulating number of cells in the peripheral blood but showed increased myeloid DC type 1 in a subset of individuals having increased anxiety and TS compared to TS without anxiety (Sarchioto et al., 2021).

When looking at inflammatory cytokines, the results agree to point toward a proinflammatory state; however, there is again inconsistency across available data. A proinflammatory cytokine TNF-α was found elevated in the serum of TS and OCD patients compared to controls (Leckman et al., 2005). Similar results were appreciated in studies investigating other tic disorders (Li et al., 2015; Yeon et al., 2017) and tic exacerbations (Parker-Athill et al., 2015). A recent meta-analysis focusing on cytokine expression in TS reported a large effect of TNF-α in the analyzed studies (Li et al., 2015). On the other

Table 15.1

Summary of the key findings from immunophysiologic studies of Tourette syndrome and tic disorders.

Immunogenetics	CNS inflammation	Peripheral inflammation
• Genetic correlation with allergies in GWAS (Thion et al., 2018) • "Monogenic" phenocopy of TS associated with *HDC* variants (Ercan-Sencicek et al., 2010) • Association of TS diagnosis with polymorphic variants of *FLT3* and *ASH1L* in GWAS • Upregulation of the expression of genes encoding proinflammatory immune effector molecules (IL-1R, TNF-α, IL-6, IL-12) in the periphery and in the striatum (Hong et al., 2004; Lennington et al., 2016) • Increased expression of FLT3 in lymphoblastoid cell lines from TS patients (Liao et al., 2022) and in cerebellum and thalamus • Hypermethylation of the dopamine D2 receptor gene (Müller-Vahl et al., 2017)	• Altered expression of immune-related and microglial activation genes in basal ganglia (Hong et al., 2004; Lennington et al., 2016) • Possible evidence of activated microglia in the basal ganglia of people with tic disorders (Kumar et al., 2015)	• Altered proportions of immune cell subtypes, especially of T-helper lymphocyte subsets (Li et al., 2015) • High variability across studies of cytokine and chemokine peripheral levels (Lamothe et al., 2021) • Possible changes pointing toward abnormal innate immune responses (Matz et al., 2012; Weidinger et al., 2014) • Preliminary evidence of raised peripheral levels of autoantibodies directed toward dopamine D2 receptors (Dale et al., 2012; Addabbo et al., 2020), but inconclusive evidence of an association with antibodies targeting striatal antigens or tissue

CNS, central nervous system; *GWAS*, genome-wide association studies; *IL*, interleukin; *TNF*, tumor necrosis factor.

hand, a cross-sectional study showed reduced levels of TNF-α (Matz et al., 2012), and one research did not show any difference (Bos-Veneman et al., 2010). Another inflammatory mediator, IL-12, has also been found elevated in the serum of tic disorder and OCD patients, in conjunction with TNF-α (Leckman et al., 2005). Similarly, IL-1ß was increased in a group of patients with TS compared to controls (Cheng et al., 2012) and in a study looking at hospitalized subjects due to exacerbation of previously diagnosed TS and OCD (Gariup et al., 2015). In that same study, the TS and OCD group displayed higher interferon-gamma (IFN-γ)-induced protein-10 levels. A research looking at patients with OCD and TS compared to TS alone and controls found an increased serum level of IL-12 and IL-2 in the TS+OCD group, even though IL-2 levels were not considered meaningful (Gabbay et al., 2009). In another investigation, there was a statistical trend toward increased IL-2 levels in patients with higher tic severity (Bos-Veneman et al., 2010). IL-6 exploration also produced conflicting results, with some studies showing increased levels in TS patients (Cheng et al., 2012), and others no difference (Gabbay et al., 2009; Matz et al., 2012) or even reduced levels (Li et al., 2015). The previously mentioned meta-analysis described highly variable effect size for IL-6 in TS patients (Li et al., 2015). One study looking at IL-8 in 58 TS patients showed a decreased serum concentration compared to controls (Li et al., 2015). Increased circulating levels of IL-17 were also reported in the serum of TS patients (Cheng et al., 2012; Yeon et al., 2017). Neopterin is an inflammatory mediator described as elevated in TS and OCD patients (Luo et al. 2004; Matz et al. 2012). It is derived from pteridine and is a proinflammatory mediator in the Th1 response. Finally, a meta-analysis focused on the comparison of the level of different cytokines between TS patients and controls and found no difference in the levels of TNF-α, IL-1ß, or IL-6 (Lamothe et al., 2021).

A dysregulated immune response may also reflect an abnormality in innate responses. This is corroborated by a study examining differences in inflammatory mediators between TS subjects and healthy controls throughout different age groups. There were lower levels of TNF-α, reduced IL-1 antagonist receptors, and monocyte-derived soluble CD14 (sCD14) in the TS population, which is indicative of a dysregulated innate response (Matz et al., 2012). Moreover, another study found an abnormal response to lipopolysaccharide (LPS) stimulation in monocytes from the peripheral blood of TS patients, with reduced TLR-4 and sCD14 expression, thus suggesting an abnormal innate response to pathogens (Weidinger et al., 2014).

Changes in the circulating levels of immunoglobulins (Ig) were identified by a very limited number of studies, thus requiring replication and more detailed exploration. Immunoglobulin E (IgE), important in allergic immune responses, was found to be higher in the sera of TS patients than in healthy controls in a single case-control study (Landau et al., 2012). In another cross-sectional study (Bos-Veneman et al., 2011), TS patients were found to have lower peripheral concentrations of IgG3, an essential mediator for complement activation, phagocytosis, and antibodies-mediated cellular cytotoxicity (Damelang et al., 2019). Finally, a single case-control study found lower levels of total IgA in the sera of patients with TS and OCD (Kawikova et al., 2010).

Autoantibodies

The role of autoantibodies in tic disorders was initially considered after the description of antistriatal antibodies in Sydenham's chorea (Husby et al., 1976) and later reported in several studies of TS patients (Kiessling et al., 1993; Church et al., 2003; Zykov et al., 2009). One research demonstrated the presence of antineuronal antibodies in subjects with tic disorders without streptococcal infection (Yeh et al., 2006). Similar studies with immunofluorescence (Martino et al., 2007) or immunoprecipitation (Yeh et al., 2012) failed to demonstrate autoantibodies-antigen complexes in TS patients. Furthermore, two studies found anti-dopamine D2 receptors (Dr) in 7%–9% of children with chronic tic disorders (Dale et al., 2012; Addabbo et al., 2020) and almost 15% seroconversion after tic exacerbation in children previously negative to Dr autoantibodies (Addabbo et al., 2020). Autoantibodies related to neurologic disorders were investigated for TS patients with negative results, including against NMDA receptors, anti-AMPA-1/2 receptors, GABA-A and GABA-B1/B2 receptors, LGI1, and CASPR2, in the sera of children (Baglioni et al., 2019) and adults (Sühs et al., 2015), CSF of adults with TS (Baumgaertel et al., 2019), and unaffected siblings of TS children (Baglioni et al., 2019).

Central nervous system inflammation

The evidence discussed so far points toward alterations in the immune system translating into a proinflammatory state. This statement is supported by postmortem TS brain transcriptomic evidence of altered cytokines related to inflammatory cell activation and migration like IL-1ß, IL-2 receptor ß, and MCP-1 (Hong et al., 2004; Morer et al., 2010). An upregulation of genes involved in microglia activation in the striatum was also described in a small study of postmortem brains of people who lived with TS (Lennington et al., 2016). This concept was also tested with positron emission tomography using a specific tracer for activated microglia named (11)C-[R]-PK1195 (PK), a ligand binding the transporter protein expressed by activated microglia. A case-control study found an increased binding of PK in the caudate nuclei of the tic disorders group, demonstrating the presence of activated microglia and, consequently, altered inflammatory response in that region (Kumar et al., 2015).

Inflammatory changes of the CSF in TS patients were also investigated. A common indicator of inflammation in the CSF is the presence of oligoclonal bands (OBs), but this marker demonstrated conflicting results in TS. One study of 21 TS patients showed the presence of CSF-specific OB (4 or more OBs) in 6 subjects (Wenzel et al., 2011). Interestingly, some patients mirrored the OBs in the CSF and serum, indicating possible systemic inflammation. A follow-up study looked at OB in a group of 20 TS patients and found the presence of bands in 4 of them (Baumgaertel et al., 2019). On the other hand, this was not confirmed by all studies: when looking specifically at the cytokines, chemokines, immunoglobulins, and inflammatory cells profile in the CSF of a small group of children with TS and ASO compared to controls, no statistical difference was reported (Pranzatelli et al., 2017).

COMORBIDITIES: CONFOUNDERS OR EPIPHENOMENA?

Considering the possible connection between the altered immune system response and tic disorders, a potential association between TS and primary immune diseases has also been explored. Several different studies connected TS with atopy. A population-based study using the Canadian Community health survey detected an increased risk of asthma in TS subjects compared to the general population (Yang et al., 2017). Another population-based study from a Taiwanese population gathered data from 845 children recently diagnosed with TS and reported a higher incidence of allergic conjunctivitis, allergic rhinitis, asthma, and atopic dermatitis in the tic disorders group compared to the healthy controls (Chang et al., 2011). Similar results connecting allergic disorders with TS were reported by smaller clinic-based studies. A study in Turkish clinic-based population showed an association of TS children with allergic rhinitis and OCD with atopic dermatitis (Yuce et al., 2014), whereas a report from a Chinese population reported an association of the diagnosis of provisional tic disorder with conjunctivitis, allergic rhinitis, and asthma in children (Chen et al., 2020). These epidemiologic and clinical results related to allergic conditions find indirect support from genomic studies that suggest a link of genes involved in histamine receptor signaling pathways with the diagnosis of TS (Fernandez et al., 2012). Finally, a large population-based study in Sweden showed a correlation between primary tic disorders and Hashimoto's thyroiditis, celiac disease, scarlet fever, type 1 diabetes mellitus, and psoriasis (Mataix-Cols et al., 2018). In addition, another study found an association between TS and Sydenham's chorea/rheumatic fever (Pérez-Vigil et al., 2016).

THE LONGSTANDING DEBATE ON THE ROLE OF GROUP A STREPTOCOCCUS IN TOURETTE SYNDROME

The discussion about the connection between immune dysfunction and movement disorders began with early reports documenting the now established role of GAS in Sydenham's chorea. Subsequent research focused on immunopathogenic mechanisms not only of Sydenham's chorea, but also of other neuropsychiatric presentations. A historical overview of this debate and of its implications for the understanding of the pathophysiology of tic disorders has been summarized by A. Cavanna in recent years (Cavanna, 2018).

The earliest studies describing the connection between infectious diseases and movement disorders date back to the 19th century, depicting chorea caused by an infectious agent. At that time, "chorea" was a general nomenclature for different movement disorders. This initial concept was abandoned for almost one century after a nosologic reorganization of movement disorders. In the 20th century, clinical and scientific interest in this topic resurfaced following the documentation of the potential connection between GAS infection and Sydenham's chorea by Taranta and Stollerman (1956). However, the link between tic disorders and infectious diseases started to be discussed only years later due to the great influence of the hereditary-degenerative etiologic hypothesis for tics (Meige and Feindel, 1902).

Over time, the infectious hypothesis gained strength due to the importance of behavioral symptoms within tic disorders in parallel with the psychiatric symptoms present in neuroinfectious diseases such as "general paresis of the insane" caused by tertiary syphilis (Kaplan, 2010) and the symptoms of encephalitis lethargica (Dickman, 2001). Therefore, at the end of the 20th century, the initial ideas of the infectious cause behind tic disorders emerged, and the terms "schizovirus" and "schizococcus" were coined to explain tic disorders pathophysiology (Torrey, 1988).

In 1995, the hypothesis of TS being an infectious-triggered condition gained more visibility with the concept of pediatric, infection-triggered, autoimmune neuropsychiatric disorders (PITANDs) (Allen et al., 1995). The idea behind an infection triggering a disarranged immune response generating neuropsychiatric symptoms got more robust with the studies connecting GAS infection and TS in a similar pattern to what was initially described in Sydenham's chorea. Evolving the initial concept of PITANDs, the debate started to include a more specific terminology called "Pediatric Autoimmune Neuropsychiatric Disorders Associated with Streptococcal infections" (PANDAS) (Swedo et al., 1998), describing the importance of this bacteria in generating a wide variety of acute-onset neuropsychiatric symptoms (Singer et al., 2012; Swedo et al., 2012). The discussion of an infectious-triggered etiology for tic disorders has been debated for almost three decades, and still, there is no consensus about the importance of this association between specialists (Macerollo and Martino, 2013; Sigra et al., 2018; Prato et al., 2021).

Several studies were performed to test the GAS/TS connection, starting with case report observation describing patients with acute onset or aggravation of tic disorders and evidence of GAS infection. This association was not clearly defined as causation after further research looking at epidemiologic (Schrag et al., 2009; Martino et al., 2021) and laboratory data (Singer et al., 2005). Hence, the scientific data accumulated in recent years do not provide robust support to the premise of PANDAS (Cavanna and Coffman, 2021). Furthermore, the negative results of immune therapy in tic disorders (Hoekstra et al., 2004) led to the consensus of not using GAS laboratory data for diagnosis, immune treatment, or even GAS prophylaxis in PANDAS or tic disorder (Gerber et al., 2009).

Taking into account the controversial clinical data, the terminology used to characterize this entity was broadened, lost the infectious agent etiologic specificity, and was redefined as "Pediatric Acute Neuropsychiatric Syndrome" (PANS)

(Swedo et al., 1998) and "Childhood Acute Neuropsychiatric Syndrome" (CANS) (Singer et al., 2012; Swedo et al., 2012). As evident from these denominations, GAS infection proof was no longer necessary, and the association with tics was considered only an accompanying feature. PANS/CANS represent an acute onset condition with movement disorders and neuropsychiatric symptoms that could be caused by a constellation of different etiologies, such as infectious, postinfectious, drug-induced, toxic, traumatic, vascular, autoimmune, hypoxic, psychogenic, idiopathic. Changing the terminology, a scientific group named PANS Research Consortium, determined to reformulate this concept, released several statements (Chang et al., 2015; Cooperstock et al., 2017; Frankovich et al., 2017; Thienemann et al., 2017), which were later followed by a group of pediatric neurologist experts (Gilbert et al., 2018). The recommendation encompassed diagnosis and management. However, the application of these recommendations has been so far inconsistent within the broad community of clinicians, and the relationship between PANS/CANS and tic disorders remains therefore an object of investigation.

INSIGHTS FROM ANIMAL STUDIES

The role of immune-based mechanisms in TS has also been the object of hypotheses tested in animal models. Different strategies have been used, including animal infection, immunization with microbial antigens or mimics, transgenic animal models, and injecting cytokines, immune mediators, or autoantibodies from TS patients (Hornig and Lipkin, 2013).

Early studies injected sera from TS patients in mice, generating motor stereotypies and episodic utterances (Hallett et al., 2000). This abnormal movement induction was even more prominent with the administration of sera rich in antineural or antinuclear antibodies (Taylor et al., 2002). Interestingly, the injection of soluble receptors for cytokines in mice, specifically soluble IL-2 and IL-6 receptors, induced abnormal stereotypic behavior, associated with an inflammatory response in the animal cortex and striatum (Patel et al., 2012; Zalcman et al., 2012). A different group reported immunoreactivity of TS patients' sera against a specific protein called hyperpolarization-activated cyclic nucleotide-gated channel 4 (HCN4; Yeh et al., 2012). HCNs are protein membrane channels involved in the cyclic activity of cells in the brain and cardiovascular system (Benarroch, 2013). HCN4, more specifically, is important in the oscillatory activity within the thalamocortical system (Zobeiri et al., 2019). Interestingly, the infusion of anti-HCN4 antibodies induced motor stereotypies in rats (Yeh et al., 2012). Moreover, a recent study used 3,3-iminodipropionitrile, a known substance used to generate TS animal models, to compare the immunologic profiles of exposed rats, exposed rats subsequently treated with haloperidol, and unexposed control animals. This study described an increase of TNF-α and IL-6 concentrations and highly increased microglia activation in the exposed, untreated group and only slight immune activation in the exposed, treated group (Zhonglin et al., 2022).

Several earlier rodent models described behavioral and immunologic effects of immunization with GAS homogenate. In mice, this approach induced abnormal rearing behavior (Hoffman et al., 2004), deficits in motor coordination, learning/memory, and social interactions (Yaddanapudi et al., 2010). In the latter study, the injection of IgG from immunized mice to unexposed recipients replicated the motor abnormalities (Yaddanapudi et al., 2010). These results were further replicated in a similar study (Lotan et al., 2014). Furthermore, different protein homogenates of GAS were used to immunize mice, including one with high content of the cell wall antigen M protein type 18, which, after being injected in mice, generated abnormal movements and compulsive behaviors that were responsive to haloperidol and paroxetine (Brimberg et al., 2012). Similar symptoms were obtained after injecting anti-GAS IgM in mice, leading to the development of head bobbing, sniffing, and intense grooming (Zhang et al., 2012).

Prenatal exposure was tested with maternal sensitization (maternal immune activation) in mice. Injecting pregnant mice with polyinosinic:polycytidylic acid (poly I:C), a viral mimic, and a TLR-3 agonist resulted in stereotypic behavior in their offspring (Malkova et al., 2012). Comparably, injecting LPS, a component of bacterial cell walls, at a central period of gestation in pregnant dams generated repetitive and stereotypic behavior in the offspring (Kirsten and Bernardi, 2017).

As discussed in the section on *Immunogenetics*, the *HDC* gene has been associated with TS pathophysiology. A study exploring *HDC* knockout mice described stereotypic tic-like behaviors that improved with antidopaminergic therapy. Another study looking at the same animal model found that the *HDC* knockout mice did not demonstrate changes in the abundance of microglia, but decreased insulin-like growth factor 1 (IGF-1) and fewer ramifications in striatal microglia, suggestive of its activated state (Frick et al., 2016). IGF-1 is especially important in neurogenesis and neuronal survival. Histamine signaling abnormalities may therefore influence the functional state of striatal microglia, which might relate to the behavioral abnormalities in the phenotype of these transgenic mice.

Finally, in line with recent studies exploring the link between gut microbiome and tic disorders, one study reported an improvement of serotonin-induced tic-like behaviors in rats after administration of the probiotic *Lactobacillus plantarum* PS128 (Liao et al., 2019).

IMMUNE-BASED THERAPIES

The amount of evidence describing an inflammatory involvement in TS supports the potential usefulness of immune therapy as part of the management options for tic disorders. Intravenous immunoglobulin (IVIG) was explored in a few small studies with inconsistent results. A small and uncontrolled study of 7 children with refractory tics and autoantibodies to caudate nucleus protein showed improved motor and vocal tics after IVIG (Zykov et al., 2009). Another study comparing IVIG, plasma exchange, and sham infusion of saline in patients with exacerbation of tics and OCD symptoms triggered by infection demonstrated a striking improvement of symptom severity with the immune therapies

(Perlmutter et al., 1999). However, a double-blind, placebo-controlled study failed to confirm an improvement when comparing IVIG and placebo (Hoekstra et al., 2004). Anecdotally, a subject with tic disorders and Guillain–Barré syndrome did not improve in tic severity after being treated with IVIG for the neuromuscular disease (Szejko et al., 2020).

Cefdinir, a β-lactam antibiotic with immunologic properties, was tried in subjects with tic disorders over 30 days. This trial did not find any statistically significant benefit from the intervention, although there was a trend toward improvement on the Yale Global Tic Scale severity subscores (Murphy et al., 2015). Pidotimod, a synthetic immunomodulating agent that promotes phagocytosis in macrophages and neutrophils, activates NK cells, promotes the proliferation of lymphocytes, and regulates the ratio of CD4+ Th/CD8+ Tc cells, has been tried in a group of TS patients with an abnormal pattern of cytokines and immune cells population with positive results, although this was an uncontrolled study (Li et al., 2015). More recently, studies on the potential benefit of fecal transplantation on tic severity were published. A case report demonstrated a trend of improvement in the Yale Global Tic Severity Scale – Total Score (YGTSS-TS) after fecal microbiota transplantation (Zhao et al., 2017). The same group later expanded the study to a sample of five patients and described a statistically significant reduction in YGTSS-TS (Zhao et al., 2020). Interestingly the symptomatic improvement was linked with an increment in Bacteroidetes abundance (Zhao et al., 2020), which was recently associated with TS in a study exploring microbial taxonomy across different psychiatric disorders (Ni et al., 2022).

CONCLUSIVE REMARKS

Like other neurodevelopmental disorders, mechanistic interactions between CNS and immune system may have an important role in the pathophysiology of primary tic disorders (summarized in Fig. 15.1). This group of neurodevelopmental disorders, however, is still lagging behind other, more investigated disorders like autism spectrum disorders, in respect to the amount and quality of evidence supporting the importance of neural-immune crosstalk in their pathogenesis. A genetic predisposition to a dysfunction in these mechanistic interactions is highly plausible, even if genomic, transcriptomic, and epigenetic studies have not yet elucidated which molecular pathways and cellular interactions are the most relevant. Both genomic and maternal genetic predisposition to immune-mediated pathology (allergies, autoimmunity) have been detected in populations of patients with tic disorders, in line with the recent observation that genes represent the main contribution to maternal inheritance in TS and other tic disorders. This notwithstanding, environmental factors, for example, exposure to tobacco smoking during pregnancy, may be involved in determining a maternal inheritance effect that might lead to immune dysfunction. Albeit a very intriguing and biologically plausible pathophysiologic aspect, the involvement of abnormally activated microglia in the origin of brain network dysfunction underlying tics is supported by relatively small clinical and pathologic studies and awaits confirmation. Immune-activating environmental factors like psychosocial stressors and infections did not confirm a significant association with the onset of tics in first-degree relatives of youth with primary tic disorders. The most consistent evidence of a pathologic involvement of immune mechanisms in TS and tic disorders is that of systemic immune dysregulation, involving both innate and adaptive pathways, which support the presence of a chronic, "low- grade," inflammatory state in these patients. At the same time, strong evidence of inflammation as a key contributor to the temporal fluctuations in severity of tics is still lacking. Moreover, the clinical and pathophysiologic characterization of PANDAS/PANS/CANS, previously reported as an "indirect" evidence of immunopathology in tic disorders, has amplified the distance between this putative acute

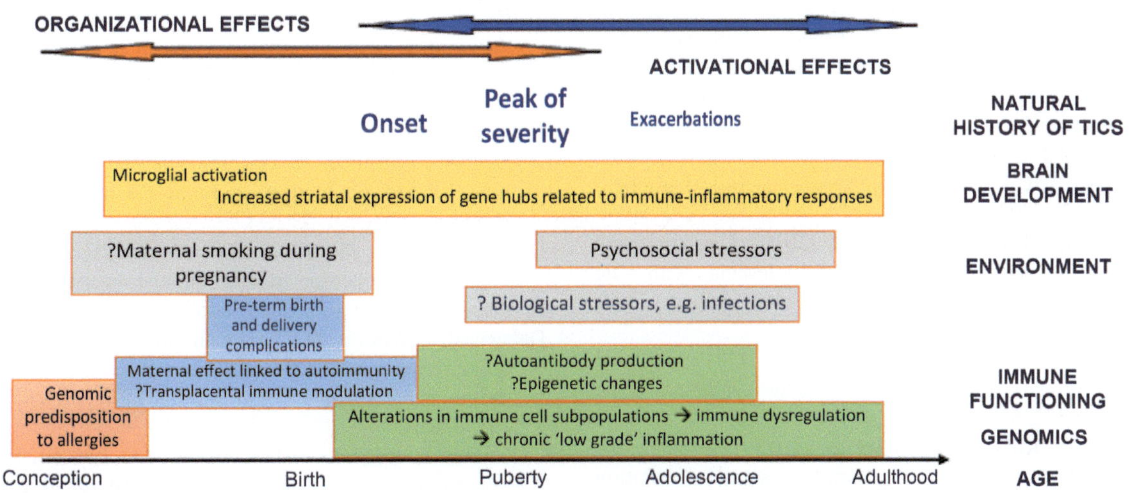

Fig. 15.1. Graphic summary of the most relevant mechanistic factors potentially at play, from conception to young adulthood, in determining dysfunctional neural-immune crosstalk in tic disorders.

encephalopathic picture – in which tics are, at best, an accessory finding – and the etiopathogenesis of TS and primary tic disorders. As a result of the pending uncertainties, there is no good evidence that immune-modulating therapies may benefit tics. The rationale to explore further immunotherapies in tic disorders needs to be corroborated by new research that should aim to illustrate the pivotal immune pathways, if existing, that are implicated in the production of tics. The increasing efforts in applying a multiomic investigative approach to the whole spectrum of neurodevelopmental disorders might help advance the knowledge of the key dysfunctions in neural-immune crosstalk in TS and related disorders.

References

Abdulkadir M, Tischfield JA, Heiman GA et al. (2022). Polygenic and environmental determinants of tics in the Avon Longitudinal Study of Parents and Children. Am J Med Genet B 192: 73–84.

Abdulkadir M, Tischfield JA, King RA et al. (2016). Pre- and perinatal complications in relation to Tourette syndrome and co-occurring obsessive-compulsive disorder and attention-deficit/hyperactivity disorder. J Psychiatr Res 82: 126.

Addabbo, Baglioni V, Schrag A et al. (2020). Anti-dopamine D2 receptor antibodies in chronic tic disorders. Devel Med Child Neurol 62: 1205–1212.

Akaltun İ, Kara T, Sertan Kara S et al. (2018). Seroprevalance anti-*Toxoplasma gondii* antibodies in children and adolescents with Tourette syndrome/chronic motor or vocal tic disorder: a case-control study. Psychiatry Res 263: 154–157.

Akdis M, Burgler S, Crameri R et al. (2011). Interleukins, from 1 to 37, and interferon-γ: receptors, functions, and roles in diseases. J Allergy Clin Immunol 127: 701–721. e70.

Allen AJ, Leonard HL, Swedo SE (1995). Case study: a new infection-triggered, autoimmune subtype of pediatric OCD and Tourette's syndrome. J Am Acad Child Adolesc Psychiatry 34: 307–311.

Alshammery S, Patel S, Jones et al. (2022). Common targetable inflammatory pathways in brain transcriptome of autism spectrum disorders and Tourette syndrome. Front Neurosci 16: 999346.

Baglioni V, Coutinho E, Menassa DA et al. (2019). Antibodies to neuronal surface proteins in Tourette syndrome: lack of evidence in a European paediatric cohort. Brain Behav Immun 81: 665–669.

Baumgaertel C, Skripuletz T, Kronenberg J et al. (2019). Immunity in Gilles de la Tourette-Syndrome: results from a cerebrospinal fluid study. Front Neurol 10: 732.

Benarroch EE (2013). HCN channels: function and clinical implications. Neurology 80: 304–310.

Bos-Veneman N, Bijzet J, Limburg PC et al. (2010). Cytokines and soluble adhesion molecules in children and adolescents with a tic disorder. Prog Neuropsychopharmacol Biol Psychiatry 34: 1390–1395.

Bos-Veneman N, Olieman R, Tobiasova Z et al. (2011). Altered immunoglobulin profiles in children with Tourette syndrome. Brain Behav Immun 25: 532–538.

Brander G, Rydell M, Kuja-Halkola R et al. (2018). Perinatal risk factors in Tourette's and chronic tic disorders: a total population sibling comparison study. Mol Psychiatry 23: 1189.

Brimberg L, Benhar I, Mascaro-Blanco A et al. (2012). Behavioral, pharmacological, and immunological abnormalities after streptococcal exposure: a novel rat model of Sydenham chorea and related neuropsychiatric disorders. Neuropsychopharmacology 37: 2076–2087.

Browne HA, Modabbernia A, Buxbaum JD et al. (2016). Prenatal maternal smoking and increased risk for Tourette syndrome and chronic tic disorders. J Am Acad Child Adolesc Psychiatry 55: 784–791.

Caine ED, Weitkamp LR, Chiverton P et al. (1985). Tourette syndrome and HLA. J Neurol Sci 69: 201–206.

Cardona F, Orefici G (2001). Group A streptococcal infections and tic disorders in an Italian pediatric population. J Pediatr 138: 71–75.

Cavanna AE (2018). Motion and emotion: the neuropsychiatry of movement disorders and epilepsy. Berlin: Springer.

Cavanna AE, Coffman KA (2021). Streptococcus and tics: another brick in the wall? Neurology 96: 560–561.

Chang K, Frankovich J, Cooperstock M et al. PANS Collaborative Consortium(2015). Clinical evaluation of youth with pediatric acute-onset neuropsychiatric syndrome (PANS): recommendations from the 2013 PANS Consensus Conference. J Child Adolesc Psychopharmacol 25: 3–13.

Chang YT, Li YF, Muo CH et al. (2011). Correlation of Tourette syndrome and allergic disease: nationwide population-based case-control study. J Dev Behav Pediatr 32: 98–102.

Chen L, Chen X, Ke N et al. (2020). Association between allergic conjunctivitis and provisional tic disorder in children. Int Ophthalmol 40: 247–253.

Cheng Y, Zheng Y, He F et al. (2012). Detection of autoantibodies and increased concentrations of interleukins in plasma from patients with Tourette's syndrome. J Mol Neurosci 48: 219–224.

Cho CE, Norman M (2013). Cesarean section and development of the immune system in the offspring. Am J Obstet Gynecol 208: 249–254.

Chou IC, Lin HC, Wang CH et al. (2010). Polymorphisms of interleukin 1 gene IL1RN are associated with Tourette syndrome. Pediatr Neurol 42: 320–324.

Church AJ, Dale RC, Lees AJ et al. (2003). Tourette's syndrome: a cross sectional study to examine the PANDAS hypothesis. J Neurol Neurosurg Psychiatry 74: 602–607.

Cibrián D, Sánchez-Madrid F (2017). CD69: from activation marker to metabolic gatekeeper. Eur J Immunol 47: 946.

Cooperstock MS, Swedo SE, Pasternack MS et al. for the PANS/PANDAS Consortium(2017). Clinical management of pediatric acute-onset neuropsychiatric syndrome: part III – treatment and prevention of infections. J Child Adolesc Psychopharmacol 27: 594–606.

Cubo E, Hortigüela M, Jorge-Roldan S et al. (2014). Prenatal and perinatal morbidity in children with tic disorders: a mainstream school-based population study in Central Spain. Tremor Other Hyperkinet Mov 4: 272.

Dale RC, Merheb V, Pillai S et al. (2012). Antibodies to surface dopamine-2 receptor in autoimmune movement and psychiatric disorders. Brain 135: 3453–3468.

Dalsgaard S, Waltoft BL, Leckman JF et al. (2015). Maternal history of autoimmune disease and later development of Tourette syndrome in offspring. J Am Acad Child Adolesc Psychiatry 54: 495–501e1.

Damelang T, Rogerson SJ, Kent SJ et al. (2019). Role of IgG3 in infectious diseases. Trends Immunol 40: 197–211.

Dickman MS (2001). von Economo encephalitis. Arch Neurol 58: 1696–1698.

Ercan-Sencicek AG, Stillman AA, Ghosh AK et al. (2010). L-Histidine decarboxylase and Tourette's syndrome. N Engl J Med 362: 1901.

Fernandez TV, Sanders SJ, Yurkiewicz IR et al. (2012). Rare copy number variants in Tourette syndrome disrupt genes in histaminergic pathways and overlap with autism. Biol Psychiatry 71: 392.

Frankovich J, Swedo S, Murphy T et al. (2017). Clinical management of pediatric acute-onset neuropsychiatric syndrome: part II – use of immunomodulatory therapies. J Child Adolesc Psychopharmacol 27: 574–593.

Frick L, Rapanelli M, Abbasi E et al. (2016). Histamine regulation of microglia: gene-environment interaction in the regulation of central nervous system inflammation. Brain Behav Immun 57: 326.

Gabbay V, Coffey BJ, Guttman LE et al. (2009). A cytokine study in children and adolescents with Tourette's disorder. Prog Neuropsychopharmacol Biol Psychiatry 33: 967.

Gariup M, Gonzalez A, Lázaro L et al. (2015). IL-8 and the innate immunity as biomarkers in acute child and adolescent psychopathology. Psychoneuroendocrinology 62: 233–242.

Gerber MA, Baltimore RS, Eaton CB et al. (2009). Prevention of rheumatic fever and diagnosis and treatment of acute Streptococcal pharyngitis: a scientific statement from the American Heart Association Rheumatic Fever, Endocarditis, and Kawasaki Disease Committee of the Council on Cardiovascular Disease in the Young, the Interdisciplinary Council on Functional Genomics and Translational Biology, and the Interdisciplinary Council on Quality of Care and Outcomes Research: endorsed by the American Academy of Pediatrics. Circulation 119: 1541–1551.

Gilbert DL, Mink JW, Singer HS (2018). A pediatric neurology perspective on Pediatric Autoimmune Neuropsychiatric Disorder Associated with Streptococcal Infection and Pediatric Acute-Onset Neuropsychiatric Syndrome. J Pediatr 199: 243–251.

Hallett JJ, Harling-Berg CJ, Knopf PM et al. (2000). Anti-striatal antibodies in Tourette syndrome cause neuronal dysfunction. J Neuroimmunol 111: 195–202.

Halvorsen, Szatkiewicz M, Mudgal J et al. (2021). Elevated common variant genetic risk for Tourette syndrome in a densely-affected pedigree. Mol Psychiatry 26: 7522–7529.

Han VX, Patel S, Jones HF et al. (2021). Maternal acute and chronic inflammation in pregnancy is associated with common neurodevelopmental disorders: a systematic review. Transl Psychiatry 11: 71.

He F, Shao X, Yi M et al. (2015). Association of IL-1α rs17561 and IL-1 RN rs315952 polymorphisms with Tourette syndrome: a family-based study. Int J Clin Exp Pathol 8: 4182.

Hoekstra PJ, Minderaa RB, Kallenberg CGM (2004). Lack of effect of intravenous immunoglobulins on tics: a double-blind placebo-controlled study. J Clin Psychiatry 65: 537–542.

Hoffman KL, Hornig M, Yaddanapudi K et al. (2004). A murine model for neuropsychiatric disorders associated with Group A β-hemolytic streptococcal infection. J Neurosci 24: 1780.

Hong JJ, Loiselle CR, Yoon DY et al. (2004). Microarray analysis in Tourette syndrome postmortem putamen. J Neurol Sci 225: 57–64.

Hornig M, Lipkin WI (2013). Immune-mediated animal models of Tourette syndrome. Neurosci Biobehav Rev 37: 1120.

Husby G, Van De Rijn I, Zabriskie JB et al. (1976). Antibodies reacting with cytoplasm of subthalamic and caudate nuclei neurons in chorea and acute rheumatic fever. J Exp Med 144: 1094.

Jones HF, Han VX, Patel S et al. (2021). Maternal autoimmunity and inflammation are associated with childhood tics and obsessive-compulsive disorder: transcriptomic data show common enriched innate immune pathways. Brain Behav Immun 94: 308–317.

Kaplan RM (2010). Syphilis, sex and psychiatry, 1789–1925: part 2. Australas Psychiatry 18: 22–27.

Kawikova I, Grady BPX, Tobiasova Z et al. (2010). Children with Tourette's syndrome may suffer immunoglobulin A dysgammaglobulinemia: preliminary report. Biol Psychiatry 67: 679.

Kawikova I, Leckman JF, Kronig H et al. (2007). Decreased numbers of regulatory T cells suggest impaired immune tolerance in children with Tourette syndrome: a preliminary study. Biol Psychiatry 61: 273–278.

Kazi JU, Rönnstrand L (2019). FMS-like tyrosine kinase 3/FLT3: from basic science to clinical implications. Physiol Rev 99: 1433–1466.

Keszler G, Kruk E, Kenezloi E et al. (2014). Association of the tumor necrosis factor-308 A/G promoter polymorphism with Tourette syndrome. Int J Immunogenet 41: 493–498.

Kiessling LS, Marcotte AC, Culpepper L (1993). Antineuronal antibodies in movement disorders. Pediatrics 92: 39–43.

Kirsten TB, Bernardi MM (2017). Prenatal lipopolysaccharide induces hypothalamic dopaminergic hypoactivity and autistic-like behaviors: repetitive self-grooming and stereotypies. Behav Brain Res 331: 25–29.

Krause D, Matz J, Weidinger E et al. (2009). Association between intracellular infectious agents and Tourette's syndrome. Eur Arch Psychiatry Clin Neurosci 260: 359–363.

Kumar A, Williams MT, Chugani HT (2015). Evaluation of basal ganglia and thalamic inflammation in children with pediatric autoimmune neuropsychiatric disorders associated with streptococcal infection and Tourette syndrome: a positron emission tomographic (PET) study using 11C-[R]-PK11195. J Child Neurol 30: 749–756.

Lajonchere C (1996). Gilles de la Tourette and the discovery of Tourette syndrome: includes a translation of his 1884 article. Arch Neurol 53: 567.

Lamothe H, Tamouza R, Hartmann A et al. (2021). Immunity and Gilles de la Tourette syndrome: a systematic review and meta-analysis of evidence for immune implications in Tourette syndrome. Eur J Neurol 28: 3187–3200.

Landau YE, Steinberg T, Richmand B et al. (2012). Involvement of immunologic and biochemical mechanisms in the pathogenesis of Tourette's syndrome. J Neural Transm (Vienna) 119: 621.

Leckman JF, Katsovich L, Kawikova I et al. (2005). Increased serum levels of interleukin-12 and tumor necrosis factor-alpha in Tourette's syndrome. Biol Psychiatry 57: 667–673.

Leivonen S, Chudal R, Joelsson P et al. (2015). Prenatal maternal smoking and Tourette syndrome: a nationwide register study. Child Psychiatry Human Dev 47: 75–82.

Lennington JB, Coppola G, Kataoka-Sasaki Y et al. (2016). Transcriptome analysis of the human striatum in Tourette syndrome. Biol Psychiatry 79: 372–382.

Leslie DL, Kozma L, Martin A et al. (2008). Neuropsychiatric disorders associated with streptococcal infection: a case-control study among privately insured children. J Am Acad Child Adolesc Psychiatry 47: 1166–1172.

Li E, Ruan Y, Chen Q et al. (2015). Streptococcal infection and immune response in children with Tourette's syndrome. Childs Nerv Syst 31: 1157–1163.

Liao C, Vuokila V, Catoire H et al. (2022). Transcriptome-wide association study reveals increased neuronal FLT3 expression is associated with Tourette's syndrome. Commun Biol 5: 289.

Liao JF, Cheng YF, Li SW et al. (2019). *Lactobacillus plantarum* PS128 ameliorates 2,5-dimethoxy-4-iodoamphetamine-induced tic-like behaviors via its influences on the microbiota–gut-brain-axis. Brain Res Bull 153: 59–73.

Lin H, Katsovich L, Ghebremichael M et al. (2007). Psychosocial stress predicts future symptom severities in children and adolescents with Tourette syndrome and/or obsessive-compulsive disorder. J Child Psychol Psychiatry 48: 157.

Lin H, Williams KA, Katsovich L et al. (2010). Streptococcal upper respiratory tract infections and psychosocial stress predict future tic and obsessive-compulsive symptom severity in children and adolescents with Tourette syndrome and/or obsessive-compulsive disorder. Biol Psychiatry 67: 684.

Lit L, Gilbert DL, Walker W et al. (2007). A subgroup of Tourette's patients overexpress specific natural killer cell genes in blood: a preliminary report. Am J Med Genet B Neuropsychiatr Genet 144B: 958–963.

Liu S, Tian M, He F et al. (2020). Mutations in ASH1L confer susceptibility to Tourette syndrome. Mol Psychiatry 25: 476–490.

Liu S, Yi M, Wang M et al. (2011). Association of IL8-251A/T, IL12B-1188A/C and TNF-α-238A/G polymorphisms with Tourette syndrome in a family-based association study in a Chinese Han population. Neurosci Lett 495: 155–158.

Liu W, Xu L, Zhang C et al. (2022). ASH1L may contribute to the risk of Tourette syndrome: combination of family-based analysis and case--control study. Brain Behav 12: e2539.

Lotan D, Benhar I, Alvarez K et al. (2014). Behavioral and neural effects of intra-striatal infusion of anti-streptococcal antibodies in rats. Brain Behav Immun 38: 249–262.

Luo F, Leckman J, Katsovich L et al. (2004). Prospective longitudinal study of children with tic disorders and/or obsessive-compulsive disorder: relationship of symptom exacerbations to newly acquired streptococcal infections. Pediatrics 113: e578–e585.

Macerollo A, Martino D (2013). Pediatric Autoimmune Neuropsychiatric Disorders Associated with Streptococcal Infections (PANDAS): an evolving concept. Tremor Other Hyperkinet Mov (NY) 3: tre-03-167-4158-7.

Malkova NV, Yu CZ, Hsiao EY et al. (2012). Maternal immune activation yields offspring displaying mouse versions of the three core symptoms of autism. Brain Behav Immun 26: 607–616.

Martino D, Chiarotti F, Buttiglione M et al. (2011). The relationship between Group A streptococcal infections and Tourette syndrome: a study on a large service-based cohort. Devel Med Child Neurol 53: 951–957.

Martino D, Defazio G, Church AJ et al. (2007). Antineuronal antibody status and phenotype analysis in Tourette's syndrome. Mov Disord 22: 1424–1429.

Martino D, Macerollo A, Leckman JF (2013). Neuroendocrine aspects of Tourette syndrome. Int Rev Neurobiol 112: 239–279.

Martino D, Schrag A, Anastasiou Z et al. (2021). Association of Group A *Streptococcus* exposure and exacerbations of chronic tic disorders: a multinational prospective cohort study. Neurology 96: e1680–e1693.

Mataix-Cols D, Frans E, Pérez-Vigil A et al. (2018). A total-population multigenerational family clustering study of autoimmune diseases in obsessive-compulsive disorder and Tourette's/chronic tic disorders. Mol Psychiatry 23: 1652.

Mathews CA, Bimson B, Lowe TL et al. (2006). Association between maternal smoking and increased symptom severity in Tourette's syndrome. Am J Psychiatry 163: 1066–1073.

Mathews CA, Scharf JM, Miller LL et al. (2014). Association between pre- and perinatal exposures and Tourette syndrome or chronic tic disorder in the ALSPAC cohort. Br J Psychiatry 204: 40.

Matz J, Krause DL, Dehning S et al. (2012). Altered monocyte activation markers in Tourette's syndrome: a case–control study. BMC Psychiatry 12: 29.

Meige H, Feindel E (1902). Les tics et leur traitement. Paris: Masson.

Mell LK, Davis RL, Owens D (2005). Association between streptococcal infection and obsessive-compulsive disorder, Tourette's syndrome, and tic disorder. Pediatrics 116: 56–60.

Melville JM, Moss TJM (2013). The immune consequences of preterm birth. Front Neurosci 7: 79.

Min S, Lee H, Park K et al. (1991). Tourette disorder and HLA typing. Yonsei Med J 32: 315–318.

Möller JC, Tackenberg B, Heinzel-Gutenbrunner M et al. (2008). Immunophenotyping in Tourette syndrome – a pilot study. Eur J Neurol 15: 749–753.

Morer A, Chae W, Henegariu O et al. (2010). Elevated expression of MCP-1, IL-2 and PTPR-N in basal ganglia of Tourette syndrome cases. Brain Behav Immun 24: 1069–1073.

Motlagh MG, Katsovich L, Thompson N et al. (2010). Severe psychosocial stress and heavy cigarette smoking during pregnancy: an examination of the pre- and perinatal risk factors associated with ADHD and Tourette syndrome. European Child Adolesc Psychiatry 19: 755–764.

Müller-Vahl KR, Loeber G, Kotsiari A et al. (2017). Gilles de la Tourette syndrome is associated with hypermethylation of the dopamine D2 receptor gene. J Psychiatr Res 86: 1–8.

Murphy T, Parker-Athill EC, Lewin AB et al. (2015). Cefdinir for recent-onset pediatric neuropsychiatric disorders: a pilot randomized trial. J Child Adolesc Psychopharmacol 25: 57.

Murphy T, Sajid M, Soto O et al. (2004). Detecting pediatric autoimmune neuropsychiatric disorders associated with streptococcus in children with obsessive-compulsive disorder and tics. Biol Psychiatry 55: 61–68.

Murphy T, Storch EA, Turner A et al. (2010). Maternal history of autoimmune disease in children presenting with tics and/or obsessive-compulsive disorder. J Neuroimmunol 229: 243.

Ni J-J, Xu Q, Yan S-S et al. (2022). Gut microbiota and psychiatric disorders: a two-sample Mendelian randomization study. Front Microbiol 12: 737197.

Orlovska S, Vestergaard CH, Bech BH et al. (2017). Association of streptococcal throat infection with mental disorders: testing key aspects of the PANDAS hypothesis in a nationwide study. JAMA Psychiatry 74: 740.

Parker-Athill EC, Ehrhart J, Tan J et al. (2015). Cytokine correlations in youth with tic disorders. J Child Adolesc Psychopharmacol 25: 86–92.

Patel A, Zhu Y, Kuzhikandathil EV et al. (2012). Soluble interleukin-6 receptor induces motor stereotypies and co-localizes with gp130 in regions linked to cortico-striato-thalamo-cortical circuits. PLoS One 7: e41623.

Pérez-Vigil A, Fernández de la Cruz L, Brander G et al. (2016). The link between autoimmune diseases and obsessive-compulsive and tic disorders: a systematic review. Neurosci Biobehav Rev 71: 542–562.

Perlmutter SJ, Leitman SF, Garvey MA et al. (1999). Therapeutic plasma exchange and intravenous immunoglobulin for obsessive-compulsive disorder and tic disorders in childhood. The Lancet 354: 1153–1158.

Peter ME, Budd RC, Desbarats J et al. (2007). The CD95 receptor: apoptosis revisited. Cell 129: 447–450.

Pranzatelli MR, Tate ED, Allison TJ (2017). Case-control, exploratory study of cerebrospinal fluid chemokines/cytokines and lymphocyte subsets in childhood Tourette syndrome with positive streptococcal markers. Cytokine 96: 49–53.

Prato A, Gulisano M, Scerbo M et al. (2021). Diagnostic approach to pediatric autoimmune neuropsychiatric disorders associated with streptococcal infections (PANDAS): a narrative review of literature data. Front Pediatr 9: 746639.

Qidwai T, Khan F (2011). Tumour necrosis factor gene polymorphism and disease prevalence. Scand J Immunol 74: 522.

Sabra S, Gratacós E, Roig MDG (2017). Smoking-induced changes in the maternal immune, endocrine, and metabolic pathways and their impact on fetal growth: a topical review. Fetal Diagn Ther 41: 241–250.

Sarchioto M, Howe F, Dumitriu IE et al. (2021). Analyses of peripheral blood dendritic cells and magnetic resonance spectroscopy support dysfunctional neuro-immune crosstalk in Tourette syndrome. Eur J Neurol 28: 1910–1921.

Schnell J, Bond M, Moll N et al. (2021). *Mycoplasma pneumoniae* IgG positivity is associated with tic severity in chronic tic disorders. Brain Behav Immun 99: 281–288.

Schoenian S, König I, Oertel W et al. (2003). HLA-DRB genotyping in Gilles de la Tourette patients and their parents. Am J Med Genet B Neuropsychiatr Genet 119B: 60–64.

Schrag A, Gilbert R, Giovannoni G et al. (2009). Streptococcal infection, Tourette syndrome, and OCD: is there a connection? Neurology 73: 1256.

Sevelsted A, Stokholm J, Bønnelykke K et al. (2015). Cesarean section and chronic immune disorders. Pediatrics 135: e92–e98.

Sigra S, Hesselmark E, Bejerot S (2018). Treatment of PANDAS and PANS: a systematic review. Neurosci Biobehav Rev 86: 51–65.

Singer HS, Gilbert DL, Wolf DS et al. (2012). Moving from PANDAS to CANS. J Pediatr 160: 725–731.

Singer HS, Hong JJ, Yoon DY et al. (2005). Serum autoantibodies do not differentiate PANDAS and Tourette syndrome from controls. Neurology 65: 1701–1707.

Sühs K-W, Skripuletz T, Pul R et al. (2015). Gilles de la Tourette syndrome is not linked to contactin-associated protein receptor 2 antibodies. Mol Brain 8: 62.

Swedo SE, Leckman JF, Rose NR (2012). From research subgroup to clinical syndrome: modifying the PANDAS criteria to describe PANS (Pediatric Acute-onset Neuropsychiatric Syndrome). Pediatr Ther 2: 1000113.

Swedo SE, Leonar HL, Garvey M et al. (1998). Pediatric autoimmune neuropsychiatric disorders associated with streptococcal infections: clinical description of the first 50 cases. Am J Psychiatry 155: 264–271.

Szejko N, Fremer C, Sühs K-W et al. (2020). Intravenous immunoglobulin treatment did not improve tics in a patient with Gilles de la Tourette syndrome and intrathecal antibody synthesis. Front Neurol 11.

Takeda K, Akira S (2015). Toll-like receptors. Curr Protoc Immunol 109: 14.12.1-14.12.10.

Taranta A, Stollerman GH (1956). The relationship of Sydenham's chorea to infection with group A streptococci. Am J Med 20: 170–175.

Taylor JR, Morshed SA, Parveen S et al. (2002). An animal model of Tourette's syndrome. Am J Psychiatry 159: 657–660.

Thienemann M, Murphy T, Leckman J et al. (2017). Clinical management of pediatric acute-onset neuropsychiatric syndrome: part I – psychiatric and behavioral interventions. J Child Adolesc Psychopharmacol 27: 566–573.

Thion MS, Ginhoux F, Garel S (2018). Microglia and early brain development: an intimate journey. Science 362: 185–189.

Tian Y, Liao IH, Zhan X et al. (2011). Exon expression and alternatively spliced genes in Tourette syndrome. Am J Med Genet 156B: 72.

Torrey EF (1988). Stalking the schizovirus. Schizophr Bull 14: 223–229.

Tsai C-S, Yang Y-H, Huang K-Y et al. (2016). Association of tic disorders and enterovirus infection: a nationwide population-based study. Medicine 95: e3347.

Tsetsos F, Yu D, Sul JH et al. (2021). Synaptic processes and immune-related pathways implicated in Tourette syndrome. Transl Psychiatry 11: 56.

Weidinger E, Krause D, Wildenauer A et al. (2014). Impaired activation of the innate immune response to bacterial challenge in Tourette syndrome. World J Biol Psychiatry 15: 453–458.

Wenzel C, Wurster U, Müller-Vahl KR (2011). Oligoclonal bands in cerebrospinal fluid in patients with Tourette's syndrome. Mov Disord 26: 343–346.

Woolfenden S, Farrar MA, Eapen V et al. (2022). Delivering paediatric precision medicine: genomic and environmental considerations along the causal pathway of childhood neurodevelopmental disorders. Dev Med Child Neurol 64: 1077–1084.

Yadav A, Saini V, Arora S (2010). MCP-1: chemoattractant with a role beyond immunity: a review. Clin Chim Acta 411: 1570–1579.

Yaddanapudi K, Hornig M, Serge R et al. (2010). Passive transfer of streptococcus-induced antibodies reproduces behavioral disturbances in a mouse model of pediatric autoimmune neuropsychiatric disorders associated with streptococcal infection. Mol Psychiatry 15: 712–726.

Yang J, Hirsch L, Osland S et al. (2017). Health status, health related behaviours and chronic health indicators in people with Tourette syndrome: a Canadian population-based study. Psychiatry Res 250: 228–233.

Yeh C, Shui HA, Chu TH et al. (2012). Hyperpolarisation-activated cyclic nucleotide channel 4 (HCN4) involvement in Tourette's syndrome autoimmunity. J Neuroimmunol 250: 18–26.

Yeh C, Wu C, Tsung H et al. (2006). Antineural antibody in patients with Tourette's syndrome and their family members. J Biomed Sci 13: 101–112.

Yeon SM, Lee JH, Kang D et al. (2017). A cytokine study of pediatric Tourette's disorder without obsessive compulsive disorder. Psychiatry Res 247: 90–96.

Yildirim Z, Karabekiroglu K, Yildiran A et al. (2020). An examination of the relationship between regulatory T cells and symptom flare-ups in children and adolescents diagnosed with chronic tic disorder and Tourette syndrome. Nord J Psychiatry 75: 18–24.

Yu D, Su JH, Tsetsos F et al. (2019). Interrogating the genetic determinants of Tourette syndrome and other tic disorders through genome-wide association studies. Am J Psychiatry 176: 217.

Yuce M, Guner S, Karabekiroglu K et al. (2014). Association of Tourette syndrome and obsessive-compulsive disorder with allergic diseases in children and adolescents: a preliminary study. Eur Rev Med Pharmacol Sci 18: 303–310.

Zalcman SS, Patel A, Mohla R et al. (2012). Soluble cytokine receptors (sIL-2Rα, sIL-2Rβ) induce subunit-specific behavioral responses and accumulate in the cerebral cortex and basal forebrain. PLoS One 7: e36316.

Zengeler K, Lukens J (2021). Innate immunity at the crossroads of healthy brain maturation and neurodevelopmental disorders. Nat Rev Immunol 21: 454–468.

Zhang D, Patel A, Zhu Y et al. (2012). Anti-streptococcus IgM antibodies induce repetitive stereotyped movements: cell activation and co-localization with Fcα/μ receptors in the striatum and motor cortex. Brain Behav Immun 26: 521–533.

Zhang T, Brander G, Isung J et al. (2022). Prenatal and early childhood infections and subsequent risk of obsessive-compulsive disorder and tic disorders: a nationwide, sibling-controlled study. Biol Psychiatry 93: 1023–1030.

Zhao H, Shi Y, Luo X et al. (2017). The effect of fecal microbiota transplantation on a child with Tourette syndrome. Case Rep Med 2017: 1–3.

Zhao HJ, Luo X, Shi YC et al. (2020). The efficacy of fecal microbiota transplantation for children with Tourette syndrome: a preliminary study. Front Psychiatry 11: 554441.

Zhonglin K, Yanhui C, Guofeng C et al. (2022). Neuroinflammation in a rat model of Tourette syndrome. Front Behav Neurosci 16: 710116.

Zobeiri M, Chaudhary R, Blaich A et al. (2019). The hyperpolarization-activated HCN4 channel is important for proper maintenance of oscillatory activity in the thalamocortical system. Cereb Cortex 29: 2291.

Zykov V, Shcherbina A, Novikova E et al. (2009). Neuroimmune aspects of the pathogenesis of Tourette's syndrome and experience in the use of immunoglobulins in children. Neurosci Behav Physiol 39: 635–638.

Handbook of Clinical Neurology, Vol. 215 (3rd series)
Tic Disorders and Tourette Syndrome
C. Ganos and Y. Worbe, Editors
https://doi.org/10.1016/B978-0-443-13554-5.00015-8

Chapter 16

Animal models of Gilles de la Tourette Syndrome and other tic disorders

YOCHEVED LOEWENSTERN[†], OREL TAHARY[†], IZHAR BAR-GAD, AND KATYA BELELOVSKY*

Leslie & Susan Goldschmied Gonda Multidisciplinary Brain Research Center, Bar-Ilan University, Ramat-Gan, Israel

Abstract

Tic disorders (TD) present considerable challenges in research due to their intricate and multifaceted nature. Their primary symptoms, tics, often co-occur with other comorbid conditions that overlap with other psychiatric and movement disorders. Tic development and expression are associated with alterations in different neurotransmitter systems, genetic factors, and environmental influences. However, the current understanding regarding the underlying causes and mechanism remains limited, with the available treatments lacking specificity and efficacy. To address these complexities, numerous animal models have been developed using diverse approaches. Some models focus on studying the neural mechanisms underlying tic expression, while others are designed to investigate the therapeutic effects on the various comorbid symptoms.

The objective of this chapter is to systematically review the animal models for TD. We will discuss the commonly used criteria for assessing the experimental model validity and evaluate the unique properties of each model based on its ability to address specific aspects of the disorder. Thus, the diverse spectrum of animal models will be established in a complementary form to enable the utilization of these models for the wide range of therapeutic and scientific purposes presented by tic disorders.

INTRODUCTION

Tic disorders (TD) are neurodevelopmental conditions characterized by the presence of simple or complex motor and vocal tics. In the case of Tourette syndrome (TS), the most extensively studied chronic TD, individuals experience multiple motor tics and at least one vocal tic for a duration of more than one year (American Psychiatric Association, 2013). Tic onset typically appears in children ages 4–7 years, peaks at ages 9–12 years, and subsides during adolescence (Leckman et al., 1998; Bloch et al., 2006; Groth et al., 2017). Symptoms tend to fluctuate and occur in "bouts" (short periods of high tic rate) that wax and wane over multiple timescales (Peterson and Leckman, 1998). Motor tics are often accompanied by sensory manifestations, including a premonitory urge preceding the tics (Leckman et al., 1993; Kwak et al., 2003) and sensory hypersensitivity, a heightened awareness of internal and/or external stimuli (Cox et al., 2018). Although tics are the hallmark of TD, up to 90% of TD patients suffer from at least one additional comorbid neuropsychiatric condition, most frequently attention-deficit hyperactivity disorder (ADHD) and obsessive-compulsive behavior/disorder (OCB/D) (Gadow et al., 2002; Hirschtritt et al., 2015).

The etiology of TD is currently unknown; nonetheless, tics are suggested to be caused by the combination of genetic predispositions and environmental triggers. TD are polygenic and highly heritable (Mataix-Cols et al., 2015; Yu et al., 2019), and genome-wide linkage studies identified candidate genes involving neuronal transmembrane genes as well as dopamine (DA) and serotonin (5-hydroxytryptamine, 5-HT) system genes (Lin et al., 2022). Various environmental factors have also been linked to TD, including prenatal environmental exposures, streptococcal infections, and immune responses (Hoekstra et al., 2013). Contextual factors such as psychosocial stressors, anxiety, and emotional tension have been found to exacerbate the expression of tics (Buse et al., 2014). Although the underlying pathophysiology of TD remains elusive, structural and

[†]These authors contributed equally.

*Correspondence to: Katya Belelovsky, Leslie & Susan Goldschmied Gonda Multidisciplinary Brain Research Center, Bar-Ilan University, Ramat-Gan, Israel. E-mail: kbelelovsky@gmail.com.

functional studies have attributed it to malfunctions in cortico-basal ganglia (CBG) pathway (Mink, 2001; Felling and Singer, 2011; Tremblay et al., 2015; Yael et al., 2015; Hsu et al., 2020).

The basal ganglia (BG) are a group of interconnected subcortical nuclei involved in motor, associative, and limbic functions (Alexander et al., 1986). Cortical and thalamic inputs are intergraded in the BG in multiple parallel, functionally defined multisynaptic pathways, and then subsequently conveyed back to frontal cortical areas via the thalamus, thus forming the CBG loop (Alexander et al., 1986; Parent and Hazrati, 1995; Lee et al., 2020). It has been hypothesized that the abnormal action inhibition in TD can be ascribed to local deficits in the CBG loop, particularly in the main input nucleus, the striatum (Albin and Mink, 2006). The striatum primarily consists of Gamma-aminobutyric acid (GABA) spiny projection neurons, modulated by different types of interneurons, such as GABAergic fast-spiking interneurons (FSIs) and cholinergic tonically active neurons (TANs) (Kita et al., 1990; Bennett and Bolam, 1994; English et al., 2012; Silberberg and Bolam, 2015). The striatum receives additional GABAergic innervation from other BG nuclei, notably the arkypallidal neurons from the downstream nucleus, the globus pallidus externus (GPe). These neurons provide feedback information from the GPe to the striatum, resulting in reciprocal activity (Mallet et al., 2012; Abdi et al., 2015). Moreover, the striatum receives various neuromodulatory input, including afferents from dopaminergic, histaminergic, and adrenergic pathways (Holmberg et al., 1999; Gerfen and Surmeier, 2011; Bolam and Ellender, 2016).

Numerous studies have demonstrated significant changes in BG structure, connectivity, and function in individuals with TD. Imaging studies have reported reduced striatal volumes in TD patients (Peterson et al., 2003; Bloch et al., 2005), although more recent investigations (Roessner et al., 2009; Forde et al., 2017; Greene et al., 2017) have failed to replicate these findings. Structural connectivity studies in TD patients have revealed widespread alterations along the CBG white matter pathways (Makki et al., 2009; Cheng et al., 2014; Müller-Vahl et al., 2014), which were found to be correlated with tic severity (Worbe et al., 2015). Likewise, several functional imaging studies examining tic generation and suppression have consistently demonstrated abnormalities in the CBG circuitry (Peterson et al., 1998; Wang et al., 2011; Neuner et al., 2014). Furthermore, multiple neuromodulator and neurotransmitter systems are altered in TD patients, most notably the dopaminergic (Singer et al., 1982, 1991; Cheon et al., 2004; Tarnok et al., 2007; Wong et al., 2008; Buse et al., 2013) and GABAergic systems (Lerner et al., 2012; Draper et al., 2014; Puts et al., 2015). Postmortem studies have revealed a substantial reduction in the cell count of FSIs and TANs in the striatum of TD patients (Kalanithi et al., 2005; Kataoka et al., 2010). Likewise, an analysis of the striatal transcriptome (Lennington et al., 2016) supports a key role for striatal interneurons in the pathogenesis.

The primary therapeutic approaches for reducing tics in TD involve behavioral interventions, which can be complemented with pharmacologic treatments such as the atypical and typical antipsychotics (which act mainly as D2-type DA receptor antagonists), α2-agonists, and anticonvulsants. However, no pharmacologic treatments can completely alleviate tics (Eddy et al., 2011; Fründt et al., 2017; Chou et al., 2023; Johnson et al., 2023). Other therapeutic options include noninvasive neuromodulation modalities (Grados et al., 2018; Dyke et al., 2019) and deep brain stimulation (DBS), targeting different brain regions across the CBG pathway, particularly for patients with severe and treatment-resistant symptoms (Israelashvili et al., 2015; Baldermann et al., 2016; Johnson et al., 2019; Wehmeyer et al., 2021). To effectively benefit TD patients, treatment considerations focus not only on its efficacy in managing tics but also on its impact on comorbid disorders (Eddy et al., 2011).

Despite the considerable accumulation of clinical data, the field of research on TD still faces numerous unanswered questions, spanning from the etiology and pathophysiology of the disorder to the clinical aspects of its symptoms and treatments. Recent advancements in genetics have made progress in identifying potential candidate genes in TD, although the precise inheritance pattern remains unclear. Furthermore, significant gaps persist in understanding the neurodevelopmental and physiologic mechanisms underlying the clinical manifestations of TD. These include factors that lead to the waxing and waning of tics, the nature of the premonitory urge and sensory deficits, the interplay with other comorbid neuropsychiatric disorders, the variability of symptomatology across individual patients, and the natural course of the disorder with the remission of symptoms over time. Limitations in current therapeutic options necessitate the exploration of more effective treatments. A deeper understanding of the underlying mechanisms can pave the way for developing novel treatments and refining existing therapies for managing tics and their associated comorbidities.

The primary objective of this chapter is to provide a comprehensive review of animal models for TD, while expanding the scope beyond what has been previously covered (Swerdlow and Sutherland, 2005; Bronfeld et al., 2013a; Hornig and Lipkin, 2013; Macrì et al., 2013; Godar et al., 2014; Nespoli et al., 2016; Yael et al., 2016; Bortolato and Pittenger, 2017). The review will address the challenges associated with modeling TD and will encompass basic approaches, including pharmacologic, genetic, and environmental models, each capturing different facets of the disorder. Furthermore, a thorough assessment of these diverse models will be presented, evaluating their validity and their ability to address specific aspects of the disorder. This analysis will highlight the strengths and limitations of each model, providing valuable insights for future research endeavors.

CHALLENGES IN MODELING TIC DISORDERS

Scientific applications of animal models

The study of pathologies, TD among them, aims to achieve two main objectives: (1) clinical, fostering the development of new and enhanced treatments for patients; (2) scientific, expanding our scientific knowledge by unraveling the underlying mechanisms responsible for the pathology in humans (LeDoux, 2005). Research conducted on human patients defines the intricate properties of the disorders: their primary symptoms, dynamics, progression, and other physiologic and functional manifestations. However, human studies suffer from significant, inherent limitations due to the inability to utilize most invasive recording methods and to perform most genetic, pharmacologic, or stimulation-based interventions. Animal models are necessary to overcome these limitations and enable a wider approach to the study of the disorders.

Humans exhibit a remarkable genetic resemblance with other mammals, such as mice with over 90% shared genome, and manifest numerous brain functional and organizational similarities with the majority of mammalian species (Krubitzer, 1995; Balleine and O'Doherty, 2010). Animal models have been pivotal in enhancing our knowledge of human physiology and the development of medical procedures. Nonetheless, interspecies differences still present severe challenges, resulting in a low conversion rate of treatments from animal studies to human patients (Markou et al., 2009).

Among the diverse species used as animal models, mice stand out as the most prevalent choice for research and modeling. They are simple to breed, easy to maintain, and can be genetically manipulated to mimic essentially any human disease or condition. Rats, which share several advantages with mice and exhibit a wider repertoire of behaviors, also serve as popular rodent models. Nonhuman primates are considerably more challenging to use but exhibit a significantly closer resemblance to human physiology and behavior compared to rodents. Various other species, including zebrafish, hamsters, pigs, chickens, and others, are employed in research; however, they are less common for modeling TD. The replication of results in animal models across multiple species reinforces the reliability of a model.

Several types of interventions are used to create animal models. Genetic manipulations can be performed by mutating one or more genes to modify proteins or induce a deficit in a manner suspected to cause pathology. Pharmacologic manipulations involve the administration of one or more active pharmaceutical ingredients (APIs) into the animal system to induce the pathology. This can be done either systematically through the bloodstream, or locally in the brain, effecting a limited area of interest. Environmental manipulations can be achieved by altering the environment to make it stressful, isolated, or enriched with stimulations, or by exposing animals to pathogens, typically viruses or antibodies, either locally or systemically. A manipulation can elicit a short-term effect lasting for minutes or hours, which we refer to as an acute manipulation. Alternatively, it can result in a long-term effect persisting for days or more, which we refer to as a chronic manipulation.

All manipulations have limitations in fully reproducing any psychiatric disorder in animals. Even models of Huntington chorea, a disorder known to be caused by a single gene mutation, has not been perfectly replicated in animals so far (Rubinsztein, 2002). Animal models are useful when they succeed in capturing some meaningful aspects of the disorder, and when results, scientific or clinical, can be translated to human patients. Estimating the effectiveness of any specific model typically involves different properties of validity (Willner, 1986; Nestler and Hyman, 2010; Kaiser and Feng, 2015).

Assessment of validity for animal models of tic disorders

The predominant framework for estimating the validity of animal models considers three properties:

- Face validity – The extent to which a model exhibits symptoms resembling those present in the disorder.
- Predictive validity – The extent to which a model is able to predict specific aspects of the disorder as seen in human patients. It is frequently used to assess the responsiveness of the model to treatments used in patients.
- Construct validity – The extent to which the neurobiologic basis of the animal model aligns with current knowledge regarding the etiology and pathophysiology of the disorder.

Face validity

The definition of face validity in TD animal models remains a subject of ongoing debate (Bortolato and Pittenger, 2017). Since TD are defined by the expression of tics, we consider tics to be the primary representation of face validity. Although the diagnostic criteria for TS include expression of motor and vocal tics (American Psychiatric Association, 2013), when assessing the face validity of TS, the presence of vocal tics may not be relevant since very few animals share our capacity for vocalization (Yartsev, 2017). Many TD models exhibit a diverse set of abnormal tic-like movements or tic-like stereotypies. However, the descriptions of these motor symptoms do not adhere to an accepted standard and are often not accompanied by a robust kinematic evaluation. The absence of standardization of evaluation criteria imposes challenges in classifying and comparing these abnormal movements (Yael et al., 2016). Therefore, in models lacking kinematic evaluation or video recordings, we refer to tic-like motor symptoms as abnormal movements.

Given that a substantial majority of individuals with TD express symptoms associated with other psychiatric disorders, primarily ADHD and OCB/D (Gadow et al., 2002; Hirschtritt et al., 2015), we consider symptoms of common comorbidities of TD in animal models as a secondary form of face validity (comorbid). These include stereotypies, repetitive behaviors, hyperactivity, obsessive grooming, spontaneous aggression, and depression. On the sensory level, TD patients experience premonitory urges preceding the tics. While the premonitory urge cannot be directly measured in animals, it is suggested to be linked to the breakdown of a sensory gating process, which can be quantified in animals. Specifically, the startle reflex, the response to sudden and relatively intense stimuli, can be inhibited if preceded by a weaker stimulus within a certain timeframe (50–300 Ms). This phenomenon is referred to as prepulse inhibition (PPI) (Davis et al., 1982) and has been found to be significantly reduced in TD patients (Swerdlow et al., 2001). Thus, a reduced PPI in animal models can also be considered a form of face validity for TD. Notably, PPI deficits are not exclusive to TD; they are also observed in various other conditions, including Huntington disease, Alzheimer, schizophrenia, and depression, among others (Swerdlow et al., 2001; Geyer, 2006). Therefore, we refer to these deficits as non-specific (NS) face validity.

Predictive validity

Predictive validity in tic models is often referred to as the efficacy of commonly used treatments in TD in alleviating symptoms in the animal models. Swerdlow and Sutherland (2005) have highlighted the challenges associated with relying solely on this aspect when using TD models. Primarily, the current drugs prescribed for TD lack specificity, as they are employed in the treatment of other psychiatric disorders and often prove to be ineffective for many patients (Eddy et al., 2011). Secondly, most of these drugs mainly target dopaminergic circuits, thereby more likely to affect models where symptoms explicitly rely on DA-related mechanisms. In addition, some antipsychotic medications such as haloperidol are shown to reduce overall animal behavior and cause immobility (Klingberg and Pickenhain, 1968; Yael et al., 2013). Using these drugs to establish predictive validity in TD animal models carries the risk of reducing the expression of most behaviors, normal and abnormal alike.

Construct validity

Robust construct validity in an animal model is achieved when it replicates the abnormalities observed in humans through a manipulation, whether it be environmental, genetic, or other factors. However, in the case of TD, the construct validity is constrained by our current limited understanding of the underlying pathophysiology and the predominantly symptomatic nature of current diagnostic parameters. As a result, a comprehensive construct-based model of TD does not currently exist.

Models may not require all aspects of validity, as different models may be designed for specific purposes and should be evaluated using distinct criteria based on their intended use. For instance, animal models with strong predictive validity are commonly used for guiding therapy and should be assessed for their sensitivity, specificity, and response to different treatments of the disorders. Construct validity is determined by findings supporting the model hypothesis, yet it is constrained by our current understanding of the underlying causes and mechanisms of the disorder. Some models with well-established construct validity may not exhibit symptoms directly comparable to human patients, such as the Dyt1 model of early-onset generalized dystonia (Dang et al., 2005, 2006). Nonetheless, models based on construct validity enable the exploration of various potential underlying causes for the disorder. Strong face validity allows the investigation of neural mechanisms related to the disorder's symptoms. However, face validity should be strictly evaluated to accurately replicate symptoms as described in human diagnostic criteria (Rose et al., 2015).

MODELS

In this section, we have classified the models according to the interventions performed on the animals. These categories consist of pharmacologic, genetic, and environment-based models.

Pharmacologic models

The pathophysiology of TD involves multiple neurotransmitters, primarily DA and GABA, but also acetylcholine, norepinephrine (NE), and 5-HT to a lesser degree. Researchers have investigated these pathways, creating various pharmacologic models. Initially, the models mostly focused on manipulating specific neurotransmitters without explicitly attempting to model TD, linking these models to other psychiatric disorders. However, abnormal movements and PPI deficits observed in these animals prompted researchers to consider them as potential models for the disorder.

Pharmacologic manipulations can be accomplished through systemic or local injections. APIs differ in affinity, with some being specific and binding to a single receptor, while others are nonspecific, binding to multiple sites and producing more diverse effects. As APIs are washed out of the system relatively quickly after injection, most pharmacologic models are acute. This section reviews the most commonly used models for TD.

GABA-based striatal disinhibition model

GABA plays a vital role as the major inhibitory neurotransmitter in the brain, effectively regulating neuronal activity. The majority of the neuronal population in the BG is GABAergic, including most of the projection neurons. The striatal disinhibition model is based on the disruption of GABA transmission within the striatum. This manipulation is thought to simulate the deficits in striatal GABAergic interneurons observed in TD (Kalanithi et al., 2005; Kataoka et al., 2010; Lerner et al., 2012). It is also theorized by the action selection model of BG function (Mink, 1996) that local disruptions in GABA function in the striatum could give rise to unwanted actions (Mink, 2001).

Researchers utilized focal striatal infusions of $GABA_A$ antagonists such as bicuculline or picrotoxin, as well as selective inhibitors of striatal FSIs like IEM-1460. Injections of IEM-1460 to rodents resulted in rest tremors but did not produce tic-like movements (Gittis et al., 2011; Oran and Bar-Gad, 2018). On the contrary, injections of $GABA_A$ antagonists led to the emergence of abnormal movements in the contralateral limbs or face of primates, rats, and mice (Tarsy et al., 1978; Crossman et al., 1984, 1988; Patel and Slater, 1988; McCairn et al., 2009; Bronfeld et al., 2013b; Pogorelov et al., 2015) and to PPI deficits (Kodsi and Swerdlow, 1995). The observed movements bear a strong resemblance to tics, displaying repetitive, brief, and nonrhythmic flexions of the limb followed by slower relaxation, seemingly independent of the ongoing actions of the animal (François et al., 2004; McCairn et al., 2009; Worbe et al., 2013; Israelashvili and Bar-Gad, 2015). In primates, a minority of the injections led to sequences of muscle activations simulating complex tics (McCairn et al., 2009), while injections into the nucleus accumbens have also induced vocal tics (McCairn et al., 2016).

In addition, bicuculline administration into nonmotor areas within the BG induced ADHD-like and OCB/D-like symptoms. In primates, it resulted in hyperactivity and stereotypy following local injections into the GPe (François et al., 2004) and nonmotor areas of the striatum (Worbe et al., 2009). A partial replication of these findings was observed in rodents, where hyperactivity was induced following injections into the nucleus accumbens (Yael et al., 2019). These results suggest a connection between disinhibition in the nonmotor regions and the emergence of comorbid symptoms in TD. These findings demonstrate a compelling effect, revealing that a similar pharmacologic manipulation in different functional regions of the BG can elicit both motor tics and their most common comorbid symptoms.

Local bicuculline infusions act acutely, eliciting effects lasting up to 2 hours (McCairn et al., 2009; Bronfeld et al., 2013b; Israelashvili and Bar-Gad, 2015). A chronic variation of this model was later developed by implanting a subcutaneous pump delivering bicuculline into dorsal striatum of rats over multiple days (Vinner et al., 2017, 2021). The chronic model was used to study the course of tics during sleep and wakefulness, demonstrating a significant reduction of tic expression during prolonged sleep (Vinner Harduf et al., 2021), consistent with observations from TD patients (Shapiro et al., 1973).

Individual motor tics were found to be correlated with the occurrence of prominent momentary transients in the local field potential (LFP), referred to as LFP spikes, in both primates and rats, during wakefulness or sleep (McCairn et al., 2009, 2013; Israelashvili and Bar-Gad, 2015; Vinner Harduf et al., 2021). It has been suggested that LFP spikes, generated by abnormal focal disinhibition in the striatum, encode tic generation (Israelashvili and Bar-Gad, 2015; Vinner Harduf et al., 2021).

This model demonstrates strong face validity, as it produces accurate tic-like movements and vocal tics, and has the ability to induce comorbid symptoms. It exhibits good construct validity based on its relation to the role of GABA in TD pathophysiology. Recent studies have provided support for the pharmacologic predictive validity by demonstrating the beneficial effects of administering GABA and partial D2 agonists in behaving mice and young rats (Pogorelov et al., 2015; Rizzo et al., 2018), and a more recent study has shown that DBS effectively mitigates these symptoms in anesthetized rats (Rusheen et al., 2023).

Dopaminergic models

Tics have long been associated with hyperfunction or imbalance of the dopaminergic system (Singer et al., 1982, 1991). The clinical use of DA agents has been reported to effectively suppress tics through dopaminergic D2 receptor antagonists (Lavenstein, 2003). Conversely, the administration of DA agonists, such as amphetamine, has been shown to exacerbate tics (Castellanos et al., 1997). Numerous nuclear imaging studies have documented aberrant DA function (Ernst et al., 1999; Singer et al., 2002; Wong et al., 2008) and several genes implicated in DA signaling were associated with TD (Comings et al., 1993, 1996; Grice et al., 1996; Rowe et al., 1998; Tarnok et al., 2007). These clinical findings led researchers to develop a dopaminergic theory of TD pathophysiology and to investigate behavioral changes in animal models following the administration of DA agonists, such as amphetamine and apomorphine. Early studies in rodents and other species have shown that systemic injections of amphetamine-induced behavioral stereotypies such as compulsive sniffing, licking or biting (Randrup and Munkvad, 1967; Russell and Pihl, 1978), motor abnormalities such a rearing and bursts of prominent locomotion (Canales and Graybiel, 2000), and impaired PPI responses (Ralph et al., 2001b). Amphetamine interventions can also be employed as a challenge in other TD animal models when the initial manipulation is insufficient to induce TD-like symptoms (Xu et al., 2015a, 2015b). Amphetamine microinjections to the ventral striatum of rats elicited orofacial dyskinesias (Kelley et al., 1988). Similarly, systemic injections of apomorphine to nonhuman primates induced orofacial dyskinesia, which was accompanied by reduced firing rate of neurons in the substantia nigra pars reticulata, one of the output structures of CBG circuit (Nevet et al., 2004). DBS or lesions in another output structure of the CBG circuit counteracted the apomorphine-induced deficient PPI, yet had inconclusive results on the apomorphine-induced motor behaviors (Lütjens et al., 2011; Posch et al., 2012). The behavioral alterations induced by stimulating DA receptors responded to antipsychotic drugs, indicating predictive validity for these substances (Del Rio and Fuentes, 1969; Mansbach et al., 1988; Arnt, 1995). In terms of face validity, this model is limited as it mostly exhibits secondary TD symptoms and comorbid conditions rather than tic-like manifestations. It is important to emphasize that these models involving DA agonist injections are not exclusive to TD and are also utilized to study other neuropsychiatric disorders, including schizophrenia, OCD, and ADHD (Rebec and Bashore, 1984; Mansbach et al., 1988; Roffman and Raskin, 1997; Szechtman et al., 1998).

A more recent protocol was developed to simulate the neurodevelopmental plasticity observed in TD (Nespoli et al., 2018). Local dopaminergic imbalance was induced in juvenile rats by unilateral injections of 6-hydroxydopamine hydrobromide in the dorsal striatum, followed by a chronic administration of D2 agonist quinpirole. This complex intervention led to altered abnormal movements that resembled tics. Decreased levels of D1 receptor RNA were observed at the lesion site, consistent with the decreased striatal D1 receptor expression seen in a human postmortem study of TD (Lennington et al., 2016), confirming the construct validity of the model. The abnormal movements were reduced by the treatment of a few agents, providing support for the predictive validity of this model.

Serotonergic model

5-HT impairment has been hypothesized to occur in TD due to its reciprocal interactions with DA (Guiard et al., 2008; de Bartolomeis et al., 2013) and its established role in OCD and other comorbidities associated with TD (Barr et al., 1993; Zohar et al., 2000). Nonetheless, imaging studies on 5-HT dysfunction in TD yielded inconclusive results. Some studies showed increased serotonin transporter binding (Wong et al., 2008), while others reported reduced binding capacity (Müller-Vahl et al., 2005). Additional inconsistent findings were reported (Leckman et al., 1995; Heinz et al., 1998; Müller-Vahl et al., 2005; Haugbøl et al., 2007; Wong et al., 2008; Nikolaus et al., 2022). To examine the role of the serotonergic system, rodent animal models utilized systemic administration of 5-HT agents, such as the 5-HT precursor or receptor agonists. These injections induced spontaneous and irregular movements, described as head shakes in mice (Corne et al., 1963; Dursun and Handley, 1996) and "wet-dog shakes" in rats (Bedard and Pycock, 1977; Yap and Taylor, 1983). These brief and repetitive abnormal movements, along with observed PPI deficits (Sipes and Geyer, 1994; Kehne et al., 1996), have been suggested to model TD (Dursun and Handley, 1996). Nonetheless, the face validity of the model is questionable due to the absence of the primary symptom. Antipsychotic drugs have been shown to inhibit 5-HT-related head shakes and wet-dog shakes, thus providing this model with predictive validity (Corne and Pickering, 1967; Dursun and Handley, 1996; Kohnomi et al., 2008).

Noradrenergic models

The involvement of NE in the pathophysiology of TD has been presumed mainly due to the beneficial therapeutic effects of the presynaptic α2-adrenergic agonist clonidine (Leckman et al., 1991; The Tourette's Syndrome Study Group, 2002; Hedderick et al., 2009), as well as its key role in comorbid symptoms such as ADHD (Arnsten et al., 2007). Nevertheless, results from lumbar fluid tests, genetic and postmortem studies did not provide supportive neurophysiologic evidence for TD impairment (Singer et al., 1990; Leckman et al., 1995; Stöber et al., 1999). In line with these findings, animal research involving pharmacologic manipulations of the NE system showed no tic-like or other abnormal behaviors in rats. They did, however, result in PPI deficits, which were subsequently reversed by clonidine (Carasso et al., 1998; Alsene et al., 2006; Swerdlow et al., 2006).

Acetylcholinergic models

Acetylcholine plays a role in controlling striatal function (Kreitzer, 2009). Postmortem brain examinations of individuals with TD revealed a reduction of approximately 50% in the density of TANs (Kataoka et al., 2010). Further support for the involvement of TANs in TD pathophysiology was obtained through the analysis of the striatal transcriptome from TD patients (Lennington et al., 2016). Based on this data, researchers developed a mice model by selectively and partially (~50%) ablating choline acetytransferase (ChAT) expressing interneurons with diphtheria toxin in the dorsal striatum of ChAT-cre transgenic mice (Xu et al., 2015b). This resulted in tic-like stereotypies following acute stress or amphetamine challenge, with no abnormalities in PPI. Since motor stereotypies have been observed in normal rats after amphetamine administration (Canales and Graybiel, 2000), it raises doubts about the impact of genetic manipulation alone on the observed symptoms. Overall, this model primarily relies on construct validity, lacking substantial evidence for face validity or any available data regarding its predictive validity.

Pharmacologic models summary

Over the years, multiple pharmacologic animal models have been employed in TD research, targeting specific impairments. The distinct attributes of each model offer opportunities to explore different aspects of the disorder, even though most do not entirely replicate the primary symptom. The GABAergic model exhibits strong face validity as it can successfully induce movements resembling tics. In recent years, many studies have utilized this model to investigate the neurophysiologic mechanisms of various aspects of TD. In contrast, other models focused on the effect of antipsychotics and additional pharmacologic agents used to treat TD, aiming to alleviate symptoms. The observed effects can be partially anticipated, as they predominantly counteract the manipulation or its byproduct. Nonetheless, these models demonstrate predictive validity, making them useful in assessing the efficacy of various pharmacologic treatments and advancing the development of novel therapeutic approaches.

The current pharmacologic models have inherent limitations, due to their acute nature, and consequently are unable to adequately address questions pertaining to the neurodevelopmental aspects of the disorder or the natural progression of symptoms over an extended period. Therefore, there is a need for alternative models that can accommodate prolonged observation periods to better understand the temporal dynamics of TD.

Transgenic models

Extensive research on TS and other TD demonstrates a strong genetic component (Price et al., 1985; State, 2010; Pauls et al.,

2014). Although most of the identified mutations are rare and no gene is known to have a major effect on TD etiology (Godar et al., 2014), multiple candidate genes were associated with the disorder (Paschou, 2013; Scharf et al., 2013; Lin et al., 2022). This led to development of several genetic models, each targeting a gene of interest and investigating its contribution to TD. Genetic manipulations involve adding, deleting, or substituting parts of the animal genetic material, using different methods such as knockout (KO), knockdown, and others. Some models reviewed are based on specific pathophysiologic hypotheses (such as the involvement of DA), while others on the serendipitous discovery of phenotypic resemblances with TD.

DAT1

The DAT protein is responsible for the reuptake of DA from the synaptic cleft. Reduced DAT1 expression and function lead to higher DA levels in the striatum. Specific genetic variations in the DAT1 gene have been associated with increased risk of developing TD (Comings et al., 1996; Rowe et al., 1998; Díaz-Anzaldúa et al., 2004; Tarnok et al., 2007; Yoon et al., 2007). DAT knockdown mutant mice express 10% normal DAT in DA neurons, whereas DAT-KO mice completely lack DAT. As a consequence, the levels of extracellular DA in the striatum of these mice are roughly twofold and fivefold compared to normal wild-type mice, respectively (Zhuang et al., 2001). These hyperdopaminergic mice show a TD-resembling comorbid behaviors, including hyperactivity, hyperlocomotion, attentional alterations, and excessive sequential stereotypy of behavioral patterns (Giros et al., 1996; Zhuang et al., 2001; Berridge et al., 2005; Fox et al., 2013). Moreover, DAT-KO mice exhibit impaired PPI responses, which can be reversed by D2 and 5-HT2A serotonin receptors antagonists and NE transporter blockers (Ralph et al., 2001a; Barr et al., 2004; Yamashita et al., 2006). The models targeting the DAT1 gene demonstrate good construct and predictive validity; however, the face validity of the model is limited by its display of secondary TD symptoms and comorbid symptoms rather than tic-like movements.

MAO-A

Monoamine oxidase A (MAO-A) enzyme participates in the metabolism of 5-HT and NE, and has an important role in the degradation of DA (Bortolato et al., 2008). Several studies have provided evidence suggesting that the MAO-A gene may serve as a genetic vulnerability for TD (Gade et al., 1998; Díaz-Anzaldúa et al., 2004). In addition, human genetic variations in MAO-A with reduced or no enzyme activity have been reported (Collins et al., 1992; Brunner et al., 1993; Sabol et al., 1998), being primarily associated with aggressive behavior. The MAO-A KO mice that have been developed, exhibit comorbid conditions related to TD, including alterations in environmental and social processing, spontaneous aggression, and some repetitive behaviors but no PPI reduction (Cases et al., 1995; Scott et al., 2008). Nevertheless, the face validity of the model is limited by the lack of any tic-like behaviors.

HDC

A rare mutation in the histidine decarboxylase (HDC) gene was found in a two-generation pedigree with multiple individuals affected by TD (Ercan-Sencicek et al., 2010). HDC encodes an enzyme that is essential for the biosynthesis of histamine (HA) from histidine. The striatum has a high density of HA receptors that modulate both dopaminergic and serotonergic neurotransmission (Haas et al., 2008).

HDC-deficient mice exhibit elevated levels of DA in the striatum, as well as information-processing deficits akin to those observed in TD patients (Baldan et al., 2014). While tic-like stereotypies are not spontaneously exhibited by these mutants, they can be elicited by the amphetamine challenge or by acute stress (Baldan et al., 2014; Xu and Li, 2015a). Similar to the ChAT-cre transgenic model (Xu et al., 2015b), the observed stereotypies in this model could be mainly attributed to these external manipulations rather than the genetic alteration itself. The absence of any tic-related movements further questions the face validity of the model. The administration of antipsychotics effectively reduces these stereotypic behaviors, thereby supporting predictive validity.

D1CT-7

D1CT-7 mice carry a transgene expressing the cholera toxin enzymatic subunit A1 that is controlled by the DA D1 receptor promoter. This alteration increases activity in specific neurons in the frontal cortex, amygdala, and other structures projecting to the striatum, thereby enhancing activity in the CBG circuitry (Campbell et al., 1999; Nordstrom and Burton, 2002). D1CT-7 mice exhibited brief and spontaneous stereotypic head or body twitches, which were typically intensified when exposed to acute stress (Nordstrom and Burton, 2002; Godar et al., 2016). These spontaneous tic-like manifestations began to appear starting from the third week of life, roughly corresponding to the age of onset of human symptoms (Godar et al., 2014). In addition, these mice showed deficits that could be compared to comorbid clinical features in patients with TD. These included locomotor hyperactivity, perseverative grooming, digging, gnawing, and leaping behaviors (Campbell et al., 1999, 2000), as well as sensorimotor gating deficits and gait impairments (Godar et al., 2016; Fowler et al., 2017; Mosher et al., 2017). Together with the high face validity, the tic-like behaviors were inhibited by antipsychotic TD therapies and D1 DA receptor antagonism, suggesting predictive validity (Swerdlow and Sutherland, 2006). However, due to the artificial nature and anatomic localization of the transgene, concerns have been raised regarding the construct validity of this model (Swerdlow and Sutherland, 2006; Godar et al., 2014).

SLITRK1

The identification of SLITRK1 as a candidate gene linked to rare, familial forms of TS (Abelson et al., 2005; O'Roak et al., 2010) led to the generation of a mouse model with a nonsense mutation in the SLITRK1 gene (Katayama et al., 2010). SLITRK1 controls neurite outgrowth and is expressed throughout the CBG pathway (Proenca et al., 2011). SLITRK1-deficient mice exhibited elevated anxiety-like behavior and a depression-like phenotype, comorbid symptoms also associated with TD. Notably, increased levels of NE were observed in the brains of these mice and the administration of clonidine alleviated the anxiety-like behavior. While this model exhibits predictive and construct validity, its face validity is limited as it does not replicate the primary symptoms of the disease.

SAPAP3

The SAPAP3 gene encodes for a postsynaptic scaffolding protein at excitatory synapses that is highly expressed in the striatum (Scannevin and Huganir, 2000). SAPAP3 KO mice were originally used as the main reference mouse model for OCD, exhibiting increased anxiety and compulsive grooming behaviors (Welch et al., 2007; Wan et al., 2014; Hadjas et al., 2019). However, with the recognition of the involvement of this gene in TD in humans (Crane et al., 2011) and its relevance to the CBG circuitry (Welch et al., 2007), the SAPAP3 KO mouse model has recently been redefined as a comorbid model encompassing a range of pathologic repetitive behaviors, including head/body twitches and excessive scratching (Lamothe et al., 2023). The head and body twitches bear a resemblance to the "wet-dog shakes," which is commonly associated with 5-HT behavioral syndrome (Bedard and Pycock, 1977; Hanson and Gorzalka, 1999), rather than representing the primary symptom of TD. Aripiprazole effectively reduced the twitches and the comorbid behaviors, affirming the predictive validity of the model.

ASH1L

The ASH1L gene is involved in regulating transcription and chromatin remodeling and is active during early brain development and neural development (An et al., 2011). In humans, a de novo mutation in ASH1L increases the risk of developing TD, leading to development of ASH1L-deficient mice (Liu et al., 2020). These mice exhibited abrupt motor movements (head jerks and body shaking), elevated self-grooming (Liu et al., 2020), anxiety, and repetitive behaviors (Yan et al., 2022). In terms of the construct validity of the model, ASH1L deficiency resulted in increased activity and DA release in the dorsal striatum (Liu et al., 2020), potentially linking the observed phenotype to neural mechanisms of TD. The exhibited abnormal movements and compulsive behaviors can be alleviated by administering haloperidol, thus providing predictive validity for the model.

Genetic models summary

TD are complex with multiple genes likely contributing to their development. No single genetic variation or mutation has been consistently linked to the disorders, and different genetic models addressed different genes linked to specific, sometimes rare mutations. While the altered phenotype is expressed from birth, most of the models do not represent the neurodevelopmental nature of the disease. Moreover, the symptoms presented are mostly secondary and comorbid. While the genetic models can provide insights into the biologic pathways and mechanisms influenced by the specific gene affected, they fail to replicate the complexity seen in TD in humans.

Environment-mediated models

Numerous studies have explored environmental factors influencing the onset and severity of TD (Singer, 2011) and related neuropsychiatric disorders in childhood (Swedo et al., 1998; Peterson et al., 2000; Goines and Ashwood, 2013; Dunn et al., 2019). Immune and infectious factors, including Group A *Streptococcus* (GAS) infections (Greenberg et al., 1998; Mell et al., 2005) and poststreptococcal autoimmune processes (Church et al., 2004), have been implicated in TD pathogenesis. Additionally, high levels of GAS-induced antibodies targeting BG epitopes have been observed (Kiessling et al., 1993; Swedo et al., 1998). Other infectious agents, such as *Mycoplasma pneumoniae* and various viruses, have also been suggested to contribute to TD development (Allen et al., 1995; Budman et al., 1997; Dale et al., 2003; Müller et al., 2004; Basheer et al., 2007). These findings led to immune-mediated animal models: antibodies exposure from TD patients, immunization with microbial antigens, immune modulator injection, and the study of transgenic animals with immune abnormalities following exposure to different stimuli (Hornig and Lipkin, 2013).

Immune mediator overexpression

Direct (focal and peripheral injections of interleukins) and indirect (maternal injections) overexpression of brain immune mediators in mice resulted in increased grooming, rearing, and stereotypic behaviors (Zalcman et al., 1998; Ponzio et al., 2007; Zalcman et al., 2012); hyperlocomotor activity (Patel et al., 2012); and PPI deficits (Smith et al., 2007). Additionally, peripherally administered interleukin were localized in brain regions associated with CBG circuits (Patel et al., 2012). Moreover, interleukin-induced excessive climbing behavior could be blocked by the DA receptor antagonists (Zalcman, 2002).

Auto-antibodies injections and microbial immunizations

Focal injections of auto-antibodies to the striatum of mice and rats induced hyperactivity, increased stereotypy, grooming, head bobbing, and episodic vocalizations (Hallett et al., 2000; Taylor et al., 2002; Smith et al., 2007; Yeh et al., 2012; Zhang et al., 2012).

Immunization with GAS or sera from GAS-immunized mice induced stereotypic movements (Hoffman et al., 2004; Yaddanapudi et al., 2010) and other motor abnormalities, and were alleviated by a D2 antagonist (Brimberg et al., 2012). Stimulating the immune system of pregnant mice by viral mimic increased locomotion and grooming and decreased PPI responses (De Miranda et al., 2010; Malkova et al., 2012).

Stress-based models

Social factors, encompassing maternal stress levels, maternal smoking and stress in the prenatal and perinatal period, significantly increase the risk of TD and symptoms severity (Leckman et al., 1990; Mathews et al., 2006; Motlagh et al., 2010). Stress-based animal models (maternal deprivation, early life stressors, social isolation in rats and primates) show stereotypies, increased locomotion, and motor abnormalities (Hall, 1998; Pryce and Feldon, 2003; Márquez-Arias et al., 2010; Rommeck et al., 2011; Hohmann et al., 2013; Yochum et al., 2014).

Environmental models summary

Animal models of environmental manipulations are based on the hypothesized etiologies and aim to maximize their construct validity by replicating the environmental factors associated with the disorder. However, it is important to emphasize that the majority of these models have not been evaluated specifically for TD. As such, their validity remains in need of thorough examination. For more comprehensive data concerning the environmental factors of TD animal models, please refer to Hornig and Lipkin (2013), Macrì et al. (2013), Godar et al. (2014), Pittenger (2015), and Nespoli et al. (2016).

SUMMARY

In this chapter, we have presented a broad scope of TD animal models (Table 16.1, Fig. 16.1). The proliferation of these models can be attributed primarily to the current lack of clarity surrounding the etiology and pathophysiology of TD. Due to their intricate nature, TD are challenging to model as they incorporate alterations in multiple neuromodulator and neurotransmitter systems. These disorders are influenced by various genes and exhibit a diverse array of symptoms with varying severities that wax and wane during the progression of the disorder and are influenced by environmental factors. Moreover, the premonitory urge, a prominent characteristic of TD, cannot be directly measured in animals. None of the existing models can fully encompass this intricate complexity.

Indeed, only a minority of TD animal models have proven successful in illuminating the pathophysiology of the disorder or aiding in the development of new effective treatments. When faced with the complexities of TD symptoms, recent models incorporating novel, advanced technologies do not present significant advantages over older pharmacologic models. Many of the models reviewed struggle in replicating motor tics, the defining symptom of TD. Instead, they rely on symptoms comorbid to TD, tic-like abnormal movements that sometimes lack clear definition and quantification, or PPI deficits that are shared with many other psychiatric disorders.

In the absence of robust face validity, some models predominantly depend on predictive validity. However, the multifactorial nature of TD and the lack of specificity in available pharmacologic treatments limit the predictive validity in TD animal models. Predictive validity is established when treatment effectively alleviates the symptoms. However, many of the symptoms presented are secondary, also associated with other psychiatric disorders, and often treated similarly. From our perspective, given that both the symptoms and treatments are not specific to TD, the utility of these models in evaluating the effects of new treatments on motor tics is considerably limited.

The construct validity of a particular model can determine its usefulness in exploring specific research questions. However, the absence of a well-established conceptual hypothesis regarding TD etiology limits the utility of a model relying solely on construct validity. Genetic models, for example, are based on gene mutations found in TD and used to explore their contribution to the neurodevelopment of the disorder. However, without the presence of the defining symptom of TD, these models may not be suitable for further scientific or clinical purposes. Similarly, environmental models of TD enable the inspection of the contribution of various environmental factors to the development of the disorder. However, their overall utility is hampered by their lack of face validity.

Animal models are assessed by their ability to translate research findings to humans. Over the years, considerable advancements have been made in modeling TD, leading to insights into the underlying neural mechanisms of tics and their interplay with comorbid disorders. However, despite notable progress, numerous questions remain unanswered due to the inadequacy of existing tools in effectively addressing these inquiries. When embarking on research endeavors, one must carefully select a suitable model tailored to the specific scientific question, while also acknowledging the inherent limitations of that chosen model. There is much potential for refining current models, as well as constructing novel ones, to enrich our understanding and foster the development of more effective treatments.

Table 16.1

Animal models of tic disorders

Animal model	Species	Acute/ chronic	Symptoms	Face validity			Construct validity	Predictive validity	References
				Primary	NS	Comorbid			
Pharmacologic									
1. GABA	Primates Rats Mice	A, C	Tics, hyperactivity, stereotypy, grooming	+	+	+	+	+	Crossman et al. (1988), Kodsi and Swerdlow (1995), François et al. (2004), McCairn et al. (2009), Worbe et al. (2009), Bronfeld et al. (2013a, 2013b), Worbe et al. (2013), Israelashvili et al. (2015), Pogorelov et al. (2015), McCairn et al. (2016), Vinner et al. (2017), Oran and Bar-Gad (2018), Rizzo et al. (2018), Yael et al. (2019), Vinner Harduf et al. (2021), Rusheen et al. (2023)
2. Dopamine	Primates Rats Mice	A, C	Abnormal movements, stereotypy, orofacial dyskinesias, PPI deficits	*	+	+	+	+	Randrup and Munkvad (1967), Del Rio and Fuentes (1969), Mansbach et al. (1988), Kelley et al. (1988), Arnt (1995), Canales and Graybiel (2000), Ralph et al. (2001a, 2001b), Nevet et al. (2004), Posch et al. (2012), Nespoli et al. (2018)
3. Serotonin	Rats Mice	A	Head shakes, "wet dog shakes", PPI deficits	–	+	+	+	+	Corne et al. (1963), Corne and Pickering (1967), Bedard and Pycock (1977), Yap and Taylor (1983), Sipes and Geyer (1994), Dursun and Handley (1996), Kehne et al. (1996), Kohnomi et al. (2008)
4. Norepinephrine	Rats	A	PPI deficits	–	+	–	?	+	Carasso et al. (1998), Alsene et al. (2006), Swerdlow et al. (2006)
5. Acetylcholine	Mice	C	Stereotypies following acute stress or amphetamine challenge	–	–	+	+	?	Xu et al. (2015b)
Genetic									
1. DAT1	Mice	C	Hyperactivity, hyperlocomotion, stereotypy, PPI deficits	–	+	+	+	+	Giros et al. (1996), Ralph et al. (2001a, 2001b), Zhuang et al. (2001), Barr et al. (2004), Berridge et al. (2005), Yamashita et al. (2006), Fox et al. (2013)
2. MAO-A	Mice	C	Spontaneous aggression, repetitive behaviors	–	–	+	+	?	Cases et al. (1995), Scott et al. (2008)
3. HDC	Mice	C	Stereotypies following acute stress or amphetamine challenge	–	?	+	+	+	Baldan et al., 2014, Xu et al. (2015a)
4. D1CT7	Mice	C	Head and body twitches, hyperactivity, perseverative grooming, early onset	+	?	+	–	+	Campbell et al. (2000), Nordstrom and Burton (2002), Godar et al. (2016)

5. SLITRK	Mice	C	Anxiety- and depression-like behaviors	–	?	+	+	+	Katayama et al. (2010)
6. SAPAP3	Mice	C	OCD-like behaviors, excessive scratching, head/body twitches	*	?	+	+	+	Welch et al. (2007), Wan et al. (2014), Lamothe et al. (2023)
7. ASH1L	Mice	C	Anxiety, repetitive behaviors, head jerks and body shakes, increased grooming	–	?	+	+	+	Liu et al. (2020), Yan et al. (2022)
Environmental									
1. Immune mediator overexpression	Mice	A, C	Hyperactivity, increased grooming and rearing, stereotypic behaviors, PPI deficits	–	+	+	+	?	Zalcman et al. (1998), Zalcman (2002), Ponzio et al. (2007), Smith et al. (2007), Patel et al. (2012), Zalcman et al. (2012)
2. Auto-antibodies injections	Mice, rats	A, C	Hyperactivity, stereotypy, grooming, head bobbing, episodic vocalizations	–	?	+	+	?	Hallett et al. (2000), Taylor et al. (2002), Smith et al. (2007), Yeh et al. (2012), Zhang et al. (2012)
3. Immunization with microbial antigens	Mice	C	Stereotypy, increased locomotion and grooming, PPI deficits	–	+	+	+	?	Hoffman et al. (2004), De Miranda et al. (2010), Yaddanapudi et al. (2010), Brimberg et al. (2012), Malkova et al. (2012)

* represents abnormal movements.? represents currently unexamined. *NS*, nonspecific; *PPI*, prepulse inhibition.

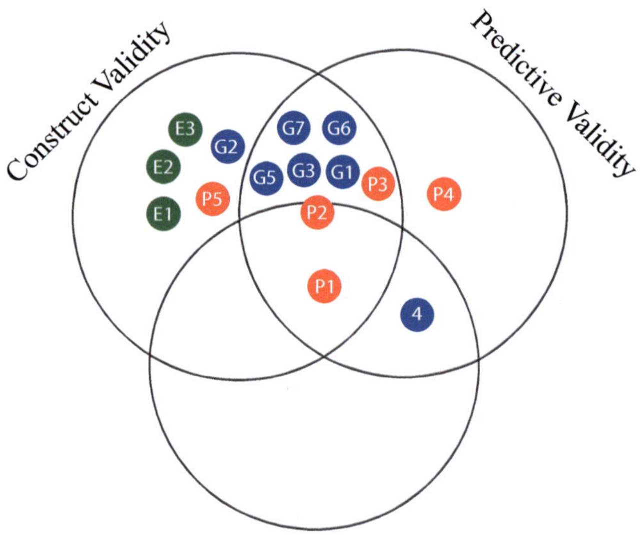

Fig. 16.1. Venn diagram presenting animal models and their shared validities. The models are categorized by the intervention type applied: pharmacologic models (red), genetic models (blue), and environmental models (green). Each model is numbered based on its order within its respective category as presented in Table 16.1.

FUNDING STATEMENT

The authors acknowledge financial support from the Israel Science Foundation (ISF) under Grant 297/18.

References

Abdi A, Mallet N, Mohamed FY et al. (2015). Prototypic and arkypallidal neurons in the dopamine-intact external globus pallidus. J Neurosci 35: 6667–6688.

Abelson JF, Kwan KY, O'Roak BJ et al. (2005). Sequence variants in SLITRK1 are associated with Tourette's syndrome. Science 310: 317–320.

Albin RL, Mink JW (2006). Recent advances in Tourette Syndrome research. Trends Neurosci 29: 175–182.

Alexander GE, DeLong MR, Strick PL (1986). Parallel organization of functionally segregated circuits linking basal ganglia and cortex. Annu Rev Neurosci 9: 357–381.

Allen AJ, Leonard HL, Swedo SE (1995). Case study: a new infection-triggered, autoimmune subtype of pediatric OCD and Tourette's syndrome. J Am Acad Child Adolesc Psychiatry 34: 307–311.

Alsene KM et al. (2006). Disruption of prepulse inhibition after stimulation of central but not peripheral α-1 adrenergic receptors. Neuropsychopharmacology 31: 2150–2161.

American Psychiatric Association (2013). Diagnostic and statistical manual of mental disorders: DSM-5, 5th ed. Washington, DC: American Psychiatric Association.

An S, Yeo KJ, Jeon YH et al. (2011). Crystal structure of the human histone methyltransferase ASH1L catalytic domain and its implications for the regulatory mechanism. J Biol Chem 286: 8369–8374.

Arnsten AF, Scahill L, Findling RL (2007). Alpha-2 adrenergic receptor agonists for the treatment of attention-deficit/hyperactivity disorder: emerging concepts from new data. J Child Adolesc Psychopharmacol 17: 393–406.

Arnt J (1995). Differential effects of classical and newer antipsychotics on the hypermotility induced by two dose levels of D-amphetamine. Eur J Pharmacol 283: 55–62.

Baldan LC, Williams KA, Williams JD et al. (2014). Histidine decarboxylase deficiency causes Tourette syndrome: parallel findings in humans and mice. Neuron 81: 77–90.

Baldermann JC, Schüller T, Huys D et al. (2016). Deep brain stimulation for Tourette-syndrome: a systematic review and meta-analysis. Brain Stimul 9: 296–304.

Balleine BW, O'Doherty JP (2010). Human and rodent homologies in action control: corticostriatal determinants of goal-directed and habitual action. Neuropsychopharmacology 35: 48–69.

Barr AM, Lehmann-Masten V, Paulus M et al. (2004). The selective serotonin-2A receptor antagonist M100907 reverses behavioral deficits in dopamine transporter knockout mice. Neuropsychopharmacology 29: 221–228.

Barr LC, Goodman WK, Price LH (1993). The serotonin hypothesis of obsessive compulsive disorder. Int Clin Psychopharmacol 8: 79–82.

Basheer SN, Wadsworth LD, Bjornson BH (2007). Anti-basal ganglia antibodies and acute movement disorder following Herpes Zoster and streptococcal infections. Eur J Paediatric Neurol 11: 104–107.

Bedard P, Pycock CJ (1977). "Wet-Dog" shake behaviour in the rat: a possible quantitative model of central 5-hydroxytryptamine activity. Neuropharmacology 16: 663–670.

Bennett BD, Bolam JP (1994). Synaptic input and output of parvalbumin-immunoreactive neurons in the neostriatum of the rat. Neuroscience 62: 707–719.

Berridge KC, Aldridge JW, Houchard KR et al. (2005). Sequential super-stereotypy of an instinctive fixed action pattern in hyper-dopaminergic mutant mice: a model of obsessive compulsive disorder and Tourette's. BMC Biol 3: 4.

Bloch MH, Peterson BS, Scahill L et al. (2006). Adulthood outcome of tic and obsessive-compulsive symptom severity in children with Tourette syndrome. Arch Pediatr Adolesc Med 160: 65–69.

Bloch MH, Leckman JF, Zhu H et al. (2005). Caudate volumes in childhood predict symptom severity in adults with Tourette syndrome. Neurology 65: 1253–1258.

Bolam JP, Ellender TJ (2016). Histamine and the striatum. Neuropharmacology 106: 74–84.

Bortolato M, Chen K, Shih JC (2008). Monoamine oxidase inactivation: from pathophysiology to therapeutics. Adv Drug Deliv Rev 60: 1527–1533.

Bortolato M, Pittenger C (2017). Modeling tics in rodents: conceptual challenges and paths forward. J Neurosci Methods 292: 12–19.

Brimberg L, Benhar I, Mascaro-Blanco A et al. (2012). Behavioral, pharmacological, and immunological abnormalities after streptococcal exposure: a novel rat model of Sydenham chorea and related neuropsychiatric disorders. Neuropsychopharmacology 37: 2076–2087.

Bronfeld M, Israelashvili M, Bar-Gad I (2013). Pharmacological animal models of Tourette syndrome. Neurosci Biobehav Rev 37: 1101–1119.

Bronfeld M, Yael D et al. (2013). Motor tics evoked by striatal disinhibition in the rat. Front Syst Neurosci 7: 50.

Brunner HG, Nelen M, Breakefield XO et al. (1993). Abnormal behavior associated with a point mutation in the structural gene for monoamine oxidase A. Science 262: 578–580.

Budman CL, Kerjakovic M, Bruun RD (1997). Viral infection and tic exacerbation. J Am Acad Child Adolesc Psychiatry 36: 162.

Buse J, Schoenefeld K, Münchau A et al. (2013). Neuromodulation in Tourette syndrome: dopamine and beyond. Neurosci Biobehav Rev 37: 1069–1084.

Buse J, Kirschbaum C, Leckman JF et al. (2014). The modulating role of stress in the onset and course of Tourette's syndrome: a review. Behav Modif 38: 184–216.

Campbell KM, de Lecea L, Severynse DM et al. (1999). OCD-like behaviors caused by a neuropotentiating transgene targeted to cortical and limbic D1+ neurons. J Neurosci 19: 5044–5053.

Campbell KM, Veldman MB, McGrath MJ et al. (2000). TS+OCD-like neuropotentiated mice are supersensitive to seizure induction. NeuroReport 11: 2335.

Canales JJ, Graybiel AM (2000). A measure of striatal function predicts motor stereotypy. Nat Neurosci 3: 377–383.

Carasso BS, Bakshi VP, Geyer MA (1998). Disruption in prepulse inhibition after alpha-1 adrenoceptor stimulation in rats. Neuropharmacology 37: 401–404.

Cases O, Seif I, Grimsby J et al. (1995). Aggressive behavior and altered amounts of brain serotonin and norepinephrine in mice lacking MAOA. Science 268: 1763–1766.

Castellanos FX, Giedd JN, Elia J et al. (1997). Controlled stimulant treatment of ADHD and comorbid Tourette's syndrome: effects of stimulant and dose. J Am Acad Child Adolesc Psychiatry 36: 589–596.

Cheng B, Braass H, Ganos C et al. (2014). Altered intrahemispheric structural connectivity in Gilles de La Tourette syndrome. NeuroImage Clin 4: 174–181.

Cheon KA, Ryu YH, Namkoong K et al. (2004). Dopamine transporter density of the basal ganglia assessed with [123I]IPT SPECT in drug-naive children with Tourette's disorder. Psychiatry Res 130: 85–95.

Chou C-Y, Agin-Liebes J, Kuo S-H (2023). Emerging therapies and recent advances for Tourette syndrome. Heliyon 9: e12874.

Church AJ, Dale RC, Giovannoni G (2004). Anti-basal ganglia antibodies: a possible diagnostic utility in idiopathic movement disorders? Arch Dis Child 89: 611–614.

Collins FA, Murphy DL, Reiss AL et al. (1992). Clinical, biochemical, and neuropsychiatric evaluation of a patient with a contiguous gene syndrome due to a microdeletion Xp11.3 including the Norrie disease locus and monoamine oxidase (MAOA and MAOB) genes. Am J Med Genet 42: 127–134.

Comings D, Muhleman D, Dietz G et al. (1993). Association between Tourette's syndrome and homozygosity at the Dopamine D3 receptor gene. Lancet 341: 906.

Comings DE, Wu S, Chiu C et al. (1996). Polygenic inheritance of Tourette syndrome, stuttering, attention deficit hyperactivity, conduct, and oppositional defiant disorder: the additive and subtractive effect of the three dopaminergic genes – DRD2, D Beta H, and DAT1. Am J Med Genet 67: 264–288.

Corne SJ, Pickering RW (1967). A possible correlation between drug-induced hallucinations in man and a behavioural response in mice. Psychopharmacologia 11: 65–78.

Corne SJ, Pickering RW, Warner BT (1963). A method for assessing the effects of drugs on the central actions of 5-hydroxytryptamine. Br J Pharmacol Chemother 20: 106–120.

Cox JH, Seri S, Cavanna AE (2018). Sensory aspects of Tourette syndrome. Neurosci Biobehav Rev 88: 170–176.

Crane J, Fagerness J, Osiecki L et al. (2011). Family-based genetic association study of DLGAP3 in Tourette syndrome. Am J Med Genet 156B: 108–114.

Crossman AR, Mitchell IJ, Sambrook MA et al. (1988). Chorea and myoclonus in the monkey induced by gamma-aminobutyric acid antagonism in the lentiform complex: the site of drug action and a hypothesis for the neural mechanisms of chorea. Brain 111: 1211–1233.

Crossman AR, Sambrook MA, Jackson A (1984). Experimental hemichorea/hemiballismus in the monkey: studies on the intracerebral site of action in a drug-induced dyskinesia. Brain 107: 579–596.

Dale RC, Church AJ, Heyman I (2003). Striatal encephalitis after Varicella Zoster infection complicated by tourettism. Mov Disord 18: 1554–1556.

Dang MT, Yokoi F, McNaught KS et al. (2005). Generation and characterization of Dyt1 DeltaGAG knock-in mouse as a model for early-onset dystonia. Exp Neurol 196: 452–463.

Dang MT, Yokoi F, Pence MA et al. (2006). Motor deficits and hyperactivity in Dyt1 knockdown mice. Neurosci Res 56: 470–474.

Davis M, Gendelman DS, Tischler MD et al. (1982). A primary acoustic startle circuit: lesion and stimulation studies. J Neurosci 2: 791–805.

de Bartolomeis A, Buonaguro EF, Iasevoli F (2013). Serotonin–glutamate and serotonin–dopamine reciprocal interactions as putative molecular targets for novel antipsychotic treatments: from receptor heterodimers to postsynaptic scaffolding and effector proteins. Psychopharmacology 225: 1–19.

De Miranda J, Yaddanapudi K, Hornig M et al. (2010). Induction of toll-like receptor 3-mediated immunity during gestation inhibits cortical neurogenesis and causes behavioral disturbances. mBio 1: e00176-10.

Del Rio J, Fuentes JA (1969). Further studies on the antagonism of stereotyped behaviour induced by amphetamine. Eur J Pharmacol 8: 73–78.

Díaz-Anzaldúa A, Joober R, Rivière JB et al. (2004). Tourette syndrome and dopaminergic genes: a family-based association study in the French Canadian Founder Population. Mol Psychiatry 9: 272–277.

Draper A, Stephenson MC, Jackson GM et al. (2014). Increased GABA contributes to enhanced control over motor excitability in Tourette syndrome. Curr Biol 24: 2343–2347.

Dunn GA, Nigg JT, Sullivan EL (2019). Neuroinflammation as a risk factor for attention deficit hyperactivity disorder. Pharmacol Biochem Behav 182: 22–34.

Dursun SM, Handley SL (1996). Similarities in the pharmacology of spontaneous and DOI-induced head-shakes suggest 5HT2A receptors are active under physiological conditions. Psychopharmacology 128: 198–205.

Dyke K, Jackson GM, Nixon E et al. (2019). Effects of single-session cathodal transcranial direct current stimulation on tic symptoms in Tourette's syndrome. Exp Brain Res 237: 2853–2863.

Eddy CM, Rickards HE, Cavanna AE (2011). Treatment strategies for tics in Tourette syndrome. Ther Adv Neurol Disord 4: 25–45.

English DF, Ibanez-Sandoval O, Stark E et al. (2012). GABAergic circuits mediate the reinforcement-related signals of striatal cholinergic interneurons. Nat Neurosci 15: 123–130.

Ercan-Sencicek AG, Stillman AA, Ghosh AK et al. (2010). L-Histidine decarboxylase and Tourette's syndrome. N Engl J Med 362: 1901–1908.

Ernst M, Zametkin AJ, Jons PH et al. (1999). High presynaptic dopaminergic activity in children with Tourette's disorder. J Am Acad Child Adolesc Psychiatry 38: 86–94.

Felling RJ, Singer HS (2011). Neurobiology of Tourette syndrome: current status and need for further investigation. J Neurosci 31: 12387–12395.

Forde NJ, Zwiers MP, Naaijen J et al. (2017). Basal ganglia structure in Tourette's disorder and/or attention-deficit/hyperactivity disorder. Mov Disord 32: 601–604.

Fowler SC, Mosher LJ, Godar SC et al. (2017). Assessment of gait and sensorimotor deficits in the D1CT-7 mouse model of Tourette syndrome. J Neurosci Methods 292: 37–44.

Fox MA, Panessiti MG, Hall FS et al. (2013). An evaluation of the serotonin system and perseverative, compulsive, stereotypical, and hyperactive behaviors in dopamine transporter (DAT) knockout mice. Psychopharmacology 227: 685–695.

François C, Grabli D, McCairn K et al. (2004). Behavioural disorders induced by external globus pallidus dysfunction in primates II. Anatomical study. Brain 127: 2055–2070.

Fründt O, Woods D, Ganos C (2017). Behavioral therapy for Tourette syndrome and chronic tic disorders. Neurol Clin Pract 7: 148–156.

Gade R, Muhleman D, Blake H et al. (1998). Correlation of length of VNTR alleles at the X-linked MAOA gene and phenotypic effect in Tourette syndrome and drug abuse. Mol Psychiatry 3: 50–60.

Gadow KD, Nolan EE, Sprafkin J et al. (2002). Tics and psychiatric comorbidity in children and adolescents. Dev Med Child Neurol 44: 330–338.

Gerfen CR, Surmeier DJ (2011). Modulation of striatal projection systems by dopamine. Annu Rev Neurosci 34: 441–466.

Geyer MA (2006). The family of sensorimotor gating disorders: comorbidities or diagnostic overlaps? Neurotox Res 10: 211–220.

Giros B, Jaber M, Jones SR et al. (1996). Hyperlocomotion and indifference to cocaine and amphetamine in mice lacking the dopamine transporter. Nature 379: 606–612.

Gittis AH, Leventhal DK, Fensterheim BA et al. (2011). Selective inhibition of striatal fast-spiking interneurons causes dyskinesias. J Neurosci 31: 15727–15731.

Godar SC, Mosher LJ, Di Giovanni G et al. (2014). Animal models of tic disorders: a translational perspective. J Neurosci Methods 238: 54–69.

Godar SC, Mosher LJ, Strathman HJ et al. (2016). The D1CT-7 mouse model of Tourette syndrome displays sensorimotor gating deficits in response to spatial confinement. Br J Pharmacology 173: 2111–2121.

Goines PE, Ashwood P (2013). Cytokine dysregulation in autism spectrum disorders (ASD): possible role of the environment. Neurotoxicol Teratol 36: 67–81.

Grados M, Huselid R, Duque-Serrano L (2018). Transcranial magnetic stimulation in Tourette syndrome: a historical perspective, its current use and the influence of comorbidities in treatment response. Brain Sci 8: 129.

Greenberg BD, Murphy DL, Swedo SE (1998). Symptom exacerbation of vocal tics and other symptoms associated with streptococcal pharyngitis in a patient with obsessive-compulsive disorder and tics. Am J Psychiatry 155: 1459–1460.

Greene DJ, Williams Iii AC, Koller JM et al. (2017). Brain structure in pediatric Tourette syndrome. Mol Psychiatry 22: 972–980.

Grice DE, Leckman JF, Pauls DL et al. (1996). Linkage disequilibrium between an allele at the dopamine D4 receptor locus and Tourette syndrome, by the transmission-disequilibrium test. Am J Hum Genet 59: 644–652.

Groth C, Mol Debes N, Rask CU et al. (2017). Course of Tourette syndrome and comorbidities in a large prospective clinical study. J Am Acad Child Adolesc Psychiatry 56: 304–312.

Guiard BP, El Mansari M, Merali Z et al. (2008). Functional interactions between dopamine, serotonin and norepinephrine neurons: an in-vivo electrophysiological study in rats with monoaminergic lesions. Int J Neuropsychopharmacol 11: 625–639.

Haas HL, Sergeeva OA, Selbach O (2008). Histamine in the nervous system. Physiol Rev 88: 1183–1241.

Hadjas LC, Lüscher C, Simmler LD (2019). Aberrant habit formation in the Sapap3-knockout mouse model of obsessive-compulsive disorder. Sci Rep 9: 12061.

Hall FS (1998). Social deprivation of neonatal, adolescent, and adult rats has distinct neurochemical and behavioral consequences. Crit Rev Neurobiol 12: 129–162.

Hallett JJ, Harling-Berg CJ, Knopf PM et al. (2000). Anti-striatal antibodies in Tourette syndrome cause neuronal dysfunction. J Neuroimmunol 111: 195–202.

Hanson LA, Gorzalka BB (1999). The influence of corticosterone on serotonergic stereotypy and sexual behavior in the female rat. Behav Brain Res 104: 27–35.

Haugbøl S, Pinborg LH, Regeur L et al. (2007). Cerebral 5-HT2A receptor binding is increased in patients with Tourette's syndrome. Int J Neuropsychopharmacol 10: 245–252.

Hedderick EF, Morris CM, Singer HS (2009). Double-blind, crossover study of clonidine and levetiracetam in Tourette syndrome. Pediatr Neurol 40: 420–425.

Heinz A, Knable MB, Wolf SS et al. (1998). Tourette's syndrome: [I-123]β-CIT SPECT correlates of vocal tic severity. Neurology 51: 1069–1074.

Hirschtritt ME, Lee PC, Pauls DL et al. (2015). Lifetime prevalence, age of risk, and genetic relationships of comorbid psychiatric disorders in Tourette syndrome. JAMA Psychiatry 72: 325–333.

Hoekstra PJ, Dietrich A, Edwards MJ et al. (2013). Environmental factors in Tourette syndrome. Neurosci Biobehav Rev 37: 1040–1049.

Hoffman KL, Hornig M, Yaddanapudi K et al. (2004). A murine model for neuropsychiatric disorders associated with Group A beta-hemolytic streptococcal infection. J Neurosci 24: 1780–1791.

Hohmann CF, Hodges A, Beard N et al. (2013). Effects of brief stress exposure during early postnatal development in Balb/CByJ mice: I. Behavioral characterization. Dev Psychobiol 55: 283–293.

Holmberg M, Scheinin M, Kurose H et al. (1999). Adrenergic A2C-receptors reside in rat striatal GABAergic projection neurons: comparison of radioligand binding and immunohistochemistry. Neuroscience 93: 1323–1333.

Hornig M, Lipkin WI (2013). Immune-mediated animal models of Tourette syndrome. Neurosci Biobehav Rev 37: 1120–1138.

Hsu CJ, Wong LC, Wang HP et al. (2020). The multimodality neuroimage findings in individuals with Tourette Syndrome. Pediatr Neonatol 61: 467–474.

Israelashvili M, Bar-Gad I (2015). Corticostriatal divergent function in determining the temporal and spatial properties of motor tics. J Neurosci 35: 16340–16351.

Israelashvili M, Loewenstern Y, Bar-Gad I (2015). Abnormal neuronal activity in Tourette syndrome and its modulation using deep brain stimulation. J Neurophysiol 114: 6–20.

Johnson KA, Fletcher PT, Servello D et al. (2019). Image-based analysis and long-term clinical outcomes of deep brain stimulation for Tourette syndrome: a multisite study. J Neurol Neurosurg Psychiatry 90: 1078–1090.

Johnson KA, Worbe Y, Foote KD et al. (2023). Tourette syndrome: clinical features, pathophysiology, and treatment. Lancet Neurol 22: 147–158.

Kaiser T, Feng G (2015). Modeling psychiatric disorders for developing effective treatments. Nat Med 21: 979–988.

Kalanithi PSA, Zheng W, Kataoka Y et al. (2005). Altered parvalbumin-positive neuron distribution in basal ganglia of individuals with Tourette syndrome. Proc Natl Acad Sci USA 102: 13307–13312.

Kataoka Y, Kalanithi PS, Grantz H et al. (2010). Decreased number of parvalbumin and cholinergic interneurons in the striatum of individuals with Tourette syndrome. J Comp Neurol 518: 277–291.

Katayama K, Yamada K, Ornthanalai VG et al. (2010). Slitrk1-deficient mice display elevated anxiety-like behavior and noradrenergic abnormalities. Mol Psychiatry 15: 177–184.

Kehne JH, McCloskey TC, Kehne JH et al. (1996). 5-HT modulation of auditory and visual sensorimotor gating: I. Effects of 5-HT releasers on sound and light prepulse inhibition in Wistar rats. Psychopharmacology 124: 95–106.

Kelley AE, Lang CG, Gauthier AM (1988). Induction of oral stereotypy following amphetamine microinjection into a discrete subregion of the striatum. Psychopharmacology 95: 556–559.

Kiessling LS, Marcotte AC, Culpepper L (1993). Antineuronal antibodies in movement disorders. Pediatrics 92: 39–43.

Kita H, Kosaka T, Heizmann CW (1990). Parvalbumin-immunoreactive neurons in the rat neostriatum: a light and electron microscopic study. Brain Res 536: 1–15.

Klingberg F, Pickenhain L (1968). Occurrence of "spindle discharges" in the rat in relation to behavior. Acta Biol Med Ger 20: 45–54.

Kodsi MH, Swerdlow NR (1995). Prepulse inhibition in the rat is regulated by ventral and caudodorsal striato-pallidal circuitry. Behav Neurosci 109: 912–928.

Kohnomi S, Suemaru K, Kawasaki H et al. (2008). Effect of aripiprazole on 5-HT2 receptor-mediated wet-dog shake responses and disruption of prepulse inhibition in rats. J Pharmacol Sci 106: 645–650.

Kreitzer AC (2009). Physiology and pharmacology of striatal neurons. Annu Rev Neurosci 32: 127–147.

Krubitzer L (1995). The organization of neocortex in mammals: are species differences really so different Trends Neurosci 18: 408–417.

Kwak C, Dat Vuong K, Jankovic J (2003). Premonitory sensory phenomenon in Tourette's syndrome. Mov Disord 18: 1530–1533.

Lamothe H, Schreiweis C, Mondragón-González LS et al. (2023). The Sapap3$^{-/-}$ mouse reconsidered as a comorbid model expressing a spectrum of pathological repetitive behaviours. Transl Psychiatry 13: 1–12.

Lavenstein BL (2003). Treatment approaches for children with Tourette's syndrome. Curr Neurol Neurosci Rep 3: 143–148.

Leckman JF, Goodman WK, Anderson GM et al. (1995). Cerebrospinal fluid biogenic amines in obsessive compulsive disorder, Tourette's syndrome, and healthy controls. Neuropsychopharmacology 12: 73–86.

Leckman JF, Hardin MT, Riddle MA et al. (1991). Clonidine treatment of Gilles de La Tourette's syndrome. Arch Gen Psychiatry 48: 324–328.

Leckman JF, Zhang H, Vitale A et al. (1998). Course of tic severity in Tourette syndrome: the first two decades. Pediatrics 102: 14–19.

Leckman JF, Dolnansky ES, Hardin MT et al. (1990). Perinatal factors in the expression of Tourette's syndrome: an exploratory study. J Am Acad Child Adolesc Psychiatry 29: 220–226.

Leckman JF, Walker DE, Cohen DJ (1993). Premonitory urges in Tourette's syndrome. Am J Psychiatry 150: 98–102.

LeDoux M (2005). Animal models and the science of movement disorders. In: Animal models of movement disorders. Academic Press, pp. 13–31.

Lee J, Wang W, Sabatini BL (2020). Anatomically segregated basal ganglia pathways allow parallel behavioral modulation. Nat Neurosci 23: 1388–1398.

Lennington JB, Coppola G, Kataoka-Sasaki Y et al. (2016). Transcriptome analysis of the human striatum in Tourette syndrome. Biol Psychiatry 79: 372–382.

Lerner A, Bagic A, Simmons JM et al. (2012). Widespread abnormality of the γ-aminobutyric acid-ergic system in Tourette syndrome. Brain 135: 1926–1936.

Lin W-D, Tsai F-J, Chou I-C (2022). Current understanding of the genetics of Tourette syndrome. Biomed J 45: 271–279.

Liu S, Tian M, He F et al. (2020). Mutations in ASH1L confer susceptibility to Tourette syndrome. Mol Psychiatry 25: 476–490.

Lütjens G, Krauss JK, Schwabe K (2011). Lesions of the entopeduncular nucleus in rats prevent apomorphine-induced deficient sensorimotor gating. Behav Brain Res 220: 281–287.

Macrì S, Onori MP, Roessner V et al. (2013). Animal models recapitulating the multifactorial origin of Tourette syndrome. Int Rev Neurobiol 112: 211–237.

Makki MI, Govindan RM, Wilson BJ et al. (2009). Altered fronto-striato-thalamic connectivity in children with Tourette syndrome assessed with diffusion tensor MRI and probabilistic fiber tracking. J Child Neurol 24: 669–678.

Malkova NV, Yu CZ, Hsiao EY et al. (2012). Maternal immune activation yields offspring displaying mouse versions of the three core symptoms of autism. Brain Behav Immun 26: 607–616.

Mallet N, Micklem BR, Henny P et al. (2012). Dichotomous organization of the external globus pallidus. Neuron 74: 1075–1086.

Mansbach RS, Geyer MA, Braff DL (1988). Dopaminergic stimulation disrupts sensorimotor gating in the rat. Psychopharmacology 94: 507–514.

Markou A, Chiamulera C, Geyer MA et al. (2009). Removing obstacles in neuroscience drug discovery: the future path for animal models. Neuropsychopharmacology 34: 74–89.

Márquez-Arias A, Santillán-Doherty AM, Arenas-Rosas RV et al. (2010). Environmental enrichment for captive stumptail macaques (*Macaca arctoides*). J Med Primatol 39: 32–40.

Mataix-Cols D, Isomura K, Pérez-Vigil A et al. (2015). Familial risks of Tourette syndrome and chronic tic disorders. a population-based cohort study. JAMA Psychiatry 72: 787–793.

Mathews CA, Bimson B, Lowe TL et al. (2006). Association between maternal smoking and increased symptom severity in Tourette's syndrome. Am J Psychiatry 163: 1066–1073.

McCairn KW, Nagai Y, Hori Y et al. (2016). A primary role for nucleus accumbens and related limbic network in vocal tics. Neuron 89: 300–307.

McCairn KW, Bronfeld M, Belelovsky K et al. (2009). The neurophysiological correlates of motor tics following focal striatal disinhibition. Brain 132: 2125–2138.

McCairn KW, Iriki A, Isoda M (2013). Global dysrhythmia of cerebro-basal ganglia–cerebellar networks underlies motor tics following striatal disinhibition. J Neurosci 33: 697–708.

Mell LK, Davis RL, Owens D (2005). Association between streptococcal infection and obsessive-compulsive disorder, Tourette's syndrome, and tic disorder. Pediatrics 116: 56–60.

Mink JW (2001). Basal ganglia dysfunction in Tourette's syndrome: a new hypothesis. Pediatr Neurol 25: 190–198.

Mink JW (1996). The basal ganglia: focused selection and inhibition of competing motor programs. Prog Neurobiol 50: 381–425.

Mosher LJ, Godar SC, Nelson M et al. (2017). Allopregnanolone mediates the exacerbation of Tourette-like responses by acute stress in mouse models. Sci Rep 7: 3348.

Motlagh MG, Katsovich L, Thompson N et al. (2010). Severe psychosocial stress and heavy cigarette smoking during pregnancy: an examination of the pre- and perinatal risk factors associated with ADHD and Tourette syndrome. Eur Child Adolesc Psychiatry 19: 755–764.

Müller N, Riedel M, Blendinger C et al. (2004). Mycoplasma pneumoniae infection and Tourette's syndrome. Psychiatry Res 129: 119–125.

Müller-Vahl KR, Szejko N, Wilke F et al. (2005). Serotonin transporter binding in Tourette syndrome. Neurosci Lett 385: 120–125.

Müller-Vahl KR, Grosskreutz J, Prell T et al. (2014). Tics are caused by alterations in prefrontal areas, thalamus and putamen, while changes in the cingulate gyrus reflect secondary compensatory mechanisms. BMC Neurosci 15: 6.

Nespoli E, Rizzo F, Boeckers TM et al. (2016). Addressing the complexity of Tourette's syndrome through the use of animal models. Front Neurosci 10: 133.

Nespoli E, Rizzo F, Boeckers T et al. (2018). Altered dopaminergic regulation of the dorsal striatum is able to induce tic-like movements in juvenile rats. PLoS One 13: e0196515.

Nestler EJ, Hyman SE (2010). Animal models of neuropsychiatric disorders. Nat Neurosci 13: 1161–1169.

Neuner I, Werner CJ, Arrubla J et al. (2014). Imaging the where and when of tic generation and resting state networks in adult Tourette patients. Front Hum Neurosci 8: 362.

Nevet A, Morris G, Saban G et al. (2004). Discharge rate of substantia nigra pars reticulata neurons is reduced in non-Parkinsonian monkeys with apomorphine-induced orofacial dyskinesia. J Neurophysiol 92: 1973–1981.

Nikolaus S, Mamlins E, Antke C et al. (2022). Boosted dopamine and blunted serotonin in Tourette syndrome – evidence from in vivo imaging studies. Rev Neurosci 33: 859–876.

Nordstrom EJ, Burton FH (2002). A transgenic model of comorbid Tourette's syndrome and obsessive-compulsive disorder circuitry. Mol Psychiatry 7: 617–625.

Oran Y, Bar-Gad I (2018). Loss of balance between striatal feedforward inhibition and corticostriatal excitation leads to tremor. J Neurosci 38: 1699–1710.

O'Roak B, Morgan TM, Fishman DO et al. (2010). Additional support for the association of SLITRK1 Var321 and Tourette syndrome. Mol Psychiatry 15: 447–450.

Parent A, Hazrati L-N (1995). Functional anatomy of the basal ganglia. I. The cortico-basal ganglia-thalamo-cortical loop. Brain Res Rev 20: 91–127.

Paschou P (2013). The genetic basis of Gilles de La Tourette syndrome. Neurosci Biobehav Rev 37: 1026–1039.

Patel A, Zhu Y, Kuzhikandathil EV et al. (2012). Soluble interleukin-6 receptor induces motor stereotypies and co-localizes with Gp130 in regions linked to cortico-striato-thalamo-cortical circuits. PLoS One 7: e41623.

Patel S, Slater P (1988). Effects of GABA compounds injected into the subpallidal regions of rat brain on nucleus accumbens evoked hyperactivity. Behav Neurosci 102: 596–600.

Pauls DL, Fernandez TV, Mathews CA et al. (2014). The inheritance of Tourette disorder: a review. J Obsessive-Compulsive Relat Disord 3: 380–385.

Peterson BS, Skudlarski P, Anderson AW et al. (1998). A functional magnetic resonance imaging study of tic suppression in Tourette syndrome. Arch Gen Psychiatry 55: 326–333.

Peterson BS, Thomas P, Kane MJ et al. (2003). Basal ganglia volumes in patients with Gilles de la Tourette Syndrome. Arch Gen Psychiatry 60: 415–424.

Peterson BS, Leckman JF, Tucker D et al. (2000). Preliminary findings of antistreptococcal antibody titers and basal ganglia volumes in tic, obsessive-compulsive, and attention-deficit/hyperactivity disorders. Arch Gen Psychiatry 57: 364–372.

Peterson BS, Leckman JF (1998). The temporal dynamics of tics in Gilles de La Tourette Syndrome. Biol Psychiatry 44: 1337–1348.

Pittenger C (2015). Chapter 47 – Animal models of Tourette syndrome and obsessive-compulsive disorder. In: LeDoux MS (Ed.), Movement disorders, 2nd edn. Boston, MA: Academic Press, pp. 747–764.

Pogorelov V, Xu M, Smith HR et al. (2015). Corticostriatal interactions in the generation of tic-like behaviors after local striatal disinhibition. Exp Neurol 265: 122–128.

Ponzio NM, Servatius R, Beck K et al. (2007). Cytokine levels during pregnancy influence immunological profiles and neurobehavioral patterns of the offspring. Ann N Y Acad Sci 1107: 118–128.

Posch DK, Schwabe K, Krauss JK et al. (2012). Deep brain stimulation of the entopeduncular nucleus in rats prevents apomorphine-induced deficient sensorimotor gating. Behav Brain Res 232: 130–136.

Price RA, Kidd KK, Cohen DJ et al. (1985). A twin study of Tourette syndrome. Arch Gen Psychiatry 42: 815–820.

Proenca CC, Gao KP, Shmelkov SV et al. (2011). Slitrks as emerging candidate genes involved in neuropsychiatric disorders. Trends Neurosci 34: 143–153.

Pryce CR, Feldon J (2003). Long-term neurobehavioural impact of the postnatal environment in rats: manipulations, effects and mediating mechanisms. Neurosci Biobehav Rev 27: 57–71.

Puts NAJ, Harris AD, Crocetti D et al. (2015). Reduced GABAergic inhibition and abnormal sensory symptoms in children with Tourette syndrome. J Neurophysiol 114: 808–817.

Ralph RJ, Paulus MP, Fumagalli F et al. (2001). Prepulse inhibition deficits and perseverative motor patterns in dopamine transporter knock-out mice: differential effects of D1 and D2 receptor antagonists. J Neurosci 21: 305–313.

Ralph RJ, Paulus MP, Geyer MA (2001). Strain-specific effects of amphetamine on prepulse inhibition and patterns of locomotor behavior in mice. J Pharmacol Exp Ther 298: 148–155.

Randrup A, Munkvad I (1967). Stereotyped activities produced by amphetamine in several animal species and man. Psychopharmacologia 11: 300–310.

Rebec GV, Bashore TR (1984). Critical issues in assessing the behavioral effects of amphetamine. Neurosci Biobehav Rev 8: 153–159.

Rizzo F, Nespoli E, Abaei A et al. (2018). Aripiprazole selectively reduces motor tics in a young animal model for Tourette's syndrome and comorbid attention deficit and hyperactivity disorder. Front Neurol 9: 59.

Roessner V, Overlack S, Baudewig J et al. (2009). No brain structure abnormalities in boys with Tourette's syndrome: a voxel-based morphometry study. Mov Disord 24: 2398–2403.

Roffman JL, Raskin LA (1997). Stereotyped behavior: effects of d-amphetamine and methylphenidate in the young rat. Pharmacol Biochem Behav 58: 1095–1102.

Rommeck I, Capitanio JP, Strand SC et al. (2011). Early social experience affects behavioral and physiological responsiveness to stressful conditions in infant rhesus macaques (*Macaca mulatta*). Am J Primatol 73: 692–701.

Rose SJ, Yu XY, Heinzer AK et al. (2015). A new knock-in mouse model of l-DOPA-responsive dystonia. Brain 138: 2987–3002.

Rowe DC, Stever C, Gard JM et al. (1998). The relation of the dopamine transporter gene (DAT1) to symptoms of internalizing disorders in children. Behav Genet 28: 215–225.

Rubinsztein DC (2002). Lessons from animal models of Huntington's disease. Trends Genet 18: 202–209.

Rusheen AE, Rojas-Cabrera J, Goyal A et al. (2023). Deep brain stimulation alleviates tics in Tourette syndrome via striatal dopamine transmission. Brain 146: 4174–4190.

Russell RL, Pihl RO (1978). The effect of dose, novelty, and exploration on amphetamine-produced stereotyped behavior. Psychopharmacology 60: 93–100.

Sabol SZ, Hu S, Hamer D (1998). A functional polymorphism in the monoamine oxidase a gene promoter. Hum Genet 103: 273–279.

Scannevin RH, Huganir RL (2000). Postsynaptic organisation and regulation of excitatory synapses. Nat Rev Neurosci 1: 133–141.

Scharf JM, Yu D, Mathews CA et al. (2013). Genome-wide association study of Tourette's syndrome. Mol Psychiatry 18: 721–728.

Scott AL, Bortolato M, Chen K et al. (2008). Novel monoamine oxidase a knock out mice with human-like spontaneous mutation. Neuroreport 19: 739–743.

Shapiro AK, Shapiro E, Wayne H (1973). Treatment of Tourette's syndrome: with haloperidol, review of 34 cases. Arch Gen Psychiatry 28: 92–97.

Silberberg G, Bolam JP (2015). Local and afferent synaptic pathways in the striatal microcircuitry. Curr Opin Neurobiol 33: 182–187.

Singer HS (2011). Tourette syndrome and other tic disorders. In: Weiner WJ and Tolosa E (Eds.), Hyperkinetic Movement Disorders. Handbook of Clinical Neurology, vol. 100. Elsevier, pp. 641–657.

Singer HS, Butler IJ, Tune LE et al. (1982). Dopaminergic dysfunction in Tourette syndrome. Ann Neurol 12: 361–366.

Singer HS, Szymanski S, Giuliano J et al. (2002). Elevated intrasynaptic dopamine release in Tourette's syndrome measured by PET. Am J Psychiatry 159: 1329–1336.

Singer HS, Hahn IH, Krowiak E et al. (1990). Tourette's syndrome: a neurochemical analysis of postmortem cortical brain tissue. Ann Neurol 27: 443–446.

Singer HS, Hahn IH, Moran TH (1991). Abnormal dopamine uptake sites in postmortem striatum from patients with Tourette's syndrome. Ann Neurol 30: 558–562.

Sipes TA, Geyer MA (1994). Multiple serotonin receptor subtypes modulate prepulse inhibition of the startle response in rats. Neuropharmacology 33: 441–448.

Smith SEP, Li J, Garbett K et al. (2007). Maternal immune activation alters fetal brain development through interleukin-6. J Neurosci 27: 10695–10702.

State MW (2010). The genetics of child psychiatric disorders: focus on autism and Tourette syndrome. Neuron 68: 254–269.

Stöber G, Hebebrand J, Cichon S et al. (1999). Tourette syndrome and the norepinephrine transporter gene: results of a systematic mutation screening. Am J Med Genet 88: 158–163.

Swedo SE, Leonard HL, Garvey M et al. (1998). Pediatric autoimmune neuropsychiatric disorders associated with streptococcal infections: clinical description of the first 50 cases. Am J Psychiatry 155: 264–271.

Swerdlow NR, Bongiovanni MJ, Tochen L et al. (2006). Separable noradrenergic and dopaminergic regulation of prepulse inhibition in rats: implications for predictive validity and Tourette syndrome. Psychopharmacology 186: 246–254.

Swerdlow NR, Karban B, Ploum Y et al. (2001). Tactile prepuff inhibition of startle in children with Tourette's syndrome: in search of an "FMRI-Friendly" startle paradigm. Biol Psychiatry 50: 578–585.

Swerdlow NR, Sutherland AN (2005). Using animal models to develop therapeutics for Tourette syndrome. Pharmacol Ther 108: 281–293.

Swerdlow NR, Sutherland AN (2006). Preclinical models relevant to Tourette syndrome. Adv Neurol 99: 69–88.

Szechtman H, Sulis W, Eilam D (1998). Quinpirole induces compulsive checking behavior in rats: a potential animal model of obsessive-compulsive disorder (OCD). Behav Neurosci 112: 1475–1485.

Tarnok Z, Ronai Z, Gervai J et al. (2007). Dopaminergic candidate genes in Tourette syndrome: association between tic severity and 3′ UTR polymorphism of the dopamine transporter gene. Am J Med Genet B Neuropsychiatr Genet 144B: 900–905.

Tarsy D, Pycock CJ, Meldrum BS et al. (1978). Focal contralateral myoclonus produced by inhibition of GABA action in the caudate nucleus of rats. Brain 101: 143–162.

Taylor JR, Morshed SA, Parveen S et al. (2002). An animal model of Tourette's syndrome. Am J Psychiatry 159: 657–660.

The Tourette's Syndrome Study Group(2002). Treatment of ADHD in children with tics: a randomized controlled trial. Neurology 58: 527–536.

Tremblay L, Worbe Y, Thobois S et al. (2015). Selective dysfunction of basal ganglia subterritories: from movement to behavioral disorders. Mov Disord 30: 1155–1170.

Vinner E, Belelovsky K, Bar-Gad I (2021). Generating acute and chronic experimental models of motor tic expression in rats. J Vis Exp 171.

Vinner E, Israelashvili M, Bar-Gad I (2017). Prolonged striatal disinhibition as a chronic animal model of tic disorders. J Neurosci Methods 292: 20–29.

Vinner Harduf E, Matzner A, Belelovsky K et al. (2021). Dissociation of tic generation from tic expression during the sleep-wake cycle. iScience 24: 102380.

Wan Y, Ade KK, Caffall Z et al. (2014). Circuit-selective striatal synaptic dysfunction in the Sapap3 knockout mouse model of obsessive-compulsive disorder. Biol Psychiatry 75: 623–630.

Wang Z, Maia TV, Marsh R et al. (2011). The neural circuits that generate tics in Gilles de la Tourette syndrome. Am J Psychiatry 168: 1326–1337.

Wehmeyer L, Schüller T, Kiess J et al. (2021). Target-specific effects of deep brain stimulation for Tourette syndrome: a systematic review and meta-analysis. Front Neurol 12: 769275.

Welch JM, Lu J, Rodriguiz R et al. (2007). Cortico-striatal synaptic defects and OCD-like behaviours in Sapap3-mutant mice. Nature 448: 894–900.

Willner P (1986). Validation criteria for animal models of human mental disorders: learned helplessness as a paradigm case. Prog Neuropsychopharmacol Biol Psychiatry 10: 677–690.

Wong DF, Brasić JR, Singer HS et al. (2008). Mechanisms of dopaminergic and serotonergic neurotransmission in Tourette syndrome: clues from an in vivo neurochemistry study with PET. Neuropsychopharmacology 33: 1239–1251.

Worbe Y, Baup N, Grabli D et al. (2009). Behavioral and movement disorders induced by local inhibitory dysfunction in primate striatum. Cereb Cortex 19: 1844–1856.

Worbe Y, Marrakchi-Kacem L, Lecomte S et al. (2015). Altered structural connectivity of cortico-striato-pallido-thalamic networks in Gilles de La Tourette syndrome. Brain 138: 472–482.

Worbe Y, Sgambato-Faure V, Epinat J et al. (2013). Towards a primate model of Gilles de La Tourette syndrome: anatomo-behavioural correlation of disorders induced by striatal dysfunction. Cortex 49: 1126–1140.

Xu M, Kobets A et al. (2015). Targeted ablation of cholinergic interneurons in the dorsolateral striatum produces behavioral manifestations of tourette syndrome. Proc Natl Acad Sci USA 112: 893–898.

Xu M, Li L et al. (2015). Histidine decarboxylase knockout mice, a genetic model of Tourette syndrome, show repetitive grooming after induced fear. Neurosci Lett 595: 50–53.

Yaddanapudi K, Hornig M, Serge R et al. (2010). Passive transfer of *Streptococcus*-induced antibodies reproduces behavioral disturbances in a mouse model of pediatric autoimmune neuropsychiatric disorders associated with streptococcal infection. Mol Psychiatry 15: 712–726.

Yael D, Tahary O, Gurovich B et al. (2019). Disinhibition of the nucleus accumbens leads to macro-scale hyperactivity consisting of micro-scale behavioral segments encoded by striatal activity. J Neurosci 39: 5897–5909.

Yael D, Zeef DH, Sand D et al. (2013). Haloperidol-induced changes in neuronal activity in the striatum of the freely moving rat. Front Syst Neurosci 7: 110.

Yael D, Israelashvili M, Bar-Gad I (2016). Animal models of Tourette syndrome—from proliferation to standardization. Front Neurosci 10: 132.

Yael D, Vinner E, Bar-Gad I (2015). Pathophysiology of tic disorders. Mov Disord 30: 1171–1178.

Yamashita M, Fukushima S, Shen HW et al. (2006). Norepinephrine transporter blockade can normalize the prepulse inhibition deficits found in dopamine transporter knockout mice. Neuropsychopharmacology 31: 2132–2139.

Yan Y, Tian M, Li M et al. (2022). ASH1L haploinsufficiency results in autistic-like phenotypes in mice and links Eph receptor gene to autism spectrum disorder. Neuron 110: 1156–1172.e9.

Yap CY, Taylor DA (1983). Involvement of 5-HT2 receptors in the wet-dog shake behaviour induced by 5-hydroxytryptophan in the rat. Neuropharmacology 22: 801–804.

Yartsev MM (2017). The Emperor's new wardrobe: rebalancing diversity of animal models in neuroscience research. Science 358: 466–469.

Yeh C-B, Shui HA, Chu TH et al. (2012). Hyperpolarisation-activated cyclic nucleotide channel 4 (HCN4) involvement in Tourette's syndrome autoimmunity. J Neuroimmunol 250: 18–26.

Yochum C, Doherty-Lyon S, Hoffman C et al. (2014). Prenatal cigarette smoke exposure causes hyperactivity and aggressive behavior: role of altered catecholamines and BDNF. Exp Neurol 254: 145–152.

Yoon DY, Rippel CA, Kobets AJ et al. (2007). Dopaminergic polymorphisms in Tourette syndrome: association with the DAT gene (SLC6A3). Am J Med Genet B Neuropsychiatr Genet 144B: 605–610.

Yu D, Sul JH, Tsetsos F et al. (2019). Interrogating the genetic determinants of Tourette syndrome and other tic disorders through genome-wide association studies. Am J Psychiatry 176: 217–227.

Zalcman S, Murray L, Dyck DG et al. (1998). Interleukin-2 and -6 induce behavioral-activating effects in mice. Brain Res 811: 111–121.

Zalcman SS (2002). Interleukin-2-induced increases in climbing behavior: inhibition by dopamine D-1 and D-2 receptor antagonists. Brain Res 944: 157–164.

Zalcman SS, Patel A, Mohla R et al. (2012). Soluble cytokine receptors (SIL-2Rα, SIL-2Rβ) induce subunit-specific behavioral responses and accumulate in the cerebral cortex and basal forebrain. PLoS One 7: e36316.

Zhang D, Patel A, Zhu Y et al. (2012). Anti-streptococcus IgM antibodies induce repetitive stereotyped movements: cell activation and co-localization with Fcα/μ receptors in the striatum and motor cortex. Brain Behav Immun 26: 521–533.

Zhuang X, Oosting RS, Jones SR et al. (2001). Hyperactivity and impaired response habituation in hyperdopaminergic mice. Proc Natl Acad Sci USA 98: 1982–1987.

Zohar J, Chopra M, Sasson Y et al. (2000). Obsessive compulsive disorder: serotonin and beyond. World J Biol Psychiatry 1: 92–100.

Handbook of Clinical Neurology, Vol. 215 (3rd series)
Tic Disorders and Tourette Syndrome
C. Ganos and Y. Worbe, Editors
https://doi.org/10.1016/B978-0-443-13554-5.00019-5

Chapter 17

Clinical neurophysiology of tics

ANNA LATORRE[1,*], MARIO MELONI[2], AND LORENZO ROCCHI[3]

[1]*Department of Clinical and Movement Neurosciences, UCL Queen Square Institute of Neurology, University College London, London, United Kingdom*

[2]*Department of Medical Sciences and Public Health, Institute of Neurology, University Hospital of Cagliari, Cagliari, Italy*

[3]*Department of Medical Sciences and Public Health, University of Cagliari, Cagliari, Italy*

Abstract

Clinical neurophysiology constitutes a relevant tool for clinicians in movement disorders since it expands their characterization beyond clinically observable features and supports their diagnosis. Although the field has yet to identify a specific neurophysiologic marker for the diagnosis of tics, neurophysiology serves as a valuable tool for clinicians in differentiating tics from other hyperkinetic disorders. Beside diagnosis, recent years have witnessed the emergence of two primary domains in which neurophysiology plays a pivotal role in the realm of tic disorders: investigating the pathophysiologic foundations of tics, in an attempt to unravel the neural mechanisms that give rise to this complex movement disorder, and the application of noninvasive brain stimulation (NIBS) techniques as an avenue for therapeutic intervention. The two domains are inherently connected, as the development of noninvasive treatment methods is grounded in the mechanisms believed to underlie tics. Although the precise neurobiology of tic disorders remains incompletely understood, growing evidence implicates dysfunction within cortical-striatal-thalamic-cortical circuits as a key contributor to their pathophysiology. In this context, NIBS has been explored as a potential therapeutic approach aimed at modulating pathologic cortical hyperexcitability in specific regions, thereby enhancing inhibitory circuit engagement and ultimately alleviating tics. An intriguing alternative to transcranial stimulation is peripheral electrical nerve stimulation, which offers a distinct yet complementary mechanism of action. For instance, studies have demonstrated that rhythmic stimulation of the median nerve within the alpha frequency band can entrain neuronal oscillations in the sensorimotor cortex, leading to a measurable reduction in tic frequency and intensity.

In this chapter, we will first present a comprehensive overview of clinical neurophysiology studies that have significantly contributed to our understanding of the pathophysiology of tics, and then we will offer a brief synopsis of the application of noninvasive stimulation for the treatment of tics, illustrating that clinical neurophysiology is a powerful tool that provides valuable insights into the neural underpinnings of tics and holds potential as adjunct therapy for tic management.

INTRODUCTION

Clinical neurophysiology constitutes a relevant tool in movement disorders since it expands their characterization beyond clinically observable features and supports their diagnosis. Its application is particularly useful in hyperkinetic disorders, which can be regular and rhythmic, like tremor; more sustained and patterned, like dystonia; brief and flowing, like chorea; abrupt and jerky, like myoclonus; or sudden, repetitive, stereotyped, and temporarily suppressible, like tics. When these characteristics are subtle and not easily discernible through clinical observation alone, neurophysiologic techniques can come to the fore.

Tics, in their phenomenological essence, are characterized by sudden, repetitive, nonrhythmic movements or vocalizations. These may be brief and rapid (clonic) or result from prolonged contractions, occasionally leading to abnormal postures (tonic/dystonic). In this context, neurophysiology serves as a valuable tool for clinicians for differentiating tics from other hyperkinetic disorders, although the field has yet to identify a specific neurophysiologic marker for the diagnosis of tics.

Beside diagnosis, recent years have witnessed the emergence of two primary domains in which neurophysiology plays a pivotal role in the realm of tic disorders, which will be the

*Correspondence to: Dr. Anna Latorre, Department of Clinical and Movement Neurosciences, UCL Queen Square Institute of Neurology, University College London, 33 Queen Square, London, WC1N 3BG. E-mail: a.latorre@ucl.ac.uk.

primary focus of this chapter. The first involves delving into the pathophysiologic foundations of tics, in an attempt to unravel the neural mechanisms that give rise to this complex movement disorder. The second revolves around the application of noninvasive brain stimulation (NIBS) techniques as an avenue for therapeutic intervention.

PATHOPHYSIOLOGY OF TICS

In this section, we will present a comprehensive overview of clinical neurophysiology studies that have significantly contributed to our understanding of the pathophysiology of tics. Evidence will be organized according either to the primary anatomic and physiologic subdivisions of the motor system that can be investigated through noninvasive electrophysiologic recordings (e.g., motor units, brainstem circuitry, cortex) or by the neurophysiologic technique used for the investigations, such as transcranial magnetic stimulation (TMS) and electroencephalography (EEG).

Motor unit involvement in tic generation

Although tics lack the flexibility of voluntary actions and are misplaced in context and time, they frequently look like voluntary movements. This resemblance extends even to electromyography (EMG), where surface recording of tics reveals burst durations typically ranging from 50 to 200 ms, or more (Fig. 17.1); these bursts can exhibit a triphasic pattern akin to that seen in voluntary ballistic movements (Hallett, 2002, 2003). In the case of dystonic or complex tics, burst durations might extend even further or manifest more intricate patterns on EMG. However, these do not possess specific features that readily distinguish them from voluntary actions or, for instance, from dystonic movements (Hallett, 2002, 2003). Nevertheless, a recent study has demonstrated that the power spectral density derived from a combination of EMG and accelerometer data, analyzed through an objective algorithm, could achieve 96.69% accuracy in differentiating tics from voluntary actions (Cernera et al., 2022).

Although techniques like EMG have not been extensively applied to investigate the behavior and properties of motor units in tics, primarily due to their similarities with voluntary movements, EMG holds potential clinical utility in distinguishing tics from other hyperkinetic disorders. For instance, unlike tics, certain types of myoclonus are characterized by very brief burst durations, often less than 50 ms, and can even manifest as negative bursts (Latorre et al., 2022). In contrast, EMG bursts in chorea have varying lengths in random muscles (Bonomo et al., 2020). These distinctions underscore the value of EMG as a diagnostic tool in clinical practice.

Brainstem dysfunction in tic disorders

Several pieces of evidence suggest that the dopaminergic system is crucially involved in the pathophysiology of tics (Ganos et al., 2021). The spontaneous blink rate is often abnormal in disorders of dopamine (Fitzpatrick et al., 2012) and the increased rate observed in patients with tics aligns with this observation (Tulen et al., 1999; Tharp et al., 2015). Notably, eye-blinking is among the most common simple tics (Nilles et al., 2023). These factors

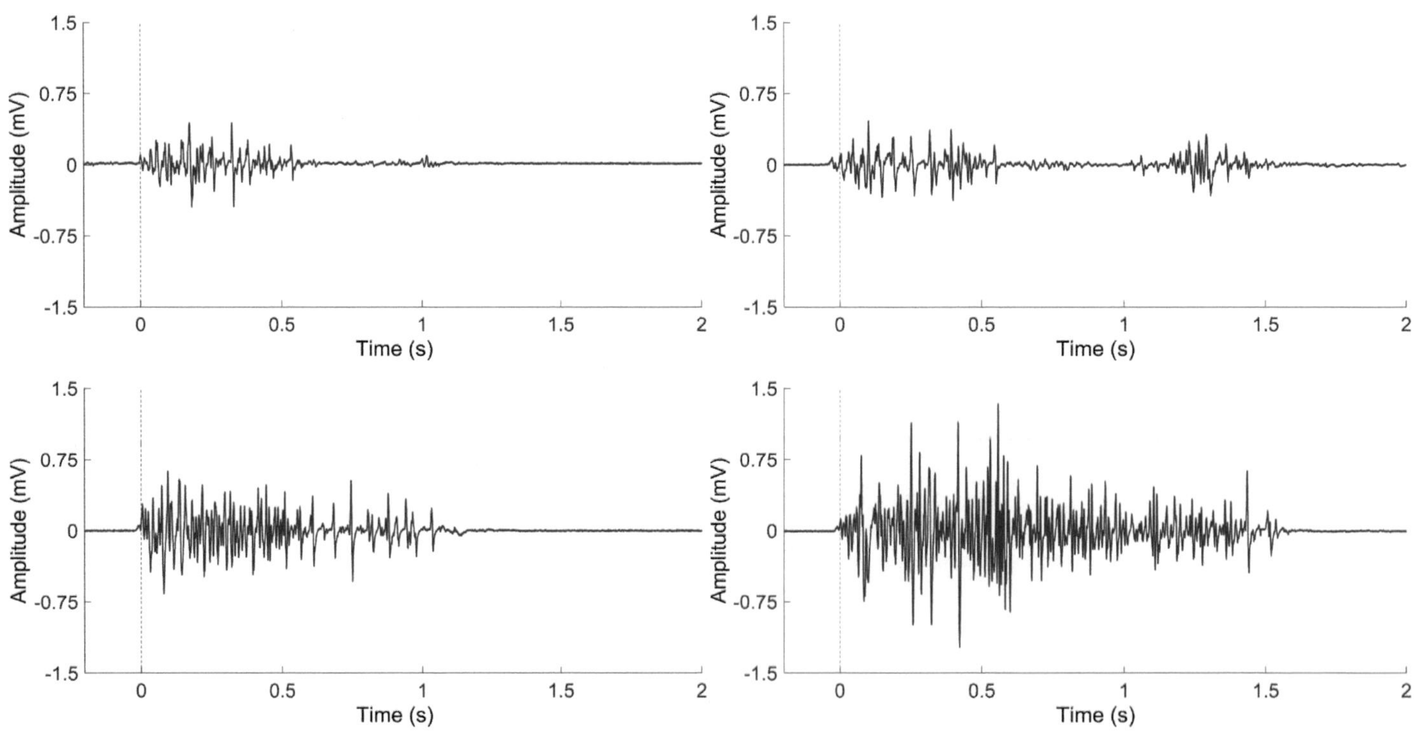

Fig. 17.1. Examples of electromyography activity related to tics.

collectively suggest a potential brainstem dysfunction in tic disorders, a notion supported by neurophysiologic investigations of the blink reflex. One accessible method for studying the blink reflex and the brainstem inhibitory circuitry responsible for its generation is the blink reflex recovery cycle (BRC). This test examines the recovery curve or excitability cycle of the blink reflex circuitry through a double pulse technique. After an initial input, the excitability of a reflex circuit tends to decrease (Aramideh et al., 2003). The BRC involves single or double electrical pulses applied at various interstimulus intervals (ISIs) over the supraorbital nerve, with the response recorded via surface EMG of the orbicularis oculi muscle. The first shock is called "conditioning" and the response to the second shock is called "test" response. Stimulation of the supraorbital nerve elicits two responses: the brief unilateral early reflex, R1, occurring at approximately 10 ms latency in the muscle ipsilateral to the stimulation site, and the longer-duration, bilateral late response, R2, with a latency of about 30 ms, corresponding to eyelid closure. In a normal recovery curve, the test R2 is typically completely abolished at ISIs shorter than 200 ms. At higher ISIs, it gradually recovers, reaching roughly 40%–50% at the 500 ms interval and 70%–90% at the 1500 ms interval. In Tourette syndrome (TS), studies have shown that the duration of R2 is prolonged (Smith and Lees, 1989; Raffaele et al., 2006), while the suppression of R2 is reduced in some individuals, irrespective of the distribution of tics (Smith and Lees, 1989). Interestingly, one study found that voluntary suppression of blinks in control subjects and both blinks and tics in patients led to a reduction in the R2 amplitude of the test stimulus (Smith and Lees, 1989). Additionally, it has been demonstrated that patients with tics exhibit no response to plasticity protocols involving the brainstem circuitry responsible for the blink reflex (Suppa et al., 2011). This could indicate compensatory mechanisms aimed at suppressing tics.

Given the evident involvement of the brainstem in tics, as evidenced by abnormal BRC in some subjects and the observation that excessive startle reflex is part of this syndrome, researchers have explored the audiogenic startle reflex in a few studies. However, results have been inconsistent, possibly due to variations in methodology for eliciting the startle reflex or the relatively small number of patients studied. While one study reported normal startle reflex responses in TS (Sachdev et al., 1997) and another noted a lack of habituation in two out of five patients (Stell et al., 1995), a more recent study found a greater amplitude of response, increased spread, and reduced habituation. This was observed alongside normal latencies of the startle response and muscle activation patterns compared to healthy controls (Gironell et al., 2000). According to the authors, this suggests a normal function of neural pathways mediating the reflex, but a disinhibition of the reflex mediated by the hyperdopaminergic state (Gironell et al., 2000).

Insights into the role of the motor cortex in tic generation by transcranial magnetic stimulation

From a pathophysiologic perspective, tics are very likely associated with abnormal processing in the basal ganglia-thalamo-cortical circuits, leading to abnormal activation of circuits responsible for voluntary movement within the primary motor cortex (M1) (Orth and Münchau, 2013). Consequently, exploring in vivo electrophysiologic aspects of the human M1 may provide valuable insights into the underlying mechanisms of tics. With this premise in mind, numerous researchers have endeavored to characterize M1 physiology in individuals with TS utilizing TMS.

TMS involves the use of coils that receive a powerful current, rapidly discharged from capacitors, thereby generating magnetic fields perpendicular to the coil's plane. These magnetic fields can depolarize cortical neurons through the induction of a secondary electric field (Hallett, 2007). In the ensuing paragraphs, we present a concise overview of the pathophysiologic findings obtained through diverse TMS protocols applied to the M1 region in TS patients.

Assessing net excitability of the primary motor cortex: resting motor threshold, active motor threshold, input/output relationship

The primary outcome measure commonly assessed following a single TMS pulse applied to the M1 is the motor evoked potential (MEP). This EMG response represents the muscle twitch triggered by M1 activation (Rothwell et al., 1987). While there are some caveats to consider (Rawji et al., 2020), the MEP amplitude is generally regarded as a reflection of the overall excitability of corticospinal tract neurons and associated interneurons (Spampinato et al., 2023). A straightforward derivative of MEP analysis involves the measurement of "thresholds," which quantify the stimulus intensities needed to elicit an MEP of a predetermined amplitude, typically ranging between 50 and 200 μV. These thresholds can be assessed both at rest (resting motor threshold, RMT) and during voluntary muscle contraction (active motor threshold, AMT) (Rossini et al., 2015). In adults with TS, RMT and AMT are generally considered normal (Ziemann et al., 1997; Orth et al., 2005a; Heise et al., 2008; Orth et al., 2008) (Fig. 17.2A and B). However, it has been observed that RMT is higher in TS compared to healthy controls, particularly in children, suggesting a developmental delay in the maturation of key brain networks (Pépés et al., 2016).

While simple thresholds provide some insight into M1 excitability at low-stimulation intensity, a more comprehensive understanding can be gained by increasing the stimulation intensity. This enables the examination of a broader population of M1 neurons and facilitates more accurate inferences about input/output (I/O) relationships (Fig. 17.2C). When a larger M1 neuron population is engaged, a consensus emerges regarding lower MEP amplitude in TS compared to healthy controls, evident in studies exploring single TMS intensities (Larsh et al., 2023) and I/O curves (Orth et al., 2008; Pépés et al., 2016). This hypoexcitability of M1 circuits has been linked to various clinical aspects of tics, including urge intensity (Larsh et al., 2023) and the severity of hand/finger, complex, and vocal tics (Orth et al., 2008).

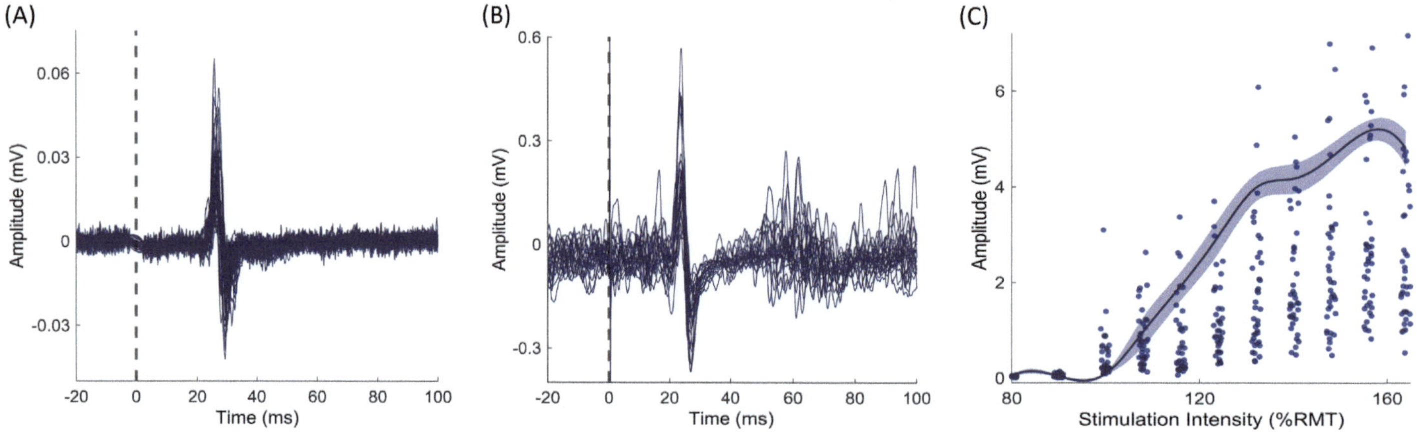

Fig. 17.2. (**A**) Resting motor threshold (RMT). (**B**) Active motor threshold (AMT). (**C**) Input/output curve. For RMT and AMT, every blue line represents a motor evoked potential (MEP), and black dashed vertical lines correspond to the transcranial magnetic stimulation pulse. In Panel **C**, the curved blue line is the interpolated average of MEPs, the shaded area is the standard error of the mean, and each dot indicates the amplitude of a single MEP.

Employing TMS over multiple points on the scalp corresponding to the M1 allows the assessment of motor cortical representations through motor mapping. An important metric obtained from this technique is the "centre of gravity," representing the location most likely to elicit the largest MEP in a specific muscle (Sondergaard et al., 2021). Notably, a study suggested that, despite substantial similarity in motor maps between TS and healthy individuals, the intensity of premonitory urges is negatively correlated with the distance between the center of gravity of facial and hand muscles and positively correlated with the area of the latter. This suggests that the loss of spatial specificity in cortical muscle representation might be associated with specific aspects of the pathophysiology of TS (Sigurdsson et al., 2020).

Assessment of facilitatory and inhibitory intracortical circuitry: SICI, LICI, CSP, ICF

One limitation of using MEPs obtained from single-pulse TMS lies in their ability to provide information solely about the overall excitability of M1 (Spampinato et al., 2023). To overcome this constraint and investigate specific M1 circuits, researchers have explored alternative TMS techniques. One approach involves applying TMS during voluntary muscle contraction to elicit the cortical silent period (CSP), an interruption of ongoing EMG activity mediated by GABAb-mediated neurotransmission. The CSP allows for the exploration of specific inhibitory inputs to corticospinal neurons (Valls-Solé et al., 1992; Hupfeld et al., 2020) (Fig. 17.3). Another method, known as paired-pulse TMS, employs two rapid-succession stimuli applied via the same coil, enabling the investigation of distinct inhibitory or excitatory intracortical circuits by varying the ISI and intensity of the conditioning stimulus.

Paradigms such as short-interval intracortical inhibition (SICI), intracortical facilitation (ICF), and long-interval intracortical inhibition (LICI) are used to probe GABAa-mediated inhibition (Kujirai et al., 1993; Di Lazzaro and Ziemann, 2013), glutamatergic facilitation (Ziemann et al., 1996; Di Lazzaro and Ziemann, 2013), and GABAb-mediated inhibition (Chen, 2004; Valls-Solé et al., 1992), respectively (Fig. 17.3). Somatosensory interactions within M1 can be probed by pairing electrical stimulation of a peripheral nerve with M1-TMS, which results in a form of MEP suppression named short-latency afferent inhibition (SAI) (Tokimura et al., 2000; Hwang et al., 2018), thought to be mediated by cholinergic neurotransmission (Ziemann et al., 2015).

When examining M1 circuits in TS, intracortical facilitatory interactions are generally considered normal (Ziemann et al., 1997; Moll et al., 1999; Heise et al., 2010; Batschelett et al., 2023), except for one study where they were increased (Orth and Rothwell, 2009). In contrast, M1 inhibition has garnered more attention, with LICI suggested to be decreased and negatively correlated with tic urges (Larsh et al., 2023). Several studies have reported a decrease in CSP duration in TS (Ziemann et al., 1997; Orth et al., 2008; Heise et al., 2010). Ziemann et al. (1997) found reduced CSP particularly in patients with distal tics involving the EMG target muscle of the hand and in those not treated with neuroleptics. However, subsequent studies only partially replicated these findings, as the CSP obtained from hand muscles was reduced regardless of tic distribution (Moll et al., 1999). The discrepancy may be attributed to age differences: further evidence suggests that CSP is increased in children irrespective of tic distribution, indicating a generalized inhibitory deficit, which later manifests only in neuronal pathways within the sensorimotor circuit controlling the muscle groups affected by tics (Moll et al., 2001, 2006).

Similar to CSP, SICI appears to serve as a reliable marker of defective inhibitory interactions in TS (Ziemann et al., 1997; Orth et al., 2005a; Heise et al., 2010), particularly in patients with distal tics (Ziemann et al., 1997). SICI has been reported to be negatively correlated with tic severity (Gilbert et al., 2004, 2005; Batschelett et al., 2023). However, some studies have reported SICI as normal (Moll et al., 1999; Batschelett et al.,

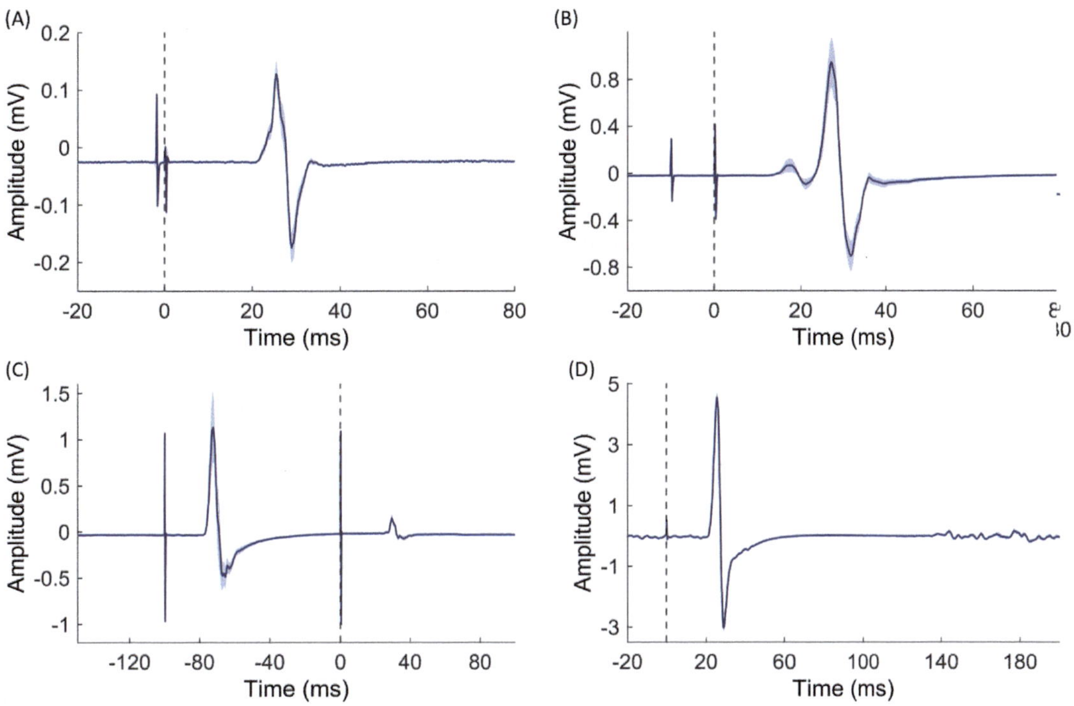

Fig. 17.3. (**A**) SICI (ISI 3 ms. TS~1 mV, CS 80% RMT). (**B**) ICF (ISI 10 ms, TS~1 mV, CS 80% RMT). (**C**) LICI (TS~1 mV, CS 120% RMT, ISI 100 ms). (**D**) CSP (TS~1 mV, EMG~15%–20% maximal voluntary contraction). Black dashed vertical lines correspond to the test TMS pulse. *CS*, conditioning stimulus; *CSP*, cortical silent period; *EMG*, electromyography; *ICF*, intracortical facilitation; *ISI*, interstimulus interval; *LICI*, long intracortical inhibition; *RMT*, resting motor threshold; *SICI*, short-interval intracortical inihibition; *TS*, test stimulus.

2023). This discrepancy may be attributed to the presence of decreased SICI only when attention-deficit/hyperactivity disorder (ADHD) coexists with TS (Moll et al., 1999; Batschelett et al., 2023), as supported by SICI reduction in ADHD, independent of TS presence (Moll et al., 1999, 2000, 2001; Batschelett et al., 2023).

Compared with healthy adults, subjects with TS have been found to have reduced SAI, and this difference is normalized by exposure to nicotine, which corresponds to amelioration of tics (Orth et al., 2005a; Orth and Rothwell, 2009). With regard to comorbidities, SAI was found to be reduced in patients with noncomplicated TS and TS associated with ADHD, but not in those with TS with obsessive-compulsive disorder (OCD) (Orth and Rothwell, 2009). Given the possible influence of sensory inputs in triggering the release of tics, reduced SAI may be a direct physiologic reflection of increased access of sensory input to motor output in TS.

A general conclusion drawn from these studies is that the gain of many M1 circuits is reduced in TS, resulting in decreased sensitivity to small changes in input from other areas. This might be an adaptive response to abnormal inputs from the basal ganglia, preventing release of unwanted movements (Orth et al., 2008). However, an important limitation of these investigations is the lack of control for the occurrence of tics, where EMG activation could potentially introduce bias in some TMS measures, such as SICI (Fong et al., 2021). There may also be times when TS patients experience no overt tics but still have sensory urges or are actively trying to suppress tics. These variations could indicate that the M1 is not consistently in the same state, influencing its responses to TMS (Orth and Münchau, 2013). Addressing these factors in an experimental setting proves challenging. To this end, Ganos et al. (2018) compared M1 physiology in TS patients when they were free to tic versus during voluntary tic inhibition (VTI). The authors observed a substantial reduction in MEP amplitude during VTI, suggesting that the previously observed reduced M1 excitability could potentially reflect state-dependent effects of VTI, either in place of or in addition to a trait reduction of excitability.

Cortical plasticity

In addition to its capability to probe M1 excitability, TMS also offers the possibility to induce cortical plasticity through repetitive transcranial magnetic stimulation (rTMS), typically employing regular or patterned trains of TMS pulses (Latorre et al., 2019; Jannati et al., 2023). When applied to the M1, the resultant changes in excitability depend on the intensity and frequency of the stimulation. For example, rTMS delivered at 1 Hz effectively suppresses MEPs (Dayan et al., 2013). Theta burst stimulation (TBS), operating by administering bursts of TMS pulses, is another method to induce plasticity. Continuous TBS (cTBS) typically leads to decreased cortical excitability (Huang et al., 2005; Nardella et al., 2014; Georgiev et al., 2016; Méndez et al., 2017), while intermittent TBS (iTBS), where rTMS bursts are interspersed with pauses, generally increases cortical excitability (Huang et al., 2005; Tremblay et al., 2017).

Paired associative stimulation (PAS) is an alternative plasticity protocol that repeatedly pairs M1 input derived from stimulation of peripheral nerves or other cortical areas with direct M1 stimulation via TMS (Stefan et al., 2000; Suppa et al., 2015a,b; Conte et al., 2016); its effects are believed to operate through spike-timing-dependent plasticity mechanisms (Wolters et al., 2003; Markram et al., 2012).

In two separate studies, TS patients exhibited no MEP modulation following both cTBS and iTBS, indicating a deficiency in long-term potentiation (LTP)-like and long-term depression (LTD)-like plasticity (Marsili et al., 2017), regardless of the presence of OCD (Suppa et al., 2014). Abnormal cortical plasticity induced by sensorimotor interactions in TS was corroborated by a study that reported a lack of MEP enhancement after PAS involving electrical nerve stimulation (Brandt et al., 2014), and by another study which suggested an excessively pronounced suppression of somatosensory evoked potentials after 1 Hz rTMS applied to the M1 hand area (Sun et al., 2020). While plasticity depending on homosynaptic M1 mechanisms and sensorimotor interactions seems deranges, plasticity involving other cortico-cortical connections may be normal in TS. This is supported by a study demonstrating normal after-effects on MEPs induced by a form of PAS involving conditioning of the supplementary motor cortex (Tübing et al., 2018).

The contribution of electroencephalography to the study of tics

In the context of hyperkinetic disorders, tics pose a unique challenge due to their potential for voluntary suppression. This distinctive feature blurs the line between voluntary and involuntary movements, leading to the introduction of the term "unvoluntary," used when an action is perceived as a voluntary response to an uncontrolled and involuntary urge to move (Cavanna and Nani, 2013). To delve into this dual nature, researchers have extensively explored movement-related cortical potentials (MRCPs) in the study of tics. MRCPs are EEG potentials that manifest around the time of movement, offering insights into the cortical activity underlying both typical and aberrant actions. To enhance the signal-to-noise ratio, the technique of EEG averaging, time-locked to EMG activity, is employed, with the onset of the movement serving as the timing event. During voluntary actions, the earliest detectable MRCP is known as the Bereitschaftspotential (BP). It exhibits a gradual negative trend, commencing approximately 2 seconds prior to movement onset. This negativity is symmetrical and broadly distributed across the scalp, with maximal amplitude at the vertex. As the movement onset approaches, roughly 400–650 ms prior, the negative slope becomes more pronounced over the central region, forming what is termed the late component. This late component reaches its peak about 100 ms after the initiation of movement (Fig. 17.4). The early component, characterized by a shallower slope, is believed to indicate movement preparation and originates from the supplementary motor area (SMA), presupplementary motor area (pre-SMA), and the lateral premotor cortex bilaterally (Utianski et al., 2016). In contrast, the late BP, marked by a steeper slope, is linked to the execution phase of the movement and has its source in the primary motor cortex and lateral premotor cortex (Utianski et al., 2016). While other postmovement MRCPs have been identified (Shibasaki et al., 1980), the BP has garnered the most attention in tic research.

In the original study conducted by Obeso et al.(1981), it was shown that tics were not preceded by BP, while mimicked tics were accompanied by premotor potentials. Following studies, however, have reported that tics can indeed be preceded by BP, albeit with a short duration falling within the range of late BP, or by normal BP (Karp et al., 1996; van der Salm et al., 2012; Panyakaew et al., 2016; Triggiani et al., 2023). It has been suggested that this differences in premotor potentials may be related to the variability of the premonitory urge (Triggiani et al., 2023).

MRCP, and in particular BP, have also been used to monitor the cerebral activity related to cognitive events. For instance, patients affected by chronic tic disorders, as opposed to controls, fail to demonstrate a relationship between motor

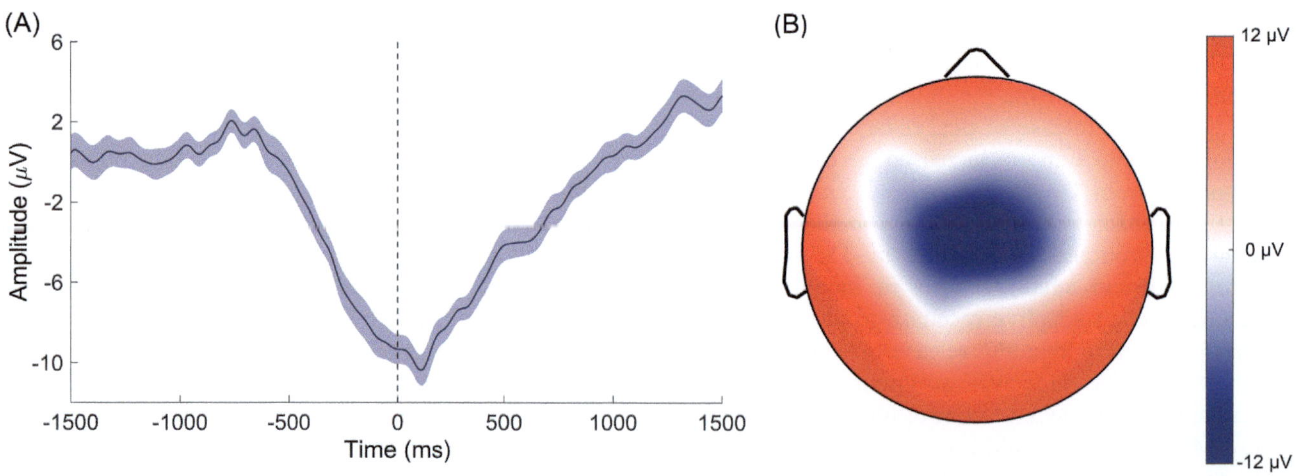

Fig. 17.4. Bereitschaftspotential. (**A**) Average signal of 50 voluntary extension movements of the right wrist, recorded at Cz. The black dashed vertical line indicates the onset of movement. (**B**) Scalp voltage topography of the same condition at time 0, showing a large vertex negativity.

output and preparation of cortical activation (i.e., BP) during a reaction time task (O'Connor et al., 2001), supporting the idea that patients with tic disorders may not be able to fine-tune cortical activation when planning and executing motor responses. Additionally, studies employing the lateralized readiness potential, which originates from generator sources in M1 (Requin and Riehle, 1995), the SMA (Rektor, 2002), and the basal ganglia (Rektor et al., 2003), have suggested that these abnormal readiness potentials may reflect a quicker retrieval of required motor programs, possibly associated with an over-activation of motor cortical areas (Thibault et al., 2009).

The BP has also been employed to investigate whether the timing of the intention to move and the perception of movement differs in individuals with tics. Studies have supported the idea that the awareness of movement lags behind the intention to move in healthy individuals (Braun et al., 2021). Interestingly, it has been observed that during tics, as well as during voluntary actions in healthy controls, the intention to move follows the onset of the BP but precedes the conscious sense of movement. This suggests that individuals with tics maintain a normal sense of agency (Triggiani et al., 2023). In this study, alongside the BP, another EEG marker of sensorimotor cortical activity during voluntary movement was utilized, namely event-related power changes in the beta frequency band (13–30 Hz), which typically occur during the planning and execution of voluntary movements. The authors found no substantial decrease in power (known as event-related desynchronization or ERD) in the beta band preceding tics, which is believed to be associated with an attempt to suppress tics (Triggiani et al., 2023). Another study revealed normal beta-band ERD before the onset of movement in individuals with TS, but an absence of mu-band ERD, which was clearly evident in control subjects at the onset of movement. Furthermore, prior to the onset of tic movements, there was no evidence of either beta- or mu-band ERD (Morera Maiquez et al., 2022). Both of these frequency bands have been linked to the initiation of voluntary movements; the beta band is associated with the disinhibition of neuronal populations related to movement, while mu-band oscillations are believed to play a crucial role in inhibiting neuronal populations that might otherwise interfere with the selection of movements. These findings have been interpreted as indicative of a dysfunction of inhibition in TS, contributing to an impaired ability to suppress neuronal populations that may compete with the processes involved in movement preparation (Morera Maiquez et al., 2022).

Abnormal patterns of beta oscillations in individuals with tics have also been observed in a magnetoencephalography study, revealing a reduction in beta power over sensorimotor regions during both premovement and actual movement phases in patients with tics while performing a go/no-go task. The authors of this study suggested that this reduction might reflect a compensatory process necessary to regulate tics and premonitory urges (Niccolai et al., 2019). However, this hypothesis was challenged by evidence indicating that abnormal beta rhythm recorded over the contralateral motor cortex during the execution of a voluntary movement returns to normal levels during tic suppression (Zapparoli et al., 2019).

Other event-related potentials have been associated with mechanisms of motor inhibition and have been explored in individuals with tics. The P3 (P300), for instance, is believed to represent specific aspects of motor inhibition. It is characterized by a frontal midline positive deflection with a peak occurring between 250 and 500 ms, and both its amplitude and latency have been implicated in response inhibition (Kok et al., 2004; Wessel and Aron, 2015). Patients with TS exhibit a smaller P300 during tasks that require participants to occasionally interrupt an ongoing response in favor of an alternative response. This suggests that patients respond more accurately than healthy controls in trials requiring an immediate response change following the inhibition of a previously prepared response. It also implies that they integrate multisensory information more efficiently during response selection processes (Brandt et al., 2017). This finding may be explained by the relative hyperconnectivity observed in fronto-striatal networks in individuals with tic disorders (Ganos et al., 2013; Worbe et al., 2015). Fronto-striatal networks play a crucial role in action selection, multicomponent behavior, goal-directed inhibition, and habitual inhibition (Jahanshahi et al., 2015); dysfunction in these networks seems to be implicated in both the generation and inhibition of tics (Worbe et al., 2015).

EEG studies have revealed compensatory mechanisms that can ameliorate this fronto-striatal dysfunction. One study demonstrated increased EEG coherence between sensorimotor areas and the prefrontal and mesial frontal cortex during the voluntary suppression of tics, as well as enhanced fronto-mesial EEG coherence during the suppression of voluntary movements in individuals with TS compared to healthy subjects. This suggests that the dynamic circuit involved in normal inhibition is over-active in TS and can be further activated acutely during the voluntary suppression of tics (Serrien et al., 2005). Building upon this research and considering that alpha oscillations exhibit high coherence over large distances (Nunez et al., 2001), a recent study investigated the global topology of whole-brain connectivity during tic suppression, with a focus on the alpha band. The authors found increased connectivity among many cortical areas during tic suppression, including the right superior frontal gyrus and the left precuneus, both of which are part of the default mode network (Morand-Beaulieu et al., 2023).

An intriguing recent application of EEG in this context involves the measurement of 1/*f* noise, also known as scale-free neural activity, and aperiodic activity. This is done to test the hypothesis suggesting that noise during sensorimotor processing is heightened in patients with TS, and that tics and urges may be attributed to the processing of neural noise and adjustments in the signal-to-noise ratio during information processing (Münchau et al., 2021). The term "1/*f* noise" is often used to describe "pink noise" or scale-free neural activity (He, 2014). It is specifically associated with the organizational patterns relevant to information processing and brain functioning (He, 2014; Voytek and Knight, 2015), which undergo modulation during sensorimotor processes (Pertermann et al.,

2019) and can be measured using EEG data. On the other hand, aperiodic oscillatory activity, which exhibits a 1/*f*-like distribution (Donoghue et al., 2020), refers to irregular activity that does not align with any specific frequency band, similar to 1/*f* activity. Studies have demonstrated an increase in 1/*f* noise and aperiodic activity during sensorimotor processing in patients with TS, potentially linked to abnormalities in GABAergic and dopaminergic transmission (Adelhöfer et al., 2021; Münchau et al., 2021). Notably, these differences observed between individuals with TS and healthy controls were driven by high-frequency oscillations rather than lower-frequency activity, which is believed to be significant in the pathophysiology of tics. This body of evidence lends support to the "neural noise account" in TS. However, the absence of a correlation between neural noise and clinical measures suggests that this phenomenon may not be directly involved in tic generation. Instead, it seems to represent a tic-independent but disease-relevant information processing mode in TS (Adelhöfer et al., 2021).

Noninvasive brain stimulation as a possible treatment avenue for tics

The treatment landscape for tics primarily encompasses pharmacologic and behavioral therapies. However, recent advancements have introduced deep brain stimulation as a promising option for patients grappling with severe, drug-resistant tics, often accompanied by conditions such as OCD (Coulombe et al., 2018). In response to the growing demand for nonpharmacologic interventions, researchers have turned to NIBS techniques. NIBS carries potential advantages, including minimal adverse effects when administered and monitored appropriately, as well as its noninvasive nature. The strategic use of NIBS in treatment aims to complement traditional approaches, particularly in cases where symptoms prove resistant to conventional therapies, or to mitigate the side-effects associated with certain medications. In the following paragraphs, we offer a brief overview of the application of NIBS for the treatment of tics or TS (detailed review in Chapter 25).

Repetitive transcranial magnetic stimulation

As discussed earlier, TMS serves various purposes, including the assessment of cortical excitability, intracortical circuits, and cortical plasticity in individuals with tics in comparison to healthy individuals. However, another intriguing application of TMS is its potential as a therapeutic tool, specifically to alleviate symptoms of tics or associated neuropsychiatric conditions.

The therapeutic effects of NIBS, like rTMS, are believed to arise from induced changes in synaptic plasticity (Latorre et al., 2019). Synaptic plasticity refers to the brain's capacity to reorganize its structure and function based on experience, ultimately resulting in behavioral changes. This process relates to adjustments in synaptic strength that either enhance or weaken the postsynaptic neural response. Additionally, synapses can undergo long-term modifications in their transmission efficiency, which can lead to persistent alterations in the morphology of neurons, particularly dendritic spines – a phenomenon known as structural plasticity (Bosch and Hayashi, 2012). Functional and structural plasticity processes occur continuously and simultaneously, giving rise to phenomena like LTP and LTD, that are believed to represent the underpinnings of learning and memory. LTP involves persistent enhancements in synaptic transmission, while LTD entails the weakening of synaptic efficacy (Lüscher and Malenka, 2012).

Numerous studies provide evidence that rTMS can induce plasticity akin to LTP and LTD in humans (Hoogendam et al., 2010). Specifically, when applied over M1, rTMS can bring about medium-term changes in the amplitude of MEPs that endure beyond the stimulation period (Chen et al., 1997). High-frequency rTMS (e.g., 10–20 Hz) tends to increase MEPs, while low-frequency rTMS (e.g., 1 Hz) reduces MEPs, reflecting heightened or diminished cortical excitability (He et al., 2020). These effects are thought to be associated with NMDA receptors, which are known to play a role in LTP and LTD processes (Huang et al., 2008; Ziemann et al., 2008; Hoogendam et al., 2010). Importantly, the effects of rTMS persist for several minutes after the stimulation period, shedding light on the plasticity and behavior of the stimulated brain region (Dayan et al., 2013).

A more recent rTMS protocol is TBS, which involves short bursts of high-frequency stimulation delivered at predefined repetition rates. TBS can be administered continuously (cTBS) or intermittently (iTBS). cTBS, with its three pulses at 50 Hz applied at 5 Hz for either 20 seconds (300 total stimuli) or 40 seconds (600 stimuli), induces MEP inhibition, whereas iTBS, comprising 22-second trains of cTBS separated by 8 seconds, leads to MEP facilitation (Huang et al., 2005). This technique enables rapid and efficient induction of neuroplasticity changes.

rTMS offers a versatile means of modulating cortical activity, and researchers have developed specific stimulation protocols designed to address the unique pathologic changes associated with various neurologic disorders. While the precise neurobiology underlying tic disorders, such as TS, remains incompletely understood, mounting evidence implicates dysfunction within cortical-striatal-thalamic-cortical circuits in their pathophysiology (Greene et al., 2015). NIBS methods, including rTMS, have been explored as a means of mitigating pathologic cortical hyperexcitability in specific regions to enhance inhibitory circuit engagement, thereby alleviating tics. The SMA has emerged as a prime target for NIBS interventions in the context of treating tics. This choice is based on findings suggesting that individuals with TS exhibit altered SMA metabolic activity (Eidelberg et al., 1997; Bloch et al., 2006) and higher concentrations of the inhibitory neurotransmitter GABA in this area (Draper et al., 2015). Additionally, SMA activity has been observed immediately prior to the occurrence of tics (Neuner et al., 2013), and there is a positive correlation between symptom severity and increase in white matter pathways connecting the striatum and thalamus with both M1 and SMA (Worbe et al., 2015). These findings support the hypothesis that normalizing SMA excitability could serve as a potential therapeutic strategy for tics.

The most common inhibitory rTMS protocol applied over SMA for the treatment of tics has been low-frequency rTMS. This protocol has yielded encouraging results in several studies. In one study by Kwon and colleagues, 10 children with TS underwent 10 sessions of 1 Hz rTMS over the SMA. This intervention led to a significant decrease in Yale Global Tic Severity Scale (YGTSS) scores, a widely used measure of tic severity; notably, these reductions in tic severity persisted at a 12-week follow-up, and there were also significant improvements in overall well-being (Kwon et al., 2011). Similarly, Le and colleagues applied the same low-frequency rTMS protocol, but in this case, the treatment spanned 20 days for 25 children with TS. Once again, there was a significant decrease in YGTSS scores, indicating a reduction in tic severity (Le et al., 2013). A study by Mantovani and colleagues reported significant reductions in YGTSS scores and Yale Brown Obsessive-Compulsive Scale scores, which measure obsessive-compulsive symptom severity, in five adults/adolescents with TS following 10 days of active 1 Hz rTMS over the SMA. Additionally, an increase in the RMT was correlated with symptom improvement and remained stable at the three-month follow-up (Mantovani et al., 2006). The same group also observed promising results in two cases involving OCD and ADHD (Mantovani et al., 2007). More recently, a study applied 15 sessions of bilateral 1 Hz rTMS over the SMA in 10 children with TS, leading to a significant reduction in YGTSS scores with good tolerability (Kahl et al., 2021). However, it is worth noting that not all studies using low-frequency rTMS have reported uniformly positive outcomes. For instance, a pilot study utilizing deep rTMS, a technology enabling deeper cortical stimulation, did not find significant changes in YGTSS scores in 12 participants with TS after 20 sessions of 1 Hz rTMS. Interestingly, in a subgroup of six patients with both TS and OCD, a significant improvement in tic severity was observed (Bloch et al., 2016).

These findings stem primarily from open-label studies, and while they appear encouraging, their validity has not been confirmed in two recent double-blind studies, where the effects of the therapy were not found to significantly surpass those of a placebo. In a randomized, double-blind, sham-controlled trial, the effects of 15 sessions (1 Hz; 30 minutes; 1800 pulses per day) of rTMS over the SMA on YGTSS scores were similar between sham and active stimulation (Landeros-Weisenberger et al., 2015). Another double-blind, sham-controlled study, utilizing 8 sessions of cTBS over the SMA for two days with a group of 12 subjects, reported no noteworthy disparity in tic severity (Wu et al., 2014).

Efforts have also ventured into exploring alternative stimulation targets, yet rTMS studies targeting prefrontal and premotor regions have, thus far, yielded underwhelming results. In three sham-controlled cross-over studies, encompassing varying frequencies and methodologies, no significant reduction in tic severity was observed (Munchau et al., 2002; Chae et al., 2004; Orth et al., 2005b), However, these investigations typically involved fewer sessions compared to those focusing on the SMA. Previous neuroimaging studies employing functional MRI (fMRI) have implicated the potential involvement of the parietal lobe in tic generation (Bohlhalter et al., 2006; Neuner et al., 2014). Consequently, a recent study delved into the potential of low-frequency bilateral rTMS over the parietal lobe as a treatment for tics. Impressively, this approach resulted in a significant amelioration of motor and phonic tics, as well as premonitory urges in individuals with TS. Following 10 consecutive days of treatment, the active group exhibited significant improvements in YGTSS scores, and these benefits persisted for at least a month (Fu et al., 2021).

Transcranial direct current stimulation

Transcranial direct current stimulation (tDCS) is a common NIBS technique that, differently from TMS, employs a low-intensity electrical current (usually within the range between 1 and 2 mA) delivered transcutaneously by two electrodes of different polarities (anode and cathode) placed on the scalp (Priori, 2003). The effect of tDCS is produced by creating a potential difference between the two electrodes, which causes a subthreshold shift of neuronal resting membrane potentials toward depolarization or hyperpolarization (Woods et al., 2016). Cortical excitability is temporarily increased in regions beneath the anode, whereas the opposite occurs in areas under the cathode (Priori et al., 2014). Although tDCS cannot directly cause neuronal firing, anodal stimulation is believed to increase net cortical excitability and to induces LTP-like effects; on the other hand, the most frequently reported effect of cathodal stimulation is a decrease in cortical excitability, possibly linked to LTD-like processes (Nitsche and Paulus, 2000; Nitsche et al., 2003). tDCS is cheaper and more accessible than TMS, and it can be administered at home. However, tDCS has been less explored in terms of therapeutic application for neurologic conditions, including TS.

To date, therapeutic use of tDCS in TS patients has involved protocols that place the cathode on the SMA, premotor, or motor cortex to reduce cortical excitability, based on the principles discussed above for TMS. For example, Carvalho and colleagues studied the effects of 10 daily sessions of cathodal tDCS over the pre-SMA in a patient with refractory TS. The authors found that tic severity decreased by 46% after tDCS treatment, and the benefit persisted for 6 months (Carvalho et al., 2015). Similar results have been shown in two other participants compared to a sham-controlled condition, with a reduction in tic severity after 5 days of cathodal tDCS positioned over the left premotor cortex (Mrakic-Sposta et al., 2008). In a recent pilot study, two patients received either sham tDCS for three weeks followed by six weeks of active cathodal tDCS (1.4 mA, 18 sessions, over SMA) or six weeks of active sessions followed by three weeks of sham sessions, with a follow-up period of three and six months. The participants experienced decreased tic and urge after receiving tDCS stimulation over SMA for 6 weeks (Eapen et al., 2017). However, these positive results have not always been confirmed. A recent sham-controlled study of 10 subjects found only a small difference in tic frequency and intensity after SMA cathodal stimulation, with no clear effect on cortical excitability changes (Dyke et al., 2019). Other areas of stimulation and protocols have been investigated, such as bilateral cathodal pre-SMA stimulation (30 minutes, 2 mA), which was tested on three subjects and showed a 34.5% reduction in tic severity after 10 sessions over 5 days in one of them (Behler et al., 2018).

Peripheral nerve stimulation

While rTMS and tDCS hold promise as treatment modalities for tics, they necessitate complex technologies and scalp-based stimulation, which may not always be practical, particularly in children. An intriguing alternative to transcranial stimulation is peripheral electrical nerve stimulation. Rhythmic stimulation of the median nerve, delivered within the alpha frequency band, has shown the ability to entrain (i.e., increase power and phase synchrony) neuronal oscillations in the sensorimotor cortex, resulting in reduced tic frequency and intensity (Morera Maiquez et al., 2020). Building on this evidence, a recent parallel double-blind study was conducted across the United Kingdom to evaluate the effectiveness of rhythmic median nerve stimulation in mitigating tics. This study employed a wearable, "watch-like" device that allowed for home-based administration of rhythmic 10-Hz median nerve stimulation over a period of 4 weeks. Active stimulation led to significantly greater and clinically meaningful reductions in tics compared to sham stimulation; this finding suggests that this home-administered device holds promise as an effective community-based treatment for tic disorders (Maiquez et al., 2023).

CONCLUSION

In conclusion, neurophysiology is a powerful tool that has provided valuable insights into the neural underpinnings of tics and holds potential as adjunct therapy for tic management. Emerging approaches, such as NIBS and peripheral nerve stimulation, offer novel and handy avenues that might change the treatment of disorders in the future. These advancements underscore the complex nature of tics and the importance of innovative strategies to improve the lives of individuals affected by this condition.

References

Adelhöfer N, Paulus T, Mückschel M et al. (2021). Increased scale-free and aperiodic neural activity during sensorimotor integration—a novel facet in Tourette syndrome. Brain Commun 3: fcab250.

Aramideh M, Cruccu G, Visser BWO (2003). Cranial nerve reflexes: anatomical pathways, recording techniques and normative data. In: Hallett M (Ed.), Handbook of clinical neurophysiology. Elsevier.

Batschelett MA, Huddleston DA, Crocetti D et al. (2023). Biomarkers of tic severity in children with Tourette syndrome: motor cortex inhibition measured with transcranial magnetic stimulation. Dev Med Child Neurol 65: 1321–1331.

Behler N, Leitner B, Mezger E et al. (2018). Cathodal tDCS over motor cortex does not improve Tourette syndrome: lessons learned from a case series. Front Behav Neurosci 12: 194.

Bloch MH, Peterson BS, Scahill L et al. (2006). Adulthood outcome of tic and obsessive-compulsive symptom severity in children with Tourette syndrome. Arch Pediatr Adolesc Med 160: 65–69.

Bloch Y, Arad S, Levkovitz Y (2016). Deep TMS add-on treatment for intractable Tourette syndrome: a feasibility study. World J Biol Psychiatry 17: 557–561.

Bohlhalter S, Goldfine A, Matteson S et al. (2006). Neural correlates of tic generation in Tourette syndrome: an event-related functional MRI study. Brain 129: 2029–2037.

Bonomo R, Latorre A, Balint B et al. (2020). Voluntary inhibitory control of chorea: a case series. Mov Disord Clin Pract 7: 308–312.

Bosch M, Hayashi Y (2012). Structural plasticity of dendritic spines. Curr Opin Neurobiol 22: 383–388.

Brandt VC, Niessen E, Ganos C et al. (2014). Altered synaptic plasticity in Tourette's syndrome and its relationship to motor skill learning. PLoS One 9: e98417.

Brandt VC, Stock AK, Münchau A et al. (2017). Evidence for enhanced multi-component behaviour in Tourette syndrome—an EEG study. Sci Rep 7: 7722.

Braun MN, Wessler J, Friese M (2021). A meta-analysis of Libet-style experiments. Neurosci Biobehav Rev 128: 182–198.

Carvalho S, Goncalves OF, Soares JM et al. (2015). Sustained effects of a neural-based intervention in a refractory case of Tourette syndrome. Brain Stimul 8: 657–659.

Cavanna AE, Nani A (2013). Tourette syndrome and consciousness of action. Tremor Other Hyperkinet Mov (NY) 3: tre-03-181-4368-1.

Cernera S, Pramanik L, Boogaart Z et al. (2022). The human tic detector: an automatic approach to tic characterization using wearable sensors. Clin Neurophysiol 134: 102–110.

Chae JH, Nahas Z, Wassermann E et al. (2004). A pilot safety study of repetitive transcranial magnetic stimulation (rTMS) in Tourette's syndrome. Cogn Behav Neurol 17: 109–117.

Chen R (2004). Interactions between inhibitory and excitatory circuits in the human motor cortex. Experimental Brain Research 1541: 1–10.

Chen R, Classen J, Gerloff C et al. (1997). Depression of motor cortex excitability by low-frequency transcranial magnetic stimulation. Neurology 48: 1398–1403.

Conte A, Li Voti P, Pontecorvo S et al. (2016). Attention-related changes in short-term cortical plasticity help to explain fatigue in multiple sclerosis. Mult Scler 22: 1359–1366.

Coulombe MA, Elkaim LM, Alotaibi NM et al. (2018). Deep brain stimulation for Gilles de la Tourette syndrome in children and youth: a meta-analysis with individual participant data. J Neurosurg Pediatr 23: 236–246.

Dayan E, Censor N, Buch ER et al. (2013). Noninvasive brain stimulation: from physiology to network dynamics and back. Nat Neurosci 16: 838–844.

Donoghue T, Haller M, Peterson EJ et al. (2020). Parameterizing neural power spectra into periodic and aperiodic components. Nat Neurosci 23: 1655–1665.

Draper A, Jude L, Jackson GM et al. (2015). Motor excitability during movement preparation in Tourette syndrome. J Neuropsychol 9: 33–44.

Dyke K, Jackson GM, Nixon E et al. (2019). Effects of single-session cathodal transcranial direct current stimulation on tic symptoms in Tourette's syndrome. Exp Brain Res 237: 2853–2863.

Eapen V, Baker R, Walter A et al. (2017). The role of transcranial direct current stimulation (tDCS) in Tourette syndrome: a review and preliminary findings. Brain Sci 7: 161.

Eidelberg D, Moeller JR, Antonini A et al. (1997). The metabolic anatomy of Tourette's syndrome. Neurology 48: 927–934.

Fitzpatrick E, Hohl N, Silburn P et al. (2012). Case-control study of blink rate in Parkinson's disease under different conditions. J Neurol 259: 739–744.

Fong PY, Spampinato D, Rocchi L et al. (2021). Two forms of short-interval intracortical inhibition in human motor cortex. Brain Stimul 14: 1340–1352.

Fu M, Wei H, Meng X et al. (2021). Effects of low-frequency repetitive transcranial magnetic stimulation of the bilateral parietal cortex in patients with Tourette syndrome. Front Neurol 12: 602830.

Ganos C, Neumann WJ, Muller-Vahl KR et al. (2021). The phenomenon of exquisite motor control in tic disorders and its pathophysiological implications. Mov Disord 36: 1308–1315.

Ganos C, Rocchi L, Latorre A et al. (2018). Motor cortical excitability during voluntary inhibition of involuntary tic movements. Mov Disord 33: 1804–1809.

Ganos C, Roessner V, Munchau A (2013). The functional anatomy of Gilles de la Tourette syndrome. Neurosci Biobehav Rev 37: 1050–1062.

Georgiev D, Rocchi L, Tocco P et al. (2016). Continuous theta burst stimulation over the dorsolateral prefrontal cortex and the pre-SMA alter drift rate and response thresholds respectively during perceptual decision-making. Brain Stimul 9: 601–608.

Gilbert DL, Bansal AS, Sethuraman G et al. (2004). Association of cortical disinhibition with tic, ADHD, and OCD severity in Tourette syndrome. Mov Disord 19: 416–425.

Gilbert DL, Sallee FR, Zhang J et al. (2005). Transcranial magnetic stimulation-evoked cortical inhibition: a consistent marker of attention-deficit/hyperactivity disorder scores in Tourette syndrome. Biol Psychiatry 57: 1597–1600.

Gironell A, Rodriguez-Fornells A, Kulisevsky J et al. (2000). Abnormalities of the acoustic startle reflex and reaction time in Gilles de la Tourette syndrome. Clin Neurophysiol 111: 1366–1371.

Greene DJ, Schlaggar BL, Black KJ (2015). Neuroimaging in Tourette syndrome: research highlights from 2014-2015. Curr Dev Disord Rep 2: 300–308.

Hallett M (2002). Chapter 58: Clinical neurophysiology of tics. In: Reisin RC, Nuwer MR, Hallett M, Medina C (Eds.), Supplements to Clinical Neurophysiology. Elsevier.

Hallett M (2003). Chapter 33: Tics. In: Hallett M (Ed.), Handbook of Clinical Neurophysiology. Elsevier.

Hallett M (2007). Transcranial magnetic stimulation: a primer. Neuron 55: 187–199.

He BJ (2014). Scale-free brain activity: past, present, and future. Trends Cogn Sci 18: 480–487.

He W, Fong PY, Leung TWH et al. (2020). Protocols of non-invasive brain stimulation for neuroplasticity induction. Neurosci Lett 719: 133437.

Heise CA, Wanschura V, Albrecht B et al. (2008). Voluntary motor drive: possible reduction in Tourette syndrome. J Neural Transm (Vienna) 115: 857–861.

Heise KF, Steven B, Liuizzi G et al. (2010). Altered modulation of intracortical excitability during movement preparation in Gilles de la Tourette syndrome. Brain 133: 580–590.

Hoogendam JM, Ramakers GM, Di Lazzaro V (2010). Physiology of repetitive transcranial magnetic stimulation of the human brain. Brain Stimul 3: 95–118.

Huang YZ, Edwards MJ, Rounis E et al. (2005). Theta burst stimulation of the human motor cortex. Neuron 45: 201–206.

Huang YZ, Rothwell JC, Edwards MJ et al. (2008). Effect of physiological activity on an NMDA-dependent form of cortical plasticity in human. Cereb Cortex 18: 563–570.

Hupfeld KE, Swanson CW, Fling BW et al. (2020). TMS-induced silent periods: a review of methods and call for consistency. J Neurosci Methods 346: 108950.

Hwang YT, Rocchi L, Hammond P et al. (2018). Effect of donepezil on transcranial magnetic stimulation parameters in Alzheimer's disease. Alzheimers Dement (NY) 4: 103–107.

Jahanshahi M, Obeso I, Rothwell JC et al. (2015). A fronto-striato-subthalamic-pallidal network for goal-directed and habitual inhibition. Nat Rev Neurosci 16: 719–732.

Jannati A, Oberman LM, Rotenberg A et al. (2023). Assessing the mechanisms of brain plasticity by transcranial magnetic stimulation. Neuropsychopharmacology 48: 191–208.

Kahl CK, Kirton A, Pringsheim T et al. (2021). Bilateral transcranial magnetic stimulation of the supplementary motor area in children with Tourette syndrome. Dev Med Child Neurol 63: 808–815.

Karp BI, Porter S, Toro C et al. (1996). Simple motor tics may be preceded by a premotor potential. J Neurol Neurosurg Psychiatry 61: 103–106.

Kok A, Ramautar JR, De Ruiter MB et al. (2004). ERP components associated with successful and unsuccessful stopping in a stop-signal task. Psychophysiology 41: 9–20.

Kujirai T, Caramia MD, Rothwell JC et al. (1993). Corticocortical inhibition in human motor cortex. J Physiol 471: 501–519.

Kwon HJ, Lim WS, Lim MH et al. (2011). 1-Hz low frequency repetitive transcranial magnetic stimulation in children with Tourette's syndrome. Neurosci Lett 492: 1–4.

Landeros-Weisenberger A, Mantovani A, Motlagh MG et al. (2015). Randomized sham-controlled double-blind trial of repetitive transcranial magnetic stimulation for adults with severe Tourette syndrome. Brain Stimul 8: 574–581.

Larsh TR, Huddleston DA, Horn PS et al. (2023). From urges to tics in children with Tourette syndrome: associations with supplementary motor area GABA and right motor cortex physiology. Cereb Cortex 33: 3922–3933.

Latorre A, Hale B, Rocchi L (2022). How do I find clues about where myoclonus is originating? Mov Disord Clin Pract 9: 721–722.

Latorre A, Rocchi L, Berardelli A et al. (2019). The use of transcranial magnetic stimulation as a treatment for movement disorders: a critical review. Mov Disord 34: 769–782.

Di Lazzaro V, Ziemann U (2013). The contribution of transcranial magnetic stimulation in the functional evaluation of microcircuits in human motor cortex. Front Neural Circuits 7: 18.

Le K, Liu L, Sun M et al. (2013). Transcranial magnetic stimulation at 1 Hertz improves clinical symptoms in children with Tourette syndrome for at least 6 months. J Clin Neurosci 20: 257–262.

Lüscher C, Malenka RC (2012). NMDA receptor-dependent long-term potentiation and long-term depression (LTP/LTD). Cold Spring Harb Perspect Biol 4: a005710.

Maiquez BM, Smith C, Dyke K et al. (2023). A double-blind, sham-controlled, trial of home-administered rhythmic 10-Hz median nerve stimulation for the reduction of tics, and suppression of the urge-to-tic, in individuals with Tourette syndrome and chronic tic disorder. J Neuropsychol 17: 540–563.

Mantovani A, Leckman JF, Grantz H et al. (2007). Repetitive transcranial magnetic stimulation of the supplementary motor area in the treatment of Tourette syndrome: report of two cases. Clin Neurophysiol 118: 2314–2315.

Mantovani A, Lisanby SH, Pieraccini F et al. (2006). Repetitive transcranial magnetic stimulation (rTMS) in the treatment of obsessive-compulsive disorder (OCD) and Tourette's syndrome (TS). Int J Neuropsychopharmacol 9: 95–100.

Markram H, Gerstner W, Sjöström et al. (2012). Spike-timing-dependent plasticity: a comprehensive overview. Front Synaptic Neurosci 4: 2.

Marsili L, Suppa A, Di Stasio F et al. (2017). BDNF and LTP-/LTD-like plasticity of the primary motor cortex in Gilles de la Tourette syndrome. Exp Brain Res 235: 841–850.

Méndez JC, Rocchi L, Jahanshahi M et al. (2017). Probing the timing network: a continuous theta burst stimulation study of temporal categorization. Neuroscience 356: 167–175.

Moll GH, Heinrich H, Gevensleben H et al. (2006). Tic distribution and inhibitory processes in the sensorimotor circuit during adolescence: a cross-sectional TMS study. Neurosci Lett 403: 96–99.

Moll GH, Heinrich H, Trott G et al. (2000). Deficient intracortical inhibition in drug-naive children with attention-deficit hyperactivity disorder is enhanced by methylphenidate. Neurosci Lett 284: 121–125.

Moll GH, Heinrich H, Trott GE et al. (2001). Children with comorbid attention-deficit-hyperactivity disorder and tic disorder: evidence for additive inhibitory deficits within the motor system. Ann Neurol 49: 393–396.

Moll GH, Wischer S, Heinrich H et al. (1999). Deficient motor control in children with tic disorder: evidence from transcranial magnetic stimulation. Neurosci Lett 272: 37–40.

Morand-Beaulieu S, Wu J, Mayes LC et al. (2023). Increased alpha-band connectivity during tic suppression in children with Tourette syndrome revealed by source electroencephalography analyses. Biol Psychiatry Cogn Neurosci Neuroimaging 8: 241–250.

Morera Maiquez BM, Jackson GM, Jackson SR (2022). Examining the neural antecedents of tics in Tourette syndrome using electroencephalography. J Neuropsychol 16: 1–20.

Morera Maiquez BM, Sigurdsson HP, Dyke K et al. (2020). Entraining movement-related brain oscillations to suppress tics in Tourette syndrome. Curr Biol 30: 2334–2342e3.

Mrakic-Sposta S, Marceglia S, Mameli F et al. (2008). Transcranial direct current stimulation in two patients with Tourette syndrome. Mov Disord 23: 2259–2261.

Munchau A, Bloem BR, Thilo KV et al. (2002). Repetitive transcranial magnetic stimulation for Tourette syndrome. Neurology 59: 1789–1791.

Münchau A, Colzato LS, Aghajaniafjei A et al. (2021). A neural noise account of Gilles de la Tourette syndrome. Neuroimage Clin 30: 102654.

Nardella A, Rocchi L, Conte A et al. (2014). Inferior parietal lobule encodes visual temporal resolution processes contributing to the critical flicker frequency threshold in humans. PLoS One 9: e98948.

Neuner I, Schneider F, Shah NJ (2013). Functional neuroanatomy of tics. Int Rev Neurobiol 112: 35–71.

Neuner I, Werner CJ, Arrubla J et al. (2014). Imaging the where and when of tic generation and resting state networks in adult Tourette patients. Front Hum Neurosci 8: 362.

Niccolai V, Korczok S, Finis J et al. (2019). A peek into premonitory urges in Tourette syndrome: temporal evolution of neurophysiological oscillatory signatures. Parkinsonism Relat Disord 65: 153–158.

Nilles C, Martino D, Fletcher J et al. (2023). Have we forgotten what tics are? A re-exploration of tic phenomenology in youth with primary tics. Mov Disord Clin Pract 10: 764–773.

Nitsche MA, Fricke K, Henschke U et al. (2003). Pharmacological modulation of cortical excitability shifts induced by transcranial direct current stimulation in humans. J Physiol 553: 293–301.

Nitsche MA, Paulus W (2000). Excitability changes induced in the human motor cortex by weak transcranial direct current stimulation. J Physiol 527: 633–639.

Nunez PL, Wingeier BM, Silberstein RB (2001). Spatial-temporal structures of human alpha rhythms: theory, microcurrent sources, multiscale measurements, and global binding of local networks. Hum Brain Mapp 13: 125–164.

O'Connor K, Lavoie ME, Robert M (2001). Preparation and motor potentials in chronic tic and Tourette syndromes. Brain Cogn 46: 224–226.

Obeso JA, Rothwell JC, Marsden CD (1981). Simple tics in Gilles de la Tourette's syndrome are not prefaced by a normal premovement EEG potential. J Neurol Neurosurg Psychiatry 44: 735–738.

Orth M, Amann B, Robertson MM et al. (2005a). Excitability of motor cortex inhibitory circuits in Tourette syndrome before and after single dose nicotine. Brain 128: 1292–1300.

Orth M, Kirby R, Richardson MP et al. (2005b). Subthreshold rTMS over pre-motor cortex has no effect on tics in patients with Gilles de la Tourette syndrome. Clin Neurophysiol 116: 764–768.

Orth M, Münchau A (2013). Transcranial magnetic stimulation studies of sensorimotor networks in Tourette syndrome. Behav Neurol 27: 57–64.

Orth M, Münchau A, Rothwell JC (2008). Corticospinal system excitability at rest is associated with tic severity in Tourette syndrome. Biol Psychiatry 64: 248–251.

Orth M, Rothwell JC (2009). Motor cortex excitability and comorbidity in Gilles de la Tourette syndrome. J Neurol Neurosurg Psychiatry 80: 29–34.

Panyakaew P, Cho HJ, Hallett M (2016). Clinical neurophysiological evaluation for simple motor tics. Clin Neurophysiol Pract 1: 33–37.

Pertermann M, Muckschel M, Adelhofer N et al. (2019). On the interrelation of 1/f neural noise and norepinephrine system activity during motor response inhibition. J Neurophysiol 121: 1633–1643.

Priori A (2003). Brain polarization in humans: a reappraisal of an old tool for prolonged non-invasive modulation of brain excitability. Clin Neurophysiol 114: 589–595.

Priori A, Ciocca M, Parazzini M et al. (2014). Transcranial cerebellar direct current stimulation and transcutaneous spinal cord direct current stimulation as innovative tools for neuroscientists. J Physiol 592: 3345–3369.

Pépés SE, Draper A, Jackson GM et al. (2016). Effects of age on motor excitability measures from children and adolescents with Tourette syndrome. Dev Cogn Neurosci 19: 78–86.

Raffaele R, Rampello L, Vecchio I et al. (2006). Blink reflex abnormalities in children with Tourette syndrome. Eur J Neurol 13: 869–873.

Rawji V, Latorre A, Sharma N et al. (2020). On the use of TMS to investigate the pathophysiology of neurodegenerative diseases. Front Neurol 11: 584664.

Rektor I (2002). Scalp-recorded Bereitschaftspotential is the result of the activity of cortical and subcortical generators – a hypothesis. Clin Neurophysiol 113: 1998–2005.

Rektor I, Kaiiovsky P, Bares M et al. (2003). A SEEG study of ERP in motor and premotor cortices and in the basal ganglia. Clin Neurophysiol 114: 463–471.

Requin J, Riehle A (1995). Neural correlates of partial transmission of sensorimotor information in the cerebral cortex. Acta Psychol (Amst) 90: 81–95.

Rossini PM, Burke D, Chen R et al. (2015). Non-invasive electrical and magnetic stimulation of the brain, spinal cord, roots and peripheral nerves: basic principles and procedures for routine clinical and research application. An updated report from an I.F.C.N. Committee. Clin Neurophysiol 126: 1071–1107.

Rothwell JC, Thompson PD, Day BL et al. (1987). Motor cortex stimulation in intact man. 1. General characteristics of EMG responses in different muscles. Brain 110: 1173–1190.

Sachdev PS, Chee KY, Aniss AM (1997). The audiogenic startle reflex in Tourette's syndrome. Biol Psychiatry 41: 796–803.

Valls-Solé J, Pascual-Leone A, Wassermann EM, Hallett M (1992). Human motor evoked responses to paired transcranial magnetic stimuli. Electroencephalography and Clinical Neurophysiology 856: 355–364.

van der Salm SM, Tijssen MA, Koelman JH et al. (2012). The Bereitschaftspotential in jerky movement disorders. J Neurol Neurosurg Psychiatry 83: 1162–1167.

Serrien DJ, Orth M, Evans AH et al. (2005). Motor inhibition in patients with Gilles de la Tourette syndrome: functional activation patterns as revealed by EEG coherence. Brain 128: 116–125.

Shibasaki H, Barrett G, Halliday E et al. (1980). Components of the movement-related cortical potential and their scalp topography. Electroencephalogr Clin Neurophysiol 49: 213–226.

Sigurdsson HP, Jackson SR, Kim S et al. (2020). A feasibility study for somatomotor cortical mapping in Tourette syndrome using neuronavigated transcranial magnetic stimulation. Cortex 129: 175–187.

Smith SJ, Lees AJ (1989). Abnormalities of the blink reflex in Gilles de la Tourette syndrome. J Neurol Neurosurg Psychiatry 52: 895–898.

Sondergaard RE, Martino D, Kiss ZHT et al. (2021). TMS motor mapping methodology and reliability: a structured review. Front Neurosci 15: 709368.

Spampinato DA, Ibanez J, Rocchi L et al. (2023). Motor potentials evoked by transcranial magnetic stimulation: interpreting a simple measure of a complex system. J Physiol 601: 2827–2851.

Stefan K, Kunesch E, Cohen LG et al. (2000). Induction of plasticity in the human motor cortex by paired associative stimulation. Brain 123: 572–584.

Stell R, Thickbroom GW, Mastaglia FL (1995). The audiogenic startle response in Tourette's syndrome. Mov Disord 10: 723–730.

Sun Y, Wei H, Lin Y et al. (2020). The suppressive effect of the motor system on the sensory system in patients with Tourette syndrome. Front Neurol 11: 855.

Suppa A, Belvisi D, Bologna M et al. (2011). Abnormal cortical and brain stem plasticity in Gilles de la Tourette syndrome. Mov Disord 26: 1703–1710.

Suppa A, Li Voti P, Rocchi L et al. (2015a). Early visuomotor integration processes induce LTP/LTD-like plasticity in the human motor cortex. Cereb Cortex 25: 703–712.

Suppa A, Marsili L, Di Stasio F et al. (2014). Cortical and brainstem plasticity in Tourette syndrome and obsessive-compulsive disorder. Mov Disord 29: 1523–1531.

Suppa A, Rocchi L, Li Voti P et al. (2015b). The photoparoxysmal response reflects abnormal early visuomotor integration in the human motor cortex. Brain Stimul 8: 1151–1161.

Tharp JA, Wendelken C, Mathews CA et al. (2015). Tourette syndrome: complementary insights from measures of cognitive control, eyeblink rate, and pupil diameter. Front Psychiatry 6: 95.

Thibault G, O'Connor KP, Stip E et al. (2009). Electrophysiological manifestations of stimulus evaluation, response inhibition and motor processing in Tourette syndrome patients. Psychiatry Res 167: 202–220.

Tokimura H, Di Lazzaro V, Tokimura Y et al. (2000). Short latency inhibition of human hand motor cortex by somatosensory input from the hand. J Physiol 523: 503–513.

Tremblay S, Hannah R, Rawji V et al. (2017). Modulation of iTBS after-effects via concurrent directional TDCS: a proof of principle study. Brain Stimul 10: 744–747.

Triggiani AI, Scheman K, Pirio Richardson S et al. (2023). Physiological and introspective antecedents of tics and movements in adults with tic disorders. Clin Neurophysiol 151: 143–150.

Tübing J, Gigla B, Brandt VC et al. (2018). Associative plasticity in supplementary motor area – motor cortex pathways in Tourette syndrome. Sci Rep 8: 11984.

Tulen JH, Azzolini M, de Vries JA et al. (1999). Quantitative study of spontaneous eye blinks and eye tics in Gilles de la Tourette's syndrome. J Neurol Neurosurg Psychiatry 67: 800–802.

Utianski RL, Caviness JN, Rubin DI et al. (2016). Movement-related potentials and event-related potentials. Clinical neurophysiology. Oxford University Press.

Valls-Solé J, Pascual-Leone A, Wassermann EM et al. (1992). Human motor evoked responses to paired transcranial magnetic stimuli. Electroencephalogr Clin Neurophysiol 85: 355–364.

Voytek B, Knight RT (2015). Dynamic network communication as a unifying neural basis for cognition, development, aging, and disease. Biol Psychiatry 77: 1089–1097.

Wessel JR, Aron AR (2015). It's not too late: the onset of the frontocentral P3 indexes successful response inhibition in the stop-signal paradigm. Psychophysiology 52: 472–480.

Wolters A, Sandbrink F, Schlottmann A et al. (2003). A temporally asymmetric Hebbian rule governing plasticity in the human motor cortex. J Neurophysiol 89: 2339–2345.

Woods AJ, Antal A, Bikson M et al. (2016). A technical guide to tDCS, and related non-invasive brain stimulation tools. Clin Neurophysiol 127: 1031–1048.

Worbe Y, Marrakchi-Kacem L, Lecomte S et al. (2015). Altered structural connectivity of cortico-striato-pallido-thalamic networks in Gilles de la Tourette syndrome. Brain 138: 472–482.

Wu SW, Maloney T, Gilbert DL et al. (2014). Functional MRI-navigated repetitive transcranial magnetic stimulation over supplementary motor area in chronic tic disorders. Brain Stimul 7: 212–218.

Zapparoli L, Macerollo A, Joyce EM et al. (2019). Voluntary tic suppression and the normalization of motor cortical beta power in Gilles de la Tourette syndrome: an EEG study. Eur J Neurosci 50: 3944–3957.

Ziemann U, Paulus W, Nitsche MA et al. (2008). Consensus: motor cortex plasticity protocols. Brain Stimul 1: 164–182.

Ziemann U, Paulus W, Rothenberger A (1997). Decreased motor inhibition in Tourette's disorder: evidence from transcranial magnetic stimulation. Am J Psychiatry 154: 1277–1284.

Ziemann U, Reis J, Schwenkreis P et al. (2015). TMS and drugs revisited 2014. Clin Neurophysiol 126: 1847–1868.

Ziemann U, Rothwell JC, Ridding MC (1996). Interaction between intracortical inhibition and facilitation in human motor cortex. J Physiol 4963: 873–881.

Handbook of Clinical Neurology, Vol. 215 (3rd series)
Tic Disorders and Tourette Syndrome
C. Ganos and Y. Worbe, Editors
https://doi.org/10.1016/B978-0-443-13554-5.00004-3

Chapter 18

Neuroimaging of tic disorders

YULIA WORBE[1,2,*] AND BASSAM AL-FATLY[3]

[1]*Paris Brain Institute (ICM), Sorbonne University, Inserm U1127, CNRS UMR7225, UM75, Movement Investigation and Therapeutics Team, Paris, France*

[2]*Department of Clinical Neurophysiology, Sorbonne University, APHP-Sorbonne University – Hospital Saint-Antoine, Paris, France*

[3]*Movement Disorders and Neuromodulation Unit, Department of Neurology with Experimental Neurology, Charité–Universitätsmedizin Berlin, Corporate Member of Freie Universität Berlin and Humboldt-Universität zu Berlin, Berlin, Germany*

Abstract

Gilles de la Tourette syndrome (GTS) is increasingly conceptualized as a neurodevelopmental network disorder. In this chapter, we have synthesized two decades of neuroimaging findings spanning gray- and white-matter morphology, structural connectivity, and functional network dynamics in GTS across childhood and adulthood. Structural MRI studies using voxel-based morphometry (VBM) have reported heterogeneous volume differences—both decreases and increases—across cortico-striatal-thalamo-cortical regions, motivating the use of meta-analytic approaches to identify more consistent loci, including cingulate, sensorimotor, thalamic, striatal, and insular regions. In contrast, cortical thickness studies more consistently indicate cortical thinning in motor/premotor and prefrontal areas, with links to tic phenomenology and premonitory urges. Diffusion and volumetric white-matter measures further implicate interhemispheric pathways and fronto-striatal projections, with some indices relating to tic severity. At the circuit level, tractography and structural covariance studies support altered connectivity within the cortico-striato-pallido-thalamo-cortical loop and highlight a complementary contribution of cerebello-cortical pathways. Functional MRI studies converge on temporally structured node engagement during tic generation—preceding activity in supplementary motor, cingulate, insular, and opercular regions, followed by sensorimotor, thalamic, and cerebellar involvement—and reveal large-scale reconfiguration of cortico-basal ganglia networks associated with tic severity and comorbid symptoms. Finally, lesion-network mapping and coordinate-based network mapping studies in secondary tics and GTS-related atrophy converge on a common tic-associated network, providing a translational framework to guide neuromodulation and to derive clinically meaningful network biomarkers for treatment targeting.

GRAY AND WHITE MATTER CHANGES

Gray matter alterations are typically assessed using voxel-based morphometry (VBM), cortical thickness measures, and morphometric analyses of cortical sulci.

Studies using VBM analysis showed heterogenous findings between youngsters and adults with Gilles de la Tourette Syndrome (GTS) and even sometimes in the same age group. For instance, one study did not find any alterations in gray matter volumes between GTS and healthy boys (Roessner et al., 2009). Several other studies within the same age range succeeded to demonstrate significant differences between GTS and healthy subjects (Garraux et al., 2006; Ludolph et al., 2006; Wittfoth et al., 2012; Debes et al., 2015; Greene et al., 2017). Reduced gray matter volume were evident in the sensorimotor cortex, the left caudate nucleus, the left postcentral, hippocampal and superior parietal gyrus, the anterior cingulate cortex, and different frontal areas including the inferior frontal gyri, ventrolateral prefrontal and orbito-frontal cortices. Areas with a greater gray matter volume in GTS patients included the midbrain, the hypothalamus, the posterior thalamus, the dorsal-lateral prefrontal regions, and bilateral ventral striatum. These conflicting results called for attempts to reconcile them into common brain locations. Such attempts benefited from the methods of neuroimaging meta-analyses. Using activation likelihood estimation (ALE), Wen and colleagues included 6 studies summing up 247 GTS patients from 7 to 30 years old and 236 healthy subjects of the same age (Wen et al., 2021). They concluded that the bilateral pulvinar thalamus, right precentral and postcentral gyri, lentiform nucleus, and hypothalamus, in addition to the left inferior parietal lobule, insula, and postcentral gyrus expressed increased gray matter

*Correspondence to: Pr. Yulia Worbe, MD, PhD, Department of Clinical Neurophysiology, Sorbonne University – Hospital Saint Antoine, Paris, France. E-mail: yulia.worbe@icm-institute.org.

volume in the included studies in GTS patients compared to healthy controls. Significant gray matter volume reduction was prominent in GTS patients in the regions of the left postcentral gyri, the anterior cingulate cortices bilaterally, and left posterior cingulate cortex. Another study using seed-based d-mapping by Wan et al. were able to include 10 eligible studies according to their criteria that investigated GTS patients and healthy subjects of age ranging from 11 to 32 years, reporting VBM results (Wan et al., 2021). The result of these meta-analysis was different from the one identified in the study using ALE (Wen et al., 2021), with the bilateral inferior frontal gyri demonstrated decreased gray matter volume, while the cerebellum, right putamen, and thalamus (specifically pulvinar nucleus) displayed increased volumes. The authors tried meta-regression analysis to find a relationship between tic severity documented in the studies included and the morphologic changes identified by their method. However, their analysis indicated that no such relationship can be inferred.

A study investigating cortical thickness changes associated with GTS in comparison to age- and sex-matched healthy children revealed decreased cortical thickness in left postcentral and superior parietal gyri (Kong et al., 2020). In addition, the left rostral anterior cingulate cortex, the right paracentral gyrus, and the right precuneus expressed cortical thinning. These findings partially agree with other studies which have been conducted on adult patients. In particular, in the study by Worbe et al., the authors described cortical thinning in the left motor and premotor cortices, prefrontal and the right ventrolateral and lateral orbitofrontal areas (Worbe et al., 2010). These same regions were priorly reported to show cortical thinning in children (Sowell et al., 2008) and adults (Fahim et al., 2010) with GTS. It is worth mentioning that the left hemispheric predominance could imply asymmetric cortical maturation that stayed until adulthood (Klöppel et al., 2010). On the contrary, VBM studies did not demonstrate such asymmetry in their results. Interestingly, the study by Worbe et al. (2010) illustrated that simple tics were associated with cortical thinning in the motor areas of the face and hand. This distribution mirrors the phenomenological distribution of simple tic expression in the orofacial and dominant hand body parts. Complex tic expression was, however, associated with cortical thinning in diffuse frontal and parietal regions, which fits well with previous studies investigating movements of increasing complexity (Catalan et al., 1998; Lehéricy et al., 2006) and learning of new motor sequences (Lehéricy et al., 2005; Doyon et al., 2009). The first study to report the relationship between cortical thickness and premonitory urges was the one by Draper et al. (2016). They reported an inverse relationship between cortical thickness in the left insula and the right sensorimotor cortex in children and teenage individuals, the same areas that Bohlhalter et al. showed to activate 2 seconds prior to tic occurrence (Bohlhalter et al., 2006; see Functional Network Alterations section). The authors have also reported reduced cortical thickness in the sensorimotor cortex, the insula, and the anterior cingulate cortex in GTS patients when compared to matched healthy individuals.

Finally, changes in white matter microstructure have been also demonstrated in GTS patients in comparison to healthy controls. For instance, children with GTS had less axial diffusivity in all five segments of the corpus callosum, which negatively associated with tic severity (Plessen et al., 2006; Wolff et al., 2016). In a large cohort of patients, Greene et al. demonstrated bilaterally reduced white matter volume projecting to the orbital and medial frontal cortices (Greene et al., 2017). Children with GTS have been shown to express microstructural changes in median and axial diffusivity of the right cingulum, anterior thalamic radiation, and the forceps minor bundles. In addition, volumetric alterations have been noticed in the same set of white matter bundles as well. A correlation has previously been shown between increased fractional anisotropy and less severe tics in the sensorimotor network (Thomalla et al., 2009).

Overall, studies addressing brain volumes showed alteration of the widespread regions in patients with GTS and mostly showed rather inconsistent results, while cortical thickness measure mostly points to the cortical thinning in children and adults with GTS.

STRUCTURAL CONNECTIVITY ALTERATIONS

Convergent evidence identified the cortico-striato-pallido-thalamo-cortical circuit as primary for the pathologic process underpinning GTS (McNaught and Mink, 2011). When talking about a circuit, we try to understand how the different hubs in this circuit are interacting. As such, anatomic connectivity can be the best descriptive measure of such interplay. The study of Worbe et al. has elegantly addressed changes in the connectivity of the aforementioned circuit in adult patients with GTS versus matched healthy subjects (Worbe et al., 2015). Increased structural connectivity between both the striatum and the thalamus on one side and the sensorimotor cortex, paracentral lobule, supplementary motor area, and parietal cortices on the other side. Intriguingly, connectivity of the motor cortex correlated positively with tic severity. Yet, other studies provided evidence about the importance of the cerebello-cortical (specifically cerebellum to frontal pole and cingulate cortex) structural covariance in the mechanism of tics (Sigurdsson et al., 2020; Jackson et al., 2021). The latter findings indicated another key structure to play a role in tic generation, which is the cerebellum and its pathways. These findings agree with computational modeling and animal studies evidence (McCairn et al., 2013; Caligiore et al., 2017). Another type of structural connectivity that was also investigated in GTS patients is interhemispheric connectivity (Plessen et al., 2006; Wolff et al., 2016). In monozygotic adult twins discordant for GTS, patients exhibited widespread alterations in interhemispheric connectivity of the corpus callosum (Cavanna et al., 2010). Additionally, another study found that this trans-callosal connectivity correlates with the severity of tics (Cheng et al., 2013). Furthermore, right to left supplementary motor cortex, structural connectivity was also shown to be increased in adult with GTS, which was also correlative with motor timing task performance, specifically timing accuracy deficit (Martino et al., 2019).

Lastly, a recent study documented that subthalamic area hyperconnectivity correlated with anxiety and impulsivity as psychiatric comorbidities in GTS (Temiz et al., 2023). The authors found, in general, an increased structural connectivity between the medial subthalamic area and limbic cortical structures like the entorhinal, insular, and temporal cortices in GTS patients versus healthy subjects. Impulsivity and anxiety scores were positively correlated with left subthalamic nucleus-insula connection, which means that the higher the impulsivity/anxiety scores, the more structural connection between these regions. This signifies the involvement of insular hyperconnectivity to the subthalamic nucleus in developing impulsivity and anxiety in GTS.

FUNCTIONAL NETWORK ALTERATIONS

Results of multiple functional magnetic resonance imaging (fMRI) studies supported the network architecture of tic generation (Wang et al., 2011). Generally, these studies can be classified based on the specific measures under investigation into the following: studies concerned with changes in spontaneous resting or event-related neuronal activities, graph theory measures, or network (resting and/or dynamic) connectivity.

Bohlhalter et al. (2006) performed an event-related fMRI experiment on 10 adult patients diagnosed with GTS. The authors investigated the time just before (2 seconds before) and at simple tic expression. The authors found that blood oxygen level-dependent (BOLD) signal was increased in the supplementary motor, anterior cingulate, insular and parietal opercular cortices in the time frame of 2 seconds before tic expression. On the contrary, the primary motor cortices and the cerebellum showed increased BOLD signal with tic onset. Another study extended the previous results by investigating the difference in BOLD signal between imitated and real tics (Wang et al., 2011). In addition, the authors reported that real tics are associated with an excess in BOLD activity in the cortico-basal ganglia circuitry, specifically in the sensorimotor cortex. Increased effective connectivity was found in the sensorimotor circuits using Granger causality. The aforementioned studies paved the way toward understanding the temporal functional changes occurring in the brain in association with tic occurrence. In order to further explain this temporal profile, Neuner et al. conducted a study on 16 GTS patients, in which they determined event-related functional changes that happen 2 and 1 second before tics, as well as at the exact time point of tic occurrence (Neuner et al., 2014). Their results indicated that 2 seconds before tics, there was an increase in activation of the supplementary motor area (SMA), the sensorimotor cortex, and parietal operculum, while, at the time-point of 1 second before tics, the anterior cingulate, putamen, insula, amygdala, cerebellum, and the extrastriatal-visual cortex exhibited activation. Tic onset demonstrated activation, which engaged the thalamus, central operculum, and the sensorimotor cortex. This finding extended the one described by Wang et al. in their prior study. In general, the preparatory engagement of the SMA prior to tics resembles the one observed in preparation of normal volitional movements (Lee et al., 1999). Another resting-state fMRI study by Cui et al. investigated the changes in amplitude of low-frequency fluctuations (ALFF) and fractional ALFF in 17 drug-naïve children with GTS in comparison to 15 matched healthy control subjects (Cui et al., 2014). Significantly decreased ALFF was observed in the posterior cingulate cortex, the precuneus, and bilateral inferior and superior parietal lobules in children with GTS together with reduced fractional ALFF in the anterior cingulate cortex, superior parietal lobule, and bilateral middle and superior frontal lobules. Furthermore, increased fractional ALFF was demonstrated in the left putamen and bilateral thalamus. These results highlighted interesting abnormalities in spontaneous functional brain activity underpinning possible mechanistic changes of GTS in childhood.

On the other hand, other studies focused on disentangling static and dynamic network characteristics that correlate with tic pathophysiology. A study by Worbe et al. constructed the sensorimotor, associative, and limbic networks of the cortico-basal ganglia circuits using 91 regions of interest (Worbe et al., 2012). They extracted graph theory measures of resting-state fMRI acquisitions in 59 adult patients diagnosed with GTS compared to age- and gender-matched control cohort of 27 adult subjects. The authors found stronger global disorganization of the cortico-basal ganglia network in GTS patients. Additionally, greater interregional interactions were observed with shorter path in patients with GTS when compared to controls. The authors aimed at characterizing network properties that could correlate with tic severity and symptoms expression. They demonstrated that the aforementioned functional abnormalities of the three cortico-basal ganglia networks correlated significantly with tic severity. In addition, only sensorimotor and associative networks expressed functional abnormalities that correlated to tic complexity. Abnormalities in the associative and limbic networks correlated with severity of obsessive-compulsive symptoms in GTS patients. These findings were pivotal to understanding the functional segregation of the main brain networks engaged in the pathophysiology and the relationship to the clinical phenomenology of tics. Another study by Ramkiran et al. explored resting-state fMRI network dysfunction using functional and effective-like connectivity in 28 adult GTS patients compared to 28 age-matched control subjects (Ramkiran et al., 2019). Additionally, graph theory measures were estimated. There was increased connectivity in the basal ganglia-cortical and the thalamo-cortical networks in GTS patients. On the contrary, functional connectivity was decreased in the cortico-cerebellar network accompanied by greater global efficiency of parallel communication through the basal ganglia, thalamus, and cerebellum. Reduced path length was also observed in multiple cortical region of interest (ROI), which is in agreement with the observation by Worbe et al. (2012). The findings hinted toward a possible lack of brain maturation as a mechanism of tics (Supekar et al., 2009).

Harnessing these dynamic brain changes, especially the ones anticipating tics, could aid in predicting tic occurrence. In this regard, Ramkiran et al. demonstrated that using advanced analytic techniques, resting-state dysfunctions of the primary motor cortex, the prefrontal basal-ganglia loop, and the amygdala-mediated visual social processing network were among the cross-validated predictors of tics (Ramkiran et al., 2023).

The results of these studies summarized above supported the notion that GTS is a network disorder that expresses spatial functional changes in specific networks in addition to temporally segregated network-node engagement in different stages of tic generation. Specific network characteristics correlated with tic severity/complexity and different network parts demonstrated changes in relation to differential clinical expression of tics and associated symptoms like obsessive-compulsive behavior.

FOCAL LESIONS AND ATROPHY SITES DEFINE A COMMON TIC-ASSOCIATED BRAIN NETWORK

Focal brain lesions of various etiologies could be linked to different secondary neurologic symptoms, including tics (Fig. 18.1). While this particular manifestation of tics is not the result of a primary brain disorder like in GTS, they may share similar share similar neuroanatomic substrates with GTS. Recently, a study by Ganos et al. looked into this in detail (Ganos et al., 2022). The authors identified 22 case reports documenting tics in association with inciting focal brain lesions. Lesion locations were traced on a common brain template (the Montreal Neurological Institute brain template). Using the method of lesion network mapping (LNM), lesion tracings were used as seed regions to derive brain-wide functional connectivity profiles in a 1000 subject normative connectome (GSP1000 Preprocessed Connectome – CohenLab Dataverse, n.d.; Thomas Yeo et al., 2011). They found that the causative lesions were connected to a common neural network involving the anterior cingulate cortex, insula, posterior putamen, thalamus (including the central median and the parafascicular nuclei), internal and external pallidum, and cerebellum. These areas represent the central hubs of the cortico-striato-pallido-thalamo-cortical circuit which underpins tic pathophysiology (Wang et al., 2011). In addition, the thalamic and pallidal areas are the

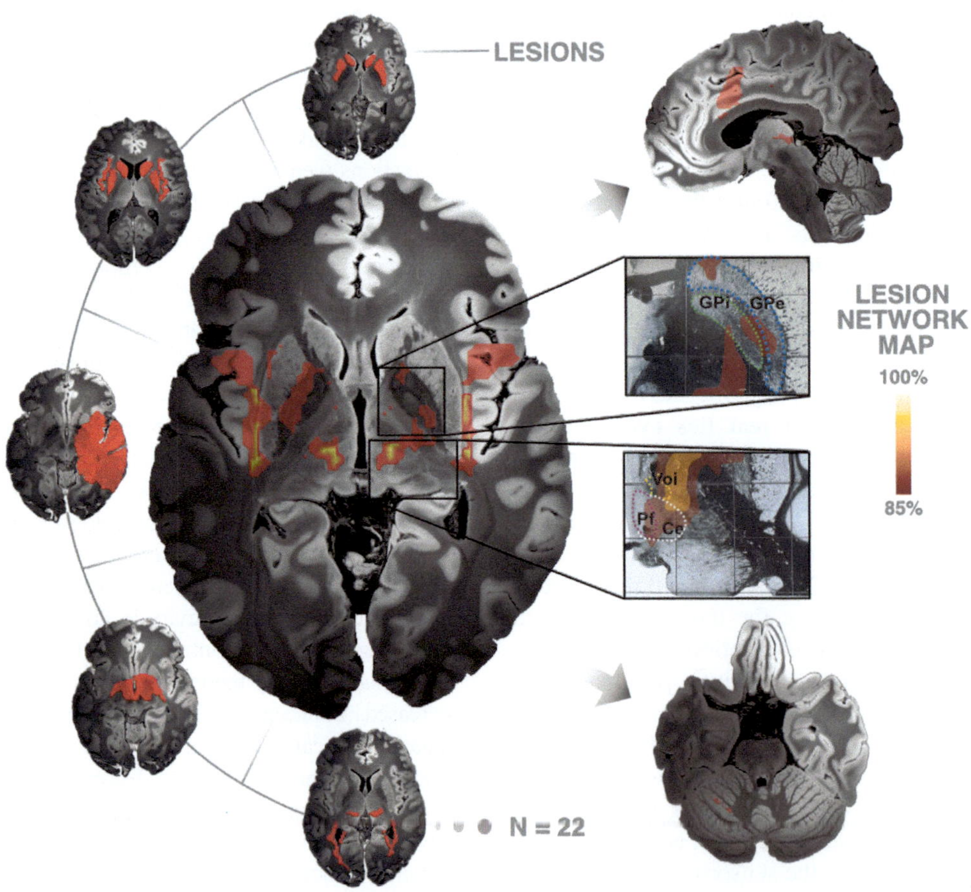

Fig. 18.1. Lesions from secondary tic cases map to a common brain network. Twenty-two case reports were identified from the literature and traced on a common brain template (MNI) in order to be used as seeds for functional connectomics analysis to derive a common lesion network map (Ganos et al. 2022). Brain regions highly connected to 85% of the lesions encompassed the thalamus, the pallidum, the putamen, the insula, anterior cingulate cortex, and the cerebellum. Specifically, the pallidal (GPi/GPe) and thalamic (Voi/Pf/CE) regions highlighted in the lesion network map agree with previously used surgical targets to treat primary GTS. ACC, anterior cingulate cortex; CB, cerebellum; CM, centromedial nucleus of the thalamus; GPe, globus pallidus externus; GPi, globus pallidus internus; Pf, parafascicular nucleus of the thalamus; Voi, ventralis oralis internus nucleus of the thalamus.

commonest deep brain stimulation (DBS) surgical target used to treat tics (Xu et al., 2020). The area of the putamen was actually specific to the occurrence of tics when to connectivity maps of other neuropsychiatric symptoms. A conjunction map (between LNM and its specificity map) was also derived and emphasized the importance of the anterior putaminal cluster for tic pathophysiology. Multiple conclusions could be inferred from this study. First, tics could be described as a "circuitopathy" (Horn and Fox, 2020). Second, a common neuroanatomic substrate represented by an interconnected functional brain network can be determined as a plausible correlate for tics. Third, this network represents a specific set of brain areas that can be translated into pathophysiologic mechanisms and therapeutic targets in tic disorders. The insula and anterior cingulate cortex, for instance, have been shown to be major players in the occurrence of premonitory urges (Tinaz et al., 2015; see Gray and White Matter Changes section). Abnormalities across the cortico-basal ganglia-thalamo-cortical pathways have already been regarded as a central in GTS pathophysiology (Worbe et al., 2010; Wang et al., 2011). The anterior putamen cluster identified in this study aligns well with the associative-limbic zone of the striatum, which represents a complex hub for information processing (Choi et al., 2012). Animal studies hinted toward the possibility of inducing tics by pharmacologically disinhibiting gamma-aminobutyric acid (GABA) ergic neurons in this particular striatal zone (McCairn et al., 2013; Tremblay et al., 2015). Additionally, pathologic evidence underlined the involvement of this region in associative-limbic more than sensorimotor interconnections in GTS patients (Kataoka et al., 2010). The anterior cingulate cortex, thalamus, insula, and putamen have been shown by previous studies to activate before the generation of tics (Bohlhalter et al., 2006; Wang et al., 2011; see Functional Network Alterations section). Importantly, pathologic studies have confirmed the role of the aforementioned regions in tic generation (Kalanithi et al., 2005; Kataoka et al., 2010).

In order to bridge the gap between secondary tics and GTS networks, another study was conducted by Zouki et al. using LNM and coordinate-based network mapping (CBNM) (Zouki et al., 2023). The second part (CBNM) entailed systematically searching the internet for studies that reported coordinates of brain atrophy in MRI/VBM studies or decreased blood flow/perfusion in positron emission tomography or single-photon emission computed tomography in patients with GTS versus healthy subjects. The results of aforementioned neuroimaging techniques are usually alterations in brain structures (atrophy or hypoperfusion) that are associated with the occurrence of tics. Noteworthy, no increase in gray matter volume or increased perfusion coordinates were included as the occurrence of such structural alteration could be interpreted as a compensatory mechanism to tics in GTS rather than neuroimaging correlates of the disorder. The study replicated the finding of the above-mentioned LNM using case reports (n=19) that largely overlapped with Ganos et al.'s study in addition to few different ones. The reason for discrepancy in the case reports included was due to a slight difference in the inclusion criteria (Ganos and Horn, 2023). Nevertheless, the study replicated the exact same network demonstrated in its preceding Ganos et al.'s study, which could be seen as a nice showcase that overcomes the neuroimaging replication crisis. Specificity mapping was also similar to Ganos et al.'s study with the addition of areas in the frontal lobe due to the use of a smaller arrays of the control groups (only cervical dystonia, parkinsonism, and Holmes tremor cases were used). Additionally, the negative (anticorrelated) part of the network was also demonstrated, which showed a cluster located majorly in the precuneus. Interestingly, the authors found that although ALE meta-analysis did not succeed to show any results in their case (in contrary to the studies mentioned in the Gray and White Matter Changes section), CBNM showed crucially a highly similar distribution of network areas (including the insula, the brainstem, the putamen, the cerebellum, and the thalamus). Negatively connected areas included the right occipital fusiform gyrus and left superior lateral occipital cortex in addition to the precuneus. The conjunction between LNM and CNBM negative networks was the precuneus, which has been suggested as a potential target for noninvasive neuromodulation techniques to interfere with tic circuit. The premise here is that anticorrelated networks could mean regions that functionally compete with the pathologically lesioned or atrophied regions and that need to be inhibited by noninvasive stimulation since they are expected to increase activity when pathologic lesions/structural alterations occur (Corp et al., 2019).

NETWORK TARGETING AS A TOOL IN TREATMENT OF TICS

The results of neuroimaging studies have led to a paradigm shift from the traditional concept of "localizational" neurology to a contemporary "network" neurology (Fox, 2018). This has mainly impacted the way with which we understand brain disorders. Furthermore, they provided a powerful translational opportunity to effectively select therapeutic targets for surgical and nonsurgical interventions. In this context, Ganos and colleagues harnessed the possibility of translating their results of LNM into a therapeutic network that can explain the variance in response after DBS in GTS patients (Ganos et al., 2022). They studied three cohorts of GTS patients from three European clinical centers who have been implanted in different DBS targets, namely the anteromedial globus pallidus internus for two cohorts, and the centromedial/parafascicular nuclei of the thalamus for one cohort. Interestingly, connectivity from the stimulation sites to the derived tic-LNM could predict clinical improvement in tics in all cohorts, as well as separately. Additionally, connectivity to the specificity and the conjunction maps were also significantly predictive. Crucially, it is important to notice that the tic-LNM is derived from case reports that have not been implanted with DBS as a therapeutic measure nor the DBS cohorts have developed tics due to focal inciting lesions. This means that the network derived from "secondary tics" cases could explain the therapeutic success of DBS in patients suffering from a "primary tic disorder," that is, GTS. The authors have also demonstrated that brain networks correlating with DBS improvement to voxel-wise connectivity metrics, of pallidal and thalamic cohorts, agreed on specific brain clusters that nicely aligned with the regions in the tic-LNM. The correlative thalamic-DBS map was also shown in another study, in which the authors used the same cohort and successfully predicted DBS-induced improvement in tics using leave-one-out cross-validation (Baldermann et al., 2022). This means that the latter map could potentially be used in subsequent investigations as an a priori DBS target to guide surgical planning instead of using a specific stereotactic coordinate. In addition, future work could incorporate the LNM to retune DBS programming in patients with tics. One particular applicability is the possible combination of finding the transcriptomics signature of the tic-LNM using resources like the Allen Institute Brain Atlas (Hawrylycz et al., 2012). The profile of expression of different neuronal receptors expressed within the tic-LNM could shed light toward the exploration of new molecular targets that may pave the way to develop new pharmacotherapeutic measures and inform future trials to treat tic. While the globus pallidus internus and the thalamus could be regarded as the most commonly targeted regions for DBS, the subthalamic nucleus has also been used as another target. In one study including 10 patients with GTS, functional connectivity of the stimulation volume to similar regions like in the aforementioned tic-LNM was demonstrated (Dai et al., 2022). Importantly, the study did not declare any predictive power of the network probably due to the low number of participants (Fig. 18.2).

Apart from that, other studies have looked at the structural and functional connectivity, using similar normative connectomic resources (Horn et al., 2017; Al-Fatly et al., 2019; Neudorfer et al., 2023), to infer a network biomarker for improvement of tics after DBS surgery. One retrospective study of 66 patients diagnosed with GTS conducted a whole-brain structural connectomic analysis in order to delineate a network of interconnected anatomic regions relevant to tic improvement subsequent to DBS (Johnson et al., 2020). The patients were implanted to the globus pallidus internus (n=34) and the centromedial nucleus of the thalamus (n=32). The structural connectomic model predicted improvement in patients implanted in the pallidum as well as in the ones implanted in the thalamus. Notably, the connectomic modeling technique yielded two different topologic profiles of pallidal- and thalamic-DBS predictive models. On the one hand, connectivity to the limbic-associative cortical networks, the cerebellum, and the caudate was positively associated with improvement in tics. On the other hand, connectivity to the sensorimotor networks

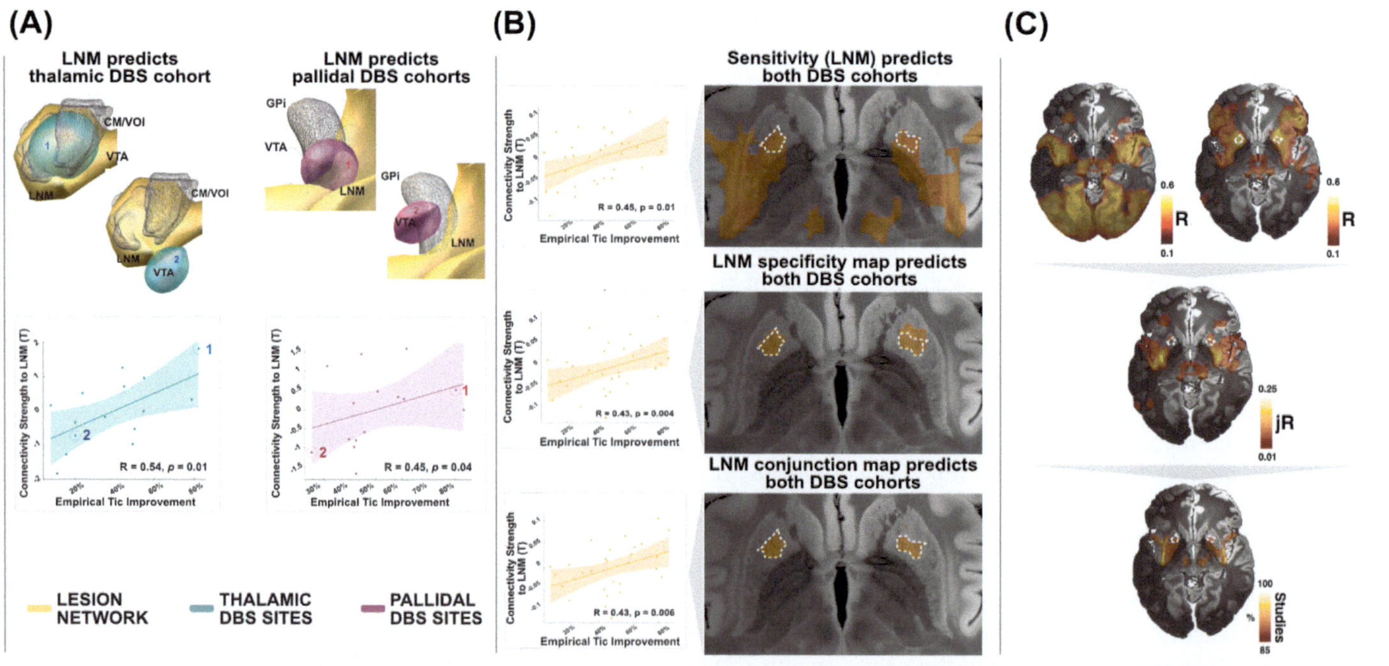

Fig. 18.2. Connectivity to lesion network map of secondary tic disorders predicts DBS improvement in primary GTS patients. A. Two cohorts of DBS were investigated in Ganos et al., namely thalamic consisting of 15 patients with GTS from Cologne and pallidal cohort comprising 15 GTS patients from Maastricht and Paris centers. Improvement in both cohorts can be predicted from the strength of connectivity of DBS electrode location to the LNM identified from the independent LNM (22 literature-based cases). **B.** Improvement of all patients from all cohorts gathered together can also be predicted from the strength of connectivity from DBS sites to clusters of sensitivity, specificity and areas conjunction of LNM. Cluster of specificity of the LNM is the result of comparing connectivity from the tic-inducing lesions to an array of different lesions causal of different symptoms. Conjunction of both sensitive and specific LNM yielded the conjunction map. **C.** Areas of beneficial DBS connectivity correlative with good improvement (red-yellow voxels) depicted in thalamic (upper left panel) and pallidal (upper right panel) cohorts agree on a common network (middle panel). The aforementioned networks also align with the LNM derived from cases of lesions causative of secondary tics and its specificity cluster (lower panel). Adapted from Ganos C, Al-Fatly B, Fischer J-F et al. (2022). A neural network for tics: insights from causal brain lesions and deep brain stimulation. Brain 145: 4385–4397. DBS, deep brain stimulation; LNM, lesion network map.

and the parietal-temporo-occipital networks had a negative impact on tic improvement after pallidal DBS. This topology has been shown to be almost inversely demonstrated in the thalamic DBS cohort. The authors also applied similar methods to demonstrate the capability of such network models to explain the response of comorbidities associated with GTS using similar network modeling. They succeeded to demonstrate specific networks that significantly predicted the response of obsessive-compulsive disorders in each DBS cohorts. The authors have then used "inverse tractography" to segment the stimulation area in each patient. Although this approach entailed a degree of circularity, it represents an elegant method that may provide a local stimulation signature to guide DBS programming and surgical targeting in patients with GTS. Another study by Johnson et al. investigated retrospective data of 35 GTS patients with pallidal DBS (Johnson et al., 2021). Pathways activation of predetermined white matter tracts and clinical information were entered in linear mixed effect models to explain tic and obsessive-compulsive behavior improvement. The study indicated that baseline severity and activation of the associative pallido-subthalamic pathway could predict tic reduction. Moreover, baseline severity and activation of the sensorimotor pallido-subthalamic pathway in addition to premotor, internal capsule and prefrontal pathways could best explain variance in obsessive-compulsive behavior improvement. Finally, two studies showed that preferential structural connectivity to the sensorimotor cortex in good responders among a small series of DBS patients implanted in different subcortical targets (*n*=7) (Andrade et al., 2020; Heiden et al., 2021). Bad/poor response to DBS therapy entailed connection to the more frontally located brain regions. Lastly, since GTS is usually manifesting during childhood, studying therapeutic networks could demand using age-matched normative neuroimaging resources (like connectomes) in clinical cohorts of pediatric age (Al-Fatly et al., 2023).

CONCLUSIONS

Different neuroimaging evidence provide a plethora of evidence that the mechanism of tic generation is linked to a specific set of interconnected brain regions. These regions express different functional changes during or before the occurrence of tics. They also undergo structural alterations that impact the clinical severity of tic manifestation. The identified neuroanatomic regions represent hubs of a distinct brain network that transfers the pathologic neural signals responsible for tic expression. Each region engages differently in specific temporal epochs signifying its role in the different stages of tic pathophysiology and generation. This network could be inferred from multimodal approaches and represents a unifying platform to describe tics as a circuitopathy. The network provides a potential target for possible therapeutic modalities to reduce tics as well as to future possible pharmacologic trials.

References

Al-Fatly B, Ewert S, Kübler D et al. (2019). Connectivity profile of thalamic deep brain stimulation to effectively treat essential tremor. Brain 142: 3086–3098.

Al-Fatly B, Giesler SJ, Oxenford S et al. (2023). Neuroimaging-based analysis of DBS outcomes in pediatric dystonia: insights from the GEPESTIM registry. Neuroimage Clin 39: 103449.

Andrade P, Heiden P, Hoevels M et al. (2020). Modulation of fibers to motor cortex during thalamic DBS in Tourette patients correlates with tic reduction. Brain Sci 10: 302.

Baldermann JC, Hennen C, Schüller T et al. (2022). Normative functional connectivity of thalamic stimulation for reducing tic severity in Tourette syndrome. Biol Psychiatry Cogn Neurosci Neuroimaging 7: 841–844.

Bohlhalter S, Goldfine A, Matteson S et al. (2006). Neural correlates of tic generation in Tourette syndrome: an event-related functional MRI study. Brain 129: 2029–2037.

Caligiore D, Mannella F, Arbib MA et al. (2017). Dysfunctions of the basal ganglia-cerebellar-thalamo-cortical system produce motor tics in Tourette syndrome. PLoS Comput Biol 13: e1005395.

Catalan MJ, Honda M, Weeks RA et al. (1998). The functional neuroanatomy of simple and complex sequential finger movements: a PET study. Brain 121: 253–264.

Cavanna AE, Stecco A, Rickards H et al. (2010). Corpus callosum abnormalities in Tourette syndrome: an MRI-DTI study of monozygotic twins. J Neurol Neurosurg Psychiatry 81: 533–535.

Cheng B, Braass H, Ganos C et al. (2013). Altered intrahemispheric structural connectivity in Gilles de la Tourette syndrome. Neuroimage Clin 4: 174–181.

Choi EY, Thomas Yeo BT, Buckner RL (2012). The organization of the human striatum estimated by intrinsic functional connectivity. J Neurophysiol 108: 2242–2263.

Corp DT, Joutsa J, Darby RR et al. (2019). Network localization of cervical dystonia based on causal brain lesions. Brain 142: 1660.

Cui Y, Jin Z, Chen X et al. (2014). Abnormal baseline brain activity in drug-naïve patients with Tourette syndrome: a resting-state fMRI study. Front Hum Neurosci 7: 62070.

Dai L, Xu W, Song Y et al. (2022). Subthalamic deep brain stimulation for refractory Gilles de la Tourette's syndrome: clinical outcome and functional connectivity. J Neurol 269: 6116–6126.

Debes N, Jeppesen S, Raghava JM et al. (2015). Longitudinal magnetic resonance imaging (MRI) analysis of the developmental changes of Tourette syndrome reveal reduced diffusion in the cortico-striato-thalamo-cortical pathways. J Child Neurol 30: 1315–1326.

Doyon J, Bellec P, Amsel R et al. (2009). Contributions of the basal ganglia and functionally related brain structures to motor learning. Behav Brain Res 199: 61–75.

Draper A, Jackson GM, Morgan PS et al. (2016). Premonitory urges are associated with decreased grey matter thickness within the insula and sensorimotor cortex in young people with Tourette syndrome. J Neuropsychol 10: 143–153.

Fahim C, Yoon U, Das S et al. (2010). Somatosensory-motor bodily representation cortical thinning in Tourette: effects of tic severity, age and gender. Cortex 46: 750–760.

Fox MD (2018). Mapping symptoms to brain networks with the human connectome. N Engl J Med 379: 2237–2245.

Ganos C, Al-Fatly B, Fischer J-F et al. (2022). A neural network for tics: insights from causal brain lesions and deep brain stimulation. Brain 145: 4385–4397.

Ganos C, Horn A (2023). Lesion-network mapping for tics: moving from replication to clinical translation. Brain Commun 5: fcad141.

Garraux G, Goldfine A, Bohlhalter S et al. (2006). Increased midbrain gray matter in Tourette's syndrome. Ann Neurol 59: 381–385.

Greene DJ, Williams AC, Koller JM et al. (2017). Brain structure in pediatric Tourette syndrome. Mol Psychiatry 22: 972–980.

GSP1000 Preprocessed Connectome – CohenLab Dataverse (n.d.) [WWW Document]. https://dataverse.harvard.edu/dataset.xhtml?persistentId=doi:10.7910/DVN/ILXIKS%27 (accessed 29.06.22).

Hawrylycz MJ, Lein ES, Guillozet-Bongaarts AL et al. (2012). An anatomically comprehensive atlas of the adult human brain transcriptome. Nature 489: 391–399.

Heiden P, Hoevels M, Bayram D et al. (2021). Connectivity patterns of deep brain stimulation targets in patients with Gilles de la Tourette Syndrome. Brain Sci 11: 1–11.

Horn A, Fox MD (2020). Opportunities of connectomic neuromodulation. Neuroimage 221: 117180.

Horn A, Reich M, Vorwerk J et al. (2017). Connectivity Predicts deep brain stimulation outcome in Parkinson disease. Ann Neurol 82: 67–78.

Jackson SR, Sigurdsson HP, Dyke K et al. (2021). The role of the cingulate cortex in the generation of motor tics and the experience of the premonitory urge-to-tic in Tourette syndrome. J Neuropsychol 15: 340–362.

Johnson KA, Duffley G, Anderson DN et al. (2020). Structural connectivity predicts clinical outcomes of deep brain stimulation for Tourette syndrome. Brain 143: 2607.

Johnson KA, Duffley G, Foltynie T et al. (2021). Basal ganglia pathways associated with therapeutic pallidal deep brain stimulation for Tourette syndrome. Biol Psychiatry Cogn Neurosci Neuroimaging 6: 961–972.

Kalanithi PSA, Zheng W, Kataoka Y et al. (2005). Altered parvalbumin-positive neuron distribution in basal ganglia of individuals with Tourette syndrome. Proc Natl Acad Sci USA 102: 13307–13312.

Kataoka Y, Kalanithi PSA, Grantz H et al. (2010). Decreased number of parvalbumin and cholinergic interneurons in the striatum of individuals with Tourette syndrome. J Comp Neurol 518: 277–291.

Klöppel S, Mangin JF, Vongerichten A et al. (2010). Nurture versus nature: long-term impact of forced right-handedness on structure of pericentral cortex and basal ganglia. J Neurosci 30: 3271–3275.

Kong L, Lv B, Wu T et al. (2020). Altered structural cerebral cortex in children with Tourette syndrome. Eur J Radiol 129: 109119.

Lee KM, Chang KH, Roh JK (1999). Subregions within the supplementary motor area activated at different stages of movement preparation and execution. Neuroimage 9: 117–123.

Lehéricy S, Bardinet E, Tremblay L et al. (2006). Motor control in basal ganglia circuits using fMRI and brain atlas approaches. Cereb Cortex 16: 149–161.

Lehéricy S, Benali H, Van De Moortele PF et al. (2005). Distinct basal ganglia territories are engaged in early and advanced motor sequence learning. Proc Natl Acad Sci USA 102: 12566–12571.

Ludolph AG, Juengling FD, Libal G et al. (2006). Grey-matter abnormalities in boys with Tourette syndrome: magnetic resonance imaging study using optimised voxel-based morphometry. Br J Psychiatry 188: 484–485.

Martino D, Hartmann A, Pelosin E et al. (2019). Motor timing in Tourette syndrome: the effect of movement lateralization and bimanual coordination. Front Neurol 10: 385.

McCairn KW, Iriki A, Isoda M (2013). Global dysrhythmia of cerebro-basal ganglia-cerebellar networks underlies motor tics following striatal disinhibition. J Neurosci 33: 697–708.

McNaught KSP, Mink JW (2011). Advances in understanding and treatment of Tourette syndrome. Nat Rev Neurol 7: 667–676.

Neudorfer C, Butenko K, Oxenford S et al. (2023). Lead-DBS v3.0: Mapping deep brain stimulation effects to local anatomy and global networks. Neuroimage 268: 119862.

Neuner I, Werner CJ, Arrubla J et al. (2014). Imaging the where and when of tic generation and resting state networks in adult Tourette patients. Front Hum Neurosci 8: 362.

Plessen KJ, Grüner R, Lundervold A et al. (2006). Reduced white matter connectivity in the corpus callosum of children with Tourette syndrome. J Child Psychol Psychiatry 47: 1013.

Ramkiran S, Heidemeyer L, Gaebler A et al. (2019). Alterations in basal ganglia-cerebello-thalamo-cortical connectivity and whole brain functional network topology in Tourette's syndrome. Neuroimage Clin 24: 101998.

Ramkiran S, Veselinović T, Dammers J et al. (2023). How brain networks tic: predicting tic severity through rs-fMRI dynamics in Tourette syndrome. Hum Brain Mapp 44: 4225.

Roessner V, Overlack S, Baudewig J et al. (2009). No brain structure abnormalities in boys with Tourette's syndrome: a voxel-based morphometry study. Mov Disord 24: 2398–2403.

Sigurdsson HP, Jackson SR, Jolley L et al. (2020). Alterations in cerebellar grey matter structure and covariance networks in young people with Tourette syndrome. Cortex 126: 1–15.

Sowell ER, Kan E, Yoshii J et al. (2008). Thinning of sensorimotor cortices in children with Tourette syndrome. Nat Neurosci 11: 637–639.

Supekar K, Musen M, Menon V (2009). Development of large-scale functional brain networks in children. PLoS Biol 7: e1000157.

Temiz G, Atkinson-Clement C, Lau B et al. (2023). Structural hyperconnectivity of the subthalamic area with limbic cortices underpins anxiety and impulsivity in Tourette syndrome. Cereb Cortex 33: 5181–5191.

Thomalla G, Siebner HR, Jonas M et al. (2009). Structural changes in the somatosensory system correlate with tic severity in Gilles de la Tourette syndrome. Brain 132: 765–777.

Thomas Yeo BT, Krienen FM, Sepulcre J et al. (2011). The organization of the human cerebral cortex estimated by intrinsic functional connectivity. J Neurophysiol 106: 1125–1165.

Tinaz S, Malone P, Hallett M et al. (2015). Role of the right dorsal anterior insula in the urge to tic in Tourette syndrome. Mov Disord 30: 1190–1197.

Tremblay L, Worbe Y, Thobois S et al. (2015). Selective dysfunction of basal ganglia subterritories: from movement to behavioral disorders. Mov Disord 30: 1155–1170.

Wan X, Zhang S, Wang W et al. (2021). Gray matter abnormalities in Tourette syndrome: a meta-analysis of voxel-based morphometry studies. Transl Psychiatry 11: 287.

Wang Z, Maia TV, Marsh R et al. (2011). The neural circuits that generate tics in Tourette's syndrome. Am J Psychiatry 168: 1326–1337.

Wen F, Yan J, Yu L et al. (2021). Grey matter abnormalities in Tourette syndrome: an activation likelihood estimation meta-analysis. BMC Psychiatry 21: 1–8.

Wittfoth M, Bornmann S, Peschel T et al. (2012). Lateral frontal cortex volume reduction in Tourette syndrome revealed by VBM. BMC Neurosci 13: 17.

Wolff N, Luehr I, Sender J et al. (2016). A DTI study on the corpus callosum of treatment-naïve boys with "pure" Tourette syndrome. Psychiatry Res Neuroimaging 247: 1–8.

Worbe Y, Gerardin E, Hartmann A et al. (2010). Distinct structural changes underpin clinical phenotypes in patients with Gilles de la Tourette syndrome. Brain 133: 3649–3660.

Worbe Y, Malherbe C, Hartmann A et al. (2012). Functional immaturity of cortico-basal ganglia networks in Gilles de la Tourette syndrome. Brain 135: 1937–1946.

Worbe Y, Marrakchi-Kacem L, Lecomte S et al. (2015). Altered structural connectivity of cortico-striato-pallido-thalamic networks in Gilles de la Tourette syndrome. Brain 138: 472.

Xu W, Zhang C, Deeb W et al. (2020). Deep brain stimulation for Tourette's syndrome. Transl Neurodegener 9: 1–19.

Zouki JJ, Ellis EG, Morrison-Ham J et al. (2023). Mapping a network for tics in Tourette syndrome using causal lesions and structural alterations. Brain Commun 5: fcad105.

Handbook of Clinical Neurology, Vol. 215 (3rd series)
Tic Disorders and Tourette Syndrome
C. Ganos and Y. Worbe, Editors
https://doi.org/10.1016/B978-0-443-13554-5.00023-7

Chapter 19

Neurotransmitter abnormalities in primary tic disorders and Tourette syndrome

YULIA WORBE[1,2,*], AND BENJAMIN PASQUEREAU[3]

[1]*Department of Clinical Neurophysiology, Sorbonne University, APHP-Sorbonne University – Hospital Saint-Antoine, Paris, France*

[2]*Paris Brain Institute (ICM), Sorbonne University, Inserm U1127, CNRS UMR7225, UM75, Movement Investigation and Therapeutics Team, Paris, France*

[3]*Institute of Cognitive Sciences–Marc Jeannerod, UMR-5229 CNRS, Bron, France*

Abstract

Tourette syndrome (TS) and primary tic disorders are increasingly understood as neurodevelopmental conditions arising from dysfunction within the cortico-basal ganglia-thalamo-cortical circuits, which govern motor, cognitive, and affective processes. This chapter reviews current knowledge on the neurochemical underpinnings of TS, drawing on evidence from neuroimaging studies – notably positron emission tomography – genetics, animal models, and neuropathology. Particular emphasis is placed on the roles of dopamine, serotonin, glutamate and γ-aminobutyric acid (GABA). Dopaminergic hyperactivity, especially involving increased D2 receptor sensitivity in the striatum, has been strongly linked to tic expression. Serotonergic dysfunction, though less consistently defined, may contribute to both tics and common comorbidities such as obsessive-compulsive behaviors, anxiety, and depression. However, the exact nature of serotonin's involvement remains unresolved, complicated by the influence of comorbidities and treatment effects. Additionally, glutamate and GABA, the brain's primary excitatory and inhibitory neurotransmitters, respectively, have emerged as crucial in modulating excitatory/inhibitory balance within the cortico-basal ganglia-thalamo-cortical loops. Glutamatergic hyperactivity and GABAergic deficits may lead to the disinhibition of motor outputs, exacerbating tic symptoms, and interact synergistically with dopaminergic circuits. Taken together, these findings underscore the multifactorial and interconnected nature of neurotransmitter abnormalities in TS. Rather than stemming from a single neurochemical deficit, TS likely arises from a complex interplay between multiple systems – dopaminergic, serotonergic, glutamatergic, GABAergic, and others – converging within dysfunctional brain networks that regulate motor and behavioral control.

NEUROCHEMICAL ABNORMALITIES IN TOURETTE SYNDROME

From neuroimaging and preclinical studies, it is commonly accepted that Tourette syndrome (TS) is associated with a pathophysiologic involvement of the cortico-basal ganglia-thalamo-cortical circuits. These interconnected loops are essential for the regulation of motor, cognitive, and affective functions. Several neurotransmitter systems that modulate these circuits have been proposed to contribute to the pathophysiology of TS, with deleterious effects on their normal operation (Singer and Wendlandt, 2001).

Among these systems, dopamine has been the most extensively studied. Hyperdopaminergic activity – particularly increased sensitivity or density of dopaminergic D2 receptors in the striatum – has been strongly associated with the manifestation of motor and vocal tics. Alongside dopamine, serotonin is also considered a key modulator of the cortico-subcortical circuits that govern movement, emotion, and cognition. This provides a compelling neurochemical basis to explain both the core symptoms of TS and its frequent psychiatric comorbidities, such as obsessive-compulsive behavior (OCB), depression, and anxiety. Over the past three decades, research has suggested that altered serotonin signaling – including disruptions in receptor activity, transporter expression, and downstream signaling cascades – may contribute to tic expression and related behavioral symptoms (Muller-Vahl, 2007). However, due to conflicting findings across clinical, pharmacologic, neuroimaging, and genetic studies, the exact relationship between serotonin dysfunction and TS remains unresolved. One of the principal challenges is determining whether

*Correspondence to: Pr. Yulia Worbe, MD, PhD, Department of Clinical Neurophysiology, Sorbonne University - Hospital Saint Antoine, Paris, France. E-mail: yulia.worbe@icm-institute.org.

serotonergic abnormalities represent a primary pathophysiologic feature of TS, or are instead secondary phenomena influenced by comorbid psychiatric conditions such as OCB or by pharmacologic interventions, including antipsychotic treatments.

Crucially, the pathophysiology of tic disorders is now understood to involve alterations in the transmission of neural signals through the cortico-basal ganglia-thalamo-cortical circuits, a process that depends on an intact and balanced neurotransmitter environment. Beyond dopamine and serotonin, increasing attention has turned to the role of glutamate and γ-aminobutyric acid (GABA), the brain's principal excitatory and inhibitory neurotransmitters, respectively.

Glutamate plays a critical role in excitatory synaptic transmission and plasticity within the cortico-basal ganglia-thalamo-cortical loops. Dysregulated glutamatergic activity – whether through excessive excitation or receptor hypersensitivity – may disturb the excitation-inhibition balance and promote abnormal motor output, potentially contributing to tic generation. Conversely, GABAergic dysfunction, particularly a reduction in GABAergic interneuron activity within the striatum and cortex, may lead to disinhibition of motor pathways, allowing inappropriate or excessive movements to emerge. These mechanisms are thought to interact with dopaminergic signaling, further exacerbating the loss of inhibitory control.

Thus, the neurochemical basis of TS likely involves multiple, interdependent neurotransmitter systems, including dopamine, serotonin, glutamate, and GABA. This is consistent with current scientific evidence supporting the involvement of a wide range of neurotransmitters and neuromodulators – such as acetylcholine and the opioid system – in the complex neurobiology of TS. Although certain neurotransmitters have been preferentially implicated, their interactions and regulatory balance within the cortico-basal ganglia-thalamo-cortical circuitry suggest that TS arises not from a single neurochemical deficit, but from a multifactorial network disruption involving excitatory, inhibitory, and modulatory systems (Singer and Wendlandt, 2001; Muller-Vahl, 2007).

GLUTAMATE AND GABA IN PRIMARY TICS AND TOURETTE SYNDROME

Dysfunction of the excitatory neurotransmitter glutamate (Glu) and the inhibitory neurotransmitter GABA has emerged as a key mechanism underlying motor disinhibition and altered sensorimotor processing in TS. The balance between glutamatergic excitation and GABAergic inhibition is crucial for maintaining proper signal flow within cortico-basal ganglia-thalamo-cortical (CBGTC) loops, and disruptions in this equilibrium are believed to contribute to the pathogenesis of tics (for further details, see also Chapter 16).

Postmortem studies provide compelling evidence of GABAergic interneuron deficits in TS. Kalanithi et al. (2005) reported increased total neuron numbers in the globus pallidus internus (GPi), alongside reduced densities of parvalbumin-positive GABAergic interneurons in the caudate and putamen, suggesting a developmental defect in inhibitory neuron migration and an imbalance in basal ganglia output pathways. Complementary findings come from peripheral biomarker research. Qian et al. (2022) measured neurotransmitter concentrations in plasma and urine of children with tic disorders and found elevated levels of norepinephrine, glutamic acid, and GABA in plasma, as well as increased normetanephrine and 5-hydroxyindoleacetic acid in urine. These markers, especially in the 8–12-year-old subgroup, showed high diagnostic accuracy (area under the curve=0.930, sensitivity=0.821, specificity=0.974), supporting the notion of systemic neurotransmitter abnormalities that reflect central nervous system dysfunction in TS.

Imaging studies have further elucidated the neurochemical alterations in vivo. Using [(11)C]flumazenil positron emission tomography (PET) imaging, Lerner et al. (2012) demonstrated reduced GABA(A) receptor binding in several key CSTC nodes – including the ventral striatum, globus pallidus, thalamus, amygdala, and right insula – alongside increased binding in the substantia nigra, periaqueductal gray, posterior cingulate cortex, and cerebellum. These findings are consistent with disinhibited thalamocortical circuits and highlight additional regions such as the insula and cerebellum that may participate in tic generation.

Mahone et al. (2018), using 7T proton magnetic resonance spectroscopy (1H MRS), found elevated glutamate levels in the premotor cortex (PMC) of children with TS compared to controls, though no group differences were observed in GABA levels. Interestingly, higher PMC glutamate was associated with improved selective motor inhibition, suggesting a possible compensatory role of glutamatergic signaling in motor control. Further exploration of GABAergic function was carried out by Puts et al. (2015), who used GABA-edited MRS to measure GABA levels in the primary sensorimotor cortex and conducted vibrotactile behavioral testing. Children with TS showed significantly reduced primary sensorimotor cortex GABA concentrations and impaired performance on tactile adaptation and detection tasks, pointing to altered sensory gating. Moreover, primary sensorimotor cortex GABA levels were negatively correlated with tic severity, and the typical correlation between GABA and sensory discrimination seen in healthy controls was absent in TS, underscoring disrupted GABAergic modulation of sensorimotor integration.

Altogether, these findings converge to support a model in which TS arises from a complex interplay of neurotransmitter system abnormalities – particularly involving GABA and glutamate – within cortico-basal ganglia-thalamo-cortical and extended motor and sensory circuits. These abnormalities likely underlie both the generation of tics and the broader spectrum of sensorimotor and behavioral dysfunctions in TS.

ROLE OF DOPAMINE IN TICS AND TOURETTE SYNDROME

Animal models and genetics

A growing body of evidence from both animal models and genetic studies strongly implicates dopaminergic dysfunction in the pathophysiology of TS. These complementary approaches shed light on the molecular and neurobiologic underpinnings of the disorder.

Animal models have proven particularly valuable for dissecting the role of dopamine in TS (see also Chapter 16). Mice generated using clustered regularly interspaced short palindromic repeats/CRISPR-associated protein 9 (CRISPR/Cas9) to carry mutations in two high-confidence TS genes, CELSR3 and WWC1, exhibited behavioral abnormalities such as sensorimotor gating deficits (impaired prepulse inhibition (PPI)), repetitive behaviors (excessive grooming and rearing), and sex-specific differences, with females showing more severe symptoms (Nasello et al., 2024). These behaviors were attenuated by aripiprazole, a D2 receptor partial agonist, suggesting a role for dopaminergic receptor signaling. Additionally, increased striatal dopamine release was observed, further implicating dopaminergic mechanisms. Further evidence comes from a mouse model carrying the W88C mutation in WWC1, which demonstrated repetitive behaviors, PPI deficits, and elevated dopamine release in the dorsal striatum (Lv et al., 2025). This mutation led to KIBRA protein degradation, disrupted the Hippo signaling pathway, and resulted in abnormal synaptic function. Importantly, restoring wild-type WWC1 reversed these deficits, confirming a causal link between the gene, dopamine signaling, and TS-relevant phenotypes. In a different line of investigation, stress-exposed D1CT-7 mice – which express a neuropotentiating transgene in D1 dopamine receptor-expressing cortical neurons – showed elevated corticosterone, exacerbated tic-like behaviors, and PPI impairments, all of which were absent in wild-type controls (Godar et al., 2016). These abnormalities were reversed by dopaminergic antagonists, indicating that both dopamine hypersensitivity and stress–dopamine interactions can exacerbate TS-related symptoms.

Complementary insights into striatal dysfunction come from O'Brien et al. (2018) who studied D1CT-7 mice with chronically potentiated dopamine D1 receptor-expressing neurons. They found that chronic glutamate output from cortical and amygdaloid regions led to increased striatal GABA, decreased D-serine, and abolished Per1 expression, indicating a state of striatal desensitization and persistent CSTC loop activation. These findings suggest that hyperactive dopaminergic-glutamatergic signaling may result in sustained output from motor circuits, supporting tic generation and maintenance.

In parallel, human genetic studies provide some evidence for the involvement of dopamine-related genes in TS.

Analysis of DNA methylation patterns revealed that hypermethylation of DRD2 correlated with tic severity, while hypomethylation of dopamine transporter (DAT) may act as a compensatory mechanism to buffer dopaminergic tone (Müller-Vahl et al., 2017). These findings point to the importance of epigenetic modulation in TS. Additional genetic associations have been reported for polymorphisms in adenosine receptor genes ADORA1 and ADORA2A, which modulate dopaminergic neurotransmission. These variants were linked not only to earlier tic onset, but also to comorbid obsessive-compulsive disorder (OCD), depression, and attention-deficit/hyperactivity disorder (ADHD) (Janik et al., 2015). Moreover, peripheral gene expression of DRD2 and SLC6A4 (serotonin transporter, SERT) was shown to correlate with tic severity, suggesting that gene activity outside the central nervous system might reflect or contribute to central neurotransmitter dysregulation (Gunther et al., 2012).

Support for dopaminergic gene involvement also comes from population studies. In a South American cohort, DRD2 polymorphisms were significantly associated with TS, with notable transmission distortions observed (Herzberg et al., 2010). In parallel, the DRD4 7-repeat allele, known to affect receptor efficiency, was over-transmitted in TS families, further implicating dopamine receptor dysregulation (Grice et al., 1996). A broader polygenic model was supported by findings that polymorphisms in DRD2, DBH (dopamine β-hydroxylase), and DAT collectively contributed to tic severity and a range of TS-related comorbidities, including ADHD and OCD (Comings et al., 1996). In contrast, investigations into the DRD1 gene yielded negative findings, as no linkage was observed, thereby refining the focus on other dopamine receptors and transporters (Gelernter et al., 1994).

PET studies of dopamine receptors and transporter studies in tics and Tourette syndrome

Additional support for dopaminergic involvement comes from PET and single-photon emission computed tomography (SPECT) neuroimaging studies of dopamine receptor and transporter function.

Denys et al. (2013) used [(11)C]raclopride PET to assess striatal D2/3 receptor availability and dopamine release following amphetamine challenge. They observed reduced D2/3 receptor binding in the putamen, and dopamine release in the ventral striatum was positively correlated with tic severity, indicating that dopaminergic reactivity may exacerbate tics. In a separate study, Wong et al. (2008) identified elevated dopamine release in the ventral striatum and increased D2 receptor binding in TS patients with comorbid OCD. In addition, abnormalities in SERT and 5-HT2A receptor binding were observed, highlighting the interplay between dopamine and serotonin systems in TS pathophysiology (see also Serotonin in Tourette Syndrome section).

Broadening this perspective, Segura and Strafella (2013) reviewed neuroimaging studies and concluded that dopaminergic dysfunction in TS extends beyond the basal ganglia, with cortical and extrastriatal regions also contributing to symptom expression.

With regard to DAT function, Cheon et al. (2004) demonstrated that drug-naive children with TD had significantly increased DAT binding in the basal ganglia, suggesting early abnormalities in dopamine reuptake regulation. In contrast, Stamenkovic et al. (2001) and Hwang et al. (2008) found no significant group differences in DAT or D2 receptor binding between TS patients and controls using SPECT. However, they emphasized that functional dopaminergic abnormalities could not be ruled out based on their imaging parameters (Stamenkovic et al., 2001; Hwang et al., 2008).

Similarly, Albin et al. (2009) used [11C]DTBZ and [11C] MP PET tracers to measure VMAT2 and DAT binding in adults with TS. They found no evidence of increased dopaminergic terminal density in the striatum, a result that contrasts with earlier imaging reports, and underscores the heterogeneity of neurobiologic findings in TS.

Together, these animals, genetic, and neuroimaging studies create a coherent picture of dopaminergic dysregulation as a core component of TS. They reveal that dopamine receptor sensitivity, striatal dopamine release, genetic variation, epigenetic modulation, receptor availability, transporter function, and cortical–subcortical interactions all converge to shape the expression and severity of TS symptoms. Nonetheless, inconsistencies between imaging findings and the variability in methods and patient populations underscore the need for standardized, longitudinal approaches to fully elucidate the dopaminergic mechanisms underlying TS.

Dopamine dysfunction in Tourette syndrome: insights from computational models and deep brain stimulation

Recent advances in computational modeling, behavioral paradigms, and neuromodulation techniques have contributed significant insight into the dopaminergic mechanisms underlying TS.

Maia and Conceição (2018) reviewed a range of dopaminergic imaging studies and proposed a unifying hypothesis of dopaminergic hyperinnervation in TS. They suggested that this abnormality accounts for the elevated tonic and phasic dopamine observed across multiple studies. In their framework, tonic dopamine heightens the expression of learned tics, while phasic dopamine enhances their reinforcement and learning, establishing a mechanistic basis for both the emergence and persistence of tics in TS. Expanding on these findings, Maia and Conceição (2017) integrated evidence from reinforcement learning and habit formation paradigms. They argued that tics function as maladaptive motor habits, driven by aberrant phasic dopamine bursts that reinforce their execution. Tonic dopamine, meanwhile, sustains the background activation required for tic performance. According to this model, antipsychotic medications may be effective because they shift striatal plasticity and reduce both the expression and reinforcement of tics through modulation of dopamine signaling. This concept of enhanced habit formation in TS is supported by empirical findings from Delorme et al. (2016), who used a three-stage instrumental learning paradigm to examine the balance between goal-directed and habitual behavior in patients with TS. In this study, unmedicated patients exhibited a marked reliance on habitual, outcome-insensitive behavioral control, which was positively correlated with tic severity. In contrast, antipsychotic-medicated patients showed intermediate performance between unmedicated patients and healthy controls. Neuroimaging using diffusion tensor imaging revealed that increased engagement in habitual responding was associated with greater structural connectivity in the right motor cortico-striatal network. Furthermore, stronger connectivity between the supplementary motor area and sensorimotor putamen predicted more severe tics in unmedicated patients. These results suggest that aberrant reinforcement signaling and structural alterations in motor circuits underlie the habitual nature of tics in TS.

Adding a systems-level dimension, Caligiore et al. (2017) developed a computational model that incorporates the basal ganglia, cerebellum, thalamus, and cortex to simulate tic generation. Their model demonstrated that dopaminergic surges, combined with external cortical input, can initiate tics. It also predicted that tic frequency varies as a function of striatal dopamine levels, and that basal ganglia-cerebellar communication plays a critical role in this process. This integrative framework supports the involvement of distributed motor circuits and suggests novel neuromodulatory targets beyond the traditional CBGTC loops.

Further evidence for dopaminergic involvement in TS comes from neurostimulation studies. Rusheen et al. (2023) demonstrated that deep brain stimulation (DBS) targeting the centromedian-parafascicular (CMPf) complex of the thalamus exerts therapeutic effects by modulating dopaminergic neurotransmission. Specifically, CMPf-DBS evoked synaptic dopamine release and elevated tonic dopamine levels in the dorsomedial striatum, ultimately leading to a reduction in tic-like behaviors. These effects were mediated through D2 receptor activation and cholinergic interneuron (CIN) pathways, providing strong evidence that DBS can alter striatal dopaminergic tone to ameliorate motor symptoms in TS.

Together, these findings underscore the importance of considering both neurochemical dynamics and circuit-level interactions in understanding TS. They also highlight how aberrant dopamine signaling, particularly the balance between tonic and phasic activity, contributes to tic development, learning, and persistence. Computational, behavioral, and interventional approaches are therefore pivotal in refining pathophysiologic models and guiding future treatment strategies (Fig. 19.1).

SEROTONIN IN TOURETTE SYNDROME

The serotonin system

Serotonin (5-hydroxytryptamine, 5-HT) is a neuromodulator widely distributed in the brain. It is synthesized from the amino acid tryptophan and its principal metabolite is the 5-hydoxyindoleacetic acid (5 HIAA). Serotonin plays a critical role in numerous physiologic and psychologic functions, including mood regulation, impulse control, adaptation to stressors and motor coordination (Lucki, 1998). Accordingly, 5-HT dysfunctions are thought to contribute to a myriad of disorders including mood disorders, OCD, migraine, obesity, and autism. Serotonin is produced from neurons of the raphe nuclei in the midbrain that project primarily to the limbic cortico-subcortical circuits (Hensler, 2006). Seven distinct classes of 5-HT receptors (5-HT1 to 5-HT7) have been described in different parts of the brain. The duration of the action of serotonin is

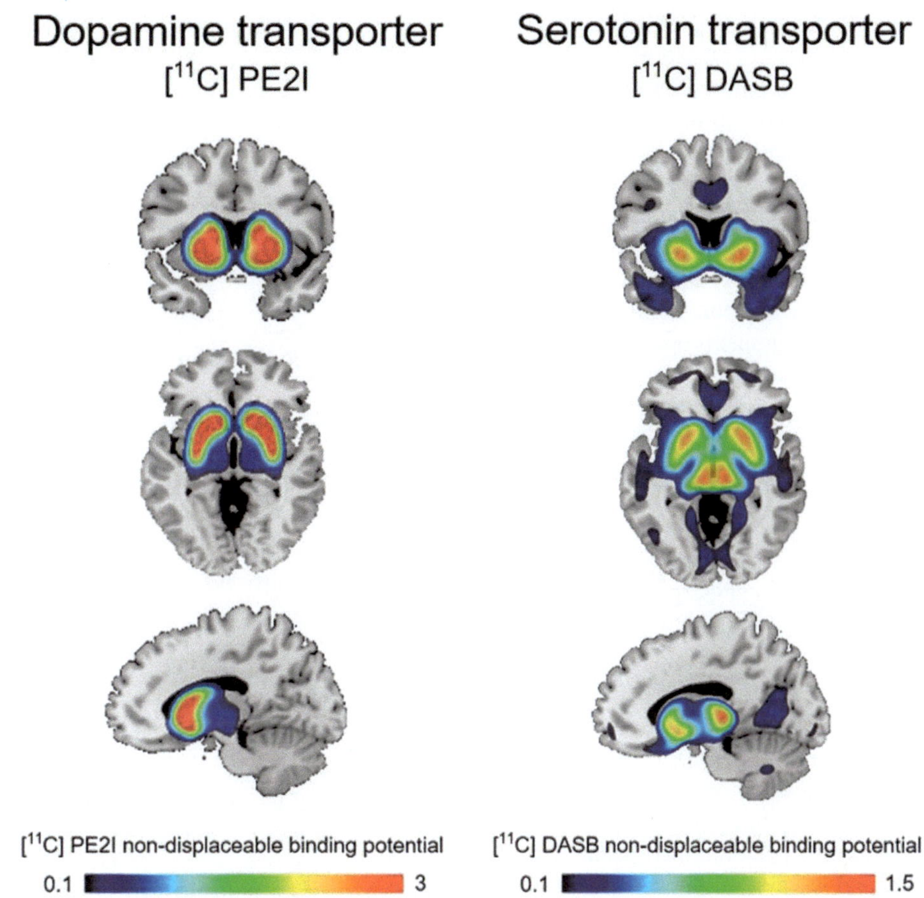

Fig. 19.1. Coronal (y=10), axial (z=−2.5), and sagittal (x=−14) views of dopaminergic and serotonergic innervation visualized by positron emission tomography in 15 healthy controls. Parametric images illustrate the average maps for [^{11}C]PE2I nondisplaceable binding potential to the dopamine transporter (DAT) and [^{11}C]DASB nondisplaceable binding potential to the serotonin transporter (SERT). Note that striatal regions are strongly co-innervated by both neurotransmitter systems.
Reproduced from Prange et al. (2022). Thanks to Elise Metereau and Stephane Thobois for sharing it.

regulated by pre-synaptic transporters (SERTs) localized on 5-HT terminals. By binding the SERT, selective serotonin reuptake inhibitors (SSRIs) are a well-known effective pharmacologic treatment for patients with OCD and mood disorders. Because most of TS patients (60%–80%) have comorbid behavioral disorders such as OCD, depression, and anxiety, and considering the therapeutic action of SSRIs, from the clinical point of view it can be assumed that abnormalities in the serotonergic system may likely occur in TS.

Treatment influencing the 5-HT system

First-line pharmacologic treatment for OCB in the setting of TS is very similar to that for OCD in isolation from it and consists primarily of SSRIs (Neri and Cardona, 2013). Although this class of serotonergic treatment has been described to be frequently effective against comorbidities in TS, a selective action of SSRI on the motor symptoms remains undefined. In a double-blind placebo-controlled trial of fluoxetine monotherapy (20 mg daily), Scahill et al. (1997) found no fluoxetine effect on tics after 8 weeks of treatment, while this treatment was beneficial in reducing comorbid OCB symptoms. In contrast, Husted et al. (2007) observed that tics were significantly reduced after 8 weeks of a fluoxetine monotherapy (up to 40 mg) in 13 OCD patients with comorbid tic disorders. Similar beneficial action of fluoxetine on tics has been reported in a very preliminary trial with fluoxetine in two TS patients (Silvestri et al., 1994), but an indirect drug effect via a stabilized emotional balance and better self-regulatory abilities may likely explain such general improvement in motor symptoms.

Antipsychotics, which are D2 receptor antagonists, are often the first-choice treatment in the therapy of tic disorders. In addition to their antidopaminergic action, these typical as well as atypical neuroleptic drugs are characterized by a complex interaction with multiple 5-HT receptors. Specifically, antipsychotics were found to interact in different ways with 5-HT receptors, that is, inhibition of 5-HT2A, 5-HT2B, 5-HT2C receptors, activation of 5-HT1A receptors, interaction with 5-HT1D, 5-HT6, and 5-HT7 receptors, and affinity for SERT (Wood et al., 2006). It is unclear whether this complex interaction of antidopaminergic drugs with 5-HT system contributes to the therapeutic efficacy of these drugs in the treatment of tics, or whether this receptor binding profile is just associated with processes, reducing the appearance of adverse side effects such as extrapyramidal and metabolic syndrome (Meltzer, 1999). Nonetheless, the atypical antipsychotics the most effective in treating tics, including risperidone and aripiprazole, have strong 5-HT2A properties suggesting a specific role for this class of receptors in regulating tics. Indeed, a small open-label study using ketanserin, a pharmacologic compound with mainly 5-HT2A antagonist properties and weak D2 antagonist properties (on the order of 200 times less than haloperidol), has also been reported effective in treating tics (Bonnier et al., 1999). Pimavanserin, a 5-HT2A receptor inverse agonist and antagonist, was also recently tested in an open-label trial over 8 weeks in 10 adult patients with TS (Billnitzer and Jankovic, 2021). A small but noteworthy decrease in tic severity (−12%) was noted on the yale global tic severity scale – total tic score (YGTSS-TTS), as well as improvements on scales assessing global clinical impression and quality of life. In addition to 5-HT2A agents, ondansetron, a selective 5-HT3 antagonist has been tested for 3 weeks as an anti-tic drug in a randomized double-blind placebo-controlled study in 30 patients with TS (Toren et al., 2005). Here again, a limited improvement was reported on tic severity via the Tourette's Syndrome Global Scale, but not by the YGTSS. Together, these findings support the potential therapeutic role of the 5-HT modulation in TS, and warrant further directed studies at 5-HT drugs.

5-HT metabolism in Tourette syndrome

Metabolic analyses such as measuring levels of 5-HT, its precursor tryptophan and its principal metabolite 5-HIAA are globally consistent with an involvement of the 5-HT system in TS pathophysiology. Compared to healthy controls, TS patients have shown lower blood levels of both 5-HT and tryptophan that treatment with clonidine partially reversed (Leckman et al., 1984; Cath et al., 2001). These findings are in line with a postmortem study in which levels of 5-HT, tryptophan, and 5-HIAA were found to be reduced in four adult TS patients (Anderson et al., 1992). The identified metabolic deficits were more marked in TS patients with the presence of OCD symptoms (Cath et al., 2001), suggesting a more selective link with comorbidities. However, the reduced levels of 5-HIAA measured in cerebral spinal fluid of TS patients were associated with the severity of the tics (Cohen et al., 1978). Administration of dextroamphetamine has shown to deteriorate the severity of tics in conjunction with 5-HIAA levels (Cohen et al., 1978), reflecting processes involving the 5-HT system in the control of motor symptoms. Overall, these analyses reflect a more or less reduction of 5-HT metabolism in TS, depending on the severity of different symptoms expressed in each patient.

5-HT genes in Tourette syndrome

TS is thought to be of polygenic etiology. Candidate gene studies suggest that the 5-HT system is involved in TS pathogenesis (Dehning et al., 2010; Moya et al., 2013a, 2013b). In a clinical trial with 98 patients, Mössner et al. (2007) found that specific variants of the enzyme responsible for 5-HT synthesis in the brain (TPH2) exert influence on TS risk. Also, studies have reported elevated expression levels of the gene for SERT (SLC6A4) in TS patients (Hildonen et al., 2021). Higher expression of SLC6A4 mRNA were detected in the blood in conjunction with the severity of TS tics (Gunther et al., 2012).

A rare SLC6A4 variant Ile425Val known to modulate SERT activity was reported to have a higher prevalence in TS individuals compared to controls (Kilic et al., 2003; Moya et al., 2013b). The SERT-linked polymorphic region (5-HTTLPR) in the promoter region immediately upstream to SLC6A4 has also been implicated in both OCD (Hu et al., 2006; Wendland et al., 2008) and TS etiology (Moya et al., 2013b). Finally, Dehning et al. (2010) have observed an association between two polymorphisms in the 5-HTR2C receptor and tic symptoms in a comparison of 87 TS patients and 311 matched controls. Together, these findings describe multiple plausible susceptibility genes for TS and further support the etiologic relevance for 5-HT system in this disorder.

5-HT imaging in Tourette syndrome

In neuroimaging, while significant effort aimed to elucidate the involvement of the dopamine in TS, much less attention has been given to the 5-HT system. Heinz et al. (1998) used SPECT and the radioligand [^{123}I][β]-carbomethoxy-3[β]-(4-iodophenyl) tropane ([^{123}I]β-CIT) to measure the presynaptic SERT in 10 adult TS patients. SERT binding potential (BP) was significantly reduced in the midbrain in TS subjects compared to normal controls, reflecting reduced transporter density, affinity, or activity in this region that could be secondary to altered levels of 5-HT. Interestingly, a negative correlation of overall tic severity and SERT availability to midbrain and thalamus was reported. Another SPECT study using the identical radioligand also found a significantly reduced binding in 12 SSRI-free TS patients compared to normal controls (Müller-Vahl et al., 2005). But this time, there was a negative correlation between SERT availability and comorbid OCD symptoms. Using PET scans and the radiotracer [^{11}C]McN5652 (McN), Wong et al. (2008) found reduction of SERT availability in different brain areas including midbrain, caudate, and putamen in 11 adult TS patients. Consistent with Müller-Vahl et al. (2005), this reduction of SERT BP was especially true for the subgroup of TS patients with OCD symptoms. Recently, however, in contrast with the majority of previous imaging studies on TS, Müller-Vahl et al. (2019) used SPECT with [^{123}I]β-CIT and found an increase of SERT availability in eight patients with TS + OCD in caudate and midbrain (SPECT), but not in TS-only or OCD-only patients. Surprisingly, these findings were also not in line with the multiple studies that report reduced SERT binding in OCD patients (Reimold et al., 2007; Zitterl et al., 2007). Nonetheless, a meta-analysis of imaging results on medication-free TS patients has summarized all available data by reporting a reduction of SERT binding in putamen, thalamus, and hypothalamus, and either an increase/decrease of BP in caudate nucleus (Nikolaus et al., 2022).

There are only two imaging studies available investigating postsynaptic 5-HT receptors in TS. Haugbøl et al. (2007) used PET and the radioligand [^{18}F]altanserin to assess 5-HT2A receptor binding in 20 medicated TS patients. Compared to control subjects, significant global elevations in [^{18}F]altanserin binding were measured in cortical regions and cingulate gyrus as well as subcortically in thalamus, caudate, and putamen, reflecting an increased density and/or affinity for 5-HT2A receptors. No specific correlation between 5-HT2A availability and comorbid OCD symptoms was found in TS patients, while another PET study (using [^{18}F]altanserin) in OCD patients without tics has demonstrated increased 5-HT2A receptor binding in caudate (Adams et al., 2005). To disentangle medication-induced effects from TS pathophysiology, Wong et al. (2008) conducted PET scans in 11 treatment-free TS patients with the radioligand [^{11}C]MDL100,907 (MDL) to measure 5-HT2A receptor binding. No significant differences were noted in 5-HT2A measures in any brain region between the whole group of TS patients and controls, suggesting that previous observations from Haugbøl et al. (2007) reflected secondary adaptations of 5-HT2A receptors to pharmacologic treatments. Although Wong et al. (2008) reports a weak elevation of 5-HT2A in TS patients depending on OCD symptoms, the limited number of patients without OCD in this study (n=2) questions the validity of this information.

In another imaging study, Behen et al. (2007) evaluated tryptophan metabolism in 26 children with TS compared to age-matched control subjects via PET scans and the radiotracer [^{11}C]methyl-L-tryptophan (AMT). Whole brain AMT uptake was globally reduced in TS condition. But, specifically, the study revealed decreased regional AMT uptake in the dorsolateral prefrontal cortex (PFC) and increased AMT uptake in caudate and thalamus of TS children. Correlations between regional AMT uptake and clinical symptoms of TS patients were also observed. Tic severity increased as function of the thalamic AMT uptake, while comorbidities such as OCD and ADHD were related to the AMT uptake in the dorsolateral PFC.

In summary, the serotonergic system including SERT, postsynaptic receptors, and the tryptophan metabolism is dysregulated in TS. These functional imaging data provide evidence for a role of serotonergic mechanisms in the pathophysiology of TS. However, to date, it remains uncertain which part of these serotonergic abnormalities is a primary cause of TS, and which one is secondary to treatments and comorbidities. Inconsistent findings are likely due to varying treatments, comorbidities, and disease severities. Investigations on entirely medication-free and larger cohorts are lacking.

5-HT in animal models

A plethora of animal models are able to mimic tic-like behaviors observed in patients suffering from TS. Among various rodent models, for instance, it has been proposed that movement abnormalities such as head shakes or twitches following administration of the selective 5-HT2A/2C receptor agonist 2,5-dimethoxy-4-iodophenyl-2-aminopropane (DOI) can be assimilated to tics in TS (Handley and Dursun, 1993; Tizabi et al., 2001). Effective drugs in the treatment of tics in patients including haloperidol (Dursun and Handley, 1996), nicotine (Tizabi et al., 2001), and donepezil (acetylcholinesterase inhibitor; Hayslett and Tizabi, 2005) have also been reported to suppress DOI-induced head shakes in mice. It has been demonstrated that DOI administration suppresses the discharge of 5-HT neurons in the dorsal raphe in conjunction with the emergence of tic-like movements, while the 5-HT2A receptor antagonist M100907 reverses this effect (Martín-Ruiz et al., 2001). Thus, DOI-induced head twitches might be mediated primarily by stimulation of 5-HT2A receptors (Hayslett and Tizabi, 2005; Swerdlow and Sutherland, 2006), supporting the etiopathologic relevance for this subtype of 5-HT receptors in tic disorders based on this animal model.

5-HT regulation of dopamine transmission

In conjunction with the potential direct link between the 5-HT system and TS symptoms, significant insights suggest that the involvement of 5-HT would be indirectly due to its interactions with the dopamine (DA) system (Steeves and Fox, 2008). Numerous findings have provided evidence regarding interactions between both 5-HT and DA systems (Soubrié et al., 1984; Schmidt et al., 1994). 5-HT neurons synapse directly onto both DA neurons in the midbrain (Hervé et al., 1987) and striatal neurons (Parent et al., 1981), positioning them to influence the DA transmission in a variety of ways. Specifically, acute and chronic systemic administration of the SERT inhibitor fluoxetine decreased DA in the striatum (Ichikawa and Meltzer, 1995) but increased DA in the PFC (Koch et al., 2002). In line with this, it has been shown that administration of SSRI could modulate the activity of DA cells in the midbrain, which would predict changes of DA in output regions (Di Mascio et al., 1998; Dremencov et al., 2009). Consequently, a potential part of the therapeutic actions of SSRIs on tics and comorbidities reported in TS could be a consequence of indirect interactions with DA neurotransmission.

The 5-HT system modulates DA transmission in a complex manner that depends on the subtypes of postsynaptic receptor involved (Alex and Pehek, 2007; Bubar and Cunningham, 2006, 2008; Di Matteo et al., 2008; Navailles and De Deurwaerdère, 2011). In particular, the 5-HT2A and 5-HT2C receptors have been investigated as likely candidates for mediating such influence on DA transmission. PET studies in human subjects with [^{11}C]raclopride have reported that administration of the 5-HT2A receptor agonists fenfluramine and psilocybin increases the release of DA in the striatum (Smith et al., 1997;

Vollenweider et al., 1999; Kuroki et al., 2003), and rodent studies have shown that the 5-HT2A and 5-HT2C receptor subtypes exert opposite effects on DA release. Specifically, 5-HT2A receptor facilitates transient release of DA in response to stimulus but does not influence tonic, basal release of DA (Schmidt et al., 1992; Gudelsky et al., 1994; Schmidt and Fadayel, 1996). Conversely, 5-HT2C receptor exerts a tonic inhibitory control on mesolimbic and nigrostriatal dopaminergic activity, suppressing stimulus-evoked phasic release of DA as well (Di Giovanni et al., 1999; Di Matteo et al., 1999; Hutson et al., 2000). Hence, given the complex cross-talk between 5-HT and DA systems, it remains challenging to determine which one of these neurotransmitters is the source of abnormalities in TS. Wong et al. (2008) have proposed that the primary dysfunction in TS could be a hypo-serotonergic state inducing secondary disturbances of the DA system via an upregulation of 5-HT2A receptors, while Steeves and Fox (2008) have suggested that DA dysfunctions could trigger 5-HT disturbances. Beyond these conflicting visions concerning the etiology of TS and the primary cause of symptoms, DA and 5-HT systems are likely not the sole factors responsible for the development of this disorder.

OTHER NEUROTRANSMITTERS IN TOURETTE SYNDROME

The role of neuroactive steroids in tic exacerbation

Steroids such as estrogens, androgens, progestogens, glucocorticoids, and mineralocorticoids, synthesized in glands like the gonads and adrenal cortices, exert their effects by modulating both cytoplasmic and membrane-bound receptors and play crucial roles in stress regulation.

In the context of tic disorders and TS, growing evidence implicates neuroactive steroids in pathophysiology. Two key neurosteroids, dehydroepiandrosterone sulfate (DHEAS) and allopregnanolone (AP), both produced by brain and gonads, have emerged as important modulators. Elevated DHEAS levels have been observed in boys with TS, aligning with the timing of adrenarche, suggesting a potential role in disease onset. AP is particularly implicated in stress-related exacerbation of tics. Both DHEAS and AP modulate GABA-A receptor activity, indicating that neurosteroid-induced alterations in GABAergic signaling could underlie tic severity and frequency (Bortolato et al., 2022; Branca and Bortolato, 2024).

In addition, multiple preclinical studies converge on the finding that the neurosteroid AP plays a key role in worsening tic-like behaviors and sensorimotor gating deficits in animal models of TS, particularly under conditions of acute stress (Branca & Bortolato, 2024). In a mouse model of partial striatal CIN depletion (mimicking TS neuropathology), mild acute stress (spatial confinement) increased grooming stereotypies and PPI deficits selectively in males. Systemic or intra-PFC administration of AP dose-dependently worsened these stress-induced behaviors. Conversely, inhibiting AP synthesis or blocking AP signaling reduced tic-like symptoms, highlighting the PFC as a key site of AP's effects (Cadeddu et al., 2023). Parallel work using the D1CT-7 TS mouse model demonstrated that acute stress elevated PFC levels of progesterone derivatives, particularly AP, and that only AP exacerbated TS-like behaviors and sensorimotor deficits (Mosher et al., 2017). These effects were reversed by haloperidol and by finasteride, a 5α-reductase inhibitor that blocks AP synthesis.

Further, administration of isoallopregnanolone (isoAP), a natural antagonist at the AP-binding site on GABA-A receptors, successfully reduced stress- and AP-induced tic-like behaviors and restored sensorimotor gating. IsoAP's efficacy was comparable to that of haloperidol and finasteride, and it was well-tolerated (Cadeddu et al., 2020).

Collectively, these findings suggest that stress exacerbates TS symptoms via AP synthesis and GABAergic disinhibition in the PFC. Modulating neurosteroid signaling, particularly targeting AP synthesis or action, may offer novel therapeutic strategies for managing symptom fluctuations in TS. These findings support the growing interest in therapeutic strategies targeting neuroactive steroids to manage tic disorders.

Acetylcholine in tics and Tourette syndrome

Emerging evidence strongly implicates striatal cholinergic dysfunction in the pathophysiology of TS.

In a pilot transcriptomic study, Tian et al. (2011) found that gene expression profiles related to GABAergic and cholinergic pathways were significantly associated with tic severity in TS patients. Specifically, they identified differential expression and alternative splicing of several key genes, including GABRA4, GABRG1, and CHRNA4, suggesting that disruptions in both GABA and acetylcholine (ACh) signaling may contribute to the clinical manifestations of tics. However, the interpretation of striatal cholinergic deficits in TS is not fully straightforward. Using [18 F]FEOBV PET imaging, Albin et al. (2017) assessed vesicular acetylcholine transporter (VAChT) expression, a marker of cholinergic terminals, in living TS patients compared to controls. Contrary to earlier postmortem findings, they found no evidence of significant cholinergic deficits in the striatum of TS patients. The authors suggest that discrepancies between postmortem and in vivo imaging results may reflect agonal or postmortem changes, medication effects, or heterogeneity within TS populations

Building on postmortem observations of reduced CINs in TS (Kataoka et al., 2010), Xu et al. (2015) used a targeted interneuron ablation strategy in mice to model this deficit. They demonstrated that selective loss of approximately 50% of CINs in the dorsolateral striatum (DLS) led to stress- and amphetamine-induced tic-like stereotypies and motor coordination deficits, without affecting sensorimotor gating. These findings establish a causal role for DLS CIN loss in generating core features of TS.

Further supporting the involvement of cholinergic dysfunction, Du et al. (2023) developed a conditional Slitrk1 knockdown (Slitrk1-KD) mouse model, targeting the dorsal striatum. Slitrk1-KD mice exhibited TS-like tic behaviors, impaired PPI, and delayed sensorimotor responses, paralleling TS phenomenology. These abnormalities were associated with reduced numbers of Slitrk1-positive CINs, diminished acetylcholine and dopamine release, increased tonic dopamine levels, and downregulated DAT expression. Notably, Slitrk1-negative CINs were found to be less excitable than Slitrk1-positive ones. These findings emphasize the pivotal role of Slitrk1 in maintaining striatal CIN activity and subsequent dopaminergic balance, thus providing a novel translational model for TS.

To better understand the dynamics of cholinergic interneuron tonically active neuron (TAN) activity, Codianni and Rubin (2023) developed a computational model of striatal TANs. Their model accurately reproduced the range of firing patterns observed experimentally, including tonic firing, surges, and pauses in response to excitatory inputs. The simulations suggest that ACh release dynamics can critically influence striatal outputs and potentially tic generation, offering a mechanistic framework for interpreting TAN behavior in TS and reinforcement learning contexts.

From a therapeutic perspective, Cadeddu et al. (2024) explored whether enhancing muscarinic receptor activity could mitigate tic-like behaviors. Using two TS mouse models (CIN-d and D1CT-7), they found that xanomeline, a M1/M4 muscarinic receptor agonist, significantly reduced tic-like and TS-related responses. This effect was specifically mediated by M4 receptors, as it was blocked by M4 (but not M1) antagonists and mimicked by the M4 positive allosteric modulator VU0467154, but not by an M1 agonist. Additionally, M4 receptor expression was found to be downregulated in CIN-d mice, and VU0467154 treatment reduced striatal neural activity (c-Fos levels). These findings position M4 receptor activation as a promising novel therapeutic strategy for TS.

Together, these studies underscore the central role of striatal CINs and muscarinic receptor signaling in the pathophysiology of TS. Disruption of cholinergic tone appears to contribute both to the emergence of tics and to the modulation of striatal output, reinforcing the importance of acetylcholine as a therapeutic target.

Targeted modulation of M4 muscarinic receptors offers a particularly promising avenue for developing more effective and better-tolerated treatments for TS, complementing traditional dopaminergic strategies. Further research is warranted to translate these findings into clinical applications and to explore how cholinergic and GABAergic dysfunctions interact in the broader neurocircuitry of tic disorders.

Endogenous opioid systems in Tourette syndrome

Using a genetic and exome-sequencing approach, Depienne et al. (2023) identified rare, possibly deleterious variants in genes related to the opioid system among TS patients, notably an excess of disruptive variants in the OPRK1 gene (encoding the κ opioid receptor). Furthermore, downregulation of the Oprk1 orthologue in zebrafish induced hyperkinetic behavior, supporting the hypothesis that altered opioid receptor function could contribute to TS.

Postmortem and cerebrospinal fluid (CSF) studies also implicate endogenous opioids. Leckman et al. (1988) reported elevated CSF concentrations of dynorphin A in TS patients, with dynorphin levels correlating with obsessive-compulsive symptom severity, although not directly with tic severity. This finding suggests a role for opioid peptides in nonmotor symptoms of TS. Additionally, prior neuropathologic work by Haber et al. (1986) demonstrated reduced dynorphin A immunoreactivity in striatal projections to the globus pallidus in a TS brain, highlighting a possible dysfunction of opioid-mediated modulation within basal ganglia circuits. From an endocrine perspective, Sandyk (1989) provided evidence for aberrant μ-opiate receptor modulation of anterior pituitary hormone release in TS patients, implying that endorphins and enkephalins are significantly involved in TS neurobiology through neuroendocrine dysregulation.

In a neurochemical study of deer mice, a model of repetitive behaviors resembling TS, Augustine et al. (2020) observed a negative correlation between repetitive behaviors and levels of endogenous opioids (leu-enkephalin and dynorphin A) in the frontal cortex. These findings suggest that reduced opioid activity could alter cortical GABAergic and glutamatergic transmission, promoting stereotyped and tic-like behaviors.

Altogether, these findings indicate that alterations in endogenous opioid signaling, particularly involving κ and μ receptors, may contribute to both motor and nonmotor symptoms of TS through complex interactions with dopaminergic, GABAergic, and neuroendocrine systems.

Several clinical trials have tested pharmacologic agents acting on the opioid system to alleviate TS symptoms.

In a small open-label study, Chappell et al. (1993) administered spiradoline, a selective κ opioid receptor agonist, to TS patients. They found that low doses of spiradoline significantly reduced motor and phonic tics, as measured by both blind video assessments and clinician ratings. Interestingly, an intermediate dose tended to worsen tics, suggesting a dose-dependent dual modulatory effect of κ receptor activation. Furthermore, TS patients displayed enhanced growth hormone secretion after spiradoline, indicating hypersensitivity of κ receptors involved in neuroendocrine regulation.

In a double-blind, randomized controlled trial, Kurlan et al. (1991) tested the effects of naltrexone, an opioid receptor antagonist, and propoxyphene, an opioid agonist, in TS patients. Naltrexone treatment was associated with a significant reduction in self-reported tic severity compared to placebo, along with improvements in attention and visuomotor planning, suggesting that blocking opioid receptors may exert beneficial effects in TS.

Collectively, both experimental and clinical studies converge on the notion that the endogenous opioid system plays a multifaceted role in TS. Dysregulation of opioid peptides (particularly dynorphin and enkephalin) and alterations in κ and μ opioid receptor signaling appear to contribute to the motor, behavioral, and neuroendocrine manifestations of TS. Moreover, pharmacologic modulation of these pathways shows promising therapeutic potential, though further large-scale, controlled studies are needed to optimize and individualize treatment strategies targeting opioid systems in TS.

Histamine neurotransmission in Tourette disorder

Histaminergic dysfunction has been increasingly implicated in the pathophysiology of TS, based on evidence from both animal models and human genetic studies.

In experimental models, Pittenger (2020) demonstrated that Hdc knockout (KO) mice, which lack histamine synthesis, display upregulation of histamine H3 receptors (H3R) in the striatum. Given the constitutive activity of H3 receptors even in the absence of ligand, this upregulation leads to abnormal signaling in striatal medium spiny neurons (MSNs), closely resembling the patterns observed after pharmacologic H3R activation in wild-type animals. Moreover, activation of H3R in Hdc KO mice accentuates these signaling abnormalities and produces behavioral stereotypies, suggesting that constitutive H3R signaling is a key driver of tic-like behaviors in this model. Expanding on these findings, Rapanelli et al. (2018) showed that H3R dysregulation in Hdc KO mice alters intracellular signaling cascades in MSNs: D1 receptor-expressing neurons exhibit elevated mitogen-activated protein kinase (MAPK) signaling at baseline, while D2 receptor-expressing neurons show reduced Akt phosphorylation. These abnormalities mirror the effects of H3R agonists in wild-type mice and support the hypothesis that overactive H3 receptor pathways underlie both molecular and behavioral disturbances associated with TS.

While these molecular findings suggest profound functional alterations, Abdurakhmanova et al. (2017) reported that Hdc KO mice do not exhibit gross morphologic or cytologic abnormalities in striatal or cortical structures. Nonetheless, these mice display behavioral impairments, particularly in exploratory behavior and habituation, implying that neurotransmitter dysregulation – rather than structural deficits – may drive the clinical features of TS in this model. Complementing these observations, Kononoff Vanhanen et al. (2016) demonstrated that H3R KO mice present with impaired sensorimotor gating, as measured by reduced PPI, another endophenotype linked to TS. Furthermore, the absence of H3 receptors disrupted dopaminergic receptor–mediated signaling in the striatum, notably affecting D1 and D2 receptor pathways, thereby reinforcing the idea that histaminergic and dopaminergic systems are closely intertwined in regulating TS-relevant behaviors.

In parallel to animal model findings, human genetic studies have also identified the HDC gene as a risk locus for TS. Karagiannidis et al. (2013), in a large multicenter study involving 520 nuclear families from Europe and Canada, found significant associations between TS and two SNPs (rs854150 and rs1894236) within the HDC region. These results support the role of histamine pathways in neuronal development and in the etiology of TS. Similarly, Ercan-Sencicek et al. (2010) reported the identification of a rare, functional mutation in the HDC gene within a two-generation TS pedigree, providing direct genetic evidence that impaired histamine biosynthesis contributes to TS and tic manifestation.

Altogether, these converging data from animal and human studies strongly support a model in which histamine deficiency – particularly through dysregulated H3 receptor signaling – plays a critical role in TS pathophysiology, affecting both motor behaviors and dopaminergic transmission. Despite this compelling body of evidence, clinical trials investigating therapies targeting histaminergic systems in TS remain to be conducted, representing a crucial next step for translating these findings into novel therapeutic strategies.

References

Abdurakhmanova S, Chary K, Kettunen M et al. (2017). Behavioral and stereological characterization of Hdc KO mice: relation to Tourette syndrome. J Comp Neurol 525: 3476–3487.

Adams KH, Hansen ES, Pinborg LH et al. (2005). Patients with obsessive-compulsive disorder have increased 5-HT2A receptor binding in the caudate nuclei. Int J Neuropsychopharmacol 8: 391–401.

Albin RL, Koeppe RA, Wernette K et al. (2009). Striatal [11C] dihydrotetrabenazine and [11C]methylphenidate binding in Tourette syndrome. Neurology 72: 1390–1396.

Albin RL, Minderovic C, Koeppe RA (2017). Normal striatal vesicular acetylcholine transporter expression in Tourette syndrome. eNeuro 4: ENEURO.0178-17.2017.

Alex KD, Pehek EA (2007). Pharmacologic mechanisms of serotonergic regulation of dopamine neurotransmission. Pharmacol Ther 113: 296–320.

Anderson GM, Pollak ES, Chatterjee D et al. (1992). Postmortem analysis of subcortical monoamines and amino acids in Tourette syndrome. Adv Neurol 58: 123–133.

Augustine F, Rajendran S, Singer HS (2020). Cortical endogenous opioids and their role in facilitating repetitive behaviors in deer mice. Behav Brain Res 379: 112317.

Behen M, Chugani HT, Juhász C et al. (2007). Abnormal brain tryptophan metabolism and clinical correlates in Tourette syndrome. Mov Disord 22: 2256–2262.

Billnitzer A, Jankovic J (2021). Pilot study to evaluate Pimavanserin for the treatment of motor and behavioral symptoms of Tourette syndrome. Mov Disord Clin Pract 8: 694–700.

Bonnier C, Nassogne MC, Evrard P (1999). Ketanserin treatment of Tourette's syndrome in children. Am J Psychiatry 156: 1122–1123.

Bortolato M, Coffey BJ, Gabbay V et al. (2022). Allopregnanolone: the missing link to explain the effects of stress on tic exacerbation? J Neuroendocrinol 34: e13022.

Branca C, Bortolato M (2024). The role of neuroactive steroids in tic disorders. Neurosci Biobehav Rev 160: 105637.

Bubar MJ, Cunningham KA (2006). Serotonin 5-HT2A and 5-HT2C receptors as potential targets for modulation of psychostimulant use and dependence. Curr Top Med Chem 6: 1971–1985.

Bubar MJ, Cunningham KA (2008). Prospects for serotonin 5-HT2R pharmacotherapy in psychostimulant abuse. Prog Brain Res 172: 319–346.

Cadeddu R, Bäckström T, Floris G et al. (2020). Isoallopregnanolone reduces tic-like behaviours in the D1CT-7 mouse model of Tourette syndrome. J Neuroendocrinol 32: e12754.

Cadeddu R, Braccagni G, Branca C et al. (2024). Activation of M(4) muscarinic receptors in the striatum reduces tic-like behaviours in two distinct murine models of Tourette syndrome. Br J Pharmacol 181: 3064–3081.

Cadeddu R, Van Zandt M, Santovito LS et al. (2023). Prefrontal allopregnanolone mediates the adverse effects of acute stress in a mouse model of tic pathophysiology. Neuropsychopharmacology 48: 1288–1299.

Caligiore D, Mannella F, Arbib MA et al. (2017). Dysfunctions of the basal ganglia-cerebellar-thalamo-cortical system produce motor tics in Tourette syndrome. PLoS Comput Biol 13: e1005395.

Cath DC, Spinhoven P, Landman AD et al. (2001). Psychopathology and personality characteristics in relation to blood serotonin in Tourette's syndrome and obsessive–compulsive disorder. J Psychopharmacol 15: 111–119.

Chappell PB, Leckman JF, Scahill LD et al. (1993). Neuroendocrine and behavioral effects of the selective kappa agonist spiradoline in Tourette's syndrome: a pilot study. Psychiatry Res 47: 267–280.

Cheon KA, Ryu YH, Namkoong K et al. (2004). Dopamine transporter density of the basal ganglia assessed with [^{123}I]IPT SPECT in drug-naive children with Tourette's disorder. Psychiatry Res 130: 85–95.

Codianni MG, Rubin JE (2023). A spiking computational model for striatal cholinergic interneurons. Brain Struct Funct 228: 589–611.

Cohen DJ, Shaywitz BA, Caparulo B et al. (1978). Chronic, multiple tics of Gilles de la Tourette's disease. CSF acid monoamine metabolites after probenecid administration. Arch Gen Psychiatry 35: 245–250.

Comings DE, Wu S, Chiu C et al. (1996). Polygenic inheritance of Tourette syndrome, stuttering, attention deficit hyperactivity, conduct, and oppositional defiant disorder: the additive and subtractive effect of the three dopaminergic genes—DRD2, DβH, and DAT1. Am J Med Genet 67: 264–288.

Dehning S, Müller N, Matz J et al. (2010). A genetic variant of HTR2C may play a role in the manifestation of Tourette syndrome. Psychiatr Genet 20: 35–38.

Delorme C, Salvador A, Valabrègue R et al. (2016). Enhanced habit formation in Gilles de la Tourette syndrome. Brain 139: 605–615.

Denys D, de Vries F, Cath D et al. (2013). Dopaminergic activity in Tourette syndrome and obsessive-compulsive disorder. Eur Neuropsychopharmacol 23: 1423–1431.

Depienne C, Ciura S, Trouillard O et al. (2023). Association of rare genetic variants in opioid receptors with Tourette syndrome. Tremor Other Hyperkinet Mov (NY) 13: 22.

Di Giovanni G, De Deurwaerdére P, Di Mascio M et al. (1999). Selective blockade of serotonin-2C/2B receptors enhances mesolimbic and mesostriatal dopaminergic function: a combined in vivo electrophysiological and microdialysis study. Neuroscience 91: 587–597.

Di Mascio M, Di Giovanni G, Di Matteo V et al. (1998). Selective serotonin reuptake inhibitors reduce the spontaneous activity of dopaminergic neurons in the ventral tegmental area. Brain Res Bull 46: 547–554.

Di Matteo V, Di Giovanni G, Di Mascio M et al. (1999). SB 242084, a selective serotonin 2C receptor antagonist, increases dopaminergic transmission in the mesolimbic system. Neuropharmacology 38: 1195–1205.

Di Matteo V, Di Giovanni G, Pierucci M et al. (2008). Serotonin control of central dopaminergic function: focus on in vivo microdialysis studies. Prog Brain Res 172: 7–44.

Dremencov E, El Mansari M, Blier P (2009). Effects of sustained serotonin reuptake inhibition on the firing of dopamine neurons in the rat ventral tegmental area. J Psychiatry Neurosci 34: 223–229.

Du JC, Chang MH, Yeh CJ et al. (2023). Pivotal role of Slitrk1 in adult striatal cholinergic neurons in mice: implication in Tourette syndrome. Ann Neurol.

Dursun SM, Handley SL (1996). Similarities in the pharmacology of spontaneous and DOI-induced head-shakes suggest 5HT 2A receptors are active under physiological conditions. Psychopharmacology 128: 198–205.

Ercan-Sencicek AG, Stillman AA, Ghosh AK et al. (2010). L-histidine decarboxylase and Tourette's syndrome. N Engl J Med 362: 1901–1908.

Gelernter J, Pauls DL, Leckman J et al. (1994). D2 dopamine receptor alleles do not influence severity of Tourette's syndrome. Results from four large kindreds. Arch Neurol 51: 397–400.

Godar SC, Mosher LJ, Strathman HJ et al. (2016). The D1CT-7 mouse model of Tourette syndrome displays sensorimotor gating deficits in response to spatial confinement. Br J Pharmacol 173: 2111–2121.

Grice DE, Leckman JF, Pauls DL et al. (1996). Linkage disequilibrium between an allele at the dopamine D4 receptor locus and Tourette syndrome, by the transmission-disequilibrium test. Am J Hum Genet 59: 644–652.

Gudelsky GA, Yamamoto BK, Nash JF (1994). Potentiation of 3,4-methylenedioxymethamphetamine-induced dopamine release and serotonin neurotoxicity by 5-HT2 receptor agonists. Eur J Pharmacol 264: 325–330.

Gunther J, Tian Y, Stamova B et al. (2012). Catecholamine-related gene expression in blood correlates with tic severity in Tourette syndrome. Psychiatry Res 200: 593–601.

Haber SN, Kowall NW, Vonsattel JP et al. (1986). Gilles de la Tourette's syndrome. A postmortem neuropathological and immunohistochemical study. J Neurol Sci 75: 225–241.

Handley SL, Dursun SM (1993). Serotonin and Tourette's syndrome: movements such as head-shakes and wet-dog shakes may model human tics. Adv Biosci 85: 235.

Haugbøl S, Pinborg LH, Regeur L et al. (2007). Cerebral 5-HT2A receptor binding is increased in patients with Tourette's syndrome. Int J Neuropsychopharmacol 10: 245–252.

Hayslett RL, Tizabi Y (2005). Effects of donepezil, nicotine and haloperidol on the central serotonergic system in mice: implications for Tourette's syndrome. Pharmacol Biochem Behav 81: 879–886.

Heinz A, Knable MB, Wolf SS et al. (1998). Tourette's syndrome: [I-123]β-CIT SPECT correlates of vocal tic severity. Neurology 51: 1069–1074.

Hensler JG (2006). Serotonergic modulation of the limbic system. Neurosci Biobehav Rev 30: 203–214.

Hervé D, Pickel VM, Joh TH et al. (1987). Serotonin axon terminals in the ventral tegmental area of the rat: fine structure and synaptic input to dopaminergic neurons. Brain Res 435: 71–83.

Herzberg I, Valencia-Duarte AV, Kay VA et al. (2010). Association of DRD2 variants and Gilles de la Tourette syndrome in a family-based sample from a South American population isolate. Psychiatr Genet 20: 179–183.

Hildonen M, Levy AM, Dahl C et al. (2021). Elevated expression of SLC6A4 encoding the serotonin transporter (SERT) in Gilles de la Tourette Syndrome. Genes (Basel) 12: 86.

Hu X-Z, Lipsky RH, Zhu G et al. (2006). Serotonin transporter promoter gain-of-function genotypes are linked to obsessive-compulsive disorder. Am J Hum Genet 78: 815–826.

Husted DS, Shapira NA, Murphy TK et al. (2007). Effect of comorbid tics on a clinically meaningful response to 8-week open-label trial of fluoxetine in obsessive compulsive disorder. J Psychiatr Res 41: 332–337.

Hutson PH, Barton CL, Jay M et al. (2000). Activation of mesolimbic dopamine function by phencyclidine is enhanced by 5-HT(2C/2B) receptor antagonists: neurochemical and behavioural studies. Neuropharmacology 39: 2318–2328.

Hwang WJ, Yao WJ, Fu YK et al. (2008). [99mTc]TRODAT-1/[^{123}I] IBZM SPECT studies of the dopaminergic system in Tourette syndrome. Psychiatry Res 162: 159–166.

Ichikawa J, Meltzer HY (1995). Effect of antidepressants on striatal and accumbens extracellular dopamine levels. Eur J Pharmacol 281: 255–261.

Janik P, Berdyński M, Safranow K et al. (2015). Association of ADORA1 rs2228079 and ADORA2A rs5751876 polymorphisms with Gilles de la Tourette syndrome in the Polish population. PLoS One 10: e0136754.

Kalanithi P, Zheng W, Kataoka Y et al. (2005). Altered parvalbumin-positive neuron distribution in basal ganglia of individuals with Tourette syndrome. Proc Natl Acad Sci USA 102: 13307–13312.

Karagiannidis I, Dehning S, Sandor P et al. (2013). Support of the histaminergic hypothesis in Tourette syndrome: association of the histamine decarboxylase gene in a large sample of families. J Med Genet 50: 760–764.

Kataoka Y, Kalanithi PS, Grantz H et al. (2010). Decreased number of parvalbumin and cholinergic interneurons in the striatum of individuals with Tourette syndrome. J Comp Neurol 518: 277–291.

Kilic F, Murphy DL, Rudnick G (2003). A human serotonin transporter mutation causes constitutive activation of transport activity. Mol Pharmacol 64: 440–446.

Koch S, Perry KW, Nelson DL et al. (2002). R-fluoxetine increases extracellular DA, NE, as well as 5-HT in rat prefrontal cortex and hypothalamus: an in vivo microdialysis and receptor binding study. Neuropsychopharmacology 27: 949–959.

Kononoff Vanhanen J, Nuutinen S, Tuominen M et al. (2016). Histamine H3 receptor regulates sensorimotor gating and dopaminergic signaling in the striatum. J Pharmacol Exp Ther 357: 264–272.

Kurlan R, Majumdar L, Deeley C et al. (1991). A controlled trial of propoxyphene and naltrexone in patients with Tourette's syndrome. Ann Neurol 30: 19–23.

Kuroki T, Meltzer HY, Ichikawa J (2003). 5-HT 2A receptor stimulation by DOI, a 5-HT 2A/2C receptor agonist, potentiates amphetamine-induced dopamine release in rat medial prefrontal cortex and nucleus accumbens. Brain Res 972: 216–221.

Leckman JF, Anderson GM, Cohen DJ et al. (1984). Whole blood serotonin and tryptophan levels in Tourette's disorder: effects of acute and chronic clonidine treatment. Life Sci 35: 2497–2503.

Leckman JF, Riddle MA, Berrettini WH et al. (1988). Elevated CSF dynorphin A [1-8] in Tourette's syndrome. Life Sci 43: 2015–2023.

Lerner A, Bagic A, Simmons JM et al. (2012). Widespread abnormality of the gamma-aminobutyric acid-ergic system in Tourette syndrome. Brain 135: 1926–1936.

Lucki I (1998). The spectrum of behaviors influenced by serotonin. Biol Psychiatry 44: 151–162.

Lv J, Liang S, Qin P et al. (2025). WWC1 mutation drives dopamine dysregulation and synaptic imbalance in Tourette's syndrome. Sci Adv 11: eadr4588.

Mahone EM, Puts NA, Edden RAE et al. (2018). GABA and glutamate in children with Tourette syndrome: a (1)H MR spectroscopy study at 7T. Psychiatry Res Neuroimaging 273: 46–53.

Maia TV, Conceição VA (2017). The roles of phasic and tonic dopamine in tic learning and expression. Biol Psychiatry 82: 401–412.

Maia TV, Conceição VA (2018). Dopaminergic disturbances in Tourette syndrome: an integrative account. Biol Psychiatry 84: 332–344.

Martín-Ruiz R, Puig MV, Celada P et al. (2001). Control of serotonergic function in medial prefrontal cortex by serotonin-2A receptors through a glutamate-dependent mechanism. J Neurosci 21: 9856–9866.

Meltzer HY (1999). The role of serotonin in antipsychotic drug action. Neuropsychopharmacology 21: 106S–115S.

Mosher LJ, Godar SC, Nelson M et al. (2017). Allopregnanolone mediates the exacerbation of Tourette-like responses by acute stress in mouse models. Sci Rep 7: 3348.

Mössner R, Müller-Vahl KR, Döring N et al. (2007). Role of the novel tryptophan hydroxylase-2 gene in Tourette syndrome. Mol Psychiatry 12: 617–619.

Moya PR, Dodman NH, Timpano KR et al. (2013a). Rare missense neuronal cadherin gene (CDH2) variants in specific obsessive-compulsive disorder and Tourette disorder phenotypes. Eur J Hum Genet 21: 850–854.

Moya PR, Wendland JR, Rubenstein LM et al. (2013b). Common and rare alleles of the serotonin transporter gene, SLC6A4, associated with Tourette's disorder. Mov Disord 28: 1263–1270.

Muller-Vahl K (2007). The pathophysiological role of the serotonergic system in Tourette syndrome. Curr Psychiatry Rev 3: 271–276.

Müller-Vahl KR, Loeber G, Kotsiari A et al. (2017). Gilles de la Tourette syndrome is associated with hypermethylation of the dopamine D2 receptor gene. J Psychiatr Res 86: 1–8.

Müller-Vahl KR, Meyer GJ, Knapp WH et al. (2005). Serotonin transporter binding in Tourette syndrome. Neurosci Lett 385: 120–125.

Müller-Vahl KR, Szejko N, Wilke F et al. (2019). Serotonin transporter binding is increased in Tourette syndrome with obsessive compulsive disorder. Sci Rep 9: 972.

Nasello C, Poppi LA, Wu J et al. (2024). Human mutations in high-confidence Tourette disorder genes affect sensorimotor behavior, reward learning, and striatal dopamine in mice. Proc Natl Acad Sci USA 121: e2307156121.

Navailles S, De Deurwaerdère P (2011). Presynaptic control of serotonin on striatal dopamine function. Psychopharmacology (Berl) 213: 213–242.

Neri V, Cardona F (2013). Clinical pharmacology of comorbid obsessive-compulsive disorder in Tourette syndrome. Int Rev Neurobiol 112: 391–414.

Nikolaus S, Mamlins E, Antke C et al. (2022). Boosted dopamine and blunted serotonin in Tourette syndrome – evidence from in vivo imaging studies. Rev Neurosci 33: 859–876.

O'Brien K, Sharrief A, Nordstrom E et al. (2018). Biochemical markers of striatal desensitization in cortical-limbic hyperglutamatergic TS- & OCD-like transgenic mice. J Chem Neuroanat 89: 11–20.

Parent A, Descarries L, Beaudet A (1981). Organization of ascending serotonin systems in the adult rat brain. A radioautographic study after intraventricular administration of [3H]5-hydroxytryptamine. Neuroscience 6: 115–138.

Pittenger C (2020). The histidine decarboxylase model of tic pathophysiology: a new focus on the histamine H(3) receptor. Br J Pharmacol 177: 570–579.

Prange S, Metereau E, Maillet A et al. (2022). Limbic serotonergic plasticity contributes to the compensation of apathy in early Parkinson's disease. Mov Disord 37: 1211–1221.

Puts NA, Harris AD, Crocetti D et al. (2015). Reduced GABAergic inhibition and abnormal sensory symptoms in children with Tourette syndrome. J Neurophysiol 114: 808–817.

Qian QQ, Tan QQ, Sun D et al. (2022). A pilot study on plasma and urine neurotransmitter levels in children with tic disorders. Brain Sci 12: 880.

Rapanelli M, Frick L, Jindachomthong K et al. (2018). Striatal signaling regulated by the H3R histamine receptor in a mouse model of tic pathophysiology. Neuroscience 392: 172–179.

Reimold M, Smolka MN, Zimmer A et al. (2007). Reduced availability of serotonin transporters in obsessive-compulsive disorder correlates with symptom severity – a [11C]DASB PET study. J Neural Transm 114: 1603–1609.

Rusheen AE, Rojas-Cabrera J, Goyal A et al. (2023). Deep brain stimulation alleviates tics in Tourette syndrome via striatal dopamine transmission. Brain 146: 4174–4190.

Sandyk R (1989). Abnormal opiate receptor functions in Tourette's syndrome. Int J Neurosci 44: 209–214.

Scahill L, Riddle MA, King RA et al. (1997). Fluoxetine has no marked effect on tic symptoms in patients with Tourette's syndrome: a double-blind placebo-controlled study. J Child Adolesc Psychopharmacol 7: 75–85.

Schmidt CJ, Fadayel GM (1996). Regional effects of MK-801 on dopamine release: effects of competitive NMDA or 5-HT2A receptor blockade. J Pharmacol Exp Ther 277: 1541–1549.

Schmidt CJ, Fadayel GM, Sullivan CK et al. (1992). 5-HT2 receptors exert a state-dependent regulation of dopaminergic function: studies with MDL 100,907 and the amphetamine analogue, 3,4-methylenedioxymethamphetamine. Eur J Pharmacol 223: 65–74.

Schmidt CJ, Sullivan CK, Fadayel GM (1994). Blockade of striatal 5-hydroxytryptamine2 receptors reduces the increase in extracellular concentrations of dopamine produced by the amphetamine analogue 3,4-methylenedioxymethamphetamine. J Neurochem 62: 1382–1389.

Segura B, Strafella AP (2013). Functional imaging of dopaminergic neurotransmission in Tourette syndrome. Int Rev Neurobiol 112: 73–93.

Silvestri R, Raffaele M, De Domenico P et al. (1994). Serotoninergic agents in the treatment of Gilles de la Tourette's syndrome. Acta Neurol (Napoli) 16: 58–63.

Singer HS, Wendlandt JT (2001). Neurochemistry and synaptic neurotransmission in Tourette syndrome. Adv Neurol 85: 163–178.

Smith GS, Dewey SL, Brodie JD et al. (1997). Serotonergic modulation of dopamine measured with [11C]raclopride and PET in normal human subjects. Am J Psychiatry 154: 490–496.

Soubrié P, Reisine TD, Glowinski J (1984). Functional aspects of serotonin transmission in the basal ganglia: a review and in vivo approach using the push-pull cannula technique. Neuroscience 13: 605–625.

Stamenkovic M, Schindler SD, Asenbaum S et al. (2001). No change in striatal dopamine re-uptake site density in psychotropic drug naive and in currently treated Tourette's disorder patients: a [(123) I]-beta-CIT SPECt-study. Eur Neuropsychopharmacol 11: 69–74.

Steeves T, Fox S Neurobiological basis of serotonin–dopamine antagonists in the treatment of Gilles de la Tourette syndrome. Prog Brain Res 172: 495–513.

Swerdlow NR, Sutherland AN (2006). Preclinical models relevant to Tourette syndrome. Adv Neurol 99: 69–88.

Tian Y, Gunther JR, Liao IH et al. (2011). GABA- and acetylcholine-related gene expression in blood correlate with tic severity and microarray evidence for alternative splicing in Tourette syndrome: a pilot study. Brain Res 1381: 228–236.

Tizabi Y, Russell LT, Johnson M, Darmani NA (2001). Nicotine attenuates DOI-induced head-twitch response in mice: implications for Tourette syndrome. Prog Neuropsychopharmacol Biol Psychiatry 25: 1445–1457.

Toren P, Weizman A, Ratner S et al. (2005). Ondansetron treatment in Tourette's disorder: a 3-week, randomized, double-blind, placebo-controlled study. J Clin Psychiatry 66: 499–503.

Vollenweider FX, Vontobel P, Hell D, Leenders KL (1999). 5-HT modulation of dopamine release in basal ganglia in psilocybin-induced psychosis in man – a PET study with [^{11}C]raclopride. Neuropsychopharmacology 20: 424–433.

Wendland JR, Moya PR, Kruse MR et al. (2008). A novel, putative gain-of-function haplotype at SLC6A4 associates with obsessive-compulsive disorder. Hum Mol Genet 17: 717–723.

Wong DF, Brašić JR, Singer HS et al. (2008). Mechanisms of dopaminergic and serotonergic neurotransmission in Tourette syndrome: clues from an in vivo neurochemistry study with PET. Neuropsychopharmacology 33: 1239–1251.

Wood M, Scott C, Clarke K et al. (2006). Pharmacological profile of antipsychotics at monoamine receptors: atypicality beyond 5-HT2A receptor blockade. CNS Neurol Disord Drug Targets 5: 445–452.

Xu M, Kobets A, Du JC et al. (2015). Targeted ablation of cholinergic interneurons in the dorsolateral striatum produces behavioral manifestations of Tourette syndrome. Proc Natl Acad Sci USA 112: 893–898.

Zitterl W, Aigner M, Stompe T et al. (2007). [^{123}I]-beta-CIT SPECT imaging shows reduced thalamus-hypothalamus serotonin transporter availability in 24 drug-free obsessive-compulsive checkers. Neuropsychopharmacology 32: 1661–1668.

Handbook of Clinical Neurology, Vol. 215 (3rd series)
Tic Disorders and Tourette Syndrome
C. Ganos and Y. Worbe, Editors
https://doi.org/10.1016/B978-0-443-13554-5.00018-3

Chapter 20

Premonitory urges and other sensory phenomena in primary tic disorders

JUAN FRANCISCO MARTIN-RODRIGUEZ[1,2,3], DANIEL MACIAS-GARCIA[1,2], AND PABLO MIR[1,2,4,*]

[1]*Servicio de Neurología y Neurofisiología Clínica, Unidad de Trastornos del Movimiento, Instituto de Biomedicina de Sevilla, (IBiS), Hospital Universitario Virgen del Rocío/CSIC/Universidad de Sevilla, Seville, Spain*

[2]*Centro de Investigación Biomédica en Red sobre Enfermedades Neurodegenerativas (CIBERNED), Madrid, Spain*

[3]*Departamento de Psicología Experimental, Facultad de Psicología, Universidad de Sevilla, Sevilla, Spain*

[4]*Departamento de Medicina, Facultad de Medicina, Universidad de Sevilla, Seville, Spain*

Abstract

Premonitory sensations, or urges, have become a central focus in the study of Gilles de la Tourette syndrome (TS). These bodily sensations, which precede tics, play a crucial role in the clinical manifestations of the syndrome. Tics are often performed to alleviate the discomfort caused by premonitory urges and provide temporary relief. Despite their importance, the causes and mechanisms of premonitory urges remain poorly understood. The severity of tics and an individual's ability to control them are closely related to the intensity and presence of these urges. Current research is actively investigating the neuroanatomic and neurophysiologic basis of premonitory urges, focusing on specific brain regions, neural pathways, and neurotransmitter systems. Understanding premonitory urges is essential for characterizing the clinical phenotype of TS and other neurodevelopmental tic disorders and for developing targeted therapies. By investigating the relationship between premonitory urges, tics, and associated clinical features – such as obsessions, compulsions, and anxiety – researchers aim to improve prognostic tools and refine treatment strategies. Ongoing investigations into these mechanisms hold great promise for improving the care and support of people with TS and, ultimately, their quality of life.

INTRODUCTION

Premonitory sensations or urges have become a central point of interest in the research on the neuroanatomic and neurophysiologic underpinnings of Gilles de la Tourette syndrome (TS). These sensations play a critical role as they offer valuable insights into key aspects of the syndrome's phenotype and possibly prognosis. The presence and strength of premonitory urges are strongly associated with the severity of tics and the individual's capacity to control them (Steinberg et al., 2010; Li et al., 2019). Premonitory urges are bodily sensations that manifest before the onset of tics. These sensations, which are perceived by most individuals with tics, are often described as uncomfortable feelings or a building tension that compels the person to respond in some way. Tics, in this context, are actions aimed at alleviating or reducing the intensity of involuntary urges. When individuals with TS engage in tic behaviors, they experience temporary relief from the discomfort associated with the premonitory urge (Rajagopal et al., 2013).

This chapter examines the complex nature of premonitory urges in primary tic disorders. We begin by reviewing the epidemiologic and clinical characteristics of these sensory phenomena, tracing the evolution of our understanding over time. We then examine the relationship between premonitory urges and common psychiatric comorbidities in TS. We then provide a comprehensive review and critical evaluation of the clinical assessment tools used to measure premonitory urges. The chapter concludes with an analysis of neuroimaging findings, neurophysiologic data, and etiologic models related to premonitory urges. Finally, we discuss the potential therapeutic implications of premonitory urges in the treatment of tic disorders.

[*]Correspondence to: Pablo Mir, Unidad de Trastornos del Movimiento. Instituto de Biomedicina de Sevilla (IBiS), Hospital Universitario Virgen del Rocío. Avda. Manuel Siurot s/n, 41013, Sevilla, España. Tel: +34 955923039, Fax: +34 955923101, E-mail: pmir@us.es.

EPIDEMIOLOGY AND CLINICAL ASPECTS

Among hyperkinetic movement disorders, tics are one of the most frequent phenotypes. The most clinically relevant primary tic disorder is TS, where tics are defined as involuntary, brief, recurrent, and stereotyped movements or vocalizations. Tics can be simple (consisting of a discrete movement or sound) or complex (involving combinations of movements and/or vocalizations). While tics are classified as *involuntary* movement disorders, early tic descriptions also noted that patients can suppress their tics through willpower for a short duration (Cruchet, 1902; Meige, 1907). Furthermore, many patients with tics experience a sensory phenomenon or premonitory urge, characterized by a growing inner tension or an urge to move, which is temporarily alleviated after the tic is expressed and typically aggravated during tic suppression. Tics are often associated with comorbidities such as obsessions, compulsions, and/or attention-deficit disorders, suggesting the additional involvement of nonmotor neuronal circuits in the pathophysiology underlying primary tic disorders.

Over the years, physicians approached primary tic disorders predominantly from a motor perspective. However, advancements in neurophysiologic techniques provided valuable insights. Specifically, in patients with TS, simple tics were differentiated from voluntary movements by observing the absence of premotor potentials before tics (Obeso et al., 1981). Initially perceived as unintended, purposeless, and involuntary movements or merely learned motor habits, tics were classically studied based on factors such as anatomic distribution, frequency, or number (Shapiro et al., 1988). Clinicians predominantly focused on external features of movements and vocalizations in patients with tic disorders, and this was reflected in the structure of the main scales used to measure the severity of tic disorders, such as the Yale Global Tic Severity Scale (Leckman et al., 1989).

However, beginning in the 1980s, there was a notable shift in attention toward the perceptions and subjective experiences of patients with TS, driven by their self-reports and a subsequent increasing body of research (Bliss, 1980; Kurlan et al., 1989; Lang, 1991; Leckman et al., 1993; Eapen et al., 1994; Kane, 1994; Kwak et al., 2003). For example, Kurlan et al. (1989) reported on three cases of TS patients, both adults and children, who experienced unpleasant sensory phenomena partially relieved by tics. In the same report, a telephone interview conducted with 34 patients with TS, aged 13–64 years, revealed that 71% experienced their tics as imminent, and 76% reported a feeling or sensation preceding the tics. Interestingly, 74% noted that this sensation was relieved after the tics occurred (Kurlan et al., 1989). Among the participants, 14 individuals with TS (41%) identified a specific focal sensory phenomenon referred to as a "sensory tic." In another study conducted at a specialized movement disorders clinic (Lang, 1991), a group of 60 TS patients and another group of 110 patients with other hyperkinetic movement disorders were asked about their subjective perception of their movements. Notably, 93% of TS patients described their movements as either entirely or largely voluntary, whereas a similar percentage of patients with nontic hyperkinetic movement disorders experienced their movements as completely involuntary. Intriguingly, the questionnaire acknowledged the term "voluntary" when tics were performed to alleviate a sensation or inner urge. These findings align with a previous work where only 24% of patients reported no sensory phenomena related to their tics (Kurlan et al., 1989).

Lang (1991) also reported that many TS patients described a characteristic and unpleasant somatic tension that builds up before their tics and is only partially relieved by their occurrence, a phenomenon that had already been noted in previous studies. In this line, Bliss (1980) clearly stated from a patient's perspective that "the movement/sound is not the sole symptom" but is preceded by "certain preliminary sensory signals and, in turn, followed by sensory impressions at the end of the action (tic)." Another first-person report argued that these sensations precipitated tics more than acting as a mere precursor, establishing a potential causal relationship from the TS patients' perspective (Kane, 1994). Anecdotal descriptions of extracorporeal sensations preceding tics in TS patients have also been documented, such as itchy sensations preceding motor tics that might manifest in other people or objects, prompting a need to scratch or touch them (Karp and Hallett, 1996). Furthermore, in a cohort of 130 patients with chronic tic disorders, Leckman et al. (1993) reported that 93% of them experienced premonitory urges, with a similar percentage perceiving their tics as fully or partially in response to these urges, consistent with previous studies (Lang, 1991). Additionally, up to 84% of the patients experienced a feeling of relief following the tic. Patients also noted that the premonitory sensation could be more bothersome than the tics themselves and described site sensitization with tactile, auditory, and/or visual stimuli (Cohen and Leckman, 1992; Leckman et al., 1993). These findings support case descriptions linking TS behaviors to both internal and external phenomena (Eapen et al., 1994), possibly related to abnormal central sensorimotor processing, altered interoceptive awareness, and atypical threshold detection of externally applied sensory stimuli (Ganos et al., 2015b; Schunke et al., 2016).

In a separate study, Kwak et al. (2003) investigated 50 patients with TS and chronic motor tics, with 92% reporting the presence of premonitory urges. Most patients (88%) described their movements as voluntary responses to involuntary sensations, with the premonitory urge intensifying if the tic was inhibited and providing transient relief after tic expression. More recently, a study with 291 adult patients confirmed the high presence of premonitory urges among chronic tic disorders (Brandt et al., 2023). The frequency and intensity of tics showed a significant correlation with the presence and strength of premonitory urges for both motor and vocal simple/complex tics (Brandt et al., 2023). Notably, the majority of patients with premonitory sensations (97%) reported experiencing temporary relief after at least one of their tics, in line with previous cohort studies (Kurlan et al., 1989; Leckman et al., 1993; Kwak et al., 2003; Essing et al., 2022) and highlighting sensory phenomena as a clinically significant aspect of TS. This sense of relief was associated with most of their tics (up to 85%), regardless of whether the tics were motor or vocal, simple or complex (Brandt et al., 2023). Interestingly, in this work, diverse qualities of premonitory urges (e.g., feeling of tension, building-up, energy, just-right experiences, discomfort, itch, and incompleteness) were described. The feeling of tension was the most frequently reported urge quality associated with different types of tics (simple or complex, motor or vocal); however, there were no differences in the perceived intensity of urges among these tics.

A specific anatomic distribution of premonitory urges has been described with some body regions identified as areas with a higher frequency of premonitory sensations (Leckman et al., 1993; Kwak et al., 2003; Ganos et al., 2012; Essing et al., 2022). A larger series subsequently revealed that premonitory urges were more commonly experienced in specific body regions, such as the head, limbs (palms, shoulders), midline abdomen, and throat, often preceding tics in the head, neck, and shoulders (Leckman et al., 1993; Kwak et al., 2003). These findings were consistent with other studies reporting similar body distributions (Ganos et al., 2012). A recent extensive multicenter study involving nearly 300 adult patients with primary tic disorders showed that patients consistently associated their premonitory urges with the anatomic regions where their tics occurred (Essing et al., 2022). Simple tics were more strongly correlated with premonitory urges than complex tics, while motor tics were more frequently associated with urges than vocal tics. However, this study focused on urges in areas with already reported tics, which implies that tics, which are not preceded by premonitory sensations, may have been underreported. This notwithstanding, the most commonly affected regions by premonitory urges for motor tics were the face/head, arms, and trunk, while for vocal tics, they were the mouth, throat, and chest. These results reinforced previous findings that the face/head is most frequently affected by premonitory urges (Kwak et al., 2003; Ganos et al., 2012). Interestingly, premonitory urges were reported to be more often localized to muscles rather than the skin or joints (Leckman et al., 1993; Kwak et al., 2003). In this line, studies evaluating the efficacy of botulinum toxin in treating tic disorders found that the treatment had some effect in reducing premonitory urge scores (Marras et al., 2001; Kwak et al., 2003).

Premonitory urges are slightly more prevalent in males, and patients typically become aware of them after the age of 10, around 3 years after the onset of tics (Leckman et al., 1993; Banaschewski et al., 2003). In adults, premonitory urge scores were positively correlated with tic severity and had the potential to impact the psychologic well-being of individuals with TS (Ganos et al., 2012; Crossley and Cavanna, 2013; Kano et al., 2015; Gu et al., 2020; Brandt et al., 2023). Notably, a cross-sectional study of 100 adult TS patients (with 50% having comorbidities like obsessive-compulsive disorder (OCD), attention-deficit hyperactivity disorder (ADHD), anxiety, or depression) indicated that the impact of premonitory urges on quality of life was greater in patients without comorbidities (Eddy and Cavanna, 2014). The Premonitory Urge for Tics Scale (PUTS) score exhibited a negative correlation with quality of life and self-reported measures of obsessive-compulsive, attention-deficit, and anxiety symptoms in the overall group, but its impact was more pronounced in uncomplicated TS. Moreover, premonitory urge scores were also found to be correlated with tic interference measured by the Yale Global Tics Severity Scale (YGTSS), supporting the notion that sensory phenomena could be equally or even more distressing than motor/vocal manifestations (Ganos et al., 2012). Subsequent studies by Openneer et al. (2020) and Brandt et al. (2023) have corroborated the adverse influence of premonitory urges on the quality of life of individuals with tic disorders.

Regarding the timing of the appearance of tics and premonitory urges in TS, the most frequent tics at the outset are often eye blinking and facial tics, which are less commonly associated with premonitory sensations (Leckman et al., 1993; Essing et al., 2022). On the other hand, head and limb tics typically manifest around 10–11 years of age, which aligns with the onset of premonitory urges (Sweet et al., 1973). In a cross-sectional study involving 251 children and adolescent patients with TS, approximately 60% of them reported experiencing both premonitory sensory phenomena and the ability to suppress their tics (Banaschewski et al., 2003). Interestingly, premonitory urges were found to be less common in younger children (under 10–14 years old), and the co-occurrence of premonitory urges and tic suppressibility did not always happen together (Ganos et al., 2012). These findings raise intriguing questions about whether the delayed awareness of premonitory urges (after 10 years old) is linked to neural maturation or the specific type of tics involved, as not all tics are associated with an urge to tics.

Since the initial descriptions of premonitory urges in chronic tic disorders, researchers have also explored the potential therapeutic implications of the tic-urge relationship (Bullen and Hemsley, 1983). Some patients have reported that being aware of their premonitory urges can aid in suppressing their tics (Cohen and Leckman, 1992; Leckman et al., 1993). In a study of 15 adult patients with Gilles de la Tourette Syndrome (GTS), Ganos et al. (2012) demonstrated that all of them reported premonitory urges and exhibited tic suppressibility; however, despite their simultaneous presence, there was no correlation between urge scores and the efficacy of tic inhibition. This may suggest that different neuronal mechanisms could underlie tic generation and tic control. In addition to an anatomic distribution of tics, a somatotopic distribution for tic inhibition was described showing differences between the two (Ganos et al., 2015b). Body parts with fewer tics were most amenable to tic inhibitory control. However, the poorly known spatial distribution of premonitory urges and its potential link with tic distribution and inhibition remains still to be solved.

The presence of premonitory urges offers valuable insights for distinguishing tics from other jerky movements. In support of this notion, Obeso et al. reported that simple motor tics did not exhibit premotor potentials compared to voluntary movements in the same individuals with TS (Obeso et al., 1981). Subsequently, another study using the same paradigm confirmed this distinction, but intriguingly, two out of five TS patients showed premotor potentials prior to their tics, with one of them experiencing premonitory urges and a subjective perception of voluntariness (Karp et al., 1996). Anecdotal case reports have also described the presence of premotor potentials preceding tics in TS patients with premonitory urges (Duggal and Nizamie, 2002); however, large EEG studies are needed (Cohen and Leckman, 1992). Of note, the presence, frequency, and intensity of premonitory urges (assessed by the PUTS) do not appear to differ between individuals with primary tic disorders and those with functional tic-like behaviors (Szejko et al., 2024).

PREMONITORY URGES AND COMORBIDITIES

Research exploring the relationship between premonitory urges and obsessive-compulsive symptoms (OCSs)/OCD has been conducted early on (Leckman et al., 1993; Miguel et al., 1997; Kwak et al., 2003). Subsequent investigations have delved further into this connection by utilizing systematic measurements of premonitory urges (Eddy and Cavanna, 2014; Openneer et al., 2020). Eddy and Cavanna (2014) found a significant correlation between the PUTS score and self-reported measures of obsessive-compulsive, attention-deficit, and anxiety symptoms. Interestingly, OCS emerged as the most influential predictor of the various PUTS scores. Additionally, in cases of complicated TS, anxiety severity seemed to influence premonitory urge scores, indicating possible differences in these phenomena depending on comorbidities (Eddy and Cavanna, 2014). A Japanese study supported these findings by demonstrating that the severity of premonitory urges was associated with the complexity of tics (measured by the YGTSS subscales) and obsessive-compulsive scores (Kano et al., 2020), consistent with previous research (Crossley and Cavanna, 2013; Gulisano et al., 2015). However, it is worth noting that not all studies have reported a correlation between PUTS scores and OCS scores (Reese et al., 2014).

In a large European study, researchers observed that the severity of PUTS increased with the age of patients, and there was a positive correlation between tic severity and OCSs within the entire cohort (Openneer et al., 2020). This correlation was particularly pronounced in patients between ages 8 and 10 years. Several separate studies have described various subtypes of premonitory sensory phenomena, including bodily and mental sensations (Miguel et al., 1997), as well as different qualities of premonitory urges (Brandt et al., 2023). Among these, the most frequently reported sensations were feelings of tension/pressure, build-up, and inner energy that needed to be relieved. Additionally, mental phenomena such as "just-right experiences" were also commonly described. Interestingly, premonitory urges such as feelings of tension, energy, or build-up were more frequently associated with comorbidities such as ADHD, depression, and anxiety. On the other hand, "just-right experiences" were more commonly reported among individuals with OCD (Table 20.1).

Patients with TS and/or OCS/OCD might exhibit differences in the phenomenology of sensations preceding repetitive behaviors. On the one hand, TS patients usually report sensory phenomena (both bodily and mental premonitory sensations), while the OCD group reports anxiety and cognitive phenomena instead of sensory experiences (Miguel et al., 1997). The presence of mental premonitory sensations (Table 20.1), specifically the "just-right" phenomenon and the feeling of incompleteness, is particularly frequent in individuals with both conditions (TS and OCD) (Miguel et al., 1997; Kwak et al., 2003). The phenomenon of "just-right" perceptions preceding motor behaviors has been extensively studied in patients with TS and OCD (Leckman et al., 1994). This type of premonitory sensation consists of repetitive actions performed until a premonitory feeling or urge is alleviated, leading the person to feel "just right." Different studies have confirmed that TS patients with comorbid obsessions/compulsions exhibited a higher prevalence of this phenomenon than those with OCD alone or TS without any OCD symptoms (Kwak et al., 2003; Neal and Cavanna, 2013). These "just-right" experiences correlated with self-reported measures of compulsivity, suggesting an intrinsic link between this type of premonitory phenomenon and obsessive-compulsive comorbidity in TS (Leckman et al., 1994). Different premonitory sensations might correlate with different repetitive behaviors, such as motor complex tics or compulsions (Rajagopal and Cavanna, 2014; Brandt et al., 2023). In a large cohort of adult patients with chronic tic disorder (Brandt et al., 2023), the presence of premonitory urges was associated with factors such as anxiety, depression, older age, and female sex. However, the intensity of premonitory urges was correlated with the presence of OCD symptoms and a younger age. Interestingly, different comorbidities (anxiety, depression, ADHD, and OCD) were associated with different qualities of premonitory urges (Table 20.1); however, the number of comorbidities did not correlate with the urge to tic's intensity but rather with the presence of these phenomena.

In a study involving 166 adolescent and adult TS patients, Worbe et al. (2010) categorized repetitive behaviors into "tic-like" and "OCD-like" behaviors (Table 20.1). The presence of repetitive behaviors was associated with more severe TS and more frequent complex tics. Those with predominantly "tic-like" behaviors performed them without anxiety and in response to an "urge." The occurrence of "just-right" phenomena was significantly higher in this group than in those with predominantly "OCD-like" behaviors, confirming previous reports of an association with TS and comorbid OCD (Miguel et al., 1997; Kwak et al., 2003). On the other hand, Brandt et al. (2023) classified "just-right" experiences as associated with OCD-like behaviors. In their cross-sectional study of 111 adults with TS, the presence of "just-right" experiences (assessed by specific questionnaires) correlated more strongly with OCD symptoms and feelings of incompleteness than with tic severity or even the specific "just-right" item on the PUTS (item 4). However, the lack of an OCD control group makes it difficult to compare these findings with the premonitory sensations associated with pure obsessions/compulsions in patients with OCD alone.

In a series of compelling studies, researchers have delved into intriguing associations between psychologic factors (personal traits), interoception, and sensory phenomena

Table 20.1

Sensory phenomena in patients with Tourette syndrome.

Type of sensory phenomena	Categories or qualities	Definition/notes
Bodily premonitory sensations Focal or generalized body sensations occurring before the repetitive behaviors	Tension	Premonitory sensation most frequently associated with tics. Correlated with levels of comorbidities (ADHD, OCS, anxiety, and depression)
	Focal energy building-up	The feeling of focal energy needed to be relieved. Correlated with levels of some comorbidities (ADHD, anxiety, and depression)
	Discomfort	Correlated with levels of anxiety and depression
	Pressure	Correlated with levels of depression
	Itch	
Mental premonitory sensations General, uncomfortable feelings or perceptions occurring before or while the repetitive behaviors are performed	Just-right experience	More frequent in patients with TS+OCD. Correlated with levels of OCS and anxiety
	Urge only	A drive/force/impulse to perform the repetitive behaviors without any obsession, fear, worry, and bodily sensation
	Energy release	A general (nonfocal) feeling of inner tension, pressure, or mental energy that builds up and needs to be discharged. Correlated with levels of OCS, anxiety, and depression
	Incompleteness	An inner sense of incompleteness, imperfection, or insufficiency (subjective feeling of discomfort). More frequent in patients with TS+OCD and correlated with levels of anxiety
	Symmetry searching rituals	Proposed as tic-like repetitive behavior
	Touching rituals	Proposed as tic-like repetitive behavior
	Counting rituals	Proposed as tic-like repetitive behavior
	Washing rituals	Proposed as OCD-like repetitive behavior
	Checking rituals	Proposed as OCD-like repetitive behavior
Others	Sensory intolerance	High levels of distress brought on by commonplace tactile sensations or sounds

TS: Gilles de la Tourette Syndrome. *OCD*: Obsessive-Compulsive Disorder. *OCS*: Obsessive-Compulsive Symptoms. *ADHD*: Attention Deficit and Hyperactivity Disorders.

among patients with OCD primary tic disorders. A noteworthy study by Lee et al. (2009) revealed a significant link between perfectionism and specific subtypes of sensory phenomena in OCD patients. The findings suggested that certain personality traits and psychologic factors could influence the type of premonitory feelings experienced by these patients. In addition, a higher proportion of self-reported OCD symptoms and tics were identified among individuals with sensory intolerance (high level of distress from everyday sounds or sensations) in a large community-based study (Taylor et al., 2014). In a separate study, Ganos et al. (2015b) observed lower interoceptive awareness among adult TS patients, which emerged as a robust predictor of premonitory urges. In this line, Schutteler and Gerlach (2023) investigated interoceptive accuracy (through different tasks such as the heartbeat counting task and the muscle tension paradigm) in children, adolescents, and young adults with TS. While no significant difference in interoceptive awareness was found between TS patients and healthy individuals, data suggested a relationship between muscular interoceptive accuracy and premonitory urges and tics.

ASSESSING PREMONITORY URGES

The presence of premonitory urges is consistently described across different populations of individuals with TS, indicating that cultural influences may not play a major role in their origin (Leckman et al., 1993; Kwak et al., 2003; Gulisano et al., 2015; Kano et al., 2015; Gu et al., 2020). Woods et al. (2005) developed PUTS, a brief self-reported questionnaire, to assess the presence and intensity of premonitory sensations in individuals with tics. Although initially designed with 10 items, psychometric analysis led to the removal of one item, resulting in a final 9-item version. The PUTS comprises six items measuring sensory/mental phenomena before the tic and four items assessing phenomena occurring just before the tic, frequency, transient relief, and tic suppressibility.

The total PUTS score (ranging from 9 to 36, with higher scores indicating more frequent premonitory urges) has shown a good correlation with overall tic severity (measured by YGTSS total score) as well as tic complexity and interference (Woods et al., 2005). Interestingly, the PUTS was found to be particularly useful in individuals older than 10 years, correlating with tic severity and complexity in this age group. However, item 10, related to the self-reported ability to temporarily suppress tics, showed a low correlation with the PUTS total score, supporting the idea that premonitory urges and tic control involve distinct neurologic mechanisms (Ganos et al., 2012).

Studies with smaller cohorts of less severe TS patients did not reveal significant correlations between PUTS and tic severity (Sutherland Owens et al., 2011; Ganos et al., 2012). However, a larger European cohort study found correlations between PUTS scores, tic severity, and comorbidities such as OCD/OCS, suggesting its validity in a wide range of ages and mild-moderate TS in terms of motor/vocal tic severity (Openneer et al., 2020). OCSs were the most correlated comorbidity with premonitory phenomena, while the correlation with measures for ADHD was weaker.

The PUTS has demonstrated good sensitivity among individuals with varying frequencies of premonitory urges, along with high internal consistency and test–retest reliability among children, adolescents, and adults from different populations. However, due to reported differences between children and adults, as well as the potential age-related effect on PUTS scores, the idea of age-specific questionnaires to assess premonitory urges has been proposed (Sutherland Owens et al., 2011; Openneer et al., 2020). The PUTS is the most widely used scale for measuring the urge to tic and has been validated in various languages and populations (Steinberg et al., 2010; Gulisano et al., 2015; Kano et al., 2015; Gu et al., 2020; Openneer et al., 2020; Forcadell et al., 2023). Recently, Baumung et al. (2021) proposed slight modifications and a revised version called PUTS-R, which consists of 24 items and includes two subscales: one measuring urge *severity* and the other urge *quality*.

In addition to the PUTS, there are other instruments available to measure premonitory phenomena, such as the University of Sao Paulo Sensory Phenomena Scale (Rosario et al., 2009) and the Not Just Right Experiences-Questionnaire Revised (Neal and Cavanna, 2013). The University of Sao Paulo Sensory Phenomena Scale has been validated in English and has good sensitivity to assess the presence of sensory phenomena, showing a reasonable correlation with PUTS scores (Sutherland Owens et al., 2011; Sampaio et al., 2014). However, sensitivity to premonitory urges in children appears to be lower than adults with tics, and University of Sao Paulo Sensory Phenomena Scale scores do not seem to correlate with tic severity or comorbidities such as obsessive-compulsive symptoms (Sutherland Owens et al., 2011; Sampaio et al., 2014).

NEUROBIOLOGY OF PREMONITORY URGES IN TIC DISORDERS

Understanding the neurobiologic basis of premonitory urges is a major challenge for researchers. Premonitory urges are covert experiential and subjective, making them difficult to detect and study. As a result, the pathophysiology of premonitory urges in tic disorders remains a key area of research that is not yet fully understood.

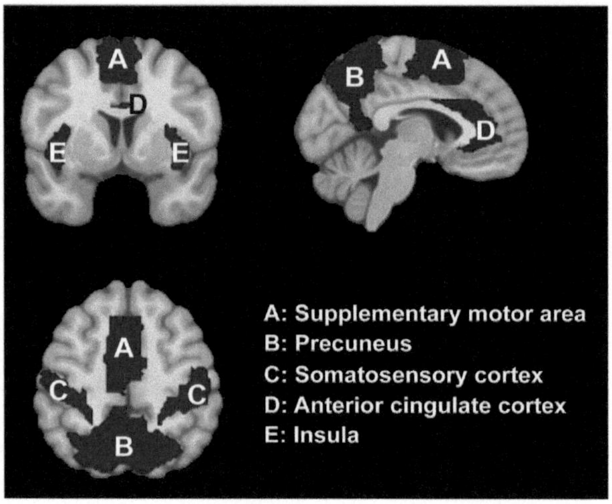

Fig. 20.1. Brain regions identified in neuroimaging studies that are commonly associated with the urge to tic.

To investigate the neural basis of premonitory urges, researchers have used a variety of neuroimaging and neurophysiology techniques, including Functional Magnetic Resonance Imaging (fMRI), Positron Emission Tomography (PET), Electroencephalograpy (EEG), or magnetoencephalography (Fig. 20.1). Several studies have highlighted that alterations in cortico-basal sensorimotor circuits may be involved in the generation of premonitory urges. For example, the supplementary motor area (Bohlhalter et al., 2006; Kataoka et al., 2010) and somatosensory cortex (Draganski et al., 2010; Wang et al., 2011) – areas previously associated with sensory phenomena in OCD—have been implicated in premonitory sensations. Notably, electric stimulation of these regions has been described to elicit the subjective sensation of a "need to move" (Fried et al., 1991).

White matter changes also appear to play a role. One study showed a negative correlation between these sensations and the integrity of the superior longitudinal fasciculus, a key tract in motor regulation and somatosensory function (Makris et al., 2005; Draganski et al., 2010). Beyond sensorimotor regions, other areas such as the insula, right ventrolateral prefrontal cortex, the medial cingulate cortex, the bilateral temporoparietal cortices, the striatum, and the thalamus have been implicated in the generation of physiologic urges to move such as blinking or itching (Leknes et al., 2007; Berman et al., 2012). These findings suggest that broad neural networks are involved in the generation and perception of premonitory urges.

Magnetoencephalography studies support the importance of sensorimotor processes in tic generation. Biermann-Ruben et al. (2012) observed that movement-evoked field amplitudes during self-paced movements were negatively correlated with tic severity and frequency, suggesting that changes in sensory feedback during voluntary movements may influence tic control. In another study, Tinaz et al. (2015) found that altered limbic input to the supplementary motor cortex, combined with changes in local network function involving GABA, may contribute to disrupted sensorimotor processing underlying tics. Indeed, emerging research indicates a distinct negative correlation between levels of bilateral supplementary motor cortex GABA and the presence, frequency, and intensity of premonitory urges; however, no such correlation has been observed with tic severity (He et al., 2022; Larsh et al., 2023).

Structural neuroimaging research has also linked paralimbic and sensorimotor changes to subjective reports of premonitory urges. Using both structural MRI and MR spectroscopy, Puts et al. (2015) reported reduced GABA in the sensorimotor cortex of children with TS. Draper et al. (2016) further found that cortical thinning in the insula and sensorimotor cortex correlated with reports of premonitory urges in young adults with TS, corroborating previous findings (Draganski et al., 2010). It is worth noting, however, that some of these findings may be explained through motion artifacts, such as the high frequency of small-amplitude head movements during scanning (Richards and Black, 2016). Collectively, these studies highlight the complex interplay between sensorimotor processing, paralimbic regions, and premonitory urges in tic disorders (Cavanna, et al., 2017).

fMRI research has been particularly informative in revealing key brain regions associated with the urge to tic, consistently pointing to the insula, cingulate cortex, and supplementary motor area. In an event-related fMRI study, Bohlhalter et al. (2006) recorded neural activation 2 seconds before tic expression and at tic onset. They found that a limbic network, including the anterior cingulate and insular cortex, supplementary motor area, and parietal operculum, was active just before tics, reflecting the subjective experience of the premonitory urge. At tic onset, sensorimotor regions such as the cerebellum and bilateral superior parietal lobule were activated. In addition, Hampson et al. (2009) showed that the supplementary motor area had a broader cross-correlation with the motor cortex during tics compared to typical intentional movements in healthy controls. Recent task-based functional connectivity studies suggest that increased activity in the precuneus, particularly under emotional conditions, may also contribute to premonitory urges (Bhikram, et al., 2021). These findings reinforce the crucial role of the supplementary motor area in tic generation while highlighting the involvement of the insula, precuneus, cingulate cortex, and other limbic regions in the urge to tic.

Although functional neuroimaging is extremely useful, its correlational nature makes it difficult to establish causality. Research based on network mapping using lesion analysis or structural changes associated with tics offers a more direct perspective on the neural circuits that drive tic symptoms. These studies consistently point to a network involving the basal ganglia, insula, anterior cingulate cortex, thalamus, and cerebellum (Ganos et al., 2022; Zouki et al., 2023). There is also evidence that some of these structures contribute to the antecedent sensations that accompany tics. Interestingly, several of these regions overlap with areas associated with normal urges such as micturition, swallowing, and blinking in both healthy individuals and patients with degenerative disorders (Hadad et al., 2024). This overlap may reflect common neural mechanisms governing both pathologic and physiologic urges (Zouki et al., 2024). There is also increasing evidence that regions outside the sensorimotor network are involved in premonitory urges. One proposed model for sensations that lead to relief-seeking actions highlights three regions: the anterior insula, which triggers the need to relieve the sensation; the anterior cingulate cortex, which selects the appropriate behavior; and a more medial insular region, which evaluates the consequences of the action and generates relief (Jackson et al., 2011). Studies in individuals with chronic TS support this model, showing increased activity in the anterior insula, which may reflect heightened interoceptive awareness, a possible explanation for premonitory urges (Ganos et al., 2015b). Furthermore, the dorsal anterior insula shows strong functional connectivity with fronto-striatal regions in TS and is positively correlated with the severity of premonitory sensations on the PUTS scale (Tinaz et al., 2015).

UNVEILING THE THERAPEUTIC SIGNIFICANCE OF PREMONITORY URGES: OPPORTUNITIES AND CHALLENGES

The identification and assessment of premonitory urges are important in managing of patients with chronic tic disorders. Premonitory urges are closely associated with tic production and have a significant impact on the severity of TS and the overall quality of life of patients (Brandt et al., 2023). Research suggests that treatments aimed at reducing premonitory urges can help control tics (Martin-Rodriguez and Mir, 2018). As a result, focusing on premonitory urges may lead to better outcomes and improved well-being for people with chronic tic disorders.

Recognizing an inner need or urge has reframed our understanding of tics as "deliberate involuntary actions." In other words, tics feel both chosen and impossible to resist. This paradox highlights new opportunities to intervene by targeting the link between urges and tic expression. Indeed, premonitory urges are central to evidence-based behavioral treatments such as habit reversal learning and comprehensive behavioral intervention for tics. These approaches train patients to recognize urges and use competing responses to prevent tic expression (Woods et al., 2008). However, less than 40% of adults with TS benefit from these treatments (Wilhelm et al., 2012), suggesting that the relationship between urges and tics may be more complex than a simple cause and effect relationship. New models are being developed to address this complexity (Friedrich et al., 2023), which may pave the way for more effective therapeutic strategies.

Despite the importance of premonitory urges, relatively little research has examined how different treatments affect them. Most data come from small case series examining medications (e.g., antidopaminergic drugs), therapies targeting the endocannabinoid system (Curtis et al., 2009; Muller-Vahl et al., 2022), or deep brain stimulation (Hashemiyoon et al., 2017; Kano et al., 2018). The limited focus on patients' internal sensory experience may partly explain why few studies have evaluated how pharmacologic therapies affect urges (Roessner et al., 2022). Some benefits have been reported with botulinum toxin injections (Marras et al., 2001;

Kwak et al., 2003; Porta et al., 2004), although the mechanism remains unclear. Antidopaminergic medications are associated with a reduced likelihood to experience urges but do not affect urgency intensity (Brandt et al., 2023). Topiramate may provide some relief (Kuo and Jimenez-Shahed, 2010; Billnitzer and Jankovic, 2020), but standardized assessments (e.g., PUTS) are often lacking, and topiramate is not considered a first-line treatment of TS (Billnitzer and Jankovic, 2020; Roessner et al., 2022). Furthermore, some treatments may reduce a patient's interoceptive awareness and thus alter the perceived intensity of urges (Ganos et al., 2015a).

Recently, nonpharmacologic approaches have also shown promise in reducing tics and the intensity of premonitory urges. For example, Morera Maiquez et al. (2020, 2023) tested indirect rhythmic electric stimulation of the sensorimotor cortex by stimulating the median nerve at EEG frequency associated with motor inhibition. This method reduced tic frequency by 31% and self-reported urge to tic by 30% during active stimulation. Iverson et al. (2023a, 2023b) reported similar results. While it remains unclear whether these improvements are sustained beyond the stimulation period and are not reflective of placebo effects, these findings suggest a promising new treatment avenue for chronic tic disorders that merits further study.

CONCLUSIONS AND PROSPECTS

Premonitory urges are a common feature of primary tic disorders. Patients often describe them as a tense or uncomfortable sensation in their muscles that drives them to perform a tic to provide brief relief. Premonitory urges are strongly associated with the presence and severity of tics, particularly those involving the face and head. Certain comorbidities also correlate with certain types of premonitory urges, and for some people, these urges can be more distressing that the tics themselves, reducing the overall quality of life. Therefore healthcare providers should routinely assess premonitory urges to provide more comprehensive support. Although premonitory urges are considered potentially beneficial for tic suppression, the relationship between tic generation and inhibition remains unclear, suggesting that different neurologic mechanisms may be involved. Further research is needed to determine whether therapies that target these internal sensations may offer new strategies for improving the quality of life of people with chronic tic disorders.

References

Banaschewski T, Woerner W, Rothenberger A (2003). Premonitory sensory phenomena and suppressibility of tics in Tourette syndrome: developmental aspects in children and adolescents. Dev Med Child Neurol 45: 700–703.

Baumung L, Muller-Vahl K, Dyke K et al. (2021). Developing the Premonitory Urges for Tic Disorders Scale-Revised (PUTS-R). J Neuropsychol 15: 129–142.

Berman BD, Horovitz SG, Morel B et al. (2012). Neural correlates of blink suppression and the buildup of a natural bodily urge. Neuroimage 59: 1441–1450.

Bhikram T, Crawley A, Arnold P et al. (2021). Neuroimaging the emotional modulation of urge inhibition in Tourette Syndrome. Cortex 135: 341–351.

Biermann-Ruben K, Miller A, Franzkowiak S et al. (2012). Increased sensory feedback in Tourette syndrome. Neuroimage 63: 119–125.

Billnitzer A, Jankovic J (2020). Current management of tics and Tourette syndrome: behavioral, pharmacologic, and surgical treatments. Neurotherapeutics 17: 1681–1693.

Bliss J (1980). Sensory experiences of Gilles de la Tourette syndrome. Arch Gen Psychiatry 37: 1343–1347.

Bohlhalter S, Goldfine A, Matteson S et al. (2006). Neural correlates of tic generation in Tourette syndrome: an event-related functional MRI study. Brain 129: 2029–2037.

Brandt V, Essing J, Jakubovski E et al. (2023). Premonitory urge and tic severity, comorbidities, and quality of life in chronic tic disorders. Mov Disord Clin Pract 10: 922–932.

Bullen JG, Hemsley DR (1983). Sensory experience as a trigger in Gilles de la Tourette's syndrome. J Behav Ther Exp Psychiatry 14: 197–201.

Cavanna AE, Black KJ, Hallett M et al. (2017). Neurobiology of the premonitory urge in Tourette's syndrome: pathophysiology and treatment implications. J Neuropsychiatry Clin Neurosci 29: 95–104.

Cohen AJ, Leckman JF (1992). Sensory phenomena associated with Gilles de la Tourette's syndrome. J Clin Psychiatry 53: 319–323.

Crossley E, Cavanna AE (2013). Sensory phenomena: clinical correlates and impact on quality of life in adult patients with Tourette syndrome. Psychiatry Res 209: 705–710.

Cruchet, R. (1902) Étude critique sur le tic convulsif et son traitement gymnastique (méthode de Brissaud et méthode de Pitres), Gounouilhou.

Curtis A, Clarke CE, Rickards HE (2009). Cannabinoids for Tourette's syndrome. Cochrane Database Syst Rev 2009: CD006565.

Draganski B, Martino D, Cavanna AE et al. (2010). Multispectral brain morphometry in Tourette syndrome persisting into adulthood. Brain 133: 3661–3675.

Draper A, Jackson GM, Morgan PS et al. (2016). Premonitory urges are associated with decreased grey matter thickness within the insula and sensorimotor cortex in young people with Tourette syndrome. J Neuropsychol 10: 143–153.

Duggal HS, Nizamie SH (2002). Bereitschaftspotential in tic disorders: a preliminary observation. Neurol India 50: 487–489.

Eapen V, Moriarty J, Robertson MM (1994). Stimulus induced behaviours in Tourette's syndrome. J Neurol Neurosurg Psychiatry 57: 853–855.

Eddy CM, Cavanna AE (2014). Premonitory urges in adults with complicated and uncomplicated Tourette syndrome. Behav Modif 38: 264–275.

Essing J, Jakubovski E, Psathakis N et al. (2022). Premonitory urges reconsidered: urge location corresponds to tic location in patients with primary tic disorders. J Mov Disord 15: 43–52.

Forcadell E, Garcia-Delgar B, Nicolau R et al. (2023). Tic disorders and premonitory urges: validation of the Spanish-language version of the Premonitory Urge for Tics Scale in children and adolescents. Neurologia (Engl Ed) 38: 319–325.

Fried I, Katz A, McCarthy G et al. (1991). Functional organization of human supplementary motor cortex studied by electrical stimulation. J Neurosci 11: 3656–3666.

Friedrich J, Rawish T, Bluschke A et al. (2023). Cognitive and neural mechanisms of behavior therapy for tics: a perception-action integration approach. Biomedicines 11: 1550.

Ganos C, Al-Fatly B, Fischer JF et al. (2022). A neural network for tics: insights from causal brain lesions and deep brain stimulation. Brain 145: 4385–4397.

Ganos C, Asmuss L, Bongert J et al. (2015). Volitional action as perceptual detection: predictors of conscious intention in adolescents with tic disorders. Cortex 64: 47–54.

Ganos C, Garrido A, Navalpotro-Gomez I et al. (2015). Premonitory urge to tic in Tourette's is associated with interoceptive awareness. Mov Disord 30: 1198–1202.

Ganos C, Kahl U, Schunke O et al. (2012). Are premonitory urges a prerequisite of tic inhibition in Gilles de la Tourette syndrome? J Neurol Neurosurg Psychiatry 83: 975–978.

Gu Y, Li Y, Cui Y (2020). Correlation between premonitory urges and tic symptoms in a Chinese population with tic disorders. Pediatr Investig 4: 86–90.

Gulisano M, Cali P, Palermo F et al. (2015). Premonitory urges in patients with Gilles de la Tourette syndrome: an Italian translation and a 7-year follow-up. J Child Adolesc Psychopharmacol 25: 810–816.

Hadad R, Mandelli ML, Rankin KP et al. 2024. Itching frequency and neuroanatomic correlates in frontotemporal lobar degeneration. JAMA Neurol. 81: 977-984.

Hampson M, Tokoglu F, King RA et al. (2009). Brain areas coactivating with motor cortex during chronic motor tics and intentional movements. Biol Psychiatry 65: 594–599.

Hashemiyoon R, Kuhn J, Visser-Vandewalle V (2017). Putting the pieces together in Gilles de la Tourette syndrome: exploring the link between clinical observations and the biological basis of dysfunction. Brain Topogr 30: 3–29.

He JL, Mikkelsen M, Huddleston DA et al. (2022). Frequency and intensity of premonitory urges-to-tic in Tourette syndrome is associated with supplementary motor area GABA+ levels. Mov Disord 37: 563–573.

Iverson AM, Arbuckle AL, Song DY et al. (2023). Median nerve stimulation for treatment of tics: a 4-week open trial with ecological momentary assessment. J Clin Med 12: 2545.

Iverson AM, Arbuckle AL, Ueda K et al. (2023). Median nerve stimulation for treatment of tics: randomized, controlled, crossover trial. J Clin Med 12: 2514.

Jackson SR, Parkinson A, Kim SY et al. (2011). On the functional anatomy of the urge-for-action. Cogn Neurosci 2: 227–243.

Kane MJ (1994). Premonitory urges as "attentional tics" in Tourette's syndrome. J Am Acad Child Adolesc Psychiatry 33: 805–808.

Kano Y, Fujio M, Kaji N et al. (2020). Changes in sensory phenomena, tics, obsessive-compulsive symptoms, and global functioning of Tourette syndrome: a follow-up after four years. Front Psychiatry 11: 619.

Kano Y, Matsuda N, Nonaka M et al. (2015). Sensory phenomena related to tics, obsessive-compulsive symptoms, and global functioning in Tourette syndrome. Compr Psychiatry 62: 141–146.

Kano Y, Matsuda N, Nonaka M et al. (2018). Sensory phenomena and obsessive-compulsive symptoms in Tourette syndrome following deep brain stimulation: two case reports. J Clin Neurosci 56: 199–201.

Karp BI, Hallett M (1996). Extracorporeal 'phantom' tics in Tourette's syndrome. Neurology 46: 38–40.

Karp BI, Porter S, Toro C et al. (1996). Simple motor tics may be preceded by a premotor potential. J Neurol Neurosurg Psychiatry 61: 103–106.

Kataoka Y, Kalanithi PS, Grantz H et al. (2010). Decreased number of parvalbumin and cholinergic interneurons in the striatum of individuals with Tourette syndrome. J Comp Neurol 518: 277–291.

Kuo SH, Jimenez-Shahed J (2010). Topiramate in treatment of tourette syndrome. Clin Neuropharmacol 33: 32–34.

Kurlan R, Lichter D, Hewitt D (1989). Sensory tics in Tourette's syndrome. Neurology 39: 731–734.

Kwak C, Dat Vuong K, Jankovic J (2003). Premonitory sensory phenomenon in Tourette's syndrome. Mov Disord 18: 1530–1533.

Lang A (1991). Patient perception of tics and other movement disorders. Neurology 41: 223–228.

Larsh TR, Huddleston DA, Horn PS et al. (2023). From urges to tics in children with Tourette syndrome: associations with supplementary motor area GABA and right motor cortex physiology. Cereb Cortex 33: 3922–3933.

Leckman JF, Riddle MA, Hardin MT et al. (1989). The Yale Global Tic Severity Scale: initial testing of a clinician-rated scale of tic severity. J Am Acad Child Adolesc Psychiatry 28: 566–573.

Leckman JF, Walker DE, Cohen DJ (1993). Premonitory urges in Tourette's syndrome. Am J Psychiatry 150: 98–102.

Leckman JF, Walker DE, Goodman WK et al. (1994). "Just right" perceptions associated with compulsive behavior in Tourette's syndrome. Am J Psychiatry 151: 675–680.

Lee JC, Prado HS, Diniz JB et al. (2009). Perfectionism and sensory phenomena: phenotypic components of obsessive-compulsive disorder. Compr Psychiatry 50: 431–436.

Leknes SG, Bantick S, Willis CM et al. (2007). Itch and motivation to scratch: an investigation of the central and peripheral correlates of allergen- and histamine-induced itch in humans. J Neurophysiol 97: 415–422.

Li Y, Wang F, Liu J et al. (2019). The correlation between the severity of premonitory urges and tic symptoms: a meta-analysis. J Child Adolesc Psychopharmacol 29: 652–658.

Makris N, Kennedy DN, McInerney S et al. (2005). Segmentation of subcomponents within the superior longitudinal fascicle in humans: a quantitative, in vivo, DT-MRI study. Cereb Cortex 15: 854–869.

Marras C, Andrews D, Sime E et al. (2001). Botulinum toxin for simple motor tics: a randomized, double-blind, controlled clinical trial. Neurology 56: 605–610.

Martin-Rodriguez JF, Mir P (2018). Automatic and voluntary motor inhibition: Intact processes for tic suppression? Mov Disord 33: 1667–1669.

Meige H (1907). Tics and their treatment. William Wood & Company.

Miguel EC, Baer L, Coffey BJ et al. (1997). Phenomenological differences appearing with repetitive behaviours in obsessive-compulsive disorder and Gilles de la Tourette's syndrome. Br J Psychiatry 170: 140–145.

Morera Maiquez B, Sigurdsson HP, Dyke K et al. (2020). Entraining movement-related brain oscillations to suppress tics in Tourette syndrome. Curr Biol 30: 2334–2342.e3.

Morera Maiquez B, Smith C, Dyke K et al. (2023). A double-blind, sham-controlled, trial of home-administered rhythmic 10-Hz median nerve stimulation for the reduction of tics, and suppression of the urge-to-tic, in individuals with Tourette syndrome and chronic tic disorder. J Neuropsychology 17: 540–563.

Muller-Vahl KR, Fremer C, Beals C et al. (2022). Endocannabinoid modulation using monoacylglycerol lipase inhibition in Tourette Syndrome: a phase 1 randomized, placebo-controlled study. Pharmacopsychiatry 55: 148–156.

Neal M, Cavanna AE (2013). "Not just right experiences" in patients with Tourette syndrome: complex motor tics or compulsions? Psychiatry Res 210: 559–563.

Obeso JA, Rothwell JC, Marsden CD (1981). Simple tics in Gilles de la Tourette's syndrome are not prefaced by a normal premovement EEG potential. J Neurol Neurosurg Psychiatry 44: 735–738.

Openneer TJC, Tarnok Z, Bognar E et al. (2020). The Premonitory Urge for Tics Scale in a large sample of children and adolescents: psychometric properties in a developmental context. An EMTICS study. Eur Child Adolesc Psychiatry 29: 1411–1424.

Porta M, Maggioni G, Ottaviani F et al. (2004). Treatment of phonic tics in patients with Tourette's syndrome using botulinum toxin type A. Neurol Sci 24: 420–423.

Puts NA, Harris AD, Crocetti D et al. (2015). Reduced GABAergic inhibition and abnormal sensory symptoms in children with Tourette syndrome. J Neurophysiol 114: 808–817.

Rajagopal S, Cavanna AE (2014). Premonitory urges and repetitive behaviours in adult patients with Tourette syndrome. Neurol Sci 35: 969–971.

Rajagopal S, Seri, Cavanna AE (2013). Premonitory urges and sensorimotor processing in Tourette syndrome. Behav Neurol 27: 65–73.

Reese HE, Scahill L, Peterson AL et al. (2014). The premonitory urge to tic: measurement, characteristics, and correlates in older adolescents and adults. Behav Ther 45: 177–186.

Richards CA, Black KJ (2016). Tourette syndrome research highlights 2015. F1000Res 5: 1493.

Roessner V, Eichele H, Stern JS et al. (2022). European clinical guidelines for Tourette syndrome and other tic disorders-version 2.0. Part III: pharmacological treatment. Eur Child Adolesc Psychiatry 31: 425–441.

Rosario MC, Prado HS, Borcato S et al. (2009). Validation of the University of Sao Paulo Sensory Phenomena Scale: initial psychometric properties. CNS Spectr 14: 315–323.

Sampaio AS, McCarthy KD, Mancuso E et al. (2014). Validation of the University of Sao Paulo's Sensory Phenomena Scale-English version. Compr Psychiatry 55: 1330–1336.

Schunke O, Grashorn W, Kahl U et al. (2016). Quantitative Sensory Testing in adults with Tourette syndrome. Parkinsonism Relat Disord 24: 132–136.

Schutteler C, Gerlach AL (2023). Metacognitions and interoceptive sensibility in the perception of premonitory urges in tic disorders across the lifespan. Z Kinder Jugendpsychiatr Psychother 51: 275–282.

Shapiro AK., Shapiro ES, Young JG et al. 1988. Gilles de la Tourette syndrome. Raven Press, Publishers.

Steinberg T, Shmuel Baruch S, Harush A et al. (2010). Tic disorders and the premonitory urge. J Neural Transm (Vienna) 117: 277–284.

Sutherland Owens AN, Miguel EC, Swerdlow NR (2011). Sensory gating scales and premonitory urges in Tourette syndrome. ScientificWorldJournal 11: 736–741.

Sweet RD, Solomon GE, Wayne H et al. (1973). Neurological features of Gilles de la Tourette's syndrome. J Neurol Neurosurg Psychiatry 36: 1–9.

Szejko N, Fletcher J, Martino D et al. (2024). Premonitory urge in patients with tics and functional tic-like behaviors. Mov Disord Clin Pract 11: 276–281.

Taylor S, Conelea CA, McKay D et al. (2014). Sensory intolerance: latent structure and psychopathologic correlates. Compr Psychiatry 55: 1279–1284.

Tinaz S, Malone P, Hallett M et al. (2015). Role of the right dorsal anterior insula in the urge to tic in Tourette syndrome. Mov Disord 30: 1190–1197.

Wang Z, Maia TV, Marsh R et al. (2011). The neural circuits that generate tics in Tourette's syndrome. Am J Psychiatry 168: 1326–1337.

Wilhelm S, Peterson AL, Piacentini J et al. (2012). Randomized trial of behavior therapy for adults with Tourette syndrome. Arch Gen Psychiatry 69: 795–803.

Woods DW, Piacentini J, Chang S et al. (2008). Managing Tourette syndrome: a behavioral intervention for children and adults therapist guide. Oxford University Press.

Woods DW, Piacentini J, Himle MB et al. (2005). Premonitory Urge for Tics Scale (PUTS): initial psychometric results and examination of the premonitory urge phenomenon in youths with Tic disorders. J Dev Behav Pediatr 26: 397–403.

Worbe Y, Mallet L, Golmard JL et al. (2010). Repetitive behaviours in patients with Gilles de la Tourette syndrome: tics, compulsions, or both? PLoS One 5: e12959.

Zouki JJ, Eapen V, Efron D et al. (2024). Functional brain networks associated with the urge for action: implications for pathological urge. Neurosci Biobehav Rev 163: 105779.

Zouki JJ, Ellis EG, Morrison-Ham J et al. (2023). Mapping a network for tics in Tourette syndrome using causal lesions and structural alterations. Brain Commun 5: fcad105.

Handbook of Clinical Neurology, Vol. 215 (3rd series)
Tic Disorders and Tourette Syndrome
C. Ganos and Y. Worbe, Editors
https://doi.org/10.1016/B978-0-443-13554-5.00007-9

Chapter 21

Inhibitory control and voluntary tic suppression

SIMON MORAND-BEAULIEU[1,2,*]

[1]*Department of Psychology, McGill University, Montréal, QC, Canada*

[2]*Department of Psychology, Université du Québec à Montréal, Montréal, QC, Canada*

Abstract

Tics, such as those seen in Tourette syndrome (TS), are brief, sudden, and involuntary movements or vocalizations. Although tics are involuntary, most individuals with TS can suppress them on command for short periods of time. This capacity, which is called voluntary tic suppression, is a hallmark feature of TS and differentiates it from most other movement disorders. Several studies have investigated inhibitory control of voluntary actions in TS, given its potential associations with both tic expression and voluntary tic suppression. However, studies on response inhibition in TS have produced quite mixed results. Furthermore, the mechanisms of voluntary tic suppression remain poorly understood. The current chapter offers an overview of the evidence related to inhibitory control of voluntary actions and voluntary tic suppression in TS. It reviews evidence from clinical, behavioral, neuropsychologic, and neurobiologic studies in order to understand to what extent these processes may be similar or different.

INTRODUCTION

The study of Tourette syndrome (TS) is quite fascinating. Though it is fundamentally a neurologic disorder, its severity is influenced by several factors that are not directly related with its neurologic origin. Fatigue, boredom, anxiety, thinking about or focusing on tics, and social situations are all factors that have been shown to increase tic severity (Iverson and Black, 2022). Furthermore, tics change across time, and have been shown to have a fractal distribution (Peterson and Leckman, 1998; Beeler et al., 2022). In essence, within an hour, there can be a few minutes where tics will be more intense. Within a day, there will likely be a few hours with more severe tics. During a week, there will for sure be days when tics seem more present than others. Same goes for months, years, and even decades. Thus, tic frequency and severity are highly variable and influenced by several factors.

Tics are often preceded or accompanied by feelings of discomfort, tension, or urge to move, which are known as premonitory urges (Cohen and Leckman, 1992; Leckman et al., 1993; Cavanna et al., 2017). These urges can be felt as focal or generalized uncomfortable sensations (Rajagopal et al., 2013). Discomfort or tension can often be temporarily relieved following the expression of a single or a series of multiple tics (Leckman et al., 2022). Premonitory urges have been recognized as a crucial feature of TS for a long time. In the early 20th century, Meige and Feindel (1902) reported the existence of an irresistible impulse preceding tic onset. An important description of premonitory urges was later given by Joseph Bliss, who reported various unpleasant sensations in certain parts of his body prompting him to tic in order to get relief from those sensations (Bliss et al., 1980). Such reports revealed that tics were not as involuntary as once thought. Tics have thus sometimes been described as "unvoluntary" (Tourette Syndrome Classification Study Group, 1993), because they can be considered as being halfway between voluntary and involuntary movements (Cavanna and Nani, 2013). Thus, if tics are not fully involuntary movements, they can be controlled to an extent. Indeed, in most individuals with TS, tics can be suppressed for short periods (Ueda et al., 2021). Voluntary tic suppression is a core feature of TS and has been the subject of several investigations, given its relevance as a mechanism leading to decreased tic severity. Researchers interested in tics and their suppression often used tasks assessing the voluntary control of actions to better understand the neurocognitive and neurobiologic correlates of voluntary tic suppression. Thus, response inhibition became one of the most studied neurocognitive processes in TS (Morand-Beaulieu et al., 2017b). Over the years, studies of volitional inhibitory control have shared different interpretations about tics and their suppression. Such views notably include that (1) tics

[*]Correspondence to: Simon Morand-Beaulieu, PhD, Department of Psychology, Université du Québec à Montréal, 100 Sherbrooke W Street, Montréal, QC H2X 3P2, Canada. E-mail: morand-beaulieu.simon@courrier.uqam.ca.

(and the inability to control them) could result from impairments in volitional inhibitory control and (2) neural processes involved in response inhibition during experimental tasks may also be involved in tic suppression. The former view is considered to be somewhat outdated nowadays (see Ganos et al., 2018b; Jackson and Jackson, 2020), and the latter has been the subject of several recent studies (e.g., Rae et al., 2020).

This chapter will first review evidence from neuropsychologic and neurobiologic studies pertaining to response inhibition in children and adults with TS. The focus will then shift to studies pertaining to clinical, cognitive, and neurobiologic features of voluntary tic suppression. The final section of the chapter focuses on whether there are shared features or mechanisms between response inhibition and voluntary tic suppression.

RESPONSE INHIBITION IN TOURETTE SYNDROME

In TS, behavioral assessment of response inhibition using neuropsychologic testing mostly began in the 1990s (e.g., Ozonoff et al., 1994; Ozonoff et al., 1998; Sherman et al., 1998), and picked up steam in the next two decades. Those studies have provided quite mixed results, as diminished (e.g., Goudriaan et al., 2005; Eddy et al., 2012), intact (e.g., Marsh et al., 2007; Roessner et al., 2008), and even enhanced (e.g., Jackson et al., 2007; Thibault et al., 2009; Jackson et al., 2011) response inhibition abilities have been reported. Several factors, such as the presence of co-occurring disorders, medication intake, or the use of different experimental tasks, may explain such differences across studies. A meta-analysis, which included 61 studies and more than 1700 individuals with TS, aimed to better understand the portrait of response inhibition in TS (Morand-Beaulieu et al., 2017a). The main finding was that both children and adults with TS showed reduced response inhibition abilities compared to individuals without TS (Fig. 21.1). Between-groups differences were evident for sentence completion tasks, Stroop tasks, circle tracing tasks, and the Continuous Performance Test (CPT). However, for other commonly used tasks, such as the Go/NoGo task or the stop-signal task, there were no significant between-group differences. Attention-deficit/hyperactivity disorder (ADHD) was found to have an important impact on response inhibition. Effects sizes for between-group comparisons with controls were twice as large in individuals with TS+ADHD, compared with those without co-occurring ADHD. However, reduced performance in response inhibition tasks was also found in individuals without clinically diagnosed co-occurring conditions. Those differences were small but could suggest that there are some difficulties with response inhibition that are inherently part of TS. However, even without a diagnosis of a co-occurring condition, some traits associated with common co-occurring conditions, such as inattention, hyperactivity, or obsessive-compulsive symptoms, may have been present and may be responsible for such differences. For instance, many studies using a 2×2 factorial design to assess the co-occurrence of TS and ADHD have reported enhanced inattention/hyperactivity symptoms in children with TS without ADHD relative to typically developing controls (Shephard et al., 2016; Openneer et al., 2020; Morand-Beaulieu et al., 2022). Thus, it is hard, in experimental studies and even more in meta-analyses, to disentangle the impact of the core symptoms of TS (i.e., tics) on neuropsychologic measures of response inhibition from that of other frequent co-occurring symptoms, such as hyperactivity or inattention.

Since the publication of that meta-analysis, response inhibition continues to be an important topic in TS research. Notably, recent studies have tried to parse out distinct aspects of cognitive control to better understand how it may be related to the pathophysiology of TS. For instance, a few studies have investigated automatic inhibition with the aid of subliminal priming. In this type of paradigm, a subliminal prime (or cue) is presented, which is followed by a supraliminal target after a short delay. That target can be either compatible or incompatible with the prime. The delay between the prime and the target is called the stimulus onset asynchrony. A short (< 100 ms) or very long (>200 ms) stimulus onset asynchrony usually improves performance (positive compatibility effect), while stimulus onset asynchronies between 100 and 200 ms typically decrease performance (negative compatibility effect). The negative compatibility effect is thought to reflect the automatic inhibition of motor responses (Eimer and Schlaghecken, 2003; Schlaghecken et al., 2008). Using such paradigm, Stenner et al. (2018) found intact automatic inhibition in adults with TS. However, they only assessed automatic inhibition at a single stimulus onset asynchrony, which does not allow investigating the full range of priming effects. This issue was notably tackled by Rawji et al. (2020), who used a masked priming task with eight different stimulus onset asynchronies ranging from 0 to 250 ms. In their study, they also used a stop-signal task in order to assess both volitional and automatic inhibition in the same sample of adults with TS. By using those two tasks, they wished to investigate processes associated with not only tic suppression but also tic generation. They found intact volitional inhibition (including both reactive and proactive inhibition) of motor responses in TS, as assessed with the stop-signal task. However, in the masked priming task, adults with TS did not show a negative compatibility effect at stimulus onset asynchronies of 100 or 150 ms, which suggests impaired automatic inhibition in those individuals. Based on those results, the authors suggested that intact volitional inhibition could be used to voluntary suppress tics, while deficits in automatic inhibition may give rise to tics. In another study, proactive and reactive inhibition in TS were further parsed using advanced metrics during the stop-signal task (Indrajeet et al., 2022). The metric representing proactive inhibition estimated the time spent waiting for a stop signal in order to improve the chances of successfully inhibiting the motor response. The metric representing reactive inhibition estimated the slowing down of preparatory activity initiated by the onset of the stop signal. Using those advanced metrics, the authors found that only reactive inhibition was impaired in adults with TS, while proactive inhibition was intact. These

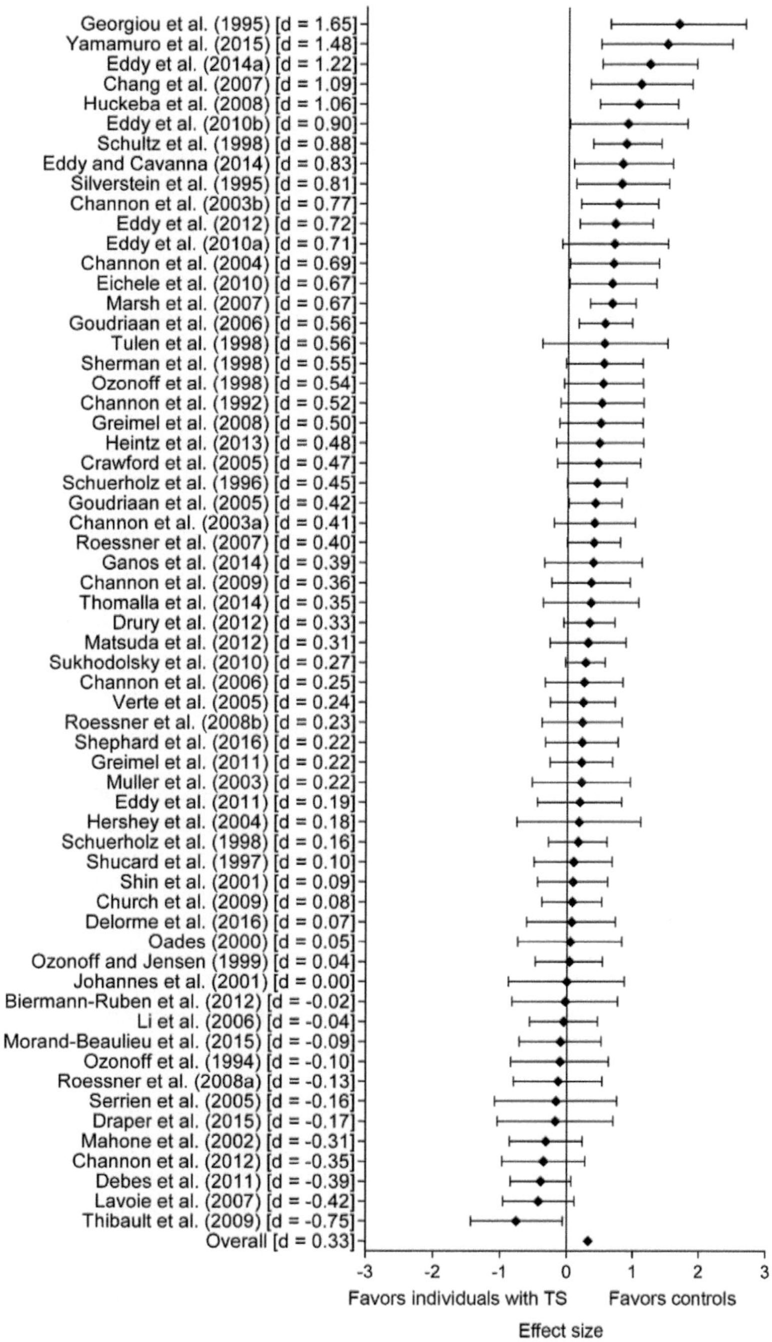

Fig. 21.1. Meta-analysis of response inhibition in TS. Individuals with TS had more inhibitory difficulties during response inhibition tasks than controls. Errors bars represent the 95% confidence interval and diamonds represent effect sizes from individual studies. *d*, Cohen's *d*; *TS*, Tourette syndrome. Adapted from Morand-Beaulieu S, Grot S, Lavoie J, et al. (2017a) The puzzling question of inhibitory control in Tourette syndrome: a meta-analysis. Neurosci Biobehav Rev 80: 240–262.

results suggest that individuals with TS do not differ from controls in their ability to postpone actions, but that the deceleration of the already initiated response appears to be reduced in participants with TS, relative to controls.

Several studies also assessed brain correlates of response inhibition in TS. Several brain regions, especially in the frontal cortex, are involved in inhibitory control. Notably, the right inferior frontal gyrus plays a major role in inhibitory control and is part, along with the presupplementary area and the subthalamic nucleus, of a cognitive control network (Aron et al., 2007a, 2007b; Aron et al., 2014; Aron et al., 2016; Obeso et al., 2013). An common occurrence in the TS neuroscience literature is a pattern of enhanced frontal activation (Johannes et al., 2001; Thibault et al., 2009; Jackson et al., 2011; Morand-Beaulieu et al., 2018; Rae et al., 2020) as revealed by electroencephalographic (EEG) (see Fig. 21.2) and functional magnetic resonance imaging (fMRI) studies, as well as increased functional connectivity in frontal regions (Serrien et al., 2005) for stimuli requiring volitional inhibitory control. This pattern has been suggested to represent compensatory processes implemented to maintain task performance, and possibly to regulate tic severity (i.e., through voluntary tic suppression; Marsh et al., 2007). Other recent studies investigating connectivity have highlighted important contributions of prefrontal regions to volitional inhibitory control in TS. For instance, Jurgiel et al. (2021) found increased effective connectivity between nodes of the fronto-parietal network in children with TS for incongruent trial of a Flanker task, which were associated with worse inhibitory performance. Similarly, worse inhibitory performance in a stop-signal task has been associated with increased connectivity between the right inferior frontal gyrus and the bilateral superior temporal gyri (Atkinson-Clement et al., 2020). That latter study also found interesting results pertaining to brain structure suggesting that worse inhibitory performance is correlated with increased gray matter signal in the inferior frontal gyrus and supplementary motor area.

Whether those increased activations and enhanced connections involving prefrontal regions represent compensatory processes allowing to maintain good inhibitory performance and possibly control tics has been debated. Jung et al. (2013) have suggested that frontal overactivations may not constitute the source of inhibitory task-control signals but would rather lead to hyperexcitability of motor areas. Relatedly, several studies have investigated how motor areas contribute to volitional inhibitory control as well as the emergence of tics. Notably, the supplementary motor area appears to be an important structure in this context. For instance, Ganos et al. (2014c) found that increased activation in the supplementary motor area during action inhibition was positively correlated with tic severity. Those results suggest that the supplementary motor area could serve as a global inhibition mechanism involved in voluntary tic suppression and action inhibition. Furthermore, the lateralized readiness potential, an event-related potential partly generated by the supplementary motor area and associated with movement preparation and execution, was found to be larger in adults with TS and correlated with increased phonic tic severity (Morand-Beaulieu et al., 2018). Also, similar to their findings regarding the inferior frontal gyrus, Atkinson-Clement et al. (2020) found worse inhibitory performance to be associated with increased gray matter signal in the supplementary motor area.

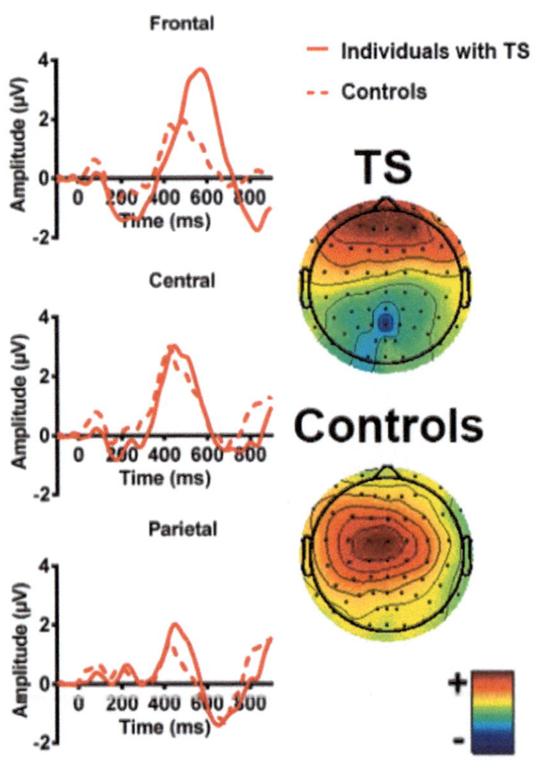

Fig. 21.2. Frontal shift of the NoGo P300 in TS. For stimuli requiring to inhibit a motor response (NoGo stimuli), adults with TS were found to have a frontal shift of the NoGo P300. In controls, the NoGo P300 was maximal over central electrodes. *ms*, milliseconds; *TS*, Tourette syndrome; *μV*, microvolts. Adapted from Morand-Beaulieu S, O'Connor KP, Blanchet PJ et al. (2018). Electrophysiological predictors of cognitive-behavioral therapy outcome in tic disorders. J Psychiatr Res 105: 113–122.

There are also interesting findings pertaining to the role of the presupplementary in response inhibition. For instance, in a recent study, Rae et al. (2023) compared internally and externally cued action and inhibition in adults with TS. They used a technique called representational similarity analyses, which allows assessing the degree of similarity in neural responses to different task conditions. In the presupplementary motor area, they found increased similarity in adults with TS relative to controls for several contrasts: externally cued action and externally cued inhibition, internally cued action and internally cued inhibition, externally cued action and internally cued action, and externally cued inhibition and internally cued inhibition. However, only the last contrast remained significant after correction for multiple comparisons. These results reveal that adults with TS may have decreased differentiation between action and inhibition in the presupplementary motor area. The authors also mentioned that this decreased differentiation could explain why increased prefrontal activation is required to inhibit actions in TS.

The aforementioned findings from neurobiologic studies highlight several brain processes involved in the inhibition of volitional actions in individuals with TS. Yet, volitional actions differ from tics in several aspects. While tics may be controlled for short period via voluntary suppression, it remains unclear whether the same processes are involved in the inhibition of voluntary actions and tics.

VOLUNTARY TIC SUPPRESSION

Phenomenology of voluntary tic suppression

As mentioned above, voluntary tic suppression is a core feature of TS. That feature distinguishes it from most other movement disorders (Ueda et al., 2021). A vast majority of individuals with TS are able to suppress their tics (Conelea et al., 2018), which however comes with the cost of a significant effort (Ganos et al., 2018b; Ueda et al., 2021). The ability to suppress tics does not seem to be influenced by co-occurring disorders such as ADHD or OCD, nor by psychotropic medication intake (Conelea et al., 2018). Voluntary tic suppression mostly occurs in situations where people dislike expressing their tics, such with unfamiliar people, at work or school, or in public places (Matsuda et al., 2016). Tic suppression abilities may be present from the very onset of tics (Greene et al., 2015) and are known to increase throughout childhood and adolescence (Banaschewski et al., 2003; Conelea et al., 2018). Voluntary tic suppression is also an important factor to consider in the developmental trajectory of TS, as reduced tic suppression abilities have been shown to predict worse tic severity one year later (Kim et al., 2019). There is a common idea that voluntary tic suppression leads to a rebound effect (i.e., an increase in tic severity following suppression), even among physicians and psychologists treating individuals with TS (Marcks et al., 2004). However, this idea is not supported by the scientific literature (Himle and Woods, 2005; Meidinger et al., 2005; Verdellen et al., 2007; Specht et al., 2013).

A better understanding of tic suppression has potential implications for treatments. However, voluntary tic suppression in itself must not be seen has viable long-term option for coping with tics. Indeed, almost all individuals with TS report feeling uncomfortable while suppressing their tics (Matsuda et al., 2016). As pointed out by Ganos et al. (2018b), it is important to distinguish voluntary tic suppression from other types of tic control, such as control through behavioral, cognitive, or pharmacologic interventions. Tics can also be reduced through the modulation of attention (i.e., some people tic less when they are concentrated) or with aging/brain maturation. For instance, tics have been shown to be reduced when someone is with strangers or a physician, and to be increased when they are home alone (Goetz et al., 2001). Such diminished tic severity in specific contexts does not necessarily equate with voluntary tic suppression, as there may not be a conscious effort to suppress tics. Although those types of tic control share some overlapping features with voluntary tic suppression, there may be specific and distinct mechanisms underlying voluntary tic suppression.

Initial reports of volitional control over the expression of tics suggested that this ability was directly related to premonitory urges (Bliss et al., 1980). However, more recent research pertaining to the relationship between premonitory urges and the ability to voluntarily suppress tics has resulted in mixed findings (Ganos et al., 2018b), suggesting that this relationship is not straightforward. Premonitory urges most likely consist of a very complex phenomenon with important variance across individuals, which might explain some discrepancies across studies. Recently, Matsuda et al. (2020) developed a new scale to identify additional aspects of tic-related sensations: rumination about sensations and awareness of emerging tics. They reported that awareness was positively associated with voluntary tic suppression abilities, suggesting that awareness of emerging signals for tics facilitates self-initiated tic suppression.

Another interesting aspect of tic suppression is its somatotopy. Indeed, prior research has shown that tic suppression, just like tic expression, follows a somatotopic pattern (Ganos et al., 2015). Tic expression is more common and more severe in body parts with the largest cortico-striatal representations, such as the face or the hands. Conversely, in body parts that are less affected by tics, which typically have smaller cortico-striatal representation, voluntary tic suppression is more effective. Thus, this suggests that tics with a stronger motor command are harder to suppress (Ganos et al., 2018b). A similar pattern has been observed with median nerve stimulation, which may have stronger effects on more complex and peripheral tics, while more simple facial tics (e.g., blinking, mouth tics, etc.) may persist during stimulation (Morera Maiquez et al., 2020). These results must however be interpreted with caution, given the novelty to this treatment as well as a poor understanding of the underlying mechanisms (Iverson et al., 2023). Finally, the somatotopy of tic suppression is also consistent with the idea of selective suppression. For instance, in some settings, like a classroom, phonic tics can be selectively suppressed while some motor tics may persist (Ganos et al., 2015). Prior research has also shown that some individuals may have specific tics that are only visible at home when they are alone and not in other contexts (Goetz et al., 2001). However, it is not clear whether such type of selective tic suppression is really voluntary and conscious, or if some individuals unconsciously express specific tics only in certain settings.

Cognitive aspects of voluntary tic suppression

For those without tics, it is hard to understand what goes through the mind of people with TS in order to voluntary suppress their tics. As mentioned by Ganos et al. (2018b), the processes involved in tic suppression can only be studied indirectly. It is however clear that this process requires an important and conscious effort (Matsuda et al., 2016). While the neurocognitive mechanisms underlying the initiation of voluntary tic suppression remain unclear, some studies have pointed toward specific cognitive functions associated with tic suppression. For instance, Specht et al. (2014) reported that verbal intelligence was correlated with the ability to suppress tics. Also, accuracy on tasks requiring attention or concentration has been shown to decrease during voluntary tic suppression (Conelea and Woods, 2008). Similarly, tic suppression is more efficient when people can focus on their tics, and less efficient when they need to divide their attention with external tasks (Herrmann et al., 2019). Those results are consistent with reports that tic suppression interferes with concentration (Wadman et al., 2016). It has been suggested that focusing on controlling tics makes it harder to give full attention to the external world (Cutler et al., 2009). Thus, if attentional and executive resources are required to effectively suppress tics, it is possible that those resources allocated to tic suppression are not available for other tasks. Alternatively, it is possible that the mechanisms that are involved and/or activated during voluntary tic suppression are incompatible with those required during tasks involving attention or executive control.

Neurobiologic correlates of voluntary tic suppression

Only a few studies have assessed the neurobiologic substrates of tic suppression. The first of such studies was conducted by Peterson et al. (1998), who contrasted fMRI activations during intervals of tic suppression and intervals of free ticcing. During tic suppression, they found both increased and decreased activity in various brain regions compared to intervals when participants were allowed to tic freely. Most subcortical areas showed reduced activation during tic suppression, except for the right caudate nucleus, which exhibited increased activity. Across the cortex, heightened activation was found in the right frontal cortex, bilateral temporal gyri, right anterior cingulate cortex, and bilateral occipital cortices, while decreased activation occurred in the right posterior cingulate cortex, bilateral hippocampus, cuneus, and left sensorimotor and inferior parietal cortices. That study, and in particular the increased right frontal activation, is often cited to support the idea that tic suppression recruits prefrontal brain regions that are also involved in response inhibition. This notion was further strengthened by further studies published afterward. Notably, frontal cortex activations during tic suppression were reported by two other fMRI studies in adults with TS. Ganos et al. (2014a) found increased regional homogeneity of the left inferior frontal gyrus during tic suppression relative to free ticcing. In another study, tic suppression mainly activated the superior prefrontal gyrus, the anterior and dorsolateral prefrontal cortex, and the dorsal anterior cingulate cortex (van der Salm et al., 2018). An EEG study investigating coherence between pairs of EEG electrodes as a proxy for functional connectivity revealed increased coherence in the alpha band over four pairs of frontal and central electrodes in nine adults with TS (Serrien et al., 2005). Enhanced coherence between those electrode pairs has been interpreted as increased functional connectivity between sensorimotor areas and the prefrontal and mesial frontal cortex. Similarly, a study by Hong et al. (2013) analyzed effective connectivity in nine children with TS, reporting increased coherence in the theta and beta bands during tic suppression. However, those two studies assessed functional connectivity between pairs of electrodes, which does not necessarily represent the functional connectivity between brain regions where those electrodes are located, due to volume conduction. This issue was tackled in a recent study assessing functional connectivity during tic suppression in the source space and in a large group of children with TS (Morand-Beaulieu et al., 2023b). Increased alpha-band functional connectivity was found during suppression (relative to free ticcing), but the brain regions involved in those connections somewhat differed from previous findings. There were three main hubs in the subnetwork showing increased functional connectivity during tic suppression: the right superior frontal gyrus, left precuneus, and left banks of the superior temporal sulcus (Fig. 21.3). Furthermore, and consistent with the idea that understanding tic suppression is relevant for behavioral treatments, functional connectivity within that subnetwork predicted reductions in vocal tic severity following comprehensive behavioral intervention for tics (CBIT) in children (Morand-Beaulieu et al., 2023a). All in all, studies assessing brain activation and functional connectivity during tic suppression revealed mixed results, with different and distributed brain areas involved in tic suppression. However, the involvement of prefrontal brain regions appears to be somewhat consistent across studies. Furthermore, there has been one study assessing the association between structural brain alterations in TS and voluntary tic suppression (Ganos et al., 2014b). Although gray matter volumes were reduced in two prefrontal structures, namely the right inferior frontal gyrus and the left frontal pole, those volume reductions were not associated with the capacity to voluntarily suppress tics.

Additionally, transcranial magnetic stimulation studies may provide interesting insights into the mechanisms of voluntary tic suppression. In a study assessing the effect of voluntary tic suppression on primary motor cortex excitability, it was discovered that voluntary tic suppression reduced corticospinal excitability, relative to free ticcing (Ganos et al., 2018a). Also, participants who were better at suppressing their tics showed larger differences in corticospinal excitability between voluntary tic suppression and free ticcing. These results, which inform us about how tic suppression modulates primary motor cortex excitability, are consistent with prior findings showing associations between reduced excitability of that region and decreased tic severity (Orth et al., 2008). Another study assessed the presupplementary motor area-mediated inhibition of the left motor cortex in children with TS (Bruce et al., 2021). It was found that less inhibition (of the left motor cortex) was associated with worse self-reported tic suppressibility, highlighting the importance of the presupplementary motor area in motor control in TS.

Finally, while the aforementioned studies investigated neural correlates of voluntary tic suppression during resting-state sessions, one study assessed the modulation of neural oscillations associated with voluntary tic suppression during the execution of a motor task (Zapparoli et al., 2019). In that study, the authors aimed to assess how tic suppression impacted the activity of the sensorimotor cortex during a motor task. In the tic-freely session, participants with TS showed reduced modulation of beta oscillations during movement execution compared with controls. When someone makes a hand movement, it typically comes with a decrease in beta oscillations over the contralateral sensorimotor cortex. This phenomenon is called beta desynchronization. In the current study, the modulation of beta oscillations during the task was negatively associated with tic severity, such that individuals with the most severe tics showed the least modulation of beta power. However, in the tic suppression condition, beta modulation during movement execution was not different across groups. These findings suggest that tic suppression may operate by normalizing sensorimotor beta oscillations.

ARE THE SAME MECHANISMS INVOLVED IN RESPONSE INHIBITION AND VOLUNTARY TIC SUPPRESSION?

By contrasting voluntary tic suppression with free ticcing, some of the studies mentioned above allow a better understanding of

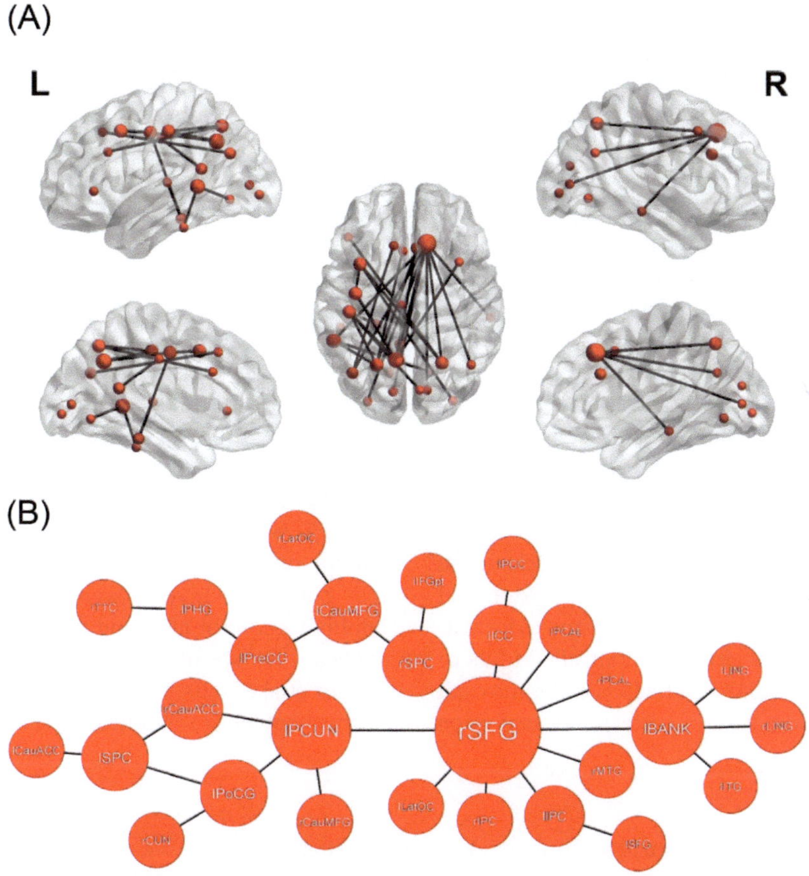

Fig. 21.3. Functional connectivity during tic suppression. (**A**) Subnetwork in which functional connectivity was higher during tic suppression relative to free ticcing. (**B**) Graphical representation of the tic suppression subnetwork. *Note*: "l" and "r" denote the lateralization of the nodes. *CauACC*, Caudal anterior cingulate cortex; *CauMFG*, caudal middle frontal gyrus; *CUN*, cuneus; *ICC*, isthmus of the cingulate cortex; *IFGpt*, pars triangularis of the inferior frontal gyrus; *IPC*, inferior parietal cortex; *ITG*, inferior temporal gyrus; *LatOC*, lateral occipital cortex; *LING*, lingual gyrus; *MTG*, middle temporal gyrus; *PCAL*, pericalcarine cortex; *PCC*, posterior cingulate cortex; *PCUN*, precuneus; *PHG*, parahippocampal gyrus; *PoCG*, postcentral gyrus; *PreCG*, precentral gyrus; *SFG*, superior frontal gyrus; *SPC*, superior parietal cortex; *TTC*, transverse temporal cortex. Reproduced from Morand-Beaulieu S, Wu J, Mayes LC, et al. (2023b) Increased alpha-band connectivity during tic suppression in children with Tourette syndrome revealed by source electroencephalography analyses. Biol Psychiatry Cogn Neurosci Neuroimaging 8: 241–250.

the brain mechanisms allowing self-driven tic control. However, they do not necessarily tell us whether tics are inhibited in the same way voluntary movements are controlled. According to the brain regions activated or the networks showing increased connectivity during tic suppression relative to rest, some of those studies suggest possible overlap between voluntary tic suppression and inhibitory control of voluntary actions (Peterson et al., 1998; Serrien et al., 2005; Ganos et al., 2014a), while others suggest mechanisms that are not so consistent with those known to be involved in the inhibition of voluntary actions (van der Salm et al., 2018; Morand-Beaulieu et al., 2023b). To really grasp the similarities and differences between tic suppression and response inhibition, studies must be designed in a way to assess both processes with the same group of participants. So far, there has only been a limited number of studies using such experimental design.

One of the earliest studies to suggest parallels between response inhibition and voluntary tic suppression was conducted by Serrien et al. (2005) who investigated patterns of EEG coherence during tic suppression and during a Go/NoGo task. In nine adults with TS, enhanced coherence within the alpha-frequency band was found across pairs of electrodes spanning the frontal cortex. In the same pairs of electrodes, TS adults' alpha coherence was increased for the NoGo trials of the Go/NoGo task relative to Go trials, while this difference was not found in control participants. It was thus concluded that the same network recruited for response inhibition during the Go/NoGo task may be engaged in voluntary tic suppression. More than a decade later, van der Salm et al. (2018) used a different approach to tackle a similar question. They compared brain activation in adults with TS suppressing ocular tics to that of non-TS controls suppressing eyeblinks. The main takeaway from that study is that different brain regions are recruited for eyeblink suppression in controls and tic suppression in individuals with TS. In controls, suppression of eyeblinks led to activation of brain regions commonly involved in response inhibition, such as the right ventrolateral prefrontal cortex and bilateral supplementary motor areas. In individuals with TS, tic suppression led to increased activation in the anterior cingulate cortex and medial prefrontal regions. Thus, those latter results contrast those of Serrien et al. (2005) and suggest that the mechanism of voluntary tic suppression may be different than that of response inhibition.

Most neurocognitive studies also reported poor concordance between response inhibition and voluntary tic suppression. For instance, neuropsychologic measures of response inhibition, including the CPT as well as Go/NoGo, Flanker, Stroop, and stop-signal tasks, were not predictive of voluntary tic suppression capacities (Ganos et al., 2014c; Specht et al., 2014; Sturm et al., 2021). Similarly, Woods et al. (2008) reported that omissions errors during the CPT, but not commission errors, were associated with voluntary tic suppression (although in a small sample of 13 children). While not directly related to tic suppression, findings from behavioral therapy trials may also be relevant to this topic. Indeed, voluntary tic suppression is sometimes thought to underlie the effectiveness of behavioral therapies for TS, such as CBIT, and both tic suppression and CBIT have been shown to share some mechanisms (Morand-Beaulieu et al., 2023a). In large randomized controlled trials of CBIT in children (Chang et al., 2018) and adults (Abramovitch et al., 2017), neuropsychologic measures of response inhibition were unrelated to CBIT outcome. However, please note that a more recent but smaller study found a link between CBIT outcome and a specific measure related to cognitive control and switching (McGuire et al., 2022). All in all, these studies suggest that the association between response inhibition measured through neuropsychologic tests and tic suppression (either voluntary of through CBIT) is not that straightforward.

CONCLUSION AND FUTURE DIRECTIONS

Over the last few decades, response inhibition has been the subject of many studies in TS, in part because of its putative role in tic suppression. Evidence supporting this hypothesis has been mixed. A possible explanation for inconsistencies across studies may be the heterogeneous profile of symptoms in individuals with TS, which complicates most studies investigating inhibitory control of voluntary actions. Another issue is that some tasks may not be sensitive enough to detect the specific difficulties in inhibitory control affecting people with TS. As shown in recent studies, very specific aspects of inhibition seem to differ in TS (Rawji et al., 2020; Indrajeet et al., 2022). The main conclusions of this chapter are outlined in Table 21.1.

Table 21.1

Summary of the literature on inhibitory control and voluntary tic suppression.

Domain	Main findings
Response inhibition	
Behavioral measures	• Mixed findings but a meta-analysis revealed mild difficulties with volitional inhibition in Tourette syndrome. • These difficulties are enhanced in those with co-occurring conditions and high tic severity. • Potential difficulties with automatic inhibition as well.
Neurobiologic correlates	• Increased frontal activation/functional connectivity during the inhibition of prepotent responses. • It is not clear whether those increased frontal activation/functional connectivity represent compensatory mechanisms.
Voluntary tic suppression	
Phenomenology	• Voluntary tic suppression may differ from other types of tic suppression, such as the reduced expression of tics in social contexts. • Some tics may be easier to voluntarily suppress than others.
Behavioral measures	• Voluntary suppressing tics may impair performance in external tasks. • Voluntary tic suppression is less efficient when focusing on external tasks.
Neurobiologic correlates	• Increased activation of several cortical regions including notably prefrontal areas during voluntary tic suppression.
Parallels between response inhibition and voluntary tic suppression	
Behavioral measures	• No associations between experimental measures of response inhibition and objective tic suppressibility.
Neurobiologic correlates	• Mixed findings, too few studies to draw conclusions.

In the end, this chapter comes to a similar conclusion to that of Kurvits and Ganos (2022). There is not, to this date, strong evidence supporting the relationship between the mechanisms of voluntary tic suppression and inhibitory control of voluntary actions. If tic suppression genuinely involves the same mechanisms as response inhibition, it is possible that lab tasks designed to assess response inhibition lack the necessary ecologic validity to correctly identify that effect. Furthermore, the emergence of tics and premonitory urges does not compare to a stream of stimuli in a Go/NoGo task. People suppressing their tics are not monitoring the emergence of premonitory urges and making the split-second decision to tic or not. Rather, they most likely enter a different neurocognitive state that may involve both concentration and relaxation. Therefore, future studies aiming to assess whether response inhibition, or any other psychologic process, are involved in tic suppression should do so by using an experimental design that parallels more closely the experience of tic suppression.

References

Abramovitch A, Hallion LS, Reese HE et al. (2017). Neurocognitive predictors of treatment response to randomized treatment in adults with tic disorders. Prog Neuropsychopharmacol Biol Psychiatry 74: 9–14.

Aron AR, Behrens TE, Smith S et al. (2007a). Triangulating a cognitive control network using diffusion-weighted magnetic resonance imaging (MRI) and functional MRI. J Neurosci 27: 3743–3752.

Aron AR, Durston S, Eagle DM et al. (2007b). Converging evidence for a fronto-basal-ganglia network for inhibitory control of action and cognition. J Neurosci 27: 11860–11864.

Aron AR, Herz DM, Brown P et al. (2016). Frontosubthalamic circuits for control of action and cognition. J Neurosci 36: 11489–11495.

Aron AR, Robbins TW, Poldrack RA (2014). Inhibition and the right inferior frontal cortex: one decade on. Trends Cogn Sci 18: 177–185.

Atkinson-Clement C, Porte C-A, de Liege A et al. (2020). Neural correlates and role of medication in reactive motor impulsivity in Tourette disorder. Cortex 125: 60–72.

Banaschewski T, Woerner W, Rothenberger A (2003). Premonitory sensory phenomena and suppressibility of tics in Tourette syndrome: developmental aspects in children and adolescents. Dev Med Child Neurol 45: 700–703.

Beeler P, Jensen NO, Kim S et al. (2022). Fractality of tics as a quantitative assessment tool for Tourette syndrome. J R Soc Interface 19: 20210742.

Bliss J, Cohen DJ, Freedman DX (1980). Sensory experiences of Gilles de la Tourette syndrome. Arch Gen Psychiatry 37: 1343–1347.

Bruce AB, Yuan W, Gilbert DL et al. (2021). Altered frontal-mediated inhibition and white matter connectivity in pediatric chronic tic disorders. Exp Brain Res 239: 955–965.

Cavanna AE, Black KJ, Hallett M et al. (2017). Neurobiology of the premonitory urge in Tourette's syndrome: pathophysiology and treatment implications. J Neuropsychiatry Clin Neurosci 29: 95–104.

Cavanna AE, Nani A (2013). Tourette syndrome and consciousness of action. Tremor Other Hyperkinet Mov (NY) 3: tre-03-181-4368-1.

Chang SW, McGuire JF, Walkup JT et al. (2018). Neurocognitive correlates of treatment response in children with Tourette's disorder. Psychiatry Res 261: 464–472.

Cohen AJ, Leckman JF (1992). Sensory phenomena associated with Gilles de la Tourette's syndrome. J Clin Psychiatry 53: 319–323.

Conelea CA, Wellen B, Woods DW et al. (2018). Patterns and predictors of tic suppressibility in youth with tic disorders. Front Psychiatry 9: 188.

Conelea CA, Woods DW (2008). Examining the impact of distraction on tic suppression in children and adolescents with Tourette syndrome. Behav Res Ther 46: 1193–1200.

Cutler D, Murphy T, Gilmour J et al. (2009). The quality of life of young people with Tourette syndrome. Child Care, Health Dev 35: 496–504.

Eddy CM, Rickards HE, Cavanna AE (2012). Executive functions in uncomplicated Tourette syndrome. Psychiatry Res 200: 46–48.

Eimer M, Schlaghecken F (2003). Response facilitation and inhibition in subliminal priming. Biol Psychol 64: 7–26.

Ganos C, Bongert J, Asmuss L et al. (2015). The somatotopy of tic inhibition: where and how much? Mov Disord 30: 1184–1189.

Ganos C, Kahl U, Brandt V et al. (2014a). The neural correlates of tic inhibition in Gilles de la Tourette syndrome. Neuropsychologia 65: 297–301.

Ganos C, Kühn S, Kahl U et al. (2014b). Prefrontal cortex volume reductions and tic inhibition are unrelated in uncomplicated GTS adults. J Psychosom Res 76: 84–87.

Ganos C, Kuhn S, Kahl U et al. (2014c). Action inhibition in Tourette syndrome. Mov Disord 29: 1532–1538.

Ganos C, Rocchi L, Latorre A et al. (2018a). Motor cortical excitability during voluntary inhibition of involuntary tic movements. Mov Disord 33: 1804–1809.

Ganos C, Rothwell J, Haggard P (2018b). Voluntary inhibitory motor control over involuntary tic movements. Mov Disord 33: 937–946.

Goetz CG, Leurgans S, Chmura TA (2001). Home alone: methods to maximize tic expression for objective videotape assessments in Gilles de la Tourette syndrome. Mov Disord 16: 693–697.

Goudriaan AE, Oosterlaan J, de Beurs E et al. (2005). Decision making in pathological gambling: a comparison between pathological gamblers, alcohol dependents, persons with Tourette syndrome, and normal controls. Brain Res Cogn Brain Res 23: 137–151.

Greene DJ, Koller JM, Robichaux-Viehoever A et al. (2015). Reward enhances tic suppression in children within months of tic disorder onset. Dev Cogn Neurosci 11: 65–74.

Herrmann K, Sprenger A, Baumung L et al. (2019). Help or hurt? How attention modulates tics under different conditions. Cortex 120: 471–482.

Himle MB, Woods DW (2005). An experimental evaluation of tic suppression and the tic rebound effect. Behav Res Ther 43: 1443–1451.

Hong HJ, Sohn H, Cha M et al. (2013). Increased frontomotor oscillations during tic suppression in children with Tourette syndrome. J Child Neurol 28: 615–624. .

Indrajeet I, Atkinson-Clement C, Worbe Y et al. (2022). Compromised reactive but intact proactive inhibitory motor control in Tourette disorder. Sci Rep 12: 2193.

Iverson AM, Arbuckle AL, Ueda K et al. (2023). Median nerve stimulation for treatment of tics: randomized, controlled, crossover trial. J Clin Med 12: 2514.

Iverson AM, Black KJ (2022). Why tic severity changes from then to now and from here to there. J Clin Med 11: 5930.

Jackson GM, Jackson SR (2020). Rethinking the nature of inhibitory control deficits in Tourette syndrome. Brain 143: 721–722.

Jackson GM, Mueller SC, Hambleton K et al. (2007). Enhanced cognitive control in Tourette Syndrome during task uncertainty. Exp Brain Res 182: 357–364.

Jackson SR, Parkinson A, Jung J et al. (2011). Compensatory neural reorganization in Tourette syndrome. Curr Biol 21: 580–585.

Johannes S, Wieringa BM, Mantey M et al. (2001). Altered inhibition of motor responses in Tourette Syndrome and Obsessive-Compulsive Disorder. Acta Neurol Scand 104: 36–43.

Jung J, Jackson SR, Parkinson A et al. (2013). Cognitive control over motor output in Tourette syndrome. Neurosci Biobehav Rev 37: 1016–1025.

Jurgiel J, Miyakoshi M, Dillon A et al. (2021). Inhibitory control in children with tic disorder: aberrant fronto-parietal network activity and connectivity. Brain Commun 3: fcab067.

Kim S, Greene DJ, Robichaux-Viehoever A et al. (2019). Tic suppression in children with recent-onset tics predicts 1-year tic outcome. J Child Neurol 34: 757–764.

Kurvits L, Ganos C (2022). Inhibitory control and tic generation. In: Martino D Leckman J (Eds.), Tourette syndrome. Oxford University Press.

Leckman JF, Bloch MH, Sukhodolsky DG et al. (2022). Phenomenology of tics and sensory urges: the self under siege. In: Martino D Leckman JF (Eds.), Tourette syndrome. Oxford University Press.

Leckman JF, Walker DE, Cohen DJ (1993). Premonitory urges in Tourette's syndrome. Am J Psychiatry 150: 98–102.

Marcks BA, Woods DW, Teng EJ et al. (2004). What do those who know, know? Investigating providers' knowledge about Tourette's Syndrome and its treatment. Cogn Behav Pract 11: 298–305.

Marsh R, Zhu H, Wang Z et al. (2007). A developmental fMRI study of self-regulatory control in Tourette's syndrome. Am J Psychiatry 164: 955–966.

Matsuda N, Kono T, Nonaka M et al. (2016). Self-initiated coping with Tourette's syndrome: effect of tic suppression on QOL. Brain Dev 38: 233–241.

Matsuda N, Nonaka M, Kono T et al. (2020). Premonitory awareness facilitates tic suppression: subscales of the Premonitory Urge for Tics Scale and a new self-report questionnaire for tic-associated sensations. Front Psychiatry 11: 592.

McGuire JF, Sturm A, Ricketts EJ et al. (2022). Cognitive control processes in behavior therapy for youth with Tourette's disorder. J Child Psychol Psychiatry 63: 296–304.

Meidinger AL, Miltenberger RG, Himle M et al. (2005). An investigation of tic suppression and the rebound effect in Tourette's disorder. Behav Modif 29: 716–745.

Meige H, Feindel ECL (1902). Les tics et leur traitement. Paris: Masson et Cie.

Morand-Beaulieu S, Crowley MJ, Grantz H et al. (2023a). Functional connectivity during tic suppression predicts reductions in vocal tics following behavior therapy in children with Tourette syndrome. Psychol Med 53: 1–8.

Morand-Beaulieu S, Grot S, Lavoie J et al. (2017a). The puzzling question of inhibitory control in Tourette syndrome: a meta-analysis. Neurosci Biobehav Rev 80: 240–262.

Morand-Beaulieu S, Leclerc JB, Valois P et al. (2017b). A review of the neuropsychological dimensions of Tourette syndrome. Brain Sci 7: 106.

Morand-Beaulieu S, O'Connor KP, Blanchet PJ et al. (2018). Electrophysiological predictors of cognitive-behavioral therapy outcome in tic disorders. J Psychiatr Res 105: 113–122.

Morand-Beaulieu S, Smith SD, Ibrahim K et al. (2022). Electrophysiological signatures of inhibitory control in children with Tourette syndrome and attention-deficit/hyperactivity disorder. Cortex 147: 157–168.

Morand-Beaulieu S, Wu J, Mayes LC et al. (2023b). Increased alpha-band connectivity during tic suppression in children with Tourette syndrome revealed by source electroencephalography analyses. Biol Psychiatry Cogn Neurosci Neuroimaging 8: 241–250.

Morera Maiquez B, Sigurdsson HP, Dyke K et al. (2020). Entraining movement-related brain oscillations to suppress tics in Tourette syndrome. Curr Biol 30: 2334–2342.e3.

Obeso I, Robles N, Muñoz-Marrón E et al. (2013). Dissociating the role of the pre-SMA in response inhibition and switching: a combined online and offline TMS approach. Front Hum Neurosci 7: 150.

Openneer TJC, Forde NJ, Akkermans SEA et al. (2020). Executive function in children with Tourette syndrome and attention-deficit/hyperactivity disorder: cross-disorder or unique impairments? Cortex 124: 176–187.

Orth M, Münchau A, Rothwell JC (2008). Corticospinal system excitability at rest is associated with tic severity in Tourette syndrome. Biol Psychiatry 64: 248–251.

Ozonoff S, Strayer DL, McMahon WM et al. (1994). Executive function abilities in autism and Tourette syndrome: an information processing approach. J Child Psychol Psychiatry 35: 1015–1032.

Ozonoff S, Strayer DL, McMahon WM et al. (1998). Inhibitory deficits in Tourette syndrome: a function of comorbidity and symptom severity. J Child Psychol Psychiatry 39: 1109–1118.

Peterson BS, Leckman JF (1998). The temporal dynamics of tics in Gilles de la Tourette syndrome. Biol Psychiatry 44: 1337–1348.

Peterson BS, Skudlarski P, Anderson AW et al. (1998). A functional magnetic resonance imaging study of tic suppression in Tourette syndrome. Arch Gen Psychiatry 55: 326–333.

Rae CL, Parkinson J, Betka S et al. (2020). Amplified engagement of prefrontal cortex during control of voluntary action in Tourette syndrome. Brain Commun 2: fcaa199.

Rae CL, Raykov P, Ambridge EM et al. (2023). Elevated representational similarity of voluntary action and inhibition in Tourette syndrome. Brain Commun 5: fcad224.

Rajagopal S, Seri S, Cavanna AE (2013). Premonitory urges and sensorimotor processing in Tourette syndrome. Behav Neurol 27: 65–73.

Rawji V, Modi S, Latorre A et al. (2020). Impaired automatic but intact volitional inhibition in primary tic disorders. Brain 143: 906–919.

Roessner V, Albrecht B, Dechent P et al. (2008). Normal response inhibition in boys with Tourette syndrome. Behav Brain Funct 4: 29.

Schlaghecken F, Rowley L, Sembi S et al. (2008). The negative compatibility effect: a case for self-inhibition. Adv Cogn Psychol 3: 227–240.

Serrien DJ, Orth M, Evans AH et al. (2005). Motor inhibition in patients with Gilles de la Tourette syndrome: functional activation patterns as revealed by EEG coherence. Brain 128: 116–125.

Shephard E, Jackson GM, Groom MJ (2016). The effects of co-occurring ADHD symptoms on electrophysiological correlates of cognitive control in young people with Tourette syndrome. J Neuropsychol 10: 223–238.

Sherman EM, Shepard L, Joschko M et al. (1998). Sustained attention and impulsivity in children with Tourette syndrome: comorbidity and confounds. J Clin Exp Neuropsychol 20: 644–657.

Specht MW, Nicotra CM, Kelly LM et al. (2014). A comparison of urge intensity and the probability of tic completion during tic freely and tic suppression conditions. Behav Modif 38: 297–318.

Specht MW, Woods DW, Nicotra CM et al. (2013). Effects of tic suppression: ability to suppress, rebound, negative reinforcement, and habituation to the premonitory urge. Behav Res Ther 51: 24–30.

Stenner M-P, Baumgaertel C, Heinze H-J et al. (2018). Intact automatic motor inhibition in patients with Tourette syndrome. Mov Disord 33: 1800–1804.

Sturm A, Ricketts EJ, McGuire JF et al. (2021). Inhibitory control in youth with Tourette's Disorder, attention-deficit/hyperactivity disorder and their combination and predictors of objective tic suppressibility. Psychiatry Res 304: 114163.

Thibault G, O'Connor KP, Stip E et al. (2009). Electrophysiological manifestations of stimulus evaluation, response inhibition and motor processing in Tourette syndrome patients. Psychiatry Res 167: 202–220.

Tourette Syndrome Classification Study Group(1993). Definitions and classification of tic disorders. Arch Neurol 50: 1013–1016.

Ueda K, Kim S, Greene DJ et al. (2021). Correlates and clinical implications of tic suppressibility. Curr Dev Disord Rep 8: 112–120.

van der Salm SMA, van der Meer JN, Cath DC et al. (2018). Distinctive tics suppression network in Gilles de la Tourette syndrome distinguished from suppression of natural urges using multimodal imaging. NeuroImage Clin 20: 783–792.

Verdellen CWJ, Hoogduin CAL, Keijsers GPJ (2007). Tic suppression in the treatment of Tourette's syndrome with exposure therapy: the rebound phenomenon reconsidered. Mov Disord 22: 1601–1606.

Wadman R, Glazebrook C, Beer C et al. (2016). Difficulties experienced by young people with Tourette syndrome in secondary school: a mixed methods description of self, parent and staff perspectives. BMC Psychiatry 16: 14.

Woods DW, Himle MB, Miltenberger RG et al. (2008). Durability, negative impact, and neuropsychological predictors of tic suppression in children with chronic tic disorder. J Abnorm Child Psychol 36: 237–245.

Zapparoli L, Macerollo A, Joyce EM et al. (2019). Voluntary tic suppression and the normalization of motor cortical beta power in Gilles de la Tourette syndrome: an EEG study. Eur J Neurosci 50: 3944–3957.

Section IV

Treatments

Handbook of Clinical Neurology, Vol. 215 (3rd series)
Tic Disorders and Tourette Syndrome
C. Ganos and Y. Worbe, Editors
https://doi.org/10.1016/B978-0-443-13554-5.00030-4

Chapter 22

Behavioral interventions for tic disorders

MICHAEL B. HIMLE[1,*], MARIKA MARKLIN[1], AND BRIANNA WELLEN[2]

[1]*Department of Psychology, University of Utah, Salt Lake City, UT, United States*

[2]*Department of Psychiatry and Behavioral Sciences, University of Minnesota, Minneapolis, MN, United States*

Abstract

Behavioral interventions are now recommended within evidence-based treatment guidelines as an alternative or adjunct to medication for the management of tic disorders (TD). The behavioral model of TD conceptualizes tics as neurologic based yet modifiable behaviors that are influenced by internal and external contextual antecedents and tic-contingent consequences. The behavioral interventions with the strongest empirical support are habit reversal training (HRT), an expanded version of HRT referred to as comprehensive behavioral intervention for tics (CBIT), and a version of exposure and response prevention (ERP) adapted from an evidence-based cognitive-behavioral intervention for obsessive-compulsive disorder. Each of these interventions involves teaching volitional tic suppression and environmental modifications aimed at reducing contextual worsening of tics. In this chapter, we review the behavioral model of TD, the core components of HRT, CBIT, and ERP, and discuss their supporting evidence base. We then briefly review several persistent and unfounded concerns and misunderstandings regarding behavior interventions for TD that inhibit adoption among some professionals and patients. Finally, despite evidence supporting their efficacy, access to behavior therapies for TD remains limited. We conclude with a review of existing and emerging approaches aimed at increasing dissemination and implementation and future directions for research.

BEHAVIORAL MODEL OF TIC DISORDERS

The behavioral model of tic disorders (TD) (see Woods et al., 2023) posits that although tics are the direct result of structural and functional abnormalities within the basal ganglia and associated cortico-striatal-thalamo-cortical circuitry (see Ganos et al., 2013), their expression and clinical course is influenced through two primary learning processes. First, studies have shown that most individuals report that tics are immediately preceded by unpleasant premonitory urges, often described as an inner tension, pressure, or discomfort that is temporarily alleviated when tics are executed (Leckman et al., 1993; Kwak et al., 2003; Brandt et al., 2016; Cavanna et al., 2017). The tic-contingent removal of aversive premonitory urges is thought to strengthen tics through automatic negative reinforcement, resulting in tics becoming more frequent, forceful, and/or complex over time (Capriotti et al., 2014; Brandt et al., 2016; Wellen et al., 2024). Second, it is well-documented that tic expression is situationally variable (Silva et al., 1995; Himle et al., 2014; Capriotti et al., 2015; Barnea et al., 2016; Barber et al., 2024). The behavioral model hypothesizes that some of this variability results, at least in part, from modifiable contextual antecedents and tic-contingent consequences (Woods et al., 2023). Specifically, within the behavioral model, contextual antecedents refer to events or environmental stimuli that precede tics and influence their likelihood, frequency, intensity, or appearance on a moment-to-moment basis (Himle et al., 2014). Several descriptive and experimental studies have shown that specific mood states, settings, persons, and activities can momentarily attenuate or exacerbate tics. For example, Silva et al. (1995) assessed 29 situational factors in 14 youth with TD and found that tics were reported to increase during periods of stress or excitement, watching television, social gatherings, fatigue, medical visits, and when socializing (among others) and decrease during focused tasks. Similarly, Barnea et al. (2016) directly observed tics in 41 youth with TD under five daily-life situations and found that tics were more frequent when they were watching television compared to when they were alone, doing homework, or talking to a stranger. Finally, Himle et al. (2014) conducted descriptive functional assessment interviews with 51 treatment-seeking youth with TD and their caregivers. Of the 11 antecedent situations assessed

*Correspondence to: Michael B. Himle, PhD, Department of Psychology, University of Utah, 380 S. 1530 E., BEHS 502, Salt Lake City, UT 84112, United States. Tel: +1 801-581-7529. E-mail: michael.himle@utah.edu.

(e.g., watching TV, during homework, in the classroom, public places, etc.), 10 were endorsed as tic-exacerbating by at least half of the sample. In contrast to antecedents, consequences are events or outcomes that are contingently delivered *following* tics that increase (i.e., reinforce) or decrease their likelihood in the future. For example, several lab-based experimental studies have demonstrated that reinforcing tic-free periods can encourage suppression and temporarily reduce tics (see Mohamed et al., 2024). Studies have also shown that tic-contingent attention (e.g., asking about tics, receiving comfort, or being told to stop ticcing) and escape or avoidance of nonpreferred tasks (e.g., being asked to leave the classroom, removal of homework demands) are associated with an increase in both tics (Himle et al., 2015; Eaton et al., 2017) and premonitory urges (Capriotti et al., 2015). Importantly, the influence of both antecedents and tic-contingent consequences have been shown to be highly individualized to each patient.

Although behavioral therapies for TD differ with respect to their specific therapeutic techniques, each involves teaching and encouraging volitional tic suppression coupled with environmental modifications to reduce the contextual worsening of tics (Woods et al., 2008b; Verdellen et al., 2011). Below, we describe the three behavioral interventions with the strongest empirical support, their rationale, and supporting evidence.

HABIT REVERSAL TRAINING

Habit reversal training (HRT) is a behavioral therapy that was initially developed to treat a broad range of impairing repetitive behaviors, including motor and vocal tics, habitual hair pulling (i.e., hair pulling disorder), excoriation, and oral-digital habits, among others (Azrin and Nunn, 1973). Applied to tics, HRT involves the sequential application of three core components on a tic-by-tic basis: awareness training, competing response training, and social support (Woods et al., 1996). Awareness training teaches the patient to identify discrete instances of the target tic and associated premonitory urges so that the patient is better able to detect when a tic has, or is about to, occur. During competing response training, the patient is taught to engage in a behavior that is physically incompatible with the tic and that can be sustained until the urge to tic subsides (typically one to three minutes). For example, for a head jerking tic, an appropriate competing response might involve gently pressing the chin toward the sternum, thereby preventing the head from moving. For vocal tics, the competing response typically involves engaging in controlled diaphragmatic breathing in a way that reverses the air flow pattern involved in producing the tic. Competing responses are designed to be inconspicuous and practical in real-world situations, allowing the individual to utilize them across various setting and activities. Social support involves teaching a supportive family member or caregiver to prompt and praise the use of competing responses and to encourage the patient to persistent in treatment. When implementing HRT, one tic is targeted each week, and the patient is instructed to practice utilizing their competing responses between sessions, especially in tic-eliciting settings or during tic-exacerbating activities. The number of sessions depends upon the number of problematic tics the patient experiences.

The rationale for HRT is centered around several premises. First, individuals are often unaware of their tics and associated premonitory urges on a moment-to-moment basis due to tic automaticity, limited interoceptive awareness, competing attentional demands, or contextual factors that undermine inhibitory control (e.g., stress, fatigue; Ganos et al., 2015; Barnea et al., 2016; Iverson and Black, 2022). Second, increased immediate awareness of tics and premonitory urges is positively associated with volitional tic suppression (Matsuda et al., 2020). Third, higher subjective ratings of urge intolerance are associated with more frequent and severe tics (Ramsey et al., 2022). And finally, tics are presumably strengthened (i.e., reinforced) by the removal of aversive premonitory urges; often referred to as the urge reduction model (Evers and van de Wetering, 1994; Wellen et al., 2024). HRT aims to improve momentary awareness of individually targeted tics and associated urges and subsequently interrupt or suppress the targeted tic. Repeatedly doing so is hypothesized to break the urge-tic-relief negative reinforcement cycle and facilitate habituation to premonitory urges, thereby diminishing tic frequency and intensity (see Woods et al., 2008b, 2023); however, the habituation mechanism has received mixed support (Houghton et al., 2017; Wellen et al., 2024). Alternative mechanisms have been proposed but have not been adequately investigated (see Essoe et al., 2021 and Woods et al., 2023 for reviews).

COMPREHENSIVE BEHAVIORAL INTERVENTION FOR TICS

Comprehensive behavioral intervention for tics (CBIT) combines HRT with psychoeducation, function-based assessment and intervention (FBAI), relaxation training, and several ancillary procedures to increase treatment engagement and compliance (Woods et al., 2008b). Psychoeducation aims to increase the patient's understanding of the phenomenology and clinical course of tics and TD, provide a rationale for the behavioral approach to treatment, reduce stigma, and address common myths and misunderstandings of behavior therapy for TD. FBAI is an individualized and systematic process for identifying and altering tic-exacerbating contextual antecedents and tic-contingent consequences, with the goal of creating a supportive "tic neutral environment." This primarily involves (a) minimizing or altering tic-exacerbating activities (e.g., reducing excessive screen time and tic-related talk); (b) reducing or eliminating tic-contingent attention (e.g., asking about tics, demands to stop ticcing) and escape (e.g., leaving the classroom, avoiding social situations, or reducing academic expectations); and (c) teaching skills to minimize tic worsening and impact when tic-exacerbating contextual factors cannot be eliminated (e.g., stress management skills, using HRT in social situations). Finally, relaxation training strategies such as diaphragmatic breathing and progressive muscle relaxation are included in the CBIT protocol because stress is a commonly reported tic exacerbator (Hoekstra et al., 2004), has been shown to undermine efforts at tic suppression (Conelea et al., 2011), and relaxation has been shown to be effective for reducing tics for some individuals in the short term (Tilling and Cavanna, 2020). CBIT is typically delivered over 8–10 weekly 1-hour sessions, but the exact number of sessions depends upon the patient's clinical presentation.

EXPOSURE AND RESPONSE PREVENTION

Exposure and response prevention (ERP) is a well-established cognitive-behavioral intervention for obsessive compulsive disorder wherein patients are taught to confront anxiety-provoking thoughts, images, or situations (exposure) while refraining from engaging compulsions and other anxiety-reducing and avoidant behaviors (see Abramowitz, 1996). Applied to tics, ERP involves having individuals suppress their tics (response prevention), thereby repeatedly "exposing" them to premonitory urges. In contrast to HRT in which tics are targeted individually and suppression is relatively brief, ERP involves suppressing all tics simultaneously for increasing periods of time, up to 2 hours (Verdellen et al., 2008, 2011). ERP is typically delivered over the course of 12, 2-hour sessions, and patients are instructed to practice suppressing tics between sessions and in the presence of tic-exacerbating stimuli and while engaged in tic-exacerbating activities (Verdellen et al., 2011). Repeatedly experiencing and tolerating premonitory urges during ERP is thought to facilitate habituation (Hoogduin et al., 1997), but the evidence supporting this mechanism has been mixed (Verdellen et al., 2008; van de Griendt et al., 2023).

EVIDENCE FOR THE EFFICACY OF BEHAVIORAL THERAPIES FOR TIC DISORDERS

Habit Reversal Training. HRT as a stand-alone intervention for TD has been extensively studied in both children and adults. Several early studies using tightly controlled single-case designs reported reductions in tic frequency ranging from 38% to 96% with treatment gains generally well-maintained for up to 12 months (see Himle et al., 2006). Several small open and randomized controlled trials (RCTs) have also shown positive results. Viefhaus et al. (2020) conducted an open trial of 16 sessions of HRT for 27 children (aged 8–19 years) with TD. Outcomes were measured at posttreatment by independent evaluators using the Yale Global Tic Severity Scale (YGTSS; Leckman et al., 1989) and the Clinician's Global Improvement Scale (CGI-I; Guy and Bonato, 1970). They reported 33% reduction in on the YGTSS total tic severity score (ES=0.89), and 70% of the sample was rated as clinical responders on the CGI-I. In a small RCT, Deckersbach et al. (2006) randomly assigned 30 adults with TD to 14 sessions of HRT or supportive psychotherapy. Outcomes were assessed with the YGTSS and CGI-I. HRT participants exhibited significantly greater reductions on the YGTSS total tic severity score compared to the control group at posttreatment (38% vs 4%, respectively) and 6-month follow-up (37% vs 4%). Sixty-seven percent of participants were rated as clinical responders on the CGI-I compared to 13% in the control group. In a nearly identical trial, Wilhelm et al. (2003) randomized 32 adults with TD to HRT or supportive psychotherapy and found a 35% reduction in YGTSS in the HRT group compared to a 1% increase in YGTSS scores in the control condition at the end of acute treatment. They found that gains were maintained at 10-month follow-up.

Comprehensive Behavioral Intervention for Tics. Studies on HRT efficacy provided an initial evidence base for the development of CBIT, which was more thoroughly tested in larger samples. Evidence for the efficacy of CBIT has been demonstrated in two large, parallel, multisite RCTS with both children and adults. In the child CBIT trial, Piacentini et al. (2010) randomized 126 children and adolescents (aged 9–17 years) with TD to eight sessions of CBIT (*n*=61) or a supportive therapy plus education control condition (*n*=65) delivered over eight sessions across a 10-week acute treatment phase. Participants who responded positively received three booster sessions and were assessed again at 6-month follow-up. The primary outcomes were changes in tic severity as measured by the YGTSS and treatment response as rated by the CGI-I, both administered by condition-blind independent evaluators. At posttreatment, participants who received CBIT showed significantly greater reductions on the YGTSS total tic score (31%) compared to the control group (14%; ES=0.68), and a significantly higher proportion of CBIT participants (52.5%) were rated as clinical responders compared to control participants (18.5%; number-needed-to-treat=3). Improvements in functional impairment and overall functioning as measured by the YGTSS Impairment Scale also favored CBIT (51% reduction vs 30% reduction for CBIT vs supportive therapy, respectively; ES=0.57). The treatment effect was not moderated by tic medication status or baseline tic severity. Adverse events were generally mild, and tic worsening was rare (1.6% in CBIT vs 6.2% in control). At 6-month follow-up, 87% of CBIT responders maintained their treatment gains.

In the parallel adult CBIT trial (Wilhelm et al., 2012), 122 adults (aged 16–69 years) with TD were randomized to CBIT (*n*=63) or the same psychoeducation plus supportive therapy control condition (*n*=59), delivered over eight sessions across a 10-week acute treatment phase. Participants who responded positively received three booster sessions and were assessed again at 6-month follow-up. At posttreatment, CBIT participants exhibited significantly greater reductions on the YGTSS total tic severity score (26%) compared to the control group (12%; ES=0.57), and a significantly higher proportion of CBIT participants (38.1%) were rated as clinical responders compared to control participants (6.8%; number-needed-to-treat=5). Improvements in functional impairment as measured by the YGTSS Impairment Scale also favored CBIT compared to supportive therapy (36% reduction vs 22% reduction, respectively; ES=0.50). At 6-month follow-up, 80% of CBIT responders maintained their treatment gains. Like the child trial, adverse events were generally mild, and tic worsening was rare and similar between groups (6.3% in CBIT vs 6.8% in control).

In a recent follow-up study, Espil et al. (2022) recontacted 80 of the 126 participants from the original CBIT child trial, an average of 11.2 years after initial treatment, to assess long-term durability of treatment gains. Participants were re-assessed on the YGTSS and CGI either in person or remotely via video conferencing. Across the entire sample, 40% of participants met criteria for partial or full remission (defined as a YGTSS total tic severity score < 14). Importantly, participants who were classified as treatment responders to CBIT in the original trial were significantly more likely to achieve remission status (67%) compared to supportive therapy responders (0%). CBIT responders also demonstrated a sustained reduction in tic severity, with large effect sizes observed at follow-up (ES=1.47). Thus, while both CBIT and supportive therapy responders showed initial improvement, only CBIT responders maintained their gains over time. Importantly, however, tic-related impairment decreased substantially at follow-up for all treatment responders, suggesting some level of adaptation or compensation over time regardless of the treatment received. The study highlights the potential of early behavioral intervention not only to produce clinically meaningful improvements but also to potentially alter the long-term clinical course of TD.

Although most children with TD develop their first tics between the ages of 5 and 9 years, until recently, CBIT had only been tested in children 8 years and older (Piacentini et al., 2010). This is likely due to studies showing that tics are highly variable and often spontaneously improve in the first few years after initial onset (Black et al., 2021), younger children have more difficulty suppressing their tics (Conelea et al., 2018), and a faulty assumption that younger children are less likely to report premonitory urges (Openneer et al., 2020). Clinical intuition might also raise concerns that younger children lack the motivation, self-awareness, or cognitive skills required to engage in the tic self-management strategies taught in CBIT, especially the HRT component. Nevertheless, some younger children are bothered by their tics, and the functional model predicts that tic-contingent reactions and consequences have the potential to worsen or attenuate tic severity starting at an early age, suggesting that the function-based component of CBIT might have value as an early intervention. Based on this rationale, Bennett et al. (2020) tested a developmentally tailored CBIT protocol for children 5–8 years of age referred to as CBIT-Junior. CBIT-Junior maintains the basic structure and therapeutic components of the original CBIT protocol but contains several developmental adaptations including fewer sessions (6 vs 8), a parent-only psychoeducation session, and gamifying and simplifying HRT by adapting it into a "Simon Says" style game in which the child is prompted to select and engage in one of several preselected tic-opposite responses. An open feasibility pilot trial with 15 children (ages 5–8) found high treatment satisfaction, reduction in tic severity on the YGTSS at posttreatment (36%) and 3-month follow-up (40%), and significant reductions on a measure of tic-related impairment. Fifty percent of the participants were rated as clinical responders on the CGI-I.

Exposure and Response Prevention. Initial support for ERP came from an initial case series by Hoogduin et al. (1997) that demonstrated tic reduction in four individuals (three adults and one child) who were treated with 10, 2-hour sessions of ERP. Across the four participants, observed tic frequency reduced by 68%–83% at the end of treatment and by 50%–100% at 2-month follow-up. This work led to a more rigorous study by Verdellen et al. (2004) who randomly assigned 43 children and adults with TD to either 12 sessions of ERP or 10 sessions of HRT. Outcomes were assessed by condition blind raters. Both treatments produced significant reductions in tic severity as measured by the YGTSS (32.8% and 18.3% for ERP and HRT, respectively), with no statistically significant differences between groups. Within-group effect sizes were large for both ERP (ES=1.42) and HRT (ES=1.06), and treatment gains were well-maintained at 3-month follow-up for both groups. ERP was associated with reductions in premonitory urges both within and between sessions. In a recent effectiveness study, Andrén et al. (2021) examined outcomes in a naturalistic clinical setting. Seventy-four youth with TD (ages 6–17) received either HRT or ERP based on therapist and patient preference. All patients received psychoeducation. Additionally, 46 received ERP, 14 received HRT, and 14 received a combination of ERP and HRT. The number of sessions was flexible based on each patient's needs, with an average of seven sessions across the sample (range 2–16). Outcomes were assessed pre- and posttreatment and again at 3-, 6-, and 12-month follow-up; however, not all patients completed all assessment timepoints. At posttreatment, 57% of patients were rated as treatment responders on the CGI-I, with an average 33% reduction on the YGTSS total tic score. Fifty-two of the original 74 patients (70%) completed the 12-month follow-up assessment. Of those, 39 (75%) were rated as treatment responders.

Meta-analyses. Meta-analytic studies also support the efficacy of behavioral therapies for TD. McGuire et al. (2014) synthesized data from 8 RCTs of HRT or CBIT encompassing 438 participants. The pooled analysis revealed a significant overall reduction in tic severity following behavioral therapy, with a medium-to-large effect size compared to both wait-list and active control conditions (SMD=0.67, 95% CI: 0.22, 1.12) and a large effect size when active control conditions were excluded (6 trials; SMD=0.94, 95% CI: 0.54, 1.34). Participants who received behavior therapy were also more likely to be rated as treatment responders on the CGI-I (OR=5.77, 95% CI: 3.23, 10.30), with a number needed to treat=3 (95% CI: 2.3, 4.2). Moderation analysis found a small negative association between clinical responses and attention-deficit/hyperactivity disorder status and small positive associations between treatment response and age and number of therapy sessions. There was no significant association between clinical response and medication status. In a more recent meta-analysis, Shou et al. (2022) examined 12 RCTs involving 536 participants and found a similar effect size favoring behavioral interventions relative to all control conditions (SMD=0.51, 95% CI: 0.22, 0.80). Analysis of the seven included studies that separately reported motor and vocal severity found a more modest effect for vocal tics (SMD=0.22, 95% CI: 0.11, 0.54) compared to motor tics (SMD=0.43, 95% CI: 0.11, 0.75).

Behavior Therapy Versus Medication. To date, only one study has directly compared behavioral to pharmacologic treatment for TD. Rizzo et al. (2018) randomized 110 children and adolescents with TD (ages 8–17 years) to eight sessions of behavioral therapy consisting of either HRT or ERP (BT), pharmacologic treatment consisting of either risperidone, aripiprazole, or pimozide (PT), or a psychoeducation control condition (PE). Outcome assessments were conducted by condition-blind raters at baseline, 10 weeks, and 3-month (for BT and PE) or 5-month (for PE) follow-up. Tic severity was evaluated using the YGTSS and CGI-I, along with additional clinical and quality-of-life measures. The study found that both BT and PT significantly reduced tic severity on the YGTSS at 10 weeks (42% and 35%, respectively) and follow-up (37% and 39%, respectively), while PE showed no significant improvement. CGI-I scores also improved significantly in the BT and PT groups, but not in the PE group. Furthermore, both BT and PT improved on several quality-of-life domains (e.g., mood, self-perception, autonomy), whereas PE did not.

CLINICAL CONSIDERATIONS

Several concerns and misunderstandings regarding the use of behavioral interventions for TD have been raised and continue to inhibit their adoption despite a lack of supporting evidence (Burd and Kerbeshian, 1987; Woods et al., 2010). The most frequently raised concerns include the potential for symptom substitution, symptom rebound, and concerns of attenuated efficacy in patients with psychiatric comorbidity (see Scahill et al., 2013). Below, we review each of these concerns and empirical studies showing them to be unfounded.

Symptom Substitution. Symptom substitution refers to the concern that suppressing tics may lead to the emergence of other, potentially more impairing, tics. While it is well-established that tics often change in appearance and bodily location over the clinical course (Leckman et al., 1998), empirical data do not support treatment-induced symptom substitution. The strongest evidence against this concern comes from a secondary analysis of the aforementioned CBIT RCTs that compared the emergence of new tics assessed weekly over the 10-week course of CBIT or supportive therapy. The analysis showed that participants exhibited an average of 1.25 new tics over the 10 weeks of treatment with no significant differences between the two intervention conditions (Peterson et al., 2016).

Symptom Rebound. Symptom rebound refers to a paradoxical worsening of tics following suppression. While it is common for patients to report subjective tic worsening following suppression (Bliss, 1980; Kane, 1994), studies employing direct observation and objective tic counts/ratings have consistently failed to find immediate or delayed symptom worsening following brief (Himle and Woods, 2005; Meidinger et al., 2005; Specht et al., 2013) or longer (Verdellen et al., 2007; Woods et al., 2008a; Müller-Vahl et al., 2014) periods of tic suppression. Some researchers have speculated that the inconsistency between patients' subjective reports and objective tic ratings is due to difficulties in reliable tic rating and/or a misjudgment resulting from contextual fluctuations (Müller-Vahl et al., 2014).

Efficacy of BT in Patient With Psychiatric Comorbidity. Psychiatric comorbidity is exceedingly common in patients with TD and can indeed present challenges in planning, ordering, and prioritizing interventions. Clinical intuition suggests that ADHD, in particular, might complicate behavioral tic management due to deficits in sustained attention, self-monitoring, inhibitory control, and/or motivation (Dopfner and Rothenberger, 2007); however, studies have not found strong moderating effects of comorbidity status nor neurocognitive functioning on treatment response to CBIT (McGuire et al., 2014; Sukhodolsky et al., 2017; Chang et al., 2018; Mazur-Laine et al., 2024). It is important to note however that most of the studies to date have excluded patients with acute comorbidity, raising questions about the generalizability of finding for patients with more severe symptoms. Nevertheless, while pharmacotherapy or adjunctive evidence-based psychotherapies may be necessary to effectively treat comorbid symptoms, behavioral interventions remain a viable and effective option for reducing tics in many patients with complex clinical profiles.

PREDICTORS OF RESPONSE TO BEHAVIOR THERAPY

In an examination of moderators of response to CBIT, Sukhodolsky et al. (2017) combined the samples from the aforementioned child and adult CBIT trials (n=248) and examined the association between treatment response and several baseline demographic and clinical characteristics. They found that age, sex, family functioning, treatment expectancy, and tic phenomenology (i.e., age of onset, tic complexity, overall severity, and premonitory urge severity) did not moderate treatment response. The presence of tic suppressing medication (no medication, α-2 agonists only, antipsychotics only, or both) was found to moderate treatment response, favoring those who were unmedicated. Importantly, although unmedicated participants showed greater reductions in tic severity and likelihood of treatment response, both medicated and unmedicated participants who received CBIT demonstrated clinically significant reductions in tic severity on the YGTSS, regardless of medication status or class (7.3, 6.5, and 8.5 point reductions for the unmedicated, α-2 agonist, and antipsychotic groups, respectively). The authors concluded that the significant interaction effect was likely the result of a better-than-expected response among those receiving concomitant supportive therapy (control intervention) and α-2 agonists. Not surprisingly, treatment compliance, specifically between-session practice "homework" compliance, has also been shown to be a significant predictor of CBIT response

(Essoe et al., 2021). Less is known about predictors and moderators of treatment response to ERP.

ACCESSIBILITY AND ALTERNATIVE DELIVERY FORMATS

Although behavioral therapies have been shown to be effective and are now widely recommended as a front-line treatment in clinical practice guidelines (Andrén et al., 2021; Pringsheim et al., 2019), accessibility remains inadequate, largely due to a lack of trained clinicians, especially outside of specialized academic centers (Wellen et al., 2023; Conelea et al., 2024). In addition to international efforts to train more therapists by organizations such as the Tourette Association of America (TAA, tourette.org) and the European Society for the Study of Tourette Syndrome (ESSTS, essts.org), several alternative delivery formats have been developed and tested for increasing access, including telemedicine, digital health interventions, and an abbreviated protocol for delivering CBIT in neurology clinics and other medical settings.

Telemedicine. The emergence of increasingly sophisticated and reliable video communication platforms and widespread adoption of telemedicine has spurred several studies investigating the efficacy and satisfaction of remote delivery of behavioral interventions for TD. In an early proof of concept study, Himle et al. (2012) randomized 20 youth with TD (ages 8–17) to standard in-person CBIT or CBIT delivered via high-speed videoconference (teleCBIT). In both conditions, patients were required to come to a specialty clinic to receive treatment. Outcomes were assessed at pre- and posttreatment using the YGTSS and CGI-I. YGTSS scores decreased in both groups (27% and 33% for in-person and telehealth, respectively) with no significant differences between groups. Eighty percent of participants in the telehealth group and 75% in the in-person group were rated as treatment responders. Patient and parent satisfaction with both delivery formats was high and did not differ between groups. In another study, Ricketts et al. (2016b) investigated teleCBIT delivered directly into patients' homes via peer-to-peer voice-over-internet protocol. Twenty youth with TD were randomized to either 10 weeks of teleCBIT via Skype or a waitlist control. Results revealed significantly greater reductions on the YGTSS in the treatment group relative to control condition (28% vs 8%, respectively). Similarly, Prato et al. (2022) randomly assigned 40 children to either ERP or HRT delivered over Skype during the COVID-19 pandemic and found significant improvements in YGTSS tic severity across both treatment delivery formats (45% and 47% for telehealth and in-person, respectively), with no significant differences between the groups. Finally, Capriotti et al. (2023) conducted a single-arm study in which treatment-seeking patients were offered teleCBIT at their usual care medical visits. Twenty-five percent of patients initiated and completed treatment (19 youth and 10 adults). Results showed a medium to large effect on YGTSS severity scores for both youth (ES=1.31) and adults (ES=0.67). About 68% of youth and 60% of adults were rated as treatment responders on the CGI-I. Treatment satisfaction was high, and 93% of sessions were free of interfering technical problems. Collectively, these results show that telehealth is a viable and effective option for patients who prefer remote delivery or do not have access to in-person treatment. However, although remote delivery offers clear advantages in accessibility, technological issues such as internet speed and potential audio-visual disruptions must be considered, especially in technological underserved areas.

Digital Health Interventions. Technology-based applications are also beginning to emerge as a potential tool for increasing access to behavioral interventions for TD. For example, in a study out of Isreal, Rachamim et al. (2022) evaluated the feasibility and initial efficacy of an internet-based self-help CBIT (iCBIT) that contains nine self-paced interactive online lessons designed to teach CBIT skills. In the pilot study, participants received minimal weekly assistance from a therapist (average 7 minutes per week), to help individualize the lessons, provide technical assistance, and encourage compliance. Forty-one youth (ages 7–18 years) and their caregivers were randomly assigned to receive either iCBIT or a waitlist control condition in a crossover design. Results showed that participants who completed iCBIT showed significantly greater tic reduction on the YGTSS compared to the control condition at posttreatment (29% vs 4%), and 64% of those who received iCBIT were rated as clinical responders on the CGI-I compared to 6% in the control group. At 6-month follow-up, YGTSS scores among those who completed iCBIT were reduced by 58%, and 95% of the sample was rated as clinical responders. In a follow-up study in Germany, Haas et al. (2022) randomly assigned 161 adults with TD to iCBIT with no therapist support (*n*=67), placebo (internet-delivered psychoeducation, *n*=70), or traditional, therapist delivered, face-to-face CBIT (F2F, *n*=24). At posttreatment, participants in the iCBIT group (10.5% reduction on the YGTSS total tic score) and F2F conditions (14.6% reduction) trended toward better treatment response relative to placebo (5.2% reduction), but the difference did not reach statistical significance. At follow-up, participants in the iCBIT showed greater improvement than placebo at both the 3-month (15.2% vs 6%, for iCBIT and placebo, respectively) and 6-month timepoints (17.6% vs 6.5%). There were no significant differences between iCBIT and F2F at any timepoint.

A similar set of studies out of Sweden (Andrén et al., 2019) and the United Kingdom (Hollis et al., 2021) tested the efficacy of ERP delivered online. The intervention, called BIP-ERP, consists of 10 online child and parent informational "chapters" designed to teach standard ERP for TD. The program is completed over the course of 10 weeks with practice between sessions. Therapists provide support through messaging on the platform (see Hall et al., 2019). Hollis et al. (2021) randomized 224 youth (ages 9–17 years) to BIP-ERP or an online psychoeducation condition. Outcomes were assessed 3- and 6-months after randomization. At the 3-month timepoint, YGTSS scores reduced in the ERP group by 16% compared to 6% in the control group (ES=0.31), and 36% were rated as clinical responders compared to 20% in the control group. At 6-month follow-up, YGTSS scores were decreased by 24% in the ERP group compared to 12% in the control group (ES=0.36). About

47% of participants in the ERP group were rated as treatment responders versus 29% in the control group.

Abbreviated CBIT. Ricketts et al. (2016a) pilot tested a streamlined version of the traditional in-person CBIT protocol that was designed to be delivered by front-line medical professionals in neurology clinics. Instead of the standard eight 60–90-minute sessions delivered over 8 weekly sessions, the streamlined protocol involved 4–8 sessions, each lasting 15–30 minutes. Sessions were delivered by nurses or physicians during usual care visits (e.g., medication visits). While the modified protocol maintained the core components of CBIT, it simplified the functional assessment and placed greater emphasis on home rather than in-session practice. This initial evaluation involving 14 children with TD (ages 9–17) found a significant reduction in tic severity on the YGTSS (17%); however, this was a more modest response compared to what has been reported in other studies (Piacentini et al., 2010). Just over half (56%) of participants were classified as treatment responders on the CGI-I. Clinicians' perceptions of feasibility were mostly favorable; however, several potential barriers were noted that need to be addressed in future research and development work (e.g., limited time even for shorter sessions, limited space, interruptions, billing and insurance reimbursement issues).

CONCLUSIONS AND FUTURE DIRECTIONS

Over the past two decades, a growing body of research has validated the efficacy of behavioral interventions for TD. Large multisite RCTs have demonstrated that CBIT and HRT lead to significant reductions in tic severity across both child and adult populations, with durable effects sustained at follow-up. ERP has also shown promise, with studies reporting comparable efficacy to HRT for reducing tics, and comparative studies suggest that both HRT and ERP may be as effective as pharmacotherapy for many patients. Empirical evidence has not supported historical concerns about negative side-effects, including symptom substitution and tic rebound. Accessibility remains inadequate, however, largely due to a lack of trained clinicians, especially for patients in underserved areas. New delivery models, including telehealth and internet-based interventions, have shown promise for expanding access to care. Several international best-practice guidelines now recommend behavioral interventions as a first-line treatment for those with bothersome tics who have access to trained therapists (Pringsheim et al., 2019; Andrén et al., 2022).

Although behavioral therapies have been shown to be effective for reducing tic, not all patients respond to treatment, and residual symptoms often persist, even among treatment responders. The fact that nearly half of patients do not substantially benefit highlights the importance of additional research to identify therapeutic elements and patient characteristics that predict clinical response, as well as potential differential outcomes for specific symptom and clinical profiles (e.g., tic topography, complexity, automaticity/compulsivity). Additional research is also needed to better understand the behavioral and neurobiologic mechanisms underlying behavioral interventions to gain insights into how to modify and improve individualized outcomes. Finally, additional comparative and combination studies with pharmacotherapy are needed to inform individualized comprehensive treatment planning.

References

Abramowitz JS (1996). Variants of exposure and response prevention in the treatment of obsessive-compulsive disorder: a meta-analysis. Behav Ther 27: 583–600.

Andrén P, Aspvall K, Fernández de la Cruz L et al. (2019). Therapist-guided and parent-guided internet-delivered behaviour therapy for paediatric Tourette's disorder: a pilot randomised controlled trial with long-term follow-up. BMJ Open 9:e024685.

Andrén P, Jakubovski E, Murphy TL et al. (2022). European clinical guidelines for Tourette syndrome and other tic disorders—version 2.0. Part II: psychological interventions. Eur Child Adolesc Psychiatry 31: 403–423.

Andrén P, Wachtmeister V, Franzé J et al. (2021). Effectiveness of behaviour therapy for children and adolescents with Tourette syndrome and chronic tic disorder in a naturalistic setting. Child Psychiatry Hum Dev 52: 739–750.

Azrin NH, Nunn RG (1973). Habit-reversal: a method of eliminating nervous habits and tics. Behav Res Ther 11: 619–628.

Barber KE, Ding Q, Espil FM et al. (2024). Contextual triggers and tic severity across life periods: a retrospective analysis in adults with tic disorders. Child Psychiatry Hum Dev. Published online.

Barnea M, Benaroya-Milshtein N, Gilboa-Sechtman E et al. (2016). Subjective versus objective measures of tic severity in Tourette syndrome: the influence of environment. Psychiatry Res 242: 204–209.

Bennett SM, Capriotti M, Bauer C et al. (2020). Development and open trial of a psychosocial intervention for young children with chronic tics: the CBIT-JR study. Behav Ther 51: 659–669.

Black KJ, Kim S, Yang NY et al. (2021). Course of tic disorders over the lifespan. Curr Dev Disord Rep 8: 121–132.

Bliss J (1980). Sensory experiences of Gilles de la Tourette syndrome. Arch Gen Psychiatry 37: 1343–1347.

Brandt VC, Beck C, Sajin V et al. (2016). Temporal relationship between premonitory urges and tics in Gilles de la Tourette syndrome. Cortex 77: 24–37.

Burd L, Kerbeshian J (1987). Treatment-generated problems associated with behavior modification in Tourette disorder. Dev Med Child Neurol 29: 831–833.

Capriotti MR, Brandt BC, Turkel JE et al. (2014). Negative reinforcement and premonitory urges in youth with Tourette syndrome: an experimental evaluation. Behav Modif 38: 276–296.

Capriotti MR, Piacentini JC, Himle MB et al. (2015). Assessing environmental consequences of ticcing in youth with chronic tic disorders: the Tic Accommodation and Reactions Scale. Child Health Care 44: 205–220.

Capriotti MR, Wellen BC, Young BN et al. (2023). Evaluating the feasibility, acceptability, and preliminary effectiveness of tele-comprehensive behavior therapy for tics (teleCBIT) for Tourette syndrome in youth and adults. J Telemed Telecare 31: 328–337.

Cavanna AE, Black KJ, Hallett M, Voon V (2017). Neurobiology of the premonitory urge in Tourette's syndrome: pathophysiology and treatment implications. J Neuropsychiatry Clin Neurosci 29: 95–104.

Chang SW, McGuire JF, Walkup JT et al. (2018). Neurocognitive correlates of treatment response in children with Tourette's disorder. Psychiatry Res 261: 464–472.

Conelea CA, Bennett S, Himle MB et al. (2024). Treating Tourette Together: an agenda for patient-centered research focused on comprehensive behavioral intervention for tics. Behav Ther 55: 263–276.

Conelea CA, Wellen B, Woods DW et al. (2018). Patterns and predictors of tic suppressibility in youth with tic disorders. Front Psychiatry 9: 188.

Conelea CA, Woods DW, Brandt BC et al. (2011). The impact of a stress induction task on tic frequencies in youth with Tourette syndrome. Behav Res Ther 49: 492–497.

Deckersbach T, Rauch S, Buhlmann U et al. (2006). Habit reversal versus supportive psychotherapy in Tourette's disorder: a randomized controlled trial and predictors of treatment response. Behav Res Ther 44: 1079–1090.

Dopfner M, Rothenberger A (2007). Behavior therapy in tic-disorders with co-existing ADHD. Eur Child Adolesc Psychiatry 16: 89–99.

Eaton CK, Jones AM, Gutierrez-Colina AM et al. (2017). The influence of environmental consequences and internalizing symptoms on children's tic severity. Child Psychiatry Hum Dev 48: 327–334.

Espil FM, Woods DW, Specht M et al. (2022). Determining the long-term effects of comprehensive behavioral intervention for tics (CBIT). J Am Acad Child Adolesc Psychiatry 61: 764–771.

Essoe JKY, Ramsey KA, Singer HS et al. (2021). Mechanisms underlying behavior therapy for Tourette's disorder. Curr Dev Disord Rep 8: 161–174.

Essoe JKY, Ricketts EJ, Ramsey KA et al. (2021). Homework adherence predicts therapeutic improvement from behavior therapy in Tourette's disorder. Behav Res Ther 140: 103844.

Evers RA, van de Wetering BJ (1994). A treatment model of motor tics based on a specific tension-reduction technique. J Behav Ther Exp Psychiatry 25: 255–260.

Ganos C, Roessner V, Munchau A (2013). The functional anatomy of Gilles de la Tourette syndrome. Neurosci Biobehav Rev 37: 1050–1062.

Ganos C, Garrido A, Navalpotro-Gomez I et al. (2015). Premonitory urge to tic in Tourette's is associated with interoceptive awareness. Mov Disord 30: 1198–1202.

Guy W, Bonato R. (Eds.). CGI: clinical global impressions, . Chevy Chase, MD: National Institute of Mental Health.

Haas M, Jakubovski E, Kunert K et al. (2022). ONLINE-TICS: internet-delivered behavioral treatment for patients with chronic tic disorders. J Clin Med 11: 250.

Hall CL, Davies EB, Andrén P et al. (2019). Investigating a therapist-guided, parent-assisted remote digital behavioural intervention for tics in children and adolescents—'Online Remote Behavioural Intervention for Tics' (ORBIT) trial: protocol of an internal pilot study and single-blind randomised controlled trial. BMJ Open 9e:027583.

Himle MB, Capriotti MR, Hayes LP et al. (2014). Variables associated with tic exacerbation in children with chronic tic disorders. Behav Modif 38: 163–183.

Himle MB, Freitag M, Walther M et al. (2012). A randomized pilot trial comparing videoconference versus face-to-face delivery of behavior therapy for childhood tic disorders. Behav Res Ther 50: 565–570.

Himle MB, Woods DW (2005). An experimental evaluation of tic suppression and the tic rebound effect. Behav Res Ther 43: 1443–1451.

Himle MB, Woods DW, Piacentini JC et al. (2006). Brief review of habit reversal training for Tourette syndrome. J Child Neurol 21: 719–725.

Hoekstra PJ, Steenhuis MP, Kallenberg CG et al. (2004). Association of small life events with self reports of tic severity in pediatric and adult tic disorder patients: a prospective longitudinal study. J Clin Psychiatry 65: 426–431.

Hollis C, Hall CL, Jones R et al. (2021). Therapist-supported online remote behavioural intervention for tics in children and adolescents in England (ORBIT): a multicentre, parallel group, single-blind, randomised controlled trial. Lancet Psychiatry 8: 871–882.

Hoogduin K, Verdellen C, Cath D (1997). Exposure and response prevention in the treatment of Gilles de la Tourette's syndrome: four case studies. Clin Psychol Psychother 4: 125–135.

Houghton DC, Capriotti MR, Scahill LD et al. (2017). Investigating habituation to premonitory urges in behavior therapy for tic disorders. Behav Ther 48: 834–846.

Iverson AM, Black KJ (2022). Why tic severity changes from then to now and from here to there. J Clin Med 11: 5930.

Kane MJ (1994). Premonitory urges as "attentional tics" in Tourette's syndrome. J Am Acad Child Adolesc Psychiatry 33: 805–808.

Kwak C, Dat Vuong K, Jankovic J (2003). Premonitory sensory phenomenon in Tourette's syndrome. Mov Disord 18: 1530–1533.

Leckman JF, Walker DE, Cohen DJ (1993). Premonitory urges in Tourette's syndrome. Am J Psychiatry 150: 98–102.

Leckman JF, Riddle MA, Hardin MT et al. (1989). The Yale Global Tic Severity Scale: initial testing of a clinician-rated scale of tic severity. J Am Acad Child Adolesc Psychiatry 28: 566–573.

Leckman JF, Zhang H, Vitale A et al. (1998). Course of tic severity in Tourette syndrome: the first two decades. Pediatrics 102: 14–19.

Matsuda N, Nonaka M, Kono T et al. (2020). Premonitory awareness facilitates tic suppression: subscales of the premonitory urge for tics scale and a new self-report questionnaire for tic-associated sensations. Front Psychiatry 11: 592.

Mazur-Laine E, Soubata H, Leclerc JB et al. (2024). Impacts of ADHD symptomology on the response to cognitive-behavioural therapy with Gilles de la Tourette syndrome patients. J Clin Med 13: 2975.

McGuire JF, Piacentini J, Brennan EA et al. (2014). A meta-analysis of behavior therapy for Tourette Syndrome. J Psychiatr Res 50: 106–112.

Meidinger AL, Miltenberger RG, Himle M et al. (2005). An investigation of tic suppression and the rebound effect in Tourette's disorder. Behav Modif 29: 716–745.

Mohamed ZA, Xue Y, Bai M et al. (2024). Efficacy of differential reinforcement of other behaviors therapy for tic disorder: a meta-analysis. BMC Neurol 24: 3.

Müller-Vahl KR, Riemann L, Bokemeyer S (2014). Tourette patients' misbelief of a tic rebound is due to overall difficulties in reliable tic rating. J Psychosom Res 76: 472–476.

Openneer TJC, Tárnok Z, Bognar E et al. (2020). The premonitory urge for tics scale in a large sample of children and adolescents: psychometric properties in a developmental context. An EMTICS study. Eur Child Adolesc Psychiatry 29: 1411–1424.

Peterson AL, McGuire JF, Wilhelm S et al. (2016). An empirical examination of symptom substitution associated with behavior therapy for Tourette's disorder. Behav Ther 47: 29–41.

Piacentini J, Woods DW, Scahill L et al. (2010). Behavior therapy for children with Tourette disorder: a randomized controlled trial. JAMA 303(19): 1929–1937.

Prato A, Maugeri N, Chiarotti F et al. (2022). A randomized controlled trial comparing videoconference vs. face-to-face delivery of behavior therapy for youths with Tourette syndrome in the time of COVID-19. Front Psychiatry 13: 862422.

Pringsheim T, Okun MS, Müller-Vahl K et al. (2019). Practice guideline recommendations summary: treatment of tics in people with Tourette syndrome and chronic tic disorders. Neurology 92: 896–906.

Rachamim L, Zimmerman-Brenner S, Rachamim O et al. (2022). Internet-based guided self-help comprehensive behavioral intervention for tics (ICBIT) for youth with tic disorders: a feasibility and effectiveness study with 6-month follow-up. Eur Child Adolesc Psychiatry 31: 275–287.

Ramsey KA, De Nadai AS, Espil FM et al. (2022). Urge intolerance predicts tic severity and impairment among adults with Tourette syndrome and chronic tic disorders. Front Psychiatry 13: 929413.

Ricketts EJ, Gilbert DL, Zinner SH et al. (2016a). Pilot testing behavior therapy for chronic tic disorders in neurology and developmental pediatrics clinics. J Child Neurol 31: 444–450.

Ricketts EJ, Goetz AR, Capriotti MR et al. (2016b). A randomized waitlist-controlled pilot trial of voice over Internet protocol-delivered behavior therapy for youth with chronic tic disorders. J Telemed Telecare 22: 153–162.

Rizzo R, Pellico A, Silvestri PR et al. (2018). A randomized controlled trial comparing behavioral, educational, and pharmacological treatments in youths with chronic tic disorder or Tourette syndrome. Front Psychiatry 9: 100.

Scahill L, Woods DW, Himle MB et al. (2013). Current controversies on the role of behavior therapy in Tourette syndrome. Mov Disord 28: 1179–1183.

Shou S, Li Y, Fan G et al. (2022). The efficacy of cognitive behavioral therapy for tic disorder: a meta-analysis and a literature review. Front Psychol 13: 851250.

Silva RR, Munoz DM, Barickman J et al. (1995). Environmental factors and related fluctuation of symptoms in children and adolescents with Tourette's disorder. J Child Psychol Psychiatry 36: 305–312.

Specht MW, Woods DW, Nicotra CM et al. (2013). Effects of tic suppression: ability to suppress, rebound, negative reinforcement, and habituation to the premonitory urge. Behav Res Ther 51: 24–30.

Sukhodolsky DG, Woods DW, Piacentini J et al. (2017). Moderators and predictors of response to behavior therapy for tics in Tourette syndrome. Neurology 88: 1029–1036.

Tilling F, Cavanna AE (2020). Relaxation therapy as a treatment for tics in patients with Tourette syndrome: a systematic literature review. Neurol Sci 41(5): 1011–1017.

van de Griendt JMTM, van den Berg NME, Verdellen CWJ et al. (2023). Working mechanisms of exposure and response prevention in the treatment of Tourette syndrome and tic disorders revisited: no evidence for within-session habituation to premonitory urges. J Clin Med 12: 7087.

Verdellen CWJ, Hoogduin CAL, Kato BS et al. (2008). Habituation of premonitory sensations during exposure and response prevention treatment in Tourette's syndrome. Behav Modif 32: 215–227.

Verdellen CW, Hoogduin CA, Keijsers GP (2007). Tic suppression in the treatment of Tourette's syndrome with exposure therapy: the rebound phenomenon reconsidered. Mov Disord 22: 1601–1606.

Verdellen CWJ, Keijsers GPJ, Cath DC et al. (2004). Exposure with response prevention versus habit reversal in Tourette's syndrome: a controlled study. Behav Res Ther 42: 501–511.

Verdellen C, van de Griendt J, Kriens S et al. (2011). Tics: therapist manual. Amsterdam, NL: Boom Publishers.

Viefhaus P, Feldhausen M, Gortz-Dorten A et al. (2020). Efficacy of habit reversal training in children with chronic tic disorders: a within-subject analysis. Behav Modif 44: 114–136.

Wellen BCM, Bootes KR, Braley EI et al. (2023). Caregiver perspectives on the health care system for tic disorders: utilization and barriers. J Dev Behav Pediatr 44: e581–e589.

Wellen BCM, Ramanujam K, Lavelle M et al. (2024). A test of the behavioral model of tic disorders using a dynamical systems framework. Behav Ther 55: 513–527.

Wilhelm S, Deckersbach T, Coffey BJ et al. (2003). Habit reversal versus supportive psychotherapy for Tourette's disorder: a randomized controlled trial. Am J Psychiatry 160: 1175–1177.

Wilhelm S, Peterson AL, Piacentini JC et al. (2012). Randomized trial of behavior therapy for adults with Tourette's disorder. Arch Gen Psychiatry 69: 795–803.

Woods DW, Conelea CA, Himle MB (2010). Behavior therapy for Tourette's disorder: utilization in a community sample and an emerging area of practice for psychologists. Prof Psychol Res Pr 41: 518–525.

Woods DW, Himle MB, Miltenberger RG et al. (2008a). Durability, negative impact, and neuropsychological predictors of tic suppression in children with chronic tic disorder. J Abnorm Child Psychol 36: 237–245.

Woods DW, Himle MB, Stiede JT et al. (2023). Behavioral interventions for children and adults with tic disorder. Annu Rev Clin Psychol 19: 233–260.

Woods DW, Miltenberger RG, Lumley VA (1996). Sequential application of major habit-reversal components to treat motor tics in children. J Appl Behav Anal 29: 483–493.

Woods DW, Piacentini JC, Chang SW et al. (2008). Managing Tourette syndrome: a behavioral intervention for children and adults. New York, NY: Oxford University Press.

Further reading

McGuire JF, Ricketts EJ, Piacentini J et al. (2015). Behavior therapy for tic disorders: an evidenced-based review and new directions for treatment research. Curr Dev Disord Rep 2: 309–317.

Handbook of Clinical Neurology, Vol. 215 (3rd series)
Tic Disorders and Tourette Syndrome
C. Ganos and Y. Worbe, Editors
https://doi.org/10.1016/B978-0-443-13554-5.00029-8

Chapter 23

Pharmacologic treatment of tics and Gilles de la Tourette syndrome

RENATA RIZZO*

Department of Clinical and Experimental Medicine, Child and Adolescent Neurology and Psychiatry, Catania University, Catania, Italy

Abstract

Treating children and adults with Gilles de la Tourette syndrome (GTS) remains a clinical challenge and typically requires a multidisciplinary approach, owing to the complexity of the disorder and the frequent presence of comorbidities such as obsessive-compulsive disorder and attention-deficit/hyperactivity disorder.

This chapter focuses on the pharmacologic management of tics in the context of GTS. It reviews key considerations in the selection of therapeutic agents from the broad range of pharmacologic options available in clinical practice. The strongest evidence supports the use of dopamine receptor antagonists, particularly atypical antipsychotics such as aripiprazole and risperidone. However, other agents – including α-2 adrenergic agonists, vesicular monoamine transporter 2 inhibitors, and topiramate – also play an important role in treatment strategies.

Pharmacologic management of GTS should be individualized, taking into account symptom severity, functional impact, and the risk–benefit profile of each therapeutic option.

CLINICAL MANAGEMENT OVERVIEW

Gilles de la Tourette syndrome (GTS) is a neurodevelopmental disorder with a reported prevalence ranging from 0.3% to 1% (Singer, 2019). Although it is not a rare condition, GTS remains under-recognized, and the broad heterogeneity of its clinical spectrum renders personalized treatment approaches both necessary and challenging (Hartmann and Worbe, 2018). Tics are the hallmark feature of GTS. While often mild or moderately bothersome, in some cases, they may be associated with self-injurious behaviors or become severely disabling.

The treatment of tics in GTS is particularly complex, as the underlying genetics, neurophysiology, and neuropathology of the disorder remain incompletely understood (see also corresponding chapters in this volume). Furthermore, the majority of patients with GTS also present with a range of comorbid neuropsychiatric conditions, which are often more impairing than the tics themselves. These include obsessive-compulsive disorder (OCD), attention-deficit/hyperactivity disorder (ADHD), anxiety and mood disorders, learning difficulties, sleep disturbances (Rizzo et al., 2017), and, in some cases, autism spectrum disorder (Gulisano et al., 2020).

Given this clinical complexity, a multidisciplinary approach is essential to address the unique needs of each patient. The clinical guidelines established by the European Society for the Study of Tourette Syndrome (ESSTS) recommend that individuals with mild tics may not require specific medical or behavioral interventions, although psychoeducation should always be provided. Psychoeducation helps patients and families understand the nature of the disorder, its course, comorbidities, and prognosis, while also preparing them for treatment if needed (Lewin et al., 2014).

Screening for psychiatric and developmental comorbidities is a critical step in management, as these conditions often have a more substantial impact on functioning than the tics themselves and may require targeted treatment. When tics do require intervention, behavioral therapies are recommended as the first-line treatment. These include habit reversal training (HRT), comprehensive behavioral intervention for tics (CBIT), and exposure with response prevention (ERP) (Andrén et al., 2022). Notably, some of these approaches can be delivered via online formats, increasing accessibility (Himle et al., 2012; Ricketts et al., 2016; Andrén et al., 2019; Hollis et al., 2021; Prato et al., 2022).

If behavioral therapy is not feasible, not accessible, or insufficiently effective, pharmacologic interventions become the next treatment option in most clinical settings and guidelines (Pringsheim et al., 2019a, b; Roessner et al., 2022). Prior to initiating medication, clinicians must engage in a comprehensive discussion with the patient and family about the potential

*Correspondence to: Renata Rizzo, Department of Clinical and Experimental Medicine, Child and Adolescent Neurology and Psychiatry, Catania University, via S. Sofia, 78 – 95123, Catania, Italy. Email: rerizzo@unict.it.

benefits, side-effects, and treatment goals to ensure alignment with the patient's needs and expectations.

The cornerstone of any therapeutic strategy remains appropriate patient education. Treatment planning should be individualized and informed by an understanding of the symptoms most disabling to the patient. This process involves shared decision-making, taking into account the patient's age, developmental stage, expectations of efficacy, treatment preferences, and practical considerations such as feasibility and tolerability.

Discrepancies between children's and parents' preferences are common. While parents may often advocate for a medication trial, children themselves may be reluctant, due to stigma, concerns about side-effects, or simply a lack of perceived need. Treatment expectations must be explored openly and managed realistically. Unrealistic expectations should be addressed, and patients and caregivers must be provided with accurate, comprehensible information.

A key consideration in GTS management is the waxing and waning nature of tics. These intraindividual fluctuations in intensity, frequency, anatomic distribution, and complexity may mislead clinicians and families alike. Therefore, extended observation is often necessary to avoid misinterpreting natural tic variability as treatment response. Regular monitoring of symptom severity using standardized rating scales at each follow-up visit can improve clinical outcomes and help guide long-term care. If symptom severity remains stable over a 6–12 month period, a medication trial may be considered, especially in the context of expected age-related improvement (Table 23.1).

We recommend that tic severity be systematically assessed at each visit using validated tools, and that treatment be adjusted as needed to optimize outcomes (Szejko et al., 2022). Quality of life should be evaluated alongside symptom severity using GTS-specific instruments when possible. These objective measures should complement clinical judgment. Adverse effects should be actively monitored through history, physical examination, and laboratory testing when indicated.

There is limited empirical guidance on the safety and efficacy of combining medications or sequencing different pharmacologic options. Often, optimal treatment is discovered only after multiple trials (Farag et al., 2015). The duration of treatment remains empirical and should be individualized based on perceived benefit and evolving side-effects (Stern, 2018). Factors such as brain maturation, growth, and the potential for tolerance must also be considered (Lemmon et al., 2015).

In recent years, several international guidelines have helped standardize care for patients with tic disorders. These include the original European guidelines published in 2011 (Roessner et al., 2011), Canadian guidelines (Steeves, 2012), the American Academy of Neurology guidelines (Pringsheim et al., 2019a), and the updated European guidelines version 2.0 (Roessner et al., 2022).

Despite some variation, there is a broad consensus across guidelines regarding the general principles of GTS management:

1. Psychoeducation should be provided routinely to all patients and caregivers.
2. Behavioral therapies are the preferred first-line treatment due to their favorable safety profile and demonstrated efficacy.
3. Pharmacologic treatment should be considered when tics remain significantly impairing despite nonpharmacologic approaches (Andrén et al., 2022; Roessner et al., 2022).

In clinical practice, several additional principles are useful. First, clinicians should be confident in the diagnosis and ensure that secondary causes of tics or functional tic-like behaviors are

Table 23.1

Common pharmacologic treatment options for Gilles de la Tourette and chronic tic disorders.

Medication	Start dosage	Therapeutic range	Side-effects
Typical antipsychotics			
• *Haloperidol*	0.5 mg	2.0–10.0 mg daily	Extrapyramidal symptoms, hyperprolactinemia, dry mouth, insomnia, dizziness, somnolence, weight gain, QTc prolongation, agitation, headache, hypotension.
• *Pimozide*	1.0 mg	2.0–10.0 mg daily	
Atypical antipsychotics			
• *Aripiprazole*	2.5 mg	2.5–30 mg daily	Somnolence, insomnia, dizziness, nausea, vomiting, headache, parkinsonism, metabolic abnormalities, type 2 diabetes, weight gain, hyperprolactinemia.
• *Risperidone*	0.25 mg	0.25–4.0 mg daily	
α-2 adrenergic agonists			
• *Clonidine*	0.05 mg	0.1–0.3 mg daily	Dizziness, orthostatic hypotension, dry mouth, constipation, insomnia, headache, agitation, nausea, vomiting.
GABAergic agents			
• *Topiramate*	25 mg	25–150 mg daily	Weight loss, paresthesias, headache, diarrhea, sedation, constipation, nausea, anxiety, memory problems, depression.
Vesicular monoamine transporters 2 (VMAT2) inhibitors			
• *Tetrabenazine*	6 mg	18–36 mg daily	Sedation, fatigue, nausea, insomnia, parkinsonism, depression, drowsiness, headaches, anxiety, akathisia, weight gain.
• *Deutetrabenazine*	6 mg	30–48 mg daily	
• *Valbenazine*	5 mg	10–50 mg daily	

excluded (Martino et al., 2023). Second, a complete diagnostic formulation that accounts for motor symptoms, psychiatric comorbidities, psychosocial stressors, and family history is essential. In cases where comorbidities like ADHD are driving academic or social dysfunction, these must be addressed alongside tic management. The presence of such comorbidities may influence the choice of pharmacologic agents and alter the prioritization of treatment targets.

Therapeutic discussions should be grounded in this comprehensive formulation. Patients and families should be informed about the expected time course of treatment, potential drug interactions, adherence requirements, and minimum duration needed to assess efficacy. Informed consent and, when applicable, patient assent should be obtained before starting pharmacologic treatment.

In cases of acute tic worsening with significant impact on quality of life, a short observation period may help determine whether the symptoms will improve spontaneously. This is especially important given the naturally fluctuating course of tic disorders (Roessner et al., 2022). In rare, treatment-refractory adult cases where multiple evidence-based interventions have failed, deep brain stimulation may be considered (Müller-Vahl et al., 2011).

Pharmacologic treatment resistance in GTS is defined by a lack of response to multiple well-established interventions, including psychotherapy, despite appropriate management of comorbid conditions (Kious et al., 2016). Defining treatment failure remains complex. Recently, Martino et al. (2023) used a Delphi consensus approach involving international GTS specialists to propose a standardized operational definition of treatment failure.

This chapter provides an overview of pharmacologic options that are available when treating tics in the context of neurodevelopmental tic disorders, such as GTS.

MECHANISMS OF ACTION OF MEDICATIONS FOR GILLES DE LA TOURETTE TREATMENT

A comprehensive discussion of the mechanisms of action for all pharmacologic treatments used in GTS is beyond the scope of this chapter. However, most medications currently used for GTS are believed to act primarily on the dopaminergic system (Mogwitz et al., 2013; Augustine and Singer, 2019). Dopamine, as one of the principal neurotransmitters involved in the cortico-striatal-thalamo-cortical circuits, is widely recognized as playing a key role in the pathophysiology of GTS.

That said, imbalances in other neurotransmitter systems have also been implicated in tic generation and modulation. These include the serotonergic, noradrenergic, glutamatergic, GABAergic, cholinergic, histaminergic, endocannabinoid, and opioid systems (Augustine and Singer, 2019; Roessner et al., 2022; Johnson et al., 2023).

According to current guidelines from both the American Academy of Neurology and the ESSTS, three major classes of drugs are considered effective in the treatment of tics in GTS:

1. Dopamine receptor antagonists, including both typical and atypical antipsychotics (e.g., haloperidol, tiapride, sulpiride, pimozide, risperidone, aripiprazole).
2. Nondopaminergic agents, such as α-2 adrenergic agonists (e.g., clonidine and guanfacine) and topiramate.
3. Vesicular monoamine transporter 2 (VMAT2) inhibitors (e.g., tetrabenazine, deutetrabenazine, valbenazine) (Pringsheim et al., 2019; Roessner et al., 2022).

OVERVIEW OF EFFICACY BY DRUG CLASS

Dopamine receptor antagonists and partial agonists in the treatment of Gilles de la Tourette

Dopamine-blocking agents (antipsychotics) are believed to help control tics by modulating abnormal dopaminergic activity within the cortico-striatal-thalamo-cortical circuits. The beneficial effects of dopamine D2 receptor blockade in GTS date back to the late 1960s, when haloperidol became the first drug approved by the US Food and Drug Administration (FDA) for the treatment of tics. Pimozide followed as the second FDA-approved antipsychotic for GTS. The efficacy of typical antipsychotics such as haloperidol and pimozide is supported both by randomized controlled trials (RCTs) and by long-standing clinical experience in managing GTS (Bloch et al., 2011; Roessner et al., 2011; Mogwitz et al., 2013). These first-generation agents have been used for several decades.

However, since 2011, no additional English-language RCTs have evaluated the use of first-generation antipsychotics for tic treatment (Billnitzer and Jankovic, 2020; Roessner et al., 2022). Haloperidol is typically initiated at 0.5 mg/day, with a usual dose range of 2–10 mg daily (Roth, 2018). Pimozide is used at 2–8 mg/day, administered once daily, with a maximum dose of 10 mg/day. Due to significant adverse effects – including parkinsonism, apathy, anhedonia, and Corrected QT Interval (QTc) prolongation – the use of haloperidol and pimozide has markedly declined (Pringsheim and Marras, 2009; Pringsheim et al., 2019b). According to ESSTS guidelines, haloperidol is now recommended only in severely affected and treatment-resistant cases (Liu et al., 2020; Roessner et al., 2022).

A 2009 Cochrane review assessed six RCTs involving pimozide in tic treatment: one compared pimozide with placebo, one with haloperidol, two with both placebo and haloperidol, and two with risperidone (Pringsheim and Marras, 2009). Pimozide was superior to placebo in three trials. When compared with haloperidol, it showed equivalent efficacy in two trials and inferior efficacy in one (Pringsheim and Marras, 2009).

Second- and third-generation antipsychotics – including aripiprazole, risperidone, olanzapine, ziprasidone, quetiapine, and paliperidone – exhibit a relatively greater affinity for serotonin 5-HT2A receptors than for dopamine D2 receptors. In addition to D2 receptor antagonism in the mesolimbic pathway, these agents also antagonize 5-HT2A receptors, indirectly modulating dopaminergic tone. Specifically, 5-HT2A blockade leads to reduced glutamate release in the ventral tegmental area, thereby indirectly decreasing dopaminergic transmission in the mesolimbic pathway (Orzelska-Górka et al., 2022).

Risperidone has demonstrated efficacy in tic reduction in six RCTs (Bruggeman et al., 2001; Dion et al., 2002; Gaffney et al.,

2002; Scahill et al., 2003; Gilbert et al., 2004; Ghanizadeh and Haghighi, 2014; Whittington et al., 2016). It is typically initiated at 0.25 mg/day, with doses ranging from 0.25 to 4.0 mg/day (Roessner et al., 2011).

Aripiprazole, a partial dopamine D2 receptor agonist, is also supported by multiple placebo-controlled RCTs in GTS (Sallee et al., 2000; Yoo et al., 2013; Sallee et al., 2017). It has shown a favorable efficacy and tolerability profile in children and adolescents with tic disorders (He et al., 2024). In a large multicenter RCT published in 2017, including 133 children and adolescents, aripiprazole significantly reduced tic severity as measured by the Yale Global Tic Severity Scale (YGTSS) (Sallee et al., 2017). Recommended starting dose is 2 mg/day, with a maximum of 10–20 mg/day depending on body weight (Sallee et al., 2017). A recent meta-analysis reported a standardized mean difference of 4.74 in favor of aripiprazole over placebo (Yang et al., 2019). ESSTS guidelines now suggest aripiprazole as a first-line treatment for GTS (Roessner et al., 2022).

A recent network meta-analysis by Farhat et al. (2023) evaluating pharmacologic treatments for GTS in children, adolescents, and young adults confirmed that antipsychotic medications remain the most efficacious interventions. Similarly, Yang et al. (2019) concluded that risperidone and aripiprazole offer the most robust evidence base among antipsychotic treatments.

Despite their efficacy, second- and third-generation antipsychotics are associated with metabolic side-effects, including weight gain, type 2 diabetes, hyperlipidemia, and hyperprolactinemia (Rizzo et al., 2012; Pozzi et al., 2016; Rafaniello et al., 2016).

Olanzapine, another atypical antipsychotic, has shown efficacy in GTS through several open-label studies (Krishnamoorthy, King, 1998; Stamenkovic et al., 2000; Budman et al., 2001; McCracken et al., 2008) and two double-blind crossover trials (Onofrj et al., 2000; Stephens et al., 2004). While it can be considered for GTS treatment, careful monitoring of metabolic side-effects is essential (Zuddas et al., 2011).

Ziprasidone and quetiapine have limited supporting evidence for tic reduction (Singer et al., 1995; de Jonge et al., 2007) and are used less frequently. Reported side-effects include sedation, drowsiness, and weight gain (McCracken et al., 2008; Zuddas et al., 2011).

Monitoring for metabolic and cardiac side-effects is essential when prescribing antipsychotics. Baseline evaluation should include ECG, weight, blood pressure, glucose, lipid profile, and prolactin levels. At follow-up, weight and blood pressure should be rechecked, alongside assessments for extrapyramidal symptoms and abnormal movements. While motor side-effects are more common with first-generation antipsychotics, they may still occur with newer agents. Tardive dyskinesia remains exceedingly rare in GTS.

Annual blood monitoring is recommended during ongoing treatment. Repeat ECGs are necessary if the patient is on pimozide or risperidone, or if the medication history suggests a risk of QTc prolongation. A retrospective cohort study showed that individuals with GTS or chronic tic disorders had a higher risk of metabolic or cardiovascular disorders compared to their siblings, controls, and the general population (Brander et al., 2019).

D1/D5 RECEPTOR ANTAGONIST ECOPIPAM

Ecopipam is a promising selective D1/D5 receptor antagonist to treat GTS. To date, four clinical trials were completed to evaluate the efficacy, safety, and tolerability of ecopipam. The first one, conducted by Gilbert et al. (2014), was an 8-week, phase II–III study. They enrolled 18 adults (> 18 years) that received 50 or 100 mg of ecopipam. At the end of the study, they reported a significant reduction of YGTSS (mean decrease of YGTSS-TTS=5.3 points), and only mild adverse events: sedation, fatigue, insomnia, anxiety, and muscle twitching (Gilbert et al., 2014). Subsequently, the results of a 4-week double blind placebo-controlled study on efficacy and safety of ecopipam in youth (7–17 years) were published by the same group. As in the previous study, the efficacy measured with YGTSS reduction was significant (mean reduction at 30 days in YGTSS-TSS =–3.2; 95% CI, –6.1 to –0.3; p=0.033), and side adverse events were mild and tolerable (Gilbert et al., 2018). Another multicenter, randomized, double-blind, placebo-controlled, phase IIb trial was also conducted with patients aged > 6 to < 18 years, with confirmed GTS and a minimum score of 20 on YGTSS-TSS (Gilbert et al., 2023). Participants were randomly assigned 1:1 to ecopipam (n=76) or placebo (n=77). Among children and adolescents with GTS, ecopipam reduced tics as measured by YGTSS to a greater extent than placebo, without observable evidence of common antipsychotic-associated side-effects (Gilbert et al., 2023). In addition, in a phase IIb randomized, double-blind, placebo-controlled clinical trial, participants were randomly assigned 1:1 to ecopipam (n =76) or placebo (n=77), and ecopipam significantly reduced both motor and phonic tics (Gilbert et al., 2024). Specifically, for motor tics, the largest reductions were in intensity scores (ecopipam – placebo LS means difference of –0.48; $p < 0.01$), followed by number, frequency, and interference (ecopipam – placebo differences ranging from –0.34 to –0.43; all $p < 0.05$); for phonic tics, the largest reduction was in complexity scores (ecopipam – placebo LS means difference of –0.48; p=0.02) (Gilbert et al., 2024).

NONDOPAMINERGIC AGENTS

Noradrenergic and serotoninergic systems appear to be involved in the pathophysiology of GTS, yet their exact role is still unclear. Noradrenergic agents such as clonidine and guanfacine are commonly used in children and adolescents, mainly in patients with GTS and ADHD. Canadian and US guidelines for evidence-based treatment of tic disorder indicated α-2 adrenergic agonists as first-line treatment (Pringsheim et al., 2012). Clonidine has shown effectiveness in several RCTs for the treatment of tics, five including a placebo control, and one comparing clonidine to levetiracetam (Goetz et al., 1987; Leckman, 1991; Singer et al., 1995; Tourette's Syndrome Study Group, 2002; Du et al., 2008; Hedderick et al., 2009).

The oral formulation of clonidine was used in five trials and the clonidine adhesive patch in one trial (Du et al., 2008). Recommended starting dose is 0.05 mg per day and titrate up slowly to a therapeutic range of 0.1–0.3 mg per day. Several randomized controlled studies have shown that clonidine was effective compared to placebo on tic (Whittington et al., 2016). No difference between clonidine and placebo was found in vocal tics (Goetz et al., 1987).

Guanfacine is another α-2 adrenergic receptor agonist that has a moderate effect in tic reduction and improving attention. The evidence in favor of the use of guanfacine for tic reduction comes from three randomized controlled studies (Scahill et al., 2001; Cummings et al., 2002; Murphy et al., 2017). All three RCTs compared guanfacine to placebo. In one small RCT in children and adolescents, 50% of the guanfacine and 22% of the placebo group suffered from coexisting ADHD (Murphy et al., 2017). One of these trials showed a significant statistical reduction in YGTSS of treated patients (Scahill et al., 2001). Weisman et al. (2013) conducted a meta-analysis on the efficacy of α-2 adrenergic agonists on tics in GTS. This meta-analysis indicated that the effect size of α-2 adrenergic agonist is much larger on children with tic plus ADHD (95% CI:0.36–101) than in individuals with tic without ADHD (95% CI: −0.06to .0.36) (Weisman et al., 2013). The most common side-effects of α-2 adrenergic agonists are postural hypotension, bradycardia, headache, initial sedation emerging within the first 2 weeks of treatment, and dry mouth (Hirota et al., 2014). Mania could be induced in children with bipolar disorder (Elbe et al., 2016). Prescribing these drugs requires a baseline ECG, with blood pressure and pulse checked at each visit and tapered withdrawal, if necessary.

Topiramate is a GABAergic agent that blocks AMPA/kainate receptors. There have been several studies regarding topiramate's efficacy in treating tics. In a pilot study, 29 patients with GTS age 7–65 years were randomized to receive either topiramate or placebo (Jankovic et al., 2010). They found a statistically significant reduction of the YGTSS score (Jankovic et al., 2010). Yang et al. (2013) in a meta-analysis evaluated 14 trials involving 1003 patients aged 7–17 years, to assess the efficacy and safety of topiramate compared to haloperidol (12 trials) and tiapride (2 trials). All except three trials showed no superiority of topiramate compared to haloperidol or tiapride. The typical dose is 25–150 mg/day. Side-effects include weight loss, paresthesias, headache, diarrhea, sedation, constipation, nausea, anxiety, and depression.

VESICULAR MONOAMINE TRANSPORTERS 2 INHIBITORS

The VMAT2 inhibitors are dopamine depleting agents that can block the transport of dopamine, norepinephrine, and serotonin from synaptic vesicles to synapses. Medications in this class include tetrabenazine, deutetrabenazine, and valbenazine. As far as we know, no RCTs were conducted on the efficacy of tetrabenazine in GTS. However, several case-studies or open-label studies were conducted and reported a good level of efficacy with reduction of tics on doses of 18–36 mg/day, improvement of the clinical global impression (Roth, 2018) and only mild side-effects such as sedation, fatigue, nausea, insomnia, parkinsonism, and drowsiness (Porta et al., 2008; Jankovic et al., 2016). Tetrabenazine may sometimes induce or worsen depression (Jankovic et al., 2016). The results of a nonrandomized, prospective study conducted with a total of 120 patients with GTS suggested that it may therefore be a good add-on option to improve the efficacy of other medications in treating tics, including serotonin reuptake inhibitors, antipsychotics, and/or tricyclic antidepressants (Porta et al., 2008). In an open-label pilot study of 23 patients (age 12–18) with GTS, deutetrabenazine was administered over 8 weeks (Jankovic et al., 2016a). Patients had a 37.6% reduction in tic severity measured by the YGTTS total tic severity score (TTS) (Jankovic et al., 2016). However, two recent RCTs examining effectiveness and safety of deutetrabenazine for the treatment of GTS in children and adolescents revealed no significant differences between deutetrabenazine or placebo for YGTSS-TSS (Coffey et al., 2021; Jankovic et al., 2021). In addition, a 54-week, global, phase III, open-label extension study of deutetrabenazine (6–48 mg daily) was conducted recently with a 2-week randomized withdrawal period (Jankovic et al., 2023). In this long-term trial, deutetrabenazine was well tolerated with low frequency of reported treatment-emergent adverse events, including headaches, somnolence, nasopharyngitis, weight increases, and anxiety (Jankovic et al., 2023). However, during the 2-week randomized withdrawal period (weeks 28–30), a ≥20% increase in YGTSS-TTS occurred in 17 (32.1%) participants in the deutetrabenazine group and 10 (40%) in the placebo group, but the observed worsening of YGTSS-TTS was not statistically significant (least squares mean difference, −0.4; p=0.78) (Jankovic et al., 2023).

Similar findings (i.e., failure to meet clinical endpoints) were reported with the use of valbenazine as well, which is an orally bioavailable VMAT2 inhibitor that was specifically studied for tics in GTS (Farber et al., 2021). Typical side-effects of valbenazine include depression, akathisia, weight gain, and parkinsonism (Niemann and Jankovic, 2019).

CANNABINOIDS

Cannabis-based medications have been explored in the past as a potential treatment option for tics in GTS. Two small placebo-controlled studies have suggested that cannabinoids may have beneficial effects in tic reduction in GTS (Müller-Vahl, 2013; Whiting et al., 2015). Moreover, several case studies documented not only tic reduction but also improvement in domains related to typical comorbidities in GTS, such as ADHD, sleep problems, self-injurious behaviors, impulsivity, obsessive-compulsive behavior (OCB), and depression (Lim et al., 2017; Müller-Vahl, 2024). To date, two RCTs have examined the efficacy and safety of cannabis in GTS. The first (NCT0324244) "Safety and efficacy of cannabis in Tourette" was a double-blind randomized placebo control, phase II, pilot study (Abi-Jaoude et al., 2023). In this study, each participant received a vaporized single 0.25 g dose of D9-tetrahydrocannabinol (THC) 10%, THC/cannabidiol (CBD) 9%/9%, CBD 13%, and placebo at 2-week intervals. Twelve adult patients affected by GTS were randomized, with nine completing the study. No statistically significant differences were reported on the Modified Rush Video-Based Tic Rating Scale for any of the cannabis products, although the THC 10% product was significantly better than placebo on the secondary outcome measures, which included the Premonitory Urge for Tics Scale, Subjective Units of Distress Scale, and Clinical Global Impression–Improvement (Abi-Jaoude et al., 2023). Also, the THC 10% product resulted in the most adverse events, including sedation, psychomotor effects, dizziness, cough, throat burning, dry mouth, feeling cold, and feeling high (Abi-Jaoude et al., 2023). The second study (NCT03087201) "CANNAbinoids in the Treatment of TICS (CANNA-TICS)" was a multicenter, randomized, double-blind, placebo controlled, parallel-group, phase IIIb trial, that examined the efficacy and safety of nabiximols in 97 adults with GTS/chronic tic disorder (CTD), aged > 18 years, randomized 2:1 to nabiximols:placebo (Müller-Vahl et al., 2023). Although a much larger number of patients in the nabiximols compared to the placebo group (21.9% vs 9.1%) met the primary efficacy endpoint, defined as a tic reduction of ≥25% according to the TTS of the YGTSS after 13 weeks of treatment, superiority of nabiximols could not be formally demonstrated (Müller-Vahl et al., 2023).

Furthermore, substantial trends for the improvement of tics, depression, and measures of quality of life were observed, mostly so in male patients with more severe tics, and patients with comorbid ADHD, thereby suggesting that these subgroups may benefit more from treatments with cannabis-based medication, while there were no serious adverse events (Müller-Vahl et al., 2023).

In a recent randomized, double-blind, placebo-controlled crossover trial conducted with 22 adults with severe GTS, participants were randomly assigned to a 6-week treatment period with escalating doses of an oral oil containing 5 mg/mL of THC and 5 mg/mL of CBD, followed by a 6-week course of placebo, or vice versa, separated by a 4-week washout period (Mosley et al., 2023). The study showed that an oral 1:1 THC:CBD formulation titrated upward over 6 weeks up to a daily dose of 20 mg of THC and 20 mg of CBD led to a significant reduction in tics as measured by the total tic score on the YGTSS, as well as a reduction in obsessive-compulsive symptoms and anxiety, without major adverse effects (Mosley et al., 2023). In addition, in a small single-dose phase Ib study (n–20), the safety and efficacy of an endocannabinoid modulator and inhibitor of the mono-acylglycerol lipase, which degrades the endocannabinoid2-arachidonoylglycerol (2-AG), Lu AG06466 (previously known as ABX-1431) was investigated (Muller-Vahl, 2018). In this pilot study, a single dose of Lu AG06466 was found to be effective in the reduction of tics and premonitory urges, with only mild to moderate associated adverse events (Muller-Vahl, 2018). Based on these preliminary results, the efficacy of the same compound was examined in a larger multicenter parallel-group randomized double-blind placebo-controlled trial on 49 adults with GTS (Müller-Vahl et al., 2021). After a 12-week treatment period, no beneficial effects on tic severity, premonitory urges, OCD, global impairment, and quality of life could be demonstrated (Müller-Vahl et al., 2021). Therefore, the role of cannabis-based medications as in the treatment of tics in GTS remains unclear.

COMPLEMENTARY ALTERNATIVE MEDICINE AND SUPPLEMENTATION

Although dietary and nutritional factors have been implicated in a variety of neurologic conditions, relatively few studies – often with contradictory results – have evaluated the efficacy of nutritional supplements in the treatment of tic disorders and other neuropsychiatric conditions in pediatric populations. Complementary and alternative therapies, such as traditional Chinese medicine (TCM) and vitamin supplementation, may offer some benefit for the treatment of GTS; however, these approaches have not been systematically studied.

In China, two commonly used TCM formulations for GTS are the 5-ling granule (5-LGr) and the Ningdong granule. The 5-LGr contains 11 ingredients, including herbal, animal, and placental components, and is thought to modulate the dopaminergic D2 receptor pathway. In the only RCT published to date, 603 children and adolescents with GTS (ages 5–18) were randomly assigned to receive either placebo, tiapride (200–400 mg/day), or 5-LGr (15–22.5 g/day). The trial demonstrated significant efficacy of 5-LGr in reducing tic severity compared to placebo (Zheng et al., 2016). However, these findings have not yet been independently replicated.

Nutritional supplementation with magnesium and vitamin B6 has also been explored. In one study, García-López et al. (2008) reported a 13.8-point reduction in the TTS on the YGTSS, following such supplementation in children with GTS. A more recent observational study found that 75% of children with GTS who used nutritional supplements experienced clinical improvement, particularly in motor and vocal tics, as well as in sleep quality and anxiety symptoms (Smith and Ludlow, 2023). Similarly, an open-label trial investigating the combined use of L-theanine and vitamin B6 in children with GTS and comorbid anxiety reported significant clinical benefits without consistent adverse effects or newly emergent symptoms (Rizzo et al., 2022).

Despite these promising findings, larger, well-designed placebo-controlled studies are needed to confirm the efficacy and safety of nutritional and complementary therapies in the treatment of GTS.

CONCLUSIONS

The clinical management of GTS is inherently complex due to its heterogeneous presentation, waxing and waning course of tics and frequent psychiatric comorbidities. While tics are the defining symptom, comorbid conditions such as ADHD, OCD, and anxiety often contribute more significantly to functional impairment and should be systematically screened and addressed.

Effective treatment begins with psychoeducation and shared decision-making, followed by behavioral interventions – such as CBIT and HRT – which are considered first-line therapies. Pharmacologic options should be considered when tics are functionally impairing, and behavioral interventions are unavailable or insufficient. Current evidence supports the use of dopamine receptor antagonists (particularly aripiprazole and risperidone), α-2 adrenergic agonists (clonidine and guanfacine), VMAT2 inhibitors, and topiramate. Emerging treatments such as ecopipam show promise, while cannabinoid-based therapies and complementary approaches, including TCM and nutritional supplementation, remain investigational.

Treatment must be individualized, guided by symptom burden, comorbidities, and patient-specific goals and preferences. The long-term nature of GTS, combined with its fluctuating severity, demands careful longitudinal monitoring using validated scales and quality of life assessments. Clinicians should remain vigilant for side-effects, particularly when using antipsychotics or polypharmacy strategies, and adhere to guideline-recommended monitoring protocols.

Despite the increasing availability of pharmacologic and behavioral interventions, treatment resistance remains a challenge in a subset of patients. For these individuals, advanced options such as deep brain stimulation may be considered.

Continued research is needed to elucidate the neurobiologic underpinnings of GTS, refine pharmacologic strategies, and explore novel treatment targets – including nondopaminergic systems and adjunctive therapies. The ongoing development of international guidelines and consensus statements is a critical step toward standardizing care and improving outcomes for individuals with GTS across the lifespan.

References

Abi-Jaoude E, Bhikram T, Parveen F et al. (2023). A double-blind, randomized, controlled crossover trial of cannabis in adults with Tourette syndrome. Cannabis cannabinoid Res 8: 835–845.

Andrén, Aspvall P, Fernández de la Cruz K et al. (2019). Therapist-guided and parent-guided internet-delivered behaviour therapy for paediatric Tourette's disorder: a pilot randomised controlled trial with long-term follow-up. BMJ open 9: e024685.

Andrén P, Jakubovski E, Murphy TL et al. (2022). European clinical guidelines for Tourette syndrome and other tic disorders-version 2.0. Part II: psychological interventions. Eur child & Adolesc psychiatry 31: 403–423.

Augustine F, Singer HS (2019). Merging the Pathophysiology and Pharmacotherapy of Tics. Tremor other hyperkinetic Mov 8: 595.

Billnitzer A, Jankovic J (2020). Current Management of Tics and Tourette Syndrome: Behavioral, Pharmacologic, and Surgical Treatments. Neurotherapeutics: J Am Soc Exp NeuroTherapeutics 17: 1681–1693.

Bloch M, State M, Pittenger C (2011). Recent advances in Tourette syndrome. Curr OpNeurol 24: 119–125.

Brander G, Isomura K, Chang Z (2019). Association of Tourette Syndrome and Chronic Tic Disorder With Metabolic and Cardiovascular Disorders. JAMA Neurol 76: 454–461.

Bruggeman R, van der Linden C, Buitelaar JK et al. (2001). Risperidone versus pimozide in Tourette's disorder: a comparative double-blind parallel-group study. J Clin psychiatry 62: 50–56.

Budman CL, Gayer A, Lesser M et al. (2001). An open-label study of the treatment efficacy of olanzapine for Tourette's disorder. J Clin psychiatry 62: 290–294.

Coffey B, Jankovic J, Claassen DO et al. (2021). Efficacy and Safety of Fixed-Dose Deutetrabenazine in Children and Adolescents for Tics Associated With Tourette Syndrome: A Randomized Clinical Trial. JAMA Netw open 4: e2129397.

Cummings DD, Singer HS, Krieger M et al. (2002). Neuropsychiatric effects of guanfacine in children with mild tourette syndrome: a pilot study. Clin Neuropharmacol 25: 325–332.

de Jonge JL, Cath DC, van Balkom AJ (2007). Quetiapine in patients with Tourette's disorder: an open-label, flexible-dose study. J Clin psychiatry 68: 1148–1150.

Dion Y, Annable L, Sandor P et al. (2002). Risperidone in the treatment of tourette syndrome: a double-blind, placebo-controlled trial. J Clin Psychopharmacol 22: 31–39.

Du YS, Li HF, Vance A et al. (2008). Randomized double-blind multicentre placebo-controlled clinical trial of the clonidine adhesive patch for the treatment of tic disorders. Australian & New Zealand Journal of Psychiatry 42: 807–813.

Elbe D, Perel-Panar C, Wicholas L (2016). Manic Reaction in a Child Induced by Guanfacine-Extended Release. J child Adolesc Psychopharmacol 26: 566–567.

Farag M, Stern JS, Simmons H et al. (2015). Serial pharmacological prescribing practices for tic management in Tourette syndrome. Hum Psychopharmacol 30: 435–441.

Farber RH, Angelov A, Kim K et al. (2021). Clinical development of valbenazine for tics associated with Tourette syndrome. Expert Rev neurotherapeutics 21: 393–404.

Farhat LC, Behling E, Landeros-Weisenberger A et al. (2023). Comparative efficacy, tolerability, and acceptability of pharmacological interventions for the treatment of children, adolescents, and young adults with Tourette's syndrome: a systematic review and network meta-analysis. Lancet Child & Adolesc Health 7: 112–126.

Gaffney GR, Perry PJ, Lund BC et al. (2002). Risperidone versus clonidine in the treatment of children and adolescents with Tourette's syndrome. J Am Acad Child Adolesc Psychiatry 41: 330–336.

García-López R, Romero-González J, Perea-Milla E et al. (2008). Estudio piloto sin grupo control del tratamiento con magnesio y vitamina B(6) del síndrome de Gilles de la Tourette en niños [An open study evaluating the efficacy and security of magnesium and vitamin B(6) as a treatment of Tourette syndrome in children]. Medicina clinica 131: 689–691.

Ghanizadeh A, Haghighi A (2014). Aripiprazole versus risperidone for treating children and adolescents with tic disorder: a randomized double blind clinical trial. Child psychiatry Hum Dev 45: 596–603.

Gilbert DL, Batterson JR, Sethuraman G et al. (2004). Tic reduction with risperidone versus pimozide in a randomized, double-blind, crossover trial. J Am Acad Child Adolesc Psychiatry 43: 206–214.

Gilbert DL, Budman CL, Singer HS et al. (2014). A D1 receptor antagonist, ecopipam, for treatment of tics in Tourette syndrome. Clin Neuropharmacol 37: 26–30.

Gilbert DL, Murphy TK, Jankovic J et al. (2018). Ecopipam, a D1 receptor antagonist, for treatment of tourette syndrome in children: A randomized, placebo-controlled crossover study. Mov disorders: Off J Mov Disord Soc 33: 1272–1280.

Gilbert DL, Dubow JS, Cunniff TM et al. (2023). Ecopipam for Tourette Syndrome: A Randomized Trial. Pediatrics 151: e2022059574.

Gilbert DL, Karkanias G, Munschauer F et al. (2024). Effect of Ecopipam, a selective dopamine-1 receptor antagonist, on tic characteristics as assessed by the YGTSS: results from the phase IIb (D1AMOND) randomized, doubleblind, placebo-controlled clinical trial in Tourette syndrome (S30.009). Neurology 102: 2664.

Goetz CG, Tanner CM, Wilson RS et al. (1987). Clonidine and Gilles de la Tourette's syndrome: double-blind study using objective rating methods. Ann Neurol 21: 307–310.

Gulisano M, Barone R, Mosa MR et al. (2020). Incidence of Autism Spectrum Disorder in Youths Affected by Gilles de la Tourette Syndrome Based on Data from a Large Single Italian Clinical Cohort. Brain Sci 10: 812.

Hartmann A, Worbe Y (2018). Tourette syndrome: clinical spectrum, mechanisms and personalized treatments. Curr OpNeurol 31: 504–509.

He F, Luo J, Huang Y et al. (2024). Randomized, double-blind, placebo-controlled trial of aripiprazole oral solution in children and adolescents with Tourette's disorder. Child Adolesc psychiatry Ment health 18: 88.

Hedderick EF, Morris CM, Singer HS (2009). Double-blind, crossover study of clonidine and levetiracetam in Tourette syndrome. Pediatric Neurol 40: 420–425.

Himle MB, Freitag M, Walther M et al. (2012). A randomized pilot trial comparing videoconference versus face-to-face delivery of behavior therapy for childhood tic disorders. Behav Res Ther 50: 565–570.

Hirota T, Schwartz S, Correll CU (2014). Alpha-2 agonists for attention-deficit/hyperactivity disorder in youth: a systematic review and meta-analysis of monotherapy and add-on trials to stimulant therapy. J Am Acad Child Adolesc Psychiatry 53: 153–173.

Hollis C, Hall CL, Jones R et al. (2021). Therapist-supported online remote behavioural intervention for tics in children and adolescents in England (ORBIT): a multicentre, parallel group, single-blind, randomised controlled trial. lancet Psychiatry 8: 871–882.

Jankovic J, Jimenez-Shahed J, Brown LW (2010). A randomised, double-blind, placebo-controlled study of topiramate in the treatment of Tourette syndrome. J neurology, neurosurgery, psychiatry 81: 70–73.

Jankovic J, Jimenez-Shahed J, Budman C et al. (2016). Deutetrabenazine in Tics Associated with Tourette Syndrome. Tremor other hyperkinetic Mov (N York, NY) 6: 422.

Jankovic J, Coffey B, Claassen DO et al. (2023). Safety and Efficacy of Long-Term Deutetrabenazine Use in Children and Adolescents with Tics Associated with Tourette Syndrome: An Open-Label Extension Study. Mov Disord Clin Pract 10: 1388–1398.

Johnson KA, Worbe Y, Foote KD et al. (2023). Tourette syndrome: clinical features, pathophysiology, and treatment. Lancet Neurol 22: 147–158.

Kious BM, Jimenez-Shahed J, Shprecher DR (2016). Treatment-refractory Tourette Syndrome. Prog Neuro-psychopharmacol & Biol psychiatry 70: 227–236.

Krishnamoorthy J, King BH (1998). Open-label olanzapine treatment in five preadolescent children. J child Adolesc Psychopharmacol 8: 107–113.

Leckman JF, Hardin MT, Riddle MA et al. (1991). Clonidine treatment of Gilles de la Tourette's syndrome. Arch Gen psychiatry 48: 324–328.

Lemmon ME, Grados M, Kline T et al. (2015). Efficacy of Glutamate Modulators in Tic Suppression: A Double-Blind, Randomized Control Trial of D-serine and Riluzole in Tourette Syndrome. Pediatric Neurol 52: 629–634.

Lewin AB, Wu MS, McGuire JF et al. (2014). Cognitive behavior therapy for obsessive-compulsive and related disorders. Psychiatr Clin North Am 37: 415–445.

Lim K, See YM, Lee J (2017). A Systematic Review of the Effectiveness of Medical Cannabis for Psychiatric, Movement and Neurodegenerative Disorders. Clin Psychopharmacol neuroscience: Off Sci J Korean Coll Neuropsychopharmacology 15: 301–312.

Liu ZS, Cui YH, Sun D et al. (2020). Current Status, Diagnosis, and Treatment Recommendation for Tic Disorders in China. Front Psychiatry 11: 774.

Martino D, Hedderly T, Murphy T et al. (2023). The spectrum of functional tic-like behaviours: Data from an international registry. Eur J Neurol 30: 334–343.

Martino D, Malaty I, Müller-Vahl K et al. (2023). Treatment failure in persistent tic disorders: an expert clinicians' consensus-based definition. Eur Child & Adolesc Psychiatry 32: 859–872.

McCracken JT, Suddath R, Chang S et al. (2008). Effectiveness and tolerability of open label olanzapine in children and adolescents with Tourette syndrome. J Child Adolesc Psychopharmacol 18: 501–508.

Mogwitz S, Buse J, Ehrlich S et al. (2013). Clinical pharmacology of dopamine-modulating agents in Tourette's syndrome. Int Rev Neurobiol 112: 281–349.

Mosley PE, Webb L, Suraev A et al. (2023). Tetrahydrocannabinol and Cannabidiol in Tourette Syndrome. NEJM Evid 2: EVIDoa2300012.

Müller-Vahl KR, Cath DC, Cavanna AE et al. (2011). European clinical guidelines for Tourette syndrome and other tic disorders. Part IV: deep brain stimulation. Eur Child & Adolesc psychiatry 20: 209–217.

Müller-Vahl KR (2013). Treatment of Tourette syndrome with cannabinoids. Behavioural Neurol 27: 119–124.

Muller-Vahl K, Jakubovski E, Carolin F et al. (2018). ABX-1431, a first-in-class endocannabinoid modulator, improves tics in adult patients with Tourette syndrome. Neurology 90: e2182–e2183.

Müller-Vahl KR, Szejko N, Saryyeva A et al. (2021). Randomized double-blind sham-controlled trial of thalamic versus GPi stimulation in patients with severe medically refractory Gilles de la Tourette syndrome. Brain Stimulation 14: 662–675.

Müller-Vahl KR, Pisarenko A, Szejko N et al. (2023). CANNA-TICS: Efficacy and safety of oral treatment with nabiximols in adults with chronic tic disorders - Results of a prospective, multicenter, randomized, double-blind, placebo controlled, phase IIIb superiority study. Psychiatry Res 323: 115135.

Müller-Vahl KR (2024). Cannabinoids in the Treatment of Selected Mental Illnesses: Practical Approach and Overview of the Literature. Pharmacopsychiatry 57: 104–114.

Murphy TK, Fernandez TV, Coffey BJ et al. (2017). Extended-Release Guanfacine Does Not Show a Large Effect on Tic Severity in Children with Chronic Tic Disorders. J Child Adolesc Psychopharmacol 27: 762–770.

Niemann N, Jankovic J (2019). Real-World Experience With VMAT2 Inhibitors. Clin Neuropharmacol 42: 37–41.

Onofrj M, Paci C, D'Andreamatteo G et al. (2000). Olanzapine in severe Gilles de la Tourette syndrome: a 52-week double-blind cross-over study vs. low-dose pimozide. J Neurol 247: 443–446.

Orzelska-Górka J, Mikulska J, Wiszniewska A et al. (2022). New Atypical Antipsychotics in the Treatment of Schizophrenia and Depression. Int J Mol Sci 23: 10624.

Porta M, Sassi M, Cavallazzi M et al. (2008). Tourette's syndrome and role of tetrabenazine: review and personal experience. Clin Drug Investigation 28: 443–459.

Pozzi M, Cattaneo D, Baldelli S et al. (2016). Therapeutic drug monitoring of second-generation antipsychotics in pediatric patients: an observational study in real-life settings. Eur J Clin pharmacology 72: 285–293.

Prato A, Maugeri N, Chiarotti F et al. (2022). A Randomized Controlled Trial Comparing Videoconference vs. Face-to-Face Delivery of Behavior Therapy for Youths With Tourette Syndrome in the Time of COVID-19. Front Psychiatry 13: 862422.

Pringsheim T, Doja A, Gorman D et al. (2012). Canadian guidelines for the evidence-based treatment of tic disorders: pharmacotherapy. Can J Psychiatry Rev Canadienne de Psychiatrie 57: 133–143.

Pringsheim T, Marras C (2009). Pimozide for tics in Tourette's syndrome. Cochrane database Syst Rev 2009: CD006996.

Pringsheim T, Okun MS, Müller-Vahl K et al. (2019a). Practice guideline recommendations summary: Treatment of tics in people with Tourette syndrome and chronic tic disorders. Neurology 92: 896–906.

Pringsheim T, Holler-Managan Y, Okun MS et al. (2019b). Comprehensive systematic review summary: Treatment of tics in people with Tourette syndrome and chronic tic disorders. Neurology 92: 907–915.

Rafaniello C, Pozzi M, Pisano S et al. (2016). Second generation antipsychotics in 'real-life' paediatric patients. Adverse drug reactions and clinical outcomes of drug switch. Expert Opin drug Saf 15(sup2): 1–8.

Ricketts EJ, Goetz AR, Capriotti MR et al. (2016). A randomized waitlist-controlled pilot trial of voice over Internet protocol-delivered behavior therapy for youth with chronic tic disorders. J Telemed Telecare 22: 153–162.

Rizzo R, Eddy CM, Calí P et al. (2012). Metabolic effects of aripiprazole and pimozide in children with Tourette syndrome. Pediatric Neurol 47: 419–422.

Rizzo R, Gulisano M, Martino D et al. (2017). Gilles de la Tourette Syndrome, Depression, Depressive Illness, and Correlates in a Child and Adolescent Population. J Child Adolesc Psychopharmacol 27: 243–249.

Rizzo R, Prato A, Scerbo M et al. (2022). Use of Nutritional Supplements Based on L-Theanine and Vitamin B6 in Children with Tourette Syndrome, with Anxiety Disorders: A Pilot Study. Nutrients 14: 852.

Roessner V, Plessen KJ, Rothenberger A et al. (2011). European clinical guidelines for Tourette syndrome and other tic disorders. Part II: pharmacological treatment. Eur Child & Adolesc psychiatry 20: 173–196.

Roessner V, Eichele H, Stern JS et al. (2022). European clinical guidelines for Tourette syndrome and other tic disorders-version 2.0. Part III: pharmacological treatment. Eur Child & Adolesc Psychiatry 31: 425–441.

Roth J (2018). The colorful spectrum of Tourette syndrome and its medical, surgical and behavioral therapies. Parkinsonism & Relat Disord 46: S75–S79.

Sallee F, Kohegyi E, Zhao J et al. (2017). Randomized, Double-Blind, Placebo-Controlled Trial Demonstrates the Efficacy and Safety of Oral Aripiprazole for the Treatment of Tourette's Disorder in Children and Adolescents. J Child Adolesc Psychopharmacol 27: 771–781.

Sallee FR, Kurlan R, Goetz CG et al. (2000). Ziprasidone treatment of children and adolescents with Tourette's syndrome: a pilot study. J Am Acad Child Adolesc Psychiatry 39: 292–299.

Scahill L, Chappell PB, Kim YS et al. (2001). A placebo-controlled study of guanfacine in the treatment of children with tic disorders and attention deficit hyperactivity disorder. Am J Psychiatry 158: 1067–1074.

Scahill L, Leckman JF, Schultz RT et al. (2003). A placebo-controlled trial of risperidone in Tourette syndrome. Neurology 60: 1130–1135.

Singer HS, Brown J, Quaskey S et al. (1995). The treatment of attention-deficit hyperactivity disorder in Tourette's syndrome: a double-blind placebo-controlled study with clonidine and desipramine. Pediatrics 95: 74–81.

Singer HS (2019). Tics and Tourette Syndrome. Contin (Minneapolis, Minn) 25: 936–958.

Smith BL, Ludlow AK (2023). Patterns of Nutritional Supplement Use in Children with Tourette Syndrome. J Diet Suppl 20: 28–43.

Stamenkovic M, Schindler SD, Aschauer HN et al. (2000). Effective open-label treatment of tourette's disorder with olanzapine. Int Clin Psychopharmacol 15: 23–28.

Stephens RJ, Bassel C, Sandor P (2004). Olanzapine in the treatment of aggression and tics in children with Tourette's syndrome–a pilot study. J Child Adolesc Psychopharmacol 14: 255–266.

Stern JS (2018). Tourette's syndrome and its borderland. Practical Neurol 18: 262–270.

Szejko N, Robinson S, Hartmann A et al. (2022). European clinical guidelines for Tourette syndrome and other tic disorders-version 2.0. Part I: assessment. Eur Child & Adolesc Psychiatry 31: 383–402.

Tourette's Syndrome Study Group (2002). Treatment of ADHD in children with tics: a randomized controlled trial. Neurology 58: 527–536.

Weisman H, Qureshi IA, Leckman JF et al. (2013). Systematic review: pharmacological treatment of tic disorders–efficacy of antipsychotic and alpha-2 adrenergic agonist agents. Neurosci Biobehav Rev 37: 1162–1171.

Whiting PF, Wolff RF, Deshpande S et al. (2015). Cannabinoids for Medical Use: A Systematic Review and Meta-analysis. JAMA 313: 2456–2473.

Whittington C, Pennant M, Kendall T et al. (2016). Practitioner Review: Treatments for Tourette syndrome in children and young people - a systematic review. J Child Psychol Psychiatr 57: 988–1004.

Yang CS, Zhang LL, Zeng LN et al. (2013). Topiramate for Tourette's syndrome in children: a meta-analysis. Pediatric Neurol 49: 344–350.

Yang C, Yi Q, Zhang L et al. (2019). Safety of aripiprazole for tics in children and adolescents: A systematic review and meta-analysis. Medicine 98: e15816.

Yoo HK, Joung YS, Lee JS et al. (2013). A multicenter, randomized, double-blind, placebo-controlled study of aripiprazole in children and adolescents with Tourette's disorder. J Clin psychiatry 74: e772–e780.

Zheng Y, Zhang ZJ, Han XM et al. (2016). A proprietary herbal medicine (5-Ling Granule) for Tourette syndrome: a randomized controlled trial. J Child Psychol Psychiatry, Allied Discip 57: 74–83.

Zuddas A, Zanni R, Usala T (2011). Second generation antipsychotics (SGAs) for non-psychotic disorders in children and adolescents: a review of the randomized controlled studies. Eur neuropsychopharmacology: J Eur Coll Neuropsychopharmacology 21: 600–620.

Handbook of Clinical Neurology, Vol. 215 (3rd series)
Tic Disorders and Tourette Syndrome
C. Ganos and Y. Worbe, Editors
https://doi.org/10.1016/B978-0-443-13554-5.00002-X

Chapter 24

Deep brain stimulation for Gilles de la Tourette syndrome

KARA A. JOHNSON[1,2,*], AYSEGUL GUNDUZ[1,2,3], AND MICHAEL S. OKUN[1,2]

[1]*Norman Fixel Institute for Neurological Diseases, University of Florida, Gainesville, FL, United States*

[2]*Department of Neurology, University of Florida, Gainesville, FL, United States*

[3]*J. Crayton Pruitt Family Department of Biomedical Engineering, University of Florida, Gainesville, FL, United States*

Abstract

Gilles de la Tourette syndrome (GTS) is a childhood-onset neurodevelopmental disorder characterized by motor and vocal tics and commonly associated with comorbid psychiatric and behavioral symptoms. Medication or behavioral therapies are effective for treating GTS in most cases; however, a proportion of individuals experience treatment-refractory tics. Deep brain stimulation (DBS) is an invasive surgical therapy reserved for select individuals with severe, treatment-refractory GTS. DBS has been targeted to several brain regions involved in the basal ganglia-thalamo-cortical networks implicated in GTS pathophysiology, including medial thalamic regions, the anteromedial (limbic) globus pallidus internus, and the posteroventrolateral (sensorimotor) globus pallidus internus, among others. Evidence from case studies, open-label studies, randomized controlled trials, and multicenter retrospective studies has suggested that DBS can be effective for improving tics, comorbidities, and quality of life. However, outcomes have been variable across individuals, and challenges remain regarding selecting individuals who are most likely to benefit from DBS, identifying effective target(s) for each individual, and determining optimal postoperative stimulation strategies. Progress has been made toward addressing these challenges through multicenter collaborations, neuroimaging, and neurophysiology; however, further prospective studies, clinical trials, and regulatory approvals will be needed to advance DBS as a more widely adopted therapy for treatment-refractory GTS.

BACKGROUND AND HISTORY

Gilles de la Tourette syndrome (GTS) is a neurodevelopmental disorder with childhood onset characterized by tics, which are involuntary, repetitive, nonrhythmic movements (motor tics) and vocalizations (phonic tics). GTS affects approximately 0.5% of the worldwide population, with the highest prevalence in children and adolescents (Jafari et al., 2022). GTS can have lasting detrimental effects on individuals' social, emotional, and physical well-being and quality of life (Eapen et al., 2016). Up to 90% of individuals with GTS also experience comorbid psychiatric or behavioral symptoms/disorders, most commonly obsessive-compulsive behavior/disorder (OCB/OCD), attention deficit (hyperactivity) disorders, and mood disorders (Hirschtritt et al., 2015). The clinical presentation of tics and comorbidities varies widely across individuals, including varying types and severities of tics (motor/phonic, simple/complex), as well as types and severities of comorbid behavioral or psychiatric symptoms. The pathophysiology of GTS remains incompletely understood; however, tics and comorbidities have been generally linked to pathologic neural activity within various brain regions of the basal ganglia-thalamo-cortical (BGTC) networks (Mink, 2001; Worbe et al., 2015; Johnson et al., 2022).

Clinical management of tics and comorbidities should be pursued when symptoms significantly impair the individual's daily functions and quality of life. First-line treatments for GTS include behavioral therapies and pharmacologic therapies, as either independent or conjunctive treatments. The majority of individuals with GTS experience sufficient tic improvement with first-line treatments; however, a subset of individuals continue to experience treatment-refractory symptoms, although the estimated proportion with persistent disabling symptoms is unclear because the criteria used to define treatment-refractory GTS vary across centers (Kious et al., 2016). Effective alternative treatments are critically needed for individuals experiencing severe, treatment-refractory symptoms.

Neurosurgical brain lesioning has historically been used for select individuals with severe, treatment-refractory GTS. Several

*Correspondence to: Kara A. Johnson, PhD, Department of Neurology, Norman Fixel Institute for Neurological Disorders, 3011 SW Williston Rd, Gainesville, FL 32608, United States. Tel: +1-352-273-9000. E-mail: kara.johnson@neurology.ufl.edu.

brain regions have been targeted with neurosurgical brain lesioning to treat GTS, including the frontal lobe (prefrontal lobotomy and frontal and limbic leucotomies), anterior cingulotomy, the thalamus, and the cerebellum (Temel and Visser-Vandewalle, 2004). The goal of neurosurgical brain lesioning was to abate pathologic neural activity in circuits involved in producing tics and behavioral disorders in GTS. Although some individuals experienced improvements in their tics or comorbid behavioral symptoms, many experienced severe and sometimes permanent side-effects (e.g., dystonia, hemiparesis, or dysarthria).

Based on promising results in essential tremor and Parkinson's disease, bilateral deep brain stimulation (DBS) was explored as a potential alternative neurosurgical therapy for treatment-refractory GTS to modulate pathologic circuits more reversibly and thus potentially avoid permanent side-effects. The goal of DBS is to deliver electric stimulation to focal brain regions to modulate pathologic neural activity within these BGTC networks and alleviate tics and comorbid behavioral symptoms in GTS. The first target used for DBS for GTS was a medial thalamic region at the intersection of the centromedian nucleus (CM), the substantia periventricularis (Spv), and the ventro-oralis-internus (Voi) (Vandewalle et al., 1999; Visser-Vandewalle et al., 2003). This target was defined based on atlases, the experience of the neurosurgeons, and previous neurosurgical lesioning studies (Hassler and Dieckmann, 1970). Since these seminal studies, several other brain regions have been investigated as targets for DBS for GTS within the BGTC networks. Although several studies have demonstrated the efficacy of DBS to treat GTS, outcomes have varied substantially across individuals, and critical unknowns remain regarding the appropriate patient population and how to select the optimal DBS target and stimulation parameters for each individual. DBS for GTS has not received regulatory approval for clinical use and thus remains an experimental therapy for carefully selected individuals and requires further investigation to become a widespread therapy.

In this chapter, we provide an overview of the current clinical approaches to DBS for treatment-refractory GTS, including patient selection, evidence of clinical efficacy for current DBS targets, and postoperative management. Additionally, we highlight new neuroimaging and neurophysiologic techniques and outline key areas for future research, which may advance DBS as a potential therapy for individuals with treatment-refractory GTS.

PATIENT SELECTION

The criteria for selecting suitable candidates to undergo DBS for treatment-refractory GTS vary across centers. A recent review expanded on previously proposed criteria (Schrock et al., 2015) and defined five "pillars" of patient selection: tic severity, quality of life, treatment refractoriness, behavioral comorbidities, and age (Martino et al., 2021) (Panel 24.1). First, a GTS diagnosis should be confirmed by an expert clinician using the latest diagnostic criteria in the *Diagnostic and Statistical Manual of Mental Disorders* (5th Edition) (DSM-V). The authors highlight the importance of using objective, validated measures (such as the Yale Global Tic Severity Scale (YGTSS) (Leckman et al., 1989) and the Modified Rush Video Rating Scale (MRVRS) (Goetz et al., 1999)) in addition to subjective measures to confirm high tic severity, tic-related impact on quality of life, and true treatment refractoriness (Martino et al., 2023), paying particular attention to comorbidities as potential confounders (Szejko et al., 2020). Early guidelines proposed an age cut-off of 18 years; however, recent guidelines instead suggest a careful evaluation of the risk-benefit profile for pediatric cases to be performed by a multidisciplinary team and a local ethics committee (Schrock et al., 2015; Martino et al., 2021). Importantly, a recent meta-analysis reported positive outcomes of GTS DBS in individuals under 21 years, with 64% of individuals showing greater than 50% improvement (Coulombe et al., 2018). Overall, given the invasiveness and time-intensive nature of DBS therapy, it is critically important to employ an interdisciplinary team of experts to carefully confirm candidacy and assess the potential risks and benefits of undergoing DBS for each individual.

SURGICAL TARGETS AND CLINICAL OUTCOMES

Several brain regions within the BGTC networks implicated in the pathophysiology of tics and GTS have been investigated as targets for DBS (Fig. 24.1). In this section, we review the rationale and the evidence supporting the clinical efficacy of each brain target, particularly focusing on randomized controlled trials (RCTs) (Table 24.1) and open-label long-term follow-up studies (Table 24.2). The summary tables include RCTs with double-blinded assessments of multiple DBS conditions and long-term follow-up studies with follow-up of at least 12 months in two or more patients. While RCTs provide randomized and blinded evaluations of the effects of DBS, open-label long-term follow-up studies are valuable for assessing real-world outcomes with several years of DBS, which is especially valuable for evaluating effectiveness in individuals whose symptoms were otherwise refractory to other behavioral and pharmacologic treatments.

Medial thalamic regions

The medial thalamus was the first reported DBS target for GTS, specifically at the intersection of the CM, the Spv, and the Voi (Vandewalle et al., 1999). The rationale was based on favorable results of thalamotomies in individuals with GTS (Hassler and Dieckmann, 1970) and the hypothesis that lesioning or applying high-frequency DBS to these thalamic nuclei would improve tics by reducing activity in corticostriatal circuits involving both the ventral (limbic) and dorsal (sensorimotor) striatum. Since the report of the original CM-Spv-Voi target, there have been several variations in thalamic DBS targeting, including the CM-parafascicular (Pf) complex (Houeto et al., 2005), a target point 2 mm anterior to the original CM-Spv-Voi target (Servello et al., 2008), the Voi, ventro-oralis-anterior, the ventro-oralis-posterior nuclei (Vernaleken et al., 2009), and the ventral anterior-ventral lateral nuclei (Huys et al., 2016).

Box 24.1 Criteria for selecting candidates for DBS.

Diagnosis of GTS

A diagnosis of GTS according to DSM-V criteria should be confirmed by an expert clinician.

Tic Severity

Tic severity should be evaluated using validated rating scales, such as the Yale Global Tic Severity Scale (YGTSS).

The suggested minimum YGTSS total tic score is 35/50 as criteria for DBS eligibility, meeting this criterion for 6–12 months before screening.

Additional modifiers of tic severity may also be considered (e.g., self-injurious behavior and tics that impart excessive risk and exceptional burden).

Quality of Life (QoL)

The impact of tics on QoL should be evaluated using disease-specific rating scales, such as the YGTSS impairment score or the GTS-QoL scale, as well as the impact of comorbidities on QoL using generic QoL rating scales.

The impact of tics versus comorbidities on QoL should be carefully compared to determine the main treatment focus.

Treatment Refractoriness

Treatment-refractory GTS is typically defined as the patient experiences severe tics that persist despite: (1) trialing at least three drug classes (α-adrenergic agonist, typical and atypical dopamine antagonist, and one from another class) and (2) trialing comprehensive behavioral intervention for tics by a trained therapist (including documentation of treatment adherence).

Definitions of therapy efficacy vary; a ≤25% decrease in YGTSS score has been proposed as a criterion of insufficient efficacy (Jeon et al., 2013).

Comorbidities may confound assessments of treatment refractoriness.

Behavioral Comorbidities

Thorough evaluation of comorbidities and their impact on QoL should be performed by a multidisciplinary team of experts.

Comorbidities should be stable (or decreasing) for 6 months before DBS screening.

Age

Individuals younger than 18 years with very severe tics may be considered.

A local ethics committee should be consulted for guidance.

Four RCTs have evaluated the clinical efficacy of thalamic DBS for GTS (Maciunas et al., 2007; Ackermans et al., 2011; Okun et al., 2013; Baldermann et al., 2021). These studies varied in target, cohort size, and study design (including duration of stimulation before evaluating its efficacy compared to the control condition), making it difficult to draw definitive conclusions across studies. Three of the studies reported a significantly higher reduction in tic severity as measured by either the YGTSS or the MRVRS on-DBS compared to the sham condition. Notably, the most recent RCT studied brief 2-day periods of randomized, double-blinded sham-controlled crossover evaluations and observed a 44% improvement on DBS compared to only 26% off DBS (Baldermann et al., 2021). Another study of CMpf DBS compared scheduled DBS, continuous DBS, and off DBS, indicating nonsignificant trends of higher tic improvement with scheduled and continuous DBS when compared to off DBS (Okun et al., 2013). Additionally, all of the studies reported significant reductions in tic severity in open-label follow-up, with the average percent improvement ranging from 18.6% to 44% across studies. Although promising, these studies were performed in relatively small patient cohorts (*N*=5–8 patients), and there was considerable variability in responses observed across individual patients.

In addition to RCTs, several open-label, long-term follow-up studies have been reported for thalamic DBS for GTS, ranging from 6 to 78 months of outcomes (Visser-Vandewalle et al., 2003; Porta et al., 2009; Ackermans et al., 2010; Porta et al., 2012; Huys et al., 2016; Servello et al., 2016; Testini et al., 2016; Dowd et al., 2017; Smeets et al., 2017; Kimura et al., 2020; Servello et al., 2020). These studies have reported average improvements in tic severity ranging from 48.5% to 86.9%, but similar to the RCTs, individual responses to DBS varied substantially. Some individuals experienced highly beneficial effects of DBS on their tics and comorbid behavioral symptoms (Ackermans et al., 2010), while others experienced debilitating side-effects or minimal benefits (Smeets et al., 2017).

RCTs in limited cohorts and several open-label long-term follow-up studies have demonstrated that DBS targeted to medial thalamic regions can be highly effective in some individuals with GTS; however, the effects of thalamic DBS on tics and behavioral and psychiatric comorbidities can be variable and warrant further investigation.

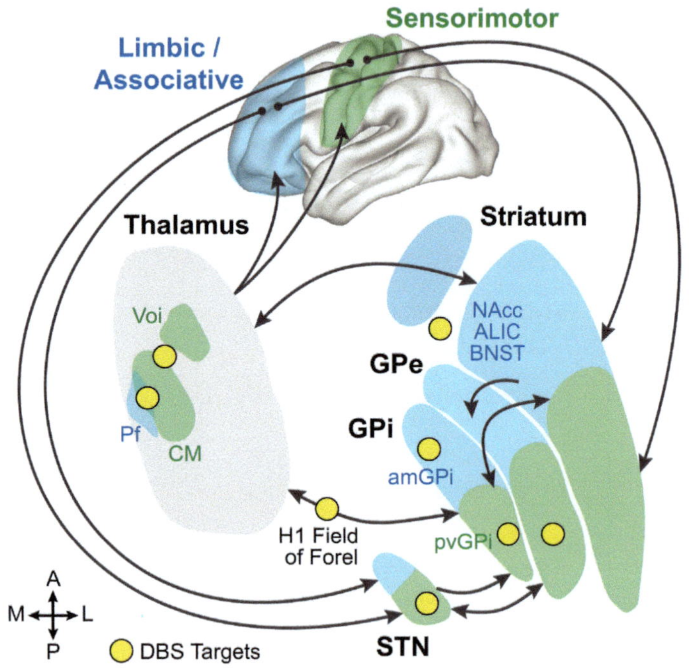

Fig. 24.1. Regions targeted with deep brain stimulation to modulate sensorimotor, associative, and limbic basal ganglia-thalamo-cortical networks implicated in Gilles de la Tourette syndrome pathophysiology. Target points and neuroanatomic structures are approximate.

Anteromedial globus pallidus internus

The anteromedial globus pallidus internus (amGPi) is one of the GTS DBS targets most commonly reported in studies (in addition to thalamic targets). The rationale is that applying DBS to the amGPi, which is an important relay for the limbic BGTC networks that play a role in GTS pathophysiology, could potentially modulate relevant pathologic neural activity (Houeto et al., 2005).

Two RCTs have been performed to evaluate the clinical efficacy of amGPi DBS compared to sham stimulation in reducing tics in GTS (Kefalopoulou et al., 2015; Welter et al., 2017). Only one of the studies found a significant improvement in tic severity on active DBS compared to sham (Kefalopoulou et al., 2015). The other study hypothesized that the lack of contrast between the active and sham conditions could be because 3 months of DBS may be insufficient to induce significant effects on tics (Welter et al., 2017). In support of this hypothesis, both studies reported significant reductions in tic severity after 8–36 months of open-label follow-up (average of 30.3% and 40.1% across individuals), along with significant improvements in quality of life and measures of depression and anxiety. Two studies have also reported outcomes of dual bilateral DBS targeted to the CMpf and amGPi, one of which was an RCT. In a single patient, DBS targeted to the CMpf and amGPi induced similar tic improvements (Houeto et al., 2005), whereas in a series in an RCT of N=3 patients, amGPi DBS induced higher tic improvements (ranging from 65% to 96%) than CMpf DBS (30% to 64%) (Welter et al., 2008).

Open-label long-term studies with up to 48 months of follow-up have also reported significant reductions in tic severity and comorbidities with amGPi DBS (Martinez-Fernandez et al., 2011; Sachdev et al., 2014; Servello et al., 2016; Smeets et al., 2016; Azimi et al., 2018; Welter et al., 2019; Servello et al., 2020). These studies have reported average percent improvement in tic severity measured with the YGTSS ranging from 38% to 69.5% across individuals, as well as significant improvements in OCD and other behavioral comorbidities.

Overall, the clinical evidence supporting the use of amGPi DBS from RCTs and open-label long-term follow-up studies has revealed it can be effective for reducing tic severity in select cases of GTS, as well as symptoms of behavioral and psychiatric comorbidities. However, mixed results from the RCTs preclude making definitive conclusions about its effectiveness and may also reflect challenges in designing clinical trials to robustly evaluate the effects of DBS in such a heterogeneous population and to account for the sometimes delayed effects of DBS on tics and psychiatric symptoms.

Posteroventrolateral globus pallidus internus

The posteroventrolateral globus pallidus internus (pvGPi) was first reported as a target for DBS for GTS in a single case report (van der Linden et al., 2002), with the rationale to modulate the motor BGTC networks via the motor subregion of the GPi to reduce tics in GTS.

Since the original report, only one RCT has been published evaluating the efficacy of pvGPi DBS; however, it was performed in N=10 patients who underwent dual bilateral targeting of the medial thalamus (CM-Voi) and the pvGPi (Müller-Vahl et al., 2021). The results indicated pvGPi DBS significantly reduced tics compared to baseline at 3 months, but neither pvGPi DBS nor CM-Voi DBS significantly reduced tics when compared to sham. Some individuals experienced positive effects of DBS; however, 50% of the study subjects discontinued DBS due to a lack of response.

Several long-term studies have been reported for pvGPi DBS for 12–99 months of follow-up. These studies have reported improvement in tics, comorbidities, and quality of life in some individuals (Dong et al., 2012; Dehning et al., 2014; Zhang et al., 2014; Servello et al., 2016; Zhang et al., 2016; Hauseux et al., 2017; Sun et al., 2021; Cui et al., 2022; Liu et al., 2022). In the subset of studies that reported tic improvement percentages, the average improvement in tics ranged from 45.7% to 57.8% across individuals (Zhang et al., 2014, 2016; Sun et al., 2021). One study investigated low-frequency stimulation (65 Hz) targeted to the pvGPi in N=6 patients with up to 48 months of follow-up and observed 4/6 individuals experienced at least a 50% reduction in tic severity as measured by the YGTSS and improvements in comorbidities and quality of life (Sun et al., 2021). However, it is unclear whether low-frequency pvGPi DBS will offer further benefits over conventional high-frequency DBS because their effects were not directly compared.

Table 24.1

Randomized controlled trials of DBS for GTS.

Target	References	Number of patients	Study design	Study design description	Time to primary outcome	Latest follow-up	Outcomes	Significant difference between active vs sham?	Main conclusions
CMpf	Maciunas et al. (2007)	5	Randomized, double-blind crossover	Patients were assessed in a double-blind randomized manner for 4 weeks, with each week spent in one of four states (unilateral each side, bilateral, or off). Patients were similarly assessed 3 months after unblinded bilateral stimulation, while repeated open programming sessions were performed.	4 weeks	3 months	*Active vs Sham*: Mean 14.7% reduction with bilateral DBS compared to sham. *Open-label*: Mean 44% YGTSS reduction compared to baseline. 3/5 patients were clearly clinically improved.	Yes	Bilateral thalamic DBS reduced tic frequency and severity in some patients with GTS.
CM-Spv-Voi	Ackermans et al. (2011)	6	Randomized, double-blind crossover	After surgery, the patients were randomly assigned to 3 months stimulation followed by 3 months OFF stimulation or vice versa. All patients then received active DBS in open-label follow-up. Assessments were performed prior to surgery and at 3, 6, and 12 months after surgery.	3 months	12 months	*Active vs Sham*: Mean 37% reduction in YGTSS on-DBS compared to off-DBS. *Open-label*: Mean 49% reduction in YGTSS compared to baseline. No significant change in OCD, ADHD, anxiety, depression, or self-injurious behavior.	Yes	Stimulation of the CM-Spv-Voi crosspoint may reduce tic severity in refractory Tourette syndrome.
CM	Okun et al. (2013)	5	Randomized, double-blind cross-over – scheduled/ cycling DBS	Patients were randomized by the statistician to receive immediate DBS activation at postoperative day 30 or delayed-start DBS activation at postoperative day 60. An experiment was performed at 6 postoperative months that used baseline data compared with acute changes in settings at 6 months (continuous, off, and scheduled).	1 h	6 months	*Active vs Sham*: YGTSS not reported. *Open-label*: Mean 18.6% reduction in YGTSS compared to baseline. Trend toward improvement in QoL.	No	This study provided safety and proof of concept that a scheduled DBS approach could improve motor and vocal tics in GTS.
CMpf-Voi	Baldermann et al. (2021)	8	Randomized, double-blind crossover	Patients underwent short randomized, double-blinded sham-controlled crossover sequences with either active or sham stimulation at both 6-month and 12-month assessments.	2 days	12 months	*Active vs Sham*: Mean 26% at 6 months and 44% at 12 months reduction in YGTSS on-DBS compared to off-DBS. *Open-label*: Significant reduction in YGTSS at 6 and 12 months compared to baseline.	Yes	Thalamic stimulation was effective in improving tic severity and overall QoL. The reduction of tic severity was primarily driven by active stimulation.
CMpf (closed-loop DBS using cortical signals)	Cagle et al. (2022)	4	Nonrandomized, double-blind crossover	Patients underwent 6 months of conventional DBS until optimal parameters were identified. Then closed-loop DBS was implemented, using cortical signals to control stimulation. At the 6-month visit, patients underwent a nonrandomized double-blind assessment to compare conventional DBS, closed-loop DBS, and off-DBS using the Modified Rush Videotape Rating Scale (MRVRS).	MRVRS video for each condition	6 months	*Active vs Sham*: Significantly lower MRVRS scores on conventional DBS and closed-loop DBS compared to off-DBS. *Open-label*: Mean 33.3% reduction in YGTSS compared to baseline. 2/4 participants had a tic improvement of > 40%.	Yes	Embedded closed-loop deep DBS was feasible, safe, and had a comparable outcome to conventional DBS for the treatment of tics.

(*Continued*)

Table 24.1 (Continued)

Target	References	Number of patients	Study design	Study design description	Time to primary outcome	Latest follow-up	Outcomes	Significant difference between active vs sham?	Main conclusions
amGPi	Kefalopoulou et al. (2015)	15	Randomized, double-blind crossover	Patients received surgery for bilateral amGPi DBS and then were randomly assigned to receive either stimulation on-first or off-first for 3 months, followed by a switch to the opposite condition for a further 3 months. All patients then received open-label active stimulation in follow-up. Note: *N*=2 patients were implanted "slightly more posteriorly" for dystonic tics.	3 months	8–36 months	*Active vs Sham*: Mean 15.3% reduction in YGTSS on-DBS vs off-DBS. *Open-label*: Mean 40.1% reduction in YGTSS and significant improvements in QoL, depression, and anxiety compared to baseline.	Yes	amGPi DBS significantly improved tics in patients with refractory GTS with an overall favorable safety profile.
amGPi	Welter et al. (2017)	19	Randomized, double-blind parallel	Patients received surgery for bilateral amGPi DBS. After 3 months, patients were randomly assigned to receive either active or sham stimulation for the subsequent 3 months in a double-blind fashion. All patients then received open-label active stimulation for the subsequent 6 months.	3 months	12 months	*Active vs Sham*: Active group – median 1.1%; sham group – median 0.0% reduction in YGTSS. 72 h OFF stimulation resulted in worsening of tics. *Open-label*: Mean 39.9% reduction in YGTSS after open-label follow-up compared to baseline.	No	amGPi DBS was effective in open-label follow-up; however, 3 months of amGPi DBS was insufficient to decrease tic severity for patients with GTS.
CMpf +amGPi	Welter et al. (2008)	3	Randomized, double-blind crossover	Patients underwent four blinded periods (each 2 months) of CMpf DBS, amGPi DBS, simultaneous CMpf+amGPi DBS, and sham DBS. Sham condition involved reducing the stimulation amplitude of combined CMpf+amGPi DBS to maintain blinding.	2 months	20–60 months	*Active vs Sham*: Motor tics returned in all patients during sham DBS. amGPi produced higher tic improvement (65%–96%) than CMpf DBS (30%–64%) or combined amGPi-CMpf DBS (43%–76%). *Open-label*: Two patients experienced longlasting benefits from DBS. One patient required parameter readjustment and continued to experience intermittent tic improvements with DBS.	–	DBS targeted to amGPi or CMpf shows promise as a treatment for GTS. amGPi DBS may result in greater tic improvement, but small sample size prevents definitive conclusions.
CM-Voi +pvGPi	Müller-Vahl et al. (2021)	10	Randomized, double-blind crossover	Patients underwent three blinded periods each lasting three months including (1) sham, (2) pvGPi DBS, and (3) CM-Voi DBS, followed by an open-label long-term follow-up.	3 months	Mean 89.9 months	*Active vs Sham*: No targets were more effective than sham. pvGPi DBS significantly reduced tics but not comorbidities. *Open-label*: No significant improvement in tics, comorbidities, or QoL. Fifty percent of patients discontinued DBS. Some individual patients benefited from CM DBS.	No	Initial effects of pvGPi DBS on tics were more favorable than CM-Voi DBS. Individual patients benefited from long-term CM-Voi DBS, but outcomes varied.

ADHD, attention-deficit hyperactivity disorder; *amGPi*, anteromedial globus pallidus internus; *CM*, centromedian nucleus; *DBS*, deep brain stimulation; *GTS*, Gilles de la Tourette syndrome; *OCD*, obsessive-compulsive disorder; *pf*, parafascicular; *pvGPi*, posteroventrolateral globus pallidus internus; *QoL*, quality of life; *Spv*, substantia periventricularis; *Voi*, ventro-oralis-internus; *YGTSS*, Yale Global Tic Severity Scale.

Table 24.2

Long-term follow-up studies of DBS for GTS.

Target	References	Number of patients	Latest follow-up	Outcome at latest follow-up	Main conclusions
CM-Spv-Voi	Visser-Vandewalle et al. (2003)	3	8–60 months	Tic reduction of 90.1% in case 1 (at 5 years), 72.2% in case 2 (at 1 year), and 82.6% in case 3 (at 8 months) via video assessment.	Chronic thalamic DBS may be an effective and safe treatment for medically intractable GTS in adult patients.
CMpf-Voa	Porta et al. (2009)	18	24 months	Mean 51.4% improvement in YGTSS at 24 months. Significant improvements in OCD, depression, anxiety, and subjective social impairment.	Bilateral thalamic DBS reduced global tic severity in patients with severe intractable GTS.
CM-Spv-Voi	Ackermans et al. (2010)	2	8–72 months	Tic improvement observed at 5 years in patient 1 (90.1%) was maintained at 10 years (92.6%). In patient 2, the tic improvement at 8 months (82%) was slightly decreased at 6 years (78%).	Bilateral thalamic DBS may provide sustained tic benefit after at least 6 years.
CMpf	Porta et al. (2012)	18	36–72 months	Mean 58.7% reduction in YGTSS at latest follow-up. Significant improvement in OCD, anxiety, and depression.	At long-term follow-up, DBS was successful for improving tics and potentially reducing obsessions, anxiety and depression.
VA/VL	Huys et al. (2016)	8	12 months	Mean 51% reduction in YGTSS at 6 and 12 months. Significant improvement in trait anxiety, QoL, and global functioning. Presurgical compulsivity, anxiety, emotional dysregulation, and inhibition appeared to be significant predictors of surgery outcome.	Stimulation of the VA/VL motor part of the thalamus may be a viable option when considering DBS for TS.
CMpf-Voi/amGPi/pvGPi/NAcc/ALIC	Servello et al. (2016)	48	48+ months	29 cases (78%) exhibited a reduction of more than 50% of the YGTSS score.	The clinical efficacy of DBS in GTS is promising but should be considered as a last resort therapeutic option in carefully selected GTS patients.
CMpf	Testini et al. (2016)	12	2–91 months	Mean 54% reduction in YGTSS total score at latest follow-up.	CMpf DBS can be a safe and effective therapy for patients with refractory GTS.
CMpf-Voi	Dowd et al. (2017)	13	6–58 months	Mean 50% improvement at latest follow-up compared to baseline.	DBS targeted to the CMpf-Voi showed long-term efficacy as a treatment for severe, refractory GTS.
CM-SPv-Voi	Smeets et al. (2017)	7	12–78 months	Mean 48.5% reduction in YGTSS at latest follow-up. Some patients experienced debilitating side-effects.	There was a disbalance of therapeutic effects and side-effects at long-term follow-up in some patients, often leading to either switching the stimulator off or new surgery with a different neuroanatomic target.
CMpf-Voi/amGPi	Servello et al. (2020)	55	48 months	A significant and stable improvement in the YGTSS and Yale-Brown Obsessive Compulsive Scale (Y-BOCS) was observed with DBS in CMpf and amGPi (% improvement not reported).	DBS is an effective therapy in treating GTS, and amGPi DBS might be superior to CMpf-Voi DBS in alleviating comorbid OCD/OCB.
CMpf	Kimura et al. (2020)	25	36 months	Mean 45.2% reduction in YGTSS tic severity score at 1 year and by 56.6% at the last follow-up.	Thalamic DBS is an effective treatment for refractory GTS, although careful indication is necessary.

(Continued)

Table 24.2 (Continued)

Target	References	Number of patients	Latest follow-up	Outcome at latest follow-up	Main conclusions
amGPi/pvGPi	Martinez-Fernandez et al. (2011)	5	3–24 months	Improvements in YGTSS scores differed following amGPi DBS vs pvGPi DBS (mean 38% vs 20%), and MRVRS scores (54% vs 37%).	In this small case series, amGPi DBS may be superior to pvGPi DBS for reducing tics in GTS.
amGPi	Sachdev et al. (2014)	17	8–46 months	12/17 were responders (> 50% reduction in YGTSS), 16/17 reported positive response. Mean 54.3% reduction in YGTSS at final follow-up.	amGPi DBS is an effective and well-tolerated treatment for a subgroup of severe TS with long-term benefits.
amGPi	Smeets et al. (2016)	5	12–38 months	Mean 69.5% reduction in YGTSS at latest follow-up.	DBS of the amGPi is effective in reducing tic severity and possibly also OCB in patients with refractory GTS without serious adverse events or side-effects.
amGPi	Azimi et al. (2018)	6	12–48 months	Total YGTSS scores and QoL were significantly improved at latest follow-up (% improvement not reported).	amGPi DBS can be an effective and well-tolerated therapy for patients with refractory GTS.
amGPi	Welter et al. (2019)	16	12–48 months	Mean 49.3% reduction in YGTSS at latest follow-up compared to baseline.	This study reported long-term follow-up of patients in previous randomized controlled trials (Welter et al. (2017)). amGPi-DBS is an effective treatment for TS patients with sustained reduction in tics and self-injurious behaviors and improvement in global functioning.
pvGPi	Dong et al. (2012)	2	12 months	Unilateral (right) pvGPi DBS reduced YGTSS scores by 58.5% (patient 1) and 53.1% (patient 2) at 12 months.	Unilateral pvGPi DBS may be a viable treatment for some patients with GTS.
pvGPi	Dehning et al. (2014)	6	12–72 months	4/6 patients were responders, who experienced improved global functioning and QoL, as well as psychosocial function.	Improvement of tics with pvGPi DBS seems to be positively correlated with improved functional outcome. Classic “tic scales” should be supplemented with GAF and GTS-QoL to assess treatment success.
pvGPi	Zhang et al. (2014)	13	13–80 months	Mean 52.1% reduction in YGTSS at latest follow-up. Improvement in QoL, 45.7%.	pvGPi DBS offers potential benefit for improving disabling tics and QoL.
pvGPi	Zhang et al. (2016)	24	12–99 months	Tics and OCD were significantly improved (mean 57.8% tic improvement, 38.6% mean OCD improvement, and 12-month follow-up), with higher reduction in motor tics than phonic tics.	pvGPi DBS alleviated tics and psychiatric comorbidities in patients with GTS and improved their QoL.
pvGPi	Hauseux et al. (2017)	3	Mean 52 months	Motor tics decreased with pvGPi in all patients, but varied over time. No significant effect on comorbidities.	There may be long-term therapeutic benefit of DBS in highly socially impaired adolescent patients with intractable GTS and severe psychiatric conditions.
pvGPi	Sun et al. (2021)	6	26–48 months	4/6 were responders (> 50% reduction in YGTSS). Median 56.6% reduction in YGTSS at final follow-up. Also reported significant improvement in OCD symptoms and QoL.	Low-frequency (65 Hz) pvGPi DBS may be an effective and practical treatment for refractory TS.

pvGPi	Cui et al. (2022)	61	Mean 73.3 months	Mean 58.1% tic improvement at final follow-up. Tics, OCD, depression, and QoL were improved at the latest follow-up compared to presurgery.	pvGPi DBS provided long-term efficacy and acceptable safety as a treatment for GTS.
pvGPi	Liu et al. (2022)	5	60+ months	Motor tics and depression were significantly improved at latest follow-up (% improvement not reported). No statistically significant improvement in vocal tics, OCD, or anxiety symptoms. Global function was mildly improved.	pvGPi DBS may be effective for motor tics in GTS patients, especially with depressive symptoms.

ALIC, anterior limb of the internal capsule; *amGPi*, anteromedial globus pallidus internus; *CM*, centromedian nucleus; *DBS*, deep brain stimulation; *GTS*, Gilles de la Tourette syndrome; OCB, obsessive-compulsive behavior; *OCD*, obsessive-compulsive disorder; *pf*, parafascicular; *pvGPi*, posteroventrolateral globus pallidus internus; *QoL*, quality of life; *Spv*, substantia periventricularis; *VA/VL*, ventral anterior-ventral lateral; *Voa*, ventro-oralis-anterior; *Voi*, ventro-oralis-internus; *Y-BOCS*, Yale-Brown Obsessive Compulsive Scale; *YGTSS*, Yale Global Tic Severity Scale.

Overall, pvGPi DBS has shown promising results for reducing tics and potentially improving comorbid symptoms in GTS. However, the paucity of RCT evidence supporting pvGPi DBS makes it challenging to draw definitive conclusions about its effectiveness or the benefits of pvGPi relative to the other targets.

Less common deep brain stimulation targets

Other brain regions implicated in motor and associative/limbic BGTC networks have been explored as DBS targets for GTS. The nucleus accumbens (NAcc) and anterior limb of the internal capsule (ALIC) (sometimes referred to as ventral capsule/ventral striatum) have been reported as targets for GTS DBS, especially in individuals with comorbid OCD (Flaherty et al., 2005; Servello et al., 2009). However, the NAcc/ALIC has waned as a GTS DBS target, likely due to unsatisfactory outcomes. Other targets include the globus pallidus externus (GPe) (Piedimonte et al., 2013; Vilela-Filho et al., 2023), the dorsal subthalamic nucleus (STN) (Martinez-Torres et al., 2009; Dai et al., 2022), the H Fields of Forel (Neudorfer et al., 2017), and the NAcc/ALIC more specifically focused on the bed nucleus of stria terminalis (Duarte-Batista et al., 2020). Evidence supporting the efficacy of these targets has been limited to single cases or case series, and there are no RCTs to date.

Clinical outcomes and adverse events in multicenter studies

Developing generalizable knowledge about GTS DBS has been challenging because a limited patient population is appropriate for GTS DBS and relatively few centers worldwide have the multidisciplinary expertise to perform GTS DBS. The Tourette Association of America International Tourette Syndrome Deep Brain Stimulation Registry and Database was developed to aggregate data across centers and enable collaborative research to study GTS DBS (Deeb et al., 2016). Using data from 185 individuals who underwent GTS DBS from 31 institutions in 10 countries, the consortium assessed the efficacy and safety of GTS DBS (Martinez-Ramirez et al., 2018). The results revealed that DBS significantly improved tics, with average improvements of 40.1% at 6 months and 45.1% at 12 months in total YGTSS scores across all brain targets. Another study investigating the long-term outcomes confirmed these findings with an average YGTSS total improvement of 46.7% at the final follow-up time point (up to 120 months) (Johnson et al., 2019). Both of these multicenter studies found no significant difference in YGTSS improvement across the most common brain targets (medial thalamic regions, amGPi, pvGPi). This finding has also been reported in another meta-analysis (Baldermann et al., 2016). Multicenter data cannot directly replace rigorous RCT evidence; however, the results have suggested that DBS improves symptoms in individuals whose symptoms are otherwise treatment-refractory, and similar to STN and GPi DBS in Parkinson disease, multiple targets within the BGTC networks can be effective for DBS in GTS.

Multicenter analysis of the adverse events reported with GTS DBS (Martinez-Ramirez et al., 2018) showed that 35.4% of individuals reported adverse event(s) in the first year of follow-up. Most adverse events were related to transitory side-effects of stimulation (e.g., paresthesias, dysarthria) without major complications. Surgery-related adverse events included infection and hemorrhage, which occurred in 2.5% and 1.3%, respectively. A more recent study using multicenter data reported an explant rate of 5.6% (15 of 269 patients); infection was the most common reason for device removal (N=7 patients), followed by discomfort from the hardware (N=3 patients), poor compliance or lack of perceived benefit (N=3 patients), formation of a cyst with associated obstructive hydrocephalus (N=1 patient), and complete resolution of tics (N=1 patient) (Deeb et al., 2020).

SURGICAL PROCEDURES

The surgical procedures for GTS DBS are similar to those used for more classical DBS indications. Some individuals with GTS may not be eligible for surgery under local anesthesia due to severe motor tics (especially involving the neck), in which case general anesthesia may be utilized. In some cases, sedation may be used to avoid general anesthesia, such as using lormetazepam and clonidine (Visser-Vandewalle et al., 2003) or propofol (van der Linden et al., 2002). Withholding anesthesia or limiting sedatives can be advantageous in some cases, as this method allows for intraoperative stimulation testing for side-effects while the patient is awake and cooperative, which can be beneficial for verifying the optimal lead location. However, since the optimal lead placement has not been established for any of the GTS DBS targets, many surgeons forgo detailed microelectrode recording or intraoperative testing. Therefore, decisions about general anesthesia versus sedation must be made on a case-by-case basis to ensure the highest likelihood of success. Once the bilateral DBS leads are implanted, the implantable pulse generator is then implanted either during the same surgery or in a separate surgery (staged procedures). Following the surgical procedures, it is crucial to verify that the leads are accurately implanted by obtaining high-resolution postoperative CT imaging and fusing it with the preoperative MRI used for targeting to visualize the contact locations relative to the target nuclei. Some centers may use a postoperative MRI for lead localization, but lead artifacts in MRI may obscure the final location of the DBS lead. Additionally, it is possible to perform unilateral DBS for GTS (Kuhn et al., 2011); however, most centers opt for bilateral placement since GTS rarely manifests with primarily unilateral tics.

POSTOPERATIVE MANAGEMENT

Following device implantation, individuals then undergo multiple follow-up clinic visits for DBS programming, during which the clinician titrates the DBS to identify stimulation parameters that reduce symptoms and minimize side-effects. Like other DBS indications, the programming process for GTS is trial-and-error and no set guidelines have been established. The clinician often begins with a "monopolar review," in which each DBS contact is evaluated separately by increasing the stimulation amplitude and asking the patient to report any changes in sensations (e.g., side-effects such as paresthesias, dysarthria, autonomic responses (Martinez-Ramirez et al., 2018)). Once the side-effect thresholds for each DBS contact are established, the clinician will then decide which contact(s) seem most promising for clinical benefit and may further refine the amplitude, pulse width, and frequency to identify the minimal amount of stimulation required to reduce symptoms. Stimulation parameters are generally identified separately for each hemisphere and then delivered simultaneously. There is no clear consensus on the most effective combination of parameters. A wide range of frequencies (60–200 Hz, although typically 100–200 Hz is used), pulse widths (60–400 μs), and amplitudes (1–10 V) have been reported in multicenter datasets (Deeb and Malaty, 2019; Johnson et al., 2019).

Identifying effective stimulation parameters in GTS can be challenging, because often DBS does not induce acute, observable effects on tics and it can take weeks to months to induce clinically meaningful changes in tic severity (Diederich et al., 2005; Kuhn et al., 2007; Kefalopoulou et al., 2015; Servello et al., 2016; Welter et al., 2017). Additionally, stimulation may need to be increased over time to maintain therapeutic benefit (Kefalopoulou et al., 2015). One study using multicenter retrospective data estimated the median time to reach a 40% improvement in YGTSS total score was 13 months (Johnson et al., 2019). This latency in response may in part be due to the laborious and time-consuming efforts to identify optimal stimulation parameters, which frequently require several post-operative clinical visits. In contrast, tic severity may worsen rapidly (within 24–72 hours) upon turning off DBS (Welter et al., 2017; Baldermann et al., 2021), which may reflect a "rebound effect" when abruptly stopping the stimulation. This phenomenon has also been reported in pallidal DBS for dystonia (Grabli et al., 2009). The mechanisms underlying the time course for the therapeutic effects of DBS in GTS remain unclear.

Since the therapeutic effects of DBS in GTS can be delayed, it is especially crucial to evaluate and document the observed symptoms and side-effects during the programming period in a rigorous and systematic way. The YGTSS has been recommended as a validated instrument to assess tic severity, which assesses symptoms over the past week (Leckman et al., 1989). The MRVRS, a video-based rating scale, can supplement the YGTSS and provide an objective measure for directly observing and quantifying tic frequency and severity (Goetz et al., 1999). It also may be helpful for individuals and/or their caregivers to maintain a symptom diary to track symptom severity at home, given the patient-specific waxing and waning nature of tics. All of these tools may collectively enable clinicians to systematically assess the relative effectiveness of different DBS programming settings, as well as to assess the overall effectiveness of DBS compared to the patient's baseline symptoms.

ADVANCES AND FUTURE DIRECTIONS

DBS can be an effective therapy for carefully selected individuals with severe, treatment-refractory GTS, and progress has been made toward understanding how to apply DBS effectively. However, several remaining challenges, unknown factors, and regulatory hurdles have limited the use of DBS in GTS.

Although general guidelines have been proposed, the patient population best suited for DBS remains unclear. Up to 90% of individuals with GTS have one or more comorbid behavioral or psychiatric conditions, and those with more severe tics tend to also have severe comorbidities that affect their quality of life (Hirschtritt et al., 2015). Several RCTs and open-label studies have reported improvements in comorbidities such as OCD, depression, anxiety, and self-injurious behavior with DBS; however, not all individuals experience lasting improvements in these symptoms, and pervasive comorbidities may impede lasting effects on tics. Given the complex interactions between tics and comorbidities in GTS, evaluating the effects of DBS on each symptom separately can be especially challenging. Recent multicenter studies and meta-analyses suggest age, tic severity, and impairment may all affect the response to DBS. However, the available data have been mixed; lower (Baldermann et al., 2016) and higher (Johnson et al., 2020b) tic severity and younger (Baldermann et al., 2016) and older (Johnson et al., 2019) age at surgery have been correlated with greater tic reduction following DBS. Further research is thus warranted to identify consistent clinical characteristics and symptom profiles most likely to respond to DBS.

It remains unclear how to identify the optimal DBS target for each patient and their specific symptom profiles. At a group level, DBS targeted to medial thalamic regions, the amGPi, or the pvGPi has been associated with similar improvements in tics and comorbidities (Baldermann et al., 2016; Martinez-Ramirez et al., 2018; Johnson et al., 2019). Notably, one recent retrospective study reported that amGPi DBS may be more effective for reducing comorbid OCD than CMpf-Voi DBS (Servello et al., 2020), but further prospective studies will be needed to confirm. Overall, DBS targeted to various nodes of the BGTC networks implicated in GTS can be effective for reducing tics. However, data-driven methods are lacking to determine which target will be most effective for each individual. Studies directly comparing the therapeutic effects of multiple targets within individuals have been limited, as implanting four or more leads (2+ bilateral targets) may not always be feasible. Other indications, such as depression (Scangos et al., 2021; Sheth et al., 2022) and chronic pain (Shirvalkar et al., 2023), have explored temporarily implanting stereo-EEG (sEEG) electrodes for stimulation and recording in multiple network nodes before permanent DBS implantation as a method to identify patient-specific stimulation targets. This approach has been applied in GTS in one case report (MacLean et al., 2021), in which sEEG testing guided bilateral DBS implantation in the pvGPi and NAcc with favorable outcomes. The use of acute testing with sEEG may be beneficial for guiding patient-specific DBS target selection and providing an opportunity to study the network effects of DBS in GTS to inform more tailored stimulation paradigms to reduce tics. Importantly, the feasibility and potential benefits and drawbacks of sEEG must be evaluated in a rigorous and systematic clinical trial to determine the future utility to guide DBS for GTS.

Once the target region has been identified, the optimal location within the GPi or medial thalamic region to stimulate with DBS to reduce tics must be identified to determine the most effective DBS lead location and stimulation parameters. Several recent neuroimaging-based studies have sought to identify the DBS "sweet spot" and brain networks to modulate with DBS in GTS (Akbarian-Tefaghi et al., 2017; Brito et al., 2019; Johnson et al., 2019; Andrade et al., 2020; Johnson et al., 2020a, 2020b; Heiden et al., 2021; Morishita et al., 2021; Baldermann et al., 2022; Ganos et al., 2022; Morishita et al., 2022). Modulation of key regions in motor networks with medial thalamic DBS (Andrade et al., 2020; Johnson et al., 2020a) and associative/limbic networks with GPi DBS (with effective stimulation modulating associative pallido-subthalamic fiber pathways in mid- to anterior GPi and into the ALIC) (Johnson et al., 2020a, 2020b) may be associated with better tic improvement. A recent study also reported a common functional network that may potentially underlie tic improvement with medial thalamic DBS and amGPi DBS (Ganos et al., 2022). These findings could potentially be used to tailor DBS targeting and stimulation parameters in order to more optimally engage networks to reduce tics.

DBS also offers a unique opportunity to invasively record neural activity via local field potentials in humans to study neurophysiologic abnormalities of GTS. Several studies point to increased low-frequency activity (1–10 Hz) in the centromedian thalamic region (Marceglia et al., 2010; Shute et al., 2016; Molina et al., 2017; Neumann et al., 2018; Cagle et al., 2020) and the amGPi (Zauber et al., 2014; Jimenez-Shahed et al., 2016; Neumann et al., 2018; Zhu et al., 2019; Wilken et al., 2021) that may be associated with tics. These findings could be used to develop new closed-loop paradigms to deliver stimulation when tic-related signals have been detected, which may improve efficacy, reduce side-effects, and prolong the battery life of the device. Initial reports of closed-loop DBS in GTS have been promising; a case study of closed-loop DBS controlled by low-frequency thalamic activity reported a 48% improvement in tic severity and a 63% longer projected battery life (Molina et al., 2017). A recent nonrandomized controlled trial used cortical signals associated with tics to drive stimulation and observed that closed-loop CMpf DBS was as effective for reducing tics as conventional DBS (Cagle et al., 2022). Next-generation DBS devices facilitate chronic recordings during DBS, which could be applied to study the effects of stimulation on neurophysiologic activity in GTS. These efforts may enable the identification of new biomarkers associated with tics and comorbidities, including individualized circadian tic patterns and behavior within a naturalistic environment.

CONCLUSIONS

Select individuals with severe, treatment-refractory GTS are candidates for DBS therapy. DBS can provide an effective alternative for reducing tics and improving quality of life. Several case reports, open-label studies, RCTs, and multicenter studies have provided convincing evidence supporting the effectiveness of DBS for GTS. However, the outcomes have been variable across individuals, and GTS DBS has not received regulatory approval for widespread clinical use. Additionally, challenges remain regarding how to apply DBS to treat GTS effectively, including identifying the subpopulation of GTS most likely to benefit from DBS, selecting the most effective target for each individual, and expediting the process of identifying therapeutic stimulation parameters. Advances in GTS DBS will require multicenter collaborations, as even expert centers implant only a handful of individuals each year. Progress has been made by using multicenter data to aggregate GTS DBS data across centers and using neuroimaging and neurophysiology to develop new approaches to improve DBS for GTS. Next-generation DBS devices and continued multidisciplinary collaboration across centers will have a high potential to advance our understanding of GTS and likely inspire novel approaches to improve the efficacy and efficiency of DBS for GTS. Finally, a worldwide effort will need to be undertaken to ensure the availability of DBS devices and refine the process for individuals to receive this therapy for the treatment of GTS and other less common disorders (Okun and Giordano, 2021).

References

Ackermans L, Duits A, Temel Y et al. (2010). Long-term outcome of thalamic deep brain stimulation in two patients with Tourette syndrome. J Neurol Neurosurg Psychiatry 81: 1068–1072.

Ackermans L, Duits A, van der Linden C et al. (2011). Double-blind clinical trial of thalamic stimulation in patients with Tourette syndrome. Brain 134: 832–844.

Akbarian-Tefaghi L, Akram H, Johansson J et al. (2017). Refining the deep brain stimulation target within the limbic globus pallidus internus for Tourette syndrome. Stereotact Funct Neurosurg 95: 251–258.

Andrade P, Heiden P, Hoevels M et al. (2020). Modulation of fibers to motor cortex during thalamic DBS in Tourette patients correlates with tic reduction. Brain Sci 10: 302.

Azimi A, Parvaresh M, Shahidi G et al. (2018). Anteromedial GPi deep brain stimulation in Tourette syndrome: the first case series from Iran. Clin Neurol Neurosurg 172: 116–119.

Baldermann JC, Hennen C, Schüller T et al. (2022). Normative functional connectivity of thalamic stimulation for reducing tic severity in Tourette syndrome. Biol Psychiatry Cogn Neurosci Neuroimaging 7: 841–844.

Baldermann JC, Kuhn J, Schüller T et al. (2021). Thalamic deep brain stimulation for Tourette syndrome: a naturalistic trial with brief randomized, double-blinded sham-controlled periods. Brain Stimul 14: 1059–1067.

Baldermann JC, Schüller T, Huys D et al. (2016). Deep brain stimulation for Tourette-syndrome: a systematic review and meta-analysis. Brain Stimul 9: 296–304.

Brito M, Teixeira MJ, Mendes MM et al. (2019). Exploring the clinical outcomes after deep brain stimulation in Tourette syndrome. J Neurol Sci 402: 48–51.

Cagle JN, Okun MS, Cernera S et al. (2022). Embedded human closed-loop deep brain stimulation for Tourette syndrome: a nonrandomized controlled trial. JAMA Neurol 79: 1064–1068.

Cagle JN, Okun MS, Opri E et al. (2020). Differentiating tic electrophysiology from voluntary movement in the human thalamocortical circuit. J Neurol Neurosurg Psychiatry 91: 533–539.

Coulombe M-A, Elkaim LM, Alotaibi NM et al. (2018). Deep brain stimulation for Gilles de la Tourette syndrome in children and youth: a meta-analysis with individual participant data. J Neurosurg 23: 236–246.

Cui Z, Wang J, Mao Z et al. (2022). Long-term efficacy, prognostic factors, and safety of deep brain stimulation in patients with refractory Tourette syndrome: a single center, single target, retrospective study. J Psychiatr Res 151: 523–530.

Dai L, Xu W, Song Y et al. (2022). Subthalamic deep brain stimulation for refractory Gilles de la Tourette's syndrome: clinical outcome and functional connectivity. J Neurol 269: 6116–6126.

Deeb W, Leentjens AFG, Mogilner AY et al. (2020). Deep brain stimulation lead removal in Tourette syndrome. Parkinsonism Relat Disord 77: 89–93.

Deeb W, Malaty I (2019). Deep brain stimulation for Tourette syndrome: potential role in the pediatric population. J Child Neurol 35: 155–165.

Deeb W, Rossi PJ, Porta M et al. (2016). The international deep brain stimulation registry and database for Gilles de la Tourette syndrome: how does it work? Front Neurosci 10: 170.

Dehning S, Leitner B, Schennach R et al. (2014). Functional outcome and quality of life in Tourette's syndrome after deep brain stimulation of the posteroventrolateral globus pallidus internus: long-term follow-up. World J Biol Psychiatry 15: 66–75.

Diederich NJ, Kalteis K, Stamenkovic M et al. (2005). Efficient internal pallidal stimulation in Gilles de la Tourette sydrome: a case report. Mov Disord 20: 1496–1499.

Dong S, Zhuang P, Zhang XH et al. (2012). Unilateral deep brain stimulation of the right globus pallidus internus in patients with Tourette's syndrome: two cases with outcomes after 1 year and a brief review of the literature. J Int Med Res 40: 2021–2028.

Dowd RS, Pourfar M, Mogilner AY (2017). Deep brain stimulation for Tourette syndrome: a single-center series. J Neurosurg 128: 596–604.

Duarte-Batista P, Coelho M, Quintas S et al. (2020). Anterior limb of internal capsule and bed nucleus of stria terminalis stimulation for Gilles de la Tourette syndrome with obsessive-compulsive disorder in adolescence: a case of success. Stereotact Funct Neurosurg 98: 95–103.

Eapen V, Cavanna AE, Robertson MM (2016). Comorbidities, social impact, and quality of life in tourette syndrome. Front Psychiatry 7: 5–10.

Flaherty AW, Williams ZM, Amirnovin R et al. (2005). Deep brain stimulation of the anterior internal capsule for the treatment of Tourette syndrome: technical case report. Neurosurgery 57: E403.

Ganos C, Al-Fatly B, Fischer J-F et al. (2022). A neural network for tics: insights from causal brain lesions and deep brain stimulation. Brain 145: 4385–4397.

Goetz CG, Pappert EJ, Louis ED et al. (1999). Advantages of a modified scoring method for the Rush video-based tic rating scale. Mov Disord 14: 502–506.

Grabli D, Ewenczyk C, Coelho-Braga M-C et al. (2009). Interruption of deep brain stimulation of the globus pallidus in primary generalized dystonia. Mov Disord 24: 2363–2369.

Hassler R, Dieckmann G (1970). Traitment stéréotaxique des tics et cris inarticulés ou coprolaliques considérés comme phénoméne d'obsession motrice au cours de la maladie de Gilles de la Tourette. Rev Neurol 123: 89–100.

Hauseux PA, Cyprien F, Cif L et al. (2017). Long-term follow-up of pallidal deep brain stimulation in teenagers with refractory Tourette syndrome and comorbid psychiatric disorders: about three cases. Eur J Paediatr Neurol 21: 214–217.

Heiden P, Hoevels M, Bayram D et al. (2021). Connectivity patterns of deep brain stimulation targets in patients with Gilles de la Tourette syndrome. Brain Sci 11: 1–11.

Hirschtritt ME, Lee PC, Pauls DL et al. (2015). Lifetime prevalence, age of risk, and genetic relationships of comorbid psychiatric disorders in tourette syndrome. JAMA Psychiatry 72: 325–333.

Houeto JL, Karachi C, Mallet L et al. (2005). Tourette's syndrome and deep brain stimulation. J Neurol Neurosurg Psychiatry 76: 992–995.

Huys D, Bartsch C, Koester P et al. (2016). Motor improvement and emotional stabilization in patients with Tourette syndrome after deep brain stimulation of the ventral anterior and ventrolateral motor part of the thalamus. Biol Psychiatry 79: 392–401.

Jafari F, Abbasi P, Rahmati M et al. (2022). Systematic review and meta-analysis of tourette syndrome prevalence; 1986 to 2022. Pediatr Neurol 137: 6–16.

Jeon S, Walkup JT, Woods DW et al. (2013). Detecting a clinically meaningful change in tic severity in Tourette syndrome: a comparison of three methods. Contemp Clin Trials 36: 414–420.

Jimenez-Shahed J, Telkes I, Viswanathan A et al. (2016). GPi oscillatory activity differentiates tics from the resting state, voluntary movements, and the unmedicated Parkinsonian state. Front Neurosci 10: 1–11.

Johnson KA, Duffley G, Anderson DN et al. (2020). Structural connectivity predicts clinical outcomes of deep brain stimulation for Tourette syndrome. Brain 143: 2607–2623.

Johnson KA, Duffley G, Foltynie T et al. (2020). Basal ganglia pathways associated with therapeutic pallidal deep brain stimulation for Tourette syndrome. Biol Psychiatry Cogn Neurosci Neuroimaging 6: 961–972.

Johnson KA, Fletcher PT, Servello D et al. (2019). Image-based analysis and long-term clinical outcomes of deep brain stimulation for Tourette syndrome: a multisite study. J Neurol Neurosurg Psychiatry 90: 1078–1090.

Johnson KA, Worbe Y, Foote KD et al. (2022). Tourette syndrome: clinical features, pathophysiology, and treatment. Lancet Neurol 22: 147–158.

Kefalopoulou Z, Zrinzo L, Jahanshahi M et al. (2015). Bilateral globus pallidus stimulation for severe Tourette's syndrome: a double-blind, randomised crossover trial. Lancet Neurol 14: 595–605.

Kimura Y, Iijima K, Takayama Y et al. (2020). Deep brain stimulation for refractory Tourette syndrome: electrode position and clinical outcome. Neurol Med Chir 61: 33–39.

Kious BM, Jimenez-Shahed J, Shprecher DR (2016). Treatment-refractory Tourette syndrome. Prog Neuropsychopharmacol Biol Psychiatry 70: 227–236.

Kuhn J, Bartsch C, Lenartz D et al. (2011). Clinical effectiveness of unilateral deep brain stimulation in Tourette syndrome. Transl Psychiatry 1: 1–4.

Kuhn J, Lenartz D, Mai JK et al. (2007). Deep brain stimulation of the nucleus accumbens and the internal capsule in therapeutically refractory Tourette-syndrome. J Neurol 254: 963–965.

Leckman JF, Riddle MA, Hardin MT et al. (1989). The Yale Global Tic Severity Scale: initial testing of a clinician-rated scale of tic severity. J Am Acad Child Adolesc Psychiatry 28: 566–573.

Liu A, Jiao Y, Zhang S et al. (2022). Improved depressive symptoms in patients with refractory Gilles de la Tourette syndrome after deep brain stimulation of posteroventral globus pallidus interna. Brain Behav 12: e2635.

Maciunas RJ, Maddux BN, Riley DE et al. (2007). Prospective randomized double-blind trial of bilateral thalamic deep brain stimulation in adults with Tourette syndrome. J Neurosurg 107: 1004–1014.

MacLean JA, Ferman D, Chu JK et al. (2021). Transient complete resolution of Tourette syndrome symptoms following personalized depth electrode placement. Brain Sci 11: 1559.

Marceglia S, Servello D, Foffani G et al. (2010). Thalamic single-unit and local field potential activity in Tourette syndrome. Mov Disord 25: 300–308.

Martinez-Fernandez R, Zrinzo L, Aviles-olmos I et al. (2011). Deep brain stimulation for Gilles de la Tourette syndrome: a case series targeting subregions of the globus pallidus internus. Mov Disord 26: 1922–1930.

Martinez-Ramirez D, Jiminez-Shahed J, Leckman JF et al. (2018). Efficacy and safety of deep brain stimulation in tourette syndrome: the International Tourette Syndrome Deep Brain Stimulation Public Database and Registry. JAMA Neurol 75: 353–359.

Martinez-Torres I, Hariz MI, Zrinzo L et al. (2009). Improvement of tics after subthalamic nucleus deep brain stimulation. Neurology 72: 1787–1790.

Martino D, Deeb W, Jimenez-Shahed J et al. (2021). The 5 pillars in Tourette syndrome deep brain stimulation patient selection. Neurology 96: 664–676.

Martino D, Malaty I, Müller-Vahl K et al. (2023). Treatment failure in persistent tic disorders: an expert clinicians' consensus-based definition. Eur Child Adolesc Psychiatry 32: 859–872.

Mink JW (2001). Basal ganglia dysfunction in Tourette's syndrome: a new hypothesis. Pediatr Neurol 25: 190–198.

Molina R, Okun MS, Shute JB et al. (2017). Report of a patient undergoing chronic responsive deep brain stimulation for Tourette syndrome: proof of concept. J Neurosurg 129: 308–314.

Morishita T, Sakai Y, Iida H et al. (2021). Neuroanatomical considerations for optimizing thalamic deep brain stimulation in Tourette syndrome. J Neurosurg 136: 1–11.

Morishita T, Sakai Y, Iida H et al. (2022). Precision mapping of thalamic deep brain stimulation lead positions associated with the microlesion effect in Tourette syndrome. Neurosurgery 93: 875–883.

Müller-Vahl KR, Szejko N, Saryyeva A et al. (2021). Randomized double-blind sham-controlled trial of thalamic versus GPi stimulation in patients with severe medically refractory Gilles de la Tourette syndrome. Brain Stimul 14: 662–675.

Neudorfer C, El Majdoub F, Hunsche S et al. (2017). Deep brain stimulation of the H Fields of Forel alleviates tics in Tourette syndrome. Front Hum Neurosci 11: 308.

Neumann W-J, Huebl J, Brücke C et al. (2018). Pallidal and thalamic neural oscillatory patterns in Tourette syndrome. Ann Neurol 84: 505–514.

Okun MS, Foote KD, Wu SS et al. (2013). A trial of scheduled deep brain stimulation for Tourette syndrome. JAMA Neurol 70: 85.

Okun MS, Giordano J (2021). Reforming the process for deep brain stimulation and neurologic device approval in rare diseases. JAMA Neurol 78: 5–6.

Piedimonte F, Andreani JCM, Piedimonte L et al. (2013). Behavioral and motor improvement after deep brain stimulation of the globus pallidus externus in a case of Tourette's syndrome. Neuromodulation 16: 55–58.

Porta M, Brambilla A, Cavanna AE et al. (2009). Thalamic deep brain stimulation for treatment-refractory Tourette syndrome: two-year outcome. Neurology 73: 1375–1380.

Porta M, Servello D, Zanaboni C et al. (2012). Deep brain stimulation for treatment of refractory Tourette syndrome: long-term follow-up. Acta Neurochir 154: 2029–2041.

Sachdev PS, Mohan A, Cannon E et al. (2014). Deep brain stimulation of the antero-medial globus pallidus interna for Tourette syndrome. PLoS One 9: e104926.

Scangos KW, Makhoul GS, Sugrue LP et al. (2021). State-dependent responses to intracranial brain stimulation in a patient with depression. Nat Med 27: 229–231.

Schrock LE, Mink JW, Woods DW et al. (2015). Tourette syndrome deep brain stimulation: a review and updated recommendations. Mov Disord 30: 448–471.

Servello D, Galbiati TF, Balestrino R et al. (2020). Deep brain stimulation for Gilles de la Tourette syndrome: toward limbic targets. Brain Sci 10: 301.

Servello D, Porta M, Sassi M et al. (2008). Deep brain stimulation in 18 patients with severe Gilles de la Tourette syndrome refractory to treatment: the surgery and stimulation. J Neurol Neurosurg Psychiatry 79: 136–142.

Servello D, Sassi M, Brambilla A et al. (2009). De novo and rescue DBS leads for refractory Tourette syndrome patients with severe comorbid OCD: a multiple case report. J Neurol 256: 1533–1539.

Servello D, Zekaj E, Saleh C et al. (2016). Deep brain stimulation in Gilles de la Tourette syndrome: what does the future hold? A cohort of 48 patients. Neurosurgery 78: 91–100.

Sheth SA, Bijanki KR, Metzger B et al. (2022). Deep brain stimulation for depression informed by intracranial recordings. Biol Psychiatry 92: 246–251.

Shirvalkar P, Prosky J, Chin G et al. (2023). First-in-human prediction of chronic pain state using intracranial neural biomarkers. Nat Neurosci 26: 1090–1099.

Shute JB, Okun MS, Opri E et al. (2016). Thalamocortical network activity enables chronic tic detection in humans with Tourette syndrome. NeuroImage Clin 12: 165–172.

Smeets AYJM, Duits AA, Leentjens AFG et al. (2017). Thalamic deep brain stimulation for refractory tourette syndrome: clinical evidence for increasing disbalance of therapeutic effects and side effects at long-term follow-up. Neuromodulation 21: 197–202.

Smeets AYJM, Duits AA, Plantinga BR et al. (2016). Deep brain stimulation of the internal globus pallidus in refractory Tourette syndrome. Clin Neurol Neurosurg 142: 54–59.

Sun F, Zhang X, Dong S et al. (2021). Effectiveness of low-frequency pallidal deep brain stimulation at 65 Hz in Tourette syndrome. Neuromodulation 25: 286–295.

Szejko N, Lombroso A, Bloch MH et al. (2020). Refractory Gilles de la Tourette syndrome—many pieces that define the puzzle. Front Neurol 11: 1–11.

Temel Y, Visser-Vandewalle V (2004). Surgery in Tourette syndrome. Mov Disord 19: 3–14.

Testini P, Zhao CZ, Stead M et al. (2016). Centromedian-parafascicular complex deep brain stimulation for Tourette syndrome: a retrospective study. Mayo Clin Proc 91: 218–225.

van der Linden C, Colle H, Vandewalle V et al. (2002). Successful treatment of tics with bilateral internal pallidum (GPi) stimulation in a 27-year-old male patient with Gilles de la Tourette's syndrome (GTS). Mov Disord 17: P1130.

Vandewalle V, van der Linden C, Groenewegen HJ et al. (1999). Stereotactic treatment of Gilles de la Tourette syndrome by high frequency stimulation of thalamus. Lancet 353: 724.

Vernaleken I, Kuhn J, Lenartz D et al. (2009). Bithalamical deep brain stimulation in Tourette syndrome is associated with reduction in dopaminergic transmission. Biol Psychiatry 66: e15–e17.

Vilela-Filho O, Souza JT, Ragazzo PC et al. (2023). Bilateral globus pallidus externus deep brain stimulation for the treatment of refractory Tourette syndrome: an open clinical trial. Neuromodulation. 27: 742–758.

Visser-Vandewalle V, Temel Y, Boon P et al. (2003). Chronic bilateral thalamic stimulation: a new therapeutic approach in intractable Tourette syndrome. Report of three cases. J Neurosurg 99: 1094–1100.

Welter M-L, Houeto J-L, Thobois S et al. (2017). Anterior pallidal deep brain stimulation for Tourette's syndrome: a randomised, double-blind, controlled trial. Lancet Neurol 16: 610–619.

Welter M-L, Houeto J-L, Worbe Y et al. (2019). Long-term effects of anterior pallidal deep brain stimulation for tourette's syndrome. Mov Disord 34: 586–588.

Welter M-L, Mallet L, Houeto JL et al. (2008). Internal pallidal and thalamic stimulation in patients with Tourette syndrome. JAMA Neurol 65: 952–957.

Wilken M, Cerquetti D, Rossi M et al. (2021). Low-frequency oscillations at the limbic globus pallidus internus seem to be associated with premonitory urges in Tourette's syndrome. Mov Disord 36: 2966–2967.

Worbe Y, Lehericy S, Hartmann A (2015). Neuroimaging of tic genesis: present status and future perspectives. Mov Disord 30: 1179–1183.

Zauber SE, Ahn S, Worth RM et al. (2014). Oscillatory neural activity of anteromedial globus pallidus internus in Tourette syndrome. Clin Neurophysiol 125: 1923–1924.

Zhang J-G, Ge Y, Stead M et al. (2014). Long-term outcome of globus pallidus internus deep brain stimulation in patients with Tourette syndrome. Mayo Clin Proc 89: 1506–1514.

Zhang X-H, Li J-Y, Zhang Y-Q et al. (2016). Deep brain stimulation of the globus pallidus internus in patients with intractable Tourette syndrome: a 1-year follow-up study. Chin Med J 129: 1022–1027.

Zhu G, Geng X, Zhang R et al. (2019). Deep brain stimulation modulates pallidal and subthalamic neural oscillations in Tourette's syndrome. Brain Behav 9: e01450.

Handbook of Clinical Neurology, Vol. 215 (3rd series)
Tic Disorders and Tourette Syndrome
C. Ganos and Y. Worbe, Editors
https://doi.org/10.1016/B978-0-443-13554-5.00012-2

Chapter 25

Noninvasive brain stimulation for Gilles de la Tourette syndrome

MAIRI S. HOULGREAVE, KATHERINE DYKE, AND STEPHEN R. JACKSON*

School of Psychology, University of Nottingham, Nottingham, United Kingdom

Abstract

This chapter explores evidence for the effectiveness of noninvasive brain stimulation (NIBS) for tic disorders and explores what the future may hold for these approaches. NIBS can be achieved through several means, including electrical stimulation of peripheral nerves and transcranial cortical stimulation using electrical or magnetic stimulation. Each of these techniques will be discussed with a focus on current evidence of effectiveness in treating tics, benefits, limitations, and current unknowns.

TRANSCRANIAL MAGNETIC STIMULATION

Background and mechanisms

Transcranial magnetic stimulation (TMS) is an approach which relies on the principles of electromagnetic induction to generate electrical currents on the cortical surface. This technique was first reported in 1985 (Barker et al., 1985) and revolutionized the field of brain stimulation, as unlike direct electrical stimulation of the scalp, TMS causes minimal, if any, discomfort. Through varying pulse intensity and frequency, TMS can be used to both measure and modulate cortical excitability and can exert influence both during (online) and following (offline) stimulation. Notably, application of repeated, patterned trains of stimulation have been shown to modulate cortical tissue beyond the stimulation period. These offline effects have been likened to long-term potentiation and long-term depression-type mechanisms (Ziemann, 2004), through which more sustained changes may be promoted via alterations in synaptic plasticity.

Much of the initial work exploring the modulatory effects of TMS in humans was conducted in the motor system, due to the ability to generate motor evoked potentials (MEPs), which can be used as an index of corticospinal excitability at the time of stimulation. MEPs are cortically driven muscle contractions that can be measured using electromyography from electrodes attached to a target muscle. Specifically, applying a TMS pulse to the motor cortex can induce a discharge predominantly of intracortical interneurons projecting to cortical motor neurons. This synchronized discharge results in descending volleys of activation from the pyramidal tracts in the brain and spinal cord to the peripheral nerve (Bestmann and Krakauer, 2015). Given enough input, this stimulation of the peripheral nerve will result in a quantifiable muscle contraction. Measuring MEPs can give insight into current levels of corticospinal excitability, the current state of the motor system during action preparation (Bestmann and Duque, 2016), and alterations in motor plasticity.

Two modulatory TMS approaches have primarily been utilized as potential interventions for reducing tics in Tourette syndrome (TS) and related tic disorders. These are repetitive (r)TMS, and a patterned form of TMS known as theta burst stimulation (TBS). Studies of neurologically healthy adults (hereafter termed neurotypical) following a single session of rTMS have demonstrated changes in motor cortical excitability indexed by alterations in MEP amplitude which can last for over 30 minutes after stimulation has ceased (e.g., Jung et al. 2008). These effects are somewhat dependent on the pattern of stimulation applied. Generally, rTMS pulses delivered at 1 Hz have been reported to reduce MEP amplitudes following stimulation, indicating reduced motor cortical excitability (e.g., Chen et al. 1997; Touge et al. 2001), whereas higher frequency stimulation (typically 5–20 Hz) can have the opposite effect (e.g., Pascual-Leone et al. 1994). For a meta-analytic review of continuous (cTBS) and intermittent (iTBS) TBS effects, see Chung et al. (2016).

It is generally assumed that rTMS effects are frequency-dependent; however, substantial individual variability in response to rTMS has been reported (e.g., Maeda et al., 2000). The same is true for TBS, which also appears to be variable both between (Hamada et al., 2012; López-Alonso et al., 2014; Vallence et al., 2015; Ozdemir et al., 2021) and

*Correspondence to: Stephen R. Jackson, PhD, School of Psychology, University of Nottingham, University Park, Nottingham, NG7 2RD, United Kingdom. Tel: +44 07974214154, E-mail: stephen.jackson@nottingham.ac.uk.

within subjects, and across different sessions (Boucher et al., 2021). While the consensual view remains that on average, TBS can produce bidirectional effects – with cTBS thought to reduce MEP amplitude following stimulation and iTBS thought to increase MEP amplitude, it is now felt by many that the field should move away from this facilitation/inhibition dichotomy in favor of alternative approaches for thinking about the effects of rTMS (Goldsworthy et al., 2021). In particular, it is now thought that the lack of reliability of rTMS to consistently modulate MEP amplitudes in a predictable direction must impact its utility as a therapy for brain health conditions.

Transcranial magnetic stimulation and tic disorders

One of the challenges for therapeutic use of modulatory TMS, is selecting an appropriate anatomic target site to stimulate. TMS has limited depth penetration (Deng et al., 2013), and while indirect stimulation of deeper regions may be possible through targeting interconnected cortical regions (e.g., Bestmann et al. 2004, 2005), a more direct target location remains preferable. One of the most used anatomic targets for protocols aiming to reduce tics in tic disorders is the supplementary motor area (SMA). This is an appealing target location due to its role in modulating descending corticospinal projections (Haber, 2003), its likely involvement in tic production (Bohlhalter et al., 2006; Hampson et al., 2009), and evidence of structural and functional alterations in this region in TS (Eidelberg et al., 1997; Worbe et al., 2015). The SMA has extensive connections to areas relating to motor control and cognitive processing (Picard and Strick, 2001), including habit formation (Voon et al., 2015), which may be altered and overactivated in TS (Delorme et al., 2016). Additionally, research using magnetic resonance spectroscopy has established relationships between levels of the inhibitory neurotransmitter γ-aminobutyric acid (GABA) within the SMA and both tics (Draper et al., 2014) and premonitory urges (He et al., 2022) in TS. Furthermore, brain imaging evidence indicates that individuals with TS have altered connectivity between SMA and primary motor cortex (M1) (Franzkowiak et al., 2012).

Together, these converging lines of evidence suggest that targeting the SMA with noninvasive brain stimulation (NIBS) approaches that are thought to reduce cortical excitability (e.g., 1 Hz rTMS or cTBS) might reduce the occurrence of tics in TS. At the time of writing, nine studies have assessed the potential of 1 Hz rTMS (seven studies) or cTBS (two studies) delivered over SMA to reduce tics. Mantovani et al. (2006) reported significant reductions in the Yale Global Tic Severity Scale (YGTSS) (Leckman et al., 1989) following 10 days of stimulation in five adults/adolescents, alongside reductions in measures of obsessive-compulsive disorder (OCD). Mantovani et al. (2007) also reported reductions in tics, and symptoms of OCD and attention-deficit hyperactivity disorder (ADHD), following 10 sessions of 1 Hz rTMS over the SMA. Subsequent studies have reported similar reductions in tic severity following rTMS delivered to the SMA (Kwon et al., 2011; Le et al., 2013; Bloch et al., 2016; Singh et al., 2018; Kahl et al., 2021) (see Dyke et al. (2021) for a review).

While the results of the studies referred to above appear promising, it is important to note that they are "open label" trials conducted without appropriate: sham controls, comparison against treatment-as-usual control, or comparisons against other interventions. Furthermore, the sample sizes in these studies are often extremely small. For these reasons, the conclusions that can be drawn from these studies are limited, and placebo effects cannot be excluded as an explanation for observed reductions in clinical symptoms. Furthermore, the results from studies that have included an appropriate sham control condition have been less positive. Thus, Landeros-Weisenberger et al. (2015) found equivalent changes in YGTSS for an active stimulation compared to a sham stimulation group after receiving 15 days of 1 Hz active versus sham rTMS. At the time of writing, a single additional sham-controlled study targeting the SMA had been published by Wu et al. (2014). In this study six participants experienced cTBS for 2 days, whilst another six experienced a sham stimulation. No significant differences in tic severity measured by YGTSS were found.

Another potential target site for stimulation is the M1, which may have reduced rates of inhibition in TS. Some of the evidence for this comes from studies using a TMS technique known as short-interval intracortical inhibition (SICI). SICI is a paired-pulse TMS protocol in which a subthreshold "conditioning" pulse is delivered over the motor cortex 1–5 ms prior to a second "test" pulse which is delivered at or above threshold. The conditioning pulse reduces the magnitude of the MEP generated in response to the test pulse compared to that generated by delivering the test pulse alone. This inhibitory effect appears to be underpinned by GABAergic interneurons (Hanajima and Ugawa, 2008; Ziemann, 2013). SICI has been reported to be reduced in adults with TS (Ziemann et al., 1997; Orth et al., 2005b, 2008; Orth and Rothwell, 2008; Heise et al., 2010) and to correlate with motor tic severity (Batschelett et al., 2023) and ADHD symptoms in both children and adults (Gilbert et al., 2004, 2005). Interestingly, SICI does not appear to be as consistently altered in younger children and adolescents; for example, both Moll et al. (1999) and, more recently, Batschelett et al. (2023) failed to find SICI differences at rest. Exactly why SICI results differ between adults and younger cohorts remains to be understood, although other general measures of motor cortical excitability also suggest the existence of age-related differences within primary motor regions (Pépés et al., 2016). Nevertheless, while M1 has been used as a target site in several therapeutic 1 Hz rTMS studies (Münchau et al., 2002; Chae et al., 2004; Orth et al., 2005a), to date all have failed to demonstrate significant reductions in YGTSS scores when compared against sham control conditions. In comparison to studies targeting the SMA, this could be taken to suggest that M1 is a poorer target for stimulation; however, these results are more in agreement with the sham-controlled SMA studies.

In summary, the current evidence that rTMS/TBS delivered to cortical motor areas produces a meaningful reduction in tics in tic disorders is unconvincing. Existing studies are limited by: inappropriately small sample sizes, lack of adequate experimental control conditions, and a lack of mechanistic exploration. Below we briefly discuss key assumptions that most often underpin rTMS/TBS interventions for tic disorders, and we suggest why these assumptions may be misplaced and outline several potential reasons for the lack of efficacy of rTMS/TBS in reducing tics in tic disorders.

Individual variability in response to repetitive transcranial magnetic stimulation/theta burst stimulation

One assumption of studies applying rTMS/TBS to reduce tics in tic disorders is that the stimulation delivered to each individual reliably and consistently produces a similar physiologic effect. Specifically, it has been assumed that rTMS/cTBS can lead to a reduction in cortical excitability within the targeted area (e.g., SMA), and for this reason "inhibitory" TMS protocols such as low-frequency (1 Hz) rTMS or cTBS were utilized. However, it is now well established that these protocols lead to considerable individual variability in the direction of any poststimulation alterations in cortical excitability (Maeda et al., 2000; Hamada et al., 2012). Thus, the same stimulation might lead to a *decrease* in cortical excitability in one individual but an *increase* in cortical excitability in another. Therefore, one explanation of the apparent lack of efficacy of NIBS for reducing tics in tic disorder might be the unreliability of these TMS protocols in producing a consistent physiologic response in individuals with tics.

Are "inhibitory" transcranial magnetic stimulation protocols the correct approach for reducing tics?

Even if we assume that "inhibitory" rTMS/cTBS could produce a reliable and consistent poststimulation reduction in cortical excitability, it is unclear to us whether this would necessarily be useful in reducing tics in tic disorders. A mechanistic assumption underpinning existing TMS interventions for reducing tics is that "inhibitory" rTMS/cTBS protocols are required to reduce cortical excitability in cortical motor areas such as the SMA or M1. But, we question whether there really is compelling evidence for "*hyper*-excitability" within these areas? Our own data (e.g., Pépés et al. (2016)) and that of others (e.g., Orth et al. (2008)) suggest that, if anything, these areas may be "*hypo*-excitable" rather than hyperexcitable. First, we have found that resting motor threshold (RMT) for TMS stimulation is increased significantly in children and adolescents with TS compared to matched controls (Pépés et al., 2016). This indicates that *increased* TMS stimulation is required to produce an MEP of a given amplitude in the TS group compared to that required to produce an equivalent MEP in neurologically healthy individuals. Note that, while this effect has not always been observed in studies of adults with TS, it is important to recognize that many of these studies often involved only very small numbers of patients and controls.

Figure 25.1 illustrates the results of a meta-analytic analysis that we have conducted of 22 published studies reporting RMT values for TMS stimulation of the first dorsal interosseous or abductor pollicis brevis muscles in groups of individuals with TS compared to groups of neurologically healthy individuals. The results of this analysis clearly indicate that RMT is increased in individuals with TS, indicating cortical motor hypo-excitability in TS.

Second, using physiologic TMS measurements, it has been demonstrated that the gain (i.e., the rate of change) of the TMS recruitment (IO) curve is significantly reduced in individuals with TS compared with healthy controls (Orth et al., 2008; Draper et al., 2014; Pépés et al., 2016) (see Fig. 25.2). However, while the gain of the TMS recruitment curve was inversely associated with tic severity scores in children with TS (Draper et al., 2014; Pépés et al., 2016) – that is, those individuals with lowest gain functions had the most severe tics – in adults with TS, tic severity was *positively* correlated with the gain of TMS IO curve (Orth et al., 2008).

Third, using physiologic TMS measures, we have demonstrated that the gain in cortical excitability ahead of volitional movements is significantly decreased in individuals with TS compared to age-matched healthy controls (Jackson et al., 2013; Draper et al., 2015; Pépés et al., 2016). Together, these studies indicate that motor cortical regions are hypo-excitable in TS compared to matched healthy controls. This finding is in direct contrast to the assumption that motor cortical regions are hyperexcitable in TS. Therefore, one explanation of the apparent lack of efficacy of NIBS for reducing tics in tic disorder might be that the use of "inhibitory" rTMS/cTBS protocols to reduce cortical excitability in cortical motor areas such as SMA or M1 might be misguided.

TRANSCRANIAL DIRECT CURRENT STIMULATION

Background and mechanisms

Transcranial direct current stimulation (tDCS) is a noninvasive brain stimulation approach that has been explored as a potential treatment for tics. Unlike TMS, which relies on the induction of electrical currents beneath the coil, tDCS involves the direct application of a low-voltage (typically between 1 and 2 mA) current flowing between a minimum of two electrodes placed on the scalp. It is important to note that these intensities are substantially lower than would be needed to induce cerebral activity (Poreisz et al., 2007).

Studies in neurotypical adults have shown that tDCS gives rise to both "online" (i.e., during stimulation) and "offline" (i.e., after stimulation has ceased) changes to cortical excitability. These appear to be underpinned by different mechanisms, with the online effects thought to be mediated by alterations in membrane potential (Stagg and Nitsche, 2011), whereas the offline affects have been linked to the operation of N-methyl-D-aspartate (NMDA) (Liebetanz et al., 2002; Nitsche et al., 2003) and α-amino-3-hydroxy-5-methyl-4-isoxazolepropionic acid (AMPA) (Stafford et al., 2018; Martins et al., 2019) receptor

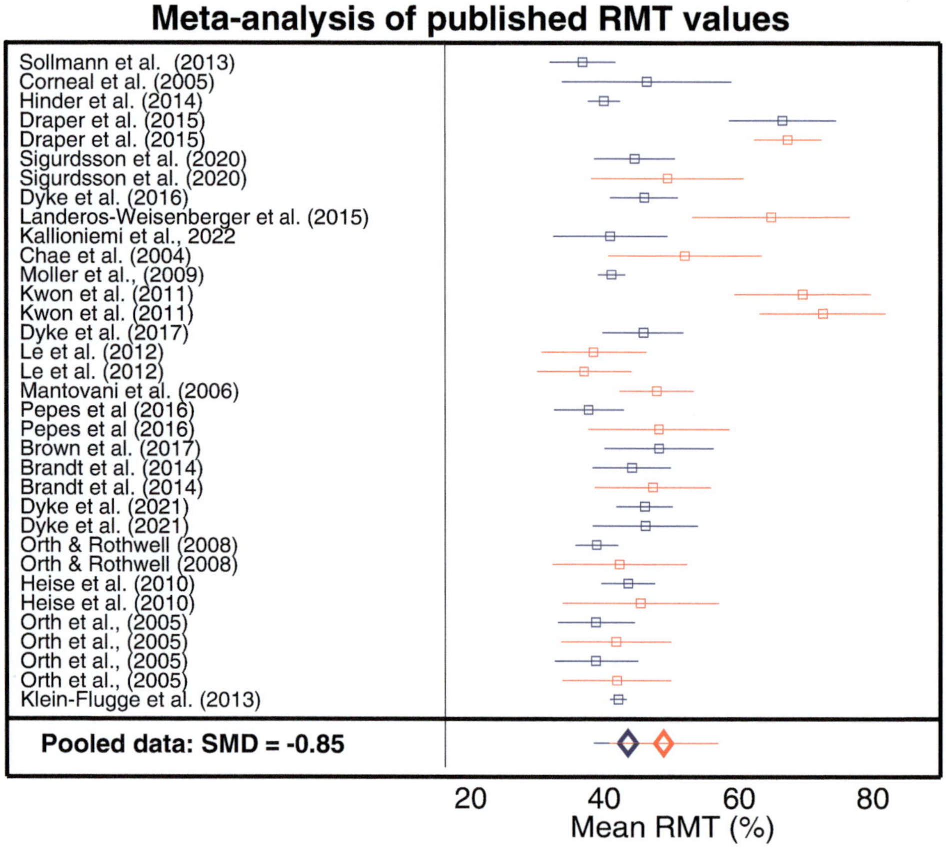

Fig. 25.1. Results of a meta-analytic analysis of published studies reporting resting motor threshold (RMT) values (mean and standard deviation (SD)) for stimulation of the first dorsal interosseous or abductor pollicis brevis muscles of the hand in groups of individuals with Tourette syndrome (TS) or neurotypical individuals. The analysis was based on 22 published studies that reported mean RMT, SD of RMT, and the number of study participants. The study was based on RMT values from 221 individuals with TS and 334 neurotypical individuals. The data illustrated are the mean RMT values (with SD error bars) reported in each study. Blue symbols represent data from neurotypical groups and red symbols data from TS groups. The standardized mean difference (SMD) between groups was computed based upon the pooled mean and SD values weighted by the number of participants within each study. The analysis revealed a large effect size (SMD) of –0.85 ($p < 0.0001$). It should be noted that the pooled participants did not differ in age (p=0.5) and that the difference between the TS and control group RMT was observed for both first dorsal interosseous (SMD=–0.54, t=–10.5, $p < 0.0001$) and abductor pollicis brevis (SMD=–1.24, t=–16.7, $p < 0.0001$) when investigated separately.

channels. Changes in levels of GABA and glutamate have also been reported following stimulation (Nitsche et al., 2004; Stagg et al., 2009). As with rTMS/TBS, it is the presence of these offline effects and their potential for facilitating longer-lasting synaptic change that has given rise to interest in tDCS as a therapeutic intervention.

tDCS can produce bidirectional effects depending on the parameters used. Specifically, cortical regions proximal to the anode have been found to *increase* in cortical excitability on average, whereas areas proximal to the cathode are reported to show *decreases* in cortical excitability. Interestingly the mechanisms underpinning these effects are more complex than simply being the reverse of one another, and the effects of cathodal stimulation are slightly less well characterized (see Stagg and Nitsche, 2011). These polarity specific effects can be convincingly demonstrated in animal studies (Creutzfeldt et al., 1962; Purpura and Mcmurtry, 1965), and have been shown to modulate MEPs in the motor cortex of humans, most notably in a seminal study by Nitsche and Paulus (2000). However, in practice the effects of tDCS are once again highly variable, and studies in neurotypical adults have shown effects to be subject to substantial intra- and intersubject variability (Wiethoff et al., 2014; Dyke et al., 2016; Horvath et al., 2016).

Transcranial direct current stimulation and tic disorder

Research into the therapeutic potential of tDCS for tic reduction has also primarily aimed to reduce cortical excitability within motor areas by placing the cathode over the targeted region. In the first published instance of this approach, Mrakic-Sposta et al. (2008), assessed the effect of five daily sessions of cathodal tDCS applied to the left premotor cortex in two participants with TS. A reduction in YGTSS scores was reported that was larger in magnitude than the effects seen using sham stimulation. Additional analysis of tics from video

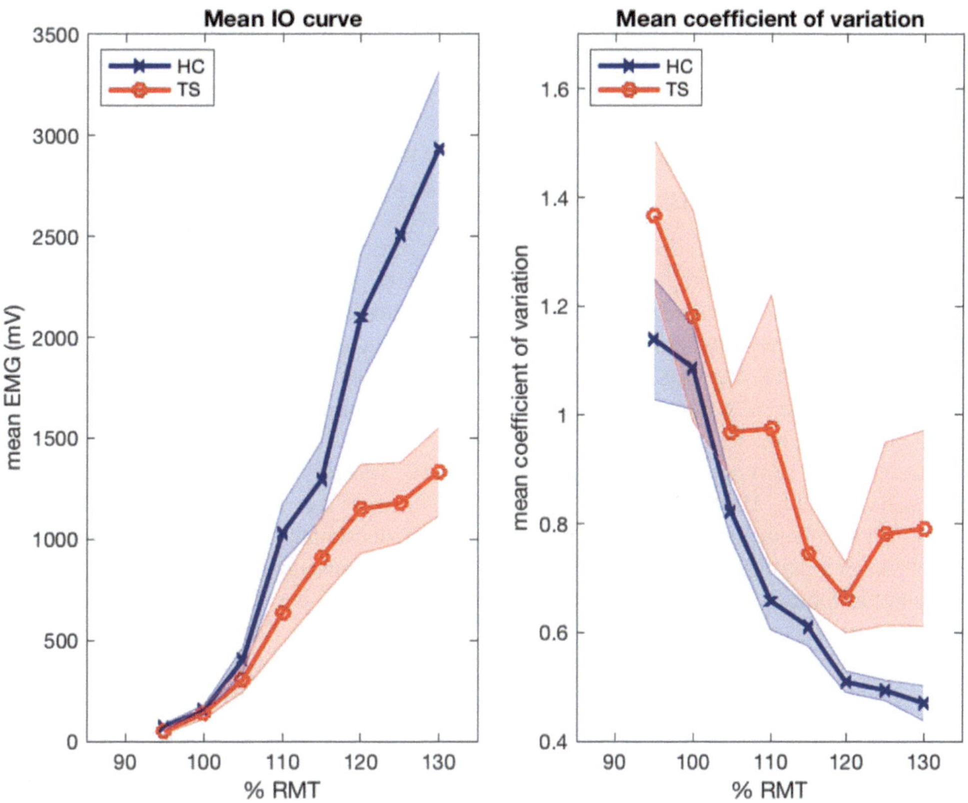

Fig. 25.2. Left panel: Illustrates TMS recruitment curves for a group of children and adolescents with Tourette syndrome (TS) and a group matched neurologically healthy controls. Inspection of this figure shows how the average gain in motor evoked potential amplitude with increasing stimulation intensity is reduced in the TS group compared to controls. Right panel: Changes in the coefficient of variation (a measure of trial-by-trial variability) with increases in stimulator output. Note that variability decreases with stimulation intensity, but at higher levels of stimulation, the TS group exhibit increased trial-by-trial variability compared to controls. Data are redrawn from Pépés SE, Draper A, Jackson GM, et al. (2016) Effects of age on motor excitability measures from children and adolescents with Tourette syndrome. Dev Cogn Neurosci 19: 78.

recording showed a substantially larger reduction in tics in comparison to sham. In later work, Eapen et al. (2017) reported reductions in urge and tic symptoms following 6 weeks of cathodal stimulation of the SMA in two participants. A sham condition was included but this always followed active stimulation and hence is difficult to interpret. In a subsequent case study, Tajadini et al. (2018) reported that 10 sessions of stimulation twice daily to motor/inferior frontal regions appeared to reduce tics, a reduction which was still present during a follow-up a year after. Additionally, another open-label case study by Carvalho et al. (2015) found that tic severity (YGTSS) was reduced by 46% following 10 sessions of cathodal tDCS over the pre-SMA. These effects were still present at a six-month follow-up, and changes in sensorimotor resting-state networks were also identified.

However, in our own work (Dyke et al., 2019), we found that a single session of cathodal tDCS delivered to the SMA marginally reduced tics immediately following stimulation in comparison to sham. However, the interaction between the two conditions was not significant, and no changes in physiologic measures of motor cortical excitability were found. It is also apparent that there can be variability in response to repeated tDCS sessions. In a study of three participants, only one showed a reduction in tic severity following 10 sessions of cathodal pre-SMA stimulation (Behler et al., 2018).

In summary, it is not possible to draw any strong conclusions about the effectiveness of tDCS as an intervention for tics owing to very small sample sizes and a general lack of adequate sham controls. At the time of writing, the largest study was conducted on 10 participants (Dyke et al., 2019), which explored the immediate impact of a single session of tDCS. As with TMS work, most studies are open label, underpowered, and lack any exploration of the underlying mechanisms that may drive tic reduction.

PERIPHERAL NERVE STIMULATION

Recently, peripheral nerve stimulation has been explored as a potential therapeutic alternative to transcranial stimulation. Peripheral nerve stimulation such as median nerve stimulation (MNS) has the advantage that it need not require patients to visit a clinic to receive treatment; rather the intervention could be delivered using a wearable therapeutic device outside of the clinic (Maiquez et al., 2023). While tDCS has been trialed in homes (Hagenacker et al., 2014; Kasschau et al., 2016; Schwippel et al., 2017), MNS would not require the patient to stay still or even within their home; instead a wearable device could be turned on in situations where tics may be increased, making the approach more adaptive and as such more useful for day-to-day life (Maiquez et al., 2023).

Background and mechanisms

One NIBS approach involves entrainment of neural oscillations (Thut et al., 2011a). Entrainment is the progressive phase synchronization of oscillations within a targeted region to an external trigger, such as the electrical/magnetic pulses delivered during NIBS (Thut et al., 2011b). For TS, the target oscillations for entrainment are those linked to the initiation of movement. During voluntary movement, oscillations in the μ-α (8–12 Hz) and β (13–30 Hz) frequency range desynchronize within the sensorimotor cortex (Pfurtscheller et al., 1996; Jurkiewicz et al., 2006). Following movement, there is an increase in synchrony in the beta band, which is known as the postmovement beta rebound. A higher synchrony/amplitude of beta oscillations is associated with a reduction in cortical excitability and a reduced probability of movement (Gilbertson et al., 2005; Engel and Fries, 2010; Jenkinson and Brown, 2011). A similar increase in μ-α synchrony is associated with inhibition of brain regions which are irrelevant to the current task (Jensen and Mazaheri, 2010; Brinkman et al., 2014; Buchholz et al., 2014; Brinkman et al., 2016). Entrainment of beta oscillations using transcranial alternating current stimulation (tACS) results in a reduction of the velocity and force of movements (Pogosyan et al., 2009; Joundi et al., 2012). While tACS is thought of as a transcranial NIBS method, there is evidence to suggest that its effects are due to stimulation of peripheral nerves (Asamoah et al., 2019). Therefore, it may be possible to entrain the sensorimotor cortex and produce changes in motor behavior indirectly via MNS.

The median nerve is a division of the brachial plexus (Murphy and Morrisonponce, 2022). This nerve provides motor innervation to the forearm and hand, in addition to sensory innervation of the medial forearm, palm, thumb, index, and middle fingers and a portion of the ring finger (Murphy and Morrisonponce, 2022). Sensory information is conveyed via the medial lemniscus pathway, where the fibers synapse in the cuneate nucleus, thalamus, and finally the primary somatosensory cortex (Passmore et al., 2014). MNS is a noninvasive technique which can manipulate neural activity through electrical stimulation of the median nerve.

Trains of 10 pulses of rhythmic MNS at a variety of frequencies (12, 19, 20 Hz) have been shown to induce increases in oscillatory amplitude and phase synchronization within the contralateral sensorimotor cortex which are specific to the frequency of stimulation, suggesting that MNS is capable of entraining sensorimotor oscillations (Maiquez et al., 2020; Morera Maiquez et al., 2020; Houlgreave et al. 2022). Arrhythmic stimulation – defined as being on average the same frequency but with a random interpulse interval – does not show these effects. An alternative explanation for MNS-induced increases in oscillatory amplitude and phase synchronization is the generation of steady-state evoked responses; however, linear modeling of a subject specific sensory evoked potentials (SEPs) demonstrated a significantly better fit for the response following arrhythmic compared to rhythmic stimulation, indicating that the presence of SEPs explains less of the variance in the rhythmic response (Houlgreave et al., 2022) and thus argues against a steady-state explanation. Furthermore, empirical mode decomposition, which decomposes a recorded signal into its components, demonstrated that in the majority of subjects, the phase-aligned frequency was not significantly different from the stimulation frequency during rhythmic trials (Quinn et al., 2021; Houlgreave et al., 2022). During sinusoidal oscillations, there is a constant frequency across phase. Therefore, in summary, these results suggest that rhythmic MNS does indeed induce entrainment of targeted cortical oscillations.

We recently explored the effects of rhythmic and arrhythmic MNS using functional magnetic resonance techniques. First, we used functional magnetic resonance spectroscopy to examine how rhythmic and arrhythmic MNS altered concentrations of neurometabolites within the contralateral sensorimotor cortex. We found that during sessions of rhythmic and arrhythmic 10 Hz MNS, glutamate increased in the contralateral sensorimotor hand area before returning to baseline as shown in yet unpublished data by Houlgreave et al. This may indicate that during MNS, regardless of stimulation pattern, there is an increase in the firing of glutamatergic neurons in the sensorimotor cortex. However, this increase is not sustained throughout stimulation, which may be caused by adaptation mechanisms in response to the repetitive nature of the stimulation. Second, we used functional magnetic resonance imaging to investigate whole brain alterations in brain activation (blood-oxygenation level-dependent signal) associated with delivering rhythmic 10 Hz MNS to the medial nerve of the right wrist (Farr et al., unpublished). Preliminary results from this study are presented in Fig. 25.3. An important initial finding is that MNS not only increases brain activation within contralateral primary and secondary cortical somatosensory areas, it also increases brain activity within the insular cortex, a brain area implicated in the experience of premonitory urge phenomena in tic disorders.

Median nerve stimulation and tics

Due to the evidence that rhythmic MNS is capable of entraining sensorimotor oscillations, the effects of 10 Hz rhythmic MNS on tic frequency and subjective ratings of the urge-to-tic were explored (Morera Maiquez et al., 2020). Tic frequency was compared in 16 participants with TS during alternating blocks of 1-minute rhythmic MNS versus no stimulation. Participants tics were recorded on video, and they were asked to continuously rate their urge-to-tic using a hand-operated slider device (Morera Maiquez et al., 2020). During periods of MNS, tic intensity and frequency were significantly reduced, compared to no stimulation (Morera Maiquez et al., 2020). There was also a strong trend for a reduction in the urge-to-tic, which has been confirmed in follow-up studies. Following the promising effects in this initial study, a double-blind, sham-controlled clinical trial was conducted, which also demonstrated a significant reduction in tic frequency

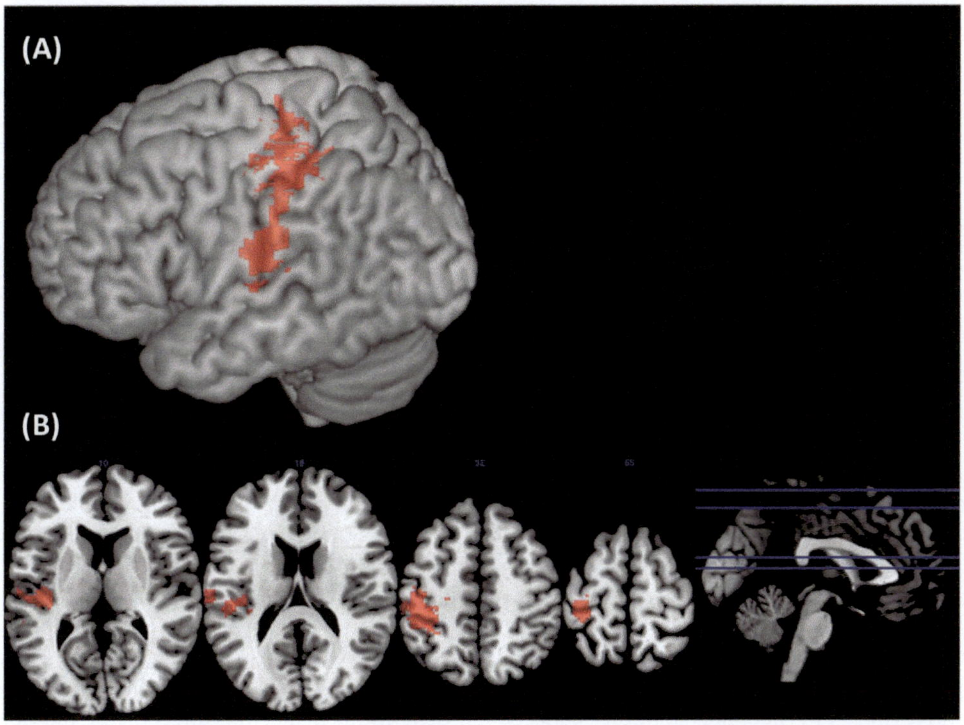

Fig. 25.3. (**A**) A rendered brain illustrating statistically significant increased blood-oxygenation level-dependent (BOLD) activation within the primary (SI) and secondary (SII) somatosensory cortical areas of the left hemisphere following rhythmic 10 Hz electrical stimulation of the median nerve. (**B**) Transverse slices at different depths illustrating that increased BOLD activation following rhythmic 10 Hz electrical stimulation of the median nerve extends to the posterior insular cortex (Farr et al., unpublished).

following rhythmic 10 Hz MNS in comparison to sham stimulation (50% of threshold) or a "treatment as usual" control group (Maiquez et al., 2023). Following the 4-week period of 14-minute daily stimulation sessions, there was also a reduction in total tic severity, suggesting that there were both offline and online benefits of stimulation. However, recent research suggests that both rhythmic and arrhythmic MNS can significantly reduce the frequency of tics (Iverson et al., 2023), suggesting that the beneficial effects might not be solely due to the entrainment of brain oscillations in the sensorimotor cortex. It should be noted that additional influences, such as attentional bias and placebo effects, could also contribute to the observed benefits reported in the Iverson et al. study. It is thought that there is a high level of "sensorimotor noise" associated with the occurrence of tics in TS, leading to a difficulty in discriminating between the signals preceding voluntary and involuntary movement (Ganos et al., 2015). One consequence of increased neural "noise" may be an increase in the variability of neural response to a given stimulus. That is, the population of neurons recruited in response to a given stimulus may be less stable and consistent in TS compared to neurotypical individuals. Therefore, it may be that increasing the synchronized firing of neuronal populations through entrainment of, or through the generation of SEPs within, the sensorimotor cortex may lead to a sustained decrease in neural variability and thus to a reduction in tics (Houlgreave et al., 2022).

Regardless of the mechanism involved, MNS has been demonstrated to result in clinically meaningful reductions in tic frequency and severity (Maiquez et al., 2023) and has the potential to be delivered in a community setting via a wearable device, providing a reduction in tics and the urge-to-tic on demand (Maiquez et al., 2023). Furthermore, the approach is suitable for children, and unlike current pharmacologic treatments, there are no side-effects.

LIMITATIONS OF CURRENT NONINVASIVE BRAIN STIMULATION APPROACHES AND FUTURE DIRECTIONS

In this section we discuss some of the challenges that we face when trying to use NIBS therapeutically for the treatment of tic disorders.

One limitation of current popular NIBS approaches, such as TMS, or transcranial electrical stimulation (tES) methods such as tDCS or tACS, is that these methods cannot be used to *directly* stimulate brain regions that may be strongly implicated in brain health conditions such as TS (e.g., the insula or cingulate cortical regions linked to the experience of premonitory urge phenomena or the generation of tics; Jackson et al., 2011; Draper et al., 2016; Jackson et al., 2020, 2021). Thus, both TMS and tES approaches are limited to stimulating brain tissue close to the external site of stimulation (i.e., the scalp) and for this reason cannot directly stimulate deeper brain regions or structures. While it can be argued that activity in deeper brain areas may be modulated *indirectly* by directly stimulating brain areas that are *connected* functionally or structurally to deep brain targets, it would seem more preferable

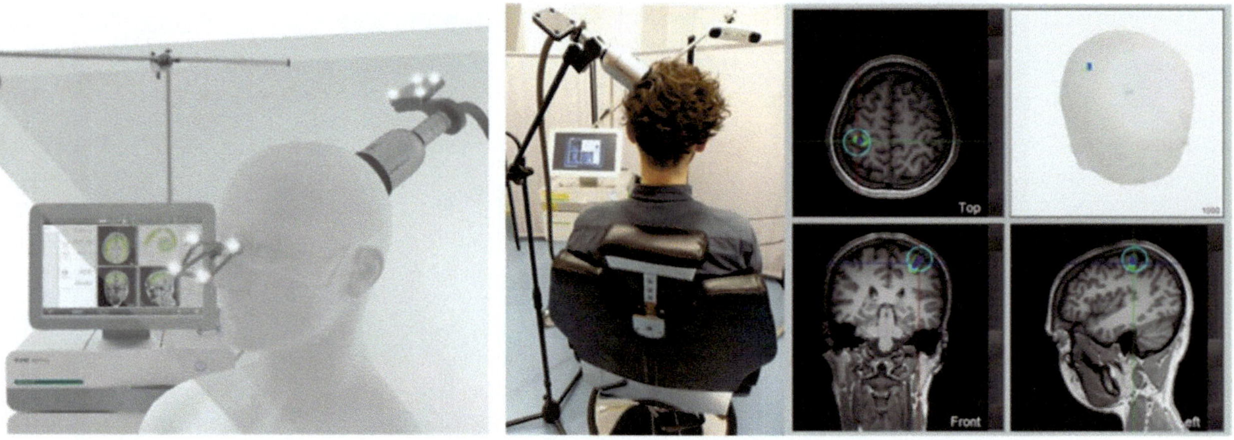

Fig. 25.4. Illustration of a focused ultrasound stimulation (fUS) system that can be used to directly stimulate deep brain areas. The left panel shows the fUS system, camera, and two position markers (one on the head and the other on the fUS system) that contribute to targeting of stimulation target. The center panel shows fUS stimulation delivery in a real research participant. The right panel illustrates how the stimulation site can be planned with respect to an individual's anatomic magnetic resonance imaging (MRI) scan using neuronavigated positioning based on an individual's MRI scan.

to be able to directly stimulate deeper brain target sites if that were feasible. Recently, this has become possible using a method known as transcranial focused ultrasound stimulation (fUS). Low-energy low-frequency fUS has been shown to be safe for use in humans. Importantly, the effects of fUS can, depending on the stimulation protocol, either increase, decrease, or interfere with neural activity within a targeted focal brain area, and these effects can last for a sustained period after stimulation has ceased. This approach has considerable potential for use both in understanding further the brain processes associated with the production of tics and in developing new forms of treatment based on effective NIBS (Fig. 25.4).

A second challenge to the therapeutic use of NIBS for the treatment of tic disorders is our limited mechanistic understanding of the effects that stimulation induces in the brains of individuals with TS. For brevity, we will confine our discussion here to two important issues linked to some common assumptions made with respect to clinical use of NIBS. First, the assumption that a one-size-fits-all approach to stimulation, in which everybody gets the same kind of stimulation without reference to that individual's current brain state, will be effective in delivering a reliable and consistent therapeutic benefit. Second, that stimulation parameters and protocols developed in neurotypical individuals (most commonly undergraduate students) will be appropriate for use in the treatment of individuals with brain health conditions (in which there is likely to be, for a number of reasons, chronic alterations in the structure and/or function of brain networks).

The effects of noninvasive brain stimulation are dependent on current brain state

There is now considerable evidence to indicate that the effects of NIBS are dependent upon the current brain state at the time of stimulation. For example, investigations of cortical metaplasticity (for review, see Müller-Dahlhaus and Ziemann, 2015) have demonstrated that the direction of plasticity induction following NIBS can be modified by preceding "priming" stimulation. Murakami et al. (2012) demonstrated that while iTBS increased motor cortical excitability (indexed by TMS-evoked MEPs), the direction of this effect could be altered by a preceding period of priming stimulation. Specifically, the pairing of identical iTBS protocols (i.e., prime → test) resulted in suppression of the nonprimed effect of iTBS (i.e., a reduction in TMS-evoked MEP amplitude). A second illustrative example comes from studies that have examined how the amplitude of TMS-evoked MEP is modulated by the timing of stimulation, relative to the phase of brain oscillations within the stimulated brain area. Brain oscillations are known to organize cortical processing and have been shown to influence perception, cognition, and behavior. In the sensorimotor cortex, α/μ-band (8–12 Hz) and β-band (13–30 Hz) oscillations are linked to the initiation of movements and alterations in movement-related cortical excitability is linked to the phase of ongoing oscillations. It is likely that this association between sensorimotor cortical excitability and the phase of movement-related brain oscillations results from time-varying bursts of synchronized firing in neuronal populations. This suggests that the effects of NIBS on TMS-evoked MEPs might be highly dependent on the timing of stimulation with respect to the phase on movement-related brain oscillations. This was examined in a recent study by Schaworonkow et al. (2019). They used real-time EEG analysis to trigger TMS-evoked MEPs during instantaneous brain-states corresponding to the phase of the μ-rhythm. TMS stimulation was delivered at different intensities on different trials permitting the researchers to examine the TMS IO curve for stimulation delivered at the positive or negative peak (trough) of the ongoing μ-rhythm. The results of this study clearly demonstrate that MEP amplitudes were modulated by the phase of the μ-rhythm, with larger MEPs associated with the negative peak leading to a leftward shift of the MEP IO curve. However, it is important to note that these effects may be frequency-dependent, as another recent study has demonstrated that while phase-dependent modulation of TMS-evoked MEP amplitude was observed for both μ and α oscillations, the relationship between MEP amplitude and oscillatory phase was opposite for μ and β oscillations (Wischnewski et al., 2022). Specifically, while the effect reported by Schaworonkow et al. (2019) was replicated, that is, TMS-evoked MEP amplitude was larger when stimulation coincided with the negative peak (trough) and rising edge of the μ-rhythm, the opposite relationship was observed for the beta oscillation. Specifically, that larger MEP amplitudes were observed when stimulation was coincided with the positive peak and falling edge of the beta oscillation (Wischnewski et al., 2022). In this context, it is of interest to note that deep brain stimulation of the globus pallidus internus (GPi), a surgical technique used to successfully reduce tics in TS, successfully reduced tic-related MEPs in an animal model of TS (Mccairn et al., 2013). Importantly, this study demonstrated that the underlying mechanism for suppression of tic-related motor activity is through the temporal locking of GPi neuronal firing with the timing of DBS stimulation pulses. Potential implications of these findings are that NIBS time-locked to brain oscillations of interest (i.e., closed-loop control of stimulation timing) might increase the efficacy and consistency of NIBS effects. By contrast, these findings may go some way to explain why open-loop NIBS (i.e., stimulation without reference to current brain state) leads to high levels of individual and between-session variability in the effects of NIBS.

CONCLUSION

The efficacy of deep brain stimulation (DBS) in reducing tics in TS is a clear proof of the concept that modulation of the brain networks involved in the genesis of tics using brain stimulation is an effective treatment for tic disorders and offers an attractive alternative to drug treatments. Nonetheless, DBS is an invasive surgical procedure that carries serious risks, is expensive, and is not considered suitable for use with children and adolescents. For these reasons, effective forms of NIBS could be transformative for individuals with tic disorders.

While current attempts to use noninvasive brain imaging approaches such as TMS or tDCS have thus far proven unconvincing, we should certainly not give up on seeking effective noninvasive approaches. First, as this chapter makes clear, aligning stimulation to coincide with ongoing brain activity will likely go a long way to improving the efficacy of stimulation and reducing the variability of its effects. Second, new forms of stimulation such as low-energy fUS and temporal interference stimulation (a novel form of noninvasive electrical stimulation based on delivering targeted overlapping high-frequency alternating currents) make the direct noninvasive stimulation of deep brain areas an exciting and highly promising new direction for TS research. Finally, new developments in neuroscience, neurotechnology, and machine learning suggest that novel sensors, coupled with wearable neuromodulation devices, may soon offer individuals with TS the possibility of effective control over tics on demand and outside of the clinic.

References

Asamoah B, Khatoun A, McLaughlin M (2019). tACS motor system effects can be caused by transcutaneous stimulation of peripheral nerves. Nat Commun 10: 1–16.

Barker AT, Jalinous R, Freeston IL (1985). Non-invasive magnetic stimulation of human motor cortex. Lancet 325: 1106–1107.

Batschelett MA, Huddleston DA, Crocetti D et al. (2023). Biomarkers of tic severity in children with Tourette syndrome: motor cortex inhibition measured with transcranial magnetic stimulation. Dev Med Child Neurol 65: 1321.

Behler N, Leitner B, Mezger E et al. (2018). Cathodal tDCS over motor cortex does not improve Tourette syndrome: lessons learned from a case series. Front Behav Neurosci 12: 359484.

Bestmann S, Baudewig J, Siebner HR et al. (2004). Functional MRI of the immediate impact of transcranial magnetic stimulation on cortical and subcortical motor circuits. Eur J Neurosci 19: 1950–1962.

Bestmann S, Baudewig J, Siebner HR et al. (2005). BOLD MRI responses to repetitive TMS over human dorsal premotor cortex. Neuroimage 28: 22–29.

Bestmann S, Duque J (2016). Transcranial magnetic stimulation: decomposing the processes underlying action preparation. Neuroscientist 22: 392–405.

Bestmann S, Krakauer JW (2015). The uses and interpretations of the motor-evoked potential for understanding behaviour. Exp Brain Res 233: 679–689.

Bloch Y, Arad S, Levkovitz Y (2016). Deep TMS add-on treatment for intractable Tourette syndrome: a feasibility study. World J Biol Psychiatry 17: 557–561.

Bohlhalter S, Goldfine A, Matteson S et al. (2006). Neural correlates of tic generation in Tourette syndrome: an event-related functional MRI study. Brain 129: 2029–2037.

Boucher PO, Ozdemir RA, Momi D et al. (2021). Sham-derived effects and the minimal reliability of theta burst stimulation. Sci Rep 11: 1–12.

Brinkman L, Stolk A, Dijkerman HC et al. (2014). Distinct roles for alpha- and beta-band oscillations during mental simulation of goal-directed actions. J Neurosci 34: 14783–14792.

Brinkman L, Stolk A, Marshall TR et al. (2016). Independent causal contributions of alpha- and beta-band oscillations during movement selection. J Neurosci 36: 8726–8733.

Buchholz VN, Jensen O, Medendorp WP (2014). Different roles of alpha and beta band oscillations in anticipatory sensorimotor gating. Front Hum Neurosci 8: 1–9.

Carvalho S, Gonçalves ÓF, Soares JM et al. (2015). Sustained effects of a neural-based intervention in a refractory case of Tourette syndrome. Brain Stimul 8: 657–659.

Chae JH, Nahas Z, Wassermann E et al. (2004). A pilot safety study of repetitive transcranial magnetic stimulation (rTMS) in Tourette's syndrome. Cognit Behav Neurol 17: 109–117.

Chen R, Classen J, Gerloff C et al. (1997). Depression of motor cortex excitability by low-frequency transcranial magnetic stimulation. Neurology 48: 1398–1403.

Chung SW, Hill AT, Rogasch NC et al. (2016). Use of theta-burst stimulation in changing excitability of motor cortex: a systematic review and meta-analysis. Neurosci Biobehav Rev 63: 43–64.

Creutzfeldt OD, Fromm GH, Kapp H (1962). Influence of transcortical d-c currents on cortical neuronal activity. Exp Neurol 5: 436–452.

Delorme C, Salvador A, Valabrègue R et al. (2016). Enhanced habit formation in Gilles de la Tourette syndrome. Brain 139: 605–615.

Deng Z-D, Lisanby SH, Peterchev AV (2013). Electric field depth–focality tradeoff in transcranial magnetic stimulation: simulation comparison of 50 coil designs. Brain Stimul 6: 1–13.
Draper A, Jackson GM, Morgan PS et al. (2016). Premonitory urges are associated with decreased grey matter thickness within the insula and sensorimotor cortex in young people with Tourette syndrome. J Neuropsychol 10: 143–153.
Draper A, Jude L, Jackson GM et al. (2015). Motor excitability during movement preparation in Tourette syndrome. J Neuropsychol 9: 33–44.
Draper A, Stephenson MC, Jackson GM et al. (2014). Increased GABA contributes to enhanced control over motor excitability in Tourette syndrome. Curr Biol 24: 2343–2347.
Dyke K, Jackson G, Jackson S (2021). Non-invasive brain stimulation as therapy: systematic review and recommendations with a focus on the treatment of Tourette syndrome. Exp Brain Res.
Dyke K, Jackson GM, Nixon E et al. (2019). Effects of single-session cathodal transcranial direct current stimulation on tic symptoms in Tourette's syndrome. Exp Brain Res 237: 2853.
Dyke K, Kim S, Jackson GM et al. (2016). Intra-subject consistency and reliability of response following 2 mA transcranial direct current stimulation. Brain Stimul 9: 819–825.
Eapen V, Baker R, Walter A et al. (2017). The role of transcranial direct current stimulation (tDCS) in Tourette syndrome: a review and preliminary findings. Brain Sci 7: 161–174.
Eidelberg D, Moeller JR, Antonini A et al. (1997). The metabolic anatomy of Tourette's syndrome. Neurology 48: 927–934.
Engel AK, Fries P (2010). Beta-band oscillations – signalling the status quo? Curr Opin Neurobiol 20: 156–165.
Franzkowiak S, Pollok B, Biermann-Ruben K et al. (2012). Motor-cortical interaction in Gilles de la Tourette syndrome. PLoS One 7: e27850.
Ganos C, Asmuss L, Bongert J et al. (2015). Volitional action as perceptual detection: predictors of conscious intention in adolescents with tic disorders. Cortex 64: 47–54.
Gilbert DL, Bansal AS, Sethuraman G et al. (2004). Association of cortical disinhibition with tic, ADHD, and OCD severity in Tourette syndrome. Mov Disord 19: 416–425.
Gilbert DL, Sallee FR, Zhang J et al. (2005). Transcranial magnetic stimulation-evoked cortical inhibition: a consistent marker of attention-deficit/hyperactivity disorder scores in Tourette Syndrome. Biol Psychiatry 57: 1597–1600.
Gilbertson T, Lalo E, Doyle L et al. (2005). Existing motor state is favored at the expense of new movement during 13-35 Hz oscillatory synchrony in the human corticospinal system. J Neurosci 25: 7771–7779.
Goldsworthy MR, Hordacre B, Rothwell JC et al. (2021). Effects of rTMS on the brain: is there value in variability? Cortex 139: 43–59.
Haber SN (2003). The primate basal ganglia: parallel and integrative networks. J Chem Neuroanat 26: 317–330.
Hagenacker T, Bude V, Naegel S et al. (2014). Patient-conducted anodal transcranial direct current stimulation of the motor cortex alleviates pain in trigeminal neuralgia. J Headache Pain 15: 78.
Hamada M, Murase N, Hasan A et al. (2012). The role of interneuron networks in driving human motor cortical plasticity. Cereb Cortex 23: 1593–1605.
Hampson M, Tokoglu F, King RA et al. (2009). Brain areas coactivating with motor cortex during chronic motor tics and intentional movements. Biol Psychiatry 65: 594–599.
Hanajima R, Ugawa Y (2008). Paired-pulse measures. Oxford handbook of transcranial stimulation. Oxford University Press.
He JL, Mikkelsen M, Huddleston DA et al. (2022). Frequency and intensity of premonitory urges-to-tic in Tourette syndrome is associated with supplementary motor area GABA+ levels. Mov Disord 37: 563–573.
Heise KF, Steven B, Liuzzi G et al. (2010). Altered modulation of intracortical excitability during movement preparation in Gilles de la Tourette syndrome. Brain 133: 580–590.
Horvath JC, Vogrin SJ, Carter O et al. (2016). Effects of a common transcranial direct current stimulation (tDCS) protocol on motor evoked potentials found to be highly variable within individuals over 9 testing sessions. Exp Brain Res 234: 2629–2642.
Houlgreave MS, Morera Maiquez B, Brookes MJ et al. (2022). The oscillatory effects of rhythmic median nerve stimulation. Neuroimage 251: 118990.
Iverson AM, Arbuckle AL, Ueda K et al. (2023). Median nerve stimulation for treatment of tics: randomized, controlled, crossover trial. J Clin Med 12: 2514.
Jackson SR, Loayza J, Crighton M et al. (2020). The role of the insula in the generation of motor tics and the experience of the premonitory urge-to-tic in Tourette syndrome. Cortex 126: 119–133.
Jackson SR, Parkinson A, Kim SY et al. (2011). On the functional anatomy of the urge-for-action. Cogn Neurosci 2: 227–243.
Jackson SR, Parkinson A, Manfredi V et al. (2013). Motor excitability is reduced prior to voluntary movements in children and adolescents with Tourette syndrome. J Neuropsychol 7: 29–44.
Jackson SR, Sigurdsson HP, Dyke K et al. (2021). The role of the cingulate cortex in the generation of motor tics and the experience of the premonitory urge-to-tic in Tourette syndrome. J Neuropsychol 15: 340–362.
Jenkinson N, Brown P (2011). New insights into the relationship between dopamine, beta oscillations and motor function. Trends Neurosci 34: 611–618.
Jensen O, Mazaheri A (2010). Shaping functional architecture by oscillatory alpha activity: gating by inhibition. Front Hum Neurosci 4: 186.
Joundi RA, Jenkinson N, Brittain JS et al. (2012). Driving oscillatory activity in the human cortex enhances motor performance. Curr Biol 22: 403–407.
Jung SH, Shin JE, Jeong YS et al. (2008). Changes in motor cortical excitability induced by high-frequency repetitive transcranial magnetic stimulation of different stimulation durations. Clin Neurophysiol 119: 71–79.
Jurkiewicz MT, Gaetz WC, Bostan AC et al. (2006). Post-movement beta rebound is generated in motor cortex: evidence from neuromagnetic recordings. Neuroimage 32: 1281–1289.
Kahl CK, Kirton A, Pringsheim T et al. (2021). Bilateral transcranial magnetic stimulation of the supplementary motor area in children with Tourette syndrome. Dev Med Child Neurol 63: 808–815.
Kasschau M, Reisner J, Sherman K et al. (2016). Transcranial direct current stimulation is feasible for remotely supervised home delivery in multiple sclerosis. Neuromodulation 19: 824–831.
Kwon HJ, Lim WS, Lim MH et al. (2011). 1-Hz low frequency repetitive transcranial magnetic stimulation in children with Tourette's syndrome. Neurosci Lett 492: 1–4.
Landeros-Weisenberger A, Mantovani A, Motlagh MG et al. (2015). Randomized sham controlled double-blind trial of repetitive

transcranial magnetic stimulation for adults with severe Tourette syndrome. Brain Stimul 8: 574–581.

Le K, Liu L, Sun M et al. (2013). Transcranial magnetic stimulation at 1 Hertz improves clinical symptoms in children with Tourette syndrome for at least 6 months. J Clin Neurosci 20: 257–262.

Leckman JF, Riddle MA, Hardin MT et al. (1989). The Yale Global Tic Severity Scale: initial testing of a clinician-rated scale of tic severity. J Am Acad Child Adolesc Psychiatry 28: 566–573.

Liebetanz D, Nitsche MA, Tergau F et al. (2002). Pharmacological approach to the mechanisms of transcranial DC-stimulation-induced after-effects of human motor cortex excitability. Brain 125: 2238–2247.

López-Alonso V, Cheeran B, Río-Rodríguez D et al. (2014). Inter-individual variability in response to non-invasive brain stimulation paradigms. Brain Stimul 7: 372–380.

Maeda F, Keenan JP, Tormos JM et al. (2000). Interindividual variability of the modulatory effects of repetitive transcranial magnetic stimulation on cortical excitability. Exp Brain Res 133: 425–430.

Maiquez BM, Jackson GM, Jackson SR (2020). Entraining movement-related brain oscillations using rhythmic median nerve stimulation. bioRxiv 2020.03.30.016097.

Maiquez BM, Smith C, Dyke K et al. (2023). A double-blind, sham-controlled, trial of home-administered rhythmic 10-Hz median nerve stimulation for the reduction of tics, and suppression of the urge-to-tic, in individuals with Tourette syndrome and chronic tic disorder. J Neuropsychol 17: 540–563.

Mantovani A, Leckman JF, Grantz H et al. (2007). Repetitive transcranial magnetic stimulation of the supplementary motor area in the treatment of Tourette syndrome: report of two cases. Clin Neurophysiol 118: 2314–2315.

Mantovani A, Lisanby SH, Pieraccini F et al. (2006). Repetitive transcranial magnetic stimulation (rTMS) in the treatment of obsessive-compulsive disorder (OCD) and Tourette's syndrome (TS). Int J Neuropsychopharmacol 9: 95–100.

Martins CW, Carla L, Rodrigues M et al. (2019). AMPA receptors are involved in prefrontal direct current stimulation effects on long-term working memory and GAP-43 expression. Behav Brain Res 362: 208–212.

Mccairn KW, Iriki A, Isoda M (2013). Deep brain stimulation reduces tic-related neural activity via temporal locking with stimulus pulses. Neurobiol Dis 33: 6581–6593.

Moll GH, Wischer S, Heinrich H et al. (1999). Deficient motor control in children with tic disorder: evidence from transcranial magnetic stimulation. Neurosci Lett 272: 37–40.

Morera Maiquez B, Sigurdsson HP, Dyke K et al. (2020). Entraining movement-related brain oscillations to suppress tics in Tourette syndrome. Curr Biol 30: 2334–2342.e3.

Mrakic-Sposta S, Marceglia S, Mameli F et al. (2008). Transcranial direct current stimulation in two patients with Tourette syndrome. Mov Disord 23: 2259–2261.

Müller-Dahlhaus F, Ziemann U (2015). Metaplasticity in human cortex. Neuroscientist 21: 185–202.

Münchau A, Bloem BR, Thilo KV et al. (2002). Repetitive transcranial magnetic stimulation for Tourette syndrome. Neurology 59: 1789–1791.

Murakami T, Müller-Dahlhaus F, Lu M-K et al. (2012). Homeostatic metaplasticity of corticospinal excitatory and intracortical inhibitory neural circuits in human motor cortex. J Physiol Neurosci 590: 5765–5781.

Murphy KA, Morrisonponce D (2022). Anatomy, shoulder and upper limb, median nerve. StatPearls.

Nitsche MA, Fricke K, Henschke U et al. (2003). Pharmacological modulation of cortical excitability shifts induced by transcranial direct current stimulation in humans. J Physiol 553: 293–301.

Nitsche MA, Liebetanz D, Schlitterlau A et al. (2004). GABAergic modulation of DC stimulation-induced motor cortex excitability shifts in humans. Eur J Neurosci 19: 2720–2726.

Nitsche MA, Paulus W (2000). Excitability changes induced in the human motor cortex by weak transcranial direct current stimulation. J Physiol 527: 633.

Orth M, Amann B, Robertson MM et al. (2005). Excitability of motor cortex inhibitory circuits in Tourette syndrome before and after single dose nicotine. Brain 128: 1292–1300.

Orth M, Kirby R, Richardson MP et al. (2005). Subthreshold rTMS over pre-motor cortex has no effect on tics in patients with Gilles de la Tourette syndrome. Clin Neurophysiol 116: 764–768.

Orth M, Münchau A, Rothwell JC (2008). Corticospinal system excitability at rest is associated with tic severity in Tourette syndrome. Biol Psychiatry 64: 248–251.

Orth M, Rothwell JC (2008). Motor cortex excitability and comorbidity in Gilles de la Tourette syndrome. J Neurol Neurosurg Psychiatry 80: 29–34.

Ozdemir RA, Boucher P, Fried PJ et al. (2021). Reproducibility of cortical response modulation induced by intermittent and continuous theta-burst stimulation of the human motor cortex. Brain Stimul 14: 949–964.

Pascual-Leone A, Valls-Sole J, Wassermann EM et al. (1994). Responses to rapid-rate transcranial magnetic stimulation of the human motor cortex. Brain 117: 847–858.

Passmore SR, Murphy B, Lee TD (2014). The origin, and application of somatosensory evoked potentials as a neurophysiological technique to investigate neuroplasticity. J Can Chiropr Assoc 58: 170–183.

Pépés SE, Draper A, Jackson GM et al. (2016). Effects of age on motor excitability measures from children and adolescents with Tourette syndrome. Dev Cogn Neurosci 19: 78.

Pfurtscheller G, Stancák A, Neuper C (1996). Post-movement beta synchronization. A correlate of an idling motor area? Electroencephalogr Clin Neurophysiol 98: 281–293.

Picard N, Strick PL (2001). Imaging the premotor areas. Curr Opin Neurobiol 11: 663–672.

Pogosyan A, Gaynor LD, Eusebio A et al. (2009). Boosting cortical activity at beta-band frequencies slows movement in humans. Curr Biol 19: 1637–1641.

Poreisz C, Boros K, Antal A et al. (2007). Safety aspects of transcranial direct current stimulation concerning healthy subjects and patients. Brain Res Bull 72: 208–214.

Purpura DP, Mcmurtry JG (1965). Intracellular activities and evoked potential changes during polarization of motor cortex. J Neurophysiol 28: 166–185.

Quinn A, Lopes-dos-Santos V, Dupret D et al. (2021). EMD: empirical mode decomposition and Hilbert-Huang spectral analyses in Python. J Open Source Softw 6: 2977.

Schaworonkow N, Triesch J, Ziemann U et al. (2019). EEG-triggered TMS reveals stronger brain state-dependent modulation of motor evoked potentials at weaker stimulation intensities. Brain Stimul 12: 110–118.

Schwippel T, Wasserka B, Fallgatter AJ et al. (2017). Safety and efficacy of long-term home treatment with transcranial direct current stimulation (tDCS) in a case of multimodal hallucinations. Brain Stimul 10: 873–874.

Singh S, Kumar S, Kumar N et al. (2018). Low-frequency repetitive transcranial magnetic stimulation for treatment of Tourette syndrome:

a naturalistic study with 3 months of follow-up. Indian J Psychol Med 40: 482.

Stafford J, Brownlow ML, Qualley A et al. (2018). AMPA receptor translocation and phosphorylation are induced by transcranial direct current stimulation in rats. Neurobiol Learn Mem 150: 36–41.

Stagg CJ, Best JG, Stephenson MC et al. (2009). Polarity-sensitive modulation of cortical neurotransmitters by transcranial stimulation. J Neurosci 29: 5202–5206.

Stagg CJ, Nitsche MA (2011). Physiological basis of transcranial direct current stimulation. Neuroscientist 17: 37–53.

Tajadini S, Mohammadi N, Tahamtan M et al. (2018). Cathodal transcranial direct current stimulation to ameliorate the frequency and severity of motor tics: a case study of Tourette syndrome. J Adv Med Sci Appl Technol 4: 21–26.

Thut G, Schyns PG, Gross J (2011). Entrainment of perceptually relevant brain oscillations by non-invasive rhythmic stimulation of the human brain. Front Psychol 2: 170.

Thut G, Veniero D, Romei V et al. (2011). Rhythmic TMS causes local entrainment of natural oscillatory signatures. Curr Biol 21: 1176–1185.

Touge T, Gerschlager W, Brown P et al. (2001). Are the after-effects of low-frequency rTMS on motor cortex excitability due to changes in the efficacy of cortical synapses? Clin Neurophysiol 112: 2138–2145.

Vallence AM, Goldsworthy MR, Hodyl NA et al. (2015). Inter- and intra-subject variability of motor cortex plasticity following continuous theta-burst stimulation. Neuroscience 304: 266–278.

Voon V, Derbyshire K, Rück C et al. (2015). Disorders of compulsivity: a common bias towards learning habits. Mol Psychiatry 20: 345–352.

Wiethoff S, Hamada M, Rothwell JC (2014). Variability in response to transcranial direct current stimulation of the motor cortex. Brain Stimul 7: 468–475.

Wischnewski M, Haigh ZJ, Shirinpour S et al. (2022). The phase of sensorimotor mu and beta oscillations has the opposite effect on corticospinal excitability. Brain Stimul 15: 1093–1100.

Worbe Y, Marrakchi-Kacem L, Lecomte S et al. (2015). Altered structural connectivity of cortico-striato-pallido-thalamic networks in Gilles de la Tourette syndrome. Brain 138: 472–482.

Wu SW, Maloney T, Gilbert DL et al. (2014). Functional MRI-navigated repetitive transcranial magnetic stimulation over supplementary motor area in chronic tic disorders. Brain Stimul 7: 212–218.

Ziemann U (2004). TMS induced plasticity in human cortex. Rev Neurosci 15: 253–266.

Ziemann U (2013). Pharmaco-transcranial magnetic stimulation studies of motor excitability. Handb Clin Neurol 116: 387–397.

Ziemann U, Paulus W, Rothenberger A (1997). Decreased motor inhibition in Tourette's disorder: evidence from transcranial magnetic stimulation. Am J Psychiatry 154: 1277–1284.

Handbook of Clinical Neurology, Vol. 215 (3rd series)
Tic Disorders and Tourette Syndrome
C. Ganos and Y. Worbe, Editors
https://doi.org/10.1016/B978-0-443-13554-5.00008-0

Chapter 26

Beneath the tip of the iceberg: treatment of neuropsychiatric comorbidities in tic disorders

ELIA ABI-JAOUDE[1,2,*], DANIEL A. GORMAN[1], AND PAUL SANDOR[2]

[1]*Department of Psychiatry, Hospital for Sick Children, University of Toronto, Toronto, ON, Canada*

[2]*Department of Psychiatry, Toronto Western Hospital – University Health Network, University of Toronto, Toronto, ON, Canada*

Abstract

While Tourette syndrome (TS) is defined by tics, comorbidities are very much part of the clinical presentation – 90% of patients having a least one associated condition, most prominently obsessive-compulsive disorder (OCD) and attention-deficit/hyperactivity disorder (ADHD), but also learning disabilities, anxiety, depression, rage outbursts, autism spectrum disorder, sensory hypersensitivities, and sleep difficulties. Moreover, the associated symptoms often are more problematic than tics, or interact with the tics such that the different symptoms exacerbate each other, adding complexity to the clinical picture. Thus, to provide optimal care to patients with TS, it is essential that clinicians consider and address associated symptoms as part of their management. In this chapter, we review the treatment of the common neuropsychiatric comorbidities in TS, focusing especially on OCD and ADHD. We describe cognitive, behavioral, and pharmacologic interventions, with the aim of making these clear and applicable for the nonspecialist clinician. The goal is for clinicians to provide holistic care that includes understanding the impact of various symptoms and collaboratively prioritizing targets for intervention while favoring conservative approaches, in order to improve the quality of life of patients with TS.

INTRODUCTION

While Tourette syndrome (TS) is defined by tics, neuropsychiatric comorbidities are strongly associated with the condition. In clinical populations, an estimated 90% of patients have at least one comorbidity (Robertson et al., 2017). The two conditions most closely associated with TS are obsessive-compulsive disorder (OCD) and attention-deficit/hyperactivity disorder (ADHD) (Gorman et al., 2010; Hirschtritt et al., 2015). In fact, most patients with TS will have symptoms of these disorders, with half meeting full criteria for OCD, half meeting full criteria for ADHD, and almost three quarters meeting criteria for either (Hirschtritt et al., 2015; Robertson et al., 2017). Other neuropsychiatric comorbidities are also common in TS, including learning disabilities, anxiety, depression, rage outbursts, autism spectrum disorder (ASD), sensory hypersensitivities, and sleep difficulties (Hirschtritt et al., 2015; Robertson et al., 2017).

Thus, TS is beyond being just a movement disorder – the presence of tics is often suggestive of difficulties in self-control across a broad range of domains, including cognitive, emotional, behavioral, and sensory (Murphy and Muter, 2012; Martino and Mink, 2013; Kurvits et al., 2020). Furthermore, it is not uncommon that the tics are not themselves problematic, but it is the other symptoms that are the most challenging for the patient (Silvestri et al., 2018). Of note, the symptoms that are most problematic can change over time (Hirschtritt et al., 2015; Groth et al., 2017). Also noteworthy is that various symptoms can interact and exacerbate one another in myriad ways, such as anxiety and sleep difficulties contributing to increase in tics and symptoms of OCD and ADHD, which in turn could worsen anxiety and sleep difficulties, resulting in a vicious cycle.

It is therefore important for the clinician to carry out a comprehensive, holistic assessment when seeing patients presenting with tics. This includes understanding the impact of various symptoms in their lives, and collaboratively developing an initial management plan that prioritizes targets for intervention. Indeed, one should be careful not to overwhelm the patient with too many interventions that are not feasible at the same time, or pursue medications liberally resulting in burdensome

[*]Correspondence to: Elia Abi-Jaoude, MD, PhD, FRCPC, Department of Psychiatry, Hospital for Sick Children, 555 University Avenue, Toronto, ON M5G 1X8, Canada. Tel: +1-416-813-7382, E-mail: elia.abi.jaoude@utoronto.ca.

polypharmacy or confusion as to the effects (positive and negative) of different agents. Rather, target symptoms should be prioritized based on severity and, even more importantly, associated impairment. Treatment plans can then be adjusted based on response and the evolution of symptoms over time. The goal should not be elimination of all symptoms, but rather mitigation of their impact on the patient's function and quality of life.

In this chapter, we review treatment of the common neuropsychiatric comorbidities in TS. We focus on the two comorbidities most closely related to TS, OCD, and ADHD, after which other comorbidities are covered more briefly. In most instances, the interventions are the same as when these conditions occur without TS. However, we highlight situations where the TS context can influence decisions about how to approach the comorbid neuropsychiatric symptoms.

OBSESSIVE-COMPULSIVE DISORDER

As discussed in Chapter 7, childhood-onset OCD is closely linked with TS: the conditions are often co-inherited; they implicate overlapping neural substrates from the cortico-striato-thalamo-cortical (CSTC) circuits (Martino and Mink, 2013); and, phenomenologically, obsessions and compulsions, similar to tics, generally involve actions and related experiences that are unwanted yet the patient feels compelled to engage in (Martino and Mink, 2013; Kurvits et al., 2020). It can sometimes be challenging to differentiate complex tics from compulsions. A useful general guide is that while tics primarily involve sensorimotor experiences, obsessive-compulsive symptoms (OCS) primarily involve cognitive and emotional processes. In other words, the urge to tic is physical, whereas the urge to perform a compulsion is mental. Still, in some cases there can be overlap, for example, a motor tic, driven by a sensory urge, needs to be carried out according to a cognitive rule, such as in response to an external stimulus or a certain number of times.

Given the phenomenological similarities between tics and OCS, there is much overlap in their first-line treatments – behavioral therapy that essentially involves progressively resisting the symptoms. As discussed in Chapter 24, the main behavioral treatments for tics are habit reversal training and exposure and response prevention (ERP), approaches that have much in common with each other. In the case of OCD, the mainstay behavioral therapy is ERP, with an added cognitive component – thus, ERP for OCD is delivered as part of cognitive behavioral therapy (CBT) (Overview | Obsessive-compulsive disorder and body dysmorphic disorder, 2005; Practice parameter for the assessment and treatment of children and adolescents with obsessive-compulsive disorder, 2012).

When it comes to pharmacotherapy, greater differences exist in the approaches to tics and OCS. Whereas the main medications used for tics are α-2 agonists and antipsychotics, the main medications for OCD are selective serotonin reuptake inhibitors (SSRIs).

In the sections that follow, we start by reviewing the importance of providing psychoeducation and support for patients with TS and OCS. We then review CBT/ERP and the use of SSRIs, clomipramine, and antipsychotic augmentation.

Psychoeducation and support

As with TS more generally, psychoeducation, reassurance, and support have a crucial role for patients presenting with comorbid OCS (Overview | Obsessive-compulsive disorder and body dysmorphic disorder, 2005). For many patients and their families, OCS can be perplexing and can engender feelings of shame, guilt, and frustration. Many patients may not realize that their experiences are common, and that they have a known neuropsychiatric condition as opposed to a unique personal oddity or moral weakness. As such, patients will often go to great lengths to keep their OCS private, sometimes even from their own families. For example, experiences of ego-dystonic obsessions (i.e., intrusive thoughts that are felt to be repugnant, distressing, unacceptable, and inconsistent with one's desires) that involve sexual themes can be very troubling for the patient. In addition, delays in task completion due to rituals, or repeated reassurance-seeking, can be experienced by others as the patient being difficult. Families often accommodate ritualistic behaviors in a well-intentioned attempt to be empathic or avert emotional outbursts, but this tends to perpetuate the patient's OCS (Przeworski and Birnkrant, 2017). Thus, making the diagnosis and explaining how common OCS are in general and especially in TS can be validating and reassuring. It is also worth providing information resources on OCS, including developmentally appropriate books and websites.

Cognitive behavioral therapy

There is extensive evidence for CBT as treatment for OCD, which has more randomized controlled trials (RCTs) and larger effect sizes compared to medication (Watson and Rees, 2008; Romanelli et al., 2014; Ivarsson et al., 2015; McGuire et al., 2015). While some of the RCTs used a waitlist control, the superiority of CBT remains even when one considers only those RCTs that used an active comparator (e.g., psychoeducation and support).

CBT is based on the idea that thoughts, feelings, and behaviors are related to and influence each other; thus, dysfunction in one of these domains contributes to dysfunction in the others, and, importantly, targeting one of these can influence the others and result in overall improvement. In CBT, as the name implies, thoughts and behaviors are targets for intervention. However, whereas CBT for other conditions, such as depression and anxiety, may include a strong cognitive component, CBT for OCS emphasizes the behavioral component in the form of ERP (Berman et al., 2017).

The cognitive behavioral model of OCS is that obsessions and related anxiety are transiently relieved by compulsive rituals, resulting in symptom maintenance through negative reinforcement (Berman et al., 2017). In other words, a behavior that eliminates a negative experience, in this case the compulsion transiently eliminating the obsession and related anxiety, is more likely to occur again. As such, engaging in ERP serves to break the cycle of reinforcement and allow for habituation to the obsession, leading to reduced symptoms. The approach involves developing a hierarchy of OCS such that patients can start with the easiest symptoms to target. In a planned and systematic manner, the patient deliberately engages in exposure to something that triggers the OCS being targeted (e.g., touching a door handle in the case of a contamination obsession) and refrains from engaging in the relieving ritual (e.g., compulsive handwashing). The exposures need to be frequent and sustained to allow for adequate habituation with diminishing need to engage in the compulsion (Berman et al., 2017). Once this is achieved for a particular OCS, the patient moves to the next target OCS, thus gradually progressing up the hierarchy to increasingly challenging OCS. Success at lower levels in the hierarchy weakens distress at higher levels. Relaxation training can be employed as needed throughout the process to help the patient tolerate the exposures.

While the focus of CBT for OCD is ERP (i.e., behavioral), addressing the cognitive aspects of OCD is also important. In addition to providing the rationale for the patient to engage in ERP, cognitive therapy aims to address the cognitive distortions that are commonly part of OCD (Berman et al., 2017; Przeworski and Birnkrant, 2017). Table 26.1 lists cognitive distortions with examples.

The distortions are addressed by being recognized, labeled, and challenged with education and Socratic questioning (Berman et al., 2017; Przeworski and Birnkrant, 2017). This also supports efforts for the patient to engage the ERP, which is where the most progress will take place.

The role of parents (in the case of a child) and other family members (such as a spouse in the case of an adult) is crucial for supporting efforts with CBT, in particular ERP. This includes consistent and firm limit setting, while remaining calm and supportive. Modeling is important, particularly as a parent may themselves have OCD. Parents should also refrain from accommodations that enable OCS in the child, especially for the symptoms being actively targeted with ERP. Indeed, family accommodation of OCS is associated with increased functional impairment (Storch et al., 2010). Parents and other family members should also recognize efforts at ERP by the patient, providing them with encouragement, praise, and even small material rewards to help reinforce their efforts. Thus, parents and other family members need to be fully engaged in the treatment along with the patient for it to be successful.

In the case of limited engagement with or benefit from CBT, an alternative approach is acceptance and commitment therapy (ACT) (Berman et al., 2017). Rather than targeting symptoms directly, ACT uses acceptance and mindfulness strategies with the goal of mitigating the impact of the

Table 26.1

Cognitive distortions.

Cognitive distortion	Description	Example
Intolerance of uncertainty	Intolerance of even the slightest possibility of a negative outcome; considered a core distortion in obsessive-compulsive disorder	Stove being left on and leading to a fire due to not engaging in ritualistic checking
Overestimation of threat	Probability and severity of a negative outcome are exaggerated	Getting a serious illness due to contamination
Overestimation of responsibility	Belief that awareness of the threat means responsibility for preventing a feared outcome	Responsibility for preventing a loved one dying in a car accident by performing a ritual
Significance of thoughts	Ascribing inordinate importance or meaning to obsessional thoughts	Ego-dystonic obsessions that involve sexually inappropriate themes
Significance of thoughts – likelihood thought–action fusion	Belief that having thoughts makes it more likely that these thoughts will lead to action	Having violent thoughts about stabbing loved ones will lead to actually stabbing them
Significance of thoughts – moral thought–action fusion	Belief that having thoughts is as bad as the action represented by these thoughts	Believing that a sexual thought about a child is as bad as being a pedophile
Need to control thoughts	Because of excessive significance ascribed to obsessive thoughts, feeling the need to control such thoughts, often leading to increase in the thoughts	Trying to fight off sexually inappropriate thoughts, leading, paradoxically, to an increase in the thoughts
Intolerance of anxiety	Belief that unless the anxiety is addressed, such as by carrying out a ritual, it will be overwhelming	Fear that unless the stove is checked, anxiety about the possibility of a fire will be intolerable or interfere with ability to function
"Just right" phenomena	Need to engage in a ritual until it feels "just right," or until a certain degree of perfection is achieved	Needing to reread a sentence until a feeling is satisfied with regard to having fully understood it
Emotional reasoning	Belief that, regardless of rational considerations, the mere presence of a fear is indicative of a serious threat	The mere presence of fear that a fire will happen indicates this is a serious threat even in the absence of evidence to support this

symptoms on one's functioning and quality of life. For further information about the practical details of ACT for OCD, as well as the evidence for its efficacy, the reader is referred to the website of the Association for Contextual Behavioral Science (https://contextualscience.org/).

Pharmacotherapy

If treatment with CBT does not result in adequate improvement in OCS, or if it is not feasible, medication can be considered, specifically SSRIs. Even when medication is pursued, it is generally worthwhile to continue CBT for several reasons: combined treatment is superior to SSRI alone, but SSRI adds little to CBT; for SSRI partial responders, adding CBT is superior to continuing SSRI; for CBT partial responders, continuing CBT is as beneficial as adding an SSRI (Romanelli et al., 2014; Ivarsson et al., 2015).

Evidence from meta-analyses indicates that all SSRIs have comparable efficacy for OCD, with small to moderate effect size (Bridge et al., 2007; Watson and Rees, 2008; Ivarsson et al., 2015; McGuire et al., 2015; Locher et al., 2017). Rather than having any specific effects on OCD-related pathophysiology, the mechanism of action of SSRIs may be through creating a degree of emotional blunting, thus decreasing the intensity of OCS-related anxiety (Langley et al., 2023; Sklivanioti Greenfield et al., 2023). This would allow the patient to better tolerate the distress from not engaging in OCD rituals, and thus better tolerate exposures, whether they be in the context of formal ERP or in the informal context of activities of daily life.

When considering medication, the harms as well as the benefits should be considered and discussed in detail with patients and families. This is the basis for informed decision-making. Furthermore, informed patients and families are better able to recognize adverse effects and bring these to the clinician's attention so they can be addressed. Common early adverse effects include gastrointestinal upset in the form of stomach upset, nausea/vomiting and diarrhea, headache, and dizziness, which tend to improve within one to two weeks. Other physical adverse effects include decreased or increased appetite, sedation, insomnia, nightmares, diaphoresis, akathisia, increased prolactin, and sexual dysfunction (desire, erection, orgasm, or ejaculation). A rare but serious adverse effect is serotonin syndrome, which is more likely to occur in patients taking multiple serotonergic agents. Psychiatric adverse effects include apathy, irritability, mood lability, agitation, akathisia, manic reaction, psychosis, and suicidality (Lexicomp, n.d.; Cartwright et al., 2016; Sharma et al., 2016; Hengartner et al., 2021). It is also important to inform patients about withdrawal symptoms when decreasing doses or discontinuing the medication. These include headache, dizziness, nausea, flu-like symptoms, agitation, anxiety, irritability, electrical sensations ("brain zaps"), mood lability, agitation, and confusion (Horowitz and Wilcock, 2022; Sørensen et al., 2022).

Should a decision be made to proceed with a trial of SSRI medication, the main options are sertraline, fluoxetine, and fluvoxamine – see Table 26.2 for dosing details. Note that fluoxetine has a relatively long half-life (around one week or longer for its active metabolite), which means that it can take longer for an adverse reaction to resolve. While citalopram and escitalopram could be reasonable options, they both have a warning related to dose-related QTc interval prolongation, which becomes especially salient should there be concomitant use of other medications with QTc-prolonging effects, such as antipsychotics. Moreover, higher SSRI doses are often required to achieve maximal benefit (Bloch et al., 2010). We do not recommend paroxetine given its higher incidence of adverse effects especially in the context of withdrawal (Boaden et al., 2020; Quilichini et al., 2022).

The dose can be increased as needed and as tolerated according to the titration guidance in Table 26.2, taking into account patient age and symptom severity. Note that the dosing titration we recommend here is somewhat more conservative than that of other references, the goal being to mitigate adverse effects, while allowing time to assess the full impact of a dose on the OCS. In terms of dosing, while higher SSRI doses are often required to achieve maximal benefit, this often comes at the expense of decreased tolerability (Bloch et al., 2010).

If there is clear evidence of benefit, the medication should be continued, typically for at least a year. After that, if symptoms are no longer a substantial active concern, and during a period of stability, the dose can be tapered gradually. Unless there is a pressing need to decrease the dose fast, it is best to be conservative with the tapering process – decreasing in small doses at intervals of at least one month, and waiting for any

Table 26.2

Selective serotonin reuptake inhibitor dosing (Lexicomp, n.d.).

Drug	Starting dose	Titration	Maximum dose
Sertraline	25–50 mg/day	Increase by 25–50 mg daily every 2–4 weeks or longer	200 mg daily
Fluvoxamine	25–50 mg/day	Increase by 25–50 mg daily every 2–4 weeks or longer	200 mg daily in children (total dose should be divided b.i.d. if > 50 mg); 300 mg in adolescents and adults (total dose should be divided b.i.d. if > 100 mg)
Fluoxetine	10–20 mg/day	Increase by 10–20 mg daily every 2–4 weeks or longer	60 mg daily in children and adolescents; 80 mg in adults

withdrawal symptoms to have resolved prior to every subsequent decrease. An advantage of SSRIs available in liquid form (e.g., fluoxetine, sertraline) is that they allow for dose tapering in very small decrements if needed.

For patients for whom a first SSRI trial fails, a trial with a second SSRI is recommended. Should this also fail, a trial with a third SSRI or clomipramine (a tricyclic antidepressant with mainly serotonergic activity) can be considered. Evidence from meta-analyses suggest that clomipramine has greater benefit compared to SSRI, though this finding is partly explained by the fact that clomipramine studies are older, and older studies generally found greater effect sizes than newer studies independent of the agent being studied. In addition, placebo-controlled RCTs of clomipramine for pediatric OCD are few in number and generally of poor quality, with only one relatively small study satisfying inclusion criteria of a Cochrane review (Ipser et al., 2009). At any rate, even if clomipramine is more efficacious than SSRIs, it is regarded as a second-line agent after at least two SSRI trials because of its adverse effect profile and the need for blood level and cardiac monitoring (given the risk of cardiac arrhythmias). Clomipramine is also highly lethal in overdose.

Another option when an SSRI trial is inadequate is augmentation with an antipsychotic drug, specifically risperidone, aripiprazole, or haloperidol, which have evidence from a number of RCTs, though all in adults. The evidence is mixed with regard to whether patients with comorbid tics respond better or worse to antipsychotic augmentation. Regardless, one should be cautious about proceeding with augmentation with antipsychotic. There is no evidence of efficacy for antipsychotic augmentation in OCD in patients who received less than 12 weeks of maximal SSRI treatment (Bloch et al., 2006). Importantly, any benefits from antipsychotics need to be weighed against the significant harms, including extrapyramidal symptoms, tardive dyskinesia, weight gain and other metabolic effects (hyperglycemia and hyperlipidemia), hyperprolactinemia, sedation, orthostasis, anticholinergic effects, QTc prolongation, seizures, neutropenia, and in the case of aripiprazole impaired impulse-control (e.g., hypersexuality, pathologic gambling, video game, and other behavioral addictions). It should also be kept in mind that fluoxetine increases risperidone, aripiprazole, and haloperidol levels through CYP 2D6 inhibition. Therefore, when one of these antipsychotics is added to fluoxetine, the dosing should be especially cautious.

ATTENTION-DEFICIT/HYPERACTIVITY DISORDER

As covered in Chapter 8, ADHD-related symptoms are commonly part of TS. Phenomenologically, whereas OCS overlap with tics from a compulsivity perspective, symptoms of inattention, hyperactivity, and impulsivity can be seen as having in common with tics the theme of self-control. Extraneous movements related to hyperactivity, in particular fidgetiness, can sometimes be confused for tics. In addition, ADHD symptoms and tics often exacerbate each other, such as tics making it harder to maintain attention, and stress from ADHD-related challenges worsening tics. Further, ADHD can make it harder to focus on behavioral therapy for tics.

ADHD has been associated with higher tic severity (Groth et al., 2019), and it often features as being associated with a negative impact on psychosocial outcomes (Debes et al., 2010; Gorman et al., 2010) and quality of life (Eddy et al., 2012) in TS. From a neurobiologic standpoint, the presence of ADHD may be indicative of CSTC loop characteristics associated with less tic control (Coffey, 2015; El Malhany et al., 2015). From a psychosocial standpoint, the challenges related to ADHD typically involve academics and social interactions, with repeated "failures" often leading to discouragement, low self-esteem, and worsening of behavioral problems. It is therefore critical to evaluate for ADHD-related challenges and provide interventions and supports to address them.

Psychosocial and behavioral intervention

A key intervention in ADHD involves the child's environment, starting with the home. Inattentive and hyperactive/impulsive challenges often necessitate extensive support from parents and other caregivers even just to carry out a child's daily routines. This often takes the form of repeated reminders and redirection that can be experienced as constant "nagging," with frequent frustration for all involved. Even when not deliberate, an implicit message is often conveyed that the child is not trying hard enough or is just "lazy." An important goal is thus to bolster a sense of self-efficacy in the child and in the parent through behavioral management and environmental interventions.

While parenting certainly involves setting and enforcing boundaries, it is important that expectations correspond to the child's capacity to abide by them, and that the child is set up for success. To this end, it is useful to view ADHD as a developmental lag in self-regulatory control. Rather than basing expectations on chronologic age, thinking about the child's actual capacity sets the stage for expectations that are more realistic, for support that is more in line with the child's developmental ability, and for teaching of self-regulation skills through productive interactions and modeling with the goal of engendering growth.

In practice, in addition to adjusting expectations, this means providing close guidance, including breaking complex tasks into smaller steps. A daily schedule with clear structure and routines can be very helpful with planning and organization. Where there is a specific pattern that presents challenges, the parent can have a discussion with the child outside of the situation about concrete steps for addressing the situation the next time it occurs. In addition, to help counterbalance the potential for frequent negative feedback, it is worthwhile for the parent or caregiver to deliberately look for opportunities for positive feedback, even for relatively minor positive action. Such feedback should be genuine, ideally immediately following the behavior ("catch them being good"), with elaboration on the positive impact of the behavior.

In a parallel process, the clinician needs to keep a positive and supportive attitude toward the parent, who is often already struggling with frustration and guilt. Parents should be encouraged to be patient and compassionate toward themselves, learn that they need not be "perfect," realize the value of "mistakes" as opportunities for growth, recognize their strengths as parents, and celebrate even the smallest of steps in a positive direction.

To optimize self-regulation capacity, it is important to ensure a healthy lifestyle. A common issue is that sleep is suboptimal, with resultant sleep deprivation compromising attention and self-control. Thus, guidance around sleep hygiene and healthy sleep routines is often necessary to ensure optimal sleep that leads to the child awaking refreshed (see section on Sleep Difficulties later in this Chapter). Regular physical activity can help with sleep, and it is also known to help with attention and related ADHD symptoms (Halperin et al., 2014; Kamp et al., 2014). A balanced and healthy diet is advisable, including avoiding artificial additives since they can adversely impact attention (Nigg et al., 2012; Sonuga-Barke et al., 2013). Excessive exposure to electronics often needs to be addressed given its multifold negative effects, including interfering with sleep, promoting a sedentary lifestyle, taking time away from homework and other productive tasks, and compromising attention and other executive functions (Swing et al., 2010; Lillard and Peterson, 2011). Not surprisingly, high-frequency digital media use has been found to be associated with emergence of symptoms of ADHD (Kudlow et al., 2025; Ra et al., 2018).

As ADHD negatively impacts classroom behavior and academic performance, school-based interventions are critical. Appropriate accommodations and behavioral strategies, often formalized in an Individualized Education Plan, can promote improved behavior and learning. The accommodations need to be tailored to the particular needs of the child, and the specific strategies that work best will vary depending on the child. For example, while one child can focus better when there is complete silence, another might focus better when there are background sounds (e.g., low volume music, white noise, sounds of nature). Common accommodations that can be helpful in the school setting include the following: assistance around workload management, organization, and planning; extra time for tests and assignments; taking tests in a smaller room with fewer distractions; assistive technology (e.g., voice-to-text software); the option for tests to be taken orally instead of in writing; and assistance with note-taking or providing notes prior to class to maximize active listening and engagement with the material being taught. It is often helpful for the clinician to provide a letter for the school, listing the diagnoses and recommending accommodations. It is also helpful for parents or caregivers to meet with the child's teachers and administrative staff to ensure everyone is on the same page with the goal of optimizing supports for the child's learning.

Another area to consider is the child's social functioning, especially how well the child is getting along with peers in school and elsewhere. ADHD symptoms, in particular hyperactivity and impulsivity, can lead to difficulties in peer relationships as the child may be experienced by peers as intrusive, annoying, or easy to provoke. These social challenges may also be exacerbated by the presence of tics, which are often a target of bullying. Where necessary, intervention by adults can help support the child and peers to ensure positive and appropriate interactions. Finally, it is also good for the child to have opportunities to take part in extracurricular activities, whether in school or outside, that build on their strengths and support their self-confidence.

Pharmacotherapy

When significant difficulties persist despite psychosocial and behavioral interventions, a stimulant medication trial is worth considering. Stimulants are the mainstay drugs used for ADHD given their large effect sizes on the symptoms (Overview | Attention deficit hyperactivity disorder, 2018). This is based on numerous clinical trials over the decades that the drugs have been in use. However, there is evidence that stimulants may be less efficacious for ADHD in the context of TS, with effect sizes closer to medium (Tourette's Syndrome Study Group, 2002; Gadow et al., 2007). Importantly, it should be noted that despite their benefits for ADHD symptoms in the short-term, the evidence has not supported improvement in long-term functional outcomes such as academic performance (Molina et al., 2009; Hinshaw et al., 2015; Swanson, 2019; Therapeutics Initiative | and [144] ADHD in adults, n.d.). Thus, rather than being viewed as a panacea, it is best when these drugs are used as part of the broad array of interventions described above.

In this context, medications that decrease ADHD symptoms can make it easier for the child to cope with expectations and demands. The effects of stimulant medications, both positive and negative, tend to be quite clear. When they work well, the patient is much better able to maintain attention and exercise self-control around behaviors. At the same time, when making decisions about starting or continuing treatment, the potential benefits need to be weighed against the harms. Adverse effects of stimulant drugs include the following: initial insomnia, decreased appetite with weight loss, stomach upset, headache, thirst, social withdrawal, irritability, agitation, aggression, anxiety, increased heart rate (mean ≤10 bpm), increased blood pressure (mean ≤5 mmHg), palpitations, tics (more on this below), body-focused repetitive behaviors (e.g., skin picking, hair pulling), psychotic/manic symptoms, growth suppression, priapism (rare), drug abuse and diversion, and possibly suicidality. In addition, there is controversy about the very rare cases of sudden cardiac death in individuals taking stimulants, and thus cardiac assessment is recommended in patients with a history of cardiac disease or family history of nonischemic cardiac disease (Perrin et al., 2008; Warren et al., 2009). Contraindications to stimulants include significant cardiac problems, hyperthyroidism, glaucoma, psychosis, and pregnancy.

Historically, it has been recommended that stimulants not be used for ADHD in the presence of TS, based largely on case reports of tics appearing following stimulant use (Lowe et al.,

1982), as well as theoretical considerations (regarding the potential role of dopamine in tics). In fact, the US Food and Drug Administration and other references still state that stimulants are contraindicated in patients with TS or even a family history of TS. However, multiple RCTs and meta-analyses have shown that, on average, stimulants do not exacerbate tics in patients with tic disorders (Bloch et al., 2009; Cohen et al., 2015; Osland et al., 2018). At the same time, stimulants have been found to worsen tics in some patients at higher doses, and considerable clinical experience supports the notion that individual patients with TS do experience tic exacerbation with stimulant medications (Bloch et al., 2012; Osland et al., 2018). Thus, in general, if the main clinical concern for a patient is ADHD symptoms and not tics, stimulants remain a good medication option given their effect sizes and ease of use.

In terms of choice of stimulant, the two types are methylphenidate and amphetamine. Many formulations of each type have been developed, with different formulations available in different countries. The formulations differ mainly in their delivery systems and corresponding durations of action, and thus the choice of formulation is based partly on the desired duration of action, considering both benefits and side-effects. In addition, while amphetamine may be a little more efficacious on average in both children and adults, methylphenidate may be a little better tolerated on average in children but not adults (Cortese et al., 2018). Another consideration is that amphetamine may be more likely to exacerbate tics, especially at higher doses (Castellanos et al., 1997). Given these factors, methylphenidate may be preferred as the first stimulant choice in individuals with TS, especially children. Nonetheless, individual responses vary considerably with respect to both efficacy and tolerability, so if an individual does not do well with methylphenidate, an amphetamine should generally be tried.

Once a stimulant is started, the dose should be increased by small increments as needed and as tolerated, at intervals of approximately one week or more, ensuring thorough assessment of benefit and harms throughout the process. Given the wide range of different stimulant formulations available, we recommend that the reader refer to the product monographs for the formulations available in their countries for more specific guidance on dosing. The use of patient/parent and teacher rating scales such as the SNAP-IV or the SWAN Rating Scale (both of which are freely available) can help with evaluation of response at each dose. In terms of medical monitoring, heart rate and blood pressure should be measured at baseline and regularly thereafter, especially following dose increases. In addition, height and weight should be plotted on a growth chart and monitored. "Drug holidays," whereby the stimulant is not taken on weekends or during summer holidays, is a common practice when the main benefits of stimulant treatment are in the area of academic performance, or when appetite suppression and weight loss are significant concerns.

When tics are not a significant concern but rather ADHD, then we would generally recommend starting with a stimulant trial in terms of pharmacotherapy given their large effect sizes on symptoms of ADHD and ease with which they can be tried. However, when both ADHD and tics are a significant concern, it is worth considering a trial with an α-2 agonist, namely clonidine or guanfacine. Interestingly, while these drugs have a small effect size for tics overall (0.3), the effect size appears to be much more robust when tics are comorbid with ADHD (0.7), based on findings from a meta-analysis (Weisman et al., 2013). Of note, it can take up to 3 months for full effect of these drugs in the treatment of tics. In terms of their efficacy for ADHD symptoms, α-2 agonists have been estimated to have a moderate effect size, with no difference in response rate between clonidine and guanfacine (Hirota et al., 2014).

Adverse effects of α-2 agonists include sedation, bradycardia, hypotension, rebound hypertension, dizziness, headache, anticholinergic effects, irritability and other emotional changes, and gastrointestinal symptoms (Cortese et al., 2018; Lexicomp, n.d.). In rare cases they can cause ECG abnormalities, and guanfacine can cause modest QTc interval prolongation. Sedation is particularly common, often limiting the use of these drugs, though it may improve after one to two months. Heart rate and blood pressure should be monitored regularly – at baseline, after each dose increase during titration, periodically during maintenance, after each dose decrease during tapering, and after discontinuation. In addition, QTc interval may need to be monitored with guanfacine based on individual risk, including concomitant use of other QTc-prolonging medications.

In terms of dosing, the usual total daily dose range for clonidine is 0.1–0.4 mg/day, and for guanfacine it is 1–4 mg/day. Immediate release formulations need to be taken in two (in the case of adults) or three (in the case of children) divided doses, while extended-release formulations are taken once daily. We would recommend conservative starting and titration doses, waiting one month prior to each dose increase, to allow time for effects to be assessed, and to mitigate adverse effects, particularly sedation. Tapering α-2 agonists should also be carried out gradually as abrupt discontinuation can lead to withdrawal symptoms, including rebound hypertension, tachycardia, arrhythmias, and anxiety. Given the risks with withdrawal, prescribers should consider the risk of nonadherence when deciding whether to prescribe an α-2 agonist.

Another pharmacologic option that can be considered for ADHD is the norepinephrine reuptake inhibitor atomoxetine. Atomoxetine has a moderate effect size for ADHD symptoms (Cunill et al., 2013; Schwartz and Correll, 2014), and it requires titration over weeks to months to bring about its full benefits. Adverse effects from atomoxetine include sedation, insomnia, headache, stomach upset, decreased appetite and weight loss, irritability, dysphoria, aggression, suicidality, psychotic/manic symptoms, increased heart rate (mean ≤10 bpm), increased blood pressure (mean ≤4 mmHg), dizziness, dry mouth, constipation, urinary hesitation and retention, dysmenorrhea, sexual dysfunction, growth suppression, rare cases of priapism, rare cases of severe liver injury, and there are case reports of life-threatening QT prolongation. As well, there have been multiple case reports of tic symptoms induced by atomoxetine (Yang et al., 2017).

Similar to stimulants, there is controversy about very rare cases of sudden cardiac death with atomoxetine, so cardiac evaluation is indicated in individuals with cardiac risk factors. Contraindications to atomoxetine include significant cardiac problems and narrow angle glaucoma.

In terms of dosing, it is recommended that atomoxetine be started at 0.5 mg/kg/day (maximum 40 mg), then increased after two weeks to 0.8 mg/kg/day (maximum 60 mg), and then increased to a "target dose" of 1.2 mg/kg/day (maximum 80 mg). If there is inadequate response and good tolerability after at least one month at the target dose, atomoxetine can be increased to an approved maximum of 1.4 mg/kg/day (maximum 100 mg). In terms of monitoring, heart rate and blood pressure should be measured at baseline and regularly thereafter, especially following dose increases; height and weight should be plotted on a growth chart; and the patient should be watched for suicidal ideation. While baseline liver enzymes and function tests are not necessary, they should be done at the first sign or symptom suggestive of liver dysfunction.

LEARNING DISABILITIES

Individuals with TS often have learning disabilities, especially when comorbid ADHD is present (Schuerholz et al., 1996; Gorman et al., 2010), and they may therefore experience academic or occupational difficulties. It is important to assess for other causes of such difficulties, including interference from tics and any comorbid conditions. However, unless other causes clearly account for the difficulties, psychoeducational testing to assess for a learning disability is indicated. Psychoeducational testing requires a referral to a psychologist with the relevant expertise, and it includes a combination of cognitive and academic testing using standardized measures.

If a learning disability or even subthreshold learning difficulties are identified, the psychologist will make recommendations to support the individual's academic or occupational functioning. The recommendations typically include accommodations to mitigate the impact of the individual's learning problems, as well as strategies for remediating them. A common example of an accommodation is extra time to complete tasks for an individual whose weaknesses in working memory and processing speed (commonly seen in ADHD) slow them down. In addition, the use of assistive technology, such as voice-to-text software, may be recommended. Remediation could involve tutoring to address an area of academic weakness, although it is important for the tutor to consider the individual's learning profile when designing a tutoring program. In some cases, a student's curriculum in one or more subjects may need to be modified to a lower grade level.

It is ideal for accommodations and other recommendations to be formalized by the school or workplace. In the school setting, this is often done in the form of a document called an "Individualized Education Plan." Having a formal document helps ensure that the recommendations are implemented consistently and on an ongoing basis, and that the school or workplace is held accountable for doing so.

ANXIETY

Anxiety is common in TS, and it frequently interacts with tics (Sapozhnikov and Vermilion, 2023). When anxiety is pervasive and severe enough to result in significant distress or impairment, including avoidance behaviors that interfere with function, it is considered to meet criteria for disorder. While the frequency of anxiety disorders in TS may be related to the theme of self-control difficulties, impacting in this case emotional regulation, there is likely a significant psychosocial contribution given the frequent challenges brought about by tics and related symptoms, including negative reactions by others.

Anxiety is a common exacerbating factor for tics, which in turn can increase anxiety, resulting in a vicious cycle. Thus, addressing anxiety can often improve tics in addition to the anxiety itself. In fact, in some cases anxiety may be the primary driver for tics, and so rather than targeting the tics directly, the focus of intervention could be on the anxiety.

Similar to what is described for OCD above, the first-line intervention for anxiety is CBT. In the case of anxiety, the cognitive aspect would include addressing cognitive distortions that result in overestimation of threat. Patients engage in charting thoughts they experience in response to significant anxiety, which they learn to identify and subsequently challenge with Socratic questioning, the goal being to arrive at more balanced and realistic thoughts. The behavioral aspects involve exposure to situations that cause anxiety, done in a stepwise fashion with the goal of achieving habituation to progressively more challenging situations. Learning relaxation strategies is also part of the process. If treatment with CBT does not result in adequate improvement or if it is not feasible, SSRI medication can be considered. Similar to OCD, the likely mechanism for improvement is through a general emotional blunting, the result of which is to decrease the intensity of anxiety (Langley et al., 2023; Sklivanioti Greenfield et al., 2023). SSRI adverse effects, dosing, and other details are generally as described above for OCD, although doses may be lower in the treatment of non-OCD anxiety.

DEPRESSION

Given the significant psychosocial adversities patients with TS often face, it is not surprising that many will experience periods of feeling down, along with negative thinking, hopelessness, associated changes in sleep and appetite, and even suicidality. While the sum of such experiences along with impact on function can be seen as consistent with a diagnosis of major depressive disorder, it is often more useful to see them in the context of the circumstances that patients are enduring. This allows for focused efforts toward addressing underlying issues, be they bullying or other difficulties at school, challenges in family interactions, difficulties with work, or other adverse psychosocial circumstances (Mynors-Wallis et al., 2000).

When the patient experiences a depressed state to a degree that is disproportionate to the psychosocial circumstances, a worthwhile starting point is to ensure healthy lifestyle, including adequate sleep (Gee et al., 2019), healthy diet (Jacka et al., 2017; Lassale et al., 2019), regular physical activity (Cooney et al., 2013; Schuch et al., 2016), and healthy social interactions (Davidson et al., 2016; Filia et al., 2021). For example, if the patient is self-isolating because they feel down, it should be explained to them that isolation can worsen negative feelings, and they should be encouraged to make efforts to go outside, interact with others, and engage in physical activity. If needed, such efforts can be pursued more formally as behavioral intervention in the context of CBT, while cognitive distortions in the form of biased negative thinking are addressed with the goal of arriving at more balanced or realistic thoughts. It should be noted that while CBT is a mainstay (Cuijpers et al., 2023), other forms of therapy can be useful in depression, including interpersonal therapy and ACT (Cuijpers et al., 2021).

In terms of antidepressant medications, in particular SSRIs, numerous RCTs show that the bulk of the improvement comes about from placebo (Kirsch et al., 2008; Fournier et al., 2010; Cipriani et al., 2018; Munkholm et al., 2019), with questionable clinical difference between drug and placebo (Leucht et al., 2013; Moncrieff and Kirsch, 2015; Hengartner and Plöderl, 2022). Nevertheless, some patients may benefit from the drug in the short-term (Stone et al., 2022), and as such, an SSRI medication trial can still be considered for major depressive disorder when psychosocial and psychotherapeutic interventions have not been adequate, if the patient is interested in such a trial after being fully informed of the possible benefits and harms, and they have not already failed such trials. The relevant information regarding SSRIs, including dosing and other details, is found in the section on OCD above, although doses may be lower compared to those used in OCD.

RAGE OUTBURSTS

Individuals with TS may also experience co-occurring episodes of sudden anger in the form of rage outbursts (Müller-Vahl et al., 2019; Conte et al., 2020). These tend to be correlated with tic severity, and with ADHD (Conte et al., 2020). The episodes involve a sudden outburst of anger that is out of proportion to the trigger and out of character for the patient. There is a sense of loss of control, and the patient may lash out verbally and even physically. This is typically followed by feelings of remorse. The prevalence of rage outbursts in patients with TS ranges from 20% to 67%, making it a significant concern for patients, families, and clinicians alike.

It is important for clinicians and families to address anger and rage in individuals with TS, as these symptoms can have a significant impact on quality of life and social functioning (Conte et al., 2020). The goal is for individuals with TS and their families to learn to better manage anger and rage symptoms and improve their overall well-being. It is useful to underscore that there are major differences between premeditated proactive aggression and impulsive reactive aggression.

Behavioral intervention

Behavioral interventions including CBT approaches and anger management have been shown to be useful in TS patients to better regulate their emotions and reduce the frequency and severity of their anger and rage outbursts (Sukhodolsky et al., 2009; Conte et al., 2020). Another useful approach, Collaborative & Proactive Solutions (formerly Collaborative Problem Solving), provides helpful strategies for parents of children with neurodevelopmental disorders leading to behavioral problems (The Explosive Child [Sixth Edition] – Ross W. Greene PhD – Paperback, n.d.). The approach is based on the premise that rage outbursts and related challenging behaviors are a result of a developmental lag in self-regulatory capacity, in particular with regard to frustration tolerance, anger control, flexibility, and problem-solving. As such, the focus is on enhancing self-regulatory skills by supporting and modeling to the child proactive and collaborative problem-solving to address common themes or patterns associated with behavioral challenges.

Pharmacologic management

Pharmacologic management of anger and rage in TS typically involves optimizing the treatment of tics and other neurodevelopmental disorders that may be present, especially ADHD. Indeed, when ADHD is present, it often drives much of the patient's rage and aggression, and these symptoms can improve substantially when the ADHD is adequately managed (Pringsheim et al., 2015a). Moreover, stimulants have a less burdensome side-effect profile than other agents such as antipsychotics and mood stabilizers. For these reasons, stimulants are considered the first-line medication for aggressive behavior in children and adolescents with comorbid ADHD (Pringsheim et al., 2015a). A systematic review of antipsychotics and mood stabilizers found moderate- to high-quality evidence that risperidone can decrease aggressive behavior in youth with oppositional defiant disorder or conduct disorder, with or without ADHD, and low or very low-quality evidence for other agents (Pringsheim et al., 2015b). Of course, all these medications are associated with considerable adverse effects, which often outweigh the benefits. RCTs of pharmacologic treatment of rage attacks in TS patients are lacking (Mol Debes, 2013). A systematic review concluded that whereas behavioral interventions for rage appear to be effective, the evidence to support the pharmacologic treatment of rage attacks is inconclusive (Conte et al., 2020).

AUTISM SPECTRUM DISORDER

Individuals with TS and comorbid ASD are likely to require support from a specialized team of individuals, particularly if the ASD is moderate to severe (Level 2 or 3). The team may include developmental pediatricians, behavioral therapists, speech-language pathologists, occupational therapists, physical therapists, and other disciplines. The overarching goal of treatment for ASD is to facilitate the acquisition of skills, remove barriers to learning, and improve the individual's functioning and well-being (Anagnostou et al., 2014). Therapeutic interventions largely involve behavioral, educational, social, and sensory-oriented approaches. No medication addresses the core symptoms of ASD (i.e., social communication deficits and restricted, repetitive patterns of behavior), but medications can be used to target associated symptoms.

Intensive behavioral interventions have focused mainly on preschool children, who are unlikely to have a diagnosis of TS given that tic onset is typically at around age 5 years. Nonetheless, these interventions can be used in individuals of any age to help with skill acquisition and address problem behaviors (Hirota and King, 2023). The most well-known form of intensive behavioral intervention is Applied Behavior Analysis, but many other forms exist as well (Lord et al., 2018). For older children and adults with ASD, a well-established educational program that uses modified behavioral techniques is TEACCH (Treatment and Education of Autistic and Related Communication-Handicapped Children) (Mesibov et al., 2005). Older children and young adults may also benefit from group social skills interventions (GSSIs), and the most widely used manual-based GSSI for ASD is UCLA PEERS (Program for the Education and Enrichment of Relational Skills) (Zheng et al., 2021). Sensory-integration approaches delivered by occupational therapists are commonly used in children with ASD, especially in North America, but evidence to support their efficacy is still limited (Weitlauf et al., 2017).

In terms of pharmacotherapy, risperidone and aripiprazole are approved by the US Food and Drug Administration for irritability and aggression in children and adolescents with ASD, and these medications may also be used to treat tics. Nonetheless, they should be used cautiously and judiciously given the significant metabolic and other side-effects associated with them. Moreover, irritability and aggression may be due to other issues, including ADHD, anxiety, depression, academic frustration, or physical discomfort (e.g. due to dental problems, constipation, or other medical problems). Individuals with ASD who present with irritability and aggression should be assessed for such underlying causes, and if present, these causes should generally be addressed with corresponding interventions before initiating an antipsychotic.

A small evidence base supports the efficacy of stimulant (methylphenidate) and nonstimulant (atomoxetine and guanfacine) medication for ADHD associated with ASD, although in general the benefits are less and the adverse effects greater compared to the use of these medications in individuals without ASD (Rodrigues et al., 2021). No RCTs support the use of SSRIs for any indication in children with ASD, and very limited controlled evidence supports the use of SSRIs for obsessive-compulsive behaviors, anxiety, and aggression in adults with ASD (Williams et al., 2013). Nonetheless, based on evidence in individuals without ASD, SSRIs may be tried cautiously for anxiety and OCS in individuals with ASD. A meta-analysis of studies in children and adults with ASD found that SSRIs did not significantly reduce restricted, repetitive behaviors, and antipsychotics were found to provide only a modest benefit (Zhou et al., 2021).

SENSORY HYPERSENSITIVITY

Sensory hypersensitivity affects at least 80% of individuals with TS (Cohen and Leckman, 1992; Belluscio et al., 2011). It is also common in conditions that are often comorbid with TS, especially OCD, ADHD, and ASD, and it tends to be more severe when these comorbidities are present (Isaacs and Riordan, 2020). Unfortunately, little research is available to guide management of sensory hypersensitivity in TS. Clinicians must therefore take a pragmatic approach that avoids doing harm and considers the individual clinical circumstances.

For example, bothersome sensory stimuli may be avoided where feasible, such as cutting out clothing tags for children who are hypersensitive to them. Adaptation of behavioral and cognitive therapy approaches is thought to be helpful as well (Cox et al., 2018). This may include exposure and desensitization strategies, although research has found that when presented with repetitive tactile stimuli, individuals with TS may not habituate normally (Puts et al., 2015). As previously noted, limited evidence supports the use of sensory-integration approaches in children with ASD (Weitlauf et al., 2017); however, it is unclear whether these approaches are helpful for sensory hypersensitivity in individuals with TS. Pharmacotherapy does not have a significant role in the management of sensory hypersensitivity, but SSRIs may help when the sensory hypersensitivity is associated with comorbid OCD or anxiety that is the primary target for medication.

SLEEP DIFFICULTIES

Sleep disturbances have been reported frequently in TS patients, both children and adults. As early as 1974, Moldofsky et al. described increased prevalence of sleep problems in an inpatient series of TS patients (Moldofsky et al., 1974), and in 1992, Allen and colleagues reported elevated prevalence of sleep problems based on a parent-completed sleep questionnaire (Allen et al., 1992). They collected data on boys, aged 7–14 years – three groups, TS-only (N=57), ADHD-only (N=21), and TS+ADHD (N=89), were compared to an age-matched control population (N=146). The complaint of "poor sleep" occurred in 26% with TS-only, 48% with ADHD-only, and 41% with TS +ADHD; all were significantly different from 10% found in controls (Allen et al., 1992). A recent a large population-based study from Sweden found the participants with tics had an elevated prevalence of insomnia (32.16%, compared to 13.70% of the tic free population) (Isomura et al., 2022), representing a 6.7-fold increased likelihood of insomnia in the population affected by tic disorders. The prevalence of sleep abnormalities detected with polysomnographic studies is somewhat lower, but still elevated. A meta-analysis of 20 polysomnographic studies found that in the TS groups, sleep efficiency was lower and sleep onset latency was higher than controls and the effect size was similar to that found in patients with primary sleep disorders (Keenan et al., 2021). Sleep efficiency was significantly reduced, and sleep onset latency was significantly increased in TS+ADHD group, though less than in TS-only group. In addition, time in bed and total sleep time were significantly higher compared with the control group. The meta-analysis found no significant differences in sleep parameters between ADHD-only and control groups (Keenan et al., 2021).

Behavioral intervention

The approach to treatment of sleep problems in the general population also applies to TS patients. Treatment of sleep disturbances should involve education and behavioral intervention, while pharmacologic treatment should be avoided unless very necessary. The patient and their family need to learn about the importance of good sleep hygiene. CBT and mindfulness techniques are seen as the first-line treatment to address insomnia. If these approaches are not sufficient, then pharmacologic treatment may be considered (Rios et al., 2019). European guidelines for the diagnosis and treatment of insomnia based on a systematic review of literature provide the general outline of intervention for insomnia (Riemann et al., 2017). According to these guidelines, to correctly diagnose insomnia and other sleep-related disorders, a clinical interview is required consisting of a sleep history supplemented by sleep questionnaires and sleep diaries. Polysomnography is an objective and comprehensive evaluation of sleep quality and quantity. Especially when specific issues are suspected (e.g., periodic limb movement disorder, sleep-related breathing disorders), polysomnographic evaluation is indicated.

CBT for insomnia usually includes a combination of psychoeducation focused on the sleep hygiene, relaxation training, stimulus control therapy, sleep restriction therapy, and cognitive therapy (Qaseem et al., 2016; Riemann et al., 2017; Hertenstein et al., 2022). It is administered by a therapist over the course of 6–10 sessions. Table 26.3 lists the things to do and to avoid as part of sleep hygiene.

Relaxation training is an important part of this approach. Behavioral strategies include restricting sleep to an appropriate time window and using the bed only for sleeping (and sex) and no other activities such as watching TV or studying. Cognitive strategies are also part of this approach and include therapeutic interventions designed to identify and change misconceptions about sleep. Mindfulness-based practices can assist in coping with psychologic distress associated with insomnia. It is not yet clear how much each of these therapeutic components contribute to the overall result (Parsons et al., 2021).

Pharmacologic agents

Use of pharmacologic agents for insomnia should be a last resort given the risks, including that of tolerance, dependence, and withdrawal. The drugs used include melatonin, benzodiazepines, and antihistamines, among others. Additionally, α-2 agonists, antipsychotics, and certain antidepressants have also been employed for this purpose. The evidence regarding safety is weak partly because most of the studies were conducted for 16 weeks or less (Rios et al., 2019). In the context of TS, pharmacologic agents used to treat tics, especially α-2 agonists and antipsychotics, tend to help with insomnia, although they may cause daytime sedation as well.

Melatonin is available over the counter and has a reputation for safety and efficacy for insomnia. Melatonin (*N*-acetyl-5-methoxytryptamine) is an indoleamine, released from the pineal gland as a neurohormone (Cipolla-Neto and Amaral, 2018). There is good rationale for using melatonin for treatment of insomnia due to its well-established effects on circadian rhythms (Cruz-Sanabria et al., 2023) and more directly its

Table 26.3

Sleep hygiene.

To do	To avoid
Have a regular bedtime routine	Excessive caffeine or use in the afternoon
Maintain the same bedtime and wake-up time	Napping later in the day (it is best to avoid daytime napping completely when trying to address insomnia)
Room environment: quiet, dark, cool	Digital screen exposure close to bedtime
If unable to sleep within 20–30 min, get out of bed and engage in relaxing	Spending excessive time (> 20–30 min) in bed while awake, whether at night or during the day
activity, returning to bed when more tired and ready to fall asleep	
Reserve the bed exclusively for sleep (and sexual activity)	"clockwatching" while trying to sleep at night
Physical exercise during the day	Physical exercise close to bedtime
Exposure to (preferably natural) light during the day	Emotional arousal or stimulating activity close to bedtime
	Eating close to bedtime
	Smoking
	Using alcohol for sleep
	Relying regularly on sedating medication for sleep

role in sleep initiation and maintenance in adults and children (Choi et al., 2022). There is limited evidence of melatonin efficacy when treating insomnia in TS patients. One study provides evidence that evening light therapy advanced the Dim Light Melatonin release in adults with TS, thus correcting a delayed release of melatonin (Ricketts et al., 2022).

A recent umbrella review found that there is statistically significant improvement in sleep latency and total sleep time with melatonin; however, it was not clear that these changes were clinically meaningful (Low et al., 2020). Moreover, the authors concluded that "existing evidence is limited by disparate methodological quality of the papers, as well as a lack of consensus regarding the type, definition and interpretation of outcome measures in the evaluation of efficacy for insomnia (Low et al., 2020)." A more recent systematic review included the outcome variables sleep onset latency, total sleep time, sleep efficiency, sleep quality, and quality of life: melatonin was only significantly effective in sleep onset latency and total sleep time in children and adolescents; it did not appear to be effective in adults (Choi et al., 2022).

The dosing of melatonin as a hypnotic treating insomnia is not well established even in the general population where usual recommendations range from 0.5 mg to 5 mg at night (Bruni et al., 2015). RCT evidence suggests that melatonin can be used as a chronobiotic, i.e., to readjust the circadian clock schedule – 0.5 mg of melatonin administered 3–4 hours before desired sleep time may advance the sleep onset by 1.5 hours on average, particularly in those patients with delayed Dim Light Melatonin onset (van Andel et al., 2021).

Existing melatonin studies based on chart reviews or questionnaires must be interpreted very cautiously because of great uncertainty regarding the melatonin content of commercially available products. In a recent study melatonin content was found to range from −83% to +478% of the labeled content. Additionally, lot-to-lot variability within a particular product varied by as much as 465% (Erland and Saxena, 2017).

It is concerning that in Canada and United States melatonin is widely sold over the counter and used as though it was a nutritional supplement. By contrast, in Europe a prescription is required. After all, melatonin is a hormone produced and release from the pineal gland with widespread effects on sleep, body temperature, circadian rhythms, reproductive system, endocrine system, and metabolism. Adverse effects include headache, dizziness, transient depression, irritability, aggression, enuresis, nausea, stomach cramps, insomnia, nightmares, hypothermia, and excessive daytime somnolence. Little is known about its long-term safety, which may include suppression of the hypothalamic pituitary axis. Therefore, it ought to be used cautiously, especially in children and preferably under medical supervision.

Clonidine, which is often used in the management of tics, has sedative effects that can help with insomnia (Jang et al., 2022); guanfacine can have similar benefits; it may be less sedating, though in the case of both drugs this is dose-dependent. Adverse effects are as already described above, including daytime somnolence and dizziness (Cortese et al., 2018), as are dosing ranges. Weighting the dosing regimen to bedtime can help with maximizing sedation at night while mitigating it during the day; however, this is not always successful and may limit continued treatment.

Antipsychotics, also used in the management of tics, tend to be sedating as well. As such, they can also help with insomnia, although as with α-2 agonists, they can cause daytime sedation as well. The potential utility of antipsychotics needs to be carefully balanced with their risk of serious harms, including neurologic, metabolic, and cardiac, as described above.

Benzodiazepines such as clonazepam and lorazepam can sometimes be used for in the short-term for insomnia and for sleep parasomnias (e.g., periodic leg movement disorder, rapid eye movement (REM) sleep behavior disorder). However, patients usually develop dependence and tolerance, and other adverse effects include cognitive problems and paradoxic disinhibition. Therefore, benzodiazepines should be used only in the short-term and with caution, especially in children and adolescents.

FINANCIAL DISCLOSURE/CONFLICT OF INTEREST

The authors have no financial conflicts of interest to disclose.

References

Allen RP, Singer HS, Brown JE et al. (1992). Sleep disorders in Tourette syndrome: a primary or unrelated problem? Pediatr Neurol 8: 275–280.

Anagnostou E, Zwaigenbaum L, Szatmari P et al. (2014). Autism spectrum disorder: advances in evidence-based practice. CMAJ 186: 509–519.

Belluscio BA, Jin L, Watters V et al. (2011). Sensory sensitivity to external stimuli in Tourette syndrome patients. Mov Disord 26: 2538–2543.

Berman NC, Schwartz R, Park J (2017). Psychological models and treatments of OCD for adults. In: The Wiley handbook of obsessive compulsive disorders. John Wiley & Sons, Ltd., pp. 223–243.

Bloch MH, Gorman DA, Abi-Jaoude E (2012). Commentaries on 'Pharmacological treatment for Attention Deficit Hyperactivity Disorder (ADHD) in children with comorbid tic disorders'. Evid Based Child Health Cochrane Rev J 7: 1231–1234.

Bloch MH, Landeros-Weisenberger A, Kelmendi B et al. (2006). A systematic review: antipsychotic augmentation with treatment refractory obsessive-compulsive disorder. [see comment][erratum appears in Mol Psychiatry. Aug118795 Rev. 49 Refs. Mol Psychiatry 11: 622–632.

Bloch MH, McGuire J, Landeros-Weisenberger A et al. (2010). Meta-analysis of the dose-response relationship of SSRI in obsessive-compulsive disorder. Mol Psychiatry 15: 850–855.

Bloch MH, Panza KE, Landeros-Weisenberger A et al. (2009). Meta-analysis: treatment of attention-deficit/hyperactivity disorder in children with comorbid tic disorders. J Am Acad Child Adolesc Psychiatry 48: 884–893.

Boaden K, Tomlinson A, Cortese S et al. (2020). Antidepressants in Children and adolescents: meta-review of efficacy, tolerability and suicidality in acute treatment. Front Psychiatry 11: 717.

Bridge JA, Iyengar S, Salary CB et al. (2007). Clinical response and risk for reported suicidal ideation and suicide attempts in pediatric antidepressant treatment: a meta-analysis of randomized controlled trials. JAMA 297: 1683–1696.

Bruni O, Alonso-Alconada D, Besag F et al. (2015). Current role of melatonin in pediatric neurology: clinical recommendations. Eur J Paediatr Neurol 19: 122–133.

Cartwright C, Gibson K, Read J et al. (2016). Long-term antidepressant use: patient perspectives of benefits and adverse effects. Patient Prefer Adherence 10: 1401–1407.

Castellanos FX, Giedd JN, Elia J et al. (1997). Controlled stimulant treatment of ADHD and comorbid Tourette's syndrome: effects of stimulant and dose. J Am Acad Child Adolesc Psychiatry 36: 589–596.

Choi K, Lee YJ, Park S et al. (2022). Efficacy of melatonin for chronic insomnia: systematic reviews and meta-analyses. Sleep Med Rev 66: 101692.

Cipolla-Neto J, Amaral FGD (2018). Melatonin as a hormone: new physiological and clinical insights. Endocr Rev 39: 990–1028.

Cipriani A, Furukawa TA, Salanti G et al. (2018). Comparative efficacy and acceptability of 21 antidepressant drugs for the acute treatment of adults with major depressive disorder: a systematic review and network meta-analysis. Lancet 391: 1357–1366.

Coffey BJ (2015). Complexities for assessment and treatment of co-occurring ADHD and tics. Curr Dev Disord Rep 2: 293–299.

Cohen AJ, Leckman JF (1992). Sensory phenomena associated with Gilles de la Tourette's syndrome. J Clin Psychiatry 53: 319–323.

Cohen SC, Mulqueen JM, Ferracioli-Oda E et al. (2015). Meta-analysis: risk of tics associated with psychostimulant use in randomized, placebo-controlled trials. J Am Acad Child Adolesc Psychiatry 54: 728–736.

Conte G, Valente F, Fioriello F et al. (2020). Rage attacks in Tourette syndrome and chronic tic disorder: a systematic review. Neurosci Biobehav Rev 119: 21–36.

Cooney GM, Dwan K, Greig CA et al. (2013). Exercise for depression. Cochrane Database Syst Rev 2013: CD004366.

Cortese S, Adamo N, Del Giovane C et al. (2018). Comparative efficacy and tolerability of medications for attention-deficit hyperactivity disorder in children, adolescents, and adults: a systematic review and network meta-analysis. Lancet Psychiatry 5: 727–738.

Cox JH, Seri S, Cavanna AE (2018). Sensory aspects of Tourette syndrome. Neurosci Biobehav Rev 88: 170–176.

Cruz-Sanabria F, Carmassi C, Bruno S et al. (2023). Melatonin as a chronobiotic with sleep-promoting properties. Curr Neuropharmacol 21: 951–987.

Cuijpers P, Miguel C, Harrer M et al. (2023). Cognitive behavior therapy vs. control conditions, other psychotherapies, pharmacotherapies and combined treatment for depression: a comprehensive meta-analysis including 409 trials with 52,702 patients. World Psychiatry 22: 105–115.

Cuijpers P, Quero S, Noma H et al. (2021). Psychotherapies for depression: a network meta-analysis covering efficacy, acceptability and long-term outcomes of all main treatment types. World Psychiatry 20: 283–293.

Cunill R, Castells X, Tobias A et al. (2013). Atomoxetine for attention deficit hyperactivity disorder in the adulthood: a meta-analysis and meta-regression. Pharmacoepidemiol Drug Saf 22: 961–969.

Davidson SK, Dowrick CF, Gunn JM (2016). Impact of functional and structural social relationships on two year depression outcomes: a multivariate analysis. J Affect Disord 193: 274–281.

Debes N, Hjalgrim H, Skov L (2010). The presence of attention-deficit hyperactivity disorder (ADHD) and obsessive-compulsive disorder worsen psychosocial and educational problems in Tourette syndrome. J Child Neurol 25: 171–181.

Eddy CM, Cavanna AE, Gulisano M et al. (2012). The effects of comorbid obsessive-compulsive disorder and attention-deficit hyperactivity disorder on quality of life in Tourette syndrome. J Neuropsychiatry Clin Neurosci 24: 458–462.

El Malhany N, Gulisano M, Rizzo R et al. (2015). Tourette syndrome and comorbid ADHD: causes and consequences. Eur J Pediatr 174: 279–288.

Erland LAE, Saxena PK (2017). Melatonin natural health products and supplements: presence of serotonin and significant variability of melatonin content. J Clin Sleep Med 13: 275–281.

Filia K, Eastwood O, Herniman S et al. (2021). Facilitating improvements in young people's social relationships to prevent or treat depression: a review of empirically supported interventions. Transl Psychiatry 11: 1–10.

Fournier JC, DeRubeis RJ, Hollon SD et al. (2010). Antidepressant drug effects and depression severity: a patient-level meta-analysis. JAMA 303: 47–53.

Gadow KD, Sverd J, Nolan EE et al. (2007). Immediate-release methylphenidate for ADHD in children with comorbid chronic multiple tic disorder. J Am Acad Child Adolesc Psychiatry 46: 840–848.

Gee B, Orchard F, Clarke E et al. (2019). The effect of non-pharmacological sleep interventions on depression symptoms: a meta-analysis of randomised controlled trials. Sleep Med Rev 43: 118–128.

Gorman DA, Thompson N, Plessen KJ et al. (2010). Psychosocial outcome and psychiatric comorbidity in older adolescents with Tourette syndrome: controlled study. Br J Psychiatry 197: 36.

Groth C, Mol Debes N, Rask CU et al. (2017). Course of Tourette syndrome and comorbidities in a large prospective clinical study. J Am Acad Child Adolesc Psychiatry 56: 304–312.

Groth C, Skov L, Lange T et al. (2019). Predictors of the clinical course of Tourette syndrome: a longitudinal study. J Child Neurol 34: 913–921.

Halperin JM, Berwid OG, O'Neill S (2014). Healthy body, healthy mind? The effectiveness of physical activity to treat ADHD in children. Child Adolesc Psychiatr Clin N Am 23: 899–936.

Hengartner MP, Amendola S, Kaminski JA et al. (2021). Suicide risk with selective serotonin reuptake inhibitors and other new-generation antidepressants in adults: a systematic review and meta-analysis of observational studies. J Epidemiol Community Health 75jech-2020-214611.

Hengartner MP, Plöderl M (2022). Estimates of the minimal important difference to evaluate the clinical significance of antidepressants in the acute treatment of moderate-to-severe depression. BMJ Evid Based Med 27: 69–73.

Hertenstein E, Trinca E, Wunderlin M et al. (2022). Cognitive behavioral therapy for insomnia in patients with mental disorders and comorbid insomnia: a systematic review and meta-analysis. Sleep Med Rev 62: 101597.

Hinshaw SP, Arnold LE MTA Cooperative Group(2015). Attention-deficit hyperactivity disorder, multimodal treatment, and longitudinal outcome: evidence, paradox, and challenge. Wiley Interdiscip Rev Cogn Sci 6: 39–52.

Hirota T, King BH (2023). Autism spectrum disorder: a review. JAMA 329: 157–168.

Hirota T, Schwartz S, Correll CU (2014). Alpha-2 agonists for attention-deficit/hyperactivity disorder in youth: a systematic review and meta-analysis of monotherapy and add-on trials to stimulant therapy. J Am Acad Child Adolesc Psychiatry 53: 153–173.

Hirschtritt ME, Lee PC, Pauls DL et al. (2015). Lifetime prevalence, age of risk, and genetic relationships of comorbid psychiatric disorders in Tourette syndrome. JAMA Psychiatry 72: 325–333.

Horowitz M, Wilcock M (2022). Newer generation antidepressants and withdrawal effects: reconsidering the role of antidepressants and helping patients to stop. Drug Ther Bull 60: 7–12.

Ipser JC, Stein DJ, Hawkridge S et al. (2009). Pharmacotherapy for anxiety disorders in children and adolescents. Cochrane Database Syst Rev 9: CD005170.

Isaacs D, Riordan H (2020). Sensory hypersensitivity in Tourette syndrome: a review. Brain Dev 42: 627–638.

Isomura K, Sidorchuk A, Sevilla-Cermeño L et al. (2022). Insomnia in Tourette syndrome and chronic tic disorder. Mov Disord 37: 392–400.

Ivarsson T, Skarphedinsson G, Kornør H et al. (2015). The place of and evidence for serotonin reuptake inhibitors (SRIs) for obsessive compulsive disorder (OCD) in children and adolescents: views based on a systematic review and meta-analysis. Psychiatry Res 227: 93–103.

Jacka FN, O'Neil A, Opie R et al. (2017). A randomised controlled trial of dietary improvement for adults with major depression (the 'SMILES' trial). BMC Med 15: 23.

Jang YJ, Choi H, Han TS et al. (2022). Effectiveness of clonidine in child and adolescent sleep disorders. Psychiatry Investig 19: 738–747.

Kamp CF, Sperlich B, Holmberg H-C (2014). Exercise reduces the symptoms of attention-deficit/hyperactivity disorder and improves social behaviour, motor skills, strength and neuropsychological parameters. Acta Paediatr 103: 709–714.

Keenan L, Sherlock C, Bramham J et al. (2021). Overlapping sleep disturbances in persistent tic disorders and attention-deficit hyperactivity disorder: a systematic review and meta-analysis of polysomnographic findings. Neurosci Biobehav Rev 126: 194–212.

Kirsch I, Deacon BJ, Huedo-Medina TB et al. (2008). Initial severity and antidepressant benefits: a meta-analysis of data submitted to the Food and Drug Administration. PLoS Med 5: e45.

Kudlow P, Treurnicht Naylor K, Abi-Jaoude E (2025). Genes and screens: attention-deficit hyperactivity disorder in the digital age. The British Journal of Psychiatry 1–3.

Kurvits L, Martino D, Ganos C (2020). Clinical features that evoke the concept of disinhibition in Tourette syndrome. Front Psychiatry 11: 21.

Langley C, Armand S, Luo Q et al. (2023). Chronic escitalopram in healthy volunteers has specific effects on reinforcement sensitivity: a double-blind, placebo-controlled semi-randomised study. Neuropsychopharmacology 48: 664–670.

Lassale C, Batty GD, Baghdadli A et al. (2019). Healthy dietary indices and risk of depressive outcomes: a systematic review and meta-analysis of observational studies. Mol Psychiatry 24: 965–986.

Leucht S, Fennema H, Engel R et al. (2013). What does the HAMD mean? J Affect Disord 148: 243–248.

Lexicomp: Evidence-Based Drug Referential Content [WWW Document] (n.d.). https://www.wolterskluwer.com/en/solutions/lexicomp (accessed 15.10.23).

Lillard AS, Peterson J (2011). The immediate impact of different types of television on young children's executive function. Pediatrics 128: 644–649.

Locher C, Koechlin H, Zion SR et al. (2017). Efficacy and safety of selective serotonin reuptake inhibitors, serotonin-norepinephrine reuptake inhibitors, and placebo for common psychiatric disorders among children and adolescents: a systematic review and meta-analysis. JAMA Psychiatry 74: 1011–1020.

Lord C, Elsabbagh M, Baird G et al. (2018). Autism spectrum disorder. Lancet 392: 508–520.

Low TL, Choo FN, Tan SM (2020). The efficacy of melatonin and melatonin agonists in insomnia – an umbrella review. J Psychiatr Res 121: 10–23.

Lowe TL, Cohen DJ, Detlor J et al. (1982). Stimulant medications precipitate Tourette's syndrome. JAMA 247: 1168–1169.

Martino D, Mink JW (2013). Tic disorders. Continuum (Minneap Minn) 19: 1287–1311.

McGuire JF, Piacentini J, Lewin AB et al. (2015). A META-Analysis of cognitive behavior therapy and medication for child obsessive-compulsive disorder: moderators of treatment efficacy, response, and remission. Depress Anxiety 32: 580–593.

Mesibov GB, Shea V, Schopler E (2005). The TEACCH approach to autism spectrum disorders. New York: Springer Science & Business Media.

Mol Debes NM (2013). Co-morbid disorders in Tourette syndrome. Behav Neurol 27: 7–14.

Moldofsky H, Tullis C, Lamon R (1974). Multiple tic syndrome (Giles de la Tourette's syndrome). J Nerv Ment Dis 159: 282–292.

Molina BSG, Hinshaw SP, Swanson JM et al. (2009). The MTA at 8 years: prospective follow-up of children treated for combined-type ADHD in a multisite study. J Am Acad Child Adolesc Psychiatry 48: 484–500.

Moncrieff J, Kirsch I (2015). Empirically derived criteria cast doubt on the clinical significance of antidepressant-placebo differences. Contemp Clin Trials 43: 60–62.

Müller-Vahl KR, Kayser L, Pisarenko A et al. (2019). The Rage Attack Questionnaire-Revised (RAQ-R): assessing rage attacks in adults with Tourette syndrome. Front Psychiatry 10: 956.

Munkholm K, Paludan-Müller AS, Boesen K (2019). Considering the methodological limitations in the evidence base of antidepressants for depression: a reanalysis of a network meta-analysis. BMJ Open 9: e024886.

Murphy T, Muter V (2012). Risk factors for comorbidity in ADHD and GTS: looking beyond a single-deficit model. Appl Neuropsychol Child 1: 129–136.

Mynors-Wallis LM, Gath DH, Day A et al. (2000). Randomised controlled trial of problem solving treatment, antidepressant medication, and combined treatment for major depression in primary care. BMJ 320: 26–30.

Nigg JT, Lewis K, Edinger T et al. (2012). Meta-analysis of attention-deficit/hyperactivity disorder or attention-deficit/hyperactivity disorder symptoms, restriction diet, and synthetic food color additives. J Am Acad Child Adolesc Psychiatry 51: 86–97.e8.

Osland ST, Steeves TD, Pringsheim T (2018). Pharmacological treatment for attention deficit hyperactivity disorder (ADHD) in children with comorbid tic disorders. Cochrane Database Syst Rev 6: CD007990.

Overview | Attention deficit hyperactivity disorder: Diagnosis and management | Guidance | NICE [WWW Document] (2018). https://www.nice.org.uk/guidance/ng87 (accessed 23.10.23).

Overview | Obsessive-compulsive disorder and body dysmorphic disorder: treatment | Guidance | NICE [WWW Document] (2005). https://www.nice.org.uk/guidance/cg31 (accessed 30.07.23).

Parsons CE, Zachariae R, Landberger C et al. (2021). How does cognitive behavioural therapy for insomnia work? A systematic review and meta-analysis of mediators of change. Clin Psychol Rev 86: 102027.

Perrin JM, Friedman RA, Knilans TK Black Box Working Group, Section on Cardiology and Cardiac Surgery (2008). Cardiovascular monitoring and stimulant drugs for attention-deficit/hyperactivity disorder. Pediatrics 122: 451–453.

Practice parameter for the assessment and treatment of children and adolescents with obsessive-compulsive disorder. J Am Acad Child Adolesc Psychiatry 51: 98–113.

Pringsheim T, Hirsch L, Gardner D et al. (2015). The pharmacological management of oppositional behaviour, conduct problems, and aggression in children and adolescents with attention-deficit hyperactivity disorder, oppositional defiant disorder, and conduct disorder: a systematic review and meta-analysis. Can J Psychiatry 60: 52–61.

Pringsheim T, Hirsch L, Gardner D et al. (2015). The pharmacological management of oppositional behaviour, conduct problems, and aggression in children and adolescents with attention-deficit hyperactivity disorder, oppositional defiant disorder, and conduct disorder: a systematic review and meta-analysis. Part 2: antipsychotics and traditional mood stabilizers. Can J Psychiatry Rev 60: 52–61.

Przeworski A, Birnkrant JM (2017). Psychological models and treatments for OCD in children. In: The Wiley handbook of obsessive compulsive disorders. John Wiley & Sons, Ltd., pp. 244–260.

Puts NAJ, Harris AD, Crocetti D et al. (2015). Reduced GABAergic inhibition and abnormal sensory symptoms in children with Tourette syndrome. J Neurophysiol 114: 808–817.

Qaseem A, Kansagara D, Forciea MA et al. (2016). Management of chronic insomnia disorder in adults: a clinical practice guideline from the American College of Physicians. Ann Intern Med 165: 125–133.

Quilichini J-B, Revet A, Garcia P et al. (2022). Comparative effects of 15 antidepressants on the risk of withdrawal syndrome: a real-world study using the WHO pharmacovigilance database. J Affect Disord 297: 189–193.

Ra CK, Cho J, Stone MD et al. (2018). Association of digital media use with subsequent symptoms of attention-deficit/hyperactivity disorder among adolescents. JAMA 320: 255–263.

Ricketts EJ, Burgess HJ, Montalbano GE et al. (2022). Morning light therapy in adults with Tourette's disorder. J Neurol 269: 399–410.

Riemann D, Baglioni C, Bassetti C et al. (2017). European guideline for the diagnosis and treatment of insomnia. J Sleep Res 26: 675–700.

Rios P, Cardoso R, Morra D et al. (2019). Comparative effectiveness and safety of pharmacological and non-pharmacological interventions for insomnia: an overview of reviews. Syst Rev 8: 281.

Robertson MM, Eapen V, Singer HS et al. (2017). Gilles de la Tourette syndrome. Nat Rev Dis Prim 3: 16097.

Rodrigues R, Lai MC, Beswick A (2021). Practitioner review: Pharmacological treatment of attention-deficit/hyperactivity disorder symptoms in children and youth with autism spectrum disorder: a systematic review and meta-analysis. J Child Psychol Psychiatry 62: 680–700.

Romanelli RJ, Wu FM, Gamba R et al. (2014). Behavioral therapy and serotonin reuptake inhibitor pharmacotherapy in the treatment of obsessive-compulsive disorder: a systematic review and meta-analysis of head-to-head randomized controlled trials. Depress Anxiety 31: 641–652.

Sapozhnikov Y, Vermilion J (2023). Co-occurring anxiety in youth with tic disorders: a review. J Child Adolesc Psychopharmacol 33: 402–408.

Schuch FB, Vancampfort D, Rosenbaum S et al. (2016). Exercise improves physical and psychological quality of life in people with depression: a meta-analysis including the evaluation of control group response. Psychiatry Res 241: 47–54.

Schuerholz LJ, Baumgardner TL, Singer HS (1996). Neuropsychological status of children with and without attention deficit hyperactivity disorder. Neurology 46: 958–965.

Schwartz S, Correll CU (2014). Efficacy and safety of atomoxetine in children and adolescents with attention-deficit/hyperactivity disorder: results from a comprehensive meta-analysis and metaregression. J Am Acad Child Adolesc Psychiatry 53: 174–187.

Sharma T, Guski LS, Freund N et al. (2016). Suicidality and aggression during antidepressant treatment: systematic review and meta-analyses based on clinical study reports. BMJ 352: i65.

Silvestri PR, Baglioni V, Cardona F et al. (2018). Self-concept and self-esteem in patients with chronic tic disorders: a systematic literature review. Eur J Paediatr Neurol 22: 749–756.

Sklivanioti Greenfield M, Wang Y, Msghina M (2023). Behavioral, cortical and autonomic effects of single-dose escitalopram on the induction and regulation of fear and disgust: comparison with single-session psychological emotion regulation with reappraisal. Front Psychiatry 13: 988893.

Sonuga-Barke EJS, Brandeis D, Cortese S et al. (2013). Nonpharmacological interventions for ADHD: systematic review and meta-analyses of randomized controlled trials of dietary and psychological treatments. Am J Psychiatry 170: 275–289.

Sørensen A, Jørgensen KJ, Munkholm K (2022). Description of antidepressant withdrawal symptoms in clinical practice guidelines on depression: a systematic review. J Affect Disord 316: 177–186.

Stone MB, Yaseen ZS, Miller BJ et al. (2022). Response to acute monotherapy for major depressive disorder in randomized, placebo controlled trials submitted to the US Food and Drug Administration: individual participant data analysis. BMJ 378: e067606.

Storch EA, Larson MJ, Muroff J et al. (2010). Predictors of functional impairment in pediatric obsessive-compulsive disorder. J Anxiety Disord 24: 275–283.

Sukhodolsky DG, Vitulano LA, Carroll DH et al. (2009). Randomized trial of anger control training for adolescents with Tourette's syndrome and disruptive behavior. J Am Acad Child Adolesc Psychiatry 48: 413–421.

Swanson JM (2019). Debate: Are stimulant medications for attention-deficit/hyperactivity disorder effective in the long term? (against). J Am Acad Child Adolesc Psychiatry 58: 936–938.

Swing EL, Gentile DA, Anderson CA et al. (2010). Television and video game exposure and the development of attention problems. Pediatrics 126: 214–221.

The Explosive Child [Sixth Edition] – Ross W. Greene PhD – Paperback [WWW Document] (n.d.). HarperCollins. https://www.harpercollins.ca/9780063092464/the-explosive-child-sixth-edition/ (accessed 29.10.23).

Therapeutics Initiative | [144] ADHD in adults (n.d.). https://www.ti.ubc.ca/2023/09/29/144-adhd-adults/ (accessed 23.10.23).

Tourette's Syndrome Study Group (2002). Treatment of ADHD in children with tics: a randomized controlled trial. Neurology 58: 527–536.

van Andel E, Bijlenga D, Vogel SWN et al. (2021). Effects of chronotherapy on circadian rhythm and ADHD symptoms in adults with attention-deficit/hyperactivity disorder and delayed sleep phase syndrome: a randomized clinical trial. Chronobiol Int 38: 260–269.

Warren AE, Hamilton RM, Bélanger SA et al. (2009). Cardiac risk assessment before the use of stimulant medications in children and youth: a joint position statement by the Canadian Paediatric Society, the Canadian Cardiovascular Society, and the Canadian Academy of Child and Adolescent Psychiatry. Can J Cardiol 25: 625–630.

Watson HJ, Rees CS (2008). Meta-analysis of randomized, controlled treatment trials for pediatric obsessive-compulsive disorder. J Child Psychol Psychiatry 49: 489–498.

Weisman H, Qureshi IA, Leckman JF et al. (2013). Systematic review: Pharmacological treatment of tic disorders – efficacy of antipsychotic and alpha-2 adrenergic agonist agents. Neurosci Biobehav Rev 37: 1162–1171.

Weitlauf AS, Sathe NA, McPheeters ML et al. (2017). Interventions targeting sensory challenges in children with autism spectrum disorder—an update. Comp Eff Rev.

Williams K, Brignell A, Randall M et al. (2013). Selective serotonin reuptake inhibitors (SSRIs) for autism spectrum disorders (ASD). Cochrane Database Syst Rev 2013: CD004677.

Yang R, Li R, Gao W et al. (2017). Tic symptoms induced by atomoxetine in treatment of ADHD: a case report and literature review. J Dev Behav Pediatr 38: 151–154.

Zheng S, Kim H, Salzman E et al. (2021). Improving social knowledge and skills among adolescents with autism: systematic review and meta-analysis of UCLA PEERS for adolescents. J Autism Dev Disord 51: 4488–4503.

Zhou MS, Nasir M, Farhat LC et al. (2021). Meta-analysis: pharmacologic treatment of restricted and repetitive behaviors in autism spectrum disorders. J Am Acad Child Adolesc Psychiatry 60: 35–45.

Section V

Lived experience, patient organizations and other useful resources

Handbook of Clinical Neurology, Vol. 215 (3rd series)
Tic Disorders and Tourette Syndrome
C. Ganos and Y. Worbe, Editors
https://doi.org/10.1016/B978-0-443-13554-5.00021-3

Chapter 27

Understanding tics and neurodevelopmental disorders through lived experience: The first-person experience and the parent perspective

BEN BROWN[1,*] AND JAN STEWART[2]

[1]*Sanford, NC, United States*

[2]*Toronto, ON, Canada*

Abstract

For many individuals with Tourette, navigating everyday judgments from strangers and peers alike can be as much of a challenge as the tics. Tourette-awareness campaigns and the rise of social media support communities have done significant work to improve the public's understanding and acceptance of tic disorders, but individuals living with it still experience unfortunate misunderstandings about their tics – sometimes in sensitive environments, like airports – which makes for a nervous base-pressure that "Touretters" can feel in their day-to-day and which can inflame their ticcing. Social difficulties, and even the fear of the them, are very much a part of the story for people with Tourette and other tic disorders. The science community has done excellent work over the years in reaching out to the Tourette community to consider those daily-life experiences and better align with what people living with Tourette would like to see improve with respect to the disorder.

The First-Person Experience

Ben Brown, Sanford, North Carolina, United States

Airports and anxiety are a classic pair. I am confident in thinking the potentials for delays and cancellations, upon the sensitivities of money, time and safety, are probably enough to make the occasional airline patron at least anxious. And perhaps it is made worse by the urgency of arriving at the airport early, after a potentially nerve-grinding parking situation, and then the gravity of airport security and the awkwardness of the important but still intimidating signs around the terminal telling all kinds of people at once, "If You See Something, Say Something." I get the purpose of the statement; I am only pointing out that it has always made me nervous, all of it, at airports, and all kinds of people can relate. Plus, I have Tourette syndrome.

A setup-and-reveal like that used to be mainstream comedy gold, but in all truth my anxiety has always spiked at places like airports, because of my Tourette, which began when I was a child with constant tics that people around me frequently viewed as disruptive. The tics have eased up in adulthood for me, now just shy of age 45, to where it is not a blatantly visible case. It is noticeable – people around me are able to see and hear my various vocalizations and face wrinkling and head-jerking, and so on – but, for whatever reason, strangers often disregard it. It has always been so easy, though, for me to imagine scenarios of airport security officers locking on to my strange, sudden flinching, or my jagged head-rocking, or my little episodes of jumpy, high-pitched sounds, all kicked up by my anxieties, and detaining me to make sure I am not intoxicated or some other risk. I mean, it is their job to investigate, right? It is on my mind every time I book a flight.

When I get to the airport, I try to play it cool and relaxed, but I am anxious about my Tourette, or at least how it might appear to bystanders, who could indeed "say something," like people did about me and my tics in various other contexts when I was a kid, in the 1980s. Situations that feel tense or anxious often inflame my tics. In a tense environment, my tics ratchet up. This a pretty common experience for people with tic disorders, citing the many conversations I have had with others diagnosed with Tourette.

This specific nightmare – involuntary tics prompting a response from airport authorities – has never played out for me in real life, even though it stays on my mind. It breaks my heart and scares me that it has indeed happened to others living with the disorder, my knowledge of which adds to my apprehensiveness. These days, I am vaguely open with my tics, whether with friends or at the workplace. But, for better or worse, from

*Correspondence to: Ben Brown, Creator, past host/producer of Tourette's Podcast, 500 N Franklin Dr., Sanford, NC, United States. Tel: +1-910-508-3552, E-mail: benbrowninput@gmail.com.

childhood, the dynamics of my Tourette were such that I developed ways of masking or "smoothing out" certain tics, to make them less conspicuous and more socially survivable. I was only able to do this with long-haul tics I have had from the beginning, like my head-rocking. I would complete the tic when it came, but I would add my own motions or dynamics to it to make it appear less obvious. That was the intent, anyway, even though it probably looked more conspicuous.

When I was growing up, one of the most frequently asked questions was whether I could hold my tics in, or refrain from them somehow. Some inquisitors were just basically curious, while others hoped I might pipe down. Every so often, I still hear the question or see someone asking on social media. Since it remains a curiosity, my experience is that restraining tics is possible for a time, but can be distressing and sometimes painful, in constantly wrestling with involuntary motions and sounds that are often hyperspecific and demanding in expression. Generally, the person I am today has a core set of tics that I have become much more open with over the years, but I find myself masking or, if possible, suppressing my tics in awkward but temporary situations, like in elevators, and feeling the grip and suspense in more sensitive or enduring environments. Efforts to completely restrain my tics can end up with them bursting out from pressure.

I know I should not have to suppress or muzzle my tics; there is nothing wrong with them, and I am not doing anything wrong – or at least that is what I know in my mind, and it is what I am getting better at believing in, thanks to the greater Tourette community's awesome advocacy and validation work over the years. But the community and I still know it is bound for tough moments. And as a middle-aged man, whose adult professional and social life has gone mostly fine with respect to Tourette, I still frequently fear and feel the pressure to conceal my tics so they do not disturb my surroundings. And I often think, cynically, that maybe that is *why* adult life has gone fine for me.

Doctors diagnosed my Tourette in 1986, when I was six years old, roughly a year after my parents began searching for answers. I was bombarding and confusing them with highly visible tics – vocalizations like snorting and movements like shrugging and head-jerking all day and night. I was their first child, and they did not know what to make of it, except it was something new on top of the behavior they previously sought a medical opinion for: I hardly ever slept, I could not focus on anything for more than a moment, I was easily upset or angered, and showed other, extra-level traits that led to an attention deficit disorder diagnosis. And now, strange tics were an out-front part of my being as my parents and I learned about Tourette in real time.

I turned in an awful grade school academic performance in the chaos of it and associated conditions, like heavy depression, sharp anxiety, and lack of sleep. I was rough socially as well. While there were loads of happy moments and fun times with friends and so on, I was also getting into fights, being disagreeable, and sustaining a bitterness, possibly connected to my self-consciousness and seemingly never-ending lines of "normal" people to convince. Although I could outwardly show excitable or upbeat behavior in fun situations with friends, I generally felt exhaustingly misunderstood.

In 1990, a bit before my 10th birthday, the local newspaper ran a story about me and my Tourette. The paper did an excellent job putting Tourette in context and conveying the basics of what the day's scientists knew. Good public education. But I was also a sensitive kid miserable on medications and doctors' efforts to shape my behavior. The attention on my Tourette felt as constant as the tics themselves. It was also isolating – heavy for a youth becoming hyperconscious of the stakes of peer acceptance.

As I entered high school, I figured it was just more practical to conceal my Tourette as much as I could and maybe avoid problems that way. I was closeted, basically, through my 20s. Later, seeing various communities gather and share on social media in a potentially more accepting world, I started to let my tics show in a natural way again in environments that felt OK. They include a few tics I have had since the beginning, such as my head movements and lumpy closed-mouth yelp, plus constant muscle clenching in my limbs. Tourette is always there.

I assume my special nervousness is a remnant from rough experiences and fears in youth, with that lived experience and layered sense of isolation. But I have also seen the situation evolve as opportunities for communication and sharing have exploded in recent decades.

Born in 1980, I was essentially first in line for the home computing and internet experience as communication and the nature of community overall were about to evolve wildly. In that decade, there were primitive online bulletin boards that people could access on computers with dial-up connections. But my main interaction with anyone in the Tourette community came through paper mail with a Tourette pen-pal my age – 10 at the time, in 1990 – named Drew. The TAA, then called the Tourette Syndrome Association, linked us up as part of a broader pen-pal program it maintained.

It wasn't until after the turn of the century that I started thinking about the ripples of that kind of peer connection in a more modern sense, with how the online Tourette community has established itself globally for awareness and support, but also for real friendships among Touretters that can revolve around any other common interests, like hobbies, with an awesome and uncommon piece of mutual background and life story.

It is no shocker that new friends would discuss mutual interests, but whether they are talking Tourette or train spotting, together in this shared context they have highlighted individual humanity and the chance to live with the same kind of respect granted to everyone else. We might tic with an amazing sense of relief in being unregarded or even understood.

That established and repeated impression of humanity has, like with other mental health focuses, placed a healthier spotlight on Tourette and how to respect people affected. Although it might still be a slow process, with misunderstandings possible around any public corner, the Tourette community's self-advocacy is improving its world.

So is the science community, whose work empowers so much of that advocacy and the quality of information to share. It has been awesome to witness the convergence of scientists and members of the Tourette and tic disorder community who believe that lived experience, how Tourette applies to peoples' lives, is an important part of the conversation: treatment options, social understanding, and other focuses ultimately angled toward quality of life. In August 2019, for instance, a diverse group of healthcare professionals, researchers, and people with Tourette gathered at a special summit called Treating Tourette Together. It essentially sought to re-center the scope of science work in accordance with what actually mattered to the community itself, to potentially re-calibrate efforts with the people they are meant to help.

Held in partnership with TAA, the Patient-Centered Outcomes Research Institute, and a board of professionals in the field, the event spanned two days in the midwest US city of Minneapolis, Minn., and I was proud to have been there for it. I got to meet with fellow Tourette community members and field professionals who flew in from various distances for moderated discussions and input sessions.

The event meant a lot to me, in seeing such a positive and esteemed group created to inform a good future with direct, diverse community-member input, a feeling that encouraged me afterward as my rideshare drove me back to the anxiety of the airport terminal.

A Parent Perspective

Jan Stewart, Toronto, Canada

I never thought I would live with fear in my own house. I grew up in a warm, loving family, started a promising career, and married a wonderful man. I always assumed my children's lives would follow in the same successful path as mine. Yet shortly after my kids were born, the cracks began to appear, and my world was turned upside down.

Aa a baby, Andrew's hands and feet moved in constant circular motion. He never played in his crib and lacked internal controls when it came to eating. As he grew, he could not maintain eye contact, became extremely anxious, and had terrible difficulty with transitions between activities. He loved to please but could not sit still or focus on any activity for a sustained period. By comparison, his little sister Ainsley was fun and mischievous, until she started school. By grade 1, her behavior had become out of control, disruptive, and rude. She could not read social cues, was careless in her work, and spent more time in the principal's office than in her classroom.

I was exhausted and frightened. I worried that Andrew would hurt himself when he started throwing his neck back repeatedly, and I spent hours applying ice. I could not stop Ainsley from chewing her tongue until it swelled and bled. They both blinked excessively, hummed, coughed, and made constant clucking sounds. But every time I approached our doctor, he reassured me the kids were just highly sensitive and would be fine, that I needed to calm down and stop being an overly vigilant AAA mother. Essentially *I* was the problem.

In hindsight, we know many of these behaviors are neurodevelopmental markers, but my husband David and I had no understanding of mental health or neurodiversity. I was perplexed by the doctor's continual admonishments, which shook my usual self-confidence. As the kids' challenges escalated, I feared we would never climb out of the abyss.

Shortly after Andrew's ninth birthday, his disorders exploded, and we hit bottom. One day, he was happily playing; the next day, he erupted into a violent two-hour meltdown. He kicked, screamed, swore, and punched holes in walls. These rages continued, virtually unabated, almost every day for months on end. Within weeks, he also started engaging in nonstop compulsive rituals driven by the fear of Ainsley being kidnapped or my husband David and me dying. He touched walls for hours, could not walk through doors for over 30 minutes and repeatedly swept the floor with his hands. And the rituals kept changing and escalating he rubbed his head against shrubs and car tires, put knives in his mouth to feel them, and even got down on all fours to gnaw the filthy subway floor. I felt like I was in the Twilight Zone.

The children were in deep distress. Andrew kept repeating, "This is not me, this is not me!" as he pleaded and pleaded for help. Ainsley was bullied because of her tics and "odd" behavior. My heart broke when I found her curled up in a ball sobbing, "Why don't I have any friends?" We just tried to get through each day, minute by minute. And although my parents had taught me to trust our doctor as the medical expert, I trusted my gut and refused to settle. I promised the kids I would never stop until we found them the right help.

Thankfully the children's symptoms had finally become so clear that they were both fully diagnosed: Andrew with Tourette syndrome, autism, obsessive-compulsive disorder (OCD), ADHD, developmental disabilities and learning disabilities, and Ainsley with Tourette syndrome, mood and anxiety disorders, ADHD and learning disabilities. Until then, my only exposure to Tourette syndrome had been from the media, which largely sensationalized those who repeatedly blurted out expletives and could not control their body movements.

Unlike many parents who understandably need time to process and absorb their children's diagnoses, however, the "labels" were such a relief to me. They validated my concerns and propelled me forward to delve into research, advocate, and find the right services and supports.

I read everything I could, watched videos, and joined Tourette Canada and other advocacy organizations. I learned that the majority of those with Tourette syndrome also have at least one co-occurring condition, that tics are involuntary and can only be temporarily suppressed, and that co-occurring conditions often require greater interventions than Tourette syndrome. We embraced life-saving medications, although we had to carefully monitor their side-effects and interactions: the children experienced everything from rapid weight gain to loss of appetite, stomach aches, nausea, shakiness, and headaches, but for us the benefits far outweighed the risks. And therapies were crucial but seemed never-ending: cognitive behavior therapy, Comprehensive Behavioral Intervention for Tics, Exposure and Response Prevention, social skills training, occupational therapy and speech therapy. Certain therapies, such as Eye Movement

Desensitization and Reprocessing, were ineffective for the kids but we never gave up.

Early on, I learned the critical importance of resetting the expectations for my life. I gradually accepted the fact that my life would not be as planned for or hoped for in many ways in order to move forward. And that included the support we could expect to receive from family members and friends. Our extended families were well-meaning but often ill-advised. An aunt told to be stricter, and a cousin who was a teacher advised us that the kids were simply attention-seeking and could stop their behaviors if they really wanted to. I learned to navigate around relatives who were critical or judgmental or those who offered un-asked-for advice. I was hurt by friends who distanced themselves and felt isolated and alone. Yet I believed I was partly at fault as I was consumed with the children's care and had no time to invest in my friendships. Instead, I turned to parent support groups – for Tourette syndrome, OCD, autism, and ADHD – and found warm, empathetic, and caring communities. We listened to one another, gave each other advice, and even laughed at everyday situations: one afternoon, Andrew and I were walking down the street and he was vocally hooting. A passer-by came running up and asked if he was OK; he looked her straight in the eye and said, "Oh, I'm fine, I'm just bird calling!" I do not know how he found that answer, but it was brilliant.

I learned additional key life lessons during this period that include the following:

Talk openly with your children. Children are naturally frightened when they do not understand what is going on in their lives, why they are wreaking havoc at home and at school. They want help, even if they do not explicitly ask for it. When Andrew told me he felt he was going crazy and wanted to die, I froze. When Ainsley wrote me a note apologizing for being a bad child, my heart sank. Yet David and I quickly picked ourselves up and explained they were not bad children and that their issues were not their fault. We did not sugar coat the challenges, but we gave them hope. And we importantly did not hide or avoid information, which led to them increasingly becoming empowered and taking part in their own treatment and advocacy. It is wonderful to see.

Embrace research. There are always new findings, information, medications, and therapeutic approaches. I still read everything I can, including on the internet as long as I verify information with experts: Tourette syndrome is not due to stress or poor parenting. Tics do not only occur in children. Of course, we all have to take a break from research from time to time, but it is important to stay current.

Insist on an integrated partnership approach with every professional involved in your child's care. The best outcomes occur when we all work together, listen and learn from one another. We parents and caregivers know our children best. When Andrew and Ainsley were growing up, their schools sent notes home every afternoon about their days which we returned the following morning with our observations, so that we were in constant communication. To this day, I ensure that the children's psychiatrists and psychologists regularly communicate with one another and with me. As difficult as it can be, if a professional does not want to collaborate with me or treat me with dignity and respect, I move on.

Be kind to your partner. This is true for all marriages and partnerships but is often magnified by the stresses and strains of raising neurodivergent children. There can be disagreements about everything from diagnoses and discipline to therapy and medications. David and I are certainly not immune from this and want to throttle each other at times. At our core, however, we know we are each trying out best, and we give each other the benefit of the doubt.

Take care of yourself. This extends far beyond eating properly, sleeping, and exercising. Reach out for the support you need, whether to healthcare professionals or support groups. Remember the importance of humor. Andrew lacks abstract thinking skills and when his teacher told him a certain school rule was set in stone, he asked where the stone was kept. Yesterday, David and I chuckled when we could not figure out if he was ticking (Tourette), stimming (autism), or engaging in a ritual (OCD). I refuel by carving out part of my life as my own, away from the children and away from David. My career has been life-saving for me: it separates me from the challenges at home, helps me rebuild my own identity, and frankly makes me a better parent. I also enjoy the treadmill at the gym, which lets me to think through my issues and regain perspective. Even when the kids are in crisis, I need breathers and either take a short, 10-minute walk by myself or close the door and soak in a warm, soothing bath.

Be kind to yourself. This is not easy for me. When I found Ainsley hiding in her closet, shaking in fear during one of Andrew's frightening meltdowns, it took me months to forgive myself: I should have protected her better. Sometimes I feel as if I have been punched in the stomach and, just as I am about to get up, another swing comes at me hard. But I also know the importance of patting myself on the back and quietly complimenting myself on my strength and resilience. That is the reason I chose Teddy Roosevelt's quote for the epigraph to my memoir *Hold on Tight*: "Believe you can and you're halfway there."

Advocate and volunteer. There are so many benefits to giving back to our communities when we have the energy and time. I have sat on many Boards of Directors and Advisory Councils, ranging from Canada's leading psychiatric hospital CAMH to Massachusetts General Hospital's Department of Neuropsychopharmacology. I currently chair Kerry's Place Autism Services, Canada's largest autism services provider, serve on the International OCD Foundation/Canada taskforce, write for many magazines, and have presented at Tourette syndrome and similar conferences. These activities have led many caregivers to reach out for help over the years, and I vividly remember being in their shoes: frightened, frozen, isolated, not knowing what to do or how, and terrified about the future. That is why I wrote *Hold on Tight* (Stewart, 2023). I give readers a brutally honest but inspirational account of what it is like to raise children with mental health and neurodevelopmental challenges from an intimate parental perspective.

Andrew and Ainsley know their lives will be endless roller coasters. Despite their challenges, however, they have grown into responsible, caring young adults who have wonderful jobs, give back to their communities and embrace life with grit, determination and optimism. Because of them, I am a more empathetic and understanding human being. Because of them, I am an ambassador of hope and perseverance. Because of them, I have found my calling in life. They are my heroes.

Further reading

Stewart J (2023). Hold on tight: a parent's journey raising children with mental illness. Toronto, ON: Barlow Publishing.

Handbook of Clinical Neurology, Vol. 215 (3rd series)
Tic Disorders and Tourette Syndrome
C. Ganos and Y. Worbe, Editors
https://doi.org/10.1016/B978-0-443-13554-5.00028-6

Chapter 28

Overview of support associations and other useful resources

SEONAID ANDERSON*

Villers Sainte Gertrude, Luxembourg, Belgium

Abstract

Support associations specifically for tic disorders, such as Tourette syndrome (TS), have the potential to profoundly impact individuals, families, and general awareness. This chapter explores the multifaceted roles of support associations, as well as their evolution – from small groups formed by individuals with a vested interest to large organizations managing multimillion-dollar research funding. Support associations face many challenges, making collaboration within and between countries essential. People with lived experience, clinicians, and researchers each bring valuable expertise to efforts aimed at improving the quality of life for those affected by tics and TS. Support associations serve as a vital nexus, connecting stakeholders and facilitating collaboration to drive meaningful progress.

THE IMPORTANCE OF SUPPORT ORGANIZATIONS AND SOME HISTORICAL BACKGROUND

Support associations play a crucial role in the healthcare ecosystem and can have a profound impact on the well-being of individuals with specific medical conditions. This chapter will cover issues around support associations specifically for tic disorders like Tourette syndrome (TS). For ease of comprehension, any tic disorder like TS, chronic tic disorder (CTD), or tics in general (but not functional tics) will be included together. It should be taken into account that tics exist in the general population at a level that may not be appreciated by clinical epidemiologic studies. That is to say, tics are generally very common; however, for someone to be diagnosed with a tic disorder, they would have to meet the diagnostic classification requirements as used by medical professionals. There will be many individuals in the population who have tics but who would not reach a diagnosis, however, they can still be greatly aided by information, advice, and support from patient associations who help individuals with tic disorders. This categorization reflects how support associations function, so someone seeking information on simple tics or chronic tics can also find support from them and not just information about tic disorders like TS. In fact, people seeking information about functional tics also often reach out to different associations for information and sometimes the two conditions coexist. During the covid pandemic, there was a large increase in people reaching out to support associations asking for support for functional tics (Malaty et al., 2022).

Support associations have multiple roles in the way they can assist someone with tics or a tic disorder, as well as their families, on their journey with the disorder, diagnosis, potential treatment and management, which can include (1) Emotional Support and Empowerment; (2) Information, Education, and Access to Resources; (3) Advocacy; (4) Networking; (5) Research and Funding; (6) Giving People a Voice in Healthcare; (7) Policy; and (8) Reducing Healthcare Disparities. The level to which a support association can provide some of these services to people with tics or TS depends on the size and resources available. Most support associations seem to follow a similar trajectory of evolution from their inception to creation as a formal association and onwards. In fact, it could be argued that each support association could be placed somewhere along a trajectory in terms of their evolution, though not all are successful, with some merging together or stopping altogether.

EVOLUTION OF A TOURETTE SYNDROME SUPPORT ASSOCIATION

TS is a neurodevelopmental condition with an onset in childhood. Therefore, the most common starting point for a support group is when parents of someone with tics or a tic disorder come together to discuss about the difficulties and challenges that they face in terms of understanding the health condition that affects their child. Dealing with a chronic illness or a medical condition such as a tic disorder can be emotionally challenging and may lead to a sense of isolation. Support associations provide a sense of community and

*Correspondence to: Seonaid Anderson, PhD, 11 Rue Du Sawheux, Villers Sainte Gertrude, 6941, Belgium. Tel: +32470859458, Email: Seonaida@hotmail.com.

understanding, allowing people to share their experiential knowledge, fears, and triumphs with others who are in similar situations. This emotional support can alleviate feelings of isolation and depression.

The early stages of a support association are often characterized by a sharing of experiences and word of mouth exchanges of information of where to seek diagnosis and treatments, either on a local or national level. Sometimes this is centered around a particular clinic or an interested medical professional. Support associations often serve as valuable sources of information and education, and they can empower people with tics or a tic disorder to take an active role in managing their health. Associations can provide resources, workshops, and seminars to help people with TS and their families understand their condition, the available treatment options, and how to manage their well-being effectively. If an association becomes an official entity (e.g., *Tourettes Action* in the United Kingdom), then often a website and social media profiles are used to share information and to connect with potential members and the wider community.

There may be face-to-face meetups for children and parents to get together and provide support for one another. It takes dedication and commitment in terms of organizing such events. It can be difficult to gain traction, as many families face challenges in terms of their child's health condition but also in terms of their ability to commit to attending meetings and then potentially becoming a volunteer in a more official sense for the support association. This early evolution is often driven by a few individuals, frequently parents, who may apply to national government or funding bodies for financial assistance to support activities such as meetups, day trips, or weekends away to provide support to families.

Once this has been established, it is often at this point that individuals may attempt to create an official support association and financial stability can become important. Anderson et al. (2022) surveyed TS support associations and found most rely on membership payments or donations as a source of financial aid. There can be many administrative hurdles in terms of formally registering an association. It requires long-term and unpaid dedication from individuals to keep this process going and create all the documents and regulations required to meet the obligations necessary to form an official support association. This dedication is often driven by the first-hand knowledge that an association can provide resources and support which can significantly improve the quality of life for people with tics or a tic disorder and their families.

Support associations are important in terms of linking up people in similar positions as well as providing information on tic disorders and the co-occurring conditions but also by directing people to healthcare professionals/healthcare services. However, cultural and political differences may mean that some people suggest that such things should be provided by the healthcare system or government in their country. Other activities, such as lobbying by support associations, are often driven by the need and desire to improve the situation of patients trying to access diagnosis and treatment. There may be a danger of the support associations filling implementation gaps and the healthcare system or government relying on the association to take care of people and raise their own funds. There seem to be few TS support associations who receive government funding to carry out their activities. In terms of fundraising, those associations without government funding should ensure a diverse portfolio for success and longevity. Some support associations may be run like a business seeking sponsorship, sales from branded goods, and investments. Smaller support associations often operate with limited budgets and rely on donations, grants, and volunteer efforts. Securing funding and sustaining financial stability can be a continuous challenge.

Support associations may also become involved in research, and in recent years it has been recognized by researchers and clinicians that closer relationships and international cooperation with support associations can improve the quality of research. In Europe, this has been the case with *The European Society for the study of Tourette Syndrome* (ESSTS) and the umbrella organization for TS support associations *Tics and Tourette Across the Globe* (TTAG). Support associations can also help advertise and increase numbers of participants with TS in research which are required for large-scale data collection "required by brain imaging and genetic studies" (Kleimaker et al., 2021). Several networks in the field of TS have developed with more of a collaboration between the international research community and support associations.

It would be fair to say that the "Tourette community" is a diverse community of people affected by TS, their families, healthcare, and education professionals and researchers. This leads, therefore, to a wide variety of views. Some individuals, in line with the neurodiversity trend of recent years, often quote their having tics and TS as their "superpower" which should be celebrated as part of their TS and neurodiversity. Some hold views that they don't "need fixed" or have treatment with medication or behavioral therapy. Some individuals would like access to a wider variety of treatments with regard to seeking help managing their TS and the challenges they encounter. There is also diversity in terms of terminology used in relation to the titles used by associations. Some suggest, for example, that the word "patient" in the "patient support association" should be dropped, as it indicates someone who is currently under medical care such as a doctor, nurse, or hospital. This may not be the case with all people with TS; they may have a diagnosis but are not currently under any healthcare supervision. Therefore, it may be advisable to not use the word patient at all but instead just the term "person with TS" or "TS support association." This also opens up the idea that the association is for anyone affected by, or wanting information about, TS and tic disorders, not just the person with the diagnosis. All types of people may benefit from contact with the association such as family/friends, healthcare professionals, or educators as well as the general public. Support associations, such as the *Tourette Association of America* and *Tourettes Action* in the United Kingdom, offer a range of support services, including helplines, support groups, and online resources to help individuals with TS and their families navigate the challenges associated with the condition.

Another issue for consideration is the use of the term Tourette Syndrome. When a support association uses only the term "Tourette" or "Tourette Syndrome" in the title of the association, this can lead to some confusion, but these associations are quick to point out that they will help anyone with tics, whether they have TS or not. In the case of functional tics, there are also specific support associations for people seeking information and support on this medical condition (e.g., the Functional Neurologic Disorder Society and FND Action). It has been suggested that using TS in the support associations title may also discourage people who still think TS only concerns swearing or that it is an extreme condition and may discourage people wanting to talk about milder tics to get in touch. Reassurance given is quick to solve these worries and whether discussing tics or tic disorders such as TS, everyone can benefit from information whether a diagnosis of TS is present or not.

LOBBYING AND PATIENTS' RIGHTS

Individuals with severe forms of TS may experience functional impairment that qualifies them for disability benefits or accommodations in certain situations such as education or employment due to the severity of their tics or associated conditions. To qualify for disability benefits or accommodations, individuals with TS may need to provide medical documentation from healthcare professionals that explains the severity of their condition and the resulting functional limitations.

As a support association develops initiatives such as lobbying may become increasingly important with activities including political campaigning to increase awareness and to advocate for patients' rights, pushing for improved access to healthcare, better insurance coverage (if applicable), and the development of new treatments or therapies. Support associations have the potential to influence healthcare policy at local, regional, and national levels and to advocate for policies that benefit people with TS. For example, support associations may also help advise individuals with TS and their access to appropriate educational accommodations and support in schools. Indeed, in many countries, individuals with TS are eligible for educational accommodations under laws that protect the rights of individuals with disabilities. These accommodations can include extended testing time in exams, flexible timetables, limiting distractions in the classroom, and access to support services. In contrast, individuals with TS in countries without a relevant support association may face greater challenges. In recent times, the digital evolution has meant more information is available globally through the internet.

FULLY FLEDGED SUPPORT ASSOCIATIONS

Creating a support association involves several steps to ensure its successful establishment and operation such as identifying the specific needs, challenges, and gaps in support for individuals affected by the condition. Continued momentum forward to a fully fledged association requires a dedicated group of individuals who are passionate about their cause. This core team will be responsible for planning, organizing, and managing the association's activities. Roles might include a president/chairperson, treasurer, secretary, and members responsible for various initiatives. Also required is a proper legal structure and registration of the association such as nonprofit organization, charity, or advocacy group. A board of directors and bylaws are needed to oversee the organization's activities as well as a governance structure, decision-making processes, and membership criteria. Ultimately, the services provided by support associations may contribute to improvements in quality of life for individuals living with tics and tic disorders. They can help individuals with TS navigate the complex healthcare system, make informed decisions, and cope with the challenges of their condition.

Over time as an association develops, it may be staffed by a paid team of staff instead of relying solely on part-time volunteers. It may be possible to facilitate networking opportunities for individuals with TS and their families to connect with healthcare professionals, researchers, and experts in the field. This can lead to valuable insights, second opinions, and access to clinical trials or new technologies. Some of the larger support associations can also provide research funding by raising money for research grants. Increasingly, there is collaboration with researchers, and more information on the development of new treatments and therapies.

In summary, support associations are essential for offering emotional support, education, advocacy, and resources that not only enhance the well-being of people with TS but also contribute to advancements in medical research and ideally improvements in healthcare systems.

AN EXAMPLE OF A FULLY FLEDGED SUPPORT ASSOCIATION

An example of one of the longest established TS support associations is that of **The Tourette Association of America (TAA)** (https://tourette.org/) formerly known as the Tourette Syndrome Association. This is a nonprofit organization in the United States dedicated to supporting individuals and families affected by TS and tic disorders. The organization's mission is to raise awareness, provide resources, offer support, and promote research into these conditions. The organization was founded in 1972 and its mission is to help affected individuals. The TAA also raises awareness among the general public, healthcare professionals, educators, and policymakers. The organization advocates for policies and practices and has education programs specifically designed for children and teens with TS. These programs aim to empower young individuals with TS and help them manage their condition effectively. TAA also has funds and promotes research, and it hosts an annual national conference. Its structure is formed of local chapters across the United States. This highlights one example of a TS support association, but recently a directory was created with an overview of the most important services and contact details of TS support associations around the world.

DIRECTORY OF TOURETTE SYNDROME PATIENT ASSOCIATIONS

The directory was created to provide an overview of the services and contact details of TS support associations around the world (Anderson et al., 2022). The aim of this was to provide information concerning available organizations in respective countries. This directory was previously included in a textbook chapter (Martino and Leckman, 2022), and was subsequently converted to an online version which also allows revisions of any of the details of the support associations (https://www.essts.org/directory).

Just as the Directory of Tourette Syndrome Patient Associations provided an overview of TS support associations around the world, Fig. 28.1 gives an overview of the geographic distribution of TS-related research publications. This information may be useful to the support associations who wish to try and create and strengthen relationships with TS researchers in their country. This figure is reproduced here with kind permission from Dr. Tara Murphy who created the map in 2022. On the map "blue" countries primarily highlight case studies primarily, whereas the "pink" countries have published larger studies.

Fig. 28.2 shows which countries (at the time of publication) are known to have TS support associations.

Included in the map of Fig. 28.2 are the current TTAG member countries. The TTAG ambassador program (indicated in red) may be helpful in supporting the creation of new support associations (https://ticsandtourette.org/join-us/become-a-ttag-ambassador). When conducting research with support associations, Anderson et al. (2022) surveyed TS support associations but was unable to get full information about support associations from some countries including Albania, Bulgaria, Caribbean, Colombia, Croatia, Ecuador, Egypt, Estonia, India, Ireland, Israel, Malta, Malaysia, Mexico, Philippines, Russian Federation, Singapore, South Africa, South Korea, and Venezuela.

ADDITIONAL CONSIDERATIONS AND CHALLENGES FOR TS SUPPORT ASSOCIATIONS

Engaging the general public and healthcare professionals

No matter which country a support association is based in, raising awareness about tics and TS remains a challenge across various social domains, from the general public to healthcare and educational professionals. Cultural differences also play a role in shaping awareness and acceptance (Stiede et al., 2021). A common theme across countries is that tics and tic disorders are often not taken seriously. They may be viewed as comical by members of the public and given low priority for treatment by government health spending. However, it has been shown that individuals with Tourette's disorder or CTDs are over four times more likely to die by suicide than the general population (Fernández de la Cruz et al., 2017). The long-term consequences of unrecognized and untreated tics extend beyond personal well-being to economic costs, as poor mental health can lead to increased healthcare expenditures.

Support associations can also play a crucial role in addressing media misrepresentations of people with tics and TS. Collado-Vázquez and Carrillo (2013) suggested that film representations of tics and Tourettes have created false beliefs and stereotyped images of the disorders. Here support associations could have a role working with media professionals to guide them in terms of creating more realistic representations of people with tics and TS.

Another significant discussion is the dominance of the medical model in light of growing neurodiversity movement. The core idea behind neurodiversity is that neurologic differences, such as autism, attention deficit hyperactivity disorder, dyslexia, and others, should be recognized and respected as a natural part of human diversity rather than treated as a "defect" or something to be "fixed." The movement also serves as a platform for neurodivergent individuals to take control of the narratives surrounding their lives and to challenge societal norms. Some within this movement view themselves as a neurologic distinct group that has historically been marginalized by the dominant "neurotypical" majority.

There are many healthcare professionals and researchers dedicated to helping individuals with TS, often becoming national specialists who drive forward awareness, training, and initiatives in what is still considered a niche field. Many of these experts collaborate closely with support associations to improve services and educate their colleagues. However, when such specialists retire or leave their profession, their expertise and advocacy efforts are often lost. Motivated professionals also face systemic challenges, including limited funding and the absence of standardized diagnostic and treatment pathways. For example, in the United Kingdom, there are no NICE (National Institute for Health and Care Excellence) guidelines for tic disorders, leading to significant regional disparities in access to diagnosis and treatment – a phenomenon often described as a "postcode lottery."

PERSONNEL FOR TOURETTE SYNDROME SUPPORT ASSOCIATIONS

Another ongoing challenge for support associations is finding volunteers or staff to run the charity. Some support associations insist that all board members have a personal relationship with tics or TS (whether themselves or a family member). In some ways, this is a good idea but in others it can be challenging when all of the team running the charity have TS and co-occurring conditions themselves which can often be a hindrance to their continued participation. In a small association, the departure or illness of a single individual can significantly disrupt daily operations and threaten the continuity of the organization. As patient associations often rely on volunteers to run programs and activities, it can be an ongoing challenge to recruit and retain dedicated volunteers as they are donating limited time and resources. Anderson et al. (2022) surveyed TS support associations and found that 58% operate without paid staff. The presence of paid employees may serve as an indicator of an association's level of establishment, with larger, long-standing organizations transitioning from being entirely volunteer-run to having dedicated staff over time.

SENSE OF COMMUNITY WITHIN A TS SUPPORT ASSOCIATION

The term "Tourette Syndrome community" is often used, but who exactly does it include? It may encompass individuals with tics or tic disorders, their families and friends, healthcare and research professionals, educators, employers, and anyone else affected by TS. This diverse group brings a wide range of motivations, opinions, and needs, making it difficult to foster a unified and engaged community within patient associations.

One of the main challenges is sustaining involvement. Many individuals and families dealing with TS already face significant stress and daily demands, leaving little time or energy for long-term engagement in a support association. While some may actively participate during particularly challenging periods, they might step away when symptoms become more manageable, leading to fluctuations in community engagement.

Additionally, individuals with TS differ greatly from one another, making generalizations difficult and sometimes unhelpful. The severity of symptoms varies widely – not only between individuals but also within a single person's lifetime. Some may experience mild tics that require little intervention, while others struggle with severe symptoms that disrupt their daily lives. This variability can create differing perspectives on what support associations should prioritize, potentially leading to fragmentation rather than cohesion.

MOVING FORWARD TOGETHER OR MOVING APART

Each country with a TS support association has a unique history of its formation and evolution. In some countries, a single national TS association represents the community, while in others, multiple smaller associations operate independently or alongside a national body. The development of these organizations can be complex, often arising when individuals feel that the main or national association does not fully meet local needs, prompting the creation of alternative groups.

This fragmentation can be seen as both a challenge and a strength. On one hand, it risks dividing the community, leading to competition for funding, membership, and public attention. It may also dilute awareness efforts by presenting multiple, sometimes conflicting, voices. On the other hand, localized support networks can better address regional needs and, in the long run, may amplify awareness efforts by increasing the number of voices advocating for TS.

A survey by Anderson et al. (2022) highlighted that some countries have multiple TS associations. Spain, for example, has seven regional associations, and while there were plans in 2019 to establish a National Federation, no progress appears to have been made as of 2024. Similar patterns exist in other European countries, including Italy, Germany, and the United Kingdom, where multiple support associations coexist.

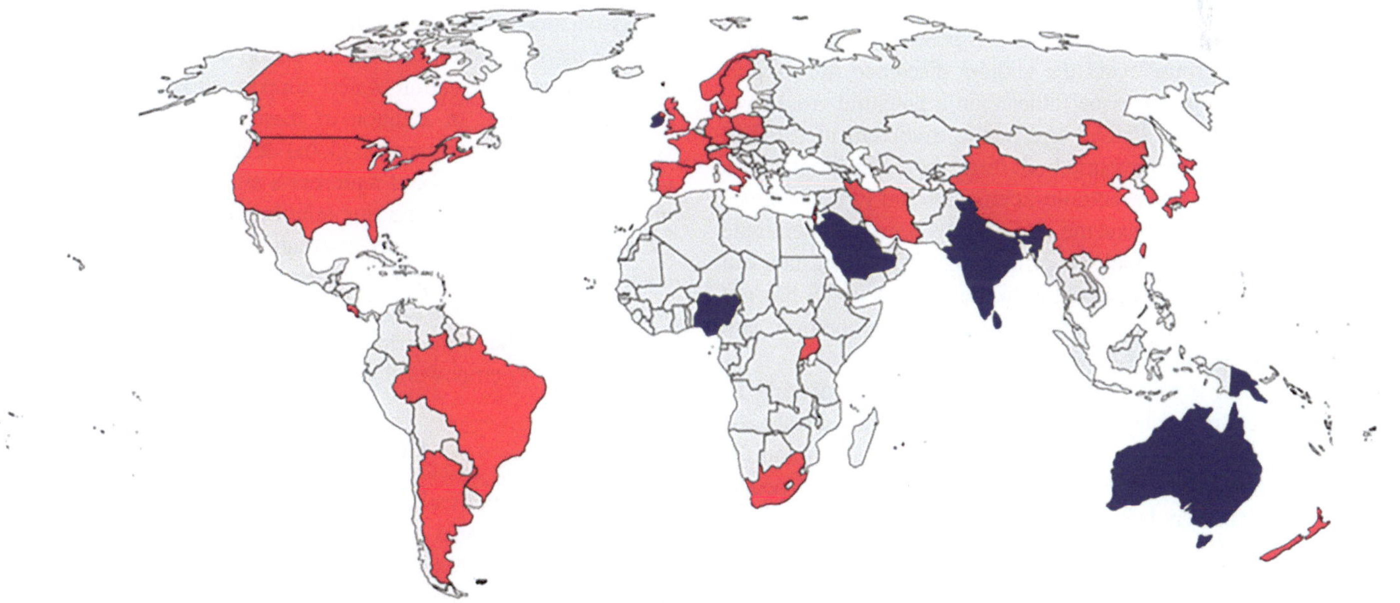

Fig. 28.1. A review of global Tourette syndrome research (Dr. Tara Murphy, 2022). Note: Map lines delineate study areas and do not necessarily depict accepted national boundaries.

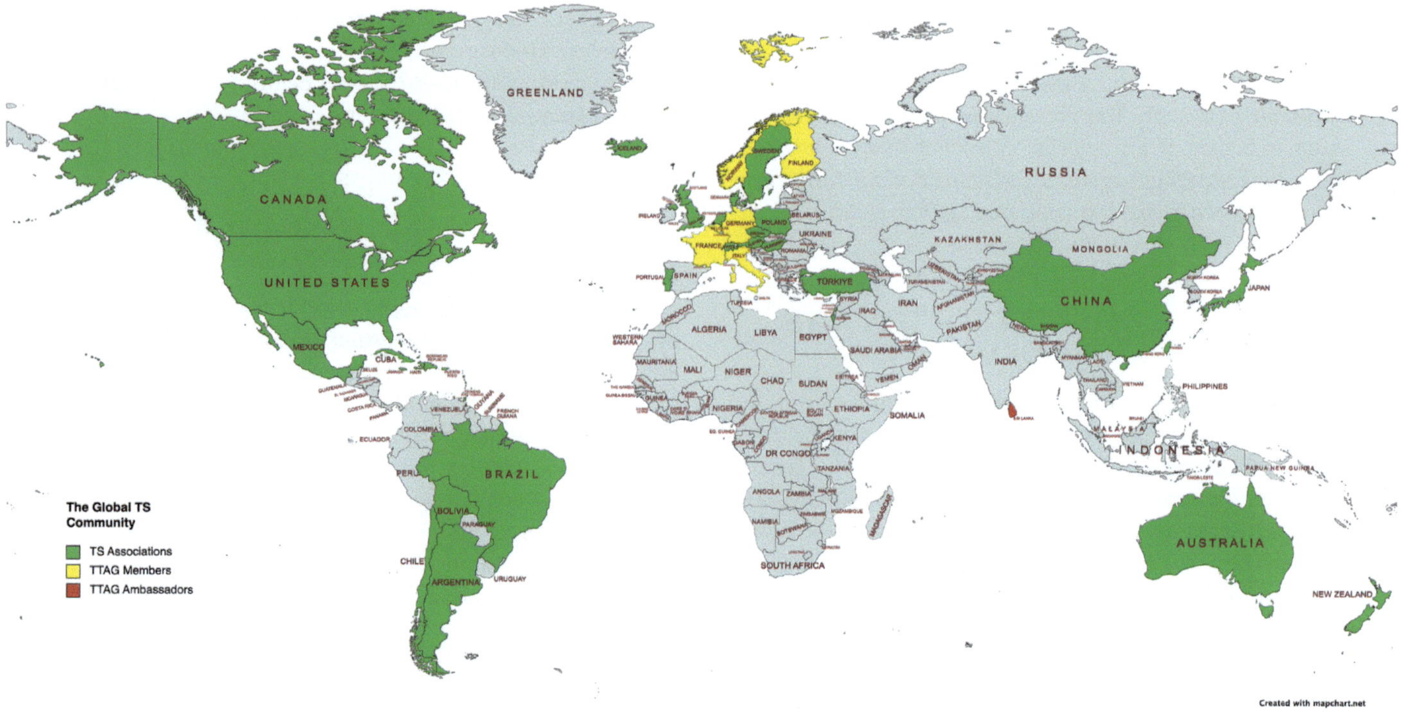

Fig. 28.2. Review of Tourette syndrome support associations (TTAG, 2023). Note: Map lines delineate study areas and do not necessarily depict accepted national boundaries.

RESEARCH AND COLLABORATION

It is an encouraging sign that there are an increasing number of published research papers in the field of tic disorders such as TS (Hartmann et al., 2023). This indicates that there is interest from researchers and clinicians, as well some research funding available in order for this research to be conducted and the research papers published. Kleimaker et al. (2021) suggested that "TS research networks should ideally be composed of international, multiprofessional groups" and the authors discussed the support associations as a separate issue rather than a potential cross-over between these groups. "Patient and public involvement (PPI) in research envisages a relationship built throughout the lifespan of a research project between academics, clinicians and PPI colleagues in order to inform, plan, execute and, in due course, disseminate and translate research" (Jackson et al., 2020). To date, an ideal definition of what PPI should entail in the context of TS has not been established; however, emerging cross-collaboration and engagement are beginning to occur.

Another aim of a well-established support association could be increasing effective collaboration with healthcare professionals and institutions and helping to bridge the gap between patients and the healthcare system. There appears to be a growing trend of people affected by tic disorders being involved in research and contributing to written materials such as this chapter and others in the reference list. This is a promising indication. Also there seems more interest in recent years from pharmacologic companies in terms of reaching out to support associations and potential collaborations. This is an issue that not all support associations may be ready for – there are pros and cons of getting involved with this type of collaboration. It potentially could lead to some much-needed funding from pharmaceutical companies for support associations, but it would need to be managed in a responsible way with clear ethical and legal guidance in place and transparency by the associations. As support associations hopefully evolve to become bigger organizations, they may also be able to fund their own research. This is already happening in some of the larger more well-established associations.

Associations on a global level conduct vital work for people affected by tic disorders. This work should of course continue and is vital. So too is the creation of new support associations in countries where none exist. In addition to these efforts within individual countries, collaborating on a global level with the international research community would represent a significant step forward for all. This is the goal of the umbrella organization TTAG to unify global efforts of TS support associations. TTAG wishes to raise awareness and provide support for associations supporting people with tics and TS worldwide. The aim is to create a "global community, where knowledge, resources, and experiences are shared freely to empower those impacted by these conditions."

CONCLUSION

With limited data available, it remains challenging to provide a comprehensive global overview of TS support associations. The most up-to-date resource with information about support associations is the previously discussed directory, yet significant gaps remain where it appears no support association exists, particularly in countries in Africa, Asia, and Oceania. This does not indicate any differences in prevalence, but disparities in diagnosis, awareness, and access to care in these geographic areas. While recent research from individual clinics in China and India suggests growing professional interest, awareness initiatives and organized support networks in these regions remain largely undocumented. Encouragingly, more researchers and academics are recognizing the role of support associations in their work (Roper et al., 2013; Anderson et al., 2022), though this is not yet universal – evident in omissions like Kleimaker et al. (2021), which overlooked long-standing patient association meetings now unified under TTAG.

There has never been a better opportunity to leverage the ongoing discourse on neurodiversity to bring tics and tic disorders into the spotlight. Rather than competing for scarce resources, such as funding and visibility, TS support associations can achieve far more by fostering collaboration within and across countries. By working together, they can strengthen advocacy efforts, share expertise, and ultimately enhance the quality of life for those affected by TS and tics. A collective approach – bringing together individuals with lived experience, clinicians, and researchers – will be the most effective way to drive meaningful progress.

ACKNOWLEDGMENTS

The author would like to thank Christos Ganos and Yulia Worbe for the invitation to write this chapter, and Dr. Tara Murphy and TTAG for their kind permission to reproduce the maps used in this chapter. The author wishes to extend appreciation and respect to all the patient associations, as well as all colleagues and people touched by TS and tics around the world. The author also extends sincere thanks to ESSTS and Anna Kanta.

USEFUL RESOURCES

ESSTS – The European Society for the study of Tourette Syndrome: https://www.essts.org/

TTAG – Tics and Tourette Across the Globe: https://ticsandtourette.org

TTAG Ambassador Program: https://ticsandtourette.org/join-us/become-a-ttag-ambassador/

Tourette Association of America: https://tourette.org

Tourettes Action (UK): https://www.tourettes-action.org.uk

References

Anderson S, Heijerman-Holtgrefe A, Hartmann A (2022). Tourette syndrome support organizations around the world. In D Martino, J Leckman (Eds.), Tourette Syndrome (2nd ed.). Oxford Academic Press.

Collado-Vázquez S, Carrillo JM (2013). Los tics y el sindrome de Tourette en la literatura, el cine y la television [Tics and Tourette syndrome in literature, cinema and television]. Rev Neurol 57: 123–133.

Fernández de la Cruz L, Rydell M, Runeson B et al. (2017). Suicide in Tourette's and chronic tic disorders. Biol Psychiatry 82: 111–118.

Hartmann A, Andrén P, Atkinson-Clément C et al. (2023). Tourette syndrome research highlights from 2022. F1000 Research 12: 826.

Jackson T, Pinnock H, Liew SM et al. (2020). Patient and public involvement in research: from tokenistic box ticking to valued team members. BMC Med 18: 79.

Kleimaker A, Kleimaker M, Behm A et al. (2021). Networks in the field of Tourette syndrome. Front Neurol 12: 624858.

Malaty IA, Anderson S, Bennett SM et al. (2022). Diagnosis and management of functional tic-like phenomena. J Clin Med 11: 6470.

Martino D, Leckman J (2022). Tourette Syndrome (2nd ed.). Oxford Academic Press.

Roper L, Hollenbeck PJ, Rickards HE (2013). Tourette syndrome support organizations around the world. In D. Martino, J. F. Leckman (Eds.), Tourette Syndrome (1st ed.). Oxford Academic Press.

Stiede JT, Woods DW, Anderson S et al. (2021). Cultural differences in reactions to tics and tic severity. Child Fam Behav Ther 43: 161–180.

Index

Note: Page numbers followed by "*f*" and "*t*" refer to figures and tables, respectively.

B

C

D

E

T